HUMAN GROWTH

3
Neurobiology and Nutrition

HUMAN GROWTH

HUMAN GROWTH

3
Neurobiology and Nutrition

Edited by

Frank Falkner

The Fels Research Institute
Wright State University School of Medicine
Yellow Springs, Ohio

and

J. M. Tanner

Institute of Child Health
London, England

PLENUM PRESS · NEW YORK AND LONDON

Library of Congress Cataloging in Publication Data

Main entry under title:

Human growth.

Includes bibliographies and index.
CONTENTS: v. 1. Principles and prenatal growth. – v. 2. Postnatal growth. – v. 3. Neurobiology and nutrition.
1. Human growth–Collected work. I. Falkner, Frank Tardrew, 1918-   II. Tanner, James Mourilyan. [DNLM: 1. Growth. 2. Gestational age. WS103 H918]
QP84.H76   612'.3   78-1440
ISBN 0-306-34463-7 (v. 3)

© 1979 Plenum Press, New York
A Division of Plenum Publishing Corporation
227 West 17th Street, New York, N.Y. 10011

Printed in the United States of America

Contributors

H. ALS
Child Development Unit
Harvard University Medical School
Boston, Massachusetts

ROBERT BALÁZS
MRC Developmental Neurobiology
   Unit
Institute of Neurology
London, England

KARL E. BERGMANN
Department of Pediatrics
J. W. Goethe University
Frankfurt/Main, West Germany

RENATE L. BERGMANN
Department of Pediatrics
J. W. Goethe University
Frankfurt/Main, West Germany

INGEBORG BRANDT
Universitatäts-Kinderklinik und
   Poliklinik
Bonn, Germany

T. B. BRAZELTON
Child Development Unit
Harvard University Medical School
Boston, Massachusetts

GEORGE F. CAHILL, JR.
Joslin Diabetes Foundation, Inc.
The Peter Bent Brigham Hospital and
   Harvard Medical School
Boston, Massachusetts

JOAQUÍN CRAVIOTO
Instituto Nacional de Ciencias y
   Tecnología de la Salud del Niño
Sistema Nacional para el Desarrollo
   Integral de la Familia (DIF)
Mexico City, Mexico

ELSA R. DELICARDIE
Rural Research and Training Center
Instituto Nacional de Ciencias y
   Tecnología de la Salud del Niño
Sistema Nacional para el Desarrollo
   Integral de la Familia (DIF)
Mexico City, Mexico

C. DREYFUS-BRISAC
Centre de Recherches de Biologie du
   Développement Foetal et Neonatal
Hópital Port Royal
Université René Descartes
Paris, France

PHYLLIS B. EVELETH
Division of Research Grants
National Institutes of Health
Bethesda, Maryland

CHRISTINE GALL
Department of Psychobiology
University of California
Irvine, California

DERRICK B. JELLIFFE
Division of Population, Family and
    International Health
School of Public Health
University of California
Los Angeles, California

E. F. PATRICE JELLIFFE
Division of Population, Family and
    International Health
School of Public Health
University of California
Los Angeles, California

PAUL D. LEWIS
Department of Histopathology
Royal Postgraduate Medical School
Hammersmith Hospital
London, England

GARY LYNCH
Department of Psychobiology
University of California
Irvine, California

PAMELA C. B. MACKINNON
University of Oxford
Oxford, England

LAURENCE MALCOLM
Health Planning and Research Unit
Christchurch, New Zealand

AMBRISH J. PATEL
MRC Developmental Neurobiology
    Unit
Institute of Neurology
London, England

T. RABINOWICZ
Division of Neuropathology
Centre Hospitalier
Universitaire Vaudois
Lausanne, Switzerland

ALDO A. ROSSINI
Joslin Diabetes Foundation, Inc.
The Peter Bent Brigham Hospital and
    Harvard Medical School
Boston, Massachusetts

F. J. SCHULTE
University of Göttingen
Göttingen, West Germany

J. P. SCOTT
Center for Research on Social
    Behavior
Bowling Green State University
Bowling Green, Ohio

J. M. TANNER
Department of Growth and
    Development
Institute of Child Health
University of London
London, England

COLWYN TREVARTHEN
University of Edinburgh
Edinburgh, Scotland

E. TRONICK
Department of Psychology
University of Massachusetts
Amherst, Massachusetts

Preface

Growth, as we conceive it, is the study of change in an organism not yet mature. Differential growth creates form: external form through growth rates which vary from one part of the body to another and one tissue to another; and internal form through the series of time-entrained events which build up in each cell the specialized complexity of its particular function. We make no distinction, then, between growth and development, and if we have not included accounts of differentiation it is simply because we had to draw a quite arbitrary line somewhere.

It is only rather recently that those involved in pediatrics and child health have come to realize that growth is the basic science peculiar to their art. It is a science which uses and incorporates the traditional disciplines of anatomy, physiology, biophysics, biochemistry, and biology. It is indeed a part of biology, and the study of human growth is a part of the curriculum of the rejuvenated science of Human Biology. What growth is not is a series of charts of height and weight. Growth standards are useful and necessary, and their construction is by no means void of intellectual challenge. They are a basic instrument in pediatric epidemiology. But they do not appear in this book, any more than clinical accounts of growth disorders.

This appears to be the first large handbook—in three volumes—devoted to Human Growth. Smaller textbooks on the subject began to appear in the late nineteenth century, some written by pediatricians and some by anthropologists. There have been magnificent mavericks like D'Arcy Thompson's *Growth and Form*. In the last five years, indeed, more texts on growth and its disorders have appeared than in all the preceding fifty (or five hundred). But our treatise sets out to cover the subject with greater breadth than earlier works.

We have refrained from dictating too closely the form of the contributions; some contributors have discussed important general issues in relatively short chapters (for example, Richard Goss, our opener, and Michael Healy); others have provided comprehensive and authoritative surveys of the current state of their fields of work (for example, Robert Balázs and his co-authors). Most contributions deal with the human, but where important advances are being made although data from the human are still lacking, we have included some basic experimental work on animals.

Inevitably there are gaps in our coverage, reflecting our private scotomata, doubtless, and sometimes our judgment that no suitable contributor in a particular field existed, or could be persuaded to write for us (the latter only in a couple of instances, however). Two chapters died on the hoof, as it were. Every reader will notice the lack of a chapter on ultrasonic studies of the growth of the fetus; the manuscript, repeatedly promised, simply failed to arrive. We had hoped, also, to include a chapter on the very rapidly evolving field of the development of the visual processes, but here also events conspired against us. We hope to repair these omissions in a second edition if one should be called for; and we solicit correspondence, too, on suggestions for other subjects.

We hope the book will be useful to pediatricians, human biologists, and all concerned with child health, and to biometrists, physiologists, and biochemists working in the field of growth. We thank heartily the contributors for their labors and their collective, and remarkable, good temper in the face of often bluntish editorial comment. No words of praise suffice for our secretaries, on whom very much of the burden has fallen. Karen Phelps, at Fels, handled all the administrative arrangements regarding what increasingly seemed like innumerable manuscripts and rumors of manuscripts, retyped huge chunks of text, and maintained an unruffled and humorous calm through the whole three years. Jan Baines, at the Institute of Child Health, somehow found time to keep track of the interactions of editors and manuscripts, and applied a gentle but insistent persuasion when any pair seemed inclinded to go their separate ways. We wish to thank also the publishers for being so uniformly helpful, and above all the contributors for the time and care they have given to making this book.

Frank Falkner
James Tanner

Yellow Springs and London

Contents

Chapter 4

Developmental Aspects of the Neuronal Control of Breathing

F. J. Schulte

Chapter 5

Ontogenesis of Brain Bioelectrical Activity and Sleep Organization in Neonates and Infants

C. Dreyfus-Brisac

Chapter 6

Sexual Differentiation of the Brain

Pamela C. B. MacKinnon

Chapter 7

Critical Periods in Organizational Processes

J. P. Scott

Chapter 8

Patterns of Early Neurological Development

Ingeborg Brandt

Chapter 9

Early Development of Neonatal and Infant Behavior

E. Tronick, H. Als, and T. B. Brazelton

VII. Nutrition

Chapter 10

Nutrition and Growth in Infancy

Renate L. Bergmann and Karl E. Bergmann

Chapter 15

Nutritional Deficiencies and Brain Development

    Robert Balázs, Paul D. Lewis, and Ambrish J. Patel

Chapter 16

Nutrition, Mental Development, and Learning

    Joaquín Cravioto and Elsa R. DeLicardie

VIII.   History of Growth Studies

Chapter 17

A Concise History of Growth Studies from Buffon to Boas

    J. M. Tanner

Contents of Volumes 1 and 2

VI
Neurobiology

1

Neuroembryology and the Development of Perception

COLWYN TREVARTHEN

1. Introduction

This review of findings in brain research is concerned with the biological causes of perception.

Stimuli, although necessary, are not sufficient for perception. Perceptual images arise in brain-cell networks of particular form, and no aspect of perception is independent of the morphological design of the cell system in which it comes about. There can be no perception without accurate and complex patterning of nerve-cell connections by growth.

Growth of the brain, like that of any other cell system, is self-regulated by physicochemical interactions of its tissue elements under the control of the genes, and it is influenced by further interaction with surrounding non-nervous tissue. However, in addition to their capacity for autonomously spinning out patterned circuits of great complexity, growing networks of brain cells also have a unique power of responding to effects transmitted to them from the environment. Perceptual mechanisms, being in direct contact with the outside world, would seem to be especially vulnerable to influence from stimuli. On the other hand, the way that the primary sensory zones categorize or sort information for consciousness and cumulative experience may be expected to be adaptively predetermined in the gene code as a consequence of strong selective pressures in evolution.

With the great advances of recent decades in neurobiology, especially in relation to the fine structure and biochemistry of developing nerve networks, there is intense interest in the relation between brain histogenesis and various kinds of

---

COLWYN TREVARTHEN • University of Edinburgh, Edinburgh, Scotland.

learning. Refined physiological and anatomical techniques add to behavioral methods of assaying effects of experimental surgery on growing brains. For the first time these analytic techniques are producing reliable direct evidence of cell changes in discrete response to patterns of stimuli, especially in newborn animals or those undergoing metamorphosis. By artificially restricting or shaping stimuli, it is possible to test the anatomical plasticity of growing perceptual mechanisms. On the other hand, the drastic methods of this research, like those of surgical interference, carry with them a danger of distorting the natural growth process beyond recognition, thereby causing a neglect of the brain's normal control of its own development when conditions are not so unbalanced. Most important, the deprivation methods used often interfere with the motor output of the brain which normally has the power to select and reject information for perception. Deprivation of stimuli is then deprivation of acts, too, and disruption of development.

In the adult, perception is not simply a matter of neural processing of input from arrays of receptors, although it is often described in this oversimplified way. It is a sensory–motor activity, and this is even more the case in developmental stages. In normally developing brains the kind of education which perception mechanisms receive from the environment is determined first by the patterned activity of motor organs, some of which, like the muscles of the lens of the eye, are adapted exclusively to the purpose of controlling sensory input. This being so, the neuroembryology of perception must include an account of the growth and development of movements that "gate" perception. In turn, movements are a property of the body and are directly dependent on body form. Therefore, from the start of what may be called *perceptogenesis* we have to look at effects of the morphogenesis of the body and the building of receptor and motor structures in relation to body form. The structures of the body are outside the nervous system.

The higher forms of intelligence elaborate on innate representations of significant objects in their world, and these templates lead them to act so that phenomena will reveal their useful attributes. Kinds of behavior become integrated together by the growing and learning brain to increase the conceptual basis for perception. Thus cognitive development enlarges the scope of perception, while memories are formed of important events and new instances that reinforce and affirm past perceptions. Cognitive concepts qualify the simpler rules of immediate perceptual control of movement, but they do not usurp them. It is necessary to look for evidence of growth in brain mechanisms that mediate these higher level developments in perception which cannot be analyzed at the level of a single nerve cell. Recent evidence from effects of brain surgery or trauma in man reveals an anatomy of consciousness with surprisingly long development and surprising rigidity of final form. Even these learning processes bear the stamp of genetic determination.

It is important to emphasize that brain development is an exceptionally protracted process. In the embryo of man the organs of the body are not yet fully differentiated and functionally mature, but although small, they have attained fairly close approximation to adult form and tissue structure by the end of this period. The brain, however, is extraordinarily undifferentiated in the embryo. All the higher cerebral structures of mammals differentiate into recognizable rudiments of the adult organs postembryonically in the fetal period. At birth the brain of man is less mature than the rest of the body, although not much smaller than the adult brain, being destined to go through exceedingly complex morphogenetic changes long after birth with little gain in bulk. It is therefore arbitrary and obfuscating to consider that neuroembryology of perception means growth of perceptual machinery "in the embryo."

As we shall see, the evidence concerning brain growth makes it virtually impossible to place an upper bound to the generation of structure in brain tissue. We cannot place its end early in the postnatal period. We have to consider that significant new development may continue throughout life. On the other hand, in the whole panoply of the brain's functional systems, those for perception seem to attain a fair degree of maturity most quickly. For this reason the most important period for the attainment of basic function in perception is early infancy.

Bearing this in mind we may divide the material for review into three developmental stages: embryo, fetus, and infant, and then consider what further developments may take place in the postinfant life of perceptual systems.

Embryo and fetus are prefunctional stages. In spite of evidence that fetuses possess some powers of psychological response to stimuli, perception of an outside world is rudimentary before birth. Even if the conditions for receiving informative stimuli were adequate inside the womb, the fetus has exceedingly restricted powers of action, and as has been said, perception is intimately dependent on controlled action. Fetuses, though they do seem to possess functional systems for categorizing vestibular, kinesthetic, and tactile (and possible olfaction and taste) stimuli, can *do* very little. They therefore perceive only to a very limited degree. In the neonate and the infant we see the most significant advances in perceptual growth as the subject develops specific adaptations to a rich environment for action. It is from recent research in this period of life that the most exciting information has come.

The main focus of this essay will be man, but it will be necessary to bring in information from many lower forms of life if human perception is to be understood. The most revealing experiments on growth of fetal and postnatal perception systems come from experiments with birds and mammals. Knowledge of the embryo phase comes partly from work with invertebrates, but mainly from studies with fish and amphibia. The embryological approach to human perception is thus also a phylogenetic one, in true Darwinian spirit. Nevertheless, caution has to be exercised, the data from forms with very different kinds of life and powers of perception requiring careful translation for comparison with the human condition. Human brain development, like human consciousness, is unique, the long postponement of free existence permitting the formulation of new adaptations that generate new kinds of perception after birth.

The following general texts contain essential information on the growth of the human embryo and fetus: Arey (1965), Barth (1953), and Hamilton *et al.* (1962). An excellent atlas of pictures of developmental stages by Böving (1965) is reproduced as an appendix to Falkner (1966) which also contains an excellent account of fetal brain development (Larroche, 1966).

2. Innate Strategy of Perceptual Systems

It is taken as a working hypothesis for this review that development of perceptual awareness and of effective voluntary movements depends upon differentiation of new refinements and new levels of integration within a scheme that is laid down before birth. In this process the environment exercises an increasing influence, selecting or validating specific forms of perceptual response from among alternatives that exist in the structures created *a priori* by growth.

Perceptual control of movements is held to obtain not so much by the growing together of initially separate sensory impressions or sensory motor arcs, but by reinforcing some of an initial excess of patterns of connection in a hierarchically structured control system, one which never loses unity while it generates large numbers of temporary states of action. Perception and movement control each other in innately specified "servo" loops. In more complex forms of action, conscious percepts, while highly responsive to what they find in the external world, are determined in large part by innate cognitive processes and by templates of memory, both generated in the mind. These mentally synthesized images, models (Craik, 1943), schemata (Bartlett, 1932; Neisser, 1976), or implicit hypotheses about reality (Gregory, 1970) may also set in motion, in the mature subject, very elaborate patterns of action whose perceptual determination is, at the outset, very remote (Bernstein, 1967; Greene, 1972).

Using Gibson's terms, if not his uncompromising ecological theory of perception (Gibson, 1966), we may classify the mechanisms for perception as a set of systems for finding information or for detecting invariants in the environmental arrays of stimuli, both "information" and "invariants" being defined with respect to possible modes of action of the brain through the body as an instrument (see Sperry, 1952). An analysis of levels of function and their anatomical correlates in fish and primate visual systems is given by Trevarthen (1968*a,b*). Ingle (1978), who also emphasizes the need to classify forms of behavior before asking about the development of perception, cites evidence for anatomically distinct brain systems for visual control of different acts in lower vertebrates. In more detail our model comprises the following complementary systems for active perception that apply to all modalities (see also Trevarthen, 1978*a*).

Body-Guidance System. This is activated when the body moves in locomotion. It must inform about the layout of surroundings and gain advance information on the distances, separations, gradients, densities, and resistances of surfaces (Gibson, 1966). Displacement and steering of the body requires widespread coordinated muscle activity. This activity must be jointly and simultaneously governed by dynamic information from the distance receptors of sight and hearing, assisted by olfaction, gustation, and pickup of radiant heat. Direct contact information about the "feel" of the relatively unresistant media of air and water, as well as the unyielding surfaces that stimulate touch, pressure, and temperature awareness, must then be integrated to the advance picture obtained by the distance receptors.

The picture of reality is obtained primitively and most unambiguously when the subject is moving about, but in mature consciousness layout and form can be seen in a single glance, without information from the subject's displacement. The world may also be seen in pictures. Many of the basic rules for perceiving by the static geometry and configuration of stimulus information may be innate, others must be learned. Their relation to the perception processes that guide movement is probably close, but it is not, in fact, known.

Because one can see, hear, smell, and feel the layout of surroundings richly when not moving, or moving very little, it is often concluded that stimuli which arise only when the body is actively moving, in the vestibular organs, joints, muscle sensors, and visceral mechanoreceptors, are in dominant, if largely unconscious, control of locomotion and shifts of posture. It is relatively simple to verify that this is not so (Gibson, 1952). Intracorporal information is integrated closely with "proprioceptive" (body sensing) information from the "teloceptive" (sensing at a distance) organs of sight, hearing, etc. (Lee, 1978; Lee and Lishman, 1975; Dichgans and Brandt, 1974). The definition of a proprioceptive (self-sensing) organ should depend on the relation of its receptor functions to the body as generator of movement. It is not dependent on the type of stimulus energy to which the receptor organ is sensitive. Both vision and audition function proprioceptively in regulating body posture, stationarity, and movement.

To perceive movement of one coherent self in relation to surroundings, the mechanism of central motor control for the whole body must receive equivalent and coincident images of space layout from the various modalities. A unified anatomical system of connections within the brain in which body movements are unambiguously and coherently represented is, therefore, essential to the primary perception of a space in which one behaves. Because the propulsive forces made by the body are directly related to its form and the geometry of skeletomuscular systems, the common field for perceptual control of movements must, in principle, be body-shaped or somatotopic in design. Guidance of locomotion (whole-body displacement) with respect to a reference outside the body requires localizing of that reference in a body-centered field representing possible movements. I call this field of control the "behavioral field" (Trevarthen, 1968*b*, 1972, 1974*b,c*, 1978*a,b*).

Stimulus effects produced by self-movement in stable surroundings must be distinguished from motion of external origin. This again may be solved by somatotopic representation or an equivalent ordered set of loci. The image of body-produced acts in such a representation may function like a background for detecting locus and direction of motion stimuli of outside origin. Such somatotopic representation would be involved in an efference–copy system (von Holst, 1954) and in a system designed to detect stimulus invariants from feedback to detect body displacement (Gibson, 1954, 1966).

Object-Locating System. Within the above common perceptual field, we distinguish this system for perception of objects by parts of the body that move separately or in combination against the objects. The class "object" is defined by reference to acts of use, and individual objects are distinguished by their kinds of use. Food is defined by the act of eating; manipulanda (tools) are defined by their handling properties, etc. Gibson (1979) calls the information for use "affordance." Correspondence must be established between advance detection of objects (e.g., by sight or odor) and their direct, final, or consummatory use. Obviously, the actual movement of "use" cannot construct object perceptions which are formed predictively (Metzger, 1974), but close equivalence of perception and act is necessary.

Since the movements for using objects are part of, or a subset of, the movements of locomotion, perception of objects must be a subset of perception of a space–time frame for locomotion. Objects are located in the body-centered behavior space. Since the different object-locating or orienting acts are to cooperate in apprehension of common physical causes (e.g., hands and eyes must together apprehend a thing as one object, not two, unless they are acting on separable parts of the thing), then not only space, but time also, must be unified within perception of objects. The different mechanisms for sensing objects must act in one frame of space and time together, and be transitively equivalent. There is now direct physiological evidence for spontaneously generated "command" activities in neurons of the parietal cortex of the monkey (Mountcastle *et al.,* 1975). These specify the aiming of eyes and/or hands toward an objective and prepare for the perceptual assimilation of touch information from contact of the objective with the fingers. They probably work in intimate association with sensory-motor programming functions of both midbrain and prefrontal cortex.

Since objects might lie anywhere in the common behavior field of the subject and still have the same properties (identity), whatever neural mechanism apprehends an object as separated from the background must have some power of independence of the locations in the primary field. This independence is the converse of the orienting ability which enables acts of object capture to be set up from many starting points in the primary field (Trevarthen, 1978*a*; Gross and Mishkin, 1977).

System for Identification and Analysis. This system, distinguishing thousands of kinds of objects and elements of which they are composed, depends on detection of their distinctive features. This is more taxing on the neural mechanism if objects are to be perceived at a distance, when the stimuli that reach the subject are attenuated. Also, many relevant features are only perceived when they are analytically teased out from the ensemble of the object's stimulation and from irrelevant details contributed by the background.

The critical motor process in identity determination is a *serial focalization* which involves deployment of discrete, high-resolution perception mechanisms that have narrowly restricted fields of detection but many neural elements (Adrian, 1946; Whitteridge, 1973). Scanning movements that regulate sampling in the general field of perceived space determine chains of focalized experience—fixations of the fovea of the eyes, setting of the frequency-response range of the ear, sniffing with the nose, touching with fingertips, tasting with the tongue. The movements of focalizing need to be precise and rapid. They must be integrated within patterns of body displacement or body transformation (body part on body part articulation). They must also be governed by perception of layout and configuration which operates quickly without movements, as in seeing a landscape by lightning flash. Seeing what things are requires quick assimilation of many pieces of information. This process would fail unless focalizing to get the right details were highly efficient and obedient to the contents of consciousness (Yarbus, 1967).

System for Modeling Properties of Object Existence. All percepts must have, in addition to spatiotemporal equivalence in different modalities, properties of use and identifying features, the further attributes of continuity, permanence, or specified change (Michotte, 1963). The system requires images in nerve activity, or readiness for activity, that persist beyond stimulation or that connect separate events in stimulation. The most general form of change to be perceived is that due to motion of objects relative to the perceiver or to each other and the solid surfaces of surroundings. The quality and dimensions of independent motion must be correctly detected in perception, if it is to serve for guidance and placing of adaptive action (Gibson, 1954). Objects are detected by their surface-bounded volumes and forms, and by qualities of color, texture, hardness, resonance, odor, taste, etc., which stay invariant in perception when the objects displace relative to the body and receptors, so object perception requires the operation of the principle of *constancy.* For vision the rules of constancy are: (1) the object keeps the same position if only the subject is displacing; (2) an object stays the same size when perceived at different distances; (3) an object stays the same form when rotated relative to the perceiver or when apprehended by perception piecemeal in different parts; (4) an object stays the same color in spite of changes in spectral composition of the light reflected from it, ambient light, or light reflected from surroundings. Objects may also deform, dissolve, or swell. These changes must be perceived through representations of their defining attributes in the brain. Finally objects must be perceived to change in perceptibility but not existence when they go into or behind masses that screen them. This requires that the perceiver know about the forms of surfaces and masses in space.

Movement-Perceiving System. A special class of motion is movement of other voluntary agents under control of their own muscular activity, and this would seem to require a specialized movement-perceiving system. A wide variety of commanding factors of biological relationship between the perceiver and other beings makes discrimination of movements a basic factor in perception of most animals. It requires detection of the special invariants that define animation (e.g., self-regeneration, rhythm, grouped bursts, purposeful aim). Attributes of the quality of animation are also important (e.g., strength/weakness, aggressiveness/friendliness, stealth/openness). Control over the perception of other beings involves communication (intersubjectivity; Trevarthen, 1975, 1977). Acts of the subject may regulate the acts of others, and perception of others includes perception of their expressions of communication.

Mechanisms for subject perception and for intersubjectivity have unique requirements which may develop (phylogenetically and ontogenetically) out of self-regulatory perception requirements. Both perception of self and perception of others seem to refer back to an image of the body as a system that is coherent and polar in design and activity. They are certainly also aided by other innately coded elements of perception which are effective in detecting persons (e.g., voice qualities, color, face configuration, etc.).

There is every reason to believe that the above classes of perceptuomotor strategy require mediation by different innate mechanisms in the brain, and this is the main justification for presenting them in summary here. They may have distinctive developmental histories. Some of the differing requirements for active information uptake for perception are reflected in anatomical specializations of the receptors; even at the point where stimuli are first received, they are segregated to serve in different levels of the process. Thus, in vision, the eye shows one set of mechanisms for picking up information for perception of surroundings (the rod system of the periphery, serving ambient vision) and another for focal detection of objects and their identifying features (the cone system of the fovea, serving focal vision) (Trevarthen, 1968*b*). Motor structures inside and surrounding the eye regulate the relative stimulation between these two receptor systems by varying the coherence and intensity of light in the retinal images, and by varying eye displacement relative to the sources of light patterns on the retina (Figure 1). Rotary movements of the eyes alternating with "fixations" favor focal vision; continuous translatory or linear displacements of the center of the eye, as in locomotion, favor ambient vision by causing large motion effects in the far-extrafoveal parts of the retinal image.

A similar analysis could be made of the perceptual appreciation of objects by the forelimbs; sense organs for mass and bulk are different from those for surface features, texture, temperature, and hardness, and they require different motor patterns to control them. Indeed this multiplicity of perceiving extends to all of the classical modalities, each of which has its own anatomical organ for focusing on information after it is located in the space of behavior.

Recent anatomical and physiological studies of the visual systems of the brain, together with clinical observations, refute the classical notion of a single autonomous geniculostriate projection system for visual awareness in man (Trevarthen and Sperry, 1973; Perenin and Jeannerod, 1975). Discovery of parallel projections each related to a distinct subsystem of the brain's sensory–motor apparatus changes the scope of developmental questions.

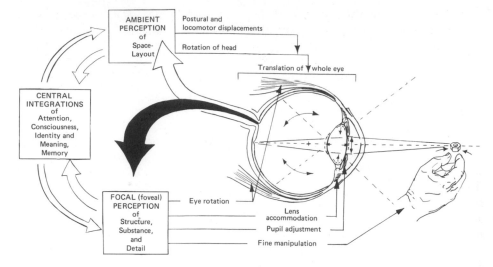

Fig. 1. The sensory–motor mechanisms of visual perception.

Not only may the different perception systems in one modality grow at different rates, but they are likely to interact with those of other modalities in the course of development in quite different ways. It is reasonable to suppose that whereas the primary space-defining general field of body action is set up at an early phase of development for all modalities jointly, focal operations of different modalities may develop at different times early in postnatal life depending on the maturation of distal motor patterns to which they are most closely allied. We may expect a series of differentiations of perceptual systems in infants in relation to their emerging patterns of action. For example, the growth of manipulation in the first year may occasion particular development in touch and vision and their association, and developments from babbling to speech would be expected to be accompanied by changes in auditory and possibly visual perception of stimuli caused by mouth movement. Writing must involve the emergence of new visuomanual and audio-manual perception devices in the brain. There is no reason to assume that such changes in perceptual differentiation and reintegration are a passive consequence of learning.

The variety of avenues for perception and the richness of their cooperative action strongly favors the view that perception involves widespread brain systems. Those parts of the brain nearest the target zones for sensory projection fibers may be assumed to be most concerned with input processing for perception, but perception cannot have any use unless its mechanisms are structured to provide for direct and versatile regulation of motor output. Advances in recent years in the comprehension of brain mechanisms of perception have come, paradoxically, from new knowledge about brain machinery for important classes of movement or "modes of action" (Trevarthen, 1978*a,b*).

In considering the development of the brain we shall attempt to bring out relationships between the program of growth of circuits and the functions outlined above. It is hoped a disciplined teleological approach will lead us to some notion of what kind of functions are present at birth in preparation for the growth of awareness in postnatal life and also to a means of distinguishing the rapidly maturing components from those with very protracted development, open to environmental influence over a large part of life's course.

3. Early Embryo

3.1. Beginnings of the Nervous System

At 2½ weeks after conception, when the human embryo is a two-layered disk less than 1 mm in diameter, the nervous system is identifiable only by its place in the bilaterally symmetrical body, the cells having the same appearance as ectoderm (Figure 2A). Surgical rearrangement studies with amphibian embryos show that neurectoderm is induced to differentiate from ectoderm by some factor from the notochord underneath it (Jacobson, 1970; Needham, 1942; Saxén and Toivonen, 1962; Waddington, 1957, 1966).

The form of the embryo body is created by ordered cell divisons and cell migrations (Gustafson and Wolpert, 1967). This seems to depend on a chemical communication system aided by close junctions at points between the walls of embryonic cells (Loewenstein, 1968). The nature of the communication is largely unknown, although biochemical and biophysical research on cell membranes and their transactions with the cytoplasm and the external media promise to provide the explanation within the next few decades (Barondes, 1970; Saxén, 1972; McMahon, 1974; McMahon and West, 1976). Biochemical experiments with social amebae, the cells of which undergo patterned movements to form bodies with definite ordered form, suggest that transmission between cells in contact controls hormone-like cyclic nucleotide molecules which regulate basic intracellular biochemistry (McMahon, 1973, 1974). The molecular or biochemical events become, in turn,

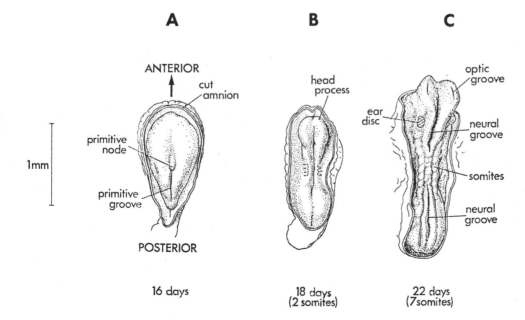

Fig. 2. Human embryos in early stages of the formation of the nervous system.

contained within patterns of cell mechanics and controlled intercellular communication (Wolpert, 1969, 1971).

Neural-plate transplant experiments with amphibian embryos show that the anteroposterior polarity of the plate is determined before the mediolateral axis, at a stage equivalent to the 2½-week-old human embryo (Roach, 1945) (Figure 2A). This order of determination of polarities remains the rule for parts of the body and nervous system through many cycles of differentiation. In the early embryo of amphibia, the organizer of cell differentiation and source of polarity for the whole body is a group of cells at the dorsal lip of the blastopore (Wolpert, 1969). These cells give rise to head mesoderm, including the anterior top of the notochord. In the embryo brain a small cluster of cells constituting the presumptive diencephalon appears to serve the same functon in marking the polarity of adjacent brain fields (Chung and Cooke, 1975).

The embryos of all vertebrates rapidly elongate when the nerve plate is formed. Midline neurectoderm cells curl by contraction of their outer (dorsal) walls to fold in a gutterlike neural groove (Figures 2B and C). At the same time, the neural crest cells at the edges of the plate multiply and rise up to form walls that turn inwards and eventually fuse. In this manner a tubular central nervous system is pinched off to lie along the middorsal line, immediately above the notochord axis of the embryo body. The neural plate is broader at the anterior end—here it forms the rudiment of the brain. Although the neural-plate cells are not visibly differentiated until later (about stage 30), they already form a mosaic of regions, each with a tendency to form a particular part of the brain (Jacobson, 1959). As the neural tube closes, mesoderm cells proliferating along the primitive groove creep outwards and forwards between ectoderm and endoderm. They form masses along either side of the notochord, and also spread forwards beyond the end of the notochord and around the place where the mouth and pharynx will form.

By three weeks the embryo has clearly demarcated head and trunk (Figure 2C). The head consists principally of brain and ectoderm, with rather diminutive pockets of muscle-forming mesoderm, but in the trunk the neural tube is flanked by large muscle plates. There is thus a gradient of increasing proportion of mesoderm to ectoderm toward the caudal end of the body. Experiments involving unsticking of embryo cells from each other, then mechanically mixing them, indicate that early neurectoderm cells are capable of growing into either brain or spinal cord. The issue is set long before nerve fibers grow, by regulatory influences from surrounding cells. If artificially combined with mixture of ectoderm and mesoderm rich in mesoderm, the neurectoderm cells differentiate into spinal structures. If in a mixture low in mesoderm, the same neurectoderm generates brain structures (Toivonen and Saxén, 1968). This, added to the evidence for generation of neural plate by influences from notochord, shows that in the early embryo the differentiating cells of the central nervous system are pulled into line with surrounding body structures. In other words, an image of the body organization is communicated to the CNS. This early event has great significance in subsequent developments by which the brain forms intimate and orderly connections with organs of the body.

3.2. Segmentation: Head and Trunk

While the neural tube is closing, cells of lateral plate mesoderm begin to aggregate into pairs of somites or muscle blocks. The chain of swellings on either side of the neural tube shows up an intrinsic division of the whole body into segments (Figures 2B and C). The original polar and bilaterally symmetric field of the body is cut by segmentation into a set of equivalent subfields (Figure 3). We do not know what the biochemical machinery is that controls the pattern formation of the body, but apparently segmentation permits it to be reiterated many times within an initially single field. Subsequent developments indicate that the segments of the nervous system form mutually compatible moduli, each impressed with a somatotopic layout, that can be grouped by formation of axonal links into equivalent subsystems, each capable of acting as an agent for the body or as one of a limited set of agents.

Where the neural tube (neurectoderm) and skin (ectoderm) join to form two separate cell layers, a row of cells separates off into the space on either side of the dorsal neural tube. These neural crest cells migrate down the sides of the neural tube to form the afferent cells of all the dorsal nerve roots, sensory cells of the auditory nerve, sympathetic ganglion cells, Schwann cell sheaths of all peripheral sensory and motor nerves, epithelial pigment cells, and probably the optic vesicles. Thus cells from the dorsal ectoderm either side of the neural plate make the greater part of the input mechanism to the brain from the viscera, the body surface, as well as the special visual, auditory, and vestibular sense organs. The olfactory sensory neurons grow from ectoderm over the anterior end of the brain. The sensory nerves from neural crest conform to the segmental pattern of the somites, making a chain of dorsal ganglia.

In the late-somite embryo the main receptor and effector suborgans of the body

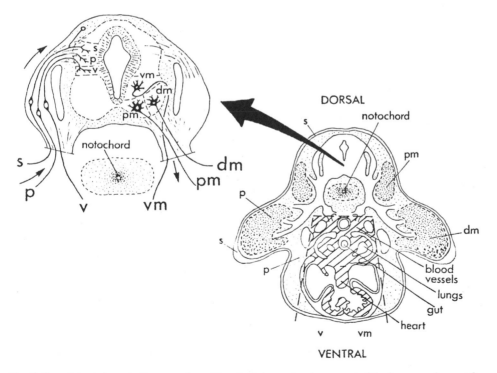

Fig. 3. Somatotopic layout of nerve cells and terminals in one trunk segment of the human embryo at five weeks. Diagrammatic sections at arm level (right), and of spinal cord (left). Arrows indicate eventual nerve conduction. s = body surface afference; p = proprioceptive afference; v = visceral afference; dm = distal motor; pm = proximal motor; vm = visceral motor.

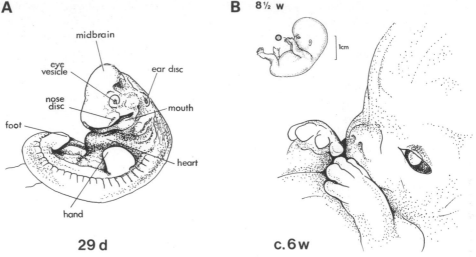

Fig. 4. Human embryos showing form of organs of perception before neuromotor activity. A: Receptor primordia in late-somite embryo. (From scanning micrograph by Lennart Nilsson.) B: Late embryo and early fetus showing coorientation of hands, eyes, and mouth. Close-up of premotor posture drawn from photograph by Lennart Nilsson.

are already making their appearance (Figure 4A). The head, representing a constellation of anterior segments with relatively small muscle primordia that are unrelated to general locomotion and in many cases specialized uniquely to control sensory information uptake, is equipped with special sense organs: nose, eyes, and ears (including vestibular organs). The mouth and tongue (formed of the appendages of anterior segments of prevertebrate ancestors) will grow into a complex organ for taking in food, while tasting and masticating it en route, and for grasping, exploring, or breaking up objects. In higher vertebrates the visceral arch musculature, as a whole, is elaborated into the muscles of the jaws and face which in man form a unique apparatus for expressive communication. The muscles of the neck, together with those turning the eyes and ears, are important in controlling sensory experience by orienting. The fore- and hindlimbs appear in the embryo as paired lobes with distal thickenings. They develop their own segmentation and in later development coopt neural sensory and motor branches from several adjacent trunk segments.

3.3. Rudimentary Sense Organs

Sense organs form from neural crest or from ectoderm next to it at the border of the neurectoderm plate. In the formation of each special receptor, for sight, olfaction, taste, or hearing, an infolding of ectoderm generates a pouch that lies against the brain. Sense cells develop in the pouch or in a cluster of neural crest cells or in a lobe of the brain that meets it. The pituitary forms in comparable manner as an ectodermal pocket which grows upwards to meet an outgrowth from the floor of the brain near the tip of the notochord. In primitive vertebrates the portion that corresponds with the anterior lobe of the hypophysis cerebri (pituitary gland) is a chemical sense organ (Sarnat and Netsky, 1974).

Receptors of the skin and of the muscles, joints, and internal organs are connected to the brain by axons that grow from spinal ganglia along the length of the body. Migration of neural crest cells to form these ganglia in a position between the ectodermal structures and central nervous system, both of which they innervate, is comparable with the formation of the optic vesicle as an outpushing from the end of the brain. Both eye and spinal ganglia will relay precise information about the locus and kind of stimulus to the brain maps by which body movements are patterned.

At 30 days the eye cup and lens, the otocyst or inner ear and an adjacent mass of neural crest that will form the vestibulocochlear ganglion, the nasal pouch and olfactory lobe of the brain, and the mouth with rudimentary tongue are all clearly identifiable. During the second 30 days, each of these structures attains a visible adaptive design with a rudimentary receptor epithelium placed in characteristic relation to the accessory structures (lens and eye chamber, semicircular canal, sacculus and cochlea of the inner ear, olfactory pouch, muscualr tongue and lips).

The distance receptors, eye and inner ear, are particularly interesting because in their geometric form they already express how they will obtain information about the shape of the outside world and the nature of external events. The geometry of eye and lens determines the optic image and how movements will transform it. The geometry of the semicircular canals determines how their cupula receptors will be excited by acceleration of the body in given directions. The sacculus is a detector oriented to regulate the correct position of the head in the gravitational field. The form of receptor mechanisms of the cochlea, which, like the central foveal structures of the retina, mature late in the fetal period, determines how the hair cells will respond differentially to sounds of different frequency aided by the pinna of the ear (Klosovskii, 1963). The shape of the turbinal bones of the nose dictates how olfactory detection may gain a space-probing capacity based on breathing (Von Békésy, 1967).

Parallel with the formation of these head organs, the perception organs of the limbs undergo closely similar changes. By the end of the first 30 days of gestation the primordia of hands and feet are clear with thickened ectoderm forming placodes in which their highly specialized receptor structures will grow (Figure 4A). The digits achieve essentially their adult appearance within the next 20 days. Thus, by the end of the second month all the specialized organs of perception have reached a close approximation to their adult forms (Figure 4B). This remarkable preparation for complex function under cerebral control takes place before the nerve fibers achieve contact between receptor cells and central nervous system, or between motoneurons of the CNS and the voluntary muscles.

3.4. Cause of Sensory Maps

The manner in which receptor links are formed between body surface and brain suggests an orderly series of inductive contacts between cell groups that are systematically rearranged by infolding or migration. When and how this specification of neuroblasts for afferent nerve circuits is achieved has been the topic of many experimental studies (Gaze, 1970; Hunt, 1975a,b; Jacobson, 1970, 1974; Meyer and Sperry, 1976). The first question concerns the programming of the primary axonal connections that grow in the late embryo. To tackle this it is necessary to make operations on receptor primordia when they are first formed, early in the embryo stage.

Most of this research has been performed on the eyes of amphibia (Figure 5).

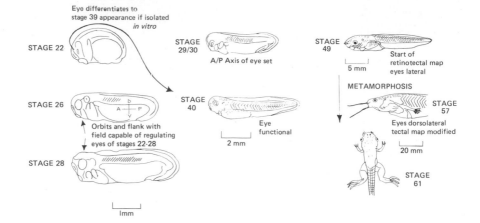

Fig. 5. *Xenopus* larvae at critical stages in the development of the visual system. (From Nieuwkoop and Faber, 1956.)

Basically, the experiment involves cutting out the eye rudiment at different stages, reimplanting it in different locations or different orientations in the embryo body, then tracing the formation of connections later in life. Earlier eye-transplant experiments (Stone, 1944; Sperry, 1945; Székely, 1957) employed orienting behavior of the adult to measure the eye–brain connections; more recent studies use direct electrophysiological mapping of optic nerve presynaptic terminals on the optic tectum of the midbrain after metamorphosis (Jacobson, 1968*a;* Gaze, 1970) and histological techniques for seeing nerve fibers and terminals (e.g., Crossland *et al.,* 1974).

If the eye vesicle of a salamander or tadpole embryo is cut out and reimplanted upside down, growth of the optic nerve to the brain leads either to normal or to partly or wholly inverted behavior in the adult. The outcome depends on the exact developmental stage at which the eye was inverted. After stage 34 in *Amblystoma* (Stone, 1960), stage 30 in *Xenopus* (Jacobson, 1968*a*), and stage 17 in *Rana pipiens* (Sharma and Hollyfield, 1974), at or just before the time the first mature neuroblasts appear at the center of the eye opposite the lens and cease to divide, each retina gains a program for specification of polarity as a unit, independent of surroundings and, as long as pieces of retina are not removed from the growing eye, this program remains to determine the directions of behavioral movements under visual guidance (Figure 5).

Jacobson (1968*b*) has shown that in the clawed frog, *Xenopus,* synthesis of DNA ceases in the neuroblast at the time it becomes somatotopically specified (stage 30), and he thought this might be related to the start of transcription of the gene code to form a protein by which the cell becomes marked as having a particular locus in the retinal array. But this is also the time at which tight permeable junctions between the embryonic retinal cells are broken and the cells become more separated and isolated biochemically (Dixon and Cronly-Dillon, 1972; Hunt, 1975*b*). More recent studies, involving culturing the eye primordium isolated from other tissues *in vitro* for several days, then reimplanting them in the body so they can form optic connections, show that axial specification starts earlier, about stage 22, when the optic vesicle contacts the ectoderm and when its cells are in tight communication, well before the first ganglion cells appear (Hunt and Jacobson,

1972, 1973a,b; Hunt, 1975a,b; Hunt and Berman, 1975). The retinal polarity is triggered in some temporary form in equilibrium with surroundings, and then an isolation process removes it from influence of cues in surrounding tissue at the time the neurons differentiate. If the eye vesicles of *Xenopus* are taken out at stage 22/23, they already have a pair of primitive axes roughly aligned with the anteroposterior (AP) and dorsoventral (DV) axes of the embryo body. Until stage 28 these axes are "replaceable" *in vivo*. They will be replaced by rotated axes within 2–6 hr if the anlagen is repositioned in the body with 180° rotation. In the 5-hr "critical period" about stage 30, the optic cup loses the capacity to undergo axial replacement first in the AP axis, then in the DV axis. However, it still retains the capacity to change one or both axes if part of the retina is cut out or if abnormal retinal combinations are made up (Berman and Hunt, 1975; Hunt and Berman, 1975).

The AP polarity of the eye is apparently fixed a few hours before the DV polarity. The sequence, the same as that seen in specification of the brain plate several stages earlier, has also been found in the specification of forelimbs and ear rudiments. That is, after polar specification of the main neural plate, individualization of special organs for perception in different modalities is followed by a second epoch of specification of axes aligned with the body. Presumably in each primordium the mechanism has much in common in the two epochs of differentiation.

Development of all vertebrate embryos follows essentially the same pattern, although the rate of development varies greatly. The stage at which the topographic axes of the retina become polarized in amphibia compares closely with that at which De Long and Coulombre (1965) and Crossland *et al.* (1974) have found that removal of retinal segments from eye cups of chick embryos results in the predicted gaps in the histologically verified projection to the optic tectum later in development. The latter study gives a more accurate picture of the process. A comparable experiment is not possible with *Xenopus* because the eye is reformed after lesions at this stage. If the operation was performed on the chick before 45 hr of incubation (stage 11 of Hamburger and Hamilton, 1951), no defect occurred. The eyes repaired themselves and generated a complete map. After 48 hr the destination of retinal quadrants on the tectum, resulting from growth and retinal axons, appears fully determined. This stage may be identified, with respect to eye development, with the human embryo at four weeks after conception (3.5 mm, Böving, 1965).

It is precisely at this stage that, in all vertebrates including man, a remarkable change is taking place in the shape of the brain (Figure 6). Known as flection, this

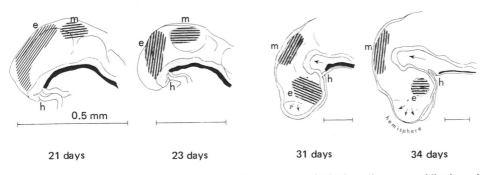

| 21 days | 23 days | 31 days | 34 days |

Fig. 6. Human brain in midsagittal section to show flection. e = retinal primordium; m = midbrain optic tectum; h = Rathke's pouch (pars anterior of hypophysis); notochord, black. (Based on Bartelmez, 1922; Bartelmez and Blout, 1954; Bartelmez and Dekaban, 1962.)

bends the front end of the brain round the tip of the notochord. It results in an approximately 180° rotation of the eye rudiment relative to the body axis and relative to the notochord. At the same time the eye changes from a rounded outpushing of the brain to a hollow spoon shape, the borders of which are closing to form the optic cup. The coincidence of flection and specification of retinal polarity makes it difficult to determine the precise causal relations between labeling of the retinal cells and their rotation as a group in the body.

Experiments with *Xenopus* show that an eye, with polarity specified after transplantation to the flank of the body, will form proper connections when the eye is transplanted without rotation to a vacant orbit. This would seem to rule out a simple interpretation of flection.

Nevertheless, the age-old problem of the inverted retinal image *is* a challenge to any nativistic theory of the visual projection. It is necessary to introduce an inverting principle to explain how the retina may be polarized within a cytochemical field common to the whole body and yet anticipate the inversion of the retinal image by optical projection of light in the eye. Either the eye map is anatomically rotated with respect to other somatotopic maps or some polarity marker (Trevarthen, 1974c), or the polarity of its somatotopic code is switched chemically. Possibly the switching occurs at the beginning of specification early in flection or experimentally every time an eye at this stage is placed in the orbital region surgically. If this is the case flection may be a concomitant or expression of the process of switching, not a cause. A diencephalic organizer found by Chung and Cooke (1975) may be responsible for repolarization. Clearly further experiments are needed on this question.

The experiments with embryo eyes of amphibia and birds prove that correspondence between points in visual space around the body and loci in the space of movements of the body are determined in general layout by some differentiation of cells in retina and tectum so they have a specified connective affinity. Whatever additional anatomical refinements must be introduced to attain the full precision of visuomotor circuits by sorting of nerve cell connections prenatally or postnatally, this basic geometric correspondence for orienting is predetermined genetically and is set up early in the formation of the embryo. Presumably the same applies to man. For correct mapping of eye to tectum, the postsynaptic tectal cells must be polarized along two axes like the ganglion cell array of the retina. Chung and Cooke (1975) have found that the source of polarity information for the tectal primordium of *Xenopus,* at least until stage 37, is in the diencephalic region near the tip of the notochord. This region regenerates if it is taken out. It may be the "organizer" of all the brain fields including the eye primordia.

It is likely that all the special perceptual systems that later come to "represent" the subject in exploration of objects and surroundings are given basic coordination with the general somatotopic field of brain and body by similar early biochemical processes. Polarization of rudimentary peripheral and central fields begins before nerve cells conduct impulses, but embryo cells are probably transmitting information actively through the epithelial sheets of body and brain. Information from eyes, ears, nose, and mouth is subsequently carried in an impulse code and mapped into one behavior field in central nerve circuits because in the middle embryo stages, corresponding to the late embryo of lower vertebrates just before the larval forms achieve active mobility, their primordia become polarized to relate to the common somatotopic map. The polarity or orientation of this map was triggered in the brain cell arrays in the early somite embryo before the neural plate closed to form the rudiment of brain and spinal cord. The formation of the first neuroblasts, immature nerve cells which divide no further, occurs in all parts of the primary nervous system at the stage when axial specificities in the optic vesicle are irreversible. This event may represent a widespread break-up of an intercellular federation of tightly connected embryonic cells. The loss of cell contacts may indeed be responsible for the observed "fixation" of polarities.

3.5. Patterning of Early Fiber Tracts in the Brain

The neural tube of a 30-day human embryo has a mitotically active ependymal layer and a surface mantle of neuroblasts. In the second half of the embryo period the neuroblasts multiply and deploy in a complex pattern of layers and clusters which clearly express a latent differentiation of types of cell and selective affinities between them. The remarkable patterning of nerve axons, when they do suddenly begin to grow out from neuroblasts and thread intricately in the brain from the 30th day onwards, appears to be a further expression of invisible affinities among the neuroblasts for chemical structure in the associated supporting cells. The mantle layer becomes covered by a mass of axons which form the marginal layer, and mantle and marginal layers are the primordia of gray and white matter in the mature central nervous system.

The pathways of growing axons demonstrate that the nerve cell projections can choose one path from among many possible ones. A first step in understanding how this may come about was obtained by Ramón y Cajal (1929), who observed a swelling at the tip of an immature nerve fiber which he called the "growth cone." Later Harrison (1907, 1910) isolated neuroblasts in tissue culture and observed activity of the growth cone and outgrowth of the axon. Subsequent research has revealed that the exceedingly fine membranous and filamentous extensions of the growth cone move actively over surfaces and may penetrate extensively between cell membranes in tissues (see Jacobson, 1970, for review). The process of filipodia formation is reversible, threads being multiplied then cut off or withdrawn to produce different net directions of progression. Growth of the axon tip thus involves an activity of the protoplasm and cell membrane similar to that seen in ameboid locomotion of protozoa. Axoplasm flows to the axon from the cell body, but there is two-way passage of material down and up the axon offering the possibility of both intrinsic direction from nucleus and cell soma to the growth tip, and back into the growth tip from media it contacts. The process offers a means of communication between growing nerve fibers and between each nerve cell and cell membrane of the medium. If the media are chemically or physically differentiated, there is a means of choice of pathway and guided growth. Alternatively, the growing nerve cells may sort each other into arrays by comparison of some property related to their locus of origin or become sorted by interaction through their effects on postsynaptic cell membranes. Self-sorting may not involve a chemical code distinguishing each cell if the axons are electrically active and their impulses are synchronized within the arrays of nerve cell bodies from which they grow. Impulse synchronization could then be used to sort them within a distant tissue (Willshaw and von der Malsburg, 1976).

The nature of the force recognized by the growth cone to lead axons selectively or to segregate axons in a population is unknown and highly controversial at the present time. That there is some selective recognition of the media in which growth occurs is proved by the criss-crossing of axon fascicles from different cell groups to make different directions of growth within one and the same medium. Fiber tracts are formed which cut across one another, as they grow, headed for quite different sites.

The directions of outgrowth of axons and the direction of movement of cells in migration may be determined intrinsically, by oriented structures of the cell wall and its contents, and by forces generated in these structures. But external factors, including hormonal agents that may penetrate widely in the tissues and cause selective effects on some cells but leave others unaffected, certainly play an important part (Jacobson, 1970).

Choice of filipodia at the growth cone is only one pattern-forming mechanism evident in early stages of development of the nerve network. A selection among alternative neural building blocks may occur between cells, those selected out dying, between branches of axons in large populations depending on their mutual interactions or on their contacts with other cells, between dendrites depending on their innervation, and between synaptic terminals once convergent populations have formed on other nerve cells or muscles. Processes of competition are essential to formation of refined patterns as in all morphogenesis (Gaze and Hope, 1976; Prestige and Willshaw, 1975; Willshaw and von der Malsburg, 1976). Loss of invisible deactivation of synapses in competition with others may, at the end of the process, be the most important means by which nerve nets are patterned and adjusted to excitatory use in later stages of development, when behavior is in active control of experience and being constantly modified by experience (Mark, 1974). All these processes depend upon communication between cells in a biochemical or biophysical language that is still unknown.

The potentiality of axon-sorting functions for ordering arrays of axons and forming refined point-to-point mapping of presynaptic onto postsynaptic sheets of cells has been emphasized by a number of authors who cast doubt on the capacity of Sperry's chemispecificity principle to carry full responsibility for the staggering precision as well as variety of nerve connections (Chung *et al.,* 1974; Székely, 1974; Prestige and Willshaw, 1975; Gaze and Hope, 1976; Hope *et al.,* 1976).

Willshaw and von der Malsburg (1976) have developed a model which, in theory, is capable of sorting the projections from one field of neurons to another with unlimited precision. It depends on an impulse synchronization according to neighborhoods in a cell array. Adjacent cells are coupled electrically, and the synchrony of activity decreases with their distance apart. Limited zones of interaction between source cells are determined with the aid of inhibition. Terminals spreading and intermingling in the target region of the CNS are reinforced or eliminated according to the synchronization of impulses in a competition that results from either presynaptic or postsynaptic interaction. Creeping or "jostling" of projections in the target region permits a continuous improvement in the accuracy of mapping until equilibrium when all sites are "filled."

The impulse patterns would presumably be effective in determining cell contacts because they result in neurotransmitters or other neurosecretions being released locally in controlled concentrations. But this model is quite different from Sperry's chemospecificity hypothesis, in detail if not in outcome, because high precision of mappings according to neighborhoods may be obtained with only a few prior cytochemical markers to determine the polarity and main layout of regions. Cells do not have to be cytochemically labeled as individuals to be given unique loci in the array. Willshaw and von der Malsburg (1976) call the polarity marking, magnification controls, etc., "boundary conditions," and they leave these fundamental morphogenetic factors unaccounted for.

Impulse coding of position according to neighborhoods is presumed to be effective in absence of environmental stimulation because nerve circuits show spontaneous electrical activity and they are formed prenatally in great refinement.

The nerve cells may be electrically active and transmitting discrete impulses in the early stages of their growth. However, the principle can obviously be extended to explain formation of additional refinements of circuitry due to synchronization of nerve cell impulse activity (or its neurosecretory products) by stimulus arrays patterned in the environment. Examples of nerve circuits which appear to be given structure by stimulation at critical times in their growth are discussed later.

Many embryonic pattern-forming processes continue to be relevant in later life of the nervous system. A large number of different mechanisms for regulating growth and maintenance of nerve networks in more mature brains have been brought to light by recent research on the effects of brain lesions and the nature of functional recovery (Guth, 1975; Jacobson, 1970; Le Vere, 1975; Lund *et al.,* 1973; Lynch *et al.,* 1973; Raisman and Field, 1973; Schneider, 1973; Schneider and Jhaveri, 1974).

4. Late Embryos and Lower Vertebrates

4.1. Growth of Optic Connections to the Brain

In most lower vertebrates the principal visual projection is a completely crossed one, the nerve fibers passing across the retina, down the optic nerve, through the chiasma to the other side of the brain, and thence in an orderly array over the surface of the optic tectum by way of two branches of each optic tract (Scalia and Fite, 1974) (Figure 7). Fibers also grow from each eye to nuclei in the

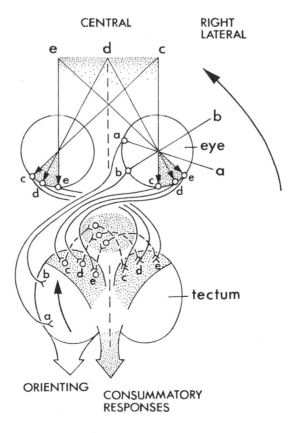

Fig. 7. Mapping of body-centered visual field on the optic tectum of vertebrates.

contralateral diencephalon and mesencephalic tegmentum. Each tectal fiber enters at a specific point to form connections at a specific depth among outer tectal cell layers. The pioneer fibers grow from the first formed ganglion cells at the center of the eye. In the human eye this occurs on about day 35 of gestation, when the embryo is 12 mm in length. Within hours a fascicle of fibers enters the optic chiasm from each optic nerve (Mann, 1964).

The precise timing of events in development of the eye and tectum and the march of projections between them has been intensively studied for *Xenopus* (Nieuwkoop and Faber, 1956; Jacobson, 1968*a,b;* Gaze and Keating, 1972; Chung *et al.,* 1974; Gaze *et al.,* 1974; Hunt, 1975*a*). At this stage, when visual axons reach the anterior end of the tectum, the number of rods and cones in the retina approximates that in the adult eye and the cell laminae of the tectum are well developed. The projection over the tectum is still growing along the posteromedial border when the tectum has completed histological differentiation. At stage 57, when the projection is near complete, the larvae escape from visual stimuli appearing above them by swimming forwards, and an optokinetic bending of the tail may be clearly observed (Mark and Feldman, 1972). Comparable growth of retinotectal connections has been traced in the chick (Cowan *et al.,* 1968; De Long and Coulombre, 1965; Crossland *et al.,* 1975).

Sperry and Hibbard (1968) have noted the extraordinary similarity in size and cell content of the retinal and tectal layers in the tadpole of *Rana,* the tectum being slightly smaller in area than the retina. Presumably this reflects a common developmental plan. Through the late embryo and larval stages of anurans and into the adult stages of fish and urodeles, both retina and optic tectum add new cells and new connections.

Formation of retinotectal connections is delayed to fetal or even postnatal stages in mammals. In the mouse almost all the neurons of the tectum originate in the 11th through 13th days of gestation, just before the optic nerve fibers reach the colliculus (De Long and Sidman, 1962). Schneider (1973) has demonstrated extensive plastic reorganization of retinocentral connections in the neonatal hamster after surgery. The hamster is born at a stage when retinotectal connections are still forming.

Gaze (1970) estimates that *Xenopus* has about 500 cells in the retina when it is specified in independence of surroundings at stage 30, and 100 times more in the adult. In *Rana* there is probably a 1000-fold increase, related to the superior visual function of this species. It is now known that retinal polarity is specified much earlier, near stage 22, and this means that a few or one precursor cell contains the trigger for specification before ganglion cells appear (Hunt, 1975*b*). Subsequently formed cells must have the retinotopic code handed on to them. Virtually all of the adult amphibian retina is formed by mitosis at the ciliary margin of the eye (Glucksman, 1940; Stone, 1959; Jacobson, 1968*b;* Straznicky and Gaze, 1971; Hollyfield, 1971), and chick and mouse eyes grow similarly (Crossland *et al.,* 1975; Sidman, 1961).

While the eye grows radially, ganglion cells being added in rings from the center outwards throughout larval and early juvenile growth, the tectum grows asymmetrically along the posterior and medial borders or "neck" (Straznicky and Gaze, 1972; Crossland *et al.,* 1975). Observation of this discrepancy in *Xenopus* has led Gaze and Keating (1972) to question the notion of a chemical encoding of affinity between ganglion cells and tectal cells as proposed by Sperry (1944, 1945, 1963, 1965). Ganglion cell terminals would appear to have to slide across a gradient in the tectum as it matures. However, the discrepancy between eye and tectum growth may disappear if allowance is made for the fact that the somatotopic axis about which loci are defined is off center in the eye. Straight ahead in the behavioral field (in front of the nose, down the sagittal axis) corresponds with the far-temporal retina. Growth of the eye about this location would appear to parallel growth of the optic tectum about its anterolateral pole from which optic fibers disperse laterally and posteriorly. Jacobson (1976) has shown that the eye of *Xenopus* does not grow uniformly in all directions, and he concludes that there is no need to postulate sliding connections on the tectum.

Crossland *et al.* (1975) have shown that, in the chick, successively established optic terminals occupy bands around the map locus that corresponds to the center of the retina, where ganglion cells first develop. They consider their findings to give strong support to the hypothesis that retinotectal connections form according to a predetermined chemoaffinity mechanism.

The first retinotectal map of *Xenopus* is diffuse and crowded in the anterior half of the available surface (Gaze and Keating, 1972; Chung *et al.*, 1974; Gaze *et al.*, 1974). The map determined electrophysiologically grows progressively more detailed and accurate after terminals have spread over the whole tectum. It is likely that this process involves an intense selection among an excess of terminal branches and among synaptic endings on tectal cells (Gaze and Hope, 1976; Prestige and Willshaw, 1975). It may also reflect progressive development of the specificities of tectal cells. An alternative hypothesis is that the correct retinal axon endings sorted within the tectum by reinforcement of synchronized impulses that encode neighborhoods in the retinal array. This is the hypothesis of Willshaw and von der Malsburg, discussed earlier (Willshaw and von der Malsburg, 1976). It is compatible with what is known of the movements of nerve cell processes during patterning of retinotectal links in *Xenopus* and in other forms.

Death of cells in retina and tectum may also eliminate misfit elements (Glucksman, 1940, 1965), but this seems to become more significant later when the projection is resculptured in metamorphosis.

Studies of the adult retinotectal system of fish and frogs discussed below show that the end product of tectal differentiation and optic innervation is refined in pattern and inflexible. This process must depend, in part, on intrinsic differentiation among tectal cells to define not only somatotopic polarities and a retinal projection density for each region, but the connective affinities of different species of nerve cell at a given locus (Maturana *et al.,* 1959; Sperry, 1951*a,b;* 1965).

4.2. Experiments on the Retinotectal Projection of the Embryo

If pieces of retina are cut out of the eyes of chick embryos before the tectum is innervated, at days four and five of incubation, the tectal projection manifests corresponding holes, observed by histological methods, in the finished projection at 12 days (De Long and Coulombre, 1965; Crossland *et al.,* 1974) (Figures 8 and 9). The same result is described below with regeneration of connections in adult goldfish (Attardi and Sperry, 1963; Meyer and Sperry, 1976). In both cases fibers made visible with silver stain or autoradiographic methods, either newly formed or newly regenerated, grew selectively to locate the appropriate tectal region.

Experiments like the above may be interpreted as showing that the retinal axons may find their way faultlessly to the one site at which they may form terminals, avoiding other vacant sites. Other preparations have apparently proved the opposite: that the early part of the process is unselective and that active competition is required to pick out the correct order in an extensive gradient of possibilities for termination of each axon (Keating, 1976). The real process seems to combine both mechanisms of sorting or regulation.

Embryo eyes may be deprived of half of their substance, then closed up to make single vesicles (Gaze, 1970; Berman and Hunt, 1975; Hunt and Berman, 1975). Alternatively double nasal or double temporal eyes carrying duplicate front or back halves of retina only may be constructed. Such eyes reform apparently normal globes, then grow neural projections which fill up the whole tectum surface in spite of apparently lacking half the population of retinal loci. This has been interpreted to show that each tectal cell has the potentiality to receive connections from cells with a wide range of retinal locus specificities (Gaze *et al.,* 1963, 1965; Strazniky *et al.,* 1971; Gaze and Keating, 1972). Such a conclusion depends on the assumption that the cells of the artificially produced eyes remain as they were at operation, and that they will represent half-eyes.

By an ingenious technique of double innervation, pairing experimental and normal eyes in duplicate innervation of the tectum, Hunt and Berman (1975) have apparently shown that this is not the case. The normal eye serves as a probe to test the specific affinities of the tectal cells. In every case the normal and experimental eyes superpose or blend their projections in register, and the map of the normal eye is normal in size and distribution over the tectal surface. It is suggested that the operated eyes regulate or undergo reformulation of the morphogenetic polarities of their cell arrays to form the retinal maps of whole eyes or double-mirror whole eyes,

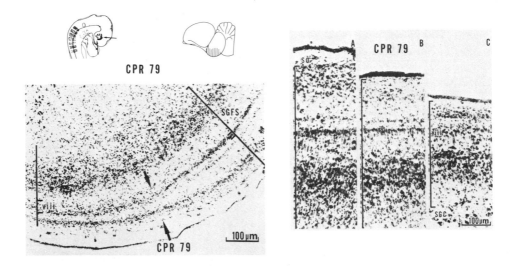

Fig. 8. Development of retinotectal projection in the chick after ablation of caudal half of the right optic cup at stage 17 (Hamburger and Hamilton, 1951), early on the third day of incubation. Chick sacrificed at the end of incubation (day 20). Retinal axons grow over the cross-hatched region (corresponding to the ablated retina and nasal visual field) and terminate on the dorsal and posterior tectum. Lower left: Photomicrograph through junction of innervated and noninnervated left optic tectum. Laminae disappear from the noninnervated stratum griseum et fibrosum superficiale (SGFS) which is reduced in thickness (to left of arrows). Right: A, normal control right tectum; B, innervated part of left tectum; C, noninnervated left tectum. The SGFS of B is reduced in thickness but has normal lamination. C lacks laminae a–f and has appearance of a tectum totally denervated by enucleation of the eye at day 3 of incubation. (From Crossland *et al.,* 1974.)

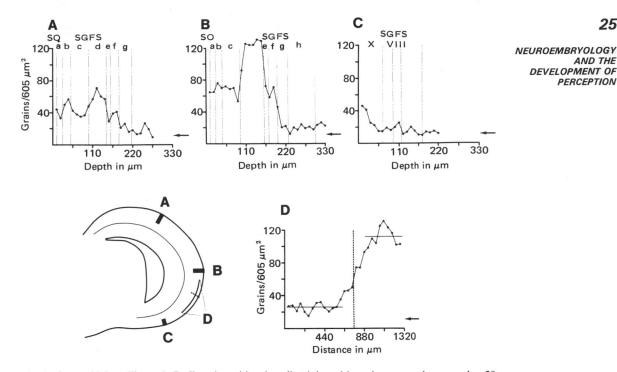

Fig. 9. Same chick as Figure 8. Radioactive tritiated proline injected into the operated eye on day 20. Embryo sacrificed 6 hr later. Radioactive protein was transported down the optic nerve fibers to the tectum. Grain density plots from the autoradiographs show distribution of labeled terminals in the innervated (A and B) and noninnervated areas (C) as well as transversely across the junction region (D). (From Crossland *et al.*, 1974.)

and these map accordingly onto the tectum. This type of regulation was also shown by Hunt and Jacobson (1973*a*). The appearance of the eye from outside is misleading as to the differentiation of the neural retina within it after mutilation. It is possible that, in these experiments, no tectal cells undergo change in locus specification. The findings do not challenge the conclusion of early regeneration studies (Sperry, 1965; Meyer and Sperry, 1976) that orderly retinotectal projection depends upon chemical tags that prescribe the formation of links between retina and tectal cells.

The mechanism by which surgically halved eyes are regulated to make a whole retina is not known, but similar phenomena occur when like surgical interference is made with the formation of the whole body at early stages of embryogenesis, and in developing limb and sense-organ primordia of insects, amphibia, and birds (Garcia-Bellido, 1972; Harrison, 1921; Spemann, 1938; Weiss, 1939; Wolpert, 1971).

4.3. Metamorphosis of the Amphibian Visual Projection: Influences from Stimuli

In seeing tadpole stages of *Xenopus* (stages 50–57), the eyes face laterally, and the most important visual response is forward flight triggered by a shadow entering the field from any direction. The highest sensitivity is in the posterior dorsal field. After metamorphosis (stages 60–65), the eyes face obliquely upwards and, the animal being an active swimmer using its limbs, the most significant part of the field is presumably anterior (Figure 5). There are correlated changes in eye shape (Jacobson, 1976). Changes observed in the tectal map (Gaze and Keating, 1972) would appear to fit these morphological adjustments in the eye and the functional changes in visual response: change in way of life from filter-feeding and tail swimming tadpole into legged adult must be associated with change in vision. A transformation in color vision of frogs at metamorphosis was found by Muntz (1962*a,b*), who recorded corresponding change in neurophysiological responses of the dicephalon to color stimulation. Pomeranz (1972; Pomeranz and Chung, 1970) has correlated ganglion cell dendritic tree anatomy and visual physiology in tadpoles and adults of frogs (*Rana*) showing that there are adaptive changes at metamorphosis in both retinal visual detectors and retinotectal connections.

How the embryogenic specification of connections between retina and tectum is made over in metamorphosis is not known, but presumably it involves loss of some components and growth of others, as in insect metamorphosis. It has been shown that the anatomy of larva and adult in a metamorphosing insect have somewhat independent genetic determination (Wigglesworth, 1954; Benzer, 1973). Within the spinal sensory–motor system of the frog and in the distal parts of the limbs a major contribution to remodeling at metamorphosis comes from programmed and selective cell death (Saunders and Fallon, 1966; Hughes, 1968; Prestige, 1970). Waves of cell death have been observed in the retina of developing eyes of *Rana* (Glucksman, 1940). Cell death is regulated by thyroid hormone (Kollros, 1968), and Levi-Montalcini (1966) has discovered a powerful active protein, nerve growth factor, which, released from an unknown source, determines the survival and growth of sympathetic ganglia in chick and mouse (see Jacobson, 1970, for a review of hormonal regulation of differentiation in the developing nervous system).

Metamorphic change in eye position of frogs enlarges the binocular visual field. During metamorphosis of *Xenopus* (stages 45–66) this change is correlated with development of a bridge of nerve fibers which projects information from the anteromedial part of each tectum via one or two synapses through the postoptic commissure to the other side of the brain (Beazley *et al.,* 1972; Keating and Gaze, 1970). This creates an ipsilateral projection (each eye projecting to the tectal lobe on the same side) which is added to the completely crossed primary projection. The intertectal pathway terminates in register with the preestablished map of the central (anterior) visual field on the other side. An accurate binocular convergence of visual information is created by which stimuli from one locus in the binocular visual field may produce excitation from both eyes at two points, one in each tectum, and a duplicate representation of the central binocular field results (Gaze and Jacobson, 1962).

Electrophysiological mapping experiments combined with surgery and manipulation of visual experience in one eye show that convergence of crossed and ipsilateral (intertectal) projections is achieved in *Xenopus* by plastic adjustment of the latter intertectal terminals under influence of visual stimulation (Gaze *et al.,* 1970; Beazley *et al.,* 1972; Keating, 1976). This central visual function is not based entirely on preprogrammed or self-regulated growth of nerve circuits.

The anatomical structure, on which binocular vision in the central area depends, appeared in these experiments to involve formation of a memory-like record of associated stimulus impressions. However, more recent evidence shows that in *Xenopus* the intertectal pathway is, after all, partly preformed. Spatiotemporal patterning of excitation by stimuli causes a further step in binocular registration between visual inputs to "calibrate" the maps (Beazley, 1975*b*). In normal development the ipsilateral projection is at first retinotopic, but it is diffuse in its distribution on the tectum. Later it becomes more precise. If one eye is removed in the embryo the intertectal projection from the remaining eye is again retinotopic but unrefined (Gaze *et al.*, 1970). The same results are obtained if the animals are reared through metamorphosis in the dark (Feldman *et al.*, 1971). The weakened, imprecise projection could result from partial atrophy of unsupported terminals. Jacobson and Hirsch (1973) found that the map to the empty ipsilateral tectum from one eye after the other is removed may be precisely patterned in the more visually developed frog *Rana*. In this species both contralateral and ipsilateral projections appear to be completely sorted without patterned stimulation.

The critical results follow inversion of one eye before growth of the ipsilateral projection, but after permanent polarization of the retina (Gaze *et al.*, 1970; Beazley, 1975*b*). This experiment confirms that exposure to light may cause formation of a correctly registered binocular map on the tectum ipsilateral to the inverted eye (contralaterally transplanted from one embryo to another) and an inverted binocular map on the other side. This would appear to prove that further sorting of connections may occur among the terminals of one intertectal pathway, even to the extent of inverting the natural tendencies for nerve connection. Possibly the neurons involved belong to a later generation of cells that are fully segregated only after they make functional contact with active neurons. They remain plastic, like the eye of *Xenopus* was at an earlier stage. There is, however, no evidence to support a similar conclusion for *Rana*, which appears to develop the binocular projection according to intrinsic coding of ganglion cells, intertectal relay cells, and tectal cells and dendrites (Hirsch and Jacobson, 1973; Jacobson and Hirsch, 1973; Scarf and Jacobson, 1974).

For functional interpretation of the binocular projections, it is important to remember that binocular stereopsis is not the only way visual circuits may measure distance. Monocular kinetic stereopsis, by motion parallax effects, is almost certainly more primitive. In swimming forms it is a reliable source of information for guidance of forward progression (Trevarthen, 1968*a*). Ingle (1976*b*) has demonstrated that frogs may have depth detection with cues available to only one eye. Evidently this is done by accommodation of the lens to vary focus.

Detection of looming and sheering effects in the field involves cointegration of stimuli at separate retinal points, and the effects must be measured in time and space. In kinetic depth perception and perception of causal relations in events, a given retinal locus may participate in many different arrays of excitation that signify different space information or different types of event. Any morphogenetic specification of nerve circuits for this kind of perceptual function, comparable, say, with the innate motion and event detectors of the static, predatory frog (Lettvin *et al.*, 1959; Ingle, 1976*a*) would be ready for adapting in evolution to specify depth detection by binocular stereopsis.

4.4. Regenerated Afferent Connections in Adult Fish and Amphibia

Whatever the influence of receptor excitation by patterned stimulation on the segregation of nerve connections in development of embryos or in metamorphosis, in the adults of lower vertebrates the primary projection from retina to optic tectum may end up with its many neural elements precisely specified so that functional connections may regenerate from eye to brain according to a fixed somatotopic code that is unresponsive to environmental stimulation. This was shown by the classical experiments of Sperry in the 1940s with frogs, salamanders, and fish (Sperry, 1943, 1944, 1945). If the eyes of these animals are rotated in the orbit, maladaptive prey-catching or swimming movements persist for months, showing no signs of the influence of learning (Figure 10). If the optic nerves from the inverted eyes are cut, new retinotectal connections develop and these reconstitute precisely mapped, but inverted and therefore totally maladaptive, responses. Regrown optic axons establish accurately selective connections on which not only orienting responses, but also color discrimination and detection of distinctive invariants in size, form, and motion of stimuli depend. They do so even if forced to wander in a tangled mess through an interval of nerve in which the orderly array of nerve sheaths is destroyed, or if routed into wrong branches of the optic nerve. To attain the place and type of cell for which they are specified the axons will make long detours through inappropriate territories of the tectum in which cells have been made vacant by denervation. The terminal branches of axons penetrate the tectum to end on the right dendrites in the optic layer and so rebuild specialized responses to different patterns of stimuli (Maturana et al., 1959).

The results of a large number of studies in which afferent nerves, central interneuronal tracts, and motor nerves were caused to regenerate proved that both cells of origin and the cells on which synapses are formed must either be specified in a cytochemical or molecular code capable of sorting them out in large numbers (Sperry, 1951a,b) or be roughly sorted in this way but precisely ordered by a self-sorting mechanism such as that proposed by Willshaw and von der Malsburg (1976). Anatomical studies or regenerated fibers suggest that the growing axon tips are influenced by attractive forces at points along their course (Attardi and Sperry, 1963; Sperry, 1963) (Figures 11B and C). However, it is not possible by these techniques alone to determine exactly how selection takes place or the limits of its precision. Electrophysiological studies suggest that fine branches of growing nerve fibers spread widely at first to create a diffuse projection (Gaze, 1970). They may sort out later to form the highly patterned pathways which are made visible by staining. On the other hand, selective Nauta staining of terminal arborizations (Roth, 1974) and autoradiographic studies (Meyer and Sperry, 1976) with goldfish and amphibia tend to confirm that the axon of a given ganglion cell may make route errors, but eventually sends terminals to only one small site in the tectum.

There are clearly important species differences; the young adult goldfish and the salamanders used in research are still undergoing growth of eye and possibly tectum, too. They may therefore respond to surgery by regulation that permits some plasticity of connections. Adult tree frogs show more refined specificity of connections which are less plastic. Attempts to resolve this question have involved removal of parts of the retina or parts of the tectum before regeneration (Gaze and Sharma, 1970; Yoon, 1971; Sharma, 1972; Gaze, 1974; Keating, 1976; Meyer and Sperry, 1976; Ingle and Dudek, 1977). In some instances, axons of part of the retina appear to become gradually expanded, spreading out over vacated territory to form a map with correctly ordered loci, or the whole retina may regenerate a compressed projection to an available half tectum (Figure 11D). These results clearly rule out a rigid one-to-one chemospecificity principle in favor of a polarized field of specificities, as in a magnet, and a competitive "fiber-sorting" process of some kind (Cook and Horder, 1974; Gaze, 1974; Gaze and Hope, 1976; Hope et al., 1976; Willshaw and von der Malsburg, 1976; Keating, 1976).

Compression of the whole retinal projection to a half tectum is reported under

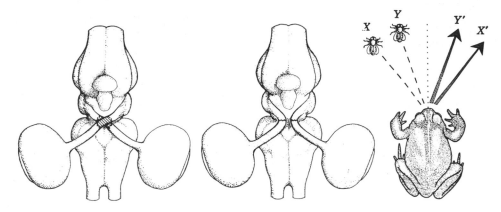

Fig. 10. Sperry's drawing of the behavioral consequences of uncrossing the optic nerves in the toad and regrowth of axons to the wrong sides of the brain. Maladaptive orienting responses prove mirror retinotopic segregation of regenerated terminals.

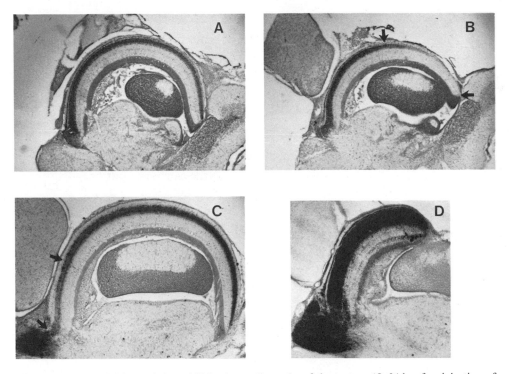

Fig. 11. Visual projections of the goldfish. Autoradiographs of the tectum 18–24 hr after injection of radioactive amino acid tritiated proline into the contralateral eye. Sections, stained in cresyl violet, in rostrocaudal plane through center of tectum. A: Normal young adult. B: Nasal half of retina ablated. Optic fibers regenerated through optic nerve crush to disorder them. Sacrificed 65 days later allowing time for terminals to reach caudal end of tectum. Terminals are sparse or absent in this region (between arrows). Plexiform layer (SGFS) collapsed. C: Temporal half of retina removed. Postoperative survival 33 days. Some aberrant endings have formed in the rostral region (between arrows), and these increase with longer survival times. This suggests a rostrocaudal gradient of affinity for all retinal fibers overlaying somatotopic positional information. D: Compressed projection at 58 days after removal of the caudal 60% of tectum and optic nerve crush. The termination layer is greatly thickened toward the caudal stump. Compare normal gradient of density in A. (From data of Dr. R. L. Meyer.)

conditions in which it is not likely that the tectal cells have died and regenerated. Yoon (1972) had made reversible compression of input to a half tectum by inserting a barrier of gelfilm across the path of regenerating axons. Later when the gelfilm becomes absorbed, the projection expands to fill the normal whole territory. Apparently expansion and compression of the retinotectal projection may persist when plastic adjustment to reversal of polarity is not possible (Sharma and Gaze, 1971; Yoon, 1973; Meyer and Sperry, 1976).

Numerous experiments with goldfish (reviewed by Gaze, 1974; Meyer and Sperry, 1976) show that regeneration of abnormal connections to tecta of fish and amphibia obey basic somatotopic specificities, but also show adjustments due to dynamic interaction between competing elements during growth. When a whole retina sends axons into a tectum reduced by about 50% by surgical ablation, the map of synapses that forms immediately is guided by somatotopic principles so that only the appropriate half of the retina connects to the reduced tectum. The fibers corresponding to the ablated tectal territory are left to crowd densely at the cut border where they probably do not form synapses (Figure 11D). A subsequent adjustment process causes the first-formed map to be gradually pushed back, the crowded fibers without proper "home" creating a competitive pressure which causes them to progressively roll back the correctly located competitors (Figure 12). The synaptic array is reorganized by a process which involves axon–axon or intersynaptic competition. After several months a topographically ordered but compressed projection is achieved, and the fish recovers a visual field without scotoma. Scott (1975) has shown that this field functions visually to permit acquisition of a conditioned respiratory response to color stimuli. The function of the field in directing orienting movements in not known. Yoon's decompression experiment (Yoon, 1972) shows that the compressed map was in an unstable condition. Observation of the behavior of regenerating axons when negotiating enforced detours or bypassing obstacles in their path suggests that an inertial growth force is guided by information in the medium or among the population of fibers, different axons reacting differently at a given site in the array (Sperry, 1965; Sperry and Hibbard, 1968). Fibers also show a strong growth preference to return to the correct laminae when they are deflected on traversing an incision scar in the tectum (Meyer and Sperry, 1976). Ingle and Dudek (1977) observed laterally displaced connections after incisions were made in the tectal afferent fibers of *Rana pipiens*.

Meyer and Sperry (1973) investigated regeneration of retinotectal connections in adult tree frogs with tectal lesions. Both behavioral and electrophysiological mapping indicated that in these animals the locus specifications are less flexible than in the goldfish. Tectal ablation causes appropriate scotomata in the visual field, and these last unchanged for at least nine months of experience of moving about and catching prey visually. Evidently the more advanced species develops more narrowly specified visuomotor circuits.

The plasticity of the retinotectal projection brought out by experiments with whole eyes growing into part tecta has been recently explored further by means of translocations, rotations, and inversions of pieces of tectum in goldfish (Yoon, 1975a) and *Xenopus* (Jacobson and Levine, 1975). The electrophysiological maps show that regenerating retinal fibers frequently terminate according to the original polarity of the rearranged tectal pieces, in spite of the discrepancy this creates in the overall map. Clearly the fibers are guided to terminate selectively with respect to polar markers in the tectal tissue. A fiber-sorting hypothesis with provision for

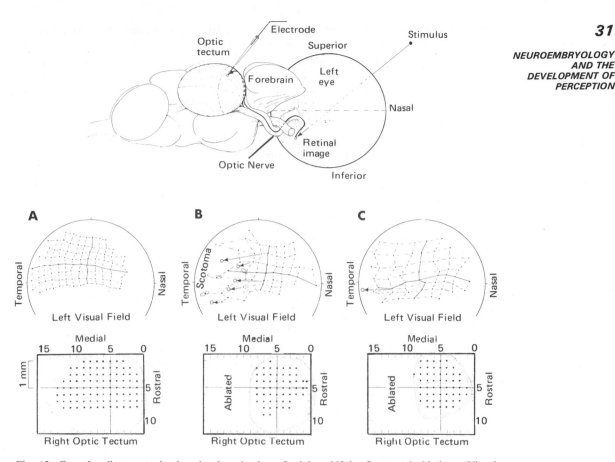

Fig. 12. Growth adjustments in the visual projection of adult goldfish after tectal ablations. Visual field plots of receptor field centers corresponding to a grid of loci on the dorsal surface of the right optic tectum. Black spots (1°–3°) presented in a standard perimeter surrounding a water-filled plastic hemisphere which reproduces the optics of the eye under water. Terminals of optic axons recorded with a glass-coated platinum–iridium microelectrode. The perimeter is centered on the optic disc. A: Normal tectum. B: Early stage of compression of regenerated tectal map at 94 days after removal of the posterior half of the tectum. Optic nerve not crushed and therefore the compression is inhomogeneous. Several loci in the temporal field have diffuse or duplicate termination reflecting development of new connections in the tectum. C: Late stage of compression of regenerated projection. Posterior half of tectum removed 214 days prior to recording. Nearly the whole superior visual field is mapped in the reduced tectal area. (From data of Dr. R. L. Meyer.)

cytochemical encoding of polarity could explain this result, but there are no conclusive experiments (Gaze and Hope, 1976; Hope *et al.*, 1976).

Plasticity observed in formation of functional connections between the skin and spinal circuits of amphibia suggests that there is a flow of specification from skin to nervous system in the direction of nerve conduction (Miner, 1956). This, in turn would favor the idea that the process of circuit selection involves transport of material down the axon by a mechanism that is linked to discharge of the action potential in the nerve membrane. Regenerating dorsal spinal roots of frogs appear to form connections widely in the skin without selection of locus or receptor type (Jacobson and Baker, 1969; Johnston *et al.*, 1975). Recovery of function after surgical disarrangement may result in this case from reorganization of the effective junctions the dorsal root cells make with spinal interneurons (Miner, 1956; Baker, 1972). According to this hypothesis, place in the skin is chemically coded and contact with the dendrites of the afferent neurons provides a means for transporting the coded label into the cord and influencing the axon of that cell to make the right connections centrally (Mark, 1975).

The above studies of epigenetic processes in visual and tactile projection pathways do not resolve the nature–nurture controversy, but they open up a new perspective on the pattern-forming process of developing nerve circuits, one in which a balance of preferences in the differentiated cytochemical and biophysical environment, created jointly by the target tissue and invading axons the way a magnetic field of force is created jointly by magnet and iron filings, determines how functional synapses will be ordered among cells and dendrites. In principle, the process may be compared to other morphogenetic processes which, by regulating interactions of element, create complex adaptive machinery in cells and tissues. But it should be remembered that the special properties of neurons allow a completely different complexity and rapidity of adjustment between intercommunicating growth elements that may be far apart in the body yet intimately affect one another. The refined circuits of the nerve net certainly use the special power of nerve impulse conduction in their processes of differentiation and in the transmission of growth response to stimuli.

The power of the growing brain circuits to regenerate and adjust in the face of drastically changed conditions is an aspect of their capacity to generate elaborate integrative functions preadaptively when they are left to grow normally. Lower vertebrates, as well as mammals, show adjustment of the preestablished nerve circuits at critical times in their growth. By this adjustment they acquire more finely adaptive perception processes with the aid of stimulus regularities from the environment (Ingle, 1978).

Mark (1974) presents an attractive thesis that all of the various instances of plasticity in formation of adaptive circuits after surgical alteration in afferent, interneuronal, and efferent links may be explained in terms of competitive selection among synapses in an initially overdispersed or overconnected network. He concludes that the genetically determined chemical labeling somehow makes possible strengthening of elements that match up functionally, and suppression or elimination of less-well-matched ones. However, the experiments cited above and others (e.g., Schneider, 1973; Hubel *et al.,* 1976) show that dynamic reorganizations, including expansions, compressions, and decompressions of projection systems, can be mediated by interaction of elements that have not yet formed synapses. Mark further proposes that an invisible deactivation of permanent synapses may be responsible for the formation of memory traces in mature brain circuits. Unfortunately, the concept of competitive synaptic inactivation without loss of synapses has not received support from recent experiments with neuromuscular preparations of lower vertebrates which it was originally proposed to explain (Scott, 1975). On the other hand, Duffey *et al.* (1976) have found evidence that suppression of vision in an unused eye of a kitten may, indeed, be due to inhibitory suppression rather than loss of connections. They reversed this inhibition to reevoke cortical responses for the unused eye by administering bicuculline with blocks inhibition by GABA (γ-aminobutyric acid). These are lively issues in current research directed to elucidate the mechanisms of pattern formation in embryonic nerve circuits and the mechanisms of memory.

5.1. Lower and Higher Vertebrate Brains

The nervous system that controls behavior of adult lower vertebrates is matched in reptiles, birds, and mammals by the system present at the end of the embryo period. The first motor, sensory, and interneuronal connections in man at this stage (Windle, 1970) may easily be compared with the brain described for the tiger salamander by Herrick (1948; Trevarthen, 1968a). Formation of this basic nerve network in a human embryo in advance of any receptor excitation suggests that the principle of a body-centered neural field for perceptual guidance of whole-body action may be retained in the growth process as a scaffold or foundation on which the dominating higher mental processes and more complex patterns of action will be formed. The latter are prepared in the fetal period by growth of additional brain structures in an unstimulating, highly controlled intrauterine environment.

Much research on brain growth makes the assumption that, in the formation of the circuits of the cerebral hemispheres and other structures that came to overmaster the primitive reflex system of the brain and spinal cord, basically the same principles apply as those which regulate nerve circuit growth in fish, salamanders, and frogs. This is supported by the universal mapping of central nuclei, tracts, and cortical fields in linked somatotopic arrays, like those of the primitive visual system. Evolutionary changes in vertebrate brains also support the concept of a basic common design which is elaborated without loss of its original layout (Sarnat and Netsky, 1974). It is likely, however, that new morphogenetic mechanisms have been evolved that are adapted to integrate brain growth to fit a course of experience.

Coghill (1929) proposed that the total pattern of behavior which integrates swimming, orienting, and feeling reactions of the salamander becomes "differentiated" into the separate mechanisms of more advanced behavior. This would leave even the highest and most elaborate functions with the imprint of their origin from the basic system. However, in the evolution of fetal forms that live and develop for some time in an egg or uterus, certain individuated sensory–motor structures or reflexes concerned with specialized parts of the body may achieve a functional isolation while the rest of the system is dormant or not yet developed (Anokhin, 1964; Gottlieb, 1973; Hamburger, 1973; Oppenheim, 1973). Movements of fetal maintenance, hatching, or birth clearly are specialized, and they are given a developmental boost relative to other behavioral structures. Adaptations to conditions arising before birth, and individuated subfunctions which emerge after birth tend to break up the inherent suprareflex unity of the system. This, while it changes the rules, does not, of course, deny the importance of Coghill's notion of intrinsic differentiation of "total pattern" nerve networks. Nor does it justify belief in an extreme Pavlovian associationist view of the development of higher intelligence.

5.2. Growth of the Brain in the Fetus

This topic has been authoritatively reviewed by Larroche (1966).

The hemispheres of man appear as thin-walled sacs from the unpaired end chamber of the brain while flection is completing at 30 days (Barthelemez and Dekaban, 1962; Hamilton et al., 1962) (Figure 6). They are small and lacking nerve connections when the first retinotectal projections and other sensory–motor circuits of brain stem and cord are established (Windle, 1970). Neuroblasts of the forebrain begin to appear just in front of the optic stalk. These later grow fibers to the hypothalamus and diencephalon. Olfactory lobe neuroblasts appear at this time. The first region to develop neuroblasts within a thickened mantle layer of the wall of the hemisphere corresponds with the corpus striatum. This is a conspicuous bipartite swelling in the wall of the cerebral vesicle by 40 days. Neurons in it produce fibers that join the caudally directed medial longitudinal tract. The thin dorsal wall (pallium) remains undifferentiated until the end of the second month when neuroblasts begin to accumulate in the primordial hippocampus and pyriform cortex (palaeopallium). The latter receives axons of secondary olfactory neurons of the olfactory bulb.

Neocortical neuroblasts are produced early in the fetus, beginning in the lateral cerebral wall and extending rapidly into the roof and medial wall above the hippocampus. Frontal and occipital poles are the last to reach this stage, which is complete in man by 65 days (50 mm). The lateral and dorsal part of the cerebral vesicle grows with the thalamus, receiving from it an influx of nonolfactory afferent axons. Thus, from its earliest fetal origins the neocortex is committed to receive sensory information relayed from the body surface and special distance receptors.

The presumptive parietal area, which is "multimodal" in the adult, is the first to show neuroblast formation, the most rapid to develop, and the first area to receive afferent axons from the thalamic neurons. The areas for the special senses of sight and hearing receive afferent connections later. Phylogenetic comparisons show that a multimodal, parietal, cortical afferent field is primitive in mammals; the unimodal, primary, afferent areas differentiate out of it in relation to increasing demands for high resolution in perception (Diamond and Hall, 1969). In the cat the first stage of form vision appears to be worked up on the parietotemporal extra-striate cortex; areas 19, 20, and 21 receive input via the midbrain and pulvinar (Sprague *et al.,* 1977).

Electrophysiological unit studies and refined anatomical techniques bring out many and varied somatotopic maps in primates, some with clearly different stimulus-analyzing properties, throughout the parietal area (Allman, 1975; Zeki, 1974). The nonspecific parietal cortex is almost certainly not an unsystematic blend of the receptor modalities, even in primitive mammals. Similar parallel organization in the visual projections of birds indicate a common anatomical scheme for all land vertebrates (Benowitz and Karten, 1976). In the adult monkey its cells are active prior to the execution of an act of orientation of hand or eyes to an object in extracorporal space. That is, they execute a "command" function directing praxic or exploratory acts and readying the processes of perceptual analysis (Mountcastle, 1975; Mountcastle *et al.,* 1975).

The cerebral hemispheres and thalamus enlarge greatly in the first fetal month, and they develop their characteristic ram's-horn shape with distinct frontal, temporal, and occipital lobes, the latter covering the midbrain by the end of this month (Larroche, 1966).

Nerve cell formation in the hemispheres occupies the 3rd to 5th month in man (Dobbing and Smart, 1974) and begins with movement of the cell bodies of neuro-epithelial cells from the region of active nuclear division near the internal limiting membrane (ventricular zone) out toward the pial surface where the neuroblasts separate off to form a dense *cortical cell plate* (Angevine, 1970; Berry and Rogers, 1966; Berry, 1974; Rakic, 1971, 1972, 1974; Sidman and Rakic, 1973). The mature neurons at any given level are relegated inwards by the migration past them of later-formed cells. The result is a series of strata, but distinct laminae only appear when the nerve cells grow axons and dendrites in the 6th–8th fetal months. Thus cells of laminae VI and V in the visual cortex, which will project to subhemispheric sites, are produced before those of layers II and I, which project intracortically or by commissures to the other hemisphere. The order of cell maturation follows this order of production, except that small Golgi type II cells produced later in ontogeny may migrate between the laminae. These cells are thought to be inhibitory in function and may contribute to postnatal memory function (Altman, 1967; Jacobson, 1974). Thalamic afferents terminate mainly in cells of layer IV.

Patterns of cell generation and migration in the growing brain have been revealed with fetal mammals by injection, into the gravid female, of tritiated amino acids, which are taken up to label cells at the time of cell division, giving a picture of "cell birthdays" (Angevine, 1970; Rakic, 1974). The concept of a highly specific spatiotemporal pattern of neuroblast dispersal during brain growth in the fetus (Coghill, 1924; Hamburger, 1948) has been strongly confirmed. Movement of cell bodies in the growing cortex is radial, producing orderly columns at right angles to the surface. These columns result from migration of cells along glial fibers that stretch radially between the two limiting membranes of the cortex (Rakic, 1972; Sidman and Rakic, 1973), and they form an anatomical unit of great significance in subsequent organization of intercellular connections (Jones, 1975; Rakic, 1976, 1977; Goldman and Nauta, 1977). They relate directly to functional components in the adult (Hubel and Wiesel, 1962; Mountcastle, 1957; Szentagothai and Arbib, 1975). Angevine (1970) has suggested that the cytochemical labeling of cortical cells that determines the formation of highly selective functional cell groups in the fetus and in postnatal maturation may follow the clonal production of cells from neuroepithelial cells, molecular tags being handed on and elaborated in the process of cell division and migration.

Some axons of thalamic and brain-stem origin arrive in the cortex early in the fetal period (Marin-Padilla, 1970; Morest, 1970), and these may determine the depth at which migrating cortical cells stop creeping outwards (Sidman and Rakic, 1973). Synapses are formed in the dendrites at the inner and outer boundaries of the cortical plate, soon after the plate begins to form (Cragg, 1972; Molliver *et al.*, 1973). Early tangential axons probably contribute important information to direct the collection of migrating cells into layers in several regions of the CNS. This exemplifies how formation of complex central neuron mechanisms involves inductive interactions between surfaces of cells brought together by migration and axon growth.

Rates of increase of DNA in the whole human brain reveal two spurts of cell multiplication (Dobbing, 1971; Dobbing and Smart, 1974) (Figure 16, page 45). The first relates to neuron production in the 12th to 20th weeks after conception (Sidman and Rakic, 1973). The second, from the 30th week to three months after birth, and continuing at a slower rate for about two years after that, is correlated with glial multiplication, dendrite growth, and myelination of axons. Gross genetic defects of cerebral formation are related to the first growth spurt which is sensitive to radiation insult and poisoning. Malnutrition of the mother during the period of glia-cell multiplication, when the child is dependent on her body for nutrition, can cause permanent deficiencies in behavior (Dobbing, 1974). It is an important social fact that both prenatal and postnatal starvation can affect psychological functions profoundly.

The intricate structure of the adult cortex at the cell level, with many distinct forms of pyramidal and nonpyramidal cells in highly ordered arrays, is determined in essentials prenatally. Most of the classical description of brain tissue structure made by the Spanish anatomist Ramón y Cajal (1909–1911) was carried out by silver staining of fetal mouse brains and the immature brains of birds. He found the immature brain clearer in structure than the adult, and his small budget limited his work to small, easily bred animals. To a remarkable degree the main cell types and the main tracts are already evident in the human brain at birth (Conel, 1939–1963). However, postnatal developments, although small in scale and adding little to brain bulk, greatly add to function of brain circuits for years after birth (Bailey and von Bonin, 1951; Conel, 1939–1963; Shkol'nik-Yarros, 1971; Yakovev and Lecours, 1967). Growth of dendrites and formation of synapses may go on until adulthood, but the relative maturity of the brain at birth differs greatly in different forms of bird or mammal. In man, the pyramidal cells of the cortex progressively spread an elaborate system of dendrites studded with synapse-bearing spines throughout childhood years. With senility these cell extensions wither and shrink as intellectual functions decline (Scheibel *et al.,* 1975).

The great commissural links between the hemispheres grow in relation to the development of the important structures they connect (Figure 15). The first is the anterior commissure between pyriform cortices and olfactory bulbs on the two sides. Soon after it the relatively small hippocampal commissure joins the hippocampal areas. The enormous neopallial commissure (corpus callosum), estimated to have 200 million fibers in the adult, first appears as a few axons passing over the hippocampal commissure at 10 weeks (60 mm). It grows long after birth, doubling in cross-sectional area from two years after birth to adulthood (Rakic and Yakovlev, 1968). The posterior end (splenium), which communicates perceptual processes of parietal and occipital regions, shows relatively greater enlargement in infancy, but less development later.

In the first postmenstrual month the occipital lobe enlarges and becomes pointed. At this stage cell proliferation throughout the cortex becomes greatly reduced and cell differentiation begins. The primary sulci are formed then, and they deepen to the 7th month. In the somatosensory cortex, trunk regions mature histologically before the limbs and forelimb areas develop in advance of those for the hindlimbs. The characteristic splitting of layer IV in the striate cortex is evident from the 8th month. Presumably it reflects proliferation of dendrites from large stellate cells destined to receive synapses from the thalamic axons.

The recently discovered anatomical asymmetries of the human cerebral cortex are detectable as early as the 5th month of gestation (Witelson and Pallie, 1973; Wada *et al.,* 1975; Le May, 1976; Chi *et al.,* 1977).

The cerebellum of mammals, and particularly of man, undergoes a protracted and elaborate development wholly comparable with that of the cerebral hemispheres and certainly significant for perceptual development (Larroche, 1966; Sidman and Rakic, 1973). The phylogenetically old flocculonodular lobes directly related to the vestibular system are the first to differentiate. The cerebellar hemispheres, which mature later, receive input from both the so-called "proprioceptive" and "exteroceptive" modalities. Even the latter are probably participating "expropioceptively" in the direct regulation of body movement. There is an important visual input to the cerebellum and pons via the posterior parietal cerebral cortex (area 18) (Glickstein and Gibson, 1976).

The complex layered histological structures of the neocerebellum, with highly specialized cell types, develop at different rates in different vertebrates, although the general order of events is the same in all (Jacobson, 1970). Precocial animals (e.g., chick, guinea pig, sheep) are born with the cerebellum at an advanced stage of maturation. It is relatively immature in helpless altricial newborns (e.g., rat, cat). In man Purkinje cells can be detected from the 24th week of gestation, and they reach their final position by 36 weeks (Larroche, 1966). Structural differentiation begins in the last weeks of pregnancy and large changes occur in the first six months after birth. Cerebellar development continues thereafter for several years. Early development of the rat cerebellum is highly sensitive to malnutrition or hormone imbalance. This is used as a model system in studies of biochemical regulation of intrauterine cerebral growth (Balázs *et al.,* 1975).

5.3. Activities and Responses of Fetuses

Spontaneous and reflex movements of fetal birds and mammals have been charted to follow the development of structures underlying movement and perception (Hamburger, 1973; Oppenheim, 1974). Human fetuses aborted at different stages have been subject to behavioral tests (Hooker, 1952; Humphrey, 1964, 1969), but these studies are rendered difficult by the exacting requirements for physiological maintenance of the fetus outside the mother, and spontaneous movements have been neglected while attention has been directed to charting reflex responses to artificial discrete stimuli. Observations of living fetuses *in utero* by X radiography and physiological studies of the amniotic fluid in which they live (Liley, 1972) give evidence that the cerebral mechanisms may control adaptive movements (heartbeat, swallowing amniotic fluid, shifts of posture) before birth. Other movements (e.g., limb extensions, thumb-sucking, respiratory movements) may be essential to modeling of structures needed only after birth.

On the whole, the most striking feature of the fetal condition is the limited range of behavior despite the exceedingly complex circuits already outlined in the brain. Neurological examination of fetuses born prematurely suggests that they are capable only of relatively weak and stereotyped movements until the last trimester of gestation, when marked changes occur (Saint-Anne Dargassies, 1966). Aborted fetuses are, of course, in very artificial and inauspicious conditions, and their normal level of activity has probably been underestimated. Nevertheless, all fetal activity seems but a specialized and restricted part of what the fetal brain circuits might reasonably have been expected to perform. Much of the cell patterns must be in prefunctional or inactivated condition. There is physiological evidence that the levels of activity of the developing central nervous circuits is regulated by inhibition which, with a delay of a few weeks after the first reflex circuits are completed, may greatly qualify the patterns of action generated in fetal motor organizer systems and their sensory control (Humphrey, 1969; Oppenheim and Reitzel, 1975). Postnatally, a degree of relative inaction and comparatively simple attentional reaction to stimuli appears to rapidly pass away with the termination of the neonatal period at 4–6 weeks.

Developments either side of term (40 weeks of gestation) indicate that this period is a clearly marked and specialized stage of cerebral maturity. Research with ungulates shows that the time of birth is under a degree of control by the fetal brain and humoral system. It is therefore probable that, under optimal conditions, human birth is regulated by the fetal brain to occur when the brain circuits have achieved a specific ''critical'' state of readiness for access to the external world. Postnatal developments in the brains of cats and monkeys, reviewed below, show that the developmental state of readiness in the brain at or soon after birth determines the influence of stimulation on the maturation of perceptual circuits.

In man, the first spontaneous movements occur at $5\frac{1}{2}$–6 weeks. (Throughout this description of fetal development, ages are expressed in weeks and in four-week lunar months since conception, that is, in conceptual rather than postmenstrual time.) The precocious appearance of touch reflexes and primary vestibular circuits (Gottlieb, 1971b), as well as the early appearance of postural reflexes and compensatory eye movements, suggest that among the first organized efferent–afferent feedback loops are those of coordination of body parts into a mechanical system capable of controlled displacement of parts. Older fetuses regulate their posture and react to positional changes caused by the mother moving her body. Finely individuated local face and hand movements also appear as early as 12–16 weeks (Humphrey, 1964, 1969). They appear to express latent neuromotor organization in circuits of brain stem and cord that are adapted to manual and oral prehension of objects. They may also be parts of innate motor patterns for communication by grimace or gesture with which man is richly endowed. At 20 weeks the human fetus may actively regulate its nutrition and protein metabolism by swallowing amniotic fluid (Liley, 1972). There is evidence of taste preference for sugar and rejection of bitter substances.

Fetal behavior could conceivably influence the development of the integrative nerve circuits through feedback stimulation of receptors which are already functioning. Self-patterning of excitation could be a part of the environment to which growing nerve circuits are adapted to respond. However, all the experimental work done so far on effects of fetal stimulation, self-produced or from outside sources, seems to point the other way. Fetal motility is shown to be patterned in independence of reafferent effects which have a significant influence only postnatally (Hamburger, 1963; Oppenheim, 1974). Oppenheim has raised a question regarding the accessibility of nerve systems, when they do become responsive, to cumulative influence from stimulation. An early form of undiscriminating habituation to repeated stimuli appears to protect the fetus from learning sensory–motor links. Then fragmentary responses to repeated stimulation become more consistent.

Virtually nothing is known about the role of prenatal self-stimulation in man, the most remarkable studies having been done with birds. Gottlieb (1971a) has shown that prehatching calls of ducklings may be one factor in the normal development of their ability to hear the mother on hatching. Precocial birds have to be born with a near-ready ability to follow the maternal assembly call which is species-specific. Vestibular, tactile, and mechanoreceptor systems may all be stimulated before birth or hatching as a result of movements of the fetus or of the mother. This could aid normal differentiation of their structure. There is, however, no evidence that it does so.

Intrauterine recordings of light and sound in the womb show that human fetuses could perceive low-frequency vibrations and sharp knocking sounds from the air round the mother, and large changes in level of light falling on the mother's belly. The loudest sounds are those originating in the mother's stomach, heartbeat, respiration, and vocal organs. Behavioral reactions of fetuses to rhythmic or sudden external sounds are frequently reported. It is not likely that such perception is significant compared to postnatal achievements in perceptual development, nor is there any evidence that prenatal experience in birds or mammals has a significant constructive role in forming specific circuits for perception. Gottlieb (1971a, 1976b) found prehatching experience to be validatory rather than constructive in birds, and the effect has limited duration.

The earliest spontaneous movements of embryos show activity cycles and bursts of movement that are rhythmical or pulsating (Hamburger, 1969; Oppenheim, 1974). It is likely that rhythmical pacemaker systems are an autogenous component of growing central networks and that they contribute an essential basis for development of perception rather than the other way about (Hamburger, 1963). Experimental studies show that while sensory feedback can modify the development and functional expression of motor patterns, it does not determine pacemaker origin (Carmichael, 1926). The importance of controlled rhythms in both normal and pathological behavior of infants is stressed by Wolff (1966, 1968). It is also a fundamental in Lashley's analysis of serial ordering processes in action and perception (Lashley, 1951) and in Bernstein's theory of human motor coordination (Bernstein, 1967).

It is important, when considering the possible significance of early fetal behavior to perceptual development and "fetal consciousness," to note that the special sense organs, having attained their basic form in the late embryo period, are physically cut off from stimulation by morphological changes in the early fetal period of neuron production (Arey, 1965; Hamilton *et al.*, 1962). The eyelids grow over the cornea to fuse at $7\frac{1}{2}$ weeks. They reopen at 6 months. The ear ossicles develop within a spongy mesoderm which remains to block transmission until the last fetal months when a cavity forms around the ossicles. The tympanic cavity remains obliterated by endodermal thickening and swelling, and is excavated shortly after birth in association with changes that accompany the onset of pulmonary respiration. From the 2nd to 6th gestational month the nostrils are closed by epithelial plugs. Fine individuated finger movements occur from the end of the 4th month, at which time the sensory structures of the skin, including hair follicles, are clearly differentiated.

The development of brain mechanisms observed through study of premature infants in the last trimester of gestation, when dendrite expansion and glial multiplication are beginning in the hemispheres, shows that there is an underlying formation of circuits throughout the CNS that follows a tightly controlled schedule in considerable independence of body weight and in resistance to the radically changed environmental conditions of hospital care (Larroche, 1967; Saint-Anne Dargassies, 1966; Schulte *et al.*, 1969). At 25 weeks the neurological and metabolic machinery of an infant is sufficiently advanced for it to be viable outside the mother in an incubator. The change to this level of competence is a sudden one, the 6-month-old fetus having achieved a characteristic state of functional maturity (Saint-Anne Dargassies, 1966).

At $7\frac{1}{2}$ months the fetus is spontaneously more alert and better coordinated in responses, waking easily to stimulation, and showing the first spontaneous eye movements with lids wide open. The "doll's eye" reflex (rotating of the eyes to oppose lateral head rotation), quick eye closure to a dazzling light, a sluggish pupillary light reflex, all attest rapid emergence of the fundamental motor controls for seeing. This is the age at which the infant may raise his hands to his face, place them in the mouth and suck them actively and rhythmically. It is the period at which exploratory, unstimulated grasping first occurs. During the 8th and 9th lunar month the infant develops muscle tone from lower to upper limbs and by the end of the 9th lunar month legs and arms lie in flection at rest and the neck may briefly support the head, but motility is less than at early stages of prematurity.

Electrical signs of maturation of the brain indicate a progressive activation of circuits from brain stem to cortex. The brain stem reticular formation, which will regulate sleep, spontaneous orienting, and consciousness after birth, exhibits patterned electrical activity first in the pons, at about 10 weeks (Bergström, 1969). The activity then advances through the midbrain reticular formation and basal ganglia, spontaneous rhythmic brain waves appearing in records from scalp electrodes at 24 weeks (Dreyfus-Brisac, 1967). At the time the fetus becomes viable, the EEG becomes more regular with bursts of theta waves (4–6 cps) that are synchronized within each hemisphere. Within a week or two this isosynchronism diminishes. At 30 weeks higher-frequency bursts become more common (10–14 cps), this being the period of most rapid development of the EEG (between 24 and 36 weeks). There are still no electroencephalic reactions to stimuli and no sleep–wake differences. Activities at this time are probably of subcortical origin.

A distinction between sleep and waking tracings appears at 37 weeks. The waking EEG is similar to that of the full-term infant but the sleep record differs, characteristics of the much younger (30-week) EEG reappearing in light sleep. The development of a distinct wakeful state may reflect maturation of the cortex or reticular formation, or both of these. The stages of EEG development form reliable indicators of conceptional age.

Electrographic responses evoked by flashes of light and sounds are readily obtained in neonates. They are large in amplitude, simplified in form, and long in latency. Thus, while demonstrating considerable complexity in cortical nerve networks, they also show that those networks are far from mature. Only slow photic driving is possible (<3/sec). Response to stimulation includes a flattening of the bioelectrical activity. Habituation has been observed before 37 weeks in respiration, body movements, and myographic activity.

Compared to adults, newborns have an immature sleep cycle in which all phases are less clearly marked. The process of maturation continues for years after birth. An awake newborn shows clear reactions of alerting to sounds, touch, changes in light, vibration, etc., and may orient to follow changes in position of sounds or visual stimuli (Peiper, 1963; Beintema, 1968). Exploratory behavior is very rudimentary in the first weeks, although dishabituation tests demonstrate that perceptual impressions of considerable refinement may be formed (see below).

5.4. Plasticity in the Fetal Mammal Brain

The Syrian hamster is born at a stage corresponding to about the 1st or 2nd fetal month of man, when the major cell migrations of the developing cortex are still taking place (Figure 13). The circuits of midbrain and thalamus are still forming. Schneider (1969) has found that adult hamsters show reciprocal defects with ablations of superior colliculus and visual cortex. Animals lacking cortex retain visual orienting, but lose pattern vision; those lacking colliculus are deficient in orienting, but they keep the ability to discriminate patterns. If the same operations are performed on newborn hamsters, there is a high degree of functional recovery (Schneider, 1973; Schneider and Jhaveri, 1974). Anatomical investigations reveal a number of compensating changes in nerve connections in the growing brain (Figures 13 and 14). These rewirings indicate that the retina, colliculus, and geniculocortical cells share a common somatotopic cytochemical code which confers a limited power of substitution in circuits between them (Schneider, 1975, 1976). Response to the code depends on competitive interaction between terminals, and patterning is greatly disturbed if sites for termination or contingents of terminals are removed

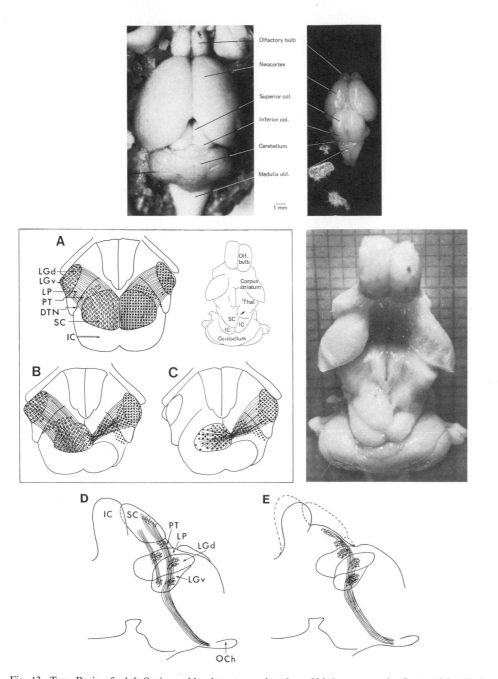

Fig. 13. Top: Brain of adult Syrian golden hamster and at day of birth, same scale. Center right: Brain stem of hamster with anomalous bundle of retinal axons crossing the midline to the anteromedial part of the left tectum after removal of surface layers of the right tectum at birth. Center left: Terminations of left (open circles) and right optic tracts (closed circles). A: Normal hamster. B: Superficial layers of right colliculus ablated at the day of birth. Anomalous connections have been formed on the left median tectum and to other nuclei along the course of the tracts. C: Similar to the previous, but with removal of the right eye at birth, denervating the left tectum and permitting extensive invasion by anomalous cross-over input from the left eye. Bottom: Reconstructed lateral views of the hamster brain. D: Normal projections. E: anomalous connections after removal of the surface layers of the superior colliculus in the neonate. LGd and LGv: dorsal and ventral nuclei of the lateral geniculate body; LP: nucleus lateralis posterior; PT: pretectal area; DTN: dorsal terminal nucleus of the accessory optic system; SC: superior colliculus; IC: inferior colliculus; OCh: optic chiasma. (From Schneider, 1973, 1976.)

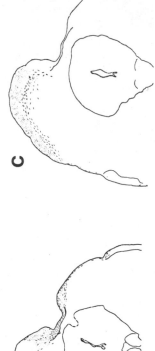

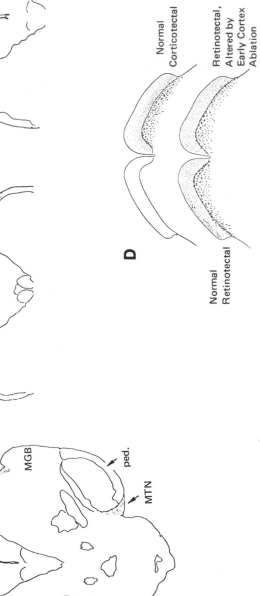

Fig. 14. Syrian golden hamster. Competitive interactions of retinal and corticofugal terminals in the tectum. Silver stained coronal sections. Degenerated axons (dashed lines) and terminals (dots). A: Surface layers of right tectum and right eye removed at birth. Extensive recrossing of fibers from the *left eye* to the left tectum (cf. Figure 13C). B: Right tectal lesion at birth, *left eye* removed in adult 5 days before sacrifice. Projection to medial zone of left tectum only (cf. Figure 13B). C: Degeneration from *right eye* removed in adult. Right tectal lesion at birth. Complementary to B (cf. Figure 13B). (From Schneider, 1973; Schneider and Jhaveri, 1974.)

surgically. Visual terminals deprived of their normal target region by ablation of the superior colliculus may even invade deafferented areas of the inferior colliculus which normally receive only auditory afference (Kalil and Schneider, 1975).

In the young colliculus the cells of deeper laminae are acceptable to retinal axons for formation of ordered connections when the superficial laminae have been removed (Figure 14). Moreover, Schneider and Jhaveri (1974) present evidence that a reduced colliculus, made small by ablation of the superficial layers near birth, may receive a correctly ordered, but compressed input of retinal axons like that produced in the goldfish by the same kind of surgery (Meyer and Sperry, 1976). Misdirected axons, terminating on inappropriate locations, were also observed. Finally, retinal axons may cross the midline to terminate in the inappropriate colliculus, and this remapping in an abnormal mirror array is facilitated if the invaded colliculus has been denervated by removal of the other eye (Figures 13 and 14). Behavioral tests (orienting to food) confirm the anatomical findings in this study which offers important evidence concerning the mechanisms for substitution, competition, and reformation of circuits after extensive damage to the still-developing brain. Functions intrinsic to the axon-producing cell and its terminal arborizations ("pruning" factors) are contrasted by Schneider (1973, 1975, 1976) with factors that operate in populations of axons seeking termination together in a potential end site (axon-ordering factors, or axon–axon interactions).

The experiments show that plasticity of the midbrain visual system extends into postembryonic stages of development in mammals. This may reflect an evolutionary change of developmental strategy to make way for incorporation of the later-formed projections from forebrain visual fields. Developmental plasticity of cerebral circuits on a small scale extends far into postnatal stages even in man but only the early fetal stages may exhibit such drastic regenerated changes as seen in the hamster (Guth, 1975).

Development of monocular dominance zones in the visual system of the monkey is completed in the superior colliculus and lateral geniculate weeks before birth; the striate cortex develops uniocular territories shortly after birth (Rakic, 1977). Throughout the visual projection, segregation of terminals from the two eyes from originally overlapping populations, although determined genetically, is sensitive to removal of one eye or reduced stimulation of one eye. Evidently there is a competitive interaction between terminals of the two eyes, and this is regulated by the relative activation of the visual cells of the two retinae (Hubel *et al.*, 1976) (Figures 19 and 20).

6. Late Fetus, Birth, and Infancy

6.1. General Features of Postnatal Brain Growth

In most mammals birth occurs around the time that cortical neurons are rapidly developing and synapses are forming over the dendrites (Figures 16 and 17). Cortical mechanisms for visual perception are more completely formed in preyed-upon species that are born above ground. Primates show a short period of postnatal maturation, and nest-born young often have poorly developed cortices at birth. Cortical development is complex and prolonged in man. Undoubtedly the first period of infancy is the most significant in the formation of working perception systems.

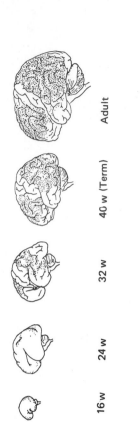

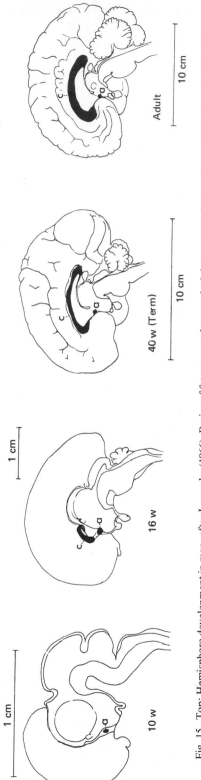

Fig. 15. Top: Hemisphere development in man, after Larroche (1966). Brains of fetuses, newborn and adult to same scale showing approximate maturation of "association" cortex (stipple). Bottom: Growth of the interhemispheric commissures before and after birth parallels developments in the "association" cortices.

Most of the secondary folds of the human cortex are formed at birth (Figure 15). The tertiary folds begin to be prominent in the latter part of the second year after birth, but they go on developing, possibly into adulthood (Larroche, 1966; Yakovlev, 1962). Sulci are patterned to reflect regional development of the cortex, and in the primary projection areas it is possible to trace a close inborn relationship between cortical gyri and the prominent afferent zones of the periphery (Adrian, 1946; Whitteridge, 1973). On the motor side, too, enlarged cortical representation correlates with variations in the complexity of motor integration for different effector organs. For example, the hand area of the highly manipulative raccoon is almost a picture of its forepaw. In man, a most important difference in the folding of the cortex in left and right hemispheres correlates with the development of a mechanism for language comprehension. This morphological asymmetry in the left parietotemporal function (Wernicke's Area) is detectable in the middle fetal stage (Galaburda *et al.,* 1978).

There is conspicuous proliferation of dendrites in early infancy (Conel, 1939– 63; Shkol'nik-Yarros, 1971), and branches of nonpyramidal "stellate" cells that receive thalamic afferents (Jones, 1975) appear more spinous in infants than later in life (Marin-Padilla, 1970; LeVay, 1973). Many synaptic contacts are formed just prior to birth, and a large number are added in the early postnatal period. Nissl granules and neurofibrils develop postnatally along with the maturation of cortical dendrites, and glia multiply to fill expanding spaces between nerve cells (Dobbing and Smart, 1974) (Figure 16). The cortex is histologically immature at birth compared to the nuclei of the central gray and brain stem. The association cortices grow more slowly than the primary projection zones and are thought to retain an embryonal plasticity throughout life (Yakovlev, 1962). Although both dendrites and synaptic arrays on them are immature at birth, recent observations with radioactive tracer staining of the newborn monkey cortex shows that the fiber systems of the cerebrum are highly organized at birth and that transcortical and commissural

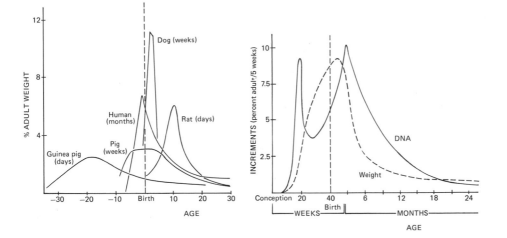

Fig. 16. Left: Brain growth rate curves by weight for various mammals. Right: Velocity curves showing incremental rates of DNA and fresh weight of whole human brain. The DNA curve has two peaks, one reflecting neuron multiplication in the midgestational period and one for glial multiplication associated with increase of brain weight, dendrite development and synaptogenesis in the first postnatal months. (After Dobbing, 1971.)

projections are distributed in a uniform system of columnar modules (Goldman and Nauta, 1977). These fragmentary observations give an indication of the complex postnatal development of refinements in brain structure.

A specialized remnant germinal layer of cells is found near the head and body of the caudate in the cerebral hemispheres of mammals. This subependymal or subventricular zone continues to proliferate in the adult rodent, generating cells that differentiate into small Golgi type II neurons after migrating considerable distances. Some consider that a majority of these cells are inhibitory and that they modify cortical response patterns, e.g., by removing unstimulated components from function. Cell production in brains after birth has been held to add to the mechanism by which learning occurs, but there is no direct evidence (Altman, 1967; Jacobson, 1970).

6.2. Synaptogenesis and Brain Function

Synapse formation has been followed in the spinal cord of birds (Foelix and Oppenheim, 1973) and monkeys (Bodian, 1968), and it relates to the development of motility and reflex response in the early fetus. In mammals there is a long period in which the early brain circuits are inactive and rather unresponsive, before the cortical fields undergo histogenesis to produce functional circuits (Hamburger, 1973; Oppenheim, 1973). Evidently the growth of inhibitory connections plays some part in the masking or inactivation of lower level structures already built (Crain, 1974; Bergström, 1969). Formation of spreading dendrites and axons, and consequent expanding of brain tissue, is followed closely by the development of vast numbers of intercellular connections, initially excitatory axondendrite contacts being formed (Bodian, 1970).

In all mammals studied there is a definite period of cortical cell branching followed by a rapid accumulation of synapses. In the sheep, a precocial form, synapse proliferation takes place about the middle of gestation (60 days) (Åström, 1967). In cats, which are born at a much earlier state of development, it is postnatal, during a period after eye-opening when visual performance is showing marked improvement.

Cragg (1975) has traced the development of synapses through the visual system of the cat for a part of the visual cortex corresponding to 60° in the periphery of the visual field (Figure 17). Optic axons reach the lateral geniculate nucleus at least 20 days before birth but synapses first appear within the near central retina only nine days before birth, and ribbon-containing (inhibitory?) bipolar synapses do not appear until birth. Few retinal receptors possess outer segments with disks at birth, but five days later disk-bearing outer segments have developed. The lateral geniculate body shows a rapid growth of synapses from five days after birth to a maximum of about 4×10^{11} synapses/cm³ at 25 days. This process precedes growth in size of geniculate neuron cell bodies by about 10 days.

Parallel with the development of retinogeniculate synapses, but delayed relative to them by about two days, geniculostriate connections grow from day seven to a maximum of 6.5×10^{11}/cm³ about day 35. This fits the time of appearance of dendritic spines on stellate neurons of the striate cortex (Adinolfi, 1971). The first-formed synapses, probably not all from visual axons, appear just above and below the optical plate of cortical cells, two weeks before birth.

The volume of gray matter per cell of visual cortex, a measure of dendrite and axon growth between the cell bodies, increases about a week ahead of the synapse population (Anker and Cragg, 1975; Cragg, 1975). Myelin starts to appear three weeks after birth (Marty and Scherrer, 1964). The average number of synapses associated with one cortical neuron rises to a peak of about 13,000 at seven weeks after birth. In the adult cat the synapses and neurons are slightly less dense in the cortex, apparently because glial cells develop. There is no evidence of loss of synapses.

Synaptic development in the cat visual pathway starts before afferent impulses can enter because the animal is not yet born and because receptor cells are immature. Cragg suggests that it may be important in all sensory pathways that the first patterns of synaptic connectivity be established in absence of stimulation. But, the main increase of synapses in the lateral geniculate nucleus and cortex takes place when the visual system is being used, in the 4th and 5th weeks (Figure 17).

The cortex of the cat is responsive to electrical shocks applied to the optic nerve at birth (Marty and Scherrer, 1964) and single-cell responses to flash stimula-

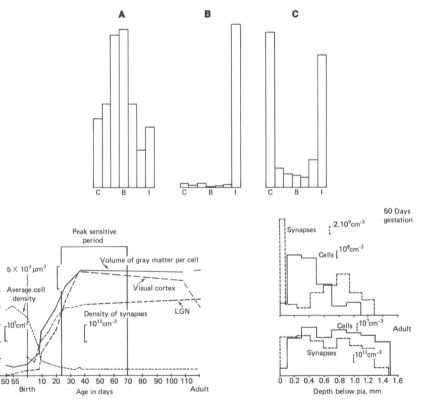

Fig. 17. Developments in the visual cortex of kittens. Top: Ocular dominance histograms, showing proportions of visual cortex units in one hemisphere driven by the contralateral eye only (C), by the ipsilateral eye only (I), by both eyes equally (B) and with intermediate degrees of dominance. A: Normal cats (Hubel and Wiesel, 1965). B: Kittens reared with the contralateral eye deprived of visual stimulation, leading to loss of connections from that eye (Wiesel and Hubel, 1965a). C: Kittens with artificial squint to prevent normal convergence of the eyes and registration of their inputs (Hubel and Wiesel, 1965). Bottom left: Formation of synapses in the visual cortex and lateral geniculate nucleus (LGN) of the cat, and decrease of cell density with spread of dendrites (Cragg, 1975). The sensitive period for formation of anomalous geniculo-striate connections with abnormal visual stimulation, according to Blakemore (1974). Bottom right: Changes in distribution of synapses in the visual cortex of the cat from 15 days before birth (50 days after gestation) to the adult (Cragg, 1975).

tion are reported at 4–6 days after birth (Huttenlocker, 1967). Cortical units recorded by Hubel and Wiesel (1963) at a week after birth were selectively responsive to direction of target displacement. Eyelids open from one to two weeks after birth, but the optic media only become fully clear by the fourth week (Barlow and Pettigrew, 1971). Evidently selective response of cortical cells for orientation of a visual target and binocular disparity develops during the 4th and 5th weeks after birth (Barlow and Pettigrew, 1971; Pettigrew, 1972) (Figure 17).

Effects of manipulation of visual experience on responses of cortical cells are described below. Stimulus-dependent changes of function occur after synapses are completely formed. They may involve selective cytochemical repression, but not necessarily anatomical loss of synapses (Mark, 1974). There is evidence that GABA-active inhibition by cells entering the cortex and developing connections postnatally may be responsible for suppression of inputs from one eye to binocular neurons after that eye has been deprived of stimulation from birth (Duffy *et al.,* 1976).

Anatomical comparison of cell developments suggest that the period of synapse formation in the cortex of cats corresponds with a period about half way through gestation in the sheep (Åström, 1967). In the monkey retinogeniculate and retinocollicular pathways appear to be fully differentiated before birth, but the striate area circuits are still developing in the first one or two months after birth (Rakic, 1976, 1977; Wiesel and Hubel, 1974) (Figure 20, page 55). Monocular deprivation experiments confirm this (Hubel *et al.,* 1977) (Figure 19, page 53). The fovea of the monkey appears to be clearly differentiated at birth (Ordy *et al.,* 1962), but it is not certain that foveal cones or their retinal connections are in a mature functional state.

In man the foveal cones are usually described as immature at birth and their end segments conspicuously elongate in the first months after birth (Mann, 1964). The soundness for the conclusion which may be based on pathological eyes has been questioned recently both by evidence of vision in the central part of the field of newborns and by doubts cast on the reliability of the anatomical data (Haith, 1978). Nevertheless, neonatal acuity is much lower than in adults. There is probably anatomical differentiation of retina as well as central visual circuits, and the fovea may develop postnatally. Striate cells undergo considerable elaboration of dendrites postnatally in the cortex of man (Conel, 1939–1963). Important postnatal changes in the synaptic mechanism of cortical circuits are suggested by the formation of dendritic spines postnatally (Conel, 1939–1963; Shkol'nik-Yarros, 1971), and by the subsequent loss of spines and other surface irregularities on dendrites later in life. It is not known how these changes relate to the development of perception, although they appear to correlate with improvements in acuity and with adaptive changes in amblyopia or astigmatism (Barlow, 1975; Fanta *et al.,* 1962; Atkinson *et al.,* 1974; Teller *et al.,* 1974; Leehey *et al.,* 1975). There appears to be a large change in visual efficiency during the first two months after birth (Haith, 1978).

6.3. Myelogenesis as a Measure of Brain Maturation

Among the most conspicuous and easily studied changes in growing brain tissue is the formation of a fatty myelin sheath on longer nerve axons. Myelin may be differentially stained and its progressive accumulation followed in nerve tracts. It used to be thought that myelin was an indicator of the start of nerve conduction. This was disproved by Langworthy (1928), who studied the movements of embryo marsupials when they crawl up the belly of the mother to enter the pouch under their own steam. At this stage the marsupial brain has no myelin.

Nevertheless, myelin deposition changes the precision and strength of transmitted patterns of excitation. Its appearance on a given axon is probably closely correlated with maturation of telodendria and dendrites of that cell. In the development of cortical areas the order of cytoarchitectonic maturation correlates well with myelogenesis in the axons on the same cells (Yakovlev, 1962). Its differential appearance in growth of the brain is undoubtedly a valuable indicator of developmental change after formation of axons and outgrowth of the main dendrite branches (Larroche, 1966; Yakovlev, 1962; Yakovlev and Lecours, 1967; Lecours, 1975).

Myelin appears in the second half of uterine life but in certain parts of the brain it has a life-long course in man after birth (Figure 18). The large afferent bundles in the brain stem myelinate well before the association fibers and reticular formation, and the hemispheres are almost devoid of myelin before birth. All the pathways of sensation and equilibration are myelinated in the midfetal stages, generally from below upward. In the cord, however, the motor roots are myelinated before the afferent roots. The vestibular root is in advance of other cranial nerves, and the white matter of the cerebellar lamellae in the flocculus is very precocious. Thus the cycle of myelination of the vestibular–cerebellar system which begins during the 6th fetal month is brief and entirely prenatal.

Marty and Scherrer (1964) have followed myelinization of the optic nerves of

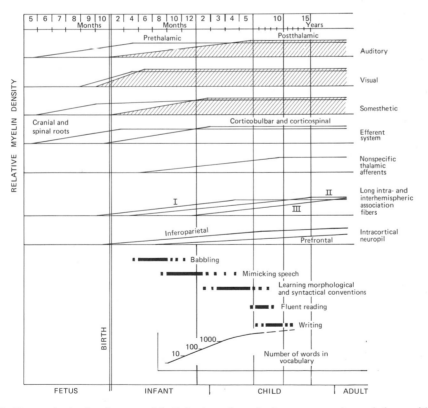

Fig. 18. Human brain development. Myelinization of cerebral systems, and correlations with the development of language functions. (After Lecours, 1975; Yakovlev and Lecours, 1967.)

the cat. In man this occurs from the chiasm toward the optic nerve and optic tract, starting in the 13th week of gestation. The thalamic optic radiations begin their cycle of myelogenesis just before term, and it is completed in the 5th postnatal month. The cycle is thus brief and postnatal. This coincides with foveal maturation and other important postnatal changes in the mechanism of vision summarized below.

The nonspecific thalamic radiations, cortical association fibers, corpus callosum, and reticular formation show the most protracted cycle of myelinogenesis, suggesting that the final process of maturation of the brain circuits, continuing long after birth, is concerned with refinement of the integrative and generative processes which derive from the core mechanisms of the brain (Figure 18). Although the special cortical sensory and motor organs have evolved into distinct cortical structures in the higher terrestrial vertebrates, and although in fetal stages they initially lag behind the parietal association cortex in maturity, late fetal and postnatal developments reverse this relationship.

The nonspecific thalamic radiations begin myelinization at about three months after birth and do not achieve maturity until at least five years later. The short tangential fibers of the intracortical nerve net in nonspecific association areas of the hemispheres begin their myelin cycle about one or two years after birth and do not complete it until several decades later. Lecours (1975) had emphasized the parallels between maturation of speech and language and the growth to maturity of postthalamic acoustic afferent pathways, intracortical neuropile in primary and secondary acoustic cortices, association pathways linked with these, and other sensory and motor structures involved in regulating sound production in speech (Figure 18). A similar analysis could be made of the development of the eye–hand system which requires convergence of excitation within the parietal cortex from visual and somesthetic areas which mature at an earlier stage.

Yakovlev and Lecours (1967) present an overview of psychological maturation in relation to the myelinogenetic cycles in different brain systems, employing Yakovlev's (1968) classification of major subdivisions of the cerebral hemispheres.

1. *Telencephalon impar* on rhinic lobe, medial and responsible primarily for visceral motility of "ablateral" organs.

2. *Telencephalon semipar* or limbic lobes, concerned with emotion, including the expressive emotional sphere of vocalization and mimicry. These are spoken of as "body-bound" movements.

3. *Telencephalon totopar* or supralimbic lobes, dealing mainly with telokinesis or effectuation (transactions through motility with objects, tools, language). This part of the brain includes the greater part of the classical frontal, temporal, parietal, and occipital lobes.

6.4. Structural Details of the Adult Sensory Pathways. Evidence of Innate Patterning.

The classical anatomy of receptors, neural projections, nerve cell bodies in different nuclear regions of the brain, and neuromotor systems is greatly enriched by the findings with modern techniques. The whole picture is of a minutely designed system. This explains why most physiologists and anatomists are more nativists than empiricists in their ideas of brain ontogeny. However, an epigenetic approach, seeking to reconcile these ways of explaining perceptual function, must accept that specific forms of elements in the perceptual mechanism may be determined by intrinsic developmental processes and yet be open to profound modification, selection, etc., by environmental effects.

The antenatal morphogenesis of receptors and effectors in the embryo and fetus of mammals commits them in highly specific ways to the functions they perform later. The same may be said for details of their histology and especially for the patterns of neural cells in receptor epithelia and projection systems which we have already discussed. Here we shall draw attention to some details of the input mechanisms of the brain that are clearly preadaptive to psychological functions.

The retina of the eye has elaborate antenatal determination of structure. In the primate the eye is nearly complete at birth, although there is a relative immaturity of the foveal system. This appears to complete its differentiation in the period shortly after birth. Other mammals have less mature eyes at birth.

Complex neural processing takes place in all vertebrate retinas and the anatomical structures underlying this are still being discovered (Rodieck, 1973). Different systems of receptor cells funnel their excitation to distinct ganglion cells that project to different parts of the brain. Kelly and Gilbert (1975) report three populations of ganglion cells of different sizes that project differentially to the lateral geniculate nucleus and superior colliculus of the cat. Peripheral and central ganglion cells of mammals receive input from different arrays of the two principal receptor types— rods and cones. The convergence of receptors onto ganglion cells and the patterning of horizontal connections between the intermediate junctions in the intervening retinal layers determine the shape of excitatory and inhibitory fields and the dynamics of stimulation for each ganglion cell (Kuffler, 1953). Spatial resolution of the visual field is determined by quantitative aspects of this convergence and lateral interaction, including the distribution and spread of inhibitory interneurons.

In monkey and man the central foveal area is rod-free and composed of narrow cones that project so that each ganglion cell represents excitation of one or a few receptors separate from their neighbors. The human foveal cones mature in the first postnatal months (Mann, 1964). Peripheral cones and rods converge in increasing numbers on each ganglion cell as one samples further toward the periphery (Polyak, 1941; Rodieck, 1973). This increasing convergence correlates closely with increased sensitivity and lowered spatial acuity. In the cat, ganglion cell density in the retina falls as receptor field size of optic nerve axons increases with eccentricity (Fisher, 1973). Peripheral and central visual functions both develop greatly in early infancy of man (Haith, 1978).

The anatomical basis of feature extraction in the central visual field is best known for the visual system of the cat and monkey, thanks principally to the work of Hubel and Wiesel (1962, 1963, 1965, 1968, 1974*a,b*). Combination of anatomical studies of intercellular connections and unit recording with microelectrodes has provided a detailed description which is still being added to. An up-to-date account of this field must emphasize the staggering detail and precision of the morphogenetically created nerve tissue. There are important correlations between psychophysical phenomena and details of the central visual anatomy. Nevertheless, doubt remains about the cause of selective connections, and there is controversy about the degree to which environmentally patterned stimulation may select, stimulate, or otherwise influence the growth process that forms particular feature extractors (Barlow, 1975). Other receptor modalities appear to have similar design, but the type of features extracted in them are at present less clear than for vision (Evans, 1974; Webster and Aitkin, 1975; Mountcastle, 1974; Jones, 1975).

The cat and monkey retinal ganglion cells have circular receptor fields with center-surround organization, and these represent different solid angles of the visual field. They project to the lateral geniculate nucleus, superior colliculus, and a number of other diencephalic and mesencephalic sites. The lateral geniculate tissue exhibits somatotopic mapping of the field (Malpeli and Baker, 1975), an orderly segregation of the input of the two eyes, and a segregation of color categories of input. The projection from geniculate neurons to the striate cortex forms an orderly map of visual field locations (Talbot and Marshall, 1941; Daniel and Whitteridge, 1961; Whitteridge, 1973) and a systematic segregation by which circular receptor fields are linked to form edge detectors of definite orientation. Each edge detector element has the property of responding to displacement in the field of a boundary between areas of differing brightness or wavelength, the direction of displacement that produces the greatest amount of excitation being at right angles to the angle of inclination of the boundary which also has an optimal length or extent. The rate of displacement of the "edge segment" or "line segment" is also a factor in producing excitation of the striate cells. It is conceivable that stationary edge detectors evolve out of velocity detectors that are insensitive to stationary stimuli. There is evidence from developments in cat, rabbit, and monkey that oriented edge detectors may arise by anatomical modification of velocity detectors.

In the visual cortex, columns or lines of adjacent cortical cells in a penetration at right angles to the cortical surface display the same angle of inclination of edge detection. In each orientation column eye predominates, and columns concerned with one eye are lined up to form bands or slabs 0.4 mm wide that divide up the cortex into meandering ribbon-like territories that look like the stripes of a zebra's back, alternate territories receiving afference predominantly from one eye (Le Vay *et al.*, 1975). The ocular dominance slabs are laid out in a clear maplike pattern of lines that intersect the vertical meridian of the somatotopic visual field at right angles and line up parallel to the horizontal meridian (Figure 19).

Tangential electrode penetrations in the visual cortex of monkeys and cats reveal a uniform mapping of units that respond to serially ordered orientation of edges (Hubel and Wiesel, 1974*a*). Evidently the pattern is complex, but there are suggestions that isoorientation columns are like monocular dominance columns, but laid out at right angles to them. In passing through about 0.8 mm of cortex there is approximately a 180° rotation in the direction of the preferred orientation of cortical cells. This hypercolumnar unit corresponds functionally and in width with two ocular dominance slabs, one for the left eye and one for the right.

Hypercolumn assemblies have remarkable anatomical uniformity over the striate cortex of the monkey, corresponding to a patch of 0.8–1 mm of cortex for all parts of the explored central 20° of the visual field (Hubel and Wiesel, 1974*b*). This territory also corresponds to one location in the visual field, the size of the region in visual degrees following the cortical magnification factor. Inside one of these unit territories individual units represent a random scattering of receptor fields within this location. The two monocular territories of a hypercolumn represent corresponding locations in the two eyes.

Finally, color detection elements, innately wired to segregate four primary colors into opponent pairs, are also distributed over the cortex in an orderly pattern (Hubel and Wiesel, 1968; Zeki, 1973). Probably the mapping of ocularity, orientation, and color intersect to permit optimal distribution of the different elements within the somatotopic map of the visual field. The functional advantages of overdispersion (orderly gridlike or alternating bandlike progression) have been discussed by Hubel and Wiesel (1974*a*).

In a vertical penetration down a line of cells through all striate cortical layers in one location in the map of the visual field, ocular dominance and orientation may remain constant. The vertical and intracortical connections seem, however, to

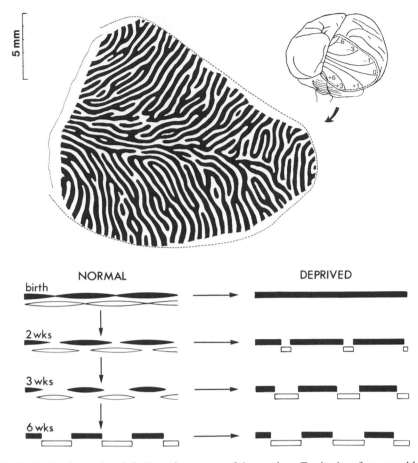

Fig. 19. Ocular dominance bands in the striate cortex of the monkey. Territories of one eye, black; the other, white. The visual field on the right occipital lobe is shown with horizontal lines 1, 3 and 6° above (+) and below (−) the horizontal meridian (0) (LeVay *et al.*, 1975). Bottom: A hypothetical scheme that might explain the effects of eye closures in newborn monkeys on segregation of optic terminals from the left and right eyes on the assumption that segregation is not complete until some weeks after birth. The width of the black and white bars represents presumed density of terminals. It is supposed that competition normally occurs between the eyes and that weaker inputs retract to make way for stronger. Strength of terminals is enhanced by stimulation, or weakened by deprivation (Hubel *et al.*, 1977).

permit combination of the primary categories into an orderly hierarchy of increasingly specific and elaborate detection elements (Hubel and Wiesel, 1962, 1968). The simplest featured detectors are cells in layer IV that receive direct input from the thalamus. Transcortical projection, from area to area of cortex, also permits integrative combination of receptor field types into higher order detectors (complex and hypercomplex fields), by means of long intracortical axons.

There are, however, many parallel, differently organized maplike projections of visual functions on the cerebral cortex of rats, cats, and monkeys (Allman, 1975; Zeki, 1974). It is probably a general feature of brains of birds and mammals for different classes of perceptual processing to be represented in a mosaic of centers that permit assembly of percepts in a number of different ways, by ascending and descending links between midbrain cortex and nuclei, thalamic nuclei and cerebral areas, as well as by transcortical "cascade" connections. Recent anatomical data for the pigeon confirm this view (Benowitz and Karten, 1976). Neither the functional significance nor the developmental history of newly discovered visual maps in areas 19, 20, and 21 of the posterior cortex are sufficiently worked out at the time of writing for a coherent interpretation.

Units in area 17 of the adult cat are apparently calibrated to respond to different binocular disparities, thus encoding depth or distance from the subject for a single stimulus that excites corresponding regions of the two eyes (Barlow *et al.*, 1967). Stereopsis units are recorded in the prestriate cortex of the monkey.

The theory of an innate anatomical system of combinations that permit progressive categorization and recombination of stimulus features is derived from the above remarkable anatomical and physiological findings. Nevertheless, the details remain incomplete, and it is not possible yet to discern the fundamental morphogenetic principles by which the network is built (see below). The more recent studies would seem to indicate that a very orderly pattern is formed by combining reiterated modular elements such as one would expect from the operation of intersecting morphogentic rules (cf. Hubel and Wiesel, 1974*a*).

Comparison of findings in a number of mammals suggests that the geniculostriate projection and parallel extrageniculate projections have the same general principles of structure in the different forms, but that the preponderance of oriented edge detection in the striate and circumstriate cortex of cat and monkey may be an adaptation associated with more refined oculomotor control and refined vision under conditions of fixation (Drager, 1975). There are also indications that the fundamental unit of detection for motion stimuli is a rate of change in visual angle, rather than in distance on the retina, which differs greatly in forms with different sized eyes. This brings up the important possibility that the fundamental unit is a morphogenetic building block of the egocentric somatotopic behavioral field (an element of behavior space). Daniels and Pettigrew (1976) propose that the extreme anatomical refinements of connections required for binocular stereopsis is a major factor in the evolution of the special characteristics of the visual cortex of cat and monkey as compared with the rabbit.

In all forms there is a bias for central binocular representation in the geniculostriate cortex and a comparatively strong representation of the peripheral field in the superior colliculi. This accords with a functional distinction between colliculus and striate cortex that has been brought out by behavioral studies of the effects of surgery to brain stem and forebrain visual centers, or split-brain surgery (Schneider, 1967, 1969; Trevarthen, 1968*b*; Trevarthen and Sperry, 1973). According to the two visions theory, midbrain visual cortex and striate cortex are complementary components in a visual mechanism which is tightly integrated through vertical connections and by convergence on a common motor system that is organized as a hierarchy of somatic elements. The functional properties of collicular units depend on projections from the cortex to the colliculus that confer form-discrimination capacity (Wickelgren and Sterling, 1969). However, at least in cats and more primitive mammals, the midbrain–pulvinar projection to areas, 19, 20, and 21 of the cortex has some form-discriminating capacity independent of the geniculostriate system (Sprague *et al.*, 1977). It is generally agreed that the midbrain cortex is inherently adapted to perception of spatial relations among stimuli and to regulation of attentional localization related to oculomotor scan (Ingle and Sprague, 1975). It is also organized to coordinate vision and audition with touch where parts of the body, such as limbs or vibrissae, project into the visual field (Drager and Hubel, 1975; Finlay *et al.*, 1978).

6.5. Postnatal Development and Adaptation of the Visual Cortex

Observation of visually naive monkeys prevented from receiving any patterned visual stimuli led to the conclusion that the above-described anatomy of retinocortical projections and intracortical connections is determined without benefit of patterned stimulation at birth (Wiesel and Hubel, 1974). This conclusion has been revised in the light of anatomical observations that show the newborn circuits to be only partly formed (Hubel *et al.*, 1977) (Figure 20). Their immediate development is largely indifferent to deprivation of patterned stimuli, however. Orientation columns and ocular dominance columns and their organization in unit hypercolumns are present after three or four weeks of visual deprivation due to suturing of both eyelids at birth. The responses of cortical cells appear as sharply tuned to their stimuli as in the adult. Precise orientation preference was observed in cortical cells

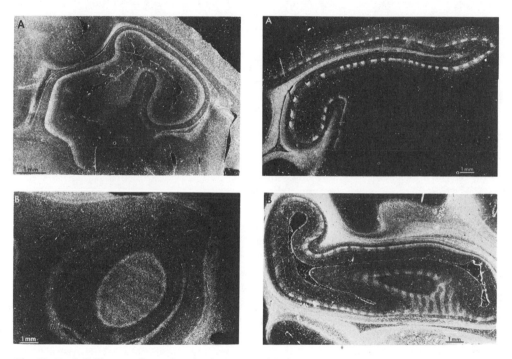

Fig. 20. Dark field autoradiographs of striate cortex following injections of radioactive fucose-proline into an eye. Left: Newborn monkey. A: Transverse section through the calcarine fissure and operculum on the right side, contralateral to the injected eye. The label is nearly continuous, with faint fluctuations in density to the right—over the dorsal calcarine stem. B: Tangential section through the operculum of the left occipital lobe. At the center, the section grazes layer IVc which shows regular banding with a periodicity of about 0.7 mm. Right: Adult monkey following right eye injections at two weeks. A: Transverse section through the right striate cortex. Labeled bands can be clearly made out in layer IVc. The white patches correspond to the labeled right eye. B: Parasagittal section through the buried calcarine cortex contralateral to the injected eye. One fold is intersected tangentially to show the layer IVc eye territories as parallel bands. The stretch of continuous label in the upper right part of the ring of layer IVc represents the monocular temporal crescent (Hubel *et al.*, 1977).

within minutes of access to stimuli, and the orderly array of responses recorded subsequently was insensitive to order of previous stimuli during the electrophysiological mapping. The monkey visual system appears to be remarkably mature in these respects soon after birth. However, precise measurements of receptor field size have not been made. It is likely that the newborn has much lower acuity than the adult. Simple complex and hypercomplex cells exist as in the adult, and orientation responses were obtained to an "on and off" of stationary slits of light, dark bars, or edges.

Binocular convergence of inputs to cortical cells is highly sensitive to visual deprivation or interocular imbalance of input near birth in the monkey. Although an accurate pattern of binocular convergence on cells of deeper layers of the cortex for corresponding locations and corresponding forms of stimulation may be observed at two days with no chance for their formation by patterned visual input, deprivation of patterned vision for a few weeks after birth, even after permitting three weeks normal access to experience first, leads to destruction of most of the binocular convergence, the majority of the cells giving responses only to one eye. Hubel *et al.* (1977) have found, by injecting one eye of a monkey with tritium-containing compounds the day after birth and following the radioactive terminals in layer IVc at one week, that the inputs from the two eyes, although already in the form of parallel bands, are only slightly segregated. Electrical recordings with tangential penetrations showed considerable overlap of inputs at 8 days, and much more advanced segregation at 30 days. They conclude that the ocular dominance columns are not fully developed until some weeks after birth. Both anatomical and electrophysiological techniques confirm that the lateral geniculate body is highly differentiated at birth, unlike the cortex (cf. Rakic, 1976, 1977). Evidently the terminals from the two eyes retract from overlapping regions by some kind of competitive repulsion to produce regular sharply defined monocular bands with centers separated by 0.4 mm (Figures 19 and 20). The periodicity must be determined by an innate morphogenetic mechanism even if the final segregation of fibers is due to a self-sorting mechanism like that proposed by Willshaw and von der Malsburg (1976). The process of segregation takes place if the monkey is reared in the dark (Wiesel and Hubel, 1974), but it is highly sensitive to monocular deprivation. Precise segregation of monocular inputs to cells in layer IVc of the striate cortex, on which, presumably, stereopsis depends, is prevented by monocular deprivation during the first postnatal month of a monkey's life (Hubel *et al.*, 1977).

These results reported by Hubel and Wiesel with the infant monkey contrast with those for kittens in which orientation detectors and binocular disparity detectors have been said to be created by coincidence of stimulation from patterned light (Blakemore, 1974; Daniels and Pettigrew, 1976). The argument concerns the degree to which the different connections, on which visual distinction of many features depends, can be determined genetically. There is no doubt that environmental stimuli may be necessary for normal circuits to develop or to be retained, but the evidence for detector formation out of less organized or unorganized and highly plastic growing circuits is less conclusive. The comparatively immature visual system of the newborn kitten becomes highly sensitive to deprivation or abnormal patterning of visual input a short time after birth when synaptic connections are forming in the visual cortex (Figure 16). Binocular disparity detection is destroyed by early postnatal deprivation of vision in one eye (Daniels and Pettigrew, 1976). It also adapts to a change in visual directions caused by placing a prism in front of one eye of a young kitten (Shlaer, 1971). But the experiments to test for educative as distinct from nonspecific supportive or stimulatory influence from the environment are difficult to perform and hard to interpret (Grobstein and Chow, 1976).

According to present information, reviewed by Barlow (1975) and Daniels and Pettigrew (1976), the more refined patterns of connection within the visual system of the cat are built with the aid of patterned stimuli which link simultaneously activated cells into permanent assemblies with strong functional commitment. Stationary edge detectors, specific for orientation, and binocularly responsive cells, by which dispartiy may be detected to permit depth detection by binocular stereopsis, are the most dependent on coherent and sufficiently varied stimulation. On the other hand, it is certain that all acquired powers of visual discrimination are set up within a highly elaborate innate ''wiring'' of nerve circuits that defines basic dimensions and categories or elements of vision and also the fundamentals of visuomotor coordination.

It is important, therefore, to distinguish each anatomical level of the visual mechanism according to its time of development in the general plan of growth. The cat and monkey differ in the maturity of their basic visual machinery at birth, but both show a progressive addition of visual analyzing components postnatally. The different components will have differing and regulated sensitivity to experience. Later developed mechanisms, even basic receptor pathways such as the foveal detector system essential for the highest resolution in central vision of the monkey or man and the central neural circuits to which it projects, may benefit from an education that is controlled by previously ''wired-in'' visuomotor circuits. Visual fixation and conjugate saccadic displacement of the direction of gaze are present at birth in man (Haith, 1978). The ''diet'' of visual detail will therefore be subject to the control of previously formed mechanisms that choose sites for fixation. They depend on an accurate and reliable motor system capable of correcting errors in convergence.

Interocular transfer of perceptual fatigue effects that lead to illusions of motion or tilt have been used by Hohman and Creutzfeldt (1975) to obtain evidence that the binocular neurons of humans are sensitive or labile over the first three years of life. In this period the experience of the two eyes is highly correlated in three-dimensional space by means of innate acts of selective visual attending. Banks *et al.* (1975) have also found that correction for congenital esotropia must be carried out before two years if it is to be permanently effective. Slater and Findlay (1975) claim evidence that binocular stereopsis is functional in the first days after birth. Other evidence for environmental effects on directionally sensitive elements in man is reviewed by Barlow (1975), but Leehey *et al.* (1975) found that the dominance of elements sensitive to horizontal and vertical contours as compared to diagonals is unlikely to be formed by pattern of retinal stimulation as claimed by some (see Daniels and Pettigrew, 1976).

It is likely that unspecified or unrestrained visual exposure of impressionable circuits would inevitably destroy their usefulness for perception. As yet, we do not have a description of the natural process of visual exploration (selective information uptake) and its part in formation of higher levels of the perceptual apparatus of the brain. At most we know a few basic looking strategies of neonates in artificial circumstances (Haith, 1978). Until we have more complete data there is little profit in speculating about the adaptive advantage of leaving circuit formation to ''instruction'' from the environment.

6.6. Changes in Visual Anatomy in Abnormal Environments and with Imbalance of Ocular Input

Many elements of the visual mechanism of young mammals are sensitive at critical stages of their development to effects of drastically reduced or distorted stimulation (Blakemore, 1974; Tees, 1976). Retina, dorsal lateral geniculate nucleus, superior colliculus, and visual cortex all show growth changes or degeneration of elements (cells, dendritic spines, or synapses) if the eyes are kept in the dark or prevented from receiving patterned stimuli by suturing the eyelids (Valverde, 1971; Blakemore, 1974; Ganz, 1978; Tees, 1974, 1976). Competitive interactions between components are revealed by monocular deprivation or by the effect of squint by which one eye is made to deviate from the normal convergence with the other eye (Garey *et al.*, 1973; Hubel and Wiesel, 1965; Hubel *et al.*, 1977; Wiesel and Hubel, 1963, 1965*a,b*). Destruction may be much greater with monocular than with binocular deprivation and the pattern of binocular convergence is destroyed by unbalanced or discordant stimulation of the two eyes (Hubel *et al.*, 1977; Guillery, 1972). Some effects of monocular deprivation or other restrictive stimulation are reversible, indicating that the structures put out of action may recover when conditions change (see Wall, 1975, for general aspects of this kind of recovery). All effects are greater if the abnormal stimulation is imposed in rapidly developing immature stages.

It is important to note that different parallel component systems of the visual apparatus may be differently sensitive to environmental effects. Further study of these differential effects may bring out the developmental relations between visual processes that analyze the stimulus information in complementary ways.

The colliculi of the cat, which show postnatal maturation immediately before that of the visual cortex (Stein *et al.*, 1973; Cragg, 1975), develop some functions normally if the visual cortex is removed (Sherman, 1972), but it is claimed that binocularity and direction selectivity of collicular cells develops under the influence of corticocollicular projections (Wickelgren, 1972). Again there appears to be functional competition for synaptic space within the colliculus tissue between retinal and striatal inputs. Anatomical evidence for this has also been obtained for the neonate hamster (Schneider, 1973) (Fig. 14).

The monkey colliculus shows ocular dominance "clumps" which become clearer prenatally during the last half of gestation (Rakic, 1976, 1977). Schneider's studies (1973, 1975, 1976) of the effects of surgery on the exceptionally immature colliculus of the neonatal hamster demonstrate the lability of early stages in differentiation and the loose anatomical commitment of first-formed terminal arborizations for their end site.

Binocularly deprived kittens show a variable degeneration of visual elements that are present in outline form at birth when the cortex is still poorly differentiated, and the deprived animals are permanently deficient in visual behavior (Wiesel and Hubel, 1963, 1965*a,b;* Ganz *et al.*, 1972; Daniels and Pettigrew, 1976). Orientation sensitivity (Barlow and Pettigrew, 1971) and binocularity (Pettigrew, 1972) are particularly sensitive to absence of visual stimulation. In the rabbit receptor fields develop after birth, but they are not sensitive to deprivation of visual experience (Grobstein *et al.*, 1973).

Monocularly deprived kittens lose binocular cells from the visual cortex completely (Wiesel and Hubel, 1965*b;* Ganz *et al.*, 1972). A period of reversed vision, exposing the eye previously deprived on its own, can partly "cure" the effects of monocular deprivation (Chow and Steward, 1972; Blakemore and Van Sluyters, 1974). There is a close correspondence between the period of maximum sensitivity to monocular deprivation and the rapid growth of synaptic terminals in the visual cortex (Cragg, 1975; Hubel and Wiesel, 1970; Blakemore, 1974; Blakemore and Van Sluyters, 1974).

When newborn monkeys have one eye removed or kept from patterned stimuli by suturing the eyelids, the imbalance of stimulation changes development, and the terminals in the striate cortex for the deprived eye occupy abnormally narrow regions while those from the other eye have larger than normal bands. The process takes between 4 and 8 weeks to complete, so it coincides with the observed maturation of ocular dominance columns and the separation of terminal arborizations (Hubel *et al.*, 1977) (Figures 19 and 20). The evidence suggests that the narrowing results from a takeover by competing terminals from the normal eye of regions normally in the territory of the terminals made weak by deprivation. Cortical dimensions do not change, and the spacing of eye dominance bands remains about 0.4 mm. The deprivation produces greater loss of territory in the hemisphere ipsilateral to the deprived eye, indicating that the crossed afference to the cortex is "stronger," and recalling the situation observed in the developing binocular visual system of *Xenopus*.

This evidence for competition between the visual circuits for the two eyes during development fits with an earlier finding that binocular deprivation in cats produces a less severe effect in cortical cell responses than predicted from monocular closures (Wiesel and Hubel, 1965*b*). There are other indications of competition between inputs from the eyes in the development of the lateral geniculate nucleus (LGN) and cortex of cats (Guillery, 1972; Guillery and Stelzner, 1970; Sherman, 1973; Sherman *et al.*, 1974). The monocular crescents of the projection are protected because they lack competing terminals from the other eye, and regions of the deprived eye corresponding to patches where retina has been removed in the other eye do not show weakened unit responses. The effect is clearly due to interaction between terminals from the two eyes, and it may depend upon a sorting principle that uses impulse synchrony to encode place of origin (Willshaw and von der Malsburg, 1976).

The population of orientation detectors in the striate cortex of the cat can be modified to conform to experience in a field of stripes all going one way. It is not certain yet how this effect may best be interpreted: as remodeling of plastic detector elements (Blakemore and Cooper, 1970; Blakemore, 1974; Pettigrew and Freeman, 1973) or selective elimination or suppression of unstimulated components (Duffy *et al.*, 1976). It may reflect reversible changes in the balance of maturation of different preestablished alternatives (Grobstein and Chow, 1976). Van Sluyters and Blakemore (1973) and Pettigrew and Freeman (1973) have produced spot detectors by raising kittens in a spotty environment. These would appear to be "created" elements since detectors of this form are rare in the normal cat cortex.

In all of the demonstrated instances of plasticity, there is no clear answer to the question about the degree to which the visual system normally achieves its patterned instruction from the visual array in which it grows. This point must be emphasized because in many textbooks and reviews extrapolations have been made from the findings to the effect that they prove visual perception of details to be acquired from the structure of the visual world or that the brain is a tabula rasa for these elements of perception. Experience may influence development of perceptual circuits in a number of different ways and have effects at different points in the program of development (Gottlieb, 1976a).

6.7. Changed Behavior with Deprivations of Experience

Animals differ greatly in their capacity to orient and aim at birth. The chicken is able to perform accurate pecks within two or three days, and Spalding (1873) showed that this act would mature without experience of trial and error. Hess (1956) showed that maturation of pecking could improve the coherence of aim even if this was displaced by prismatic bending of vision, proving that feedback or failure to hit did not guide the maturation process. More recent studies with nonmammals are reviewed by Ingle (1978). Cats are born blind and weak but after a month of development show rapid attainment of full visual control over movements. Monkeys and humans have some orienting capacity at birth and this competence progressively improves. Deprivation experiments reveal a complex interaction of maturation, environmental facilitation, and environmental formation of structure in the building of visuomotor systems (Ganz, 1975; Tees, 1976).

Walking and steering around objects is profoundly deficient in cats reared in total darkness. Although they still possess efficient space perception by auditory and tactile cues, exposure to vision causes total disruption of their locomotion. Similar disorientation and confusion is reported for humans when their vision is restored by removal of cataracts (Von Senden, 1932; Gregory and Wallis, 1963). Cats may show good recovery from effects of rearing in the dark, and monkeys deprived for less than 10 weeks recover so quickly (in 0–4 days) that one must conclude that they possess innate ability to orient in space which is destroyed by longer deprivation (Fantz, 1965; Wilson and Riesen, 1966). Response to approach of an object (looming) is absent after rearing in the dark, but it recovers after exposure to the effects of locomotion in light arrays. Human infants show a complex defensive reaction to looming of a shadow from two weeks of age (Ball and Tronick, 1971; Bower, 1974). They possess limited abilities at birth to orient vision in space and coordinate sound with seeing to do so (Haith, 1978; Mendelson and Haith, 1977).

Aiming of the limbs toward surfaces for purposes of support, or toward objects of prehension, develops after birth, and it is sensitive to deprivation of feedback from vision of limb movement. Extending of the paws to an approaching surface develops in the first postnatal month in kittens but does not depend on experience with patterned light (Hein *et al.*, 1970a,b). Deprivation in darkness causes loss of depth detection, but recovery is often rapid in a patterned light environment (Riesen and Aarons, 1959; Walk and Gibson, 1961; Wiesel and Hubel, 1965b; Baxter, 1966; Wilson and Riesen, 1966; Ganz and Fitch, 1968; Hein, 1970; Tees, 1974). The rapidity of recovery proves that the underlying competence (specific anatomical organization of structure) may be present without experience of visual information.

The perception of structural subdivision of surfaces, and corresponding orienting of the paws to left or right for accurate placing on projections, has a different maturation in the cat from triggered extension in the general direction of the approaching surface (Hein and Held, 1967). This compares with the monkey in which anatomical and behavioral studies show two separate systems, one for general ballistic aiming involving the proximal musculature in adjustment to target position, and the other concerned with local structure and aiming of the palm and fingers (distal musculature) under tactual and foveal control (Kuypers, 1973; Brinkman and Kuypers, 1973).

Reaching matures more rapidly in monkeys than in humans, but it appears to undergo the same succession of stages in a compressed schedule. Film and TV records prove that coordinated reaching toward objects in the visual field is present in outline in human neonates and that the complex sequencing of distal and proximal movements of reaching and grasping is innately programmed (Trevarthen, 1974*b,d*). However, arm extension is stereotyped and inadequate in extent for the first four months (McGraw, 1943). Observations of institutionalized infants that have much less than normal opportunity to reach and grasp objects show that deprivation can retard development of eye–hand coordination (White *et al.*, 1964).

Proprioceptively guided arm extension and sustained aim are achieved by infants before the development of manipulation with fine finger movement under central visual control. In the monkey correctly directed reaching matures in a few days after birth, but again, accurate grasping takes a longer period to mature. These developments correlate with anatomical changes in the cortical projection to spinal motor centers. The direct corticomotorneuronal projection which is concerned with refined motor control of the hand and dependent on tactile and visual guidance grows postnatally in primates (Lawrence and Kuypers, 1965). It shows a slower cycle of maturation in apes than in monkeys and is yet slower to mature in man (Kuypers, 1973). It is probable that the efferent mechanisms controlling arm and hand movement mature in relation to (probably slightly in advance of) the mechanisms for visual perception of details of form and space. In human infants accommodation, on which efficient fixation and visual resolution of detail depends, attains a primary level of maturity at the same time as visual guidance of the hand and manipulation of objects under visual control first becomes effective (McGraw, 1943; Gesell *et al.*, 1949; Haynes *et al.*, 1965; Twitchell, 1965).

Perception of depth in control of locomotion has been examined using the "visual cliff" in which information about a drop is restricted to the visual modality and dependent principally on motion parallax effects generated by eye displacement (Walk and Gibson, 1961). Visual deprivation, again, leads to temporary loss rather than destruction of this visuomotor coordination (Ganz, 1975; Tees, 1976). In the rat maturational and experimental factors in development of depth perception have been demonstrated by Tees (1974).

Perception of objects in motion and of the stable world when the subject is in movement or being displaced is helped by tracking movements of the eyes. Tracking of a complete visual array (as in compensating for body displacement) is different from pursuit of an object moving relative to the general array, but both need periodic displacement of gaze to other orientations. Recentering or redirection of gaze against motion information has very different requirements from the following movements. Pursuit of objects going behind other objects or making complex, hard-to-follow paths of displacement requires predictive capture with saccadic (ballistic) jumps that are aimed with respect to the probable future location of the target. All these different aspects of visual orienting need to be distinguished in studies of their maturation postnatally.

Since eye movements, and all other movements displacing receptors, cause change in stimulation like that which happens when things displace in the world, the brain must have some means of distinguishing what is self-produced from what is of external origin. In most cases there are distinctive features of self-produced receptor displacement in the array of stimulation, so, as Gibson (1966) has pointed out, it is unnecessary to postulate a reafference mechanism (von Holst, 1954; Teuber, 1960) in all instances.

Optokinetic tracking of moving stripes in a large surrounding field is innate in kittens, but the resolution of the system improves greatly over the first month. In monkeys and humans optokinetic tests show that visual acuity for moving striations improves more slowly. Smooth tracking of objects against a stable background is absent in infants for several months after birth in which only saccadic pursuit is observed (Haith, 1978), but smooth optokinetic following of a striped array may be obtained in neonates (McGinnis, 1930) and is present in monkeys on the day of birth (Ordy *et al.,* 1962). From the outset head and eye are coordinated in pursuit movements of infants, although the head rotations become stronger over the first month at the same time that the infant develops strategies for predicting the appearance of objects moving against complex backgrounds and behind surfaces (Mundy-Castle and Anglin, 1973; Bower, 1974; Haith and Campos, 1977). [Ingle (1978) cites evidence that in more primitive vertebrates thalamocortical circuits are essential to perception of edges that define gaps or holes.] Changes in head mobility suggest that maturation of neck muscles and their activity appears to parallel that of the proximal forelimb muscles (Trevarthen, 1974*d*). Deprivation studies with animals confirm that pursuit movements are innately determined, but sensitivity of the different components to postnatal experience has not been closely studied. Vital-Durand and Jeannerod (1974) and Flandrin and Jeannerod (1975) distinguish two levels in optokinetic response in kittens, the more precise function, which adjusts to velocity of the stripe rotation, depending more on postnatal experience and deriving input from the striate cortex.

Jeannerod (1978) presents a review of evidence that different levels in an innate hierarchy of visuomotor control mature at different times and show differing degrees of sensitivity to deprivation effects. Earlier developed ballistic orienting, used within a field of locations specified in the peripheral visual field to one level of precision, is less susceptible to deprivation effects than a later developed guided orienting function that is capable of a much higher precision. Jeannerod separately invokes the foveal (geniculostriate) visual projection and the peripheral (colliculo-pulvinar–extrastriate) projection. Robinson and Fish (1974) use this distinction to explain effects of depriving one eye of a cat of visuomotor experience for the first 18 months of life. The disadvantaged eye lacked photopic vision in the central part of the visual field and exhibited strabismus or squint. Visual orienting in low light, however, remained (see also Hein and Diamond, 1971). Kittens kept in the dark or in stroboscopic light fail to develop directional sensitivity and normal ocular dominance in units of the superior colliculus which are partly dependent on input from the striate cortex for their differentiation (Flandrin and Jeannerod, 1975).

Loss of the striate cortex in man leaves a very weak visuomotor function that has been overlooked until recently (Perenin and Jeannerod, 1975; Pöppel *et al.,* 1973; Weiskrantz *et al.,* 1974). In spite of a total deficiency in form vision due to occipital lesions, the deprived parts of the visual field still possess discrimination of direction and of the occurrence of events in light. Destriate monkeys recover vision of space and motion (ambient vision) (Humphrey, 1974; Pasik and Pasik, 1971). It is possible that the vision of space remaining after loss of the geniculostriate component is based on prenatally developed connections via the colliculi. Bronson's argument that human newborns have ambient (subcortical) vision and grow foveal (focal or cortical) vision later is, however, not well supported by developmental findings and is probably not a useful simplification (Bronson, 1974). The late developing extrastriate areas of the visual cortex send fibers to thalamic nuclei and the superior colliculi, and these probably contribute to ambient vision. Undoubtedly both systems are present but immature at birth. Peripheral vision is limited in newborns as in central acuity (Haith, 1978).

Restriction in perception of the body parts and of their joint use, and reduction in the active control of a subject in gaining experience produce severe defects of vision (Held and Hein, 1963; Hein and Held, 1967; Held and Bauer, 1967). This has led Held to propose that visual adaptation and visual maturation require exercise of self-produced effects and convergence of feedback from acts with some internal record in the brain of the "expected" or "likely" consequences of each act generated (Held and Freeman, 1963). It is possible, however, that what is missing is not a correlation between stimuli and an internal signal of acts performed, but a motor control of access to stimuli for parts of the perceptual mechanism that discriminate inside contexts established by the innate general orienting abilities. There may be very different experiences of form in stimuli in the experimental and control groups of animals and differences in the consequences (reinforcements) of acts performed (Taub, 1968; Ganz, 1974, 1975). All of these difficulties of interpretation apply to the experiment of Held and Hein (1963) which showed, by elegant use of machine by which one kitten transported a confined, near-immobile partner round a striped visual environment, that inactivity in early life can lead to pronounced, but temporary, deficiencies in visuomotor performance, in spite of non-self-produced visual stimulation.

Kittens deprived of vision of their paws, by wearing an opaque paper collar when in light, developed paw extension to an approaching surface, but failed to orient the paws to projections from the edge. Visual guidance of the limbs was also deformed in attempts to hit at objects (Held and Hein, 1963; Hein and Held, 1967; Hein *et al.*, 1970*a,b;* Hein and Diamond, 1971). There is rapid recovery of good eye–paw coordination after the normal conditions of visual control of movement are permitted. It appears that the remarkable specificity of effects of depriving perceptual monitoring of acts during early development may be due to suppression of parts of a mechanism that is innately structured in general design, but in which component parts mature in rivalry.

Experiments by Held and Bauer (1976, 1974) with infant monkeys gave results similar to those for the kittens. Deprived of vision of their arms and hands, monkeys lost normal eye–hand coordination and developed abnormal attention to their hands as objects of visual interest. But after one month or rearing without vision of arms and hands, recovery of normal function was rapid. It is important to consider the possibility that in this experiment an imbalance was caused by competition between the complementary proximal and distal mechanisms governing reach and grasp of objects. Compulsive hand regard looks like an exaggerated form of interest in manipulation under visual control, but without an object. This behavior is highly active in normal 1-month-old monkeys, but it is usually applied to an object. A different kind of deprivation procedure, allowing practice of aimed reaching under touch guidance during the period that the animals were prevented from seeing their hands, supports the conclusion that hand watching is reduced if nonvisual orienting of the arm is not allowed to degenerate (Walk and Bond, 1971). This study appears to invoke the distinction between ballistic reaching and visually guided fine grasping. Only the latter showed marked deficiency after deprivation of sight of arm and hand. However, Bauer and Held (1975) report that the practice in guided reaching outside visual control has limited application over the visual field, being progressively less effective for directions further away from the trained direction.

The above studies do not give unequivocal support to the notion that reaching movements are mapped by vision of the hands moving in visual space in early infancy. Their extension to infants reared in institutions fails to prove that human eye–hand coordination is acquired by learning (White *et al.,* 1964). They do, however, indicate that exercise plays a significant role in the maturation of visuomotor control and that full calibration of eye–hand coordination requires seeing the hand in action. Bower (1974) reports observations that human infants gain predictive adjustment of muscle contraction to the weight of a seen object in the second year. He claims that guidance of reaching by audition, by vision, and by mechanoproprioception mature at different times in a somewhat competitive interaction that is guided by experience obtained in the first two years. Fine manipulation shows developmental improvement for years after this.

Development of the mechanisms for space vision and visuomotor coordinations could, conceivably, be based largely on an innate program. The demonstrated influence of the environment might be only facilitative and not directive. In the case of skills or discriminations that must become closely adapted to unpredicted events in the environment, and that are consequently adapted to be modified by change in the features of the environment, as well as reformulations of perception to accommodate change in dimensions of body parts through growth, there must be a neural mechanism for storing effects of stimulation and modifying the processes of perception accordingly. There is increasing evidence that the mechanisms of neural maturation and of such adaptations to experience are closely related. Even newborns have motor functions that control their visual stimulation, and this probably regulates development of visual circuits in the brain. Haith (1978) suggests that eye movements are adapted to generate maximum cortical neural excitation by seeking the best patterns of stimulus change on the retina.

6.8. Learning and Modification of Perception: Relation to Brain Growth

Visual stimuli leave highly detailed impressions that last for some short time in the perceptual mechanism (Neisser, 1967). These temporary images are constantly replaced, but in the process a selection of information is retained. Remarkably detailed, but incomplete, memories may remain to affect later behavior. Evidently there is a process, or a series of processes, that transfers temporary impressions into a more permanent store which, moreover, may be ready for use in appropriate situations, even though the right matching event does not take place. Categorization of impressions confers meaning and application to related but new impressions. In other words, retained effects of experience generalize to other effects and become part of the process of perceiving.

In infancy and childhood there is an elaborate development of mental schemata that confer increasingly powerful understanding of objects and perception of their properties and uses. A major change occurs toward the end of the first year of life when many psychologists consider a human being first possesses imagistic representation and thought. But this is only the beginning of a process that continues throughout the years of childhood affecting all aspects of psychological function. The acknowledged master of this domain is Piaget (1953, 1967, 1970), whose studies have been taken up and elaborated by developmental psychologists prodigiously since the 1960s. Unfortunately there is almost no correlation between this developmental psychology and sciences of brain and growth. We have indeed very little real knowledge of the neural mechanisms of the critical events in perception or the formation of images and memories. It helps to consider the advantages of coding complex arrays of stimuli according to a set of prewired features. But doing this leaves unclear how features in stimuli become associated together in perception of forms and things, and how they may be variously interpreted or ignored by the mind (Neisser, 1976). Presumably many rules for generalization and perceptual integration in terms of phenomenal reality are derived from built-in structures of the brain, even though they may only attain full maturity through use.

This view is held by Mark (1974), who proposes a model to link concepts of brain histogenesis with a cell-process theory of memory. Findings in experiments with early postnatal visual imprinting of chicks to peck-objects provide support for this analysis. The one-trial learning tests, with different kinds of chemical blocking agents to modify nerve processes, demonstrate that excitation of cells may produce short-term effects (less than three hours) due to changes in the distribution of ions over the cell membrane under control of the sodium pump mechanism, and that these effects may be converted into more lasting traces by protein synthesis functions that are coupled to the sodium pump (Mark and Watts, 1971; Watts and Mark, 1971; Gibbs *et al.,* 1973). Mark proposed that synthesized protein, made after enhanced activity of the sodium pump has brought into a nerve cell the amino acids required, may be involved in the interneuronal recognition process and therefore influence synaptic connectivity. Selective formation of permanent impressions of stimuli results from the distributions of activity in cell connections left uncommitted by development.

Excited neurons change cation concentrations in surrounding media and they may transport ions over long distances. They also secrete chemical transmitters. Both cations and neurotransmitters have strong effects on the metabolism of cyclic nucleotides which, in turn, are agents in the regulation of cell development and differentiation (McMahon, 1974). By such links, a model may be formed that could implement formation of new nerve circuits by stimulating or altering processes of nerve cell differentiation and the formation or strengthening of synaptic links.

Unfortunately, we do not know if such processes actually occur when nerve cell growth is being modified by stimulation, or if they occur during learning. Nevertheless, even by a cautious estimate, current biochemical reasoning makes it likely that the physicochemical basis for an anatomical change in memory will turn out to be closely related to the processes that regulate formation of circuits in development from the early embryo on. The proposal that nerve circuits sort themselves spontaneously during development by a process of competitive interaction of impulses, or products of impulses, in overconnected arrays of nerve terminals (Willshaw and von der Malsburg, 1976) offers interesting possibilities for the development of Mark's theory of an active neural mechanism of memory.

6.9. Functional Effects Following Brain Surgery to Monkeys during Post-natal Development

Goldman (1971, 1972, 1976; Goldman *et al.,* 1970) has demonstrated changes in developing psychological functions of young monkeys following surgery to dorsolateral and orbital regions of the frontal cortex, and to related subcortical nuclei, at different times after birth. In part, these changes are comparable with adjustments in the distribution of visual terminals between alternative central sites on a competitive basis. The compensatory formation of neural circuits is influenced by presence or absence of components within a functional system, as by relative stimulation of components at critical stages of their growth. However, more recent results show that compensation is limited.

Goldman removes areas that in the adult are essential for certain psychological functions of perception and memory. In the six months between 12 and 18 months after birth, effects of removal of orbitofrontal cortex, head of the caudate nucleus, and dorsomedial nucleus of the thalamus are as selective and severe as those caused by the same ablations in young adults, but lesions of the dorsolateral cortex do not produce the same effects, apparently because this cortex is relatively immature and not performing the function as in the adult (Goldman, 1976). If the latter group of animals are retested at 2 years of age, when the brain is more mature, they show severe deficits on the critical delayed response tests. Orbitofrontal lesions fail to have the usual effect on monkeys much earlier, at $2\frac{1}{2}$ months of age. This area matures earlier than the dorsolateral area, and effects of its ablation appear by 1 year of age.

Differences between very young and older monkeys in the effects of cerebral lesions thus appear to be related to differing maturation rates of the removed areas. They are not easily interpreted in terms of compensatory changes and functional redistribution in the young brains. A powerful effect of an inflexible developmental program, continuing long after birth, is indicated.

6.10. Growth of Perception in Infants and Children

It is not possible in the context of this review to cover information on the progressive changes in perception in infancy and childhood and their assessment by psychologists. There is, in fact, no consensus on the order of appearance of processes involved or on their origins. As reviewed, for example, by E. J. Gibson (1969), theories of perceptual development cover a wide terrain of possibilities.*

The empiricists leave biological processes of development a small role to play. Those that emphasize cognition, imagination, inference, attentional readiness, or problem-solving with insight, place perception subordinate to the development of mental skills and strategies for coping with the world. Those that do allow innate determination in the growth of perceptuomotor structures *per se* (Koffka, 1931; Werner, 1961; Gesell *et al.,* 1934) are handicapped by the insufficiency of our knowledge of the neural circuits specifically concerned with processing peripheral information. The Russian physiologically oriented school (Zaporozhets, 1965) attributes the process of development to Pavlovian conditioning and practice rather than to growth and differentiation of generative or interpretive structures adapted to perception.

The most comprehensive theory of psychological development is that of Piaget (1953, 1967), who conceives the structures of intelligence, in the form of schemata for acts of reasoning, to grow by a self-regulating process of internal equilibration while keeping an active assimilatory relation to external conditions. Peripheral uptake of information onto schemata that determine acts of exploration or praxis gives rise to more and more powerful mental representations. The process of perception, therefore, develops in increasing dependency on thought or reasoning. His concept of early perceptual development from the "reflexes" of the sensory–motor period of infancy, while biological or embryogenic in form, does not offer

---

*The first volume of a new series on Perception and Perceptual Development, edited by Walk and Pick (1978), appeared while this review was in press. Departing from the proposals on perceptual learning of Eleanor Gibson (1969), it includes articles on the effects of environmental conditions, including cultural factors, pictorial representation, and language; on human perceptual development from infancy; and reports of recent physiological and anatomical effects of restricted experience in early life in animals.

precise directives for a neurobiological study. His great contribution lies in defining terms for a program of systematic growth of logical operations, and especially the development of concepts pertaining to the "object" as a psychological creation. He describes the mind as progressively achieving freedom from an innate tendency to perceive situations egocentrically and to become committed to one object of interest. This is overcome by a spontaneous process of decentration, which generates more complete representations of the conditions for versatile behavior. In other words, perceptions are part of immediate and direct self-regulation early in development, and then increasingly part of an extended model of reality in which the self may have multiple places or in which several selves (subjects) are interacting through communication.

In E. J. Gibson's view, which follows the perceptual theory of J. J. Gibson (1966), the problem of perceptual growth is one of the formation of increasingly efficient skills for selective uptake of information from an information-rich environment, and not one of associative learning. In her account there are, again, no clear landmarks from which to plan a neurobiological interpretation. Children gain in skill at seeing what is in the world, taking longer to learn more subtle distinctions and combinations in experience. They gain in perceptual skills as they gain in cognitive mastery and are unable to carry out the distinctively human task of perceiving meaning in coded stimuli, built of pictures, in words, or in symbolic objects without cultural training. Nevertheless, they are amazingly precocious in attempting acts on objects and in actually achieving communication with persons. How this comes about is a major biological mystery.

Bower (1974) presents a unique analysis of the development of intelligence in infants based on experiments which show attainment of steps in the development of the object concept much earlier than proposed by Piaget. Marked changes in perceptual abilities in the first two years are explained by Bower as the consequence of an epigenetic process of mutual adjustment between rules for perceiving and due to differentiation of specific distinctions within abstract formulations of goals for interest. Direct perception by vision, hearing, or touch is portrayed as limited by growth of body parts which change certain metrical dimensions affecting information uptake, and by insufficiency of motor coordination. Each perceptual modality is dependent on exercise for development, and associative affinities between modalities, while prepared for by antenatal growth of a common orienting field, are developed postnatally and are highly sensitive to disuse or exercise.

Lack of information on growth changes in perception may be due to the fact that it achieves efficiency very fast. Perception may mature in a few months as Bower proposes. There is physiological evidence that the human peripheral visual system undergoes rapid maturation in the first six months in correlation with the mastery of orienting and prehension (Pirchio *et al.*, 1978). Or the difficulty may come from the tendency for psychologists to consider perception in rational or performative terms related to adult function that do not permit one to see growth in the abstract information-selecting processes that could be identified with growth of particular brain parts.

Concepts of a monotonically ascending perceptual power, by learning from simple reactions to complex integrations, are seriously upset by recent demonstrations that young infants are capable of categorical judgments for many subtle and complex patterns of stimuli, and that they perceive within several modalities together in one preformed space of attending and acting (Haith and Campos, 1977). Perceptual discrimination begins earlier than it should if it is a practiced skill, and the findings argue strongly for an autonomous and largely prenatal formation of a general psychological process for perception of space and objects, independent of skillful action and, to a large degree, independent of cognitive achievements that require behavioral exploration of objects and their relations.

Observations of the selectivity in orienting and alerting reactions of infants, and of their deliberate control of head-turning, sucking, or limb movements in conditioning tests, reveal that within the first few months they are able to perceive sights, sounds, etc., over which they have no direct locomotor, manipulative, or articulatory control (e.g., Fantz, 1963; Papousek, 1961, 1969; Sequeland and DeLucia, 1969; Lipsitt, 1969; Eisenberg, 1975; Bower, 1974; Walk and Gibson, 1961; McKenzie and Day, 1972; Eimas *et al.*, 1971; Trehub, 1973; Bornstein, 1975). They can select, integrate, and categorize stimuli from objects, perceive the distance, direction, size, displacement, orientation, and color of an object, and detect the supramodal or amodal connection between its separate indications in receptors of eye, ear, hand, and mouth. They are also able to follow the passage of an object behind a screen and anticipate its reappearance. Even more complex perceptions must underlie their special interests in the movements of persons and especially in their acts of communication. All these highly complex perceptual integrations are made in some degree while movements of orienting and focusing of receptors, although already well-patterned, are definitely immature, and in absence of controlled locomotion or object prehension. As soon as they can crawl, infants show perception of the drop distances of horizontal surfaces from their eyes (Walk and Gibson, 1961; Tees, 1976).

Evidently the growth of the brain before birth permits differentiation of a neural field for specifying phenomenal perceptions. The characteristics of this field, when revealed in infancy, suggest that it is not composed of independent, unimodal image-forming devices. A coherent space for behavior is defined and in this stimulus distinctions are made in many equivalent perceptual subfields that process stimuli in distinct ways.

The reactions of infants 2–3 months old to signals brought under their motor control by artificial devices show that perception is linked to generation of movement at a highly elaborate level soon after birth (Papousek, 1969; Lipsitt, 1969; Kalnins and Bruner, 1973).

Bower (1974) has emphasized certain developmental transformations, including some apparent regressions, in the abilities of infants to perceive by different sense modalities. Correlated with developments in action, such as mastery of reaching to grasp objects which occurs at 4–5 months, or the development of standing and walking, these perceptual changes are not monotonically progressive, and they cannot be explained as cumulative learning effects. They argue for competitive interaction between self-organizing cerebral components. These are prepared as functional alternatives or complements for the neural regulation of acts by pick-up of perceptual information. Observations with blind, deaf, or armless infants show that processes directed to elicit or receive input through the missing organs grow for a time as if they were to be used. Then, at a stage at which they would normally mature with the benefit of exercise and stimulation, they decline (Drever, 1955; Axelrod, 1959; Lenneberg, 1967; Freedman, 1964; Fraiberg, 1974, 1976). The effects may reflect overgrowth by intact systems and may, conversely, lead to deficiencies of other forms of mental activity that are not solely dependent on the defective channel.

All aspects of psychological maturation in man are integrated with interpersonal processes that permit perception and action to be constantly extended and enriched, not only by the subject on his or her own, but also by joint activity with others who can act as agents, and also show and teach. It is not surprising, therefore, to find that there is a strong innate predisposition to cooperative, interpersonal experience, although this topic has been remarkably neglected in research. The process of sharing experience of surroundings and things becomes systematic in the second half of the first year of life, long before speech, and early enough to affect the basic maturation of object perception, although whether it actually does so or not is open to investigation. It is latent in the exceedingly complex person-perceiving functions and expressive actions of infants 2–3 months of age (Trevarthen, 1975, 1977).

Of all the achievements of infants brought to light by recent observational research, those pertaining to recognition of and communication with persons are the most remarkable. That they permit visual identification of faces and hands as organs of expression, and visual and auditory perception of expressive movements of many kinds in the first months after birth, is proved by imitation tests (Maratos, 1973; Uzgiris, 1974). Infants are selectively responsive to the sight of human faces and to human voice sounds (Frantz, 1963; Lewis, 1969; Wolff, 1969), and they categorize speech sounds in basic phonological forms (Eimas *et al.*, 1971; Trehub, 1973; Eisenberg, 1975). Tests of cognitive achievement demonstrate levels of proficiency more strongly or earlier when human objects, like the mother, are used (Décarie, 1965).

Evidently infants perceive people preferentially. To do this, they must have an innate knowledge of them as different from other events of the environment. The rapid development of game-playing, appropriate signaling of interest and orientation, and reciprocal "conversation"-like performance in exchanges with people, that are regulated by both participants, show that the brain of an infant is adapted to receive and give experience in exchange with others (Brazelton *et al.*, 1975; Schaffer, 1977). This process undoubtedly expresses an organization of the human brain for cooperative intelligence, but the neural systems involved in perceiving persons and what they do are virtually unimaginable with present knowledge.

6.11. Development of Asymmetry of Mental Functions in the Human Brain

Evidence that the cerebral hemispheres of man are inherently specialized to perform very different psychological functions came over 100 years ago from the discovery of relationships between hemispheric site of brain injury and particular neurological or psychological abnormalities (Vinken and Bruyn, 1969). Complex mental processes were implicated that bear no obvious relation to the embryogenetic somatotopic mapping of each half of the body and each half of body-centered perceptual space in the opposite half of the brain. Although the neuroanatomical basis for the lateralized functions is just beginning to be revealed (Galaburda *et al.*, 1978), these high-level psychological functions offer the best indirect evidence for protracted postnatal differentiation of adaptive cerebral mechanisms in man.

In a majority of cases, disorders of speaking, writing, or understanding of language, as well as inability to perform analytical thought or serial reasoning as in mathematical calculation, follow lesions in the left hemisphere. Accumulated evidence of many kinds leads to the conclusion that, in the adult population, nearly 90% are genotypic right-handers, and of these over 99% have left-hemisphere dominance for language. Of the approximately 10% of people who are left-handed or with mixed handedness, about 56% have left-hemisphere language dominance (Levy, 1974, 1976).

Lesions of the right hemisphere may be diagnosed by demonstration of disorders in visuospatial perception, in body image, and in recognition of complex perceptual "gestalts," including such diverse entities as faces and musical chords. Recent interpretations take the view that the hemispheres develop different kinds of perceptual processing in an orderly topography of functional regions, the right being specialized for parallel or simultaneous synthesis of images signifying layout or configuration, the left performing an analytic function dependent on brain processes that specify sequences or chains of relationship between identified elements.

Understanding of the brain mechanisms of these complementary psychological processes of the hemispheres has been greatly advanced by detailed study of consciousness and reasoning in a few epileptic patients in whom the cerebral hemispheres have been surgically disconnected to control the spread of severe seizures that were unresponsive to drugs (Sperry *et al.,* 1969; Sperry, 1970, 1974; Nebes, 1972; Levy, 1974; Trevarthen, 1974*a,b*). An elaborate series of tests with the split-brain patients show that certain perceptual processes, especially generation of images of the sounds of words on which comprehension and, above all, production of speech or writing depends, are confined to the left hemisphere (Gazzaniga and Sperry, 1967; Levy and Trevarthen, 1977; Zaidel, 1978). The ability visually to perceive whole forms from fragmentary stimuli, or the unanalyzed configuration of complex terms seen, heard, or felt as whole identities simultaneously, are much more easily performed by the right hemisphere (Nebes, 1974; Levy *et al.,* 1972; Levy and Trevarthen, 1977). Findings with patients with lateralized brain injury have thus been confirmed and clarified.

Comparable asymmetries are also brought out in normal subjects by psychophysical tests which measure the rate and efficiency of perception with different kinds of stimuli confined to the left or right visual field or hand (White, 1969). The effects observed, although consistent, are weak and require statistical methods for their demonstration. To test lateralization of auditory perception, conflicting stimuli are presented to the two ears (dichotic stimulation), and this procedure causes the bilateral attention processes, that normally mask asymmetries at the periphery, to be partially blocked. Hemispheric differences in perception of sounds become apparent as differences in hearing by the two ears.

There is no doubt that hemispheric lateralization of mental function is under some heritable control, and it has been suggested that a relatively simple one- or two-gene model may determine both handedness (or hemispheric hand control) and hemispheric dominance for language (Annett, 1964; Levy and Nagylaki, 1972; Levy, 1974, 1976). Corballis and Beale (1976) accept a theory of Morgan (1976) that lateralization of brain functions may be a result of an asymmetry of a cytoplasmic factor in the egg cell. Performance of identical twins on intelligence tests compared to the performance of nonidentical subjects matched for age indicates that visuospatial functions associated with the right hemisphere are also under genetic control. Reading disability (dyslexia) in schoolchildren correlates with indistinct manual dominance and confusions of left–right sense indicative of incomplete lateralization of functions in the cerebral hemispheres (Corballis and Beale, 1976). It has been suggested that performance differences of adolescents on tests of verbal and spatial abilities are related to maturation rates more strongly than to sex. Children maturing late were also more lateralized for speech than those maturing early. Sex differences in mental abilities may, therefore, be due to differences in cerebral functional development correlated with differential rates of biological maturation (Waber, 1976).

The effects of one-sided brain injury at different ages show that the functions of the hemispheres become progressively more distinct over a period of years after birth (Lenneberg, 1967; Hécaen, 1976). Brain injury before or at birth rarely produces distinctive language disorders later. Growth of language may be retarded, but the degree of retardation has not seemed to correlate with the site of lesion, if the injury occurs before the age of 4 (Basser, 1962; Kraschen, 1973). Symptoms of language disorder are claimed to occur with right-hemisphere lesions only in children less than 5 years old, unless they are genetically predisposed to center language control bilaterally or in the right hemisphere (Kraschen, 1973, 1975; Hécaen, 1976). Woods and Teuber (1978) suggest that the apparent involvement of the right hemisphere in language is, for most cases, an artifact due to postoperative infection of the remaining left hemisphere, which was common in cases before the adequate antibiotic treatment was available. With injuries in early childhood there is a progressive clarification of the initial clinical effects with age of trauma, and this correlates with mastery of language in normal development (Lenneberg, 1967). Injury to or removal of the left hemisphere of children in the first few years causes the child to become mute or to have disordered perception of speech. The effect of partial damage depends on the focus of lesion, whether toward frontal parts of the hemisphere (mutism), or exclusively in the temporal area (auditory perceptual defect) (Hécaen, 1976). This pattern of localization of function is like that of adult, but not exactly the same, in correlation with less-well-developed capacities of the child to form words or express ideas in language. Thus, brain-injured young children do not exhibit logorrhea, and they rarely show paraphasia.

Usually a brain-injured child recovers the appropriate level of language function completely inside a few weeks or a year, even after dominant (left) hemisphere lesions of a size that would produce permanent aphasia in an adult. Permanent effects of early trauma to the undeveloped speech control mechanisms of the young child can only be revealed by rigorous testing.

Tests with adults who have had early childhood brain lesions prove that the functional recovery may involve unusual development of language in the right hemisphere (Milner, 1974) and this may, in turn, protect their verbal functions at a later date against epileptic foci developing in the left hemisphere (Landsdell, 1962, 1969). The neural circuits of the right hemisphere of a young child must, therefore, possess the potentiality for speech perception and generation of speech, although in many left handers and nearly all right handers the same circuits would lose this capacity in development. Even so, the specialization of the left hemisphere for generative language and the right for visuospatial and constructional functions has definitely begun at birth and is reflected in more subtle psychological defects of patients who have had brain damage before or near birth (Woods and Teuber, 1973, 1978; Rudel and Denkla, 1974; Rudel *et al.*, 1974).

The anatomical foundations for asymmetrical functions of the hemisphere are now known to be established before birth. The hidden upper surface of the temporal lobe inside the Sylvian fissure of the left hemisphere (planum temporale), corresponding to Wernicke's area or the auditory association cortex, is enlarged in the majority of adult brains (Geschwind and Levitsky, 1968; Geschwind, 1974; Galaburda *et al.*, 1978). This area is also enlarged in most fetuses and newborns (Teszner *et al.*, 1972; Witelson and Pallie, 1973; Wada *et al.*, 1975). Damage to it in an adult causes severe defects in perception of speech and writing. There are also enlargements of the parietal operculum on the left side above the Sylvian fissure, and a comparable asymmetry has been found in the brain of Neanderthal man and in the orangutan (LeMay, 1976).

Psychological tests of young babies, mentioned above, show that perceptual categorization of speech sounds into entities which correspond with consonants and syllables in the production of mature speakers occurs very early in infancy, probably from birth (Eimas *et al.,* 1971; Trehub, 1973; Fodor *et al.,* 1975). Furthermore, EEG records have revealed distinctive evoked potentials for speech sounds in the left hemisphere of neonates as well as adults, as compared to the potentials evoked by melodies or noise (Molfese *et al.,* 1975). These findings strongly support the view that complex perceptual mechanisms, specialized to regulate speech, are innate in man and functioning in infancy at some level.

The condition of language in the normal nondominant right hemisphere of right-handers has been brought out most clearly by studies of cases in which the left hemisphere has been removed after language is developed, and in split-brain subjects with hemispheres disconnected after the early teens (Smith, 1966; Gazzaniga and Sperry, 1967; Zaidel, 1978). The isolated or disconnected right hemisphere appears to have the vocabulary for comprehension of all grammatical classes of words of about a 15-year-old, but it is very deficient in syntactic generative functions and is unable to formulate the acoustic image for a word to denote an object identified by nonlinguistic means. It may only speak a few automatic words or short phrases. Evidently there is accumulation of a lexicon throughout adulthood within the "minor" hemisphere, but by about 5 years of age the capacities to generate speech and to comprehend or recall context-free speech sequences more than one or two words in length are located entirely on the left side. Basic to this class of defect due to development of a strong lateralization of language is loss by the right hemisphere of the function which may evoke the sound of a word for an object from the visual or tactile image of the object, or from its remembered identity (Levy and Trevarthen, 1977).

Recent clinical and experimental data suggest that perception of speech as an adjunct to the process of language generation is lateralized in the dominant hemisphere within the first 5 years after birth. After this, language development is funneled into the left hemisphere, possibly as a secondary effect but probably by development of different neural systems with longer cycles of differentiation. Receptive semantic function continues to develop in the minor hemisphere, possibly at a progressively reduced rate relative to the left hemisphere (Zaidel, 1978). Relocation of speech synthesizing functions in the right hemispheres when the left is injured early in life may be supposed to involve a neural transformation within developing circuits analogous to that which permits optic terminals from a normally stimulated eye to invade the territory of a stimulus-deprived eye in layer IV of the striate cortex of the newborn monkey (Hubel *et al.,* 1977).

Any competitive interaction between territories in the neocortex of the hemispheres for representation of higher psychological processes and resultant asymmetrically patterned mechanisms must involve transmission of information through the corpus callosum, unlike the interactions of the striate cortex input which may occur directly between terminal arborizations of lateral geniculate cells representing the two eyes. If the conditions of interhemisphere communication are changed while the hemispheres are actively developing, then differentiation of speech-regulating and visuospatial mechanisms is imperfect, and the right hemisphere retains language functions it would normally lose. Conversely, injury or loss to the right hemisphere spoils the development of language receptive functions by forcing left hemisphere circuits to be occupied more than is normal with visuospatial operations. It may be presumed that learning of language, and practice of performative skills that depend largely on a strong inventory of visual identities and on resolution of visuospatial confusions inherent in the variety of layouts which the visual world presents, would together favor hemisphere differentiation towards an optimal dual-hemisphere system which is highly asymmetric.

Presence of the commissures would appear to be necessary for hemispheric complementarity to be completed. This conclusion is supported by the results of tests on a patient with congenital absence of the corpus callosum—a malformation of the brain that comes about in the late embryoric period (Saul and Sperry, 1968). This subject showed none of the split-brain effects and no evidence of hemispheric specialization of function. It is possible that each hemisphere carried, besides the mechanisms necessary to generate speech, full bilateral representation of the visual field. If so, the development of mirror complementary striate area projections is a consequence of competition between left and right field projections that is mediated through the neocortical commissures. Less likely is the possibility of an abnormal linking of striate areas via an enlarged anterior commissure or through the thalamus or midbrain. A higher than average verbal IQ associated with a "mild right hemisphere syndrome" (Sperry, 1974) may indicate that invasion of the minor hemisphere by language functions had preempted circuits that would otherwise have mediated the visuospatial processes. This accords with Levy's view of the evolution and development of lateralized language functions in competition with more primitive visuospatial abilities (Levy, 1969).

The timetable of development for mechanisms of language revealed by the study of the effects of brain injury at different ages correlates well with our meager knowledge of maturation of cortical and subcortical circuits for audition and for motor control of speech (Lenneberg, 1967). Developments after the age of 5 years in posterior "association" cortex tally with the optimal period for acquisition of the highly complex intermodal skills of reading and writing. The main anatomical finding, principally from studies of myelinization of tracts and intracortical axons, and their relation to developments in language abilities are summarized in Figure 18, based on Lecours (1975).

It should be noted that this method of portraying language development fails to indicate the role played by innate structures that define the deep grammatical forms of human communication which are already evident in the behavior of infants, before words are used to give wealth and precision to messages of intent and understanding (Bruner, 1975; Trevarthen, 1974*d*, 1977; Halliday, 1975). Once again we are led to consider the contrasting roles in development of brain systems that either: (1) process afferent information according to invariants in it that define layouts, events, objects, structures, and substances; (2) coordinate movement; or (3) relate present experience to an accumulated knowledge and understanding of what the world has to offer. All of these three kinds of brain functions affect perception, but only the first is directly and exclusively perceptual.

7. Conclusions

Present knowledge of brain development does not give us a neural explanation of the process of perception, nor determine, once and for all, where the contributions of inheritance end and experience takes over. Nevertheless the idea that perceptual mechanisms begin in the early embryo to unfold an innate specification appropriate to their role in regulating behavior of a moving body is clearly supported. This alone requires radical revision of psychological theories based only on experiments with perception in observing, rather than acting, adults. It is incompatible with those theories in which the concept of learning displaces interest in problems of development.

The main categories of perceptual activity in control of action, outlined in Section 2, appear to emerge as the brain grows in the embryo and fetus by a process akin to morphogenesis of metabolic and skeletomuscular machinery of the body, but are much more complex than these. An elaborate behavioral field of perceptual activities is created prefunctionally. It is related to the polarity and symmetry of the body and its hierarchy of systems of movement, and it is active in most aspects within a year of birth.

After birth, development of the perceptual mechanisms continues, and environmental patterning of circuits in a memory system of great structural complexity builds a progressively more detailed model of the meanings and distinctions of the external world which is the context and goal for voluntary behavior and consciousness. Perceptuocognitive skills continue to grow throughout life and to add to the power of perception to detect information in surroundings. This growth owes as much to neural differentiation and self-sorting of nerve circuits as to the impress of experience on what is differentiating.

On the basis of present knowledge, the growth of the human perceptual systems may be sketched as follows:

For two weeks after conception, the embryo body is a communicating system of cells in which a pattern of biochemical control of unknown molecular constitution defines a coherent symmetry of parts. In this symmetry, the nervous system is generated as a set of cells with specific positional relations both to the potential body surface and the internal effector organs, including the skeletomuscular organs of body motility. At the interface between nervous system and body surface, along the presumptive dorsal midline of the body, a bisymmetric set of receptors differentiates out at the time the body subdivides into segments and after a clear polarity has been established in the third week. An anterior head region is formed, poor in mesoderm and rich in complexes of ectoderm and neurectoderm that go to form special sense organs for perception of things at a distance. The trunk, in contrast, is rich in muscles derived from lateral plate mesoderm, and here the neural crest cells aggregate in a set of segmental primordia of somatic and visceral afferent connectives that will map the inside and the outside of the body into the brain.

By the end of the third week, each receptor primordium gains polarity markers, also of unknown constitution, that reflect the polarity of the body, and this representation is initially in equilibrium with a general biochemical or biophysical field or other topography communicated throughout the body. Eventually, about the end of the fourth week, each receptor achieves an autonomous imprint of the somatic polarities by which all its cells, including those subsequently formed, will be ordered within the whole array of the receptor epithelium.

Neuroblast migrations and outgrowth of nerve axons in the embryo of one month show that the cells and axon extensions within the central nervous system carry elements of the cytochemical code of the body field and that these may guide progress of the growth cones in the highly structured medium through which they proliferate to sort correct neurofilaments out of a mixed population. The cells either grow elongated axons or migrate from location to location in the brain. In a few days, new, highly patterned cell contacts are formed over long distances. For this orderly movement of cell extensions to occur, territories of the neuraxis must, like the primordia of receptors, be tagged so that groups of neuroblasts possess positional data related to the body axes. A selective affinity of receptor cell arrays for central target-cell arrays is shown by the precise routing of terminals when afferents grow into the central sites. How this is achieved remains uncertain. Presumably a paired chain of central target zones for afferent axons was defined by a hidden molecular differentiation which spread along the dorsal wall of the neural tube no later than the third or fourth week, when the receptor primordia were being specified. However, such a chemical labeling is probably not capable of explaining the precisely ordered topography of projections within any one receptor system.

Axon growth in the embryo is exploratory and highly responsive to the tissue medium through which axons advance. It also contributes substantially to the cell–cell communication by which nerve cell migrations and differentiations are patterned. This enables histogenesis of the nervous system to be self-regulating. Mobile extensions of the axon interact with the matrix and with each other. The local scattering of terminals in newly grown projections at end sites, and in terminals that regenerate after afferent nerves have been cut, shows that selective attachment between nerve cells is achieved progressively, through a sorting process. This confers a certain plasticity and adaptability in "systems matching" of the brain's networks, a plasticity which, in normal development, is an essential aspect of the mechanism of differentiation. Aberrant patterns of connection or groupings of cells may form, temporally, in normal brain development, but if surgical operations delete or add parts to populations of axon terminals or to groups of end cells, highly unstable and disorderly circuit layouts may be produced. These pathological nerve nets may gain the power to transmit excitation and cause altered and confused perceptual control of behavior.

The experiments with nerve networks in active development show, however, that in the normal program of growth, in which nerve cell systems are in balanced interaction, the process of cell-matching specifies reliable circuits which express in great detail a genetically preprogrammed set of instructions for wiring of receptor fields to the brain and for transmitting information by several ordered central steps to the effector neurons which excite the muscles to form patterns of movement. Nerve network patterning specifies intricate chains of behavioral coordination without benefit of experience or, in experimental cases, in opposition to experience. The development involves elaborate interaction between the growth products, especially the fine terminals of nerve axons which are formed in great excess before selection process is completed.

The 4- to 8-week embryo of man, in spite of its many patterned nerve circuits, is nearly inert. It is succeeded at two months by the fetus in which, over an initial three-month period, very large numbers of nerve cells are created, and new central mechanisms are intercalated into or superimposed on the wiring of the embryo brain. Fetal brain development involves intense competitive interaction between generations of neuroblasts and generations of circuits. The process, seen in a more elementary form in the metamorphosis of a frog, specifies a behavioral system of hierarchical design. Among other innovations, a more complex and more versatile perception machinery is built, one which, to judge from the evolution of intelligence in land vertebrates, is capable of much more complex attentional selection and more elaborate focal analysis by converging modalities in the detection and identification of objects.

The young fetal mammal brain, as observed in the baby hamster, shows equilibrium between factions of cells and between rival masses of axons converging on terminal nuclear sites. The result is a greatly enriched process of differentiation in nerve tissue. The dynamic plasticity of the process revealed by surgery to interacting neural groups does not challenge the notion of genetic cytochemical specification of nerve circuits. It rather suggests that the epigenetic process of cytodifferentiation by cell–cell interaction that began in the embryo is carried out at increasingly complex levels throughout the self-regulating process of brain growth.

Before refined circuits are consolidated and functional, neural networks formed in the fetus permit formation of a pattern of information needed to differentiate a very large number of cell species in organized arrays and to specify connective affinities between them. Indeed, such an immensely complex sorting process would appear to explain the long fetal period of development, in which the brain grows in almost total freedom from outwardly directed acts or transactions with an external world.

Centrally generated nerve excitation and then afferent excitation from receptors may affect any part of the nerve net as it is formed. This allows morphogenesis of integrative neural systems that can generate patterned central impulse activity. Nerve impulses regulate nerve cell growth. Excited nerves transmit ions, nerve transmitters, hormones, amino acids, and other morphogenetically active substances which influence cell differentiation. But, in most of the fetal period the influence of stimuli originating external to the body is kept to a minimum, by neural inhibition of brain transmission, and by morphological inactivation of receptors. The human fetus until the sixth month of gestation appears to be elaborately protected from patterned stimuli, even those which penetrate the womb. The greater part of brain system growth before birth is probably autonomous and independent of sensory stimulation.

In birds at least, stimuli available before hatching, emitted by the fetus and by outside sources, may aid differentiation of certain discriminators for perception. Imprinting in newly hatched birds is a very rapid version of this process which can occur only for a short time after hatching. It may have some human equivalent in the final stages of gestation and in early infancy.

In the newborn baby, cortical circuits in sensory areas undergoing final steps in development, with expansion of dendrites and formation of synapses, are shaped by natural patterns of stimulation. Two kinds of stimulus effect have been found in studies of growing cortical visual receptor units of kittens and newborn monkeys. In one, coordination of the eyes as duplicate space-sensing organs with different points of view permits definition of their joint use to measure distance (binocular stereopsis by disparity detection). The evidence does not prove that the binocular disparity detectors are created by concurrent stimulation of the two eyes; clearly there is complex prior specification of conjugate orienting of the eyes and of binocular convergence in the cortical circuits. But it does show that differentiation of a refined disparity mechanism requires undistorted visual experience when the primary binocular circuits are reaching completion, to sustain those units which respond discretely to appropriately small differences of visual stimulation.

A second process of selective growth of corrections may be involved in formation of high-acuity form-resolving elements, by weeding out more scattered cortical connections to leave geometrically ordered arrays. Units that respond to given features of stimuli, and that are suitable as selective filters for general application in form perception, are apparently reinforced soon after birth by this process. Again, the known anatomy favors the view that there are genetically determined categories of circuit behind the formation of feature detectors by experience. There could be no other explanation for the systematic ordering of orientation and ocular dominance columns reported for the monkey or for the complex multiple maps of the visual field in all mammals.

The plasticity in the high-resolution geniculostriate system may be adapted to counteract unpredictable malformations or changing dimensions of parts in the receptor mechanisms outside the brain, as well as defects in nerve connections, so that the eye may become as reliable as possible a transmitter of environmental information. For man, the only clear direct evidence for lasting postnatal changes in the basic neural elements of perception due to experience concerns adaptations to malformation of the optics of the eyes (astigmatism) or to failure of muscular control of one eye (strabismus or squint).

Some neurophysiologists and some psychologists of more empiricist persuasion have tended recently to jump, from evidence which shows that cortical circuits for line and shape detection can be deformed by closing one eye of a kitten at birth or restricting the neonatal kitten to a strongly patterned environment, to the conclusion that elements for contour perception are imprinted by experience on a featureless nerve matrix. On the evidence, this is not justified. The analogy for the cortical detector cell from which recordings are made, and its patterned galaxy of input, would appear to be an apple tree, the dynamic growth of which, in a dependably structured environment with normal patterns of gravity, light, air, and mechanical and metabolic support from the soil, will make a "normal" representative of its species and variety. The same self-regulating growth mechanism, if mutilated correctly and at the right state of growth, will make an espalier, flat against the wall with parallel branches. As with striate cortex cells, the environmental influence on the tree is more apparent where the treatment is extreme. Left alone again, the growth will work to reestablish a normal tree, but cannot reverse the effect of pruning completely, critical steps in development having been passed. In the normal complexity of the environment the innate forces of the tree gain clearer expression. Their adaptive nature, as they act to control the environment rather than submit to it, is then obvious.

The influence of psychobiological ideas on infant psychology is seen in the increasing recognition that the human infant is innately equipped for complex psychological performance besides learning, including perception of objects in three-dimensional space, detection of their motion, shape, size, and color, and recognition of human beings by their appearance and voice sounds. Although the brain of an infant is highly immature and certainly impressionable, environmental influences have their effects in an already specified system that is capable of categorizing stimuli from the outset, then regulating intake of stimulation and rejecting influences which are potentially destructive to its development, by highly controlled patterns of movement that may be produced spontaneously. The Pavlovian model of the brain composed of a large number of parallel elementary reflexes generating only motor responses is demonstrated to be inadequate, and infancy is seen as a highly active, self-regulating phase of psychological development with specific needs. The demonstrated learning abilities of young infants are a product of their innate powers of perceiving and their ability to regulate experience according to what they perceive.

The basic networks for perception are laid down in early infancy, permitting accurate discrimination of objects in space from many and varied signs and stimuli. They mature with great rapidity, probably inside the first year. The process is highly and specifically dependent on exercise or stimulation, but its limits of adaptation and detailed potentialities are preset during fetal brain growth. The perceptual systems clearly undergo a sequence of phases of differentiation. There is an intense period of differentiation of the fundamental cortical circuitry of seeing in the first two months fed by the new input of patterned stimulation. By four months an infant has sufficient motor coordination to regulate axial and proximal movements to shift attention and to begin prehension. After the growth of controlled focus on local information about object structure and identity, a 9- or 10-month-old begins to show evidence of problem-solving and searching behind appearances.

A great complexity of brain circuits is created postnatally by many cycles of differentiation in various regions of the CNS. Details of this process are unknown, but clearly the cerebral "association" neocortex and the cerebellum have long maturational cycles that take decades to complete. In the course of this process they are self-protected against "insult," but malnutrition, hormonal disturbance, and factors such as stress or emotional deprivation can have serious consequences for brain development. In their formation many complementary and interacting components mutually adjust to form progressively more elaborate solutions to the problem of behavioral control. The different modalities of perception mature through a competitive interaction within a common phenomenological and performative field. In the same genetically regulated growth program, a blind, deaf, or limbless child may achieve compensation, but always with some loss as preprogrammed combinations of psychological function fail to take place.

Of the processes by which infants regulate the making of their higher cognitive and perceptual skills, communication or joint enterprise with other persons is apparently the most elaborate. It is established through generation of expressive signals by a mechanism that communicates psychological states while detecting complementary signs of responsive attention from others. Communication develops into the process of education by which children acquire the conventions and knowledge of their culture. It is increasingly clear that this process has an innate basis in brain development and that it grows throughout childhood.

The higher mental functions develop asymmetrically in the cerebral hemispheres during childhood. Most information on this process concerns the development of language, but visuospatial abilities evidently follow a comparable pattern of maturation in the right hemisphere in most boys and girls. In infancy and early childhood cortical tissues become steadily more elaborate and more committed to perceptual, cognitive, and memory functions that differentiate the hemispheres. The process is genetically regulated, but it involves dynamic equilibrium between competing brain circuits, mediated in part by the neocortical commissures. Each of the higher psychological functions, as categorized by intelligence tests, perceptual puzzles, or language mastery tests, is regulated by cooperative activity of many hierarchically related cerebral subsystems. Some of these attain maturity quickly, while others remain unfinished and still adaptable for years or decades after birth. Recovery from brain trauma in different areas follows a different course depending on the age of onset of the damage. The course of maturation is changed in parts that are still undergoing tissue growth and differentiation.

As a rule, perceptual specializations of the hemispheres appear to mature early and firmly to establish asymmetry of function by the end of nursery school age. Children over five are undergoing brain development that affects their more elaborate intellectual functions that are related to perception in the way that a carpenter relates his knowledge and skills to his tools, each of which is designed to do one operation well, but which can, in extremity, be used "wrongly" to achieve an enforced solution by alternative means.

In childhood perceptual development involves creation of a complex system of cerebral tools and rules for awareness under strong sociocultural control, and the whole range of brain growth and nerve circuit differentiation is relevant to it. However, it seems clear that the primary perceptual capacities are shaped before birth, as are the principal abstract and purposeful psychological operations which perception serves. Perception of objects in relation to purposeful acts, and of people as a potential source of knowledge and cooperative enterprise function early in life. With the addition of powerful differentiating effects from experience, these innate psychological skills find the way to virtually unlimited development in skill of perception applied to particular tasks.

It is possible, in these terms, to sketch a tenuous argument linking the cell biochemistry and biophysics of brain growth at one end of our knowledge with human psychological functions of object awareness, imagery, and intersubjectivity at the other end. But it would be dishonest to leave the impression that we see any more than the beginning of an adequate psychobiological theory of perception, one which explains psychological activities in biological terms, but leaves the psychological or mental nature of their development intact. It is clear that future research will have to attend equally to the psychological and cerebral aspects of mental growth if this goal is to be achieved.

ACKNOWLEDGMENTS

This review was written while the author was traveling. He is most indebted for help with typing and bibliographic work to the Psychology Department and Medical School, the University of Auckland, and the Biology Division, California Institute of Technology. Doctors Beryl McKenzie, Ron Meyer, Gerald Schneider, and many others gave invaluable help with the recent literature. I am also grateful to Doctors Simon LeVay and Ron Meyer for permission to illustrate unpublished research or research in press.

8. References

Adinolfi, A. M., 1971, The ultrastructure of synaptic junctions in developing cerebral cortex, *Anat. Rec.* **169**:226.

Adrian, E. D., 1946, *The Physical Basis of Perception,* Oxford University Press, London.

Allman, J. M., 1977, Evolution of the visual system in the early primates, in: *Progress in Psychobiology and Physiological Psychology,* Vol. 7 (J. M. Sprague, ed.), pp. 1–53, Academic Press, New York.

Altman, J., 1967, Postnatal growth and differentiation of the mammalian brain, with implications for a morphological theory of memory, in: *The Neurosciences. A Study Program* (G. Quarton, T. Melnechuk, and F. O. Schmitt, eds.), pp. 723–743, Rockefeller University Press, New York.

Angevine, J. B., Jr., 1970, Critical cellular events in the shaping of neural centers, in: *The Neurosciences. Second Study Program* (F. O. Schmitt and T. Melnechuk, eds.), pp. 62–72, Rockefeller University Press, New York.

Anker, R. L., and Cragg, B. G., 1975, Development of the extrinsic connections of the visual cortex of the cat, *J. Comp. Neurol.* **154**:29–42.

Annett, M. A., 1964, A model of the inheritance of handedness and cerebral dominance, *Nature* **204**:59–60.

Anokhin, P. K., 1964, Systemogenesis as a general regulator of brain development, *Prog. Brain Res.* **9:**54–86.

Arey, L. B., 1965, *Developmental Anatomy,* 7th ed., Saunders, Philadelphia.

Åström, K. E., 1967, On the early development of the isocortex in fetal sheep, *Prog. Brain Res.* **26:**1–59.

Atkinson, J., Braddick, O., and Braddick, F., 1974, Acuity and contrast sensitivity of infant vision, *Nature* **247:**403–404.

Attardi, D. G., and Sperry, R. W., 1963, Preferential selection of central pathways by regenerating optic fibers, *Exp. Neurol.* **7:**46–64.

Axelrod, S., 1959, *The Effects of Early Blindness: Performance of Blind and Sighted Children on Tactile and Auditory Tasks* (Research Series, No. 7), American Foundation for the Blind, New York.

Bailey, P., and von Bonin, G., 1951, *The Isocortex of Man,* University of Illinois Press, Urbana.

Baker, R. E., 1972, Biochemical specification versus specific regrowth in the innervation of skin grafts in anurans, *Nature (London), New Biol.* **230:**235–237.

Baker, F. H., Grigg, P., and Von Noorden, G. K., 1974, Effects of visual deprivation and strabismus on the response of neurons in the visual cortex of the monkey, including studies on the striate and prestriate cortex in the normal animal, *Brain Res.* **66:**185–208.

Balász, R., 1974, Influence of metabolic factors on brain development, *Br. Med. Bull.* **30:**126–134.

Balázs, R., Lewis, P. D. and Patel, A. J., 1975, Effects of metabolic factors on brain development, in: *Growth and Development of the Brain* (M. A. B. Brazier, ed.), pp. 83–115, Raven Press, New York.

Ball, W., and Tronick, E., 1971, Infant responses to impending collision: Optical and real, *Science* **171:**818–820.

Banks, M. S., Aslin, R. N., and Letson, R. D., 1975, Critical period for the development of human binocular vision, *Science* **190:**675–677.

Barlow, H. B., 1975, Visual experience and cortical development, *Nature* **258:**199–204.

Barlow, H. B., and Pettigrew, J. D., 1971, Lack of specificity of neurons in the visual cortex of young kittens, *J. Physiol. (London)* **218:**8P–100P.

Barlow, H. B., Blakemore, C., and Pettigrew, J. D., 1967, The neural mechanism of binocular depth discrimination, *J. Physiol. (London)* **193:**327–342.

Barondes, S. H., 1970, Brain glycomacromolecules in interneuronal recognition, in: *The Neurosciences: A Second Study Program* (F. O. Schmitt, G. C. Quarton, and T. Melnechuk, eds.), pp. 744–760, Rockefeller University Press, New York.

Bartelmez, G. W., 1922, The origin of the otic and optic primordia in man, *J. Comp. Neurol.* **34:**201–232.

Bartelmez, G. W., and Blount, M. P., 1954, The formation of neural crest from the primary optic vesicle in man, *Carnegie Inst. Wash. Publ. 603, Contribs. Embryol.* **35:**55–71.

Bartelmez, G. W., and Dekaban, A. S., 1962, The early development of the human brain, *Carnegie Inst. Wash. Publ. 621, Contribs. Embryol.* **37:**15–32 (not including plates).

Barth, L. G., 1953, *Embryology,* Dryden Press, New York.

Bartlett, R. C., 1932, *Remembering,* Cambridge University Press, Cambridge.

Basser, L. S., 1962, Hemiplegia of early onset and the faculty of speech with special reference to the effect of hemispherectomy, *Brain* **85:**427–460.

Bateson, B. P. G., and Wainwright, A. A. P., 1972, The effects of prior exposure to light on the imprinting process in domestic chicks, *Behaviour* **42:**279–290.

Bauer, J., and Held, R., 1975, Comparison of visual guided reaching in normal and deprived infant monkeys, *J. Exp. Psychol. Anim. Behav. Proc.* **1:**298–308.

Baxter, B. L., 1966, Effect of visual deprivation during postnatal maturation on the electroencephalogram of the cat, *Exp. Neurol.* **14:**224–237.

Beazley, L. D., 1975a, Factors determining decussation at the chiasma by developing retinotectal fibers in *Xenopus, Exp. Brain Res.* **23:**491–504.

Beazley, L. D., 1975b, Development of intertectal neuronal connections in *Xenopus*. The effects of contralateral transposition of the eye and of eye removal, *Exp. Brain Res.* **23:**505–518.

Beazley, L. D., Keating, M. J., and Gaze, R. M., 1972, The appearance, during development, of responses in the optic tectum following visual stimulation of the ipsilateral eye in *Xenopus laevis, Vis. Res.* **12:**407–410.

Beintema, D. J., 1968, *A Neurological Study of Newborn Infants,* Spastics International and Heinemann, London.

Benowitz, L. I., and Karten, H. J., 1976, Organization of the tectofugal visual pathway in the pigeon: A retrograde transport study, *J. Comp. Neurol.* **167:**503–520.

Benzer, S., 1973, Genetic dissection of behavior, *Sci. Am.* **229:**24–37.

Bergström, R. M., 1969, Electrical parameters of the brain during ontogeny, in: *Brain and Early*

Behavior: Development in the Fetus and Infant, (R. J. Robinson, ed.), pp. 15–41, Academic Press, New York.

Berman, N., and Hunt, R. K., 1975, Visual projections to the optic tecta in *Xenopus* after partial extirpation of the embryonic eye, *J. Comp. Neurol.* **162**:23–42.

Bernstein, N., 1967, *The Coordination and Regulation of Movements,* Pergamon, Oxford.

Berry, M., 1974, Development of neocortex of the rat, in: *Studies in the Development of Behavior and the Nervous System,* Vol. 2 (G. Gottlieb, ed.), pp. 19–115, Academic Press, New York.

Berry, M., and Rogers, A. W., 1966, Histogenesis of mammalian neocortex, in: *Evolution of the Forebrain* (R. Hassler and H. Stephan, eds.), pp. 197–205, Thieme, Stuttgart.

Blakemore, C., 1974, Development of functional connexions in the mammalian visual system, *Br. Med. Bull.* **30**:152–157.

Blakemore, C. B., and Cooper, G. F., 1970, Development of the brain depends on the visual environment, *Nature (London)* **228**:477–478.

Blakemore, C., and Van Sluyters, R. C., 1974, Reversal of the physiological effects of monocular deprivation in kittens. Further evidence for a sensitive period, *J. Physiol. (London)* **237**:195–216.

Bodian, D., 1966, Development of fine structure of spinal cord in monkey fetuses. I. The motoneuron neuropil at the time of onset of reflex activity, *Bull. Johns Hopkins Hosp.* **119**:129–149.

Bodian D., 1968, Development of fine structure of spinal cord in monkey fetuses. II. Pre-reflex period to period of long intersegmental reflexes, *J. Comp. Neurol.* **133**:113–166.

Bodian, D., 1970, A model of synaptic and behavioral ontogeny, in: *The Neurosciences: Second Study Program* (F. O. Schmitt and T. Melnechuk, eds.), pp. 129–140, Rockefeller University Press, New York.

Bornstein, M. H., 1975, Qualities of colour vision in infancy, *J. Exp. Child Psychol.* **19**:401–419.

Böving, B. G., 1965, Anatomy of reproduction, in: *Obstetrics,* 13th ed. (J. P. Greenhill, ed.), pp. 8–23, Saunders, Philadelphia.

Bower, T. G. R., 1974, *Development in Infancy,* Freeman, San Francisco.

Boycott, B. B., and Wässle, H., 1974, The morphological types of ganglion cells of the domestic cat retina, *J. Physiol.* **240**:397–419.

Brazelton, T. B., Tronick, E., Adamson, L., Als, H., and Wise, S., 1975, Early mother-infant reciprocity, in: *Parent-Infant Interaction,* Ciba Foundation Symposium, No. 33, New Series (M. O'Connor, ed.), pp. 137–154, Elsevier-Excerpta Medica–North Holland, New York.

Brinkman, J., and Kuypers, H. G. J. M., 1973, Cerebral control of contralateral and ipsilateral arm, hand and finger movements in the split-brain rhesus monkey, *Brain* **96**:653–674.

Bronson, G., 1974, The postnatal growth of visual capacity, *Child Dev.* **45**:873–890.

Bruner, J. S., 1975, Ontogenesis of speech acts, *J. Child Lang.* **2**:1–19.

Carmichael, L., 1926, The development of behavior in vertebrates experimentally removed from the influence of external stimulation, *Psychol. Rev.* **33**:51–58.

Chi, J. G., Dooling, E. C., and Gillies, F. H., 1977, Gyral development of the human brain, *Ann. Neurol.* **1**:86–93.

Chow, K. L., and Steward, D. L., 1972, Reversal of structural and functional effects of long-term visual deprivation in cats, *Exp. Neurol.* **34**:409–433.

Chung, S. H., and Cooke, J., 1975, Polarity of structure and of ordered nerve connections in the developing amphibian brain, *Nature* **258**:126–132.

Chung, S. H., Keating, M. H., and Bliss, T. V. P., 1974, Functional synaptic relations during the development of the retino-tectal projection in amphibians, *Proc. R. Soc. London, Ser. B* **187**:449–459.

Coghill, G. E., 1924, Correlated anatomical and physiological studies of the growth of the nervous system of *Amphibia.* IV. Rates of proliferation and differentiation in the central nervous system of *Amblystoma, J. Comp. Neurol.* **37**:71–120.

Coghill, E. G., 1929, *Anatomy and the Problem of Behavior,* Cambridge University Press, London.

Conel, J. Le Roy, 1939–1963, *The Postnatal Development of the Human Cerebral Cortex,* Vols. I–VI, Harvard University Press, Cambridge.

Cook, J. E., and Horder, T. J., 1974, Interactions between optic fibers in their regeneration to specific sites in the goldfish tectum, *J. Physiol.* **241**:89–90.

Corballis, M. C., and Beale, I. L., 1976, *The Psychology of Left and Right,* Erlbaum, Hillsdale, New Jersey.

Cotman, C. W., and Banker, G. A., 1974, The making of a synapse, in: *Reviews of Neuroscience,* Vol. 1 (S. Ehrenpreis and I. J. Kopen, eds.), Raven Press, New York.

Cowan, W. M., 1973, Neuronal death as a regulative mechanism in the control of cell number in the nervous system, in: *Development and Aging in the Nervous System* (M. Rockstein, ed.), pp. 19–41, Academic Press, New York.

Cowan, W. M., Martin, A. H., and Wenger, E., 1968, Mitotic patterns in the optic tectum of the chick during normal development and after early removal of the optic vesicle, *J. Exp. Zool.* **169:**71–92.

Cragg, B. G., 1972, The development of synapses in the cat visual cortex, *Invest. Ophthalmol.* **11:**377–385.

Cragg, B. G., 1975, The development of synapses in the visual system of the cat, *J. Comp. Neurol.* **160:**147–166.

Craik, K. J. W., 1943, *The Nature of Explanation,* Cambridge University Press, London.

Crain, S. M., 1974, Tissue culture models of developing brain functions, in: *Studies on the Development of Behavior and the Nervous System,* Vol. 2, Aspects of Neurogenesis (G. Gottlieb, ed.), pp. 69–114, Academic Press, New York.

Crossland, W. J., Cowan, W. M., Rogers, L. A., and Kelly, J. P., 1974, The specification of the retinotectal projection in the chick, *J. Comp. Neurol.* **155:**127–164.

Crossland, W. J., Cowan, W. M., and Rogers, L. A., 1975, Studies on the development of the chick optic tectum. IV. An autoradiographic study of the development of retinotectal connections, *Brain Res.* **91:**1–23.

Daniel, P. M., and Whitteridge, D., 1961, The representation of the visual field on the cerebral cortex in monkeys, *J. Physiol. (London)* **159:**203–221.

Daniels, J. D., and Pettigrew, J. D., 1976, Development of neuronal responses in the visual system of cats, in: *Studies on the Development of Behavior and the Nervous System,* Vol. 3 (G. Gottlieb, ed.), pp. 195–232, Academic Press, New York.

Décarie, T. G., 1965, *Intelligence and Affectivity in Early Childhood,* International Universities Press, New York.

De Long, G. R., and Coulombre, A. J., 1965, Development of the retinotectal projection in the chick embryo, *Exp. Neurol.* **13:**351–363.

De Long, G. R., and Sidman, R. L., 1962, Effects of eye removal at birth on histogenesis of the mouse superior colliculus: An autoradiographic analysis with tritiated thymidine, *J. Comp. Neurol.* **118:**205–224.

Diamond, I. T., and Hall, W. C., 1969, Evolution of the neo-cortex, *Science* **164:**251–262.

Dichgans, J., and Brandt, Th., 1974, The psychophysics of visually induced perception of self-motion and tilt, in: *The Neurosciences: Third Study Program* (F. O. Schmitt and F. G. Worden, eds.), pp. 123–129, The MIT Press, Cambridge, Massachusetts.

Dixon, J. S., and Cronly-Dillon, J. R., 1972, The fine structure of the developing retina in *Xenopus laevis,* *J. Embryol. Exp. Morphol.* **28:**659–666.

Dobbing, J., 1971, Undernutrition and the developing brain: The use of animal models to elucidate the human problem, in: *Normal and Abnormal Development of Brain and Behavior* (G. B. A. Stoelinga and J. J. Van der Werffen Bosch, eds.), pp. 20–30, Williams and Wilkins, Baltimore.

Dobbing, J., 1974, The later development of the central nervous system and its vulnerability, in: *Scientific Foundations of Paediatrics* (J. A. Davis and J. Dobbing, eds.), pp. 565–577, Heinemann Medical Books, London.

Dobbing, J., and Smart, J. L., 1974, Vulnerability of developing brain and behaviour, *Br. Med. Bull.* **30:**164–168.

Drager, U. C., 1975, Receptive fields of single cells and topography in mouse visual cortex, *J. Comp. Neurol.* **160:**269–290.

Drager, U. C., and Hubel, D. H., 1975, Physiology of visual cells in mouse superior colliculus and correlation with somatosensory and auditory input, *Nature* **253:**203–204.

Drever, J., 1955, Early learning and the perception of space, *Am. J. Psychol.* **68:**605–614.

Dreyfus-Brisac, C., 1967, Ontogénèse du sommeil chez le prémature humain: Étude polygraphique, in: *Regional Development of the Brain in Early Life* (A. Minkowski, ed.), pp. 437–457, Blackwell, Oxford.

Duffy, F. H., Snodgrass, S. R., Burchfiel, J. L., and Conway, J. L., 1976, Bicuculline reversal of deprivation amblyopia in the cat, *Nature* **260:**256–257.

Eimas, P., Sigueland, E., Jusczyr, P., and Vigorito, J., 1971, Speech perception in infants, *Science* **171:**303–306.

Eisenberg, R. B., 1975, *Auditory Competence in Early Life: The Roots of Communicative Behavior,* University Park Press, Baltimore.

Evans, E. F., 1974, Neural processes for the detection of acoustic patterns and for sound localization, in: *The Neurosciences: Third Study Program* (F. O. Schmitt and G. Quarton, eds.), pp. 131–145, The MIT Press, Cambridge, Massachusetts.

Falkner, F., 1966, *Human Development,* Saunders, Philadelphia.

Fantz, R. L., 1963, Pattern vision in newborn infants, *Science* **140:**296–297.

Fantz, R. L., 1965, Ontogeny of perception, in: *Behavior of Non-human Primates: Modern Research Trends,* Vol. II (A. M. Schrier, H. F. Harlow, and F. Stollnitz, eds.), pp. 365–403, Academic Press, New York.

Fantz, R. L., Ordy, J. M., and Udelf, M. S., 1962, Maturation of pattern vision in infants during the first six months, *J. Comp. Physiol. Psychol.* **55:**907–917.

Feldman, J. D., and Gaze, R. M., 1975, The retinotectal projection from half eyes in *Xenopus, J. Comp. Neurol.* **162:**13–22.

Finlay, B. L., Schneps, S. E., Wilson, K. G., and Schneider, G. E., 1978, Topography of visual and somatosensory projections to the superior colliculus of the golden hamster, *Brain Res.* **142:**223–235.

Fisher, B., 1973, Overlap of receptive field centers and representation of the visual field in the cat's optic tract, *Vis. Res.* **13:**2113–2120.

Flandrin, J. M., and Jeannerod, M., 1975, Superior colliculus: Environmental influences on the development of directional responses in the kitten, *Brain Res.* **89:**348–352.

Fodor, J. A., Garrett, M. F., and Brill, S. L., 1975, Pi Ka pu: The perception of speech sounds by prelinguistic infants, *Percept. Psychophys.* **18:**74–78.

Foelix, R. F., and Oppenheim, R. W., 1973, Synaptogenesis in the avian embryo: Ultrastructure and possible behavioral correlates, in: *Studies on the Development of Behavior and the Nervous System,* Vol. 1 (G. Gottlieb, ed.), pp. 104–139, Academic Press, New York.

Fraiberg, S., 1974, Blind infants and their mothers: An examination of the sign system, in: *The Effect of the Infant on Its Caregiver* (M. Lewis and L. A. Rosenblum, eds.), pp. 215–232, Wiley, New York.

Fraiberg, S., 1976, Development of human attachment in infants blind from birth, *Merrill–Palmer Q.* **21:**315–334.

Freedman, D. G., 1964, Smiling in blind infants and the issue of innate vs. acquired, *J. Child Psychol. Psychiatr.* **5:**171–184.

Galaburda, A. M., Le May, M., Kemper, T. L., and Geschwind, N., 1978, Right–left asymmetries in the brain, *Science* **199:**852–856.

Ganz, L., 1975, Orientation in visual space by neonates and its modification by visual deprivation, in: *The Developmental Neuropsychology of Sensory Deprivation* (A. H. Riesen, ed.), pp. 169–210, Academic Press, New York.

Ganz, L., 1978, Sensory deprivation and visual discrimination, in: *Handbook of Sensory Physiology,* Vol. VIII (H. Teuber, ed.), Springer-Verlag, Berlin (in press).

Ganz, L., and Fitch, M., 1968, The effect of visual deprivation on perceptual behavior, *Exp. Neurol.* **22:**638–660.

Ganz, L., Fitch, M., and Satterberg, J. A., 1968, The selective effect of visual deprivation on receptive field shape determined neurophysiologically, *Exp. Neurol.* **22:**614–637.

Ganz, L., Hirsch, H. V. B., and Bliss-Tieman, S., 1972, The nature of perceptual deficits in visually deprived cats, *Brain Res.* **44:**547–568.

Garcia-Bellido, A., 1972, Pattern formation in imaginal discs, *Res. Prob. Cell Diff.* **5:**59–91.

Garey, L. J., Fisken, R. A., and Powell, T. P. S., 1973, Effects of experimental deafferentiation on cells in the lateral geniculate nucleus of the cat, *Brain Res.* **52:**363–369.

Gaze, R. M., 1970, *Formation of Nerve Connections,* Academic Press, New York.

Gaze, R. M., 1974, Neuronal specificity, *Br. Med. Bull.* **30:**116–121.

Gaze, R. M., and Hope, R. A., 1976, The formation of continuously ordered mappings, *Prog. Brain Res.* **45:**327–357.

Gaze, R. M., and Jacobson, M., 1962, The projection of the binocular visual field on the optic tecta of the frog, *Q. J. Exp. Physiol.* **47:**273–280.

Gaze, R. M., and Keating, M. J., 1972, The visual system and "neuronal specificity," *Nature (London)* **237:**375–379.

Gaze, R. M., and Sharma, S. C., 1970, Axial differences in the reinnervation of the goldfish optic tectum by regenerating optic nerve fibers, *Exp. Brain Res.* **10:**171–181.

Gaze, R. M., Jacobson, M., and Szekely, G., 1963, The retinotectal projection in *Xenopus* with compound eyes, *J. Physiol. (London)* **165:**484–499.

Gaze, R. M., Jacobson, M., and Szekely, G., 1965, On the formation of connexions by compound eyes in *Xenopus, J. Physiol. (London)* **176:**409–417.

Gaze, R. M., Keating, M. J., Szekely, G., and Beazley, L., 1970, Binocular interaction in the formation of specific intertectal neuronal connections, *Proc. R. Soc. London, Ser. B* **175:**107–147.

Gaze, R. M., Keating, M. J., and Chung, S. H., 1974, The evolution of the retinotectal map during development in *Xenopus, Proc. R. Soc. London, Ser. B* **185:**301–330.

Gazzaniga, M. S., and Sperry, R. W., 1967, Language after section of the cerebral commissures, *Brain* **90:**131–148.

Geschwind, N., 1974, The anatomical basis of hemispheric differentiation, in: *Hemisphere Function in the Human Brain* (S. J. Dimond and J. G. Beaumont, eds.), Elek Science, London.

Geschwind, N., and Levitsky, W., 1968, Human brain, left–right asymmetries in temporal speech regions, *Science* **161:**186–187.

Gesell, A. L., Thompson, H., and Amatruda, C. S., 1934, *Infant Behavior: Its Genesis and Growth,* McGraw-Hill, New York.

Gesell, A., Ilg, F. L., and Bullis, G. E., 1949, *Vision: Its Development in Infant and Child,* Hamish Hamilton, London.

Gibbs, M. E., Jeffrey, P. L., Austin, L., and Mark, R. F., 1973, Separate biochemical actions of inhibitors of short- and long-term memory, *Pharmacol. Biochem. Behav.* **1:**693–701.

Gibson, E. J., 1969, *Principles of Perceptual Learning and Development,* Appleton-Century-Crofts, New York.

Gibson, J. J., 1952, The relation between visual and postural determinants of the phenomenal vertical, *Psychol. Rev.* **59:**370–375.

Gibson, J. J., 1954, The visual perception of object motion and subjective movement, *Psychol. Rev.* **61:**304–314.

Gibson, J. J., 1966, *The Senses Considered as Perceptual Systems,* Houghton Mifflin, Boston.

Gibson, J. J., 1979, *An Ecological Approach to Visual Perception,* Houghton Mifflin, Boston (in preparation).

Glickstein, M., and Gibson, A. R., 1976, Visual cells in the pons of the brain, *Sci. Am.* **234:**90–98.

Glucksmann, A., 1940, The development and differentiation of the tadpole eye, *Br. J. Ophthalmol.* **24:**153–178.

Glucksmann, A., 1965, Cell death in normal development, *Arch. Biol.* **76:**419–437.

Goldman, P. S., 1971, Functional development of the prefrontal cortex in early life and the problem of neuronal plasticity, *Exp. Neurol.* **32:**366–387.

Goldman, P. S., 1972, Developmental determinants of cortical plasticity, *Acta Neurobiol. Exp.* **32:**495–511.

Goldman, P. S., 1976, An alternative to developmental plasticity: Heterology of CNS structures in infants and adults, in: *Plasticity and Recovery of Function in the Central Nervous System* (D. G. Stein, J. J. Rosen, and N. Butters, eds.), pp. 149–174, Academic Press, New York.

Goldman, P. S., and Nauta, W. J. H., 1977, Columnar distribution of cortico-cortical fibers in the frontal association, limbic and motor cortex of the developing rhesus monkey, *Brain Res.* **122:**393–413.

Goldman, P. S., Rosvold, H. E., and Mishkin, M., 1970, Evidence for behavioral impairment following prefrontal lobectomy in the infant monkey, *J. Comp. Physiol. Psychol.* **70:**454–463.

Gottlieb, G., 1971*a, Development of Species Identification in Birds: An Inquiry into the Prenatal Determinants of Perception,* University of Chicago Press, Chicago.

Gottlieb, G., 1971*b,* Ontogenesis of sensory function in birds and mammals, in: *Biopsychology of Development* (E. Tobach, L. R. Aronson, and E. Shaw, eds.), pp. 67–128, Academic Press, New York.

Gottlieb, G., 1973, Introduction to behavioral embryology, in: *Studies on the Development of Behavior and the Nervous System,* Vol. 1, Behavioral Embryology (G. Gottlieb, ed.), pp. 3–45, Academic Press, New York.

Gottlieb, G., 1976*a,* The roles of experience in the development of behavior and the nervous system, in: *Neural and Behavioral Specificity, Studies on the Development of Behavior and the Nervous System,* Vol. 3 (G. Gottlieb, ed.), pp. 25–54, Academic Press, New York.

Gottlieb, G., 1976*b,* Early development of species-specific auditory perception in birds, in: *Neural and Behavioral Specificity, Studies on the Development of Behavior and the Nervous System,* Vol. 3 (G. Gottlieb, ed.), pp. 237–280, Academic Press, New York.

Graybiel, A. M., 1974, Studies on the anatomical organization of posterior association cortex, in: *The Neurosciences: Third Study Program* (F. O. Schmitt and F. G. Worden, eds.), pp. 5–19, MIT Press, Cambridge, Massachusetts.

Greene, P. H., 1972, Problems of organization of motor systems; in: *Progress in Theoretical Biology,* Vol. II (R. Rosen and F. M. Snell, eds.), pp. 303–338, Academic Press, New York.

Gregory, R. L., 1970, *The Intelligent Eye,* Weidenfeld, London.

Gregory, R. L., and Wallis, J. C., 1963, Recovery from early blindness, a case study, *Exp. Psychol. Soc. Monog., No. 2,* Cambridge.

Grobstein, P., and Chow, K. L., 1976, Receptive field organization in the mammalian visual cortex: The role of individual experience in development, in: *Neural and Behavioral Specificity, Studies on the*

Development of Behavior and the Nervous System, Vol. 3 (G. Gottlieb, ed.), pp. 155–193, Academic Press, New York.

Grobstein, P., Chow, K. L., Spear, P. D., and Mathers, L. H., 1973, Development of rabbit visual cortex: Late appearance of a class of receptive fields, *Science* **180:**1185–1187.

Gross, C. G., and Mishkin, M., 1977, The neural basis of stimulus equivalence across retinal translation, in: *Lateralization in the Nervous System* (S. R. Harnand, ed.), pp. 109–121, Academic Press, New York.

Guillery, R. W., 1972, Binocular competition in the control of geniculate cell growth, *J. Comp. Neurol.* **144:**117–130.

Guillery, R. W., and Stelzner, D. J., 1970, The differential effects of unilateral lid closure upon the monocular and binocular segments of the dorsal lateral geniculate nucleus of the cat: A new interpretation, *J. Comp. Neurol.* **143:**73–100.

Gustafson, T., and Wolpert, L., 1967, Cellular movement and contact in sea urchin morphogenesis, *Biol. Rev. Cambridge Philos. Soc.* **42:**442–498.

Guth, L., 1975, History of central nervous system regeneration research, *Exp. Neurol.* **48** (3, part 2):3–15.

Haith, M. M., 1978, Visual competence in early infancy, in: *Handbook of Sensory Physiology,* Vol. VIII (R. Held, R. Leibowitz, and H. L. Teuber, eds.), Springer-Verlag, Berlin (in preparation).

Haith, M. H., and Campos, J. J., 1977, Human infancy, *Annu. Rev. Psychol.* **28:**251–293.

Halliday, M. A. K., 1975, *Learning How to Mean,* Arnold, London.

Hamburger, V., 1948, The mitotic patterns in the spinal cord of the chick embryo and their relation to histogenetic processes, *J. Comp. Neurol.* **88:**221–284.

Hamburger, V., 1963, Some aspects of the embryology of behavior, *Q. Rev. Biol.* **38:**342–365.

Hamburger, V., 1969, Origins of integrated behavior, in: *The Emergence of Order in Developing Systems* (M. Locke, ed.), pp. 251–271, Academic Press, New York.

Hamburger, V., 1973, Anatomical and physiological basis of embryonic motility in birds and mammals, in: *Studies on the Development of Behavior and the Nervous System,* Vol. 1, Behavioral Embryology (G. Gottlieb, ed.), pp. 51–76, Academic Press, New York.

Hamburger, V., and Hamilton, H. L., 1951, A series of normal stages in the development of the chick embryo, *J. Morphol.* **88:**49–92.

Hamburger, V., Wenger, E., and Oppenheim, R., 1966, Motility in the chick embryo in absence of sensory input, *J. Exp. Zool.* **162:**133–160.

Hamilton, W. J., Boyd, J. D., and Mossman, H. W., 1962, *Human Embryology,* 3rd ed., Heffer, Cambridge.

Harrison, R. G., 1907, Observations on the living developing nerve fiber, *Anat. Rec.* **1:**116–118.

Harrison, R. G., 1910, The outgrowth of the nerve fiber as a mode of protoplasmic movement, *J. Exp. Zool.* **9:**787–848.

Harrison, R. G., 1921, On relations of symmetry in transplanted limbs, *J. Exp. Zool.* **32:**1–136.

Haynes, H., White, B. L., and Held, R., 1965, Visual accommodation in human infants, *Science* **148:**528–530.

Hécaen, H., 1976, Acquired aphasia in children and the ontogenesis of hemispheric functional specification, *Brain Lang.* **3:**114–134.

Hein, A., 1970, Recovering spatial motor coordination after visual cortex lesions, in: *Perception and Its Disorders* (D. Hamburg, ed.), Research Publications of the Association for Research in Nervous Mental Disease, Vol. 48, pp. 163–185, Williams and Wilkins, Baltimore.

Hein, A., and Diamond, R. M., 1971, Independence of cat's scotopic and photopic systems in acquiring control of visually guided behavior, *J. Comp. Physiol. Psychol.* **76:**31–38.

Hein, A., and Held, R., 1967, Dissociation of the visual placing response into elicited and guided components, *Science* **158:**390–392.

Hein, A., Gower, E. C., and Diamond, R. M., 1970a, Exposure requirements for developing the triggered component of the visual-placing response, *J. Comp. Physiol. Psychol.* **73:**188–192.

Hein, A., Held, R., and Gower, E. C., 1970b, Development and segmentation of visually controlled movement by selective exposure during rearing, *J. Comp. Physiol. Psychol.* **73:**181–187.

Held, R., and Bauer, J. A., 1967, Visual guided reaching in infant monkeys after restricted rearing, *Science* **155:**718–720.

Held, R., and Bauer, J. A., 1974, Development of sensorially guided reaching in infant monkeys, *Brain Res.* **71:**265–271.

Held, R., and Freedman, S. J., 1963, Plasticity in human sensorimotor control, *Science* **142:**455–462.

Held, R., and Hein, A., 1963, Movement-produced stimulation in the development of visually guided behavior, *J. Comp. Physiol. Psychol.* **56:**872–876.

Herrick, C. J., 1948, *The Brain of the Tiger Salamander,* University of Chicago Press, Chicago.

Hess, E. H., 1956, Space perception in the chick, *Sci. Am.* **195:**71–80.

Hirsch, H. V. B., and Jacobson, M., 1973, Development and maintenance of connectivity in the visual system of the frog. II. The effect of eye removal, *Brain Res.* **49:**67–74.

Hohman, A., and Creutzfeldt, O. D., 1975, Squint and development of binocularity in humans, *Nature* **254:**613–614.

Hollyfield, J. G., 1971, Differential growth of the neural retina in *Xenopus laevis* larvae, *Dev. Biol.* **24:**264–286.

Hooker, D., 1952, *The Prenatal Origin of Behavior,* University of Kansas Press, Lawrence, Kansas.

Hope, R. A., Hammond, B. J., and Gaze, R. M., 1976, The arrow model: Retinotectal specificity and map formation in the goldfish visual system, *Proc. R. Soc. London, Ser. B* **194:**447–466.

Hubel, D. H., and Wiesel, T. N., 1962, Receptive fields, binocular interaction and functional architecture in the cat's visual cortex, *J. Physiol. (London)* **160:**106–154.

Hubel, D. H., and Wiesel, T. N., 1963, Receptive fields of cells in striate cortex of very young, visually inexperienced kittens, *J. Neurophysiol.* **26:**994–1002.

Hubel, D. H., and Wiesel, T. N., 1965, Binocular interaction in the striate cortex of kittens reared with artificial squint, *J. Neurophysiol.* **28:**1041–1059.

Hubel, D. H., and Wiesel, T. N., 1968, Receptive fields and functional architecture of monkey striate cortex, *J. Physiol. (London)* **195:**215–243.

Hubel, D. H., and Wiesel, T. N., 1970, The period of susceptibility to the physiological effects of unilateral eye closure in kittens, *J. Physiol. (London)* **206:**419–436.

Hubel, D. H., and Wiesel, T. N., 1974a, Sequence, regularity and geometry of orientation columns in the monkey striate cortex, *J. Comp. Neurol.* **158:**267–294.

Hubel, D. H., and Wiesel, T. N., 1974b, Uniformity of monkey striate cortex: A parallel relationship between field size scatter and magnification factor, *J. Comp. Neurol.* **158:**295–306.

Hubel, D. H., LeVay, S., and Wiesel, T. N., 1975, Mode of termination of retinotectal fibres in macaque monkey: An autoradiographic study, *Brain Res.* **96:**25–40.

Hubel, D. H., Wiesel, T. N., and LeVay, S., 1977, Plasticity of ocular dominance columns in monkey striate cortex, *Phil. Trans. Roy. Soc. London, Ser. B* **278:**131–163.

Hughes, A. F. W., 1968, *Aspects of Neural Ontogeny,* Academic Press, New York.

Humphrey, N., 1974, Vision in monkey without striate cortex: A case study, *Perception* **3:**241–255.

Humphrey, T., 1964, Some correlations between the appearance of human fetal reflexes and the development of the nervous system, *Prog. Brain Res.* **4:**93–135.

Humphrey, T., 1969, Postnatal repetition of human prenatal activity sequences with some suggestions of their neuroanatomical basis, in: *Brain and Early Behavior: Development in the Fetus and Infant* (R. J. Robinson, ed.), pp. 43–84, Academic Press, New York.

Hunt, R. K., 1975a, The cell cycle, cell lineage, and neuronal specificity, in: *The Cell Cycle and Cell Differentiation* (H. Holtzer and J. Reinart, eds.), pp. 43–62, Springer-Verlag, Berlin.

Hunt, R. K., 1975b, Developmental programming for retinotectal patterns, in: *Ciba Foundation Symposium on Cell Patterning, New Series* **29:**131–159.

Hunt, R. K., and Berman, N. J., 1975, Patterning of neuronal locus specificities in the retinal ganglion cells after partial extirpation of the embryonic eye, *J. Comp. Neurol.* **162:**43–70.

Hunt, R. K., and Jacobson, M., 1972, Development and stability of positional information in *Xenopus* retinal ganglion cells, *Proc. Natl. Acad. Sci. U.S.A.* **69:**780–783.

Hunt, R. K., and Jacobson, M., 1973a, Specification of positional information in retinal ganglion cells of *Xenopus:* Assays for analysis of the unspecified state, *Proc. Natl. Acad. Sci. U.S.A.* **70:**507–511.

Hunt, R. K., and Jacobson, M., 1973b, Neuronal specificity revisited, *Curr. Top. Dev. Biol.* **8:**202–259.

Huttenlocker, P. R., 1967, Development of cortical neuronal activity in the neonatal cat, *Exp. Neurol.* **17:**247–262.

Ingle, D., 1973, Two visual systems in the frog, *Science* **181:**1053–1055.

Ingle, D., 1976a, Behavioral correlates of central vision in anurans, in: *Frog Neurobiology* (R. Llinas and W. Precht, eds.), pp. 435–451, Springer-Verlag, Berlin.

Ingle, D., 1976b, Spatial vision in anurans, in: *The Amphibian Visual System* (K. Fite, ed.), pp. 119–140, Academic Press, New York.

Ingle, D., 1978, Visual behavior development in non-mammalian vertebrates, in: *Handbook of Sensory Physiology,* Vol. IX (M. Jacobson, ed.), Springer-Verlag, Berlin (in press).

Ingle, D., and Dudek, A., 1977, Aberrant retino-tectal projections in the frog, *Exper. Neurol.* **55:**567–582.

Ingle, D., and Sprague, J. M. (eds.), 1975, *Sensorimotor Function of the Midbrain Tectum* (Neurosciences Research Program Bulletin, Vol. 13), pp. 169–288, Neurosciences Research Program, Cambridge, Massachusetts.

Jacobson, C.-O., 1959, The localization of the presumptive cerebral regions in the neural plate of the axolotl larva, *J. Embryol. Exp. Morphol.* **7:**1–21.

Jacobson, M., 1968*a*, Development of neuronal specificity in retinal ganglion cells of *Xenopus*, *Dev. Biol.* **17:**202–218.

Jacobson, M., 1968*b*, Cessation of DNA synthesis in retinal ganglion cells correlated with the time of specification of their central connections, *Dev. Biol.* **17:**219–232.

Jacobson, M., 1970, *Developmental Neurobiology,* Holt, New York.

Jacobson, M., 1974, Premature specification of the retina in embryonic *Xenopus* eyes treated with ionophore X537A, *Science* **191:**288–289.

Jacobson, M., 1976, Histogenesis of retina in the clawed frog with implications for the pattern of development of retino-tectal connections, *Brain Res.* **103:**541–545.

Jacobson, M., and Baker, R. E., 1969, Development of neuronal connections with skin grafts in frogs: Behavioural and electrophysiological studies, *J. Comp. Neurol.* **137:**121–142.

Jacobson, M., and Hirsch, H. V. B., 1973, Development and maintenance of connectivity in the visual system of the frog. I. The effect of eye rotation and visual deprivation, *Brain Res.* **49:**47–65.

Jacobson, M., and Hunt, R. K., 1973, The origins of nerve-cell specificity, *Sci. Am.* **228**(2):26–35.

Jacobson, M., and Levine, R. L., 1975, Stability of implanted duplicate tectal positional markers serving as target for optic axons in adult frogs, *Brain Res.* **92:**468–471.

Jeannerod, M., 1978, Two steps in visuo-motor development, *Proceedings of the O. E. C. D. Conference on "Dips in Learning," St. Paul de Vence, March, 1975* (in preparation).

Johnston, B. T., Schrameck, J. E., and Mark, R. F., 1975, Reinnervation of axolotl limbs. II. Sensory nerves, *Proc. R. Soc., London, Ser. B* **190:**59–75.

Jones, E. G., 1975, Varieties and distribution of non-pyramidal cells in the somatosensory cortex of the squirrel monkey, *J. Comp. Neurol.* **160:**205–268.

Kalil, R. E., and Schneider, G. E., 1975, Abnormal synaptic connections of the optic tract in the thalamus after midbrain lesions in newborn hamsters, *Brain Res.* **100:**690–698.

Kalnins, I. V., and Bruner, J. S., 1973, The coordination of visual observation and instrumental behavior in early infancy, *Perception* **2:**307–314.

Keating, M. J., 1976, The formation of visual neuronal connections: an appraisal of the present status of the theory of "neuronal specificity," in: *Neural and Behavioral Specificity, Studies on the Development of Behavior and the Nervous System,* Vol. 3 (G. Gottlieb, ed.), pp. 59–110, Academic Press, New York.

Keating, M. J., and Gaze, R. M., 1970, Rigidity and plasticity in the amphibian visual system, *Brain Behav. Evol.* **3:**102–120.

Kellog, W. N., 1962, Sonar system of the blind, *Science* **137:**399–404.

Kelly, J. P., and Gilbert, C. D., 1975, The projections of different morphological types of ganglion cells in the cat retina, *J. Comp. Neurol.* **163:**65–80.

Kimura, D., 1963, Speech lateralization in young children as determined by an auditory test, *J. Comp. Physiol. Psychol.* **56:**899–902.

Klosovskii, B. N., 1963, *The Development of the Brain and Its Disturbance by Harmful Factors,* Pergamon, Oxford.

Koffka, K., 1931, *The Growth of the Mind,* Harcourt, New York.

Kollros, J. J., 1968, Endocrine influences in neural development, in: *Growth of the Nervous System* (G. E. W. Wolstenholme and M. O'Connor, eds.), pp. 179–192, Churchill, London.

Kraschen, S., 1973, Lateralization of language learning and the critical period. Some new evidence, *Lang. Learn.* **23:**63–74.

Kraschen, S., 1975, The critical period for language acquisition and its possible bases, in: *Developmental Psycholinguistics and Communication Disorders* (D. R. Aaronson and R. W. Rieber, eds.), pp. 211–224, Vol. 263, Annals of the New York Academy of Science, New York.

Kuffler, S. W., 1953, Discharge patterns and functional organization of mammalian retina, *J. Neurophysiol.* **16:**37–68.

Kühn, A., 1971, *Lectures on Developmental Physiology* (transl. R. Milkman), Springer-Verlag, New York.

Kuypers, H. G. J. M., 1973, The anatomical organisation of the descending pathways and their contributions of motor control, especially in primates, in: *New Developments in E. M. G. and Clinical Neurophysiology* (T. E. Desmedt, ed.), pp. 38–68, Vol. 3, Karger, Basel.

Landsdell, H., 1962, Laterality of verbal intelligence in the brain, *Science* **135:**922–923.

Landsdell, H., 1969, Verbal and non-verbal factors in right hemisphere speech, *J. Comp. Physiol. Psychol.* **69:**734–738.

Langworthy, O. R., 1928, The behavior of pouch young opossums correlated with myelinization of tracts, *J. Comp. Neurol.* **46:**201–240.

Larroche, J.-C., 1966, The development of the central nervous system during intrauterine life, in: *Human Development* (F. Falkner, ed.), pp. 257–276, Saunders, Philadelphia.

Larroche, J.-C., 1967, Maturation morphologique du système nerveux central: Ses rapports avec le developpement pondéral du foetus et son age gestationnel, in: *Regional Development of the Brain in Early Life* (A. Minkowski, ed.), pp. 247–256, Blackwell, Oxford.

Lashley, K. S., 1951, The problem of serial order in behaviour, in: *Cerebral Mechanisms in Behaviour* (L. A. Jeffress, ed.), pp. 112–136, Wiley, New York.

Lawrence, D. G., and Kuypers, H. G. J. M., 1965, Pyramidal and non-pyramidal pathways in monkeys: Anatomical and functional correlation, *Science* **148:**973–975.

Lecours, A. R., 1975, Myelogenetic correlates of the development of speech and language, in: *Foundations of Language Development: A Multidisciplinary Approach* (E. H. Lenneberg and E. Lenneberg, eds.), pp. 75–94, University Publishers, New York.

Lee, D. N., and Lishman, J. R., 1975, Visual proprioceptive control of stance, *J. Hum. Movement Studies* **1:**87–95.

Lee, D. N., 1978, The functions of vision, in: *Modes of Perceiving and Processing Information* (H. L. Pick and E. Saltzman, eds.), pp. 195–170, Erlbaum, Hillsdale, N.J.

Leehey, S. C., Moskowitz-Cook, A., Brill, S., and Held, R., 1975, Orientational anisotropy in infant vision, *Science* **190:**900–902.

LeMay, M., 1976, Morphological cerebral asymmetries of modern man, fossil man, and nonhuman primate, *Ann. N.Y. Acad. Sci.* **280:**349–366.

Lenneberg, E. H., 1967, *Biological Foundations of Language,* Wiley, New York.

Lettvin, J. Y., Maturana, H. R., McCulloch, W. S., and Pitts, W. H., 1959, What the frog's eye tells the frog's brain, *Proc. I.R.E.* **47:**1940–1951.

LeVay, S., 1973, Synaptic patterns in the visual cortex of the cat and monkey: Electron microscopy of Golgi preparations, *J. Comp. Neurol.* **150:**53–86.

LeVay, S., Hubel, D. H., and Wiesel, T. N., 1975, The pattern of ocular dominance columns in macaque visual cortex revealed by a reduced silver stain, *J. Comp. Neurol.* **159:**559–576.

LeVere, T. E., 1975, Neural stability, sparing and behavioral recovery following brain damage, *Psychol. Rev.* **82:**344–358.

Levi-Montalcini, R., 1966, The nerve growth factor: Its mode of action on sensory and sympathetic nerve cells, *Harvey Lect.* **60:**217–259.

Levy, J., 1969, Possible basis for the evolution of lateral specialization of the human brain, *Nature* **224:**614–615.

Levy, J., 1974, Psychobiological implications of bilateral asymmetry, in: *Hemisphere Function in the Human Brain* (S. J. Dimond and J. G. Beaumont, eds.), pp. 121–183, Paul Elek (Scientific Books), London.

Levy, J., 1976, A review of evidence for a genetic component in the determination of handedness, *Behav. Genet.* **6:**429–453.

Levy, J., and Nagylaki, T., 1972, A model for the genetics of handedness, *Genetics* **72:**117–128.

Levy, J., and Trevarthen, C., 1977, Perceptual, semantic and phonetic aspects of elementary language processes in split-brain patients, *Brain* **100:**105–118.

Levy, J., Trevarthen, C., and Sperry, R. W., 1972, Perception of bilateral chimeric figures following hemispheric deconnection, *Brain* **95:**61–78.

Levy-Agresti, J., and Sperry, R. W., 1968, Differential perceptual capacities in major and minor hemispheres, *Proc. Natl. Acad. Sci. U.S.A.* **61:**1151.

Lewis, M., 1969, Infants' responses to facial stimuli during the first year of life, *Dev. Psychol.* **1:**75–86.

Liberman, A. M., 1974, The specialization of the language hemisphere, in: *The Neurosciences. Third Study Program* (F. O. Schmitt and F. G. Worden, eds.), pp. 43–56, MIT Press, Cambridge, Massachusetts.

Liley, A. W., 1972, Disorders of amniotic fluid, in: *Pathophysiology of Gestational Disorders,* Vol. 2., Fetal–Placental Maternal Disorders (N. S. Assali, ed.), pp. 157–206, Academic Press, New York.

Lipsitt, L. P., 1969, Learning capacities of the human infant, in: *Brain and Early Behaviour: Development in the Fetus and Infant* (R. J. Robinson, ed.), pp. 227–245, Academic Press, New York.

Loewenstein, W. R., 1968, Communication through cell junctions. Implications in growth control and differentiation, *Dev. Biol. Suppl.* **2**:151–183.

Lund, J. S., and Boothe, R. G., 1975, Interlaminar connections and pyramidal neuron organisation in the visual cortex, area 17, of the macaque monkey, *J. Comp. Neurol.* **159**:305–334.

Lund, R. D., Cunningham, T. J., and Lund, J. S., 1973, Modified optic projections after unilateral eye removal in young rats, *Brain Behav. Evol.* **8**:51–72.

Lynch, G. S., Mosko, S., Parks, T., and Cotman, C. W., 1973, Relocation and hyperdevelopment of the dentate gyrus commissural system after entorhinal lesions in immature rats, *Brain Res.* **50**:174–178.

Lynch, G., Rose, G., Gall, C., and Cotman, C. W., 1975, The response of the dentate gyrus to partial deafferentation, in: *Golgi Centennial Symposium, Proceedings* (M. Santini, ed.), pp. 305–317, Raven Press, New York.

Malpeli, J. G., and Baker, H., 1975, The representation of the visual field in the lateral geniculate nucleus of *Macaca mulatta*, *J. Comp. Neurol.* **161**:569–594.

Mann, I., 1964, *The Development of the Human Eye*, Grune & Stratton, New York.

Maratos, O., 1973, The Origin and Development of Imitation in the First Six Months of Life, PhD thesis, University of Geneva.

Marie, P., 1922, Existe-t-il chez l'homme des centres préfermés ou innés du langage?, in: *Questions Neurologiques d'Actualité*, pp. 527–551, Masson, Paris.

Marin-Padilla, M., 1970, Prenatal and early post-natal ontogenesis of the cerebral cortex (neocortex) of the cat *(Felis domestica):* A Golgi study I. The primordial neocortical organization, *Z. Anat. Entwicklungsgesch.* **134**:117–145.

Mark, R. F., 1974, *Memory and Nerve-Cell Connections*, Oxford University Press, London.

Mark, R. F., 1975, Topography and topology in functional recovery of regenerated sensory and motor systems, in: *Ciba Symposium on Cell Patterning, No. 29 (New Series)*, pp. 289–313, American Elsevier, New York.

Mark, R. F., and Feldman, J., 1972, Binocular interaction in the development of optokinetic reflexes in tadpoles of *Xenopus laevis*, *Invest. Ophthalmol.* **11**:402–410.

Mark, R. F., and Watts, M. E., 1971, Drug inhibitions of memory formation in chickens. I. Long-term memory, *Proc. R. Soc. London, Ser. B* **178**:439–454.

Marty, R., and Scherrer, J., 1964, Critères de maturation des systèmes afférents corticaux, *Prog. Brain Res.* **4**:222–234.

Maturana, H. R., Lettvin, J. Y., McCulloch, W. S., and Pitts, W. H., 1959, Evidence the cut optic nerve fibers in a frog regenerate to their proper places in the tectum, *Science* **130**:1709–1710.

Maturana, H. R., Lettvin, J. Y., McCulloch, W. S., and Pitts, W. H., 1960, Anatomy and physiology of vision in the frog *(Rana pipiens)*, *J. Neurophysiol.* **43**:129–175.

McGinnis, J. M., 1930, Eye movements and optic nystagmus in early infancy, *Genet. Psych. Monogr.* **8**:321–430.

McGraw, M. B., 1943, *The Neuromuscular Maturation of the Human Newborn*, Columbia University Press, New York.

McKenzie, B. F., and Day, R. H., 1972, Object distance as a determinant of visual fixation in early infancy, *Science* **178**:1108–1110.

McMahon, D., 1973, A cell-contact model for cellular position determinations in development, *Proc. Natl. Acad. Sci. U.S.A.* **70**:2396–2400.

MaMahon, D., 1974, Chemical messengers in development: A hypothesis, *Science* **185**:1012–1021.

McMahon, D., and West, C., 1976, Transduction of positional information during development, in: *Cell Surface Interactions in Embryogenesis* (G. Poste and G. Nicolson, eds.), pp. 449–493, Elsevier North-Holland, New York.

Mendelson, M. J., and Haith, M. M., 1977, The relation between audition and vision in the human newborn, *Monogr. Soc. Res. Child Dev.* **41**:1–72.

Metzger, W., 1974, Conscious perception and action, in: *Handbook of Perception*, Vol. 1 (E. C. Carterette and M. P. Friedman, eds.), pp. 109–122, Academic Press, New York.

Meyer, R. L., and Sperry, R. W., 1973, Test for neuroplasticity in the anuran retinotectal system, *Exp. Neurol.* **40**:525–539.

Meyer, R. L., and Sperry, R. W., 1974, Explanatory models for neuroplasticity in retinotectal connections, in: *Plasticity and Recovery of Function in the Central Nervous System* (D. G. Stein, J. J. Rosen, and N. Butters, eds.), pp. 45–63, Academic Press, New York.

Meyer, R. L., and Sperry, R. W., 1976, Retinotectal specificity: Chemospecificity theory, in: *Neural and Behavioral Specificity, Studies on the Development of Behavior and the Nervous System*, Vol. 3 (G. Gottlieb, ed.), pp. 111–149, Academic Press, New York.

Michotte, A., 1963, *The Perception of Causality* (translated by T. R. Miles and E. Miles), Methuen, London.

Milner, B., 1974, Interhemispheric differences and psychological processes, *B. Med. Bull.* **27:**272–277.

Miner, N., 1956, Integumental specification of sensory fibers in the development of cutaneous local sign, *J. Comp. Neurol.* **105:**161–170.

Molfese, D. L., Freeman, R. B., and Palermo, D. S., 1975, The ontogeny of brain lateralization for speech and nonspeech stimuli, *Brain Lang.* **2:**356–368.

Molliver, M., Kostovic, I., and Van der Loos, H., 1973, The development of synapses in cerebral cortex of the human fetus, *Brain Res.* **50:**403–407.

Morest, D. K., 1970, A study of neurogenesis in the forebrain of opossum pouch young, *Z. Anat. Entwicklungsgesch.* **130:**265–305.

Morgan, M. J., 1976, Embryology and the inheritance of asymmetry, in: *Lateralization of the Nervous System* (S. R. Harnand, R. W. Doty, L. Goldstein, J. Jaynes, and G. Krautheimer, eds.), pp. 173–194, Academic Press, New York.

Mountcastle, V. B., 1957, Modality and topographic properties of single neurones of cats' somatic sensory cortex, *J. Neurophysiol.* **20:**408–434.

Mountcastle, V. B., 1974, Neural mechanisms in somesthesia, in: *Medical Physiology* (V. B. Mountcastle, ed.), Vol. 1, 13th ed., pp. 307–347, C. V. Mosby, St. Louis.

Mountcastle, V. B., 1975, The view from within. Pathways to the study of perception, *Johns Hopkins Med. J.* **136:**109–131.

Mountcastle, V. B., Lynch, J. C., Georgopolous, A., Sakata, H., and Acuna, C., 1975, Posterior parietal association cortex of the rhesus monkey: Command functions for operations within extrapersonal space, *J. Neurophysiol.* **38:**871–908.

Mundy-Castle, A., and Anglin, J., 1973, Looking strategies in infants, in: *The Competent Infant: Research and Commentary* (L. Stone, H. Smith, and L. Murphy, eds.), pp. 713–718, Basic Books, New York.

Muntz, W. R. A., 1962a, Microelectrode recordings from the diencephalon of the frog *(Rana pipiens)* and a blue-sensitive system, *J. Neurophysiol.* **25:**699–711.

Muntz, W. R. A., 1962b, Effectiveness of different colours of light in releasing positive phototactic behaviour of frogs, and a possible function of the retinal projection to the diencephalon, *J. Neurophysiol.* **25:**712–720.

Nebes, R. D., 1972, Dominance of the minor hemisphere in commissurotomized man on a test of figural unification, *Brain* **95:**633–638.

Nebes, R. D., 1974, Dominance of the minor hemisphere in commissurotomized man for the perception of part-whole relationships, in: *Hemispheric Disconnection and Cerebral Function* (M. Kinsbourne and W. L. Smith, eds.), pp. 155–164, Charles C Thomas, Springfield, Illinois.

Needham, J., 1942, *Biochemistry and Morphogenesis,* Cambridge University Press, London.

Neisser, U., 1967, *Cognitive Psychology,* Appleton-Century-Crofts, New York.

Neisser, U., 1976, *Cognition and Reality,* Freemans, San Francisco.

Nieuwkoop, P. D., and Faber, J., 1956, *Normal table of Xenopus laevis (Daudin),* North Holland, Amsterdam.

Oppenheim, R. W., 1973, Prehatching and hatching behavior: Comparative and physiological considerations, in: *Studies in the Development of Behavior and the Nervous System,* Vol. 1, Behavioral Embryology (G. Gottlieb, ed.), pp. 164–236, Academic Press, New York.

Oppenheim, R. W., 1974, The ontogeny of behavior in the chick embryo, in: *Advances in the Study of Behavior* (D. H. Lehrman *et al.,* eds.), Vol 5, pp. 133–172, Academic Press, New York.

Oppenheim, R. W., and Reitzel, J., 1975, Ontogeny of behavioral sensitivity to strychnine in the chick embryo: evidence for the early onset of CNS inhibition, *Brain Behav. and Evol.* **11:**130–159.

Ordy, J. M., Massopust, L. C., and Wolin, L. R., 1962, Postnatal development of the retina, ERG, and acuity in the rhesus monkey, *Exp. Neurol.* **5:**364–382.

Papousek, H., 1961, Conditioned head rotation reflexes in the first six months of life, *Acta Pediatr.* **50:**565–576.

Papousek, H., 1969, Individual variability in learned responses in human infants, in: *Brain and Early Behaviour: Development in Fetus and Infant* (R. J. Robinson, ed.), pp. 251–266, Academic Press, New York.

Pasik, T., and Pasik, P., 1971, The visual world of monkeys deprived of striate cortex: Effective stimulus parameters and the importance of the accessory optic system, *Vision Res., Suppl. 3,* 419–435.

Peiper, A., 1963, *Cerebral Function in Infancy and Childhood,* Consultants Bureau, New York.

Perenin, M. T., and Jeannerod, M., 1975, Residual vision in cortically blind hemifields, *Neuropsychologia* **13:**1–7.

Pettigrew, J. D., 1972, The importance of early visual experience for neurons of the developing geniculostriate system, *Invest. Ophthalmol.* **11**:386–394.

Pettigrew, J. D., and Freeman, R. D., 1973, Visual experience without lines: Effect on developing cortical neurons, *Science* **182**:599–601.

Piaget, J., 1953, *The Origins of Intelligence in Children,* Routledge and Kegan Paul, London (original French edition, 1936).

Piaget, J., 1967, *Biologie et Connaissance,* Gallimard, Paris.

Piaget, J., 1970, Piaget's theory, in: *Carmichael's Manual of Child Psychology* (P. H. Mussen, ed.), pp. 703–732, Wiley, New York.

Pirchio, M., Spinelli, D., Fiorentini, A., and Maffei, L., 1978, Infant contrast sensitivity evaluated by evoked potentials, *Brain Res.* **141**:179–184.

Polyak, S. L., 1941, *The Vertebrate Visual System,* The University of Chicago Press, Chicago.

Pomeranz, B., 1972, Metamorphosis of frog vision: Changes in ganglion cell physiology and anatomy, *Exp. Neurol.* **34**:187–199.

Pomeranz, B., and Chung, S. H., 1970, Dendritic-tree anatomy codes form vision physiology in tadpole retina, *Science* **170**:983–984.

Pöppel, E., Held, R., and Frost, D., 1973, Residual function after brain wounds involving the central visual pathways in man, *Nature* **243**:295–296.

Prestige, M. C., 1970, Differentiation, degeneration, and the role of the periphery: Quantitative considerations, in: *The Neurosciences, Second Study Program* (F. O. Schmitt, ed.), pp. 73–82, Rockefeller University Press, New York.

Prestige, M. C., and Willshaw, D. J., 1975, On the role of competition in the formation of patterned neural connections, *Proc. R. Soc. London, Ser. B* **190**:77–98.

Raisman, G., and Field, P. M., 1973, A quantitative investigation of the development of collateral reinnervation after partial deafferentation of the septal nuclei, *Brain Res.* **50**:241–264.

Rakic, P., 1971, Neuron–glia relationship during granule cell migration in developing cerebellar cortex. A golgi and electronmicroscopic study in *Macacus rhesus, J. Comp. Neurol.* **141**:283–312.

Rakic, P., 1972, Mode of cell migration to the superficial layers of fetal monkey neocortex, *J. Comp. Neurol.* **145**:61–84.

Rakic, P., 1974, Neurons in rhesus monkey visual cortex: Systematic relation between time of origin and eventual disposition, *Science* **183**:425–427.

Rakic, P., 1976, Prenatal genesis of connections subserving ocular dominance in the rhesus monkey, *Nature* **261**:467–471.

Rakic, P., 1977, Prenatal development of the visual system in the rhesus monkey, *Phil. Trans. R. Soc. London,* **278**:245–260.

Rakic, P., and Yakovlev, P. I., 1968, Development of the corpus callosum and cavum septi in man, *J. Comp. Neurol.* **132**:45–72.

Ramón y Cajal, S., 1909–1911, *Histologie du Système Nerveux de l'Homme et des Vertébrés* (L. Azoulay, transl.), 2 vols., A. Maloine, Paris (Reprinted: Consejo Superior de Investigaciones Científicas, Madrid, 1952 and 1955).

Ramón y Cajal, S., 1929, *Étude sur la Neurogenèse de quelques Vertébrés,* Madrid. (Reprinted: *Studies on Vertebrate Neurogenesis,* L. Guth, transl., Charles C Thomas, Springfield, Illinois.)

Riesen, A. H., and Aarons, L., 1959, Visual movement and intensity discrimination in cats after early deprivation of pattern vision, *J. Comp. Physiol. Psychol.* **52**:142–149.

Roach, F. C., 1945, Differentiation of the central nervous system after axial reversals of the medullary plate of *Amblystoma, J. Exp. Zool.* **99**:53–77.

Robinson, J. S., and Fish, S. E., 1974, A cat's form experienced but visuo-motor deprived eye lacks focal vision, *Dev. Psychobiol.* **7**:331–342.

Rodieck, R. W., 1973, *The Vertebrate Retina: Principles of Structure and Function,* W. H. Freeman, San Francisco.

Roth, R. L., 1974, Retinotopic organization of goldfish optic nerve and tract, *Anat. Rec.* **178**:453.

Rudel, R., and Denkla, M. B., 1974, Relation of forward and backward digit repetition to neurological impairment in children with learning disabilities, *Neuropsychologia* **12**:109–118.

Rudel, R., Teuber, H. L., and Twitchell, T. E., 1974, Levels of impairment of sensorimotor early damage, *Neuropsychologia* **12**:95–108.

Saint-Anne Dargassies, S., 1966, Neurological maturation of the premature infant of 28–41 weeks gestational age, in: *Human Development* (F. Falkner, ed.), pp. 306–325, Saunders, Philadelphia.

Sarnat, H. B., and Netsky, M. G., 1974, *Evolution of the Nervous System,* Tavistock, Oxford.

Saul, R., and Sperry, R. W., 1968, Absence of commissurotomy symptoms with agenesis of the corpus callosum, *Neurology* **18**:307.

Saunders, J. W., Jr., and Fallon, J. F., 1966, Cell death in morphogenesis, in: *Major Problems in Developmental Biology* (M. Locke, ed.), pp. 289–314, Academic Press, New York.

Saxén, L., 1972, Interactive mechanisms in morphogenesis, in: *Tissue Interactions in Carcinogenesis* (D. Tarin, ed.), pp. 49–80, Academic Press, London.

Saxén, L., and Toivonen, S., 1962, *Primary Embryonic Induction.* Academic Press, New York.

Scalia, F., and Fite, K., 1974, A retinotopic analysis of the central connections of the optic nerve in the frog, *J. Comp. Neurol.* **158:**455–478.

Scarf, B., and Jacobson, M., 1974, Development of binocularly driven single units in frogs raised with asymmetrical visual stimulation, *Exp. Neurol.* **42:**669–686.

Schaffer, H. R. (ed.), 1977, *Studies in Mother–Infant Interaction: The Loch Lomond Symposium,* Academic Press, London.

Scheibel, M. E., Lindsay, R. D., Tomiyasu, U., and Scheibel, A. B., 1975, Dendritic changes in aging human cortex, *Anat. Rec.* **181:**471.

Schneider, G. E., 1967, Contrasting visuomotor functions of tectum and cortex in the golden hamster, *Psychol. Forsch.* **31:**52–62.

Schneider, G. E., 1969, Two visual systems: Brain mechanisms for localization and discrimination are dissociated by tectal and cortical lesions, *Science* **163:**895–902.

Schneider, G. E., 1970, Mechanisms of functional recovery following lesions of visual cortex or superior colliculus in neonate and adult hamsters, *Brain Behav. Evol.* **3:**295–323.

Schneider, G. E., 1973, Early lesions of superior colliculus: factors affecting the formation of abnormal retinal projections, *Brain Behav. Evol.* **8:**73–109.

Schneider, G. E., 1975, Growth of abnormal axonal connections after brain lesions, in: *Outcome of Severe Damage to the Central Nervous System, CIBA Found. Symp. New Series,* Vol. 34, pp. 56–59, Elsevier, Amsterdam.

Schneider, G. E., 1976, Growth of abnormal neural connections following focal brain lesions: Constraining factors and functional effects, in: *Neurosurgical Treatment in Psychiatry* (W. H. Sweet, S. Obrador, and J. G. Martin-Rodrigues, eds.), pp. 5–26, University Park Press, Baltimore.

Schneider, G. E., and Jhaveri, S. R., 1974, Neuroanatomical correlates of spared or altered function after brain lesions in the newborn hamster, in: *Plasticity and Recovery of Function in the Central Nervous System* (D. G. Stein, J. J. Rosen, and N. Butters, eds.), pp. 65–109, Academic Press, New York.

Schulte, F. J., Linke, I., Michaelis, R., and Nolte, R., 1969, Excitation, inhibition, and impulse conduction in spinal motoneurones of preterm, term and small-for-dates newborn infants, in: *Brain and Early Behavior: Development in the Fetus and Infant* (R. J. Robinson, ed.), pp. 87–114, Academic Press, New York.

Scott, M. Y., 1975, Functional capacity of compressed retinotectal projection in goldfish, *Anat. Rec.* **181:**474.

Scott, S. A., 1975, Persistence of foreign innervation on reinnervated goldfish extraocular muscles, *Science* **189:**644–646.

Sequeland, E. R., and DeLucia, C. A., 1969, Visual reinforcement of non-nutritive sucking in human infants, *Science* **165:**1144–1146.

Sharma, S. C., 1972, Reformation of retinotectal projections after various tectal ablations in adult goldfish, *Exp. Neurol.* **34:**171–182.

Sharma, S. C., and Gaze, R. M., 1971, The retinotopic organization of visual responses from tectal reimplants in adult goldfish, *Arch. Ital. Biol.* **109:**357–366.

Sharma, S. C., and Hollyfield, J. G., 1974, Specification of retinal central connections in *Rana pipiens* before the appearance of the first post-mitotic ganglion cells, *J. Comp. Neurol.* **155:**395–408.

Sherman, S. M., 1972, Development of interocular alignment in cats, *Brain Res.* **37:**187–198.

Sherman, S. M., 1973, Visual field defects in monocularly and binocularly deprived cats, *Brain Res.* **49:**25–45.

Sherman, S. M., Guillery, R. W., Kaas, J. H., and Sanderson, K. J., 1974, Behavioral, electrophysiological and morphological studies of binocular competition in the development of the geniculo-cortical pathways of cats, *J. Comp. Neurol.* **158:**1–18.

Shkol'nik-Yarros, E. G., 1971, *Neurones and Interneuronal Connections of the Central Visual System* (B. Haigh, transl.), Plenum Press, New York.

Shlaer, R., 1971, Shift in binocular disparity causes compensating change in the cortical structure of kittens, *Science* **173:**638–641.

Sidman, R. L., 1961, Histogenesis of mouse retina studied with thymidine-3H, in: *Structure of the Eye* (G. K. Smelser, ed.), pp. 487–505, Academic Press, New York.

Sidman, R. L., and Rakic, P., 1973, Neuronal migration, with special reference to developing human brain: A review, *Brain Res.* **62:**1–35.

Slater, A. M., and Findlay, J. M., 1975, Binocular fixation in the newborn baby, *J. Exp. Child Psychol.* **20**:248–273.

Smith, A., 1966, Speech and other functions after left (dominant) hemispherectomy, *J. Neurol. Neurosurg. Psychiatry* **29**:467–471.

Spalding, D. A., 1873, Instinct with original observations on young animals, *Macmillan's Magazine* **27**:282–293 (reprinted in *Br. J. Anim. Behav.* **2**:2–11).

Spemann, H., 1938, *Embryonic Development and Induction*, Yale University Press, New Haven.

Sperry, R. W., 1943, Visuomotor coordination in the newt *(Triturus viridescens)* after regeneration of the optic nerves, *J. Comp. Neurol.* **79**:33–55.

Sperry, R. W., 1944, Optic nerve regeneration with return of vision in anurans, *J. Neurophysiol* **7**:57–69.

Sperry, R. W., 1945, Restoration of vision after crossing of optic nerves and after contralateral transposition of the eye, *J. Neurophysiol.* **8**:15–28.

Sperry, R. W., 1951a, Mechanisms of neural maturation, in: *Handbook of Experimental Psychology* (S. S. Stevens, ed.), pp. 236–280, Wiley, New York.

Sperry, R. W., 1951b, Regulative factors in orderly growth of neural circuits, *Growth Symp.* **10**:63–87.

Sperry, R. W., 1952, Neurology and the mind-brain problem, *Am. Sci.* **40**:291–312.

Sperry, R. W., 1963, Chemoaffinity in the orderly growth of nerve fiber patterns and connections, *Proc. Natl. Acad. Sci. U.S.A.* **50**:703–710.

Sperry, R. W., 1965, Embryogenesis of behavioral nerve nets, in: *Organogenesis* (R. L. De Haan and H. Ursprung, eds.), pp. 161–186, Holt, New York.

Sperry, R. W., 1970, Perception in the absence of the neocortical commissures, *Res. Publ. Assoc. Res. Nerv. Ment. Dis.* **48**:123–138.

Sperry, R. W., 1974, Lateral specialization in the surgically separated hemispheres, in: *The Neurosciences: Third Study Program* (F. O. Schmitt and F. G. Worden, eds.), pp. 5–20, MIT Press, Cambridge, Massachusetts.

Sperry, R. W., and Hibbard, E., 1968, Regulative factors in the orderly growth of retino-tectal connections, in: *Growth of the Nervous System* (G. E. W. Wolstenholme and M. O'Connor, eds.), pp. 41–52, Churchill, London.

Sperry, R. W., Gazzaniga, M. S., and Bogen, J. E., 1969, Interhemispheric relationships: The neocortical commissures; syndromes of hemisphere deconnection, in: *Handbook of Clinical Neurology,* Vol. 4 (P. J. Vinken and G. W. Bruyn, eds.), pp. 273–290, North Holland, Amsterdam.

Sprague, J. M., Levy, J., DiBerardino, A., and Berlucchi, G., 1977, Visual cortical areas mediating form discrimination in the cat, *J. Comp. Neurol.* **172**:441–488.

Stein, B. E., Labos, E., and Kruger, L., 1973, Sequence of changes in properties of neurones of superior colliculus of the kitten during maturation, *J. Neurophysiol.* **36**:667–679.

Stone, L. S., 1944, Functional polarization in retinal development and its reestablishment in regenerated retinae of rotated eyes, *Proc. Soc. Exp. Biol. Med.* **57**:13–14.

Stone, L. S., 1959, Experiments testing the capacity of iris to regenerate neural retina in eyes of adult newts, *J. Exp. Zool.* **142**:285–308.

Stone, L. S., 1960, Polarization of the retina and development of vision, *J. Exp. Zool.* **145**:85–93.

Straznicky, K., and Gaze, R. M., 1971, The growth of the retina in *Xenopus laevis:* An autoradiographic study, *J. Embryol. Exp. Morphol.* **26**:67–79.

Straznicky, K., and Gaze, R. M., 1972, The development of the tectum in *Xenopus laevis:* An autoradiographic study, *J. Embryol. Exp. Morphol.* **28**:87–115.

Straznicky, K., Gaze, R. M., and Keating, M. J., 1971, The retinotectal projections after uncrossing the optic chiasma in *Xenopus* with one compound eye, *J. Embryol. Exp. Morphol.* **26**:523–542.

Székely, G., 1954, Untersuchung der Entwicklung optischer Reflex mechanismem an Amphibien larven, *Acta Physiol. Acad. Sci. Hung.* **6**(Suppl. 18).

Székely, G., 1957, Regulationstendenzen in der Ausbildung der "Funktionellen Spezifität" der Retinoanlage bei *Triturus vulgaris, Arch. Entw. Org.* **150**:48–60.

Székely, G., 1974, Problems of neuronal specificity in the development of some behavioural patterns in amphibia, in: *Aspects of Neurogenesis* (G. Gottlieb, ed.), pp. 115–150, Academic Press, New York.

Szentagothai, M. J., and Arbib, M. A., 1975, The module concept in cerebral cortex architecture, *Brain Res.* **95**: 475–496.

Talbot, S. A., and Marshall, W. H., 1941, Physiological studies on neural mechanisms of visual localization and discrimination, *Am. J. Ophthalmol.* **24**:1255–1263.

Taub, E., 1968, Prism compensation as a learning phenomenon: A phylogenetic comparison, in: *The Neuropsychology of Spatially Oriented Behavior* (S. F. Freedman, ed.), pp. 173–192, Dorsey Press, Homewood, Illinois

Tees, R. C., 1974, Effect of visual deprivation on development of depth perception in the rat, *J. Comp. Physiol. Psychol.* **86:**300–308.

Tees, R. C., 1976, Perceptual development in mammals, in: *Neural and Behavioral Specificity, Studies on the Development of Behavior and the Nervous System,* Vol. 3 (G. Gottlieb, ed.), pp. 281–326, Academic Press, New York.

Teller, D. Y., Mase, R., Baton, R., and Regai, D., 1974, Visual acuity for vertical and diagonal gratings in human infants, *Vision Res.* **14:**1433–1439.

Teszner, D., Tzavaras, A., Gruner, J., and Hécaen, H., 1972, L'asymétrie droite-gauche du planum temporale: àpropos de l'étude anatomique de 100 cerveaux, *Rev. Neurol.* **126:**444–449.

Teuber, H.-L., 1960, Perception, in: *Handbook of Physiology. Section 1. Neurophysiology* (J. Field, H. W. Magoun, and V. E. Hall, eds.), Vol. III, pp. 1595–1668, American Physiological Society, Washington, D.C.

Toivonen, S., and Saxén, L., 1968, Morphogenetic interaction of presumptive neural and mesodermal cells mixed in different ratios, *Science* **159:**539–540.

Trehub, S. E., 1973, Infants' sensitivity to vowel and tonal contrasts, *Dev. Psychol.* **9:**91–96.

Trevarthen, C., 1968a, Vision in fish: The origins of the visual frame for action in vertebrates, in: *The Central Nervous System and Fish Behavior* (D. Ingle, ed.), pp. 61–94, University of Chicago Press, Chicago.

Trevarthen, C., 1968b, Two mechanisms of vision in primates, *Psych. Forsch.* **31:**299–337.

Trevarthen, C., 1972, Brain bisymmetry and the role of the corpus callosum in behavior and conscious experience, in: *Cerebral Interhemispheric Relations* (J. Cernacek and F. Podovinsky, eds.), pp. 319–333, Slovak Academy of Sciences, Bratislava.

Trevarthen, C., 1973, Behavioral embryology, in: *Handbook of Perception* (E. C. Carterette and M. P. Friedman, eds.), Vol. 3, pp. 89–117, Academic Press, New York.

Trevarthen, C., 1974a, Analysis of cerebral activities that generate and regulate consciousness in commissurotomy patients, in: *Hemisphere Function in the Human Brain* (S. J. Dimond and J. G. Beaumont, eds.), pp. 235–263, Paul Elek, London.

Trevarthen, C., 1974b, L'action dans l'espace et la perception de l'espace: mécanismes cérébraux de base, in: *De l'Espace Corporel à l'Espace Ecologique* (F. Bresson, ed.), pp. 65–80, Presses Universitaires de France, Paris.

Trevarthen, C., 1974c, Cerebral embryology and the split brain, in: *Hemispheric Disconnection and Cerebral Function* (M. Kinsbourne and W. L. Smith, eds.), pp. 208–236, Charles C Thomas, Springfield, Illinois.

Trevarthen, C., 1974d, The psychobiology of speech development, in: *Language and Brain: Developmental Aspects* (E. Lenneberg, ed.), Neurosciences Research Program Bulletin, **12:**570–585.

Trevarthen, C., 1975, Early attempts at speech, in: *Child Alive* (R. Lewin, ed.), pp. 57–74, Maurice Temple Smith, London.

Trevarthen, C., 1977, Descriptive analyses of infant communication behavior, in: *Studies in Mother–Infant Interaction: The Loch Lomond Symposium* (H. R. Schaffer, ed.), pp. 227–270, Academic Press, London.

Trevarthen, C., 1978a, Modes of perceiving and modes of acting, in: *Modes of Perceiving and Processing Information* (H. Pick and E. Saltzman, eds.), pp. 99–136, U.S. Social Science Research Council.

Trevarthen, C., 1978b, Manipulative strategies of baboons and the origins of cerebral asymmetry, in: *The Asymmetrical Function of the Brain* (M. Kinsbourne, ed.), pp. 329–391, Cambridge University Press, London.

Trevarthen, C., and Sperry, R. W., 1973, Perceptual unity of the ambient visual field in human commissurotomy patients, *Brain* **96:**547–570.

Twitchell, T. E., 1965, The automatic grasping responses of infants, *Neuropsychologia* **3:**247–259.

Uzgiris, I. C., 1974, Patterns of vocal and gestural imitation in infants, in: *The Competent Infant* (L. J. Stone, H. T. Smith, and L. B. Murphy, eds.), pp. 599–604, Tavistock, London.

Valverde, F., 1971, Rate and extent of recovery from dark rearing in the visual cortex of the mouse, *Brain Res.* **33:**1–11.

Van Sluyters, R. C., and Blakemore, C., 1973, Experimental creation of unusual neuronal properties in visual cortex of kitten, *Nature (London)* **246:**506–508.

Vinken, P. J., and Bruyn, R. W. (eds.), 1969, *Handbook of Clinical Neurology:* Vol. 3, *Disorders of Higher Nervous Activity;* Vol. 4, *Disorders of Speech Perception and Symbolic Behavior,* North Holland, Amsterdam.

Vital-Durand, F., and Jeannerod, M., 1974, Maturation of the optokinetic response: Genetic and environmental factors, *Brain Res.* **71:**249–257.

Von Békésy, G., 1967, *Sensory Inhibition,* Princeton University Press, Princeton, New Jersey.

von Holst, E., 1954, Relations between the central nervous system and the peripheral organs, *Br. J. Anim. Behav.* **2:**89–94.

Von Senden, M., 1960, *Space and Sight* (translated by P. Heath), Methuen, London.

Waber, D. P., 1976, Sex differences in cognition: A function of maturation rate? *Science* **192:**572–573.

Wada, J. A., Clarke, R., and Hamm, A., 1975. Cerebral hemispheric asymmetry in humans: Cortical speech zones in 100 adult and 100 infant brains, *Arch. Neurol.* **32:**239–246.

Waddington, C. H., 1957, *The Strategy of the Genes,* Allen and Unwin, London.

Waddington, C. H., 1966, *Principles of Development and Differentiation,* Macmillan, New York.

Walk, R. D., and Bond, E. K., 1971, The development of visually guided reaching in monkeys reared without sight of the hands, *Psychon. Sci.* **23:**115–116.

Walk, R. D., and Gibson, E. J., 1961, A comparative and analytic study of visual depth perception, *Psychol. Monogr.* **75:**Whole No. 519.

Walk, R. D., and Pick, H. L., Jr. 1978, *Perception and Experience*, Plenum Press, New York.

Wall, P. D., 1975, Signs of plasticity and reconnection in spinal cord damage, in: *Outcome of Severe Damage to the Central Nervous System,* CIBA Foundation Symposium, New Series 34, Elsevier, Amsterdam.

Watson, J. S., 1973, Smiling, cooing and "The Game," *Merrill-Palmer Q.* **18:**323–339.

Watts, M. E., and Mark, R. F., 1971, Drug inhibition of memory formation in chickens. II. Short-term memory, *Proc. R. Soc. London, Ser. B* **178:**455–464.

Webster, W. R., and Aitkin, L. M., 1975, Central auditory processing, in: *Handbook of Psychobiology* (M. S. Gazzaniga and C. Blakemore, eds.), pp. 325–364, Academic Press, New York.

Weiskrantz, L., Warrington, E. K., Saunders, M. P., and Marshall, J., 1974, Visual capacity in the hemianopic field following a restricted occipital ablation, *Brain* **97:**709–728.

Weiss, P., 1936, Selectivity controlling the central-peripheral relations in the nervous system, *Biol. Rev. Cambridge Philos. Soc.* **11:**494–531.

Weiss, P., 1939, *Principles of Development,* Holt, Rinehart and Winston, New York.

Weiss, P., 1941, Self-differentiation of the basic patterns of coordination, *Comp. Psychol. Monogr.* **17:**1–96.

Werner, H., 1961, *Comparative Psychology of Mental Development* (revised ed.), Science Editions, New York.

Wetzel, A. B., Thompson, V. E., Horel, J. A., and Meyer, P. M., 1965, Some consequences of perinatal lesions of the visual cortex in the cat, *Psychon. Sci.* **3:**381–382.

White, M. J., 1969, Laterality differences in perception: A review, *Psychol. Bull.* **72:**387–405.

White, B. L., Castle, P., and Held, R., 1964, Observations on the development of visually-directed reaching, *Child Dev.* **35:**349–364.

Whitteridge, D., 1973, Visual projections to the cortex, in: *Handbook of Sensory Physiology* (R. Jung, ed.), Vol. VII/3/B, pp. 247–268, Springer, Berlin.

Wickelgren, B., 1972, Some effects of visual deprivation on the cat superior colliculus, *Invest. Ophthalmol.* **11:**460–467.

Wickelgren, B., and Sterling, P., 1969, Influence of visual cortex on receptive fields in the superior colliculus of the cat, *J. Neurophysiol.* **32:**16–23.

Wiesel, T. N., and Hubel, D. H., 1963, Single-cell responses in striate cortex of kittens deprived of vision in one eye, *J. Neurophysiol.* **26:**1003–1017.

Wiesel, T. N., and Hubel, D. H., 1965*a,* Comparison of the effects of unilateral and bilateral eye closure on cortical unit response in kittens, *J. Neurophysiol.* **28:**1029–1040.

Wiesel, T. N., and Hubel, D. H., 1965*b,* Extent of recovery from the effects of visual deprivation in kittens, *J. Neurophysiol.* **28:**1060–1072.

Wiesel, T. N., and Hubel, D. H., 1974, Ordered arrangement of orientation columns in monkeys lacking visual experience, *J. Comp. Neurol.* **158:**307–318.

Wigglesworth, V. B., 1954, *The Physiology of Insect Metamorphosis,* Cambridge University Press, Cambridge.

Willshaw, D. J., and von der Malsburg, C., 1976, How patterned neural connections can be set up by self-organization, *Proc. R. Soc. London, Ser. B* **194:**431–445.

Wilson, P. D., and Riesen, A. H., 1966, Visual development in rhesus monkeys neonatally deprived of patterned light, *J. Comp. Physiol. Psychol.* **61:**87–95.

Windle, W. F., 1970, Development of neural elements in human embryos of four to seven weeks gestation, *Exp. Neurol. Suppl.* **5:**44–83.

Witelson, S. F., and Pallie, W., 1973, Left hemisphere specialization for language in the newborn: neuroanatomical evidence of asymmetry, *Brain* **96:**641–646.

Wolff, P. H., 1966, The causes, controls and organization of behavior in the neonate, Psychological Issues, Monograph Series, Vol. 5, No. 1, Monogr. 17, International Universities Press, New York.

Wolff, P. H., 1968, Stereotypic behavior and development, *Can. Psychol.* **9**:474–484.

Wolff, P. H., 1969, The natural history of crying and other vocalizations in early infancy, in: *Determinants of Infant Behaviour* (B. M. Foss, ed.), Vol. IV, pp. 81–110, Methuen, London.

Wolpert, L., 1969, Positional information and the spatial pattern of cellular differentiation, *J. Theor. Biol.* **25**:1–47.

Wolpert, L., 1971, Positional information and pattern formation, *Curr. Top. Dev. Biol.* **6**:183–224.

Woods, B. T., and Teuber, H.-L., 1973, Early onset of complementary specialization of cerebral hemispheres in man, *Trans. Am. Neurol. Assoc.* **98**:113–115.

Woods, B. T., and Teuber, H.-L., 1978, Changing patterns of childhood aphasia, *Ann. Neurol.* (in press).

Yakovlev, P. I., 1962, Morphological criteria of growth and maturation of the nervous system in man, in: *Mental Retardation, Research Publications of the Association for Research on Nervous and Mental Diseases,* Vol. 39, pp. 3–46, Association for Research on Nervous and Mental Diseases, New York.

Yakovlev, P. I., 1968, Telencephalon 'impar,' 'semi-par' and 'totopar.' Morphogenetic, tectogenetic and architectonic definitions, *Int. J. Neurol.* **6**:245–265.

Yakovlev, P. I., and Lecours, A. R., 1967, The myelogenetic cycles of regional maturation of the brain, in: *Regional Development of the Brain in Early Life* (A. Minkowski, ed.), pp. 3–70, Blackwell, Oxford.

Yarbus, A. L., 1967, *Eye Movements and Vision,* Plenum Press, New York.

Yoon, M. G., 1971, Reorganization of retinotectal projection following surgical operations on the optic tectum in goldfish, *Exp. Neurol.* **33**:395–411.

Yoon, M. G., 1972, Reversibility of the reorganization of retinotectal projection in goldfish, *Exp. Neurol.* **35**:565–577.

Yoon, M. G., 1973, Retention of the original topographic polarity by the 180° rotated tectum reimplant in young adult goldfish, *J. Physiol. (London)* **233**:575–588.

Yoon, M. G., 1975*a*, Readjustment of retinotectal projection following reimplantation of a rotated or inverted tectal tissue in adult goldfish, *J. Physiol. (London)* **252**:137–158.

Yoon, M. G., 1975*b*, Effects of post-operative visual environments on reorganization of retinotectal projection in goldfish, *J. Physiol. (London)* **246**:693–694.

Zaidel, E., 1978, Auditory language comprehension in the right hemisphere following cerebral commissurotomy and hemispherectomy: A comparison with child language and aphasia, in: *Language Acquisition and Language Breakdown: Parallels and Divergences* (A. Caramazza and E. B. Zurif, eds.), pp. 229–275, Johns Hopkins University Press, Baltimore.

Zaporozhets, A. V., 1965, The development of perception in the preschool child, in: *European Research in Child Development* (P. H. Mussen, ed.), *Monogr. Soc. Res. Child Devel.* **30**(Ser. No. 100):82–101.

Zeki, S. M., 1973, Colour coding in Rhesus monkey prestriate cortex, *Brain Res.* **53**:422–427.

Zeki, S. M., 1974, The mosaic organization of the visual cortex in the monkey, in: *Essays on the Nervous System—Festschrift for J. Z. Young* (R. Bellairs and E. G. Grey, eds.), pp. 327–343, Oxford University Press, London.

2

The Differentiate Maturation of the Human Cerebral Cortex

T. RABINOWICZ

1. Introduction

"Conel's work has provided us with an extraordinarily valuable picture of the development of the cerebral cortex from birth to two years. Before birth, our information is scanty and qualitative, and after two years it is practically nonexistent" (Tanner, 1961). This was written 16 years ago. In the meantime, Conel (1963, 1967) published his atlases of the cerebral cortex of children of 4 and 6 years, and we completed his last unfinished atlas of the child of 8 years. We also obtained some preliminary data on the brain at 10 years and at 22 years. We completed our atlas on the cerebral cortex of the premature baby of 8 months gestational age and now have some information on the cortex of premature babies of 7 and 6 months. Together with the data available from von Economo and Koskinas (1925) on the adult cerebral cortex, it is now possible to get at least a first idea on the maturation of the cerebral cortex from the premature infant of 6 months gestation, through 10 years postnatal age and to adulthood on almost all the cortical areas.

Although we (Rabinowicz, 1964, 1967*a,b*) used the same criteria for maturation as Conel, in the present chapter we will only use the following: (1) the total width of the cortex; (2) the number of neurons per unit of volume; and (3) the histological structure on cresyl violet stained sections. Some information was also taken from our work with Leuba and Heumann on the quantitative cytoarchitectonic development of the cerebral cortex of mice.

T. RABINOWICZ • Division of Neuropathology, Centre Hospitalier, Universitaire Vaudois, Lausanne, Switzerland.

2. Material and Methods

Cases are selected from children without any brain lesion with a complete autopsy and a medical record showing a normal development as well as a normal school record. More clinical data are given in each of Conel's atlases. At the following ages only one or two cases were studied: premature infants of the 6th and 7th month gestational age and cases of 10 and 22 years. From the premature infants of the 8th gestational month up to the child of 8 years, counts were performed on four to nine cases mostly on the left hemisphere, sometimes on both. Moreover, four to six other cases were studied with nonquantitative methods, i.e., Cajal and Golgi–Cox impregnations and myelin stainings. For each age we have thus studied between eight and nine cases and sometimes more. Each measurement (of cortical thickness or of cell dimension) was repeated 30 times, or on 30 cells of each type in each area. Numbering of cell densities was repeated 30 times in each layer of each area in all the cases studied by Conel. In our cases, and in the premature infants, cell countings are repeated 10 times. The results of each case are presented as simple arithmetic means of the 30 or 10 countings done in each area. Great care is taken to cut out blocks in exactly the same orientation (which depends on the location and size of each area) for all of the 50 studied areas. The whole histological procedure is carefully standardized. Further technical details are given in Conel (1939, pp. 3–4) and in Rabinowicz (1967a).

The techniques were exactly the same for experimental material (mice), but here, more precise stereological methods could be used: A correction of Abercrombie was made, nuclei were counted instead of whole cells on cresyl violet-stained sections of 25 μm thickness after paraffin embedding. Statistical methods for the correction of cell densities for each area on a minimum of 10 animals were also used. The evolution of the cell densities of each layer was studied at each age, as well as the whole cortical thicknesses.

3. Frontal Lobe

3.1. Precentral Gyrus, FA Gamma

The precentral gyrus was studied at different levels, together with the postcentral gyrus.

Thickness of the Cerebral Cortex (Figure 1). We show here the evolution of the cortical thickness at the level commanding the muscles of the trunk. In very young infants (premature up to full-term infants), this block comprises also the areas commanding the movements of the arm and hand. The area of the hand was studied separately in children from 1 month on up to 22 years.

At first glance, the curve of the development of the area commanding the muscles of the trunk shows an increase which is not a regular one. In this area the increase of the whole depth of the cortex is rapid until 6 months postnatal, with the exception of the period of one month before, and at, birth. From 6 months to about 2 years, this area does not show an increase in thickness that reappears only around 4 years. A sharper period of growth resumes between 8 and 10 years. We do not have any data between 10 years and the adult (which is around 40 years) whose values were given by von Economo and Koskinas (1925) in their atlas. Still, there seems to be a slow increase in thickness between 10 years and the adult values.

If one compares this evolution with that of the other areas, one notices first that

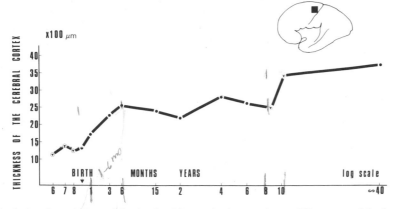

Fig. 1. Evolution through time of the depth of the cerebral cortex in area FA gamma of the frontal lobe, measured in micra (average of 10 to 30 measurements done on the wall of the circonvolution, in the middle of the area) on cresyl violet-stained, paraffin-embedded, 25-μm-thick sections. Similar techniques were employed in making the measurements which served as the basis for the other figures in this chapter which portray the evolution of cerebral cortical thickness and neuronal density through time. Time expressed on a logarithmic scale.

this is one of the thickest areas in the human cerebral cortex and also that this is one of the earliest areas to mature in the isocortical regions. On the same curve shown in Figure 1, between 6 months and 8 years the increase in thickness is a very slow one, as if most of the activities which have to be commanded by this area have been attained initially at 6 months. A second, more rapidly growing period of development is shown here after 8 years, with an almost adult value attained around 10 years.

Neuronal Density (Figure 2). The neuronal density is discussed here only for the 2nd and 5th layers, which have been taken as examples for two different types of layers. In the 2nd layer the decrease in cell density is extraordinarily fast between the premature infant of 6 months gestation and the full-term infant. At 6 months of fetal life, the cell density is more than 5000 per unit volume, 3000 at the 7th month, 2000 at the 8th month, and less than 400 at birth. This considerable decrease in cell density is a general phenomenon during the development of the cerebral cortex and is greater in the 2nd layer than in almost any other layer of the other isocortical areas. After birth, the evolution of cell density goes at a much slower pace, with density reaching almost that of an adult at around two years.

The decrease in cell density can be seen not only in the cortex, but also in the various nuclei of the brain stem, the basal ganglia, and the cerebellum (Hamburger, 1975). The reduction in the number of neurons per unit volume gives good quantitative information on how rapidly maturation is progressing (Rabinowicz, 1967*b*, 1974, 1976). The 5th layer of the same area shows a much slower decrease in neuronal density from 6 postmenstrual months to birth. In the 6-month-old premature, the cell density in the 5th layer is about 800 cells, decreasing to around 200 pyramidal cells at full-term birth. As we know from Yakovlev (1962), the 5th layer is a much earlier maturing layer (in the precentral gyrus) than all the other layers, and some future Betz cells are recognizable as early as the 6th gestational month in the premature infant. At that time, these cells are located mostly in the 5th layer, but some of them can be seen in the 6th, and even in the 4th, layer. In the 5th layer, an almost adult level of cell density is shown at around 6 months postnatally. Thus, an important difference exists between the 2nd and the 5th layers in the speed of

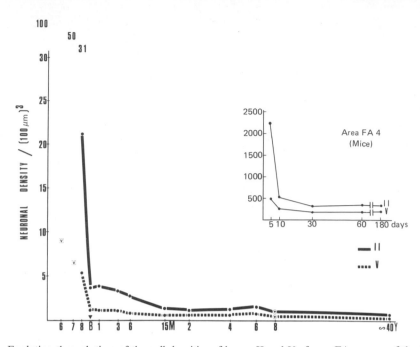

Fig. 2. Evolution through time of the cell densities of layers II and V of area FA gamma of the frontal lobe. Comparison with the evolution of the same area (4) in mice.

maturation in the same cortical area. The 2nd layer is mature only around the 15th month to the 24th month after birth. The small graph in Figure 2 shows that the phenomenon is exactly the same in the corresponding area of the cerebral cortex in mice (Leuba *et al.*, 1977).

Later during development, the graph of the neuronal density (Figure 2) shows a period, around 6 years, in which the cell density seems to be higher in both the 2nd and 5th layers. In mice one sees that around 60 days there is also a slight increase in the cell density (Leuba *et al.*, 1977). It is most probable that there is no real increase in the number of neurons, as we know that the multiplication of neuroblasts ceases around the 6th prenatal month. As we have shown, using the data obtained on mice (Heumann *et al.*, 1977) and humans (Rabinowicz, 1976), this increase in density is most probably due in part to the decrease in size of pyramidal cells. In children around 6 years, the pyramidal cells of Betz are not as long from base to apex of the cell body as they are before that age, nor as long as they will become later on. Although this was seen for Betz cells (Rabinowicz, 1976), we assume that some of the other pyramidal cells follow more or less the same pattern (Rabinowicz *et al.*, 1977).

A comparison between the graphs of neuronal density and those of cortical thickness shows that there is a rather good correlation between the thickness of the cortex and the values of cell densities found at the same point for a given age. Still, the association of relatively higher cell density and lower thickness at 6 years is most probably due to the fact that the whole cortical thickness is less, thus packing together the same amount of cells in less volume.

Histological Evolution (Figure 3). The picture of the cerebral cortex of FA gamma at the level of the arm and hand is shown in Figure 3 at 8 months prenatally, in the full-term newborn, the 15-month-old child, and the child of 6 years. The

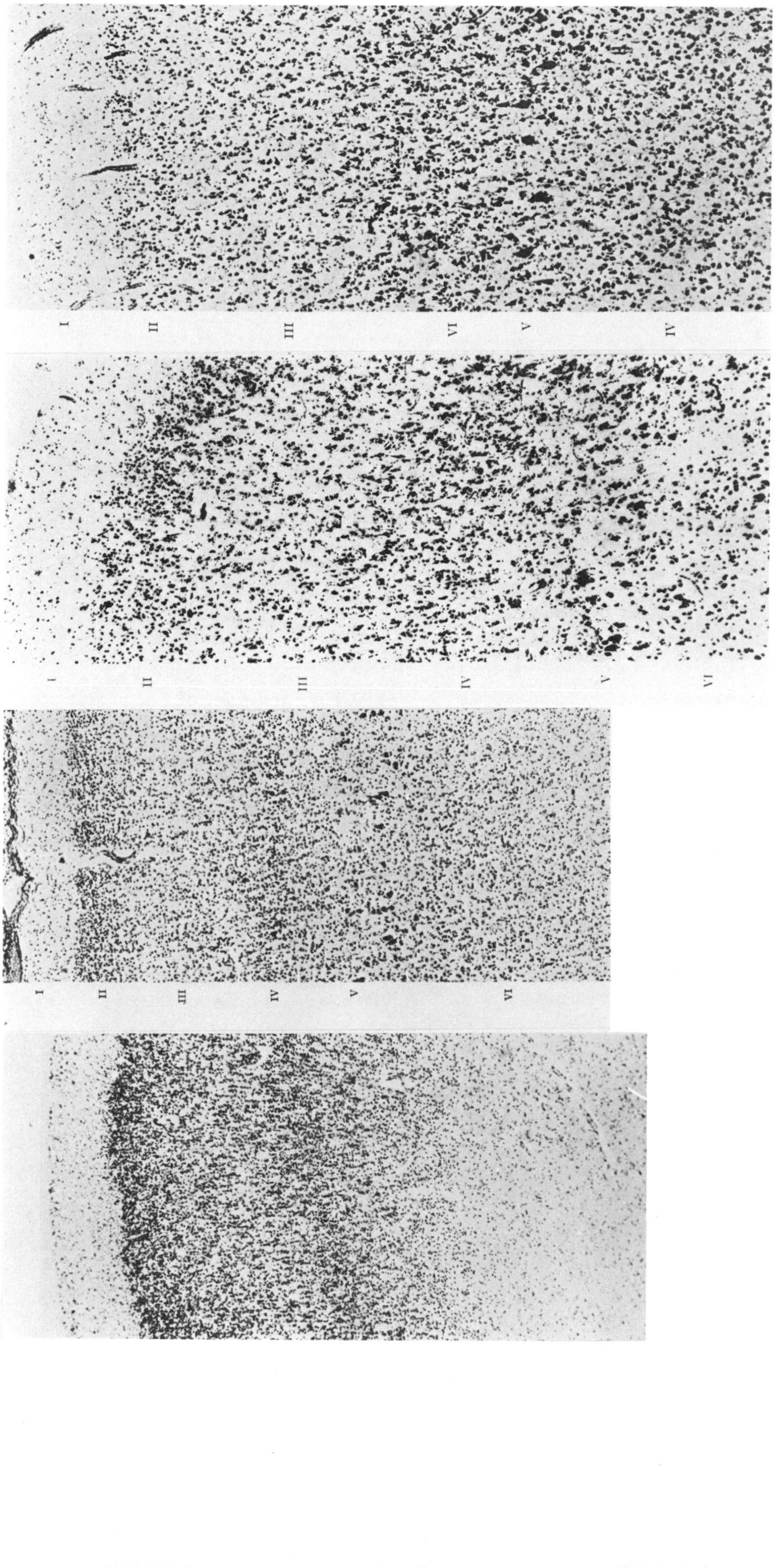

Fig. 3. Histological evolution of area FA gamma, hand. Cresyl violet, × 50. From left to right: premature of 8 months, full-term newborn, child of 15 months, child of 6 years.

increase in thickness is obvious between 8 months prenatally and the full-term newborn. Great differences exist, however, between these two stages of development: In the 8-month premature infant, the cell density is much higher mostly in the 2nd layer, the columns between cells are narrower, Betz cells are smaller, and the space around each cell is less. One also sees that in the full-term newborn the first and second layers do not increase in thickness, the 3rd layer increases slightly, and most of the increase is in the 5th and 6th layers. Between full-term birth and 15 months the differences are considerable. The whole thickness is not now present in the picture and the increase in thickness occurs to all the six layers, but more so to the pyramidal layers (3rd, 5th, and 6th layers) than to the granular layers (2nd and 4th). Between 15 months and 6 years the differences in thickness are no longer so important, but the space around each cell becomes much greater, and the size of the Betz cells becomes quite different: They are broader and shorter at 6 years than at 15 months.

Cresyl violet stainings show that Nissl bodies are already present at birth in the Betz cells, although one month earlier there is only some powdery cresyl violet-positive dust in the whole cytoplasm of these Betz cells (Rabinowicz, 1964). At 6 years, Nissl bodies are well shown, the biggest ones being at the periphery of the cell and the smaller ones around the nucleus. The evolution of the size of the Betz cells has been shown by Rabinowicz (1976).

3.2. Frontal Polar Area FE

Thickness of the Cerebral Cortex (Figure 4). In this area there is a rather rapid increase in thickness from the 6th prenatal month to the first month postnatally, except in the 8th prenatal month. From the age of 1 month this area shows a rather steady increase in thickness up to 4 years, followed by lower values from 6 to 22 years. The lower value for 10 years was obtained in one case only and thus needs confirmation. Between 6 and 22 years this area shows, even on preliminary data, a rather great variability. This is noteworthy because, at the adult level, von Economo and Koskinas (1925) also found a great variability in thickness among his cases. This is evidenced by the presence of two different adult values: one which corresponds to cases with thicknesses less than at 6 years; the other, with an increase

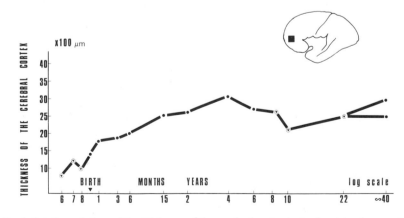

Fig. 4. Evolution through time of the thickness of the cerebral cortex in the frontal polar area (FE) of the frontal lobe.

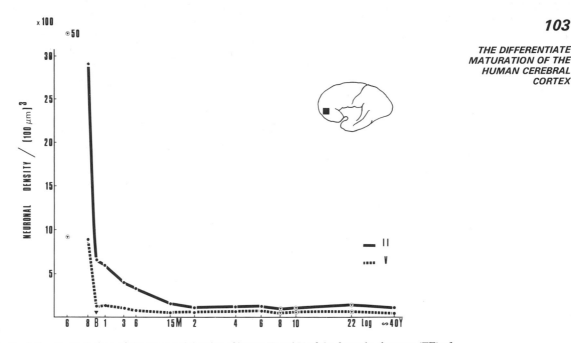

Fig. 5. Evolution through time of the neuronal density of layers II and V of the frontal polar area (FE) of the frontal lobe.

going back to the maximum value which is seen at 4 years. Generally, this area shows a progressive increase in thickness up to 4 years and a period between 6 and 22 years during which there is almost no increase at all. On the other hand, differences between adults have been so important that von Economo and Koskinas (1925) did not give a sole mean value, but had to subdivide their cases into two groups: one showing, and the other not showing an increase.

Since the frontal pole is considered to be concerned with various elaborate mental functions, the presence of a very irregular curve between 4 and 22 years may be related to great individual variability which seems to persist even at an average age of 40 years.

Compared to the precentral motor area, FE appears to be thinner, sometimes much thinner; the reasons for this fact are unclear at present.

Neuronal Density (Figure 5). At 6 months prenatally, the cell densities of both the 2nd and the 5th layers of this area are almost identical to those of the precentral motor area. However, the decrease in cell density occurs less rapidly in both layers. The cell density is around 2800 in the 2nd layer in the 8th prenatal month, and around 800 at the same time in the 5th layer. Between the 8th prenatal month and birth, the density rapidly decreases. In the 2nd layer, however, the cell density is higher and thus at birth the frontal pole is more immature than the premotor area: around 700 cells, compared to about 400 in FA gamma trunk. On the other hand, the 5th layer shows almost the same density (200) as in the precentral area, furnishing another example of differences in maturation between one layer and another in the same area.

For the 2nd layer the decrease is a more rapid one in FE than in FA gamma, but reaches, also around 2 years, the adult level; the 5th layer reaches this point one year earlier. The same phenomenon of slightly higher cell density appears at 6 years, as in the precentral gyrus, but we should not draw conclusions from the data at the 22nd year because they have been obtained from only one case. Generally speaking, area FE shows a much slower maturation of its 2nd layer compared to the precentral gyrus, while the 5th layer shows about the same speed of maturation in both areas.

As a whole, this area shows a progressive growth until 4 years. After 10 years, individual variability appears, resulting later in two groups of individuals: one without any further increase in thickness (or perhaps function?), and the other, with an increase (von Economo and Koskinas, 1925). It is intriguing also that in the whole period between 4 years and adulthood, the period of schooling, there appears little further growth. Clearly, we need more information from more cases.

3.3. Area FCBm

Thickness of the Cerebral Cortex (Figure 6). Area FCBm corresponds to the motor area of speech and seems to show three periods of growth in terms of total thickness of the cortex. The first period is a rapidly increasing one, from the 6th prenatal month to 1 month postnatally, with a decrease in the 8th prenatal month. The second period shows a slowly increasing thickness from 1 month to 2 years. The third period shows a rather rapid increase between the 2nd and 4th years, and one can consider that at around 4 years adult thickness is attained.

The evolution of this area corresponds rather well to the development of motor speech abilities as described in children. We will see later that the posterior and inferior speech centers have a quite different evolution.

Neuronal Density (Figure 7). If we compare the cell densities of the 2nd layer in the 8th prenatal month between areas FA gamma, FCBm, and FE, one notices that the precentral gyrus has at that time the lowest number of cells in the 2nd layer (2200), while the less-well-developed area FE shows 2800. The area FCBm is in between with 2500 cells per unit volume. These differences correspond well to the fact that the precentral gyrus is the earliest developed area of the frontal lobe, while FE is one of the latest.

At birth, the neuronal density of the FCBm area is also in between that of the precentral motor and the frontal polar areas.

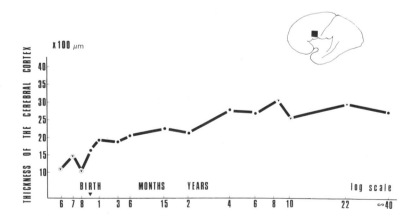

Fig. 6. Evolution through time of the thickness of the left cerebral cortex in area FCBm of the frontal lobe.

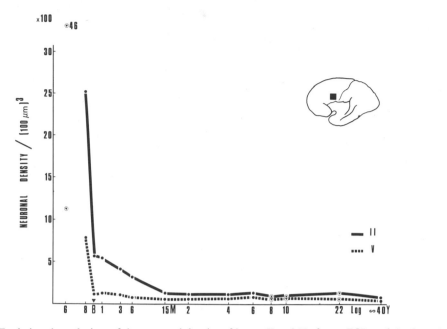

Fig. 7. Evolution through time of the neuronal density of layers II and V of area FCBm of the frontal lobe.

It is remarkable to see that some sort of topographical predeterminism is quantitatively noticeable at least 3 months before birth in the human frontal cortex. The decrease in cell density of area FCBm shows that at around 15 months it is almost at the density level of the adult in the 2nd layer, while this level is reached at 6 months for the 5th layer. For the latter, the decrease in cell density also shows a velocity which is in between FA gamma and FE. Here we can see the same phenomenon at 6 years with a slight apparent increase in cell density both for the 2nd and the 5th layer—already noted in the discussion on the FA gamma trunk area.

Histology (Figure 8). The motor center for speech is shown in Figure 8, in the 8th prenatal month, the full-term newborn, the child of 2 years, and the 6-year-old child. The histological picture corresponds quite well to our graphs both of the evolution of the thickness of the cortex, as well as of the neuronal density. The rather great difference in thickness of the cortex between the 8th prenatal month and the full-term newborn is clearly demonstrated, as well as the differences in cell densities. The difference is even more striking between the full-term newborn and the child of 2 years: Not only is the thickness of the cortex at 2 years almost double that at birth, but also the density is much lower, meaning that the dendrites developed at a considerable rate during those first 2 years after birth. This is well demonstrated with the Golgi–Cox impregnations shown by Conel (1939, 1941, 1947, 1951, 1955, 1959). On closer scrutiny, there are some quite important, if less visible, differences between the cortex at 2 years and at 6 years. In the latter, the increase in space around the cells is a clearly visible indication that dendrites were still growing during that time. Even the 1st layer shows some noticeable increase in thickness, as does the 2nd layer.

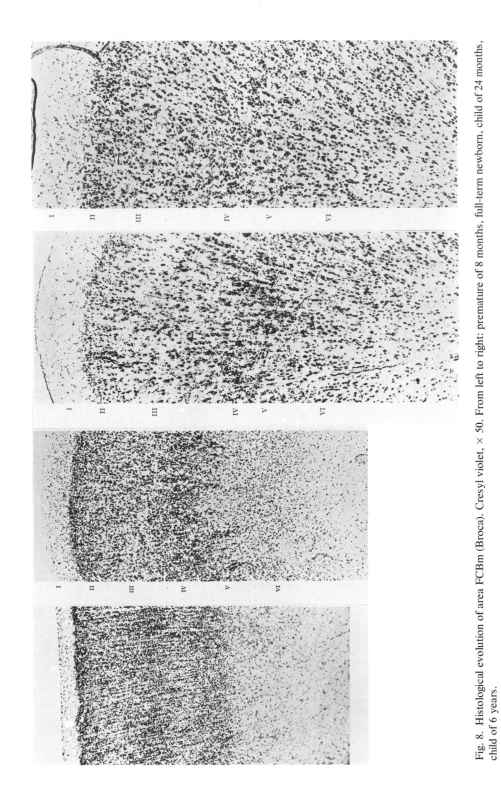

Fig. 8. Histological evolution of area FCBm (Broca). Cresyl violet, × 50. From left to right: premature of 8 months, full-term newborn, child of 24 months, child of 6 years.

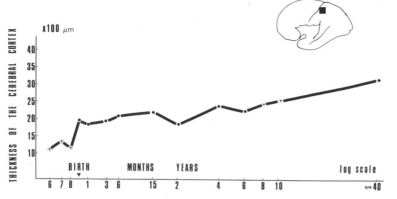

Fig. 9. Evolution through time of the thickness of the cerebral cortex in area PC of the parietal lobe.

4. Parietal Lobe

4.1. Area PC, Trunk

Thickness of the Cerebral Cortex (Figure 9). The general pattern of growth of this area is relatively simple. The increase in thickness is a rapid one from the 6th prenatal month to the full-term newborn, except around the 8th prenatal month. From that time, the increase in thickness is slow, but continuous, until adulthood. Between birth and adulthood, thickness nearly doubles, with only a slower increase around 2 and 6 years. Compared to the growth of the motor area at the same level, the differences appear rather striking: The motor area has a more rapid, but much more irregular, growth and is about 20% thicker. In the motor area, the growth is much more rapid and lasts longer between the 8th prenatal month and the 6th postnatal month, compared with the postcentral gyrus where the period of rapid growth stops at birth.

If area PC is related to the complex development of motricity, there must be differences between the motor abilities, which are developing much more rapidly until the age of 6 months, and the sensory abilities which are developing much more slowly from birth. Also noticeable is the very rapid growth of this area during the prenatal months. Is there perhaps a tendency for the PC area not to develop concomitantly with motricity?

Neuronal Density (Figure 10). The postcentral area at the level of the trunk shows, as in other areas, a great difference between the rate of maturation of the 2nd layer and that of the 5th—the former being, as usual, slower. The 2nd layer reaches an almost adult neuronal density only at 15 months. As in other areas, the cell density decreases very rapidly before birth. However, around 6 years, there is an apparent increase in cell density. Compared with the motor area at the same level, the differences are striking only for the 2nd layer, while the 5th shows almost the same patterns of development.

4.2. Area PF (Posterior Speech Center, Wernicke)

Thickness of the Left Cerebral Cortex (Figure 11). In this area the cerebral cortex shows a slight decrease in thickness from the 6th prenatal month to the 8th

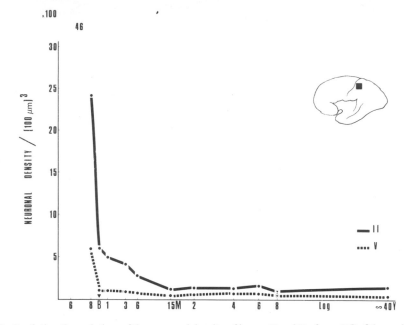

Fig. 10. Evolution through time of the neuronal density of layers II and V of area PC of the parietal lobe.

prenatal month. From birth until 15 months, the increase in thickness is rather rapid, and from that point, the average thickness does not increase very much until about 22 years. At adult levels, however, the total thickness is greater, while between 15 months and 22 years the variability is notable.

This curve reflects rather well the fact that the homogeneity of our material is very poor and that variability between individuals is certainly high from one age to another, explaining the continuous irregularity of the curve between 15 months and 10 years. The curve shows, also, despite the irregularity, an increase at adulthood. As will be later described, another speech center, the inferior center (TA), shows the same phenomenon even more markedly.

Neuronal Density (Figure 12). Area PF behaves rather similarly to area FCBm (Broca) for the 2nd and the 5th layers from birth to 15 months. Area PF shows a

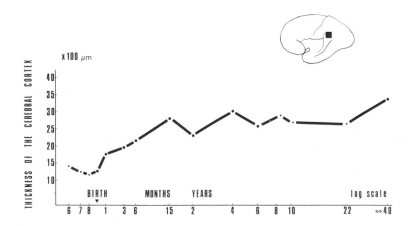

Fig. 11. Evolution through time of the thickness of the cerebral cortex in area PF of the parietal lobe.

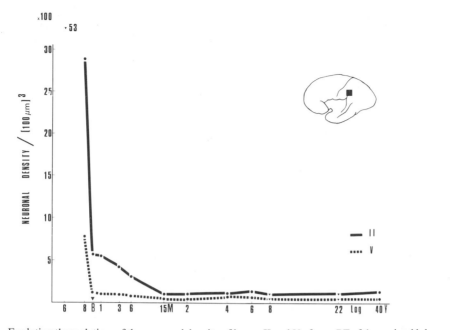

Fig. 12. Evolution through time of the neuronal density of layers II and V of area PF of the parietal lobe.

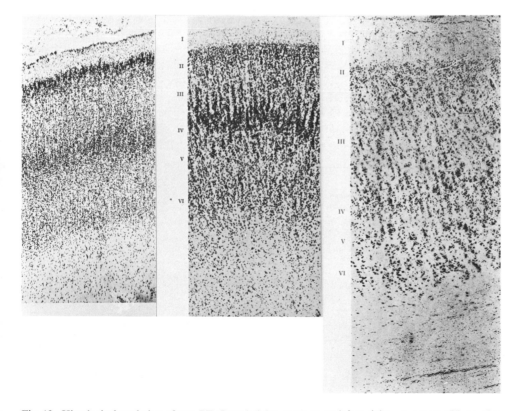

Fig. 13. Histological evolution of area PF. Cresyl violet, × 50. From left to right: premature of 8 months, full-term newborn, child 4 years.

much higher cell density in the 2nd layer in the premature, showing this layer to be less developed prenatally. At the 15th month, cell density corresponds to an almost adult level, but the great variability between individual cases is not revealed by the averaging of results used. As in the other areas, the period around 6 years corresponds to a relative increase in cell density.

Histology (Figure 13). The area is shown here at the 8th prenatal month, the full-term newborn, and at 4 years. The pictures show that there is almost no increase in thickness between the 8th prenatal month and the full-term newborn. The latter still shows a high density of cells, but the space between columns (the ''mean intercolumnar space''), is clearly increasing—a feature which is quite well developed at 4 years. As seen in other areas, the cell density of the 2nd layer is rapidly decreasing, while the thickness of the 1st, 2nd, and 3rd layers is increasing, as is the thickness of the 5th and even more so the 6th layers. Some big pyramidal cells are well developed at 4 years, while at birth they are hardly seen at all.

Not only is this area a rather variable one depending on the individual, but also the differences in degrees of maturation between layers are considerable. Along with a differential growth in thickness of layers, there is also an important increase in the distance between cells, indicating an active dendritic development.

5. Temporal Lobe

5.1. Area TA (Anterior Lower Speech Center)

Cortical Thickness (Figure 14). The left side of this area shows a slight increase in thickness between the 6th and 8th prenatal months. After birth, the increase is still a slow one up to 2 years. Between 2 and 22 years, the curve presents a striking variability at the different ages, which shows, even more clearly than in area PF, how great individual variability can be.

The posterior, as well as the anterior, midtemporal speech cortex (TA) both show much variability and inhomogeneity in the cases used for the quantitative study. This is also reflected in the histology, in which one can see that the cortex of the 15-month-old child shown here (Figure 16) is thicker than that of the 6-year-old.

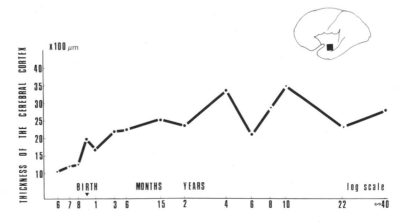

Fig. 14. Evolution through time of the thickness of the left cerebral cortex in area TA of the temporal lobe.

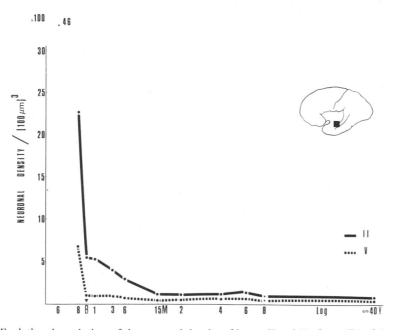

Fig. 15. Evolution through time of the neuronal density of layers II and V of area TA of the temporal lobe.

Neuronal Density (Figure 15). The cell density in the 2nd layer is relatively lower at the 8th prenatal month than in many other brain areas, but at birth the density is still quite high and then decreases rather rapidly until 15 months postnatally. The usual differences in maturation between the 2nd and the 5th layer are rather marked in area TA. The 5th layer is not far from its definitive density after 6 months. As elsewhere at 6 years, we find a relative increase in cell density corresponding to a decrease in the thickness of the cortex at that age.

Histology (Figure 16). Area TA is shown here at the 8th prenatal month, in the full-term newborn, at 15 months, and at 6 years. The decrease in cell density appears clearly if one compares the cortex of the newborn with the cortex at the 8th prenatal month. This important decrease in cell density corresponds only to a relatively moderate increase in thickness of the cortex, showing that, most probably, there has been a loss of cells. This has been effectively described by Hamburger (1975) in the brain stem of mice, and by Heumann *et al.* (1978) in the cerebral cortex of mice. Our pictures show that at 15 months the cell density is about the same as at 6 years or even slightly less: In this area, the 3rd, 5th, and 6th layers increase most in volume in the full-term newborn and up to 6 years.

5.2. Area TC (Auditory Projection Area)

Thickness of Area TC (Figure 17). This area shows a rather rapid increase in thickness until birth, and a slower one until around 15 months. Between 15 months and 2 years there is a decrease in thickness, while between 2 and 8 years there is a second period of increase ending at 10 years after a rapid growth phase. The adult values are reached at 10 years. The number of cases is low, but generally speaking,

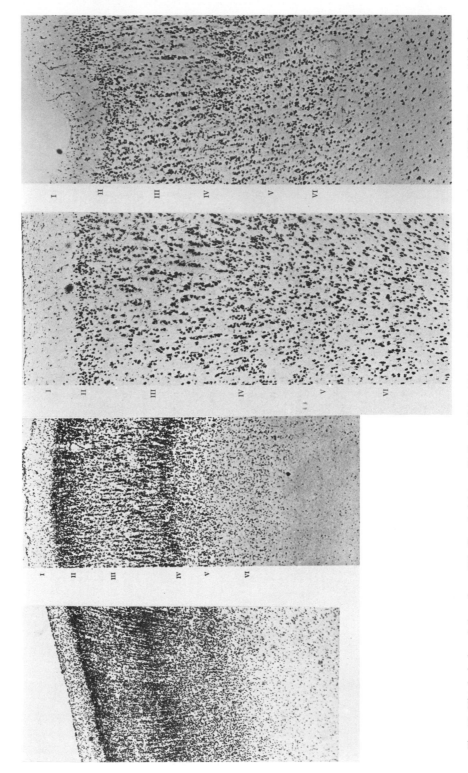

Fig. 16. Histological evolution of area TA. Cresyl violet, × 50. From left to right: premature of 8 months, full-term newborn, child of 15 months, child of 6 years.

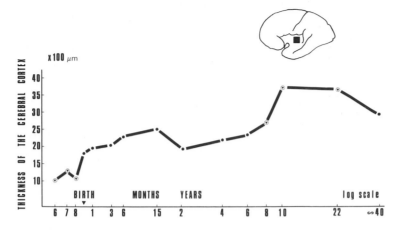

Fig. 17. Evolution through time of the thickness of the cerebral cortex of area TC of the temporal lobe.

one can say that this is a rather steadily growing area in which the adult thickness is attained between 8 and 10 years.

Interestingly enough, the adult level of 10 years is also present in the two speech centers TA and PF, and there is a turning point of musical abilities between 8 and 10 years.

Neuronal Density (Figure 18). This area shows a considerable decrease in cell density for the 2nd layer between the 6th prenatal month and birth. The same is true for the 5th layer. For the latter, however, the cell density for adult values is found around 6 months, if not even earlier, while for the 2nd layer it is attained between 15

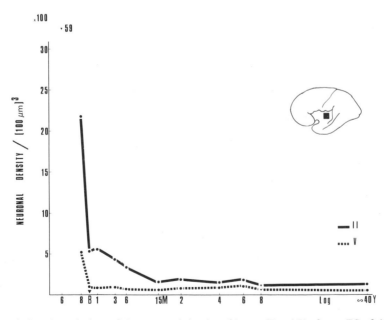

Fig. 18. Evolution through time of the neuronal density of layers II and V of area TC of the temporal lobe.

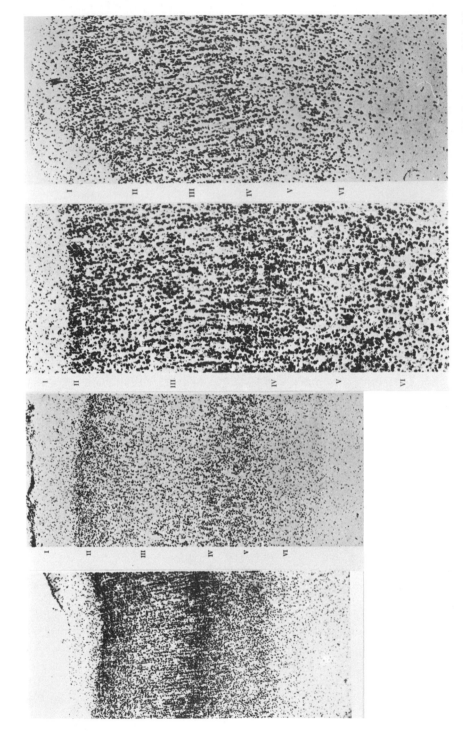

Fig. 19. Histological evolution of area TC. Cresyl violet, × 50. From left to right: premature of 8 months, full-term newborn, child of 15 months, child of 6 years.

months and 8 years. Thus, a quantitative analysis of the cytoarchitectonics shows that there must be some inequalities in the functional abilities of this area due to great differences in the speed of maturation between the 2nd and the 5th layer.

Histology (Figure 19). Area TC is illustrated here with pictures taken at the 8th prenatal month, of the full-term newborn, and of children of 15 months and 6 years old. There is a striking difference between the 8th prenatal month and the full-term newborn. The 3rd layer of the full-term newborn has considerably increased, not only by the diminution of its cell density, but also by the development of the distance between cells. Once more, the child of 15 months has, in this series of pictures, a much thicker cortex than the child of 6 years.

6. Occipital Lobe

6.1. Area OC, Calcarine, Visual Projection Area

Thickness of the Cortex (Figure 20). The thickness of the visual primary projection area shows a slight decrease between the 6th and the 8th prenatal months, and a rapid increase between the 8th prenatal month and the child 1 month old. From 1 month to 3 months there does not seem to be an increase, but there is one between 3 and 6 months. Remarkably enough, the thickness attained at 6 months is very near that found by von Economo and Koskinas (1925) in the adult— except for another increase at 8 years. One may consider that this area is most probably functioning very early in childhood. This possibility is emphasized if one studies the Golgi–Cox impregnations which have been made in the premature infant (Rabinowicz, 1964), and in children from birth to 8 years (Conel, 1939, 1941, 1947, 1951, 1955, 1959, 1963, 1967). Another feature of this area is the small variability which is expressed by an almost straight line from 6 months to adulthood. This phenomenon can be checked by looking at the pictures (Figure 22) given here: there are almost no differences in cell densities and overall structure between the cortex of a child of 6 months and that of the child of 6 years.

Neuronal Density (Figure 21). Despite its early development, area OC shows a difference between the maturation of the 2nd layer and the maturation of the 5th.

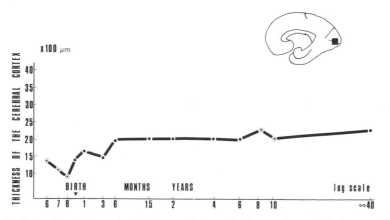

Fig. 20. Evolution through time of the thickness of the cerebral cortex in area OC of the occipital lobe.

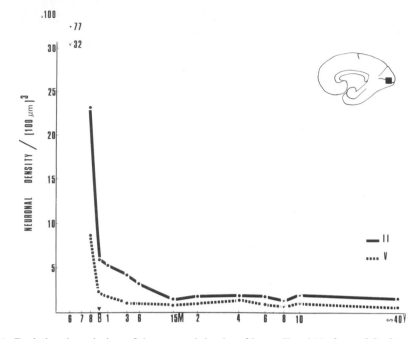

Fig. 21. Evolution through time of the neuronal density of layers II and V of area OC of the occipital lobe.

For the 2nd layer, the decrease in cell density is a considerable one, going from 7700 per unit volume in the 6th prenatal month, to about 600 at birth—probably the most rapid decrease known in the cerebral cortex. From birth to the age of about 15 months the decrease is slower, and probably the maturation of the 2nd layer is almost completed at 15 months. Meanwhile, the 5th layer attains adult values as early as 3 months. Here also we have a rapid decrease in cell density for the 5th layer, and attention should be drawn to the fact that this layer is mature earlier than any other area we have studied to date. On Golgi–Cox impregnations there is still a clear difference in the degree of maturation of the pyramidal cells of the 4th B layer, which are much slower in developing than those of the 6th or the 3rd layers (Rabinowicz, 1964).

Histology (Figure 22). Area OC shows an important increase in thickness during the last prenatal month, when all the layers are easily recognizable. It is also interesting to note that our study of the 6th, 7th, and 8th prenatal months shows that the limits between OC and OB (Limes OB gamma) appear only at the 7th prenatal month, while at that same time, the different layers of OC are easily recognizable. Our picture also shows a considerable decrease in cell densities, as well as an increase in size of the pyramidal cells in comparison between the full-term newborn and the child of 6 months. Comparing the latter with the child of 6 years shows that the increase in the size of the cells is slight and not enough to change significantly the total cortical thickness.

6.2. Area OA, Visual Association Area

Cortical Thickness (Figure 23). This area shows a remarkable evolution in comparison to the evolution of OC. OA shows an almost steady and continuous

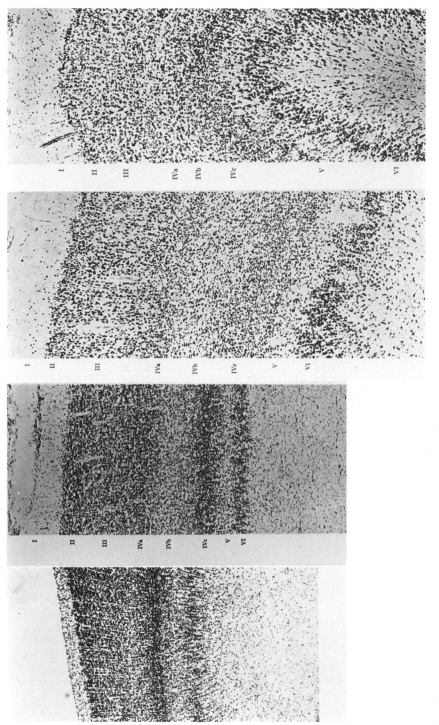

Fig. 22. Histological evolution of area OC. Cresyl violet, × 50. From left to right: premature of 8 months, full-term newborn, child of 6 months, child of 6 years.

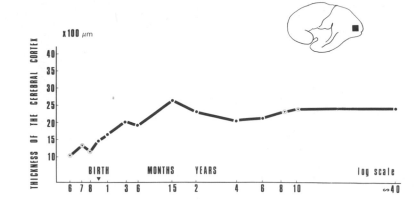

Fig. 23. Evolution through time of the thickness of the cerebral cortex in area OA of the occipital lobe.

increase in thickness up to 15 months, at which time the thickness is slightly greater than in the adult. From 15 months, there is a slow decrease with a minimum around 4 years, followed by a slow increase plateauing for the second time with the adult value at 10 years. Compared to OC, it is obvious that this area reaches an approximate adult level more than 9 months later than the primary visual cortex. This is not surprising as we know that area OA is related to visual understanding. In this area we also find a rather simple pattern of development at least with regard to the total thickness of the cortex. Thus, there seems to be some parallelism between the evolution of the primary visual cortex and that of area OA.

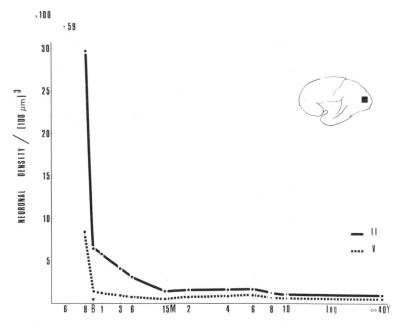

Fig. 24. Evolution through time of the neuronal density of layers II and V of area OA of the occipital lobe.

Neuronal Density (Figure 24). Here also the differences in maturation between layers 2 and 5 are quite striking, but layer 5 seems to be slightly later reaching the adult level than layer 5 in area OC, while for layer 2 the evolution is very similar after birth to that seen in layer 2 of OC. The cell density, however, is slightly lower in the 6 prenatal months for both the 2nd and the 5th layers.

7. Limbic Lobe

7.1. Anterior Limbic Area, LA

Thickness (Figure 25). The anterior limbic area shows a more irregular growth in that it is slow in the last gestational trimester and then very rapid during the last month before birth at full term. From then on, there is almost no increase in thickness until 6 months, and then a slow progressive growth until 4 years, with a clear decrease at 6 years, reaching adult thickness around 8 years. This area shows a very rapid growth spurt 1 month before to 1 month after birth, with a maximum at birth followed by a slow increase up to 4 years.

Neuronal Density (Figure 26). The anterior limbic area is remarkable for its very early maturation in terms of cell density. The density of the 2nd layer is around 3600 in the 6th prenatal month, decreasing rapidly to less than 400 at birth, thus showing a relatively rapid prenatal maturation. The 2nd layer is also irregular in maturation between 3 and 6 months, and reaches adult values at 15 months as elsewhere. For the 5th layer, the cell density of the premature infant is also very low, and the adult level is attained late, at 15 months, thus showing little difference between the maturation of the 2nd and of the 5th layer, contrary to that seen in many other areas.

7.2. Posterior Limbic Area, LC

Cortical Thickness (Figure 27). This area, like the anterior limbic area, shows a very irregular pattern of development with a rapid growth spurt around birth,

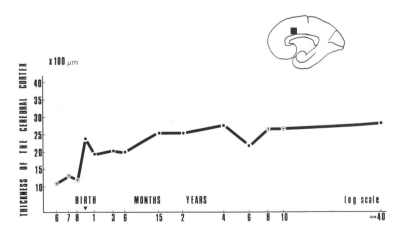

Fig. 25. Evolution through time of the thickness of the cerebral cortex in area LA of the limbic lobe.

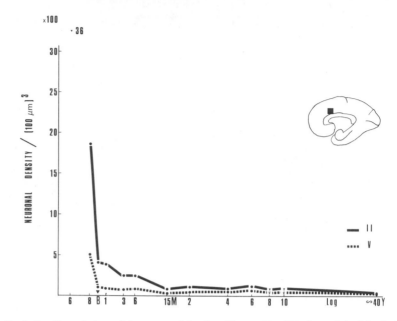

Fig. 26. Evolution through time of the neuronal density of layers II and V of area LA of the limbic lobe.

followed by a decrease in thickness at the first month and a rapid increase up to 15 months. At 15 months, the adult thickness level is attained, but decreases once more at 2 years. Then the increase resumes slowly up to 10 years when it reaches adult level. There is some similarity in the growth of both areas LA and LC from the 6th prenatal month to 15 months. LC presents one of the rare examples of a slight increase in thickness at the age of 6 years; this is not the case for LA.

Neuronal Density (Figure 28). Here the differences from the anterior limbic area LA are considerable: The cell density at the 6th prenatal month is much greater for the 2nd layer in LC than in LA, so maturation is slower. At birth, the cell density is still greater in this layer in LC than in LA, and the decrease is more rapid in LC

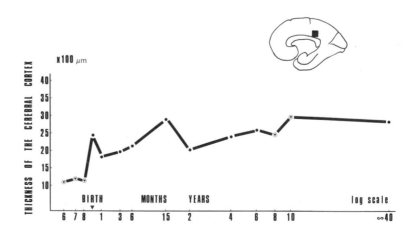

Fig. 27. Evolution through time of the thickness of the cerebral cortex in area LC of the limbic lobe.

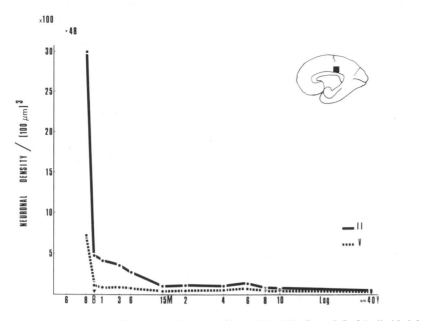

Fig. 28. Evolution through time of the neuronal density of layers II and V of area LC of the limbic lobe.

than in LA. Fifteen months is, however, also the point at which a lower density is attained. For the 5th layer the difference is not so marked, except that the cell density is greater in the 8th prenatal month in area LC than in LA, and, from birth, the evolution of the cell density of the 5th layer is almost exactly the same for LC as for LA.

If, as is generally accepted, the limbic system is associated with emotional activities and mechanisms of memory, the data provided by our study would indicate that some activity could already exist as early as 15 and 24 months. Both the curves of cortical thickness and of neuronal densities shows that more adult patterns are attained between 8 and 10 years.

8. General Considerations

Although the study of the evolution of the cortical thickness is still in its early stages, some general rules have already emerged: Besides the fact that each cerebral lobe has its own rate of development, each area in each lobe also has its own developmental rate. Moreover, each area shows differences in the rate of development of each of its layers.

The evolution of cortical thickness shows important differences, depending on the location of the area and most probably also upon individual factors. A brain growth spurt (Dobbing, 1974) exists in humans and probably reaches its peak before the 6th prenatal month, if not earlier.

Some areas are quite homogeneous in their development with no very great differences between one individual and another at a given age. This is the case in the primary visual area and in the motor center of speech. In contrast, other areas show great differences between individuals of the same age, giving thus a very irregular curve of development with important peaks and troughs. The posterior and antero-inferior speech centers are in this category.

It is also noticeable that around 6 years in most of our areas there is a decrease in the total thickness and an increase in the number of nerve cells in the different layers. This occurs also in mice around the age of 60 days. The fact that the Betz cells show an evolution of cell dimension during their development and, in particular, show at 6 years a decrease in length and an increase in width (Rabinowicz, 1976) might well correspond, at least partly, to that decrease of the total cortical thickness, provided we assume that many of the pyramidal cells of the cerebral cortex are going through the same evolution of their length and width. Moreover, the increased cell density at 6 years suggests a decrease in the dendritic pattern between the cells (Rabinowicz *et al.,* 1977) at that age.

There seems indeed to be some relationship between the functional development of the human cerebral cortex and the evolution of neuronal density, or cortical thickness and of thickness of the layers.

In almost all the areas the cell density decreases very rapidly until birth. From birth to between 3 and 6 months, the decrease is slower and it ceases at about 15 months.

In the same area differences exist in maturation between the layers, and one has to deduce that a very important moment in cortical maturation appears to be the period between 15 and 24 months, a period at which almost all the layers reach, for the first time, a similar state of maturation. Another important moment in cortical maturation seems to be the period between 6 and 8 years when a remodeling of the cortex (cortical thickness, number of neurons and dendrites) takes place (Rabinowicz *et al.,* 1977).

ACKNOWLEDGMENTS

The author is indebted to Mrs. G. Leuba, DS, and Mr. D. Heumann, BS, for their help and stimulating discussions; to Mrs. J. McD.-C. Petetot and Mrs. M.-L. Meier for histoquantitative and technical assistance; to Mr. B. Maurer for photographical work; and to Miss M. Laufer for typing the manuscript.

This work was supported by U.S . Public Health Service Grants M-156-C4-C6, HD 00326-07-08, and M 151 (from late Dr. J.-L. Conel); also by Swiss National Science Foundation Grants 364.171 and 4.434-074.

9. References

Conel, J. LeRoy, 1939, *The Postnatal Development of the Human Cerebral Cortex, Vol. I, The Cortex of the Newborn,* Harvard University Press, Cambridge, Massachusetts.

Conel, J. LeRoy, 1941, *The Postnatal Development of the Human Cerebral Cortex, Vol. II, The Cortex of the One-Month Infant,* Harvard University Press, Cambridge, Massachusetts.

Conel, J. LeRoy, 1947, *The Postnatal Development of the Human Cerebral Cortex, Vol. III, The Cortex of the Three-Month Infant,* Harvard University Press, Cambridge, Massachusetts.

Conel, J. LeRoy, 1951, *The Postnatal Development of the Human Cerebral Cortex, Vol. IV, The Cortex of the Six-Month Infant,* Harvard University Press, Cambridge, Massachusetts.

Conel, J. LeRoy, 1955, *The Postnatal Development of the Human Cerebral Cortex, Vol. V, The Cortex of the Fifteen-Month Infant,* Harvard University Press, Cambridge, Massachusetts.

Conel, J. LeRoy, 1959, *The Postnatal Development of the Human Cerebral Cortex, Vol. VI, The Cortex of the Twenty-Four-Month Infant,* Harvard University Press, Cambridge, Massachusetts.

Conel, J. LeRoy, 1963, *The Postnatal Development of the Human Cerebral Cortex, Vol. VII, The Cortex of the Four-Year Child,* Harvard University Press, Cambridge, Massachusetts.

Conel, J. LeRoy, 1967, *The Postnatal Development of the Human Cerebral Cortex, Vol. VIII, The Cortex of the Six-Year Child,* Harvard University Press, Cambridge, Massachusetts.

Conel, J. LeRoy, and Rabinowicz, T., 1978, *The Postnatal Development of the Human Cerebral Cortex, Vol. IX, The Cortex of the Eight-Year Child* (unpublished manuscript).

Dobbing, J., 1974, The later growth of the brain and its vulnerability, *Pediatrics* **53:**2–6.

Hamburger, V., 1975, Cell death in the development of the lateral motor column of the chick embryo, *J. Comp. Neurol.* **160:**535–546.

Heumann, D., Leuba, G., and Rabinowicz, T., 1977, postnatal development of the mouse cerebral neocortex. II. Quantitative cytoarchitectonics of visual and auditory areas, *J. Hirnforsch.* **18:**483–500.

Heumann, D., Leuba, G., and Rabinowicz, T., 1978, Postnatal development of the mouse cerebral neocortex. IV. Evolution of the total cortical volume of the population of neurons and glial cells *J. Hirnforsch.* **19:**411–419.

Leuba, G., Heumann, D., and Rabinowicz, T., 1977, Postnatal development of the mouse cerebral neocortex. I. Quantitative cytoarchitectonics of some motor and sensory areas, *J. Hirnforsch.* **18:**461–481.

Rabinowicz, T., 1964, The cerebral cortex of the premature infant of the 8th month, *Prog. Brain Res.* **4:**39–86.

Rabinowicz, T., 1967*a,* Techniques for the establishment of an atlas of the cerebral cortex of the premature, in: *Regional Development of the Brain in Early Life* (A. Minkowski, ed.), pp. 71–89, Blackwell, Oxford.

Rabinowicz, T., 1967*b,* Quantitative appraisal of the cerebral cortex of the premature of 8 months, in: *Regional Development of the Brain in Early Life* (A. Minkowski, ed.), pp. 91–118, Blackwell, Oxford.

Rabinowicz, T., 1974, Some aspects of the maturation of the human cerebral cortex, *Mod. Probl. Paediatr.* **13:**44–56.

Rabinowicz, T., 1976, Morphological features of the developing brain, in: *Brain Dysfunction in Infantile Febrile Convulsions* (M. A. B. Brazier and F. Coceani, eds.), pp. 1–23, II IBRO Symposium, Raven Press, New York.

Rabinowicz, T., Leuba, G., and Heumann, D., 1977, Morphologic maturation of the brain: A quantitative study, in: *Brain, Fetal and Infant* (S. R. Berenberg, ed.), pp. 28–53, s'Gravenhage.

Tanner, J. M., 1961, *Education and Physical Growth,* University of London Press, London.

von Economo, C., and Koskinas, G. N., 1925, *Die Cytoarchitektonik der Hirnrinde des Erwachsenen Menschen,* Springer, Wien.

Yakovlev, P. I., 1962, Morphological criteria of growth and maturation of the nervous system in man, *Ment. Retard.* **39:**3–46.

3

Organization and Reorganization in the Central Nervous System: Evolving Concepts of Brain Plasticity

GARY LYNCH and CHRISTINE GALL

1. Introduction

"Plasticity" is a concept which is necessary to any theory of brain function. The most complex products of the brain, the ongoing "cognitive" behaviors of the individual, are remarkable for their flexibility, their capacity for reorganization in the face of changing circumstances. Since behavior is characterized by its adaptability, it follows that the neural machinery which creates it must possess analogous features. But what in neurobiological terms is the property of the brain which allows it this plasticity? Suggested answers to this question have come from all the branches of the neurosciences. The idea most commonly advanced is that changes in the effectiveness of synaptic transmission are responsible for phenomena such as learning and memory. Physiological research, much of it quite recent, has shown that lasting changes can be created in monosynaptic systems by very brief trains of repetitive stimulation (Bliss and Lomo, 1973). Neurochemical studies have indicated that some of the subcellular systems related to the transmitter and its actions are modifiable, and this certainly provides a means through which modification in the operation of neural circuits could be achieved. The present chapter deals with still another type of mechanism by which the brain might gain its flexibility,

GARY LYNCH and CHRISTINE GALL • Department of Psychobiology, University of California, Irvine, California.

specifically, that it is capable of modifying its very structure. This idea, which is quite old, has become the subject of intense interest in recent years as newer methods (and the more routine use of some more traditional procedures) have allowed anatomists to develop a clearer picture of the fine structure of neural tissue. Studies done in the last decade indicate that the microanatomy of the neuron, its dendritic ramifications and axonal arborization, can be greatly modified and that under some circumstances the brain is capable of generating entirely new circuitry.

In the following pages, we will review selected portions of this literature, discussing first anatomical plasticity in neonates and then turning to the adult brain. This division is dictated by the long-held suspicion of psychologists that the storage of experience somehow changes with development, as well as by the common (although by no means universal) observation that immature brains recover much more completely following damage than do those of adults. These two types of findings have suggested to many workers that the young brain is more plastic than that of the adult, and it is well that we bear this in mind while considering the anatomical data. We shall also separate the literature in terms of the type of manipulation used to uncover a particular plastic property. This is done so as to emphasize that the review will cover *capabilities* of neural tissue for change—in only a few instances have attempts been made to demonstrate that these capabilities are actually expressed during the normal operation of the brain or as part of any repair process.

2. Effects of Experience and Environment on the "Molar" Composition of the Brain

Before discussing the plasticity of the constituents of the nervous system, it is appropriate that we briefly summarize the evidence that the physical size of brain areas is mutable. Several studies have demonstrated that the size of particular brain regions can be influenced by the degree of environmental complexity experienced by the animal. Generally, these experiments involve raising the subjects, most frequently rodents, either in an enriched environment with exposure to a variety of complex or novel stimuli or in a relatively impoverished environment such as isolated housing in a laboratory cage. Differential housing of this sort, for a period of several weeks from the time of weaning, results in consistent differences in the gross dimensions of several brain areas which persist as long as differential housing is maintained (Rosenzweig *et al.*, 1972; Bennett, 1976). The brains of the environmentally enriched animals have been found to be larger; one of the more frequent indexes of this difference is a thickening of the cerebral cortex, particularly in the occipital region (Diamond, 1967). These effects do not appear to be attributable to differential visual experience alone for they have been replicated with enucleated animals (Krech *et al.*, 1963). Conversely, pure sensory deprivation causes gross dimensional changes in the area related to the blocked modality. Visual deprivation by eyelid suture at birth has been shown to cause subnormal volume in both the lateral geniculate nucleus and striate cortex in comparison to similarly housed, sighted littermates (Gyllensten *et al.*, 1964; Fifkova, 1970). The origins of these changes are obscure but seem to involve an increase in the volume of the cells which constitute the affected region (see p. 128). Which type of cellular element is primarily affected and the interrelationships between blood vessels, glia, and neurons that produce the final total volumetric increase or decrease are yet to be analyzed.

3.1. Experimental Manipulation of Size, Shape, and Orientation of Dendritic Trees

Fundamental to studies on the plastic properties of immature dendrites is the question of the extent to which dendritic morphology is fixed by the cell's genetic program. The alternate perspective might ask to what extent extrinsic influences determine the morphology of developing dendrites, i.e., the extent to which their morphology is "plastic." Investigation of this latter question has involved the use of a wide variety of experimental circumstances ranging from severe afferent deprivation to the more indirect influences of differential experience. In general, Golgi studies of immature dendrites reveal a loss, or nondevelopment, of spines following removal of their normal supply of axonal input (Globus and Scheibel, 1967; Globus, 1975). Experiments of this sort suggest that the presence of a presynaptic component is essential for the development and/or maintenance of spines on immature dendrites. Although this seems to be the "rule" in studies involving forebrain structures, an exception has been demonstrated in the cerebellum. The cerebellar Purkinje cell is densely innervated by the parallel fibers (axons of the cerebellar granule cell) which terminate on spines of the more distal dendritic branches. The granule cells germinate postnatally from an external germinal layer that covers the immature cerebellum. This arrangement leaves the granule cell population vulnerable to selective destruction by treatments such as X irradiation and viral infection, treatments which leave the Purkinje cells intact. Under these circumstances the Purkinje cells never receive their normal parallel fiber input. As a result, their dendritic growth is retarded in length and tertiary branching, but spines develop nonetheless (Altman and Anderson, 1972; Hirano et al., 1972). Therefore, in this case spines develop and persist without presynaptic contact. With further maturation, the Purkinje cell dendrites of the agranular cerebellum assume a reverse-normal orientation. Rather then directing themselves toward the cerebellar surface they curve inward, toward the now dominant afferent, the climbing fibers (Herndon and Oster-Granet, 1975). Although the presence of the parallel fibers seems to be essential to the full expression of normal dendritic morphology, the dendritic spines, unlike those of several forebrain systems, are not entirely dependent upon direct interaction with a presynaptic element.

Studies of sensory deprivation suggest that the *activity* of the presynaptic component, the afferent of the dendrite under study, can also influence dendritic morphology. In an analysis of the development of pyramidal cells in the visual cortex, Valverde (1971; Valverde and Ruiz Marcos, 1969) described the growth of three classes of dendritic spines. One group followed a normal pattern of development despite visual deprivation throughout the maturation period. The second class of spines appeared when light was first supplied some days after eye opening. The third group of spines developed normally only if light was supplied at the time of eye opening, thus demonstrating dependence upon a critical period for the expression of the spine morphology.

These and several other studies indicate that the development of CNS dendrites is dependent not merely on the presence of afferent contact but is in a morphological sense responsive to the activity of those afferents. One might, therefore, question whether heightened or specialized afferent activity might also shape the morphology of the target dendrite. This idea is indirectly supported by a study of Ryugo et al. (1975), in which it was found that enucleation or whisker removal in newborn rats, producing sensory deprivation of either visual cortex or specific areas of somatosensory cortex, caused an increase in spine density on pyramidal cells of the adjacent auditory cortex. It was suggested that hyperdevelopment of the auditory system represented anatomical adjustment to a heightened dependence upon audition possibly involving increased physiological activity within this system.

The question of the influence of afferent activity in molding the morphology of immature dendrites has also been addressed in much more subtle circumstances, approximating more closely the experiential range normally encountered by the animal. The influence of environmental complexity on dendritic morphology was first studied by Holloway (1966). He found that stellate neurons in layer II visual cortex of animals reared in enriched environments displayed a higher degree of dendritic branching than did those of animals reared in isolation; subsequently other investigators have replicated and extended these results for several classes of cortical neurons (Greenough and Volkman, 1973; Globus *et al.,* 1973). Similar results were obtained in Golgi studies of animals submitted to stressful stimulation during early postnatal life (Schapiro and Vulkovich, 1970). These morphological differences have only been found in those areas where differential environment has been shown to influence cortical thickness: the most consistent differences being reported for occipital cortex, while in frontolateral cortex, where no thickness effects have been observed, dendritic changes were also not found (Greenough *et al.,* 1973). Although these observations indicate a much larger dendritic surface area for the affected cortical neurons, Globus (1975) and Greenough (1976) have failed to find increased dendritic field volume for cells where increased dendritic branching is observed, suggesting that glial rather than dendritic growth is responsible for the cortical thickness effect.

3.2. Growth of Aberrant Fiber Tracts and Expanded Terminal Fields after Early Brain Damage

Phenomena which have been accepted as demonstrations of axonal plasticity in immature animals generally involve the development of aberrant circuitry as a consequence of lesions placed within a still-developing system. As in many of the dendritic studies, plasticity in these situations may not involve a change from one morphology to another but, rather, the production of an anatomical arrangement which differs from that which normally unfolds during development. Although demonstrations of this sort do not inform us of the expressed role of neuronal plasticity within the intact organism, they provide information on the capacity of axons to deviate from what might have been thought to be a genetically programmed morphology. Plasticity of this sort has been demonstrated in axonal systems following removal of their target structure, removal of neighboring axonal projections, and transection of the subject fibers directly.

The effect of target structure removal on the development of an ingrowing population of axons has been studied in several systems, including the retinal efferents to the superior colliculus following unilateral ablation of the colliculus (Schneider, 1970). Axons of the contralateral retina which would normally innervate this structure were found to terminate more heavily in the ventral lateral geniculate, another target of these fibers which lies proximal to the lesion. In addition, fibers were found to extend their normal terminal field to include the medial portion of the intact superior colliculus, an area not normally receiving any input from the ipsilateral retina. Therefore, in this and similar studies on the olfactory system (Devor and Schneider, 1975) it appears that some axons will continue growing, if their normal targets are removed, until a termination zone is found. It has been theorized that the increased density of innervation observed in structures proximal to the lesion is a compensatory development following a principle of "conservation of the total axonal arborization." The essence of this idea is that an axon is genetically programmed to produce a certain number of branches and terminals and, if it cannot achieve this number at its normal target sites, it will develop them in some other region.

It is of interest that the development of aberrant connections between retinal axons and the superior colliculus of the ipsilateral side alters the termination of the normal retinal afferents to that structure. Retinal fibers normally terminate over the entire superficial area of the contralateral superior colliculus. In the reorganized system, where ipsilateral retinal fibers occupy the most medial areas of the intact superior colliculus, contralateral retinal fibers do not include the medial regions in their terminal fields. The afferent fields seem to exclude each other, possibly by competition for available synaptic sites.

Ramón y Cajal (1959) was the first to report the effects of axonal transection in the immature central nervous system. He found that undercutting areas of neocortex just above the white matter, involving transection of the efferent axons beyond their first point of collateral branching, caused a proliferation of collaterals proximal to the lesion which deflected back into the undercut region. Similar observations were made by Pupura and Hausepian (1961), who suggested, as a result of electrophysiological analysis of such undercut cortical regions, that these collateral branches form functional synapses within the local cortical area, thus generating aberrant functional circuits. These observations, together with the possible example of target removal studies, indicate that in several instances transection of immature axons may trigger increased collateralization of their surviving branches.

Abnormal axonal growth also takes place in intact fiber systems following the removal of adjacent input systems. These cases are particularly informative because they do not involve direct injury to the fibers of interest; damage is restricted to the neighbors of these axons. Partial deafferentation of target regions has been shown to increase the extent of innervation by the remaining intact afferents in several systems (Lund and Lund, 1971; Lynch *et al.*, 1972), in some cases involving the spread of the remaining afferents into regions they would not normally innervate (Lynch *et al.*, 1973*b*; Zimmer, 1973, 1974). More dramatic demonstrations of the plasticity of developing axons are provided by those cases where early lesions result in not only expanded terminal fields but entirely new fiber tracts. Following complete or partial unilateral retinal lesions, fibers of the intact retina have been found to grow into the deafferented region of the ipsilateral superior colliculus, a structure they do not normally innervate (Lund *et al.*, 1973). Following unilateral hemispherectomy, which removes the normal crossed cortical–spinal tract to the contralateral spinal cord, Hicks and D'Amato demonstrated the development of a small uncrossed cortical–spinal tract from the intact hemisphere (Hicks and D'Amato, 1970).

One of the best studied examples of an afferent's hyperdevelopment following removal of one of its neighbors involves the so-called commissural and associational projections of the dentate gyrus of the rat hippocampus. These two inputs share the dendrites of the granule cells (the neuronal cell type of the dentate) with a third afferent, the projections of the entorhinal cortex, in an arrangement which is remarkable for its orderliness. As shown in Figure 1, the entorhinal fibers densely innervate the outer 75% of the dentate, while the inner one quarter is the exclusive preserve of the commissural–associational axons (there are other inputs to the dentate, but they generate only a very small fraction of the terminal population of the granule cells). The two collections of fiber systems arrive in the dentate gyrus at about the same time during the first week of postnatal life and very rapidly set up the laminated pattern seen in the adult (Loy *et al.,* 1977). This is accomplished while new granule cells are still being formed and before any of their dendrites have reached even half of their adult length. If the entorhinal cortex is removed in neonatal rats, the commissural–associational fibers hyperdevelop and expand their innervation to include nearly all of the granule cell dendritic tree (Figure 1). Thus it appears that these two afferents possess the capacity for a much greater growth than they normally exhibit and that this capacity is restrained by neighboring developing fiber systems. These effects can be obtained for at least two weeks postnatal, but at three weeks the response of the commissural–associational system to the removal of the entorhinal projections changes; the two inner projections still send axons into the territory which has been denervated, but they do not arborize nearly so densely as they did after earlier lesions (Figure 1) (Lynch *et al.,* 1973*b,c*; Zimmer, 1973). (The situation for the fully mature rat will be discussed below.) This reduced proliferative response might be due to any of several variables. Quite conceivably, the commissural–associational axons have lost some of their capacity for growth, due to genetic programming or because they have formed a large population of terminals by the time the space is made available. It is also conceivable that nonneuronal elements play a role. Glia cells are added to the dentate gyrus in ever increasing numbers during the first week of life and at some point shortly thereafter might begin to take a very active role in the response of a brain region to removal of a primary afferent. Understanding of the mechanisms which are respon-

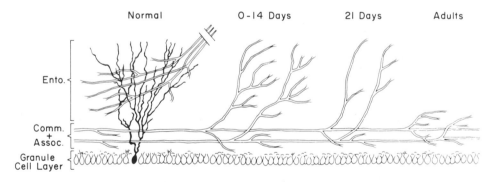

Fig. 1. Schematic illustration of the extent of the postlesion axonal growth by the commissural and associational fibers following removal of the entorhinal cortex at different ages. At the extreme left, the "normal" distribution of afferents to the dentate gyrus granule cell is illustrated. The normal distribution is sharply laminated with the entorhinal cortex afferents terminating within the distal 75% of the molecular layer and the commissural–associational fibers occupying the proximal 25%. If the ipsilateral entorhinal afferents are removed prior to postnatal (PN) day 14, the commissural–associational fibers grow to occupy the entire molecular layer. Following removal of the entorhinal fibers on PN day 21, the commissural–associational afferents are still observed to occupy the full extent of the molecular layer, but the density obtained in the more distal region is far less than is achieved with the earlier lesion. The growth observed following removal of the entorhinal fibers in the adult is limited to an expansion of the inner associational–commissural plexus into a 50-μm zone of the adjacent deafferented territory.

sible for the restrained behavior of the commissural–associational system at three weeks compared to their reaction at two weeks would surely provide important clues about the development of growth-regulatory influences of the brain.

Having shown that at least under these circumstances (removal of targets or neighboring inputs) radically aberrant circuitry will form in the brain, it becomes important to ask whether such circuits really operate. It appears that the answer is yes. Schneider (1970), in his work with the golden hamster, used a simple behavioral test and found that his animals acted as though their new retinal projections were functional. Similar studies by Hicks and D'Amato indicate the aberrant cortical–spinal circuits are also operative. More direct evidence has been obtained using neurophysiological methods in the studies of hyperdevelopment in the hippocampus. In the normal rat, electrical stimulation of the commissural projection produces monosynaptic field potentials which are restricted to the inner dendritic tree of the granule cells; following entorhinal lesions at 10 days of age, these potentials can be readily recorded from even the most distal portions of the dendrite (Lynch *et al.*, 1973*a*). This strongly suggests that the axonal projections which expand into the outer dendritic zones form competent synaptic connections.

4. Anatomical Plasticity in the Adult Brain

4.1. Changes in the Detailed Anatomy of Dendrites following Various Treatments

Deafferentation of mature dendrites has been shown to initiate plastic changes similar to those seen in the developing animal. Removal of afferents to the spines of cells throughout the neuroaxis results in the apparent loss of those spines, at least as examined in Golgi-impregnated mataral (Kemp and Powell, 1971; White and Westrum, 1964). Both light- and electron-microscopic studies have shown that adult spinal motor neurons exhibit a loss of dendritic field and the formation of varicosities in the primary and secondary dendrites following severe afferent deprivation (Bernstein and Bernstein, 1973). Partial deafferentation of the granule cells of the hippocampal dentate gyrus, by lesions of the entorhinal cortex, results in a severe decrease in the spine density of the affected dendritic zone (Parnavelas *et al.*, 1974). These hippocampal dendritic regions become repopulated with spines over time following the lesion, an event which coincides with reinnervation of the dendrites by healthy presynaptic elements (Matthews *et al.*, 1976; Lee *et al.*, 1977).

The Purkinje cells of the cerebellum normally develop two morphologically distinct types of spines. The large central dendrites, innervated by the climbing fibers, exhibit short broad spines, whereas the distal tertiary branches are densely covered by longer, thinner spines known to be contacted by the parallel fibers. Sotelo and co-workers (1975) have studied the effects of climbing-fiber removal in the adult animal by selective chemical lesions of the inferior olive. Following such lesions the primary and secondary branches have been shown to be reinnervated by parallel fibers. Concurrently, the climbing-fiber-type spine of the primary and secondary dendrites is lost in favor of the development of the longer, thinner parallel-fiber-type spine which ultimately attains a density in these regions which exceeds that seen prior to the lesion.

Therefore, in these studies of dendritic morphology following deafferentation, mature dendrites have demonstrated the capacity to lose and develop dendritic spines in coordination with the loss and redevelopment of presynaptic contact. In the case of the cerebellum the postsynaptic specialization reflects the change of the type of presynaptic element.

There appear to have been no attempts to manipulate the morphology of mature dendrites through sensory deprivation, but the effects of repetitive monosynaptic stimulation have been studied. Electrical stimulation of the neocortex of the adult cat has been reported to cause an increase in dendritic branching, dendritic field, and spine density of the pyramidal cells of the contralateral homotopic region receiving innervation from the stimulated area (Ruttledge *et al.,* 1974; Ruttledge, 1976). The interpretation of these observations, made on Golgi material, is open to dispute. One possibility is that hyperactivity of existing axodendritic synapses led to the induction of spines; alternatively, it is conceivable that the effect is due to an increase in the number of presynaptic elements, leading to an increase in the number of spines. With the latter interpretation the increase in spine density represents a secondary result of the stimulation, with the primary effect being upon the stimulated axon population. It is also possible that the stimulation altered the staining properties of the postsynaptic neurons such that they became more clearly defined by the Golgi method. If this were to be the case, dendritic elements which were not detected in the normal situation might have become apparent following stimulation. A possibility of this sort, while probably remote, serves as a cautionary note in interpreting the many experiments cited above which used the Golgi stain to reconstruct dendritic architecture after experimental manipulations. It is evident that further work using histological methods other than the Golgi stain, particularly electron microscopy, is badly needed. Despite these reservations, the experiments using repetitive stimulation are most exciting since they suggest that fully mature dendrites retain extraordinary plasticity, even in the absence of pathological conditions.

Although gross volumetric influences of differential housing have been observed in adult animals, dendritic responsiveness to this treatment instituted during maturity has not been analyzed. Rather, in the adult the interest in experimental influences upon dendritic morphology has centered around specific training situations. The notion that training can cause lasting changes in cortical dendritic morphology has been supported by data from several laboratories. Perhaps the most dramatic demonstration was obtained by Ruttledge (1976) in a classical conditioning paradigm. Stimulation to the suprasylvian gyrus was followed after a brief interval by a foot shock, creating a classical conditioning situation. The stimulated cortex and the contralateral efferent gyrus were both analyzed in Golgi preparations. As discussed immediately above, the stimulation alone, in the presence of nonassociated foot shock, causes dendritic development in the target cortex. In addition to these stimulation effects, a significant increase in the number of spines on vertical and oblique branches of the contralateral pyramidal cells was noted only in trained animals. These results indicate that some aspect of the training experience causes dendritic development that is in addition to the adjustments produced by repeated monosynaptic activation alone.

A very different training situation used by Greenough (1976) has provided data leading to similar conclusions. In this study adult rats were trained on Hebb Williams maze problems each day for 26 days. Rapid Golgi analysis of neocortical area 17 on these animals and littermate controls indicated that a significant increase in apical dendritic branching of pyramical cells within layers 3 and 4 had occurred in the former group. This study, as well as that of Ruttledge, suggests that experience can initiate elaboration of the dendrites of adult cortical neurones.

4.2. Induction of Axonal Sprouting, Terminal Proliferation, and New Circuitry by Discrete Lesions

It will be recalled that a substantial body of evidence indicates that removal of one afferent to an immature brain region will cause that region's remaining afferents to begin growing new branches and connections. There is also reason to believe that comparable phenomena occur in the adult brain. Several studies have shown that a given fiber projection will increase the density of its terminal field in partially deafferented brain areas. Goodman and Horel (1966) found that elimination of the cortical projections to the lateral geniculate nucleus resulted in an apparent proliferation of the terminals of the optic tract in the geniculate. Subsequent studies using several different anatomical methods have reported comparable effects in the colliculus (Stenevi *et al.*, 1972), septum (Moore *et al.*, 1971), and hippocampus (Lynch *et al.*, 1972; Steward *et al.*, 1973). Electron-microscopic experiments have provided evidence that the number of synaptic connections generated by one afferent can increase dramatically following removal of its neighbors; this effect has been obtained in the brain stem (Westrum and Black, 1971), colliculus (Lund and Lund, 1971), septum (Raisman, 1969; Raisman and Field, 1973), and hippocampus (Matthews *et al.*, 1976). There are problems inherent in this type of experiment which indicate the need for caution in interpreting the results. Deafferentation causes considerable shrinkage in the target area (Raisman and Field, 1973; Lynch *et al.*, 1975a), and this fact alone will produce an apparent increase in terminal and synaptic density. That is, the residual population of intact afferents will be compressed into a smaller space, resulting in an apparent increase in density but an increase that does not require any growth response. The light-microscopic experiments are further plagued by a problem which was discussed in regard to dendritic changes after experimental manipulation: the residual elements may be rendered more "visible" to the histological stain. Deafferentation could produce local (chemical) changes which alter the biochemistry of intact terminals in such a way as to increase the probability that they will be detected by the particular tracing method being used. It is also conceivable that the residual terminals might increase in size but not number; this also would tend to increase their likelihood of being counted as an ending. The electron-microscopic experiments typically count synapses, not boutons—if the number of synapses per bouton were to increase in the deafferented area, this could be misinterpreted as an increase in bouton number.

Another type of axonal growth has been reported to occur in the deafferented dentate gyrus of the rat. It will be recalled that the commissural–associational axons expand their field of termination to include the entire granule cell dendritic tree after its more distal regions have been deafferented by a lesion of the entorhinal cortex in the neonate. After the same lesion in the adult, these two fiber systems expand outwards, but not nearly to the degree that they do in the immature brain (Lynch *et al.*, 1973c, 1976). Studies using both light- and electron-microscopic techniques have revealed that this growth begins on the 5th or 6th day after the lesion (Lynch *et*

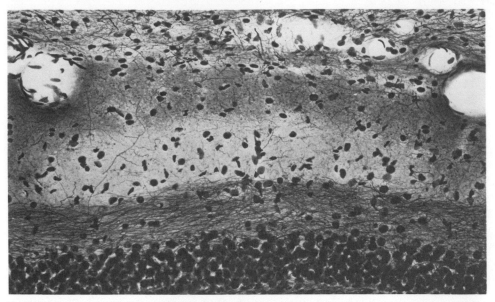

Fig. 2. Photomicrograph of a portion of the molecular layer of the rat dentate gyrus impregnated with the Holmes stain for normal fibers. The plexus of commissural–associational axons can be seen coursing in a discrete lamina close to the granule cells at the bottom of the frame. The outer 75% of the molecular layer is virtually axon-free in this animal which received an ipsilateral entorhinal cortex lesion 2 days prior to sacrifice.

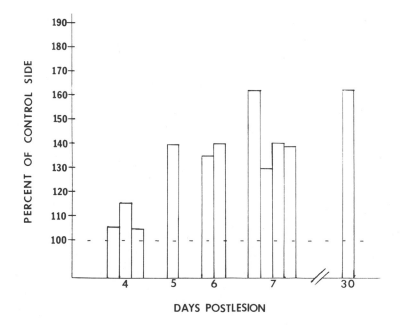

Fig. 3. Bar graph of measurements of the commissural–associational fiber plexus ipsilateral to an entorhinal cortex lesion expressed as a percentage of the plexus width within the contralateral, non-deafferented, dentate gyrus.

al., 1977; Lee *et al.*, 1977) (Figures 2 and 3) and is followed 2–3 days later by physiological changes which suggest that functional synapses have been formed (West *et al.*, 1975). These data provide strong evidence that intact axons in the mature brain retain the capacity to grow new branches, although it is evident that for unknown reasons this process is considerably reduced from that seen in younger animals.

4.3. Proliferation, Migration, and Hypertrophy of Glial Cells at Deafferented Brain Sites

The capacity of the nonneural elements of the brain to undergo plastic changes through the entire life of the animal is rarely discussed under the heading of "brain plasticity." Nonetheless, the glia cells comprise a major portion of the brain's cell population and, if one must be neurocentristic, they undoubtedly constitute a critical influence in the neuroenvironment. In almost all circumstances in which neural plasticity has been observed, glia cells have been found to undergo concurrent morphological transformations. The participation of glia cells may, therefore, be integral to a larger process of which neuroplasticity is also a part. Any attempt to evaluate this idea requires data on the temporal correlation of glial and neuronal events; the following paragraphs will review a recent attempt of this sort that used the rat hippocampus as a model system.

Starting at about 24 hr after removal of the entorhinal cortex, the astrocytes in the deafferented dentate gyrus begin to hypertrophy, an effect which reaches its climax 48–72 hr later (i.e., 3–4 days postlesion). At this point the astrocytes have greatly enlarged cell bodies, longer thicker processes, and strangely enough, processes which appear to be aligned with each other (Figure 4). Electron-microscopic analyses (Matthews *et al.*, 1976; Lee *et al.*, 1977) have shown that the astrocytes are actively engaged in phagocytosis of the degenerating terminals during the period of hypertrophy. Between 96 and 120 hr postlesion, the astroglia begin to atrophy, an effect that continues for several days thereafter. At no time was any increase in the total number of astrocytes in the dentate gyrus detected, although their distribution in the molecular layer was changed. That is, the number of astroglia in the inner molecular layer declined while the population of the outer (deafferented) zone increased. The time courses and magnitude of the two effects were nearly identical, leading to the hypothesis that the astrocytes were migrating from the inner to the outer molecular layer (see Rose *et al.*, 1976, for further discussion).

The microglia of the deafferented dentate gyrus display a very different type of behavior. It should be understood that these ubiquitous cells have been a source of controversy since their discovery by del Rio Hortega (1932), in some measure because of their similarity to oligodendrocytes. The studies to be reviewed below used autoradiographic methods, and it is not possible to be completely certain that the oligodendrocytes did not make some contribution to the results; in fact, it is likely they did.

The microglia are scattered randomly throughout the normal hippocampus and, as shown in autoradiographic experiments using labeled thymidine (which marks cells that are synthesizing DNA and hence dividing), some subgroup of their population is dividing in the naive animal. At 20 hr postlesion, but not before, the number of dividing microglia cells increases dramatically, an effect that continues for some 40–60 hr. Furthermore, these mitotic cells are found scattered throughout

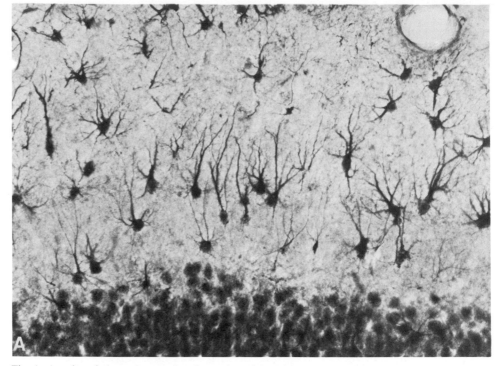

Fig. 4. A series of photomicrographs of a portion of the adult rat dentate gyrus molecular layer stained with Cajal's gold sublimate method for astrocytes. Pictured are equivalent sections from a naive animal (A) and animals sacrificed 4 (B) and 5 (C) days after lesion of the ipsilateral entorhinal cortex. In each panel the granule cell layer can be seen in the bottom of the frame, and the fissure representing the outer limit of the molecular layer lies near the top. The upper, more distal, 75% of the molecular layer is partially deafferented following such a lesion.

the hippocampus, even in areas well removed from degeneration or deafferentation. At some point in the period of cell division or shortly afterwards, the microglia begin to migrate toward the deafferented zones. This is evident from studies in which the rats were given either 6 hr, 4 days, or 8 days of survival after a brief pulse of thymidine. In the former group the labeled cells are randomly located while in the latter, longer survival group, they are concentrated in the zones of degeneration. Labeled cells are also found accumulated at sites that might logically serve as barriers to migration, e.g., the densely packed row of granule cells (Gall *et al.,* 1978).

Both microglia and astroglia undergo dramatic responses to deafferentation, and significant aspects of their responses (e.g., hypertrophy, proliferation) are concluded prior to the onset of any sprouting response (Figure 5). This leads naturally to the hypothesis that the earlier glial events cause the axonal growth, either by removing obstacles that normally retard this growth or by actively stimulating it (Lynch, 1976). This hypothesis is rendered more plausible by the several *in vitro* studies which have suggested that glial cells release nerve-growth-promoting substances (e.g., Varon and Saier, 1975, for a review); this point will be returned to in the Discussion. Finally, it is worth noting that the functional signifi-cance of the rather extraordinary behavior of the microglia remains a complete

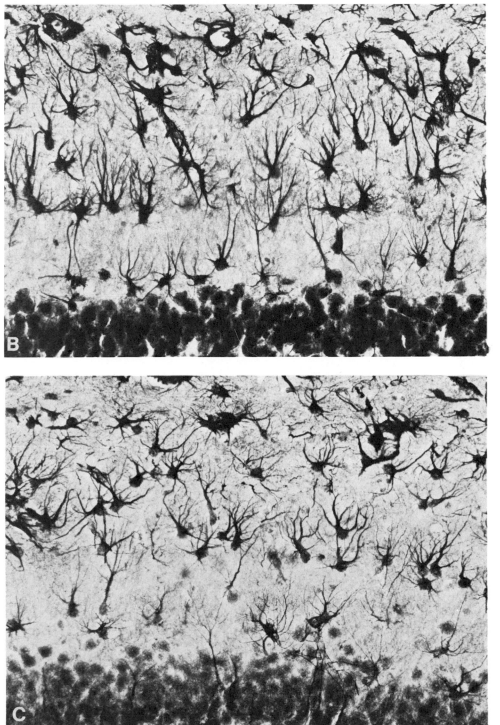

Fig. 4. (*Continued*)

**GARY LYNCH and
CHRISTINE GALL**

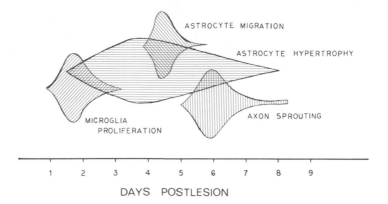

Fig. 5. A graphical illustration of the temporal position of various cellular events occurring in the dentate gyrus of the adult rat following partial deafferentation by lesion of the ipsilateral entorhinal cortex. The height of each "area" indicates the magnitude of the dynamic process involved relative to its own maxima.

mystery. The hypertrophy of the astrocyte is undoubtedly associated with their phagocytic role but the consequences of the proliferation and migration of the microglia are entirely unknown.

5. Discussion

These studies have established that the anatomical structure of the brain is plastic in that it can show a remarkable degree of reorganization under certain circumstances. Furthermore, this capacity for change is not restricted to a single cell type or to portions of their anatomy but instead seems to be a feature of all the constituents of the neuropil. As might have been expected from the clinical literature, the most dramatic examples of nervous system flexibility were found in immature animals. Dendritic development is quite clearly dependent upon an unhindered afferent supply as well as the behavioral history of the organism, while the axons of neonates prove to be capable of radically expanding their projection fields. Abundant evidence was also found for anatomical plasticity in the adult brain, although it was not so obvious as in the case of the neonates. Evidently, the intact, mature axon possesses the capacity to form new, functional synaptic connections in partially deafferented sites, and it is likely that they are able to grow new collateral branches under these circumstances. There is also a smaller selection of data which suggests that dendrites are capable of considerable morphological adjustment in response to the loss of input, to synaptic stimulation, and possibly even to different behavioral experiences in adults. Finally, the glial cells are capable of remarkable transformations, including mitosis and migration over considerable distances. These glial effects have been studied most carefully following damage to the nervous system but, as with the dendrites, there is good reason to believe that the anatomy of the glial cell is strongly influenced by environmental events.

These results have significance for several branches of the neurosciences. They suggest that CNS development is not so genetically "hard-wired" as might have been imagined; that is, that the final architecture of the brain reflects interactions between its neuronal and glial constituents as well as between the organism and its environment. The data on the growth of axonal projections and dendritic ramifications indicate that the final shape of a neuron as well as the targets of its outputs are regulated by the cell's history, by what it encountered and what happened during the encounters. More pertinent to the theme of this review, the still-evolving concept of anatomical plasticity provides a possible (and at best partial) explanation for the flexibility and adaptability of behavior. Since the brain can, under some circumstances, change its physical circuitry, is it not possible that it uses this property to encode memories and to evolve the "personality" of the organism? If we accept the idea that the environment of the neonate can influence the number of dendritic branches and spines, it would hardly be surprising that early experiences might produce profound and lasting behavioral consequences. Similarly, the growth of new connections between neural elements could very well account for the extraordinary persistence of certain memories, whether formed in immature or adult brains. The problem, of course, is to find some way of testing these ideas. The primary difficulty has been alluded to earlier: how can one realistically measure the consequences of training on brain anatomy? The majority of the studies cited above used lesions, stimulation, or rather extreme environmental circumstances to induce changes in the brain. It will be a formidable task to go from this level of analysis to finding the presumably subtler effects that would accompany an ongoing process of brain reorganization.

Ultimately, it may be necessary to search first for the mechanisms which are responsible for anatomical plasticity under the extreme circumstances in which it is easily detected and then measure the activity of these mechanisms under more typical circumstances. While there has been little work searching for biochemical processes associated with axonal sprouting, dendritic growth, etc., findings from other disciplines suggest features that the substrates of plasticity might well possess. In these final paragraphs we will consider some of this evidence (which is admittedly circumstantial in the present context) and advance the hypothesis that brain microanatomy is regulated by an intercellular exchange of materials or factors that promote and suppress growth; plasticity then could be viewed to be a perturbation in this exchange process.

This hypothesis is suggested by several lines of evidence, including the following:

1. A number of tissue culture studies have shown that neurons and glia contain certain materials which strongly affect one another's morphology and biochemical differentiation. In cultures of dissociated sensory and sympathetic ganglion it has been demonstrated that neurons cultured by themselves require media supplements of nerve growth factor (NGF) to survive but are free of this "requirement" if cultured in the presence of nonneural cells from the same structure (Varon and Saier, 1975). In fact, NGF can be shown to influence the differentiation and survival of these neurons *in vitro* only if the competence of the glia support is restricted. This glial maintenance of neuronal viability can be blocked by the addition of submaxillary NGF antibody, suggesting that the glia generate an NGF-like substance which has a tonic influence on local neurons, a relationship that may exist *in vivo* as well as *in vitro*.

In similar experiments neurotransmitter synthesis was monitored in superior cervical ganglion (SCG) neurons cultured with and without SCG nonneural cells. The presence of these cells had no influence on catecholamine synthesis but caused a 120-fold increase in acetylcholine synthesis (Patterson and Chun, 1974). These results suggest that glial cells in the superior cervical ganglion have an active influence on the neurotransmitter synthesis of the local neuronal population.

Neurons seem to influence the behavior of co-cultured glia as well. In culture studies of both peripheral and central nervous tissue, the presence of the neural element seems to determine the rate (McCarthy and Partlow, 1976) and extent (Wood and Bunge, 1975) of growth by the nonneural cells.

These studies strongly suggest that local trophic influences between neurons and glia may play a major role in the growth process of both cell types. Unfortunately, few comparable studies have been made of similar relationships in CNS tissue (Sensenbrenner and Mandel, 1974), but one suspects that such demonstrations only await further refinement of *in vitro* techniques.

2. The brain contains a variety of loosely bound chemicals which produce pronounced morphological changes in dissociated cell preparations. The application of brain "extracts," generally the soluble protein fraction from homogenized central neural tissue, to a variety of culture systems has been used to detect the presence of substances active in inducing morphological maturation of astrocytes in culture (Lim *et al.*, 1973; Lim and Mitsunobo, 1974), morphological and biochemical differentiation of central neurons in culture (Sensenbrenner *et al.*, 1972), and proliferation of CNS nonneural cells (Lim *et al.*, 1973). Although release of these substances *in vivo* has not been demonstrated, these studies encourage the notion that there are trophic substances which are released and diffuse through tissue to mediate growth processes within the brain.

3. The exchange of materials from neuron to neuron (and probably neuron to glia as well) has been documented. Related to this, axons and dendrites are thought to possess uptake and transport mechanisms which allow them to sample their environments and to rapidly send biochemical messages to the somata and hence to the genetic machinery of the cell (Lynch *et al.*, 1975*b*; Smith and Kreutzberg, 1976).

4. There are studies suggesting that sprouting in the peripheral nervous system and the glial response to deafferentation are influenced by diffusible materials emitted by degenerating or intact axons. In an early study, Hoffman and Springell (1951) found that extracts of sciatic nerve would cause intact axon fibers to sprout small branches; he felt that this substance, which he called "neurocletin," was released by Schwann cells during axonal degeneration, diffused to adjacent intact fibers, and induced in them the sprouting response. More recently, Aguilar *et al.* (1973) found that application of colchicine (an inhibitor of axonal transport) to one nerve produced neurophysiological evidence of sprouting by neighboring nerves. They suggest that the cell body of a neuron is constantly producing inhibitory materials which are sent to the axon terminal where they are released and suppress an ongoing growth response by surrounding fibers. We have already discussed the evidence which suggests that degeneration in the brain causes the release of a triggering substance which induces some glial cells to divide and is used by these cells as a guidance mechanism to reach the zones of degeneration.

To summarize: neurons and glia in culture produce and react to growth substances. The brain also produces materials which are stimulants for morphological transformations, and cells in the brain possess highly efficient channels for chemical exchange between its cells. These are separate lines of evidence, and it remains for future research to integrate them into a unified hypothesis, that is, to show that the transforming factors produced by brain are identical to those generated in tissue culture systems and that these factors are among the chemicals exchanged by brain cells. Research to date certainly supports the conclusion that nervous tissue possesses a variety of factors and the means for delivering these to specific target cells. In light of this, it is reasonable to suppose that the momentary microanatomy of the brain represents the balance of the actions of several diverse chemical factors originating from, and acting upon, several cell types. Disturbances in this balance would therefore be expected to cause morphological changes. These disturbances need not be drastic; there is evidence that the release of at least one potentially potent triggering material (adenosine) is influenced by synaptic activity (Schubert *et al.,* 1976).

This analysis of the possible mechanisms responsible for anatomical plasticity suggests an approach which may lead to a resolution of the problem of whether morphological change is used by the brain to achieve behavioral flexibility. If microanatomy is controlled by chemical factors, as suggested in the preceding paragraphs, then it may be possible to isolate and identify these materials. Such an advance might very well pave the way to experiments in which critical and identifiable biochemical events were followed throughout the course of learning, recovery of function, etc., or used in an effort to interfere with, or stimulate, these processes.

ACKNOWLEDGMENTS

This work was supported by NSF grant BNS76-17370, NIH grant NS11589-03, career development award to Gary Lynch, and a grant from the Spencer Foundation.

6. References

Aguilar, C. E., Bisby, M. A., Cooper, E., and Diamond, J., 1973, Evidence that axoplasmic transport of trophic factors is involved in the regulation of peripheral nerve fields in salamanders, *J. Physiol. (London)* **234:**449–464.

Altman, J., and Anderson, W. J., 1972, Experimental reorganization of the cerebellar cortex. I. Morphological effects of elimination of all microneurons with prolonged x-radiation started at birth, *J. Comp. Neurol.* **146:**355–406.

Bennett, E. L., 1976, Cerebral effects of differential experience and training, in: *Neural Mechanisms of Learning and Memory* (M. R. Rosenzweig and E. L. Bennett, eds.), MIT Press, Cambridge, Massachusetts. pp. 279–287.

Bernstein, M., and Bernstein, J., 1973, Regeneration of axons and synaptic complex formation rostral to the site of hemisection in the spinal cord of the monkey, *Int. J. Neurosci.* **5:**15–26.

Bliss, T., and Lomo, T., 1973, Long-lasting potentiation of synaptic transmission in the dentate area of the anaesthetized rabbit following stimulation of the perforant path, *J. Physiol.* **232:** 331–356.

Devor, M., and Schneider, G. E., 1975, Neuroanatomical plasticity: The principle of conservation of total axonal arborization, in: *Aspects of Neural Plasticity/Plasticité Nerveuse* (F. Vital-Durand and M. Jeannerod, eds.), Vol. 43, pp. 191–200, *INSERM,* Paris.

del Rio Hortega, P. Microglia, 1932, in: *Cytology and Cellular Pathology of the Nervous System,* Vol. 2 (W. Penfield, ed.), pp. 483–534, Hoeber, New York.

Diamond, M. C., 1967, Extensive cortical depth measurements and neuron size increases in the cortex of environmentally enriched rats, *J. Comp. Neurol.* **131:**357–364.

Fifkova, E., 1970, The effect of unilateral deprivation on visual centers in rats, *J. Comp. Neurol.* **140:**431–438.

Gall, C., Rose, G., Cotman, C., and Lynch, G., 1978, Proliferative and migratory activities of glial cells in the partially deafferented hippocampus, *J. Comp. Neurol.* (in press).

Globus, A., 1975, Brain morphology as a function of presynaptic morphology and activity, in: *The Developmental Neuropsychology of Sensory Deprivation* (A. Riesen, ed.), pp. 9–91, Academic Press, New York.

Globus, A., and Scheibel, A. B., 1967, Synaptic loci on parietal cortical neurons: Terminations of corpus callosum fibers, *Science* **156**:1127–1129.

Globus, A., Rosenzweig, M. R., Bennett, E. L., and Diamond, M. C. 1973, Effects of differential experience on dendritic spine counts in rat cerebral cortex, *J. Comp. Physiol. Psychol.* **82**:175–181.

Goodman, D. C., and Horel, J. A., 1966, Sprouting of optic tract projections in the brain stem of the rat, *J. Comp. Neurol.* **127**:71–88.

Greenough, W. T., 1976, Enduring brain effects of differential experience and training, in: *Neural Mechanisms of Learning and Memory* (M. R. Rosenzweig and E. L. Bennett, eds.), pp. 255–278, MIT Press, Cambridge, Massachusetts.

Greenough, W. T., and Volkman, F. R., 1973, Pattern of dendritic branching in occipital cortex of rats reared in complex environments, *Exp. Neurol.* **40**:491–504.

Greenough, W. T., Volkman, F. R., and Juraska, J. M., 1973, Effects of rearing complexity on dendritic branching in frontolateral and temporal cortex of the rat, *Exp. Neurol.* **41**:371–378.

Gyllensten, L., Malmfors, T., and Narrlin, M. L., 1964, Effect of visual deprivation on the optic centers of growing and adult mice, *Comp. Neurol.* **122**:79–90.

Herndon, R., and Oster-Granet, M., 1975, Effect of granule cell destruction on development and maintenance of the Purkinje cell dendrite, in: *Physiology and Pathology of Dendrites* (G. W. Kreutzberg, ed.), pp. 361–379, Raven Press, New York.

Hicks, S., and D'Amato, C., 1970, Motor-sensory and visual behavior after hemispherectomy in newborn and mature rats, *Exp. Neurol.* **29**:416–438.

Hirano, A., Dembitzer, H. M., and Jones, M., 1972, An electron microscopic study of cycasin induced cerebellar alterations, *J. Neuropathol. Exp. Neurol.* **31**:113–125.

Hoffman, H., and Springell, P. 1951, An attempt at the chemical identification of ''neuroclectin'' (the substance evoking axon sprouting), *Aust. J. Exp. Biol. Med. Sci.* **29**:417–424.

Holloway, R. L., Jr., 1966, Dendritic branching: Some preliminary results of training and complexity in rat visual cortex, *Brain Res.* **2**:393–396.

Kemp, J. M., and Powell, T. P. S., 1971, The termination of fibers from the cerebral cortex and thalamus upon dendritic spines in the caudate nucleus: A study with the Golgi method, *Philos. Trans. R. Soc. London, Ser. B* **262**:429–439.

Krech, P., Rosenzweig, M., and Bennett, E., 1963, Effects of complex environment and blindness on rat brain, *Arch. Neurol.* **8**:403–412.

Lee, K., Stanford, E., Cotman, C., and Lynch, G., 1977, Ultrastructural evidence for bouton sprouting in the dentate gyrus of adult rats, *Exp. Brain Res.* **29**:475–485.

Lim, R., and Mitsunobu, K., 1974, Brain cells in culture: Morphological transformation by a protein, *Science* **185**:63–66.

Lim, R., Mitsunobu, K., and Li, W., 1973, Maturation stimulating effect of brain extracts and dibutyryl cyclic AMP on dissociated embryonic brain cells in culture, *Exp. Cell Res.* **79**:243–246.

Loy, R., Lynch, G., and Cotman, C., 1977, Development of afferent lamination in the fascia dentata of the rat, *Brain Res.* **121**:229–243.

Lund, R. D., and Lund, J. S., 1971, Synaptic adjustment after deafferentation of the superior colliculus of the rat, *Science* **171**: 804–807.

Lund, R. D., Cunningham, T. S., and Lund, J. S., 1973, Modified optic projections after unilateral eye removal in young rats, *Brain, Behav. Evol.* **8**:51–72.

Lynch, G., 1976, Neuronal and glial responses to the destruction of input: The ''deafferentation syndrome,'' in: *Cerebrovascular Diseases* (P. Scheinberg, ed.), pp. 209–227, Raven Press, New York.

Lynch, G., Matthews, D. A., Mosko, S., Parks, T., and Cotman, C. W., 1972, Induced acetylcholinesterase-rich layer in rat dentate gyrus following entorhinal lesions, *Brain Res.* **42**:311–318.

Lynch, G., Deadwyler, S., and Cotman, C. W., 1973a, Postlesion axonal growth produces permanent functional connections, *Science* **180**:1364–1366.

Lynch, G., Mosko, S., Parks, T., and Cotman, C., 1973b, Relocation and hyperdevelopment of the dentate gyrus commissural system after entorhinal lesions in immature rats, *Brain Res.* **50**:174–178.

Lynch, G., Stanfield, B., and Cotman, C. W., 1973c, Developmental differences in postlesion axonal growth in the hippocampus, *Brain Res.* **59**:155–168.

Lynch, G., Rose, G., Gall, C., and Cotman, C. W., 1975a, The response of the dentate gyrus to partial deafferentation, in: *The Golgi Centennial Symposium* (M. Santini, ed.), pp. 305–317, Raven Press, New York.

Lynch, G., Smith, R., Browning, M., and Deadwyler, S., 1975*b,* Evidence for bidirectional dendritic transport of horseradish peroxidase, in: *Physiology and Pathology of Dendrites* (G. Kreutzberg, ed.), pp. 297–313, Raven Press, New York.

Lynch, G., Gall, C., Rose, G., and Cotman, C. W., 1976, Changes in the distribution of the dentate gyrus associational system after unilateral and bilateral entorhinal lesions in adult rats, *Brain Res.* **110:**57–71.

Lynch, G., Gall, C., and Cotman, C., 1977, Temporal parameters of axon "sprouting" in the brain of the adult rat. *Exp. Neurol.* **54:**179–183.

Matthews, D. A., Cotman, C., and Lynch, G., 1976, An electron microscopic study of lesion induced synaptogenesis in the dentate gyrus of the adult rat. II: Reappearance of morphologically normal synaptic contacts, *Brain Res.* **115:**23–41.

McCarthy, K. D., and Partlow, L. M., 1976, Neuronal stimulation of ^{3}H-thymidine incorporation by primary cultures of highly purified nonneuronal cells, *Brain Res.* **114:**415–426.

Monard, D., Soloman, F., Kentsch, M., and Gysin, R., 1973, Glia induced morphological differentiation in neuroblastoma cells, *Proc. Natl. Acad. Sci. U.S.A.* **70:**1682–1687.

Moore, R. Y., Bjorklund, A., and Stenevi, U., 1971, Plastic changes in the adrenergic innervation of the rat septal area in response to denervation, *Brain Res.* **33:**13–35.

Patterson, P., and Chun, L. L. Y., 1974, The influence of nonneural cells on catacholamine and acetylcholine synthesis and accumulation in cultures of dissociated sympathetic neurons, *Proc. Natl. Acad. Sci. U.S.A.* **71:**3607–3610.

Parnavelas, J., Lynch, G., Brecha, N., Cotman, C., and Globus, A., 1974, Spine loss and regrowth in the hippocampus following deafferentation, *Nature* **248:**71–73.

Purpura, D. P., and Hausepian, E. M., 1961, Morphological and physiological properties of chronically isolated immature neocortex, *Exp. Neurol.* **4:**377–401.

Raisman, G., 1969, Neuronal plasticity in the septal nuclei of the adult rat, *Brain Res.* **14:**25–48.

Raisman, G., and Field, P., 1973, A quantitative investigation of the development of collateral regeneration after partial deafferentation of the septal nuclei, *Brain Res.* **50:**241–264.

Ramón y Cajal, S., 1959, *Degeneration and Regeneration of the Nervous System* (R. M. May, trans.), reprinted by Hafner, New York.

Rose, G., Lynch, G., and Cotman, C., 1976, Hypertrophy and redistribution of astrocytes in the deafferented dentate gyrus, *Brain Res. Bull.* **1:**87–93.

Rosenzweig, M. R., Bennett, E. L., and Diamond, M. C., 1972, Chemical and anatomical plasticity of brain: Replications and extensions, in: *Macromolecules and Behavior,* 2nd ed. (J. Gaito, ed.), pp. 205–277, Appleton-Century-Crofts, New York.

Ruttledge, L. T., 1976, Synaptogenesis: Effects of synaptic use, in: *Neural Mechanisms of Learning and Memory* (M. R. Rosenzweig and E. L. Bennett, eds.), pp. 329–339, MIT Press, Cambridge, Massachusetts.

Ruttledge, L. T., Wright, C., and Duncan, J., 1974, Morphological changes in pyramidal cells of mammalian neocortex associated with increased use, *Exp. Neurol.* **44:**209–228.

Ryugo, D., Ryugo, R., Globus, A., and Killacky, H., 1975, Increased spine density in auditory cortex following visual or somatic deafferentation, *Brain Res.* **90:**143–146.

Schapiro, S., and Vulkovich, K. R., 1970, Early experience effects upon cortical dendrites. A proposed model for development, *Science* **1967:**292–294.

Schneider, G. E., 1970, Mechanisms of functional recovery following lesions of visual cortex or superior colliculus in neonate and adult hamsters, *Brain, Behav. Evol.* **3:**295–323.

Schubert, P., Lee, K., West, M., Deadwyler, S., and Lynch, G., 1976, Stimulation-dependent release of ^{3}H-adenosine derivatives from central axon terminals to target neurons, *Nature* **260:**541–545.

Sensenbrenner, M., and Mandel, P., 1974, Behavior and neuroblasts in the presence of glial cells, fibroblasts and meningeal cells in culture, *Exp. Cell Res.* **87:**159–167.

Sensenbrenner, M., Springer, N., Booker, J., and Mandel, P., 1972, Histochemical studies during the differentiation of dissociated nerve cells cultivated in the presence of brain extracts, *Neurobiology* **2:**49–60.

Smith, B., and Kreutzberg, G. (eds.), 1976, *Neurosci. Res. Prog. Bull.,* **14:**211–453.

Sotelo, C., Hillman, D., Zamora, A., and Llinas, R., 1975, Climbing fiber deafferentation: Its action on Purkinje cell dendritic spines, *Brain Res.* **98:**574–581.

Stenevi, U., Bjorklund, A., and Moore, R. Y., 1972, Growth of intact central adrenergic axons in the denervated lateral geniculate body, *Exp. Neurol.* **35:**290–299.

Steward, O., Cotman, C. W., and Lynch, G., 1973, Re-establishment of electrophysiologically functional entorhinal cortical input to the dentate gyrus deafferented by ipsilateral entorhinal lesions: Innervation by the contralateral entorhinal cortex, *Exp. Brain Res.* **18:**396–414.

Valverde, F., 1971, Rate and extent of recovery from dark rearing in the visual cortex of the mouse, *Brain Res.* **33**:1–11.

Valverde, F., and Ruiz Marcos, A., 1969, Dendritic spines in the visual cortex of the mouse. Introduction to a mathematical model, *Exp. Brain Res.* **8**:269–283.

Varon, S., and Saier, M., 1975, Culture techniques and glial-neuronal interrelationships *in vitro, Exp. Neurol.* **48**:135–162.

West, J., Deadwyler, S., Cotman, C., and Lynch, G., 1975, Time-dependent changes in commissural field potentials in the dentate gyrus following lesions of the entorhinal cortex in adult rats, *Brain Res.* **97**:215–233.

Westrum, L., and Black, R., 1971, Fine structural aspects of the synaptic organization of the spinal trigeminal nucleus (pars interpolaris) of the cat, *Brain Res.* **25**:265–288.

White, L. E., Jr., and Westrum, L. E., 1964, Dendritic spine changes in prepyriform cortex following olfactory bulb lesions—rat, Golgi method, *Anat. Rec.* **148**:410–411.

Wood, P., and Bunge, R., 1975, Evidence that sensory axons are mitogenic for Schwann cells, *Nature* **256**:662–664.

Zimmer, J., 1973, Extended commissural and ipsilateral projections in postnatally dentorhinated hippocampus and fascia dentata demonstrated in rats by silver impregnation, *Brain Res.* **64**:293–311.

Zimmer, J., 1974, Proximity as a factor in the regulation of aberrant growth in postnatally deafferented fascia dentata, *Brain Res.* **72**:137–142.

4

Developmental Aspects of the Neuronal Control of Breathing

F. J. SCHULTE

1. Brain Stem Respiratory Neurons

Neuronal respiratory control centers have been located in the medulla in a long series of experiments using ablation, transection, and electrical and chemical stimulation techniques. Baumgarten and his co-workers were able to demonstrate aggregates of both inspiratory and expiratory neurons in an area of the lateral reticular substance (Baumgarten *et al.,* 1957, 1959; Haber *et al.,* 1957; Merill, 1970). In cats and dogs their exact location is from the Vth nerve nucleus down to a few millimeters beyond the obex, dorsal and dorsomedial from the nucleus ambiguus (Figure 1). Even when the peripheral afferent inflow from the lungs, the rib cage, and the respiratory muscles is abandoned by complete muscle paralysis and vagal nerve section, these "respiratory neurons" maintain their firing pattern either together (inspiratory neurons) or exactly alternating (expiratory neurons) with phrenic nerve activity (Figure 2).

It is apparent from intracellular microelectrode recordings (Figure 3) that the fluctuation of the membrane potential of respiratory neurons during one respiratory cycle is responsible for rhythmic breathing (Baumgarten *et al.,* 1960; Salmoiraghi and Baumgarten, 1961). The membrane potential continuously declines until threshold depolarization is reached. This gives rise to a burst of action potentials, the duration of which is limited by a steady increment of the membrane potential causing respiratory neuron inhibition. It is now very likely, that, via afferent impulses, a continuous depolarization pressure is maintained on the inspiratory neurons that is periodically interrupted by several inhibitory mechanisms:

F. J. SCHULTE • University of Göttingen, Göttingen, West Germany.

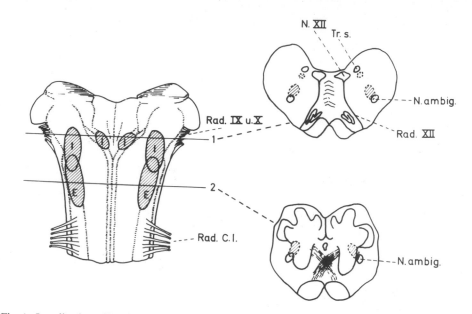

Fig. 1. Localization of inspiratory (I) and expiratory (E) neurons in the brain stem reticular formation in relation to the Tractus solitarius (Tr. s.), the N. ambiguus (N. ambig.), and the XIIth nerve nucleus (N. XII).

postexcitatory hyperpolarization due to an increase of potassium permeability; accommodation; postsynaptic inhibitory influences, such as vagal afferents from lung stretch receptors (Hering and Breuer, 1808); and inhibitory collaterals from antagonistic expiratory neurons. The role of recurrent inhibition, equivalent to the spinal cord Renshaw mechanism, is still controversial.

For our understanding of respiratory failure in immature infants, it is of particular importance that rhythmic activity is probably not an inherent characteristic of brain stem respiratory neurons but must be maintained through continuous

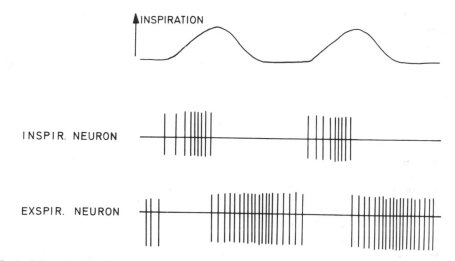

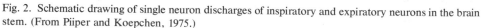

Fig. 2. Schematic drawing of single neuron discharges of inspiratory and expiratory neurons in the brain stem. (From Piiper and Koepchen, 1975.)

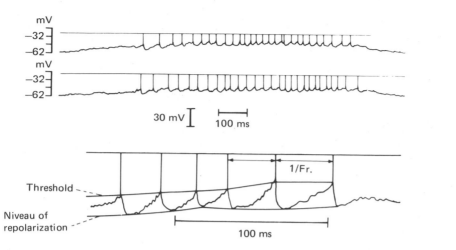

Fig. 3. Intracellular recording from an inspiratory neuron in the brain stem. For explanation see text. (From Baumgarten *et al.*, 1960.)

postsynaptic excitatory drive, i.e., depolarization pressure. Furthermore, there is some evidence in the literature that excitatory synapses are predominantly located on dendritic arborizations far away from the body of the respiratory neuron (Richter, 1974). Unfortunately, the developmental time course of dendritic arborization and the formation of synapses on brain stem respiratory neurons is not yet known. From extensive studies on the development of the cerebral cortex, however, we have to assume that immature respiratory neurons are deficient in dendrites (Purpura, 1975), which implies a lack of postsynaptic excitatory drive. This must be one important cause of the immature infant's difficulties in initiating and maintaining rhythmic breathing.

2. Chemoreceptor Regulation of Breathing

The chemical composition of blood and cerebrospinal fluid strongly influences the activity of brain stem respiratory neurons. Arterial chemoreceptors in the carotid sinus and the aortic arch (Heymens and Heymans, cited in Heymans and Neil, 1958) are activated by hypoxia, hypercapnia, and acidosis (Heymans and Neil, 1958). Central chemoreceptive areas along the ventrolateral surface of the medulla in close contact with the cerebrospinal fluid (Loeschcke, 1974) are activated by acidosis and, probably, by hypercapnia (Piiper and Koepchen, 1975).

The fundamental importance of chemoreceptor response in respiratory regulation of human neonates, both preterm and term, has received ample experimental support. In fetal lambs made hypoxic, the initiation of respiratory movements as well as an increased chemoreceptor afferent impulse activity in the sinus nerve have been demonstrated (Cross *et al.*, 1953). The sensitivity of the chemoreceptors, however, is apparently set at a lower level (Purves and Biscoe, 1966). Intrauterine P_{O_2} and P_{CO_2} blood values are thus being tolerated without vigorous fetal breathing attempts, which they would certainly cause were the same fetus born. The setting of the chemoreceptors apparently changes quite suddenly after birth, and this may be explained by an increase in sympathetic activity after cord clamping.

It is known from the exhaustive morphological studies of Boyd (1961) that in human fetuses the peripheral arterial chemoreceptors are mature. A great wealth of data is available on ventilatory responses of preterm and full-term infants to changes in blood gas concentrations, which indicates a functional significance of chemoreceptors even in immature babies, particulary regarding their response to CO_2 (Cross and Oppé, 1953; Avery *et al.*, 1963; Olinsky *et al.*, 1974; Bryan *et al.*, 1974; Bodegård, 1974, 1975). In contrast to the adult man, exposure to 4% CO_2 causes an immediate rise in respiratory frequency rather than an increase in tidal volume, the latter being prevented by powerful inhibitory influences of lung stretch receptors on brain stem respiratory neurons.

3. Mechanoreceptor Regulation of Breathing

3.1. Hering–Breuer Inflation Reflex

Inhibition of respiratory activity via vagal afferent fibers in response to distention of the lungs is called the Hering–Breuer inflation reflex (Hering and Breuer, l868).Whereas in the adult human, the Hering–Breuer reflex is weak or absent, it plays an important role in animals, including newborn cats and rabbits. After vagotomy, newborn kittens and rabbits develop periodic or gasping breathing patterns (Schwieler, l968).

The Hering–Breuer inflation reflex is present in full-term newborn infants but decreases soon after birth (Cross *et al.*, 1960). At a conceptional age of 32 weeks the reflex was found to be weak, but its strength increases to its maximum at around 38 weeks conceptional age (Bodegård *et al.*, l969).

Data obtained in newborn infants with both airway occlusion and inflation, together with exposure to 4% CO_2, give strong support to the assumption that the Hering–Breuer inflation reflex regulates respiratory depth and frequency in the perinatal period, particularly in preterm infants (Olinsky *et al.*, 1974; Bryan *et al.*, 1974; Bodegård, 1974, 1975). In contrast to the adult man, exposure to 4% CO_2 causes an immediate rise in respiratory frequency rather than an increase in tidal volume, the latter being prevented by powerful inhibitory influences of lung stretch receptors on brain stem respiratory neurons.

Thus, it seems that in the perinatal period brain stem respiratory neurons depend heavily on the periodic afferent inflow of inhibitory vagal influences to establish rhythmic inspiratory activity. Bodegård (1974) has speculated that in the very immature infant (about 32 weeks) periodic breathing and apnea may occur as a result of the weakness of the Hering–Breuer reflex, this resembling the situation mentioned above in vagotomized kittens.

3.2. Intercostal Muscle Stretch Reflexes

The existence of monosynaptic thoracic wall muscle reflexes mediated by intercostal muscle spindle afferent fibers has been clearly demonstrated (Campbell and Howell, 1962; Sears, 1965; Euler, 1966) (Figure 4). The main function of the intercostal muscle stretch reflex response to an increased airway load is to provide stability to the rib cage, thus preventing the thoracic wall from collapsing during diaphragmatic inspiratory contraction (Shannon and Zechmann, 1972).

Muscle tone as well as the strength of monosynaptic stretch reflexes are dependent upon the activity of muscle spindles and their afferent gamma motoneuron innervation (Granit, 1970). There is both morphological and functional evidence

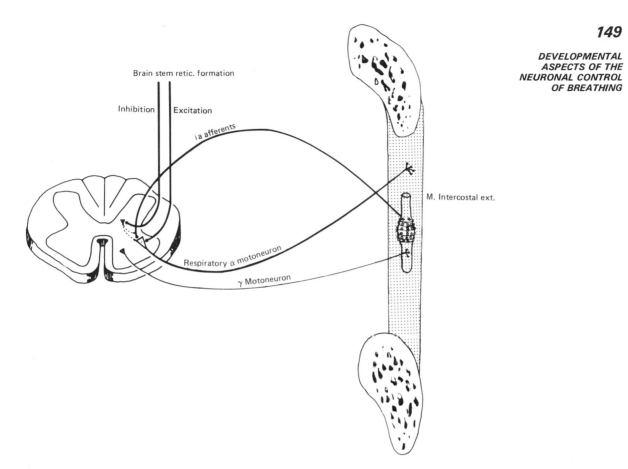

Fig. 4. Innervation of intercostal muscle. Gamma motoneurons innervate intrafusal muscle fibers of muscle spindles, thereby increasing the sensitivity of the stretch receptor. Ia afferent fibers from the muscle spindles activate alpha motoneurons, thereby increasing thoracic wall stability during diaphragmatic contraction. During REM or active sleep alpha motoneurons and/or Ia afferents are inhibited.

that tonic drive from the gamma motoneuron–muscle spindle loop to spinal alpha motoneurons is poor in the immature human infant (Skoglund, 1966; Schulte *et al.*, 1968, 1969*a,b;* Schulte, 1974).

Quite in accordance with our extensive knowledge about poor spinal motoneuron activity and weak tonic stretch reflexes in the preterm infant, the strength of intercostal muscle stretch reflexes increases in human infants gradually from 30 to 42 weeks conceptional age (Bodegård and Schwieler, 1971). The tendency of the preterm infant's chest wall to collapse during inspiration and the poor synchronization between thoracic wall extension and diaphragmatic contraction seems at least in part to be due to a lack of tonic gamma motoneuron–muscle spindle activity within the spinal motor system of intercostal muscles.

4. Patterns of Fetal and Neonatal Respiration

As early as 1888 (Ahlfeld, 1888), fetal breathing movements were observed on the abdominal wall of the pregnant woman. Only during the last few years, however, has it been firmly established that intrauterine respiratory movements are a normal phenomenon in both sheep and man (Dawes, 1974). Such movements have been clearly demonstrated in the human fetus by means of an ultrasound A scan system as early as 11 weeks of gestation (Boddy and Robinson, 1971; Boddy and Mantell, 1973). The movements are episodic, irregular, and of short duration.

In fetal lambs, observed after delivery into a saline bath, respiratory movements occurred together with rapid eye movements and with a low-voltage fast-activity EEG, indicating REM (rapid eye movement) sleep (Dawes *et al.*, 1972).

In the full-term newborn infant the breathing pattern is regular during non-REM or quiet sleep and irregular during REM or active sleep (Figure 5), with a significant correlation between the EEG amplitude and the oscillation of the respiratory rate (Prechtl *et al.*, 1968; Hathorn, 1975). Thus, at this stage of CNS maturation, coordination of breathing patterns with different behavioral states is well developed, provided that the infant is normal. Under various nonoptimal pre- and perinatal conditions, however, the dissociation of behavioral factors, particularly abnormal respiratory patterns that can mimic immaturity is one of the most sensitive indicators of deviant CNS function (Prechtl *et al.*, 1969; Schulte *et al.*, 1969*b*, 1971; Dreyfus-Brisac *et al.*, 1970; Schulte, 1974).

In very immature preterm infants of 24–27 weeks conceptional age, the brain stem is not yet able to coordinate and maintain the patterns of either active or quiet sleep and regular respiration occurs only sporadically (Dreyfus-Brisac, 1968). With increasing conceptional age, periods of muscular quiescence and discontinuous "tracé alternant" EEG cluster together with regular respiration. Periods of rapid eye movements, phasic muscular twitchings, and continuous bioelectric brain activity, however, coincide with irregular respiration and heart rate, thus indicating a progression in the organization of behavior (Dreyfus-Brisac, 1970).

Periodic breathing and the frequent occurrence of apneic spells is one of the main characteristics of the respiratory pattern in preterm infants (Daily *et al.*, 1969; Parmelee *et al.*, 1972; Fenner *et al.*, 1973; Avery, 1974). There is a significant predominance of apneic episodes during active sleep (column 6 of Table I). They occur more often, last longer, and are more frequently connected with bradycardia during active than during quiet sleep (Gabriel *et al.*, 1976). These findings can be explained by the well-known inhibitory influences acting on spinal motoneurons and primary afferents during active sleep (Pompejano, 1966, 1969). Apneic spells, spinal motoneuron inhibition, and cardiac arrhythmias (Figure 6) are neurophysiological processes which tend to occur together, and they are facilitated by neurophysiological mechanisms underlying active sleep (Schulte *et al.*, 1976).

REM sleep appears to entail the simultaneous occurrence of at least three distinct processes (Dement, 1972):

1. *Tonic motor inhibition* at the site of the ultimate central nervous system outlet, ensured by spinal motoneuron hyperpolarization. REM sleep is a time of profound motor paralysis—otherwise our dreams would have serious consequences.

2. *Phasic activity*. Paradoxically, at the same time when spinal motoneuron inhibition prevents almost all central nervous system activity from reaching the musculature, the brain produces enormous bursts of activity which give rise to a series of short-duration events like rapid eye movements, muscular twitchings, sudden changes in pupil diameter, penile erection, and the PGO (pontine–geniculate–occipital) spikes.

3. *Central nervous system arousal*. In active sleep the brain resembles the waking or aroused brain in many aspects—particularly in the area of nonspecific

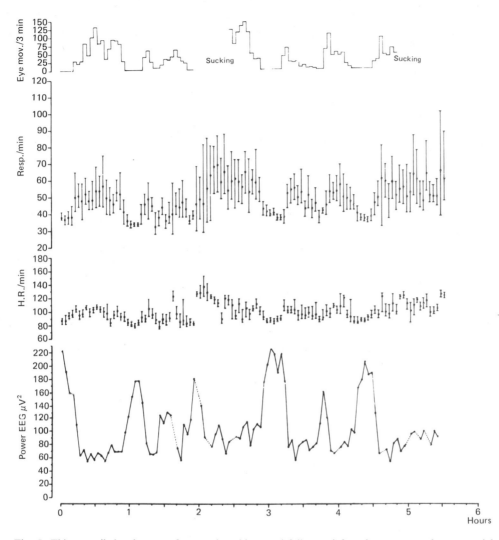

Fig. 5. This compiled polygram of a two-day-old normal full-term infant demonstrates the sequential patterns of the variables during a 5½-hr recording. The converted medians and quartile values of respiration and heart beat, the number of eye movements and the mean square voltage of the EEG are given for each 3-min epoch. During periods of REM (rapid eye movement) respiration and heart rate are irregular and the EEG is of low voltage. During quiet sleep (no eye movements) respiration and heart rate are regular and the EEG is of high voltage. (From Prechtl *et al.*, 1969.)

Table 1. Number of Apneic Spells of Different Durations in Relation to Different Sleep States[a,b]

Sleep state	Apneic spells (N)				$\dfrac{\text{Actual}}{\text{Estimated}}$ – ratio of a.sp.	Records (N)
	Total	<20 sec	20–39 sec	≥40 sec		
AS	390	294	74	22	1.29 ± 0.17 S.D.	19
QS	29	25	4	—	0.42 ± 0.22 S.D.	11
UDS	74	51	13	10	1.03 ± 0.46 S.D.	15

[a]From Gabriel *et al.* (1976).
[b]Abbreviations: AS, active sleep; QS, quiet sleep; UDS, undifferentiated sleep; a.sp., apneic spell.

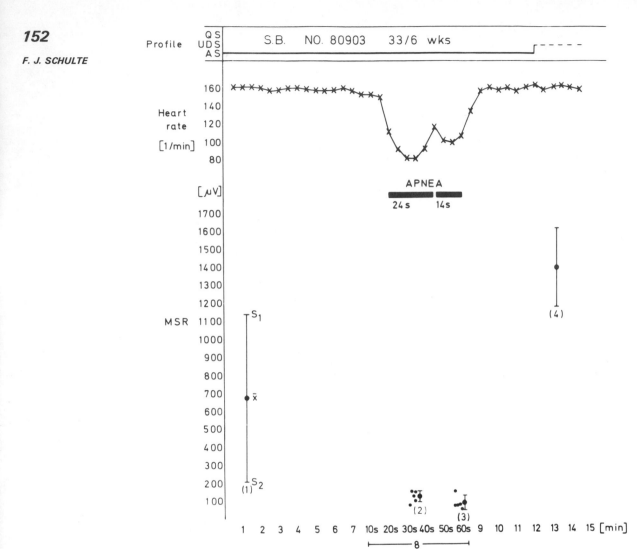

Fig. 6. During apneic spells monosynaptic reflex responses (MSR) are greatly depressed. (From Schulte *et al.*, 1979.)

measures of central nervous system activity levels like EEG, brain temperature, and cerebral blood flow—though not in awareness.

It is only with these three active-sleep brain mechanisms in mind that we can understand why during active sleep the preterm infant is breathing faster one minute and stops breathing altogether the next. In light of the above-mentioned REM sleep concept it is now understood why the first respiratory movements of the fetus *in utero* can be detected during active sleep. In active sleep there is continuous competition between strong internal central nervous system arousal and sustained spinal motoneuron inhibition. Tonic inhibition prevents most of the phasic events within the brain from leaving the central nervous system. In preterm infants there is remarkable fluctuation of tonic inhibitory processes during active sleep (Schulte *et al.,* 1979). This explains why, particularly in the very immature infant, a considerable amount of phasic activity escapes from the central nervous system, giving rise to an enormous wealth of twitchings and even to respiratory movements *in utero,* and why at other times the same baby, during active sleep, is absolutely flaccid, motionless, and even apneic.

In the preterm infant the inhibition of respiratory neurons during REM sleep becomes particularly effective: The more immature the infant, the more time it spends in active sleep and only few phases can be regarded as clear-cut quiet sleep. From 32 weeks conceptional age to 3 months after term, the percentage of quiet sleep increases from about 20% to about 60% of total sleep time (Roffwarg *et al.,* 1966; Parmelee *et al.,* 1967). Furthermore, during the same maturational period total sleep time decreases from over 18 to about 14 hr per day. Thus, the time the baby is obliged to overcome REM sleep inhibition of spinal respiratory motoneurons becomes much less with increasing age and maturity.

In newborn infants, REM sleep inhibitory influences on respiratory spinal motoneurons and their afferents not only affect respiratory rate and rhythm, but also rib cage stability and synchronization of intercostal muscle and diaphragm activity. Bryan and Bryan (1974) have shown that it is mainly during active sleep that the rib cage loses its stability, and frequently thoracic retractions, rather than expansions, can be observed during inspiration (Bodegård, 1976). This paradoxical breathing pattern occurs predominantly in preterm infants when the poorly developed tonic intercostal muscle stretch reflexes can easily be abolished during long-lasting periods of REM sleep by strong descending inhibition on the spinal motoneuron excitatory afferents from muscle spindles. Thus, inspiratory diaphragmatic contractions cause the rib cage to collapse since it is not counteracted by an increment of reflex tone in the intercostal muscles.

5. Brain Mechanisms Responsible for the Preterm Infant's Difficulties in Breathing: A Summary

1. One of the most important morphological features of central nervous system immaturity is the lack of dendritic arborization and axodendritic synaptic connections. Thus, we have to assume that synaptic excitatory drive on respiratory neurons in both brain stem and spinal cord is much weaker in preterm than in term infants.

2. P_{O_2} chemoreceptor drive is vulnerable in the preterm infant. Hypoxia produces only transient hyperapnea and can even act as a respiratory depressant when the baby is cool.

3. In the neonatal period inhibitory influences on lung stretch receptors (Hering–Breuer reflex) are important for the establishment of rhythmic inspiratory neuron activity with their normal "stop-and-go" pattern. This reflex is poorly developed at 32 weeks conceptional age.

4. The preterm infant spends most of the time in active (REM) sleep, with strong descending inhibitory influences on respiratory neuron and excitatory afferents from muscle spindles, thus causing apnea, due to spinal motoneuron arrest, and rib cage instability with thoracic/abdominal desynchronization, due to a loss of intercostal reflex muscle tone.

6. References

Ahlfeld, B., 1888, Über bisher noch nicht beschriebene intrauterine Bewegungen des Kindes, *Verh. Dtsch. Ges. Gynäkol.* **2**:203.

Avery, M. E., 1974, The lung and its disorders in the newborn infant, in: *Major Problems in Clinical Pediatrics,* Vol. 1, W. B. Saunders, Philadelphia.

Avery, M. E., Chernick, V., Dutton, R. E., and Permutt, S., 1963, Ventilatory response to inspired carbon dioxide in infants and adults, *J. Appl. Physiol.* **18**:895.

Baumgarten, R. v., Baumgarten, A. v., and Schafer, K. P., 1957, Beitrag zur Lokalisationsfrage bulboreticulärer respiratorischer Neurone der Katze, *Pfluegers Arch. Ges. Physiol.* **264**:217.

Baumgarten, R. v., Koepchen, H. P., and Aranda, L., 1959, Untersuchungen zur Lokalisation der bulbären Kreislaufzentren. 1. Mitt. Funktionelle Organisation der Vagus-Wurzeln, *Verh. Dtsch. Ges. Kreislaufforsch.* **25**:170.

Baumgarten, R. v., Balthasar, K., and Koepchen, H. P., 1960, Über ein Substrate atmungsrhythmischer Erregungsbildung im Rautenhirn der Katze, *Pfluegers Arch. Ges. Physiol.* **270**:504.

Boddy, K., and Mantell, C., 1973, Human foetal breathing *in utero, J. Physiol.* **231**:105.

Boddy, K., and Robinson, J., 1971, External method for detection of fetal breathing *in utero, Lancet* **2**:1231.

Bodegård, G., 1974, On mechanoreceptor regulation of the breathing in newborn babies. Thesis, Karolinska Institutet, Stockholm, Sweden.

Bodegård, G., 1975, Control of respiration in newborn babies. III. Developmental changes of respiratory depth and rate responses to CO_2, *Acta Paediatr. Scand.* **64**:684.

Bodegård, G., 1976, Control of respiration in newborn babies. IV. Rib cage stability and respiratory regulation, *Acta Paediatr. Scand.* **65**:257.

Bodegård, G., and Schwieler, G. H., 1971, Control of respiration in newborn babies. II. The development of the thoracic reflex response to an added respiratory load, *Acta Paediatr. Scand.* **60**:181.

Bodegård, G., Schwieler, G. H., Skoglund, S., and Zetterström, R., 1969, Control of respiration in newborn babies. I. The development of the Hering–Breuer inflation reflex, *Acta Paediatr. Scand.* **58**:567.

Boyd, J. D., 1961, The inferior aorticopulmonary glomus, *Br. Med. Bull.* **17**:79.

Brady, J. P., and Ceruti, E., 1966, Chemoreceptor reflexes in the newborn infant: Effect of varying degrees of hypoxia on heart rate and ventilation in a warm environment, *J. Physiol. (London)* **184**:631.

Bryan, A. C., and Bryan, M. H., 1974, Respiratory control in newborn infants. Read before the Congress for Perinatal Medicine, Berlin.

Bryan, M. H., Kirkpatrick, S. M. L., and Bryan, A. C., 1974, *Control of Frequency in Neonates,* Society of Pediatric Research, Washington.

Campbell, E. J. M., and Howell, J. B. L., 1962, Proprioceptive control of breathing, in: *Ciba Foundation Symposium on Pulmonary Structure and Function* (A. V. S. de Reuk and M. O'Conner, eds.), Churchill, London.

Ceruti, E., 1966, Chemoreceptor reflexes in the newborn infant: Effect of cooling in the response to hypoxia, *Pediatrics* **37**:556.

Cross, K. W., and Oppé, T. E., 1953, The effect of inhalation of high and low concentrations of oxygen on the respiration of the premature infant, *J. Physiol. (London)* **117**:38.

Cross, K.W., Hooper, J.M.D., and Oppé, T. E., 1953, The effect of inhalation of carbon dioxide in air on the respiration of the fullterm and premature infant, *J. Physiol. (London)* **122**:264.

Cross, K. W., Klaus, M., Tooley, W., and Weisser, K., 1960, The response of the newborn baby to inflation of the lungs, *J. Physiol. (London)* **151**:551.

Daily, W. I. R., Klaus, M., and Meyer, H. B. P., 1969, Apnea in premature infants monitoring incidence, heart rate changes, and an effect of environmental temperature, *Pediatrics* **43**:510.

Dawes, G. S., 1974, Breathing before birth in animals and man, *N. Engl. J. Med.* **290**:557.

Dawes, G. S., Fox, H. E., Leduc, B. M., Liggins, G. C., and Richards, R. T., 1972, Respiratory movements and rapid eye movement sleep in the foetal lamb, *J. Physiol.* **220**:119.

Dement, W. C., 1972, Sleep deprivation and the organization of behavioral states, in: *Sleep and the Maturing Nervous System* (C. D. Clemente, D. P. Purpura, and F. E. Mayer, eds.), pp. 321–364, Academic Press, New York.

Dreyfus-Brisac, C., 1968, Sleep ontogenesis in early human prematurity from 24 to 27 weeks of conceptional age, *Dev. Psychobiol.* **1**:162.

Dreyfus-Brisac, C., 1970, Ontogenesis of sleep in human prematures after 32 weeks of conceptional age, *Dev. Psychobiol.* **3**:91.

Dreyfus-Brisac, C., Monod, N., Parmelee, A. H., Prechtl, H. F. R., and Schulte, F. J., 1970, For what reasons should the pediatrician follow the rapidly expanding literature on sleep? *Neuropaediatric* **1**:349.

Euler, C. v., 1966, The control of respiratory movements, in: *Breathlessness* (J. B. L. Howell, ed.), Blackwell, Oxford.

Fenner, A., Schalk, U., Hoenicke, H., Wendenburg, A., and Roehling, T., 1973, Periodic breathing in premature and neonatal babies: Incidence, breathing pattern, respiratory gas tension, response to changes in the composition of ambient air, *Pediatr. Res.* **7**:174.

Gabriel, M., Albani, M., and Schulte, F. J., 1976, Apneic spells and sleep states in preterm infants, *Pediatrics* **57**:142.

Granit, R., 1970, *The Basis of Motor Control,* Academic Press, London.

Haber, E., Kohn, K. W., Ngai, S. H., Holaday, D. A., and Wang, S. C., 1957, Localization of spontaneous respiratory neuronal activities in the medulla oblongata of the cat: A new location of the expiratory center, *Am. J. Physiol.* **190**:350.

Hathorn, M. K. S., 1975, Analysis of the rhythm of infantile breathing, *Br. Med. Bull.* **31**:8.

Hering, E., and Breuer, J., 1868, Die Selbststeuerung der Atmung durch den nervus Vagus, *Sitzber. Akad. Wiss. Wien* **57**:672.

Heymans, C., and Neil, E., 1958, *Reflexogenic Areas of the Cardiovascular System,* J. & A. Churchill, London.

Loeschke, H. H., 1974, Regulation of the extracellular pH of the brain by the lung ventilation, in: *Central Rhythmic and Regulation* (W. Umbach and H. P. Koepchen, eds.), p. 85, Hippokrates, Stuttgart.

Merill, E. G., 1970, The lateral respiratory neurones of the medulla: Their association with nucleus ambiguus, N. retroambigualis, the spinal accessory nucleus and the spinal cord, *Brain Res.* **24**:11.

Olinsky, A., Bryan, M. H., and Bryan, A. C., 1974, Influence of lung inflation on respiratory control in neonates, *J. Appl. Physiol.* **36**:426.

Parmelee, A. H., Wenner, W. H., Akiyama, Y., Schultz, M., and Stern, E., 1967, Sleep states in premature infants, *Dev. Med. Child Neurol.* **9**:70.

Parmelee, A. H., Stern, E., and Harris, M. A., 1972, Maturation of respiration in premature and young infants, *Neuropaediatrie* **3**:294.

Piiper, J., and Koepchen, H. P., 1975, *Atmung,* 2nd ed., Urban & Schwarzenberg, Munich.

Pompejano, O., 1966, Muscular afferents and motor control during sleep. in: *Muscular Afferents and Motor Control, Nobel Symposium I* (R. Granit, ed.,), p. 415, Almqvist & Wiksell, Stockholm.

Pompejano, O., 1969, Sleep mechanisms, in: *Basic Mechanisms of the Epilepsy* (H. H. Jasper, A. A. Ward, and A. Pope, eds.), p. 453, Little, Brown, Boston.

Prechtl, H. F. R., Akiyama, Y., Zinkin, P., and Grant, D. K., 1968, in: *Studies in Infancy, Clinics in Developmental Medicine 27* (R. M. Keith and M. Bax, eds.), Heinemann Medical, London.

Prechtl, H. F. R., Weinmann, H., and Akiyama, Y., 1969, Organization of physiological parameters in normal and neurologically abnormal infants, *Neuropaediatrie* **1**:101.

Purpura, D. P., 1975, Morphogenesis of visual cortex in the preterm infant, in: *Growth and Development of the Brain* (M. A. B. Brazier, ed.), Raven Press, New York.

Purves, M. J., and Biscoe, T. J., 1966, Development of chemoreceptor activity, *Br. Med. Bull.* **22**:56–60.

Richter, D. W., 1974, Analyse der rhythmischen Aktivität respiratorischer Neurone in der Medulla oblongata, Habilitationsschrift, Munich.

Roffwarg, H. P., Muzio, I. N., and Dement, W. C., 1966, Ontogenetic development of the human sleep-dream cycle, *Science* **152**:604.

Salmoiraghi, E. C., and Baumgarten, R. v., 1961, Intracellular potentials from respiratory neurones in brain stem of cat and mechanism of rhythmic respiration, *J. Neurophysiol.* **24**:203.

Schulte, F. J., 1974, Neurological development of the neonate, in: *Scientific Foundations of Paediatrics* (J. A. Davis and J. Dobbing, eds.), p. 587, Heinemann Medical, London.

Schulte, F. J., Linke, I., Michaelis, R., and Nolte, R., 1968, Electromyographic evaluation of the moro reflex in preterm, term and small-for-dates infants, *Dev. Psychobiol.* **1**:41.

Schulte, F. J., Linke, I., Michaelis, R., and Nolte, R., 1969*a*, Excitation, inhibition, and impulse conduction in spinal motoneurones of preterm, term and small for dates newborn infants, in: *Brain and Early Behaviour* (R. J. Robinson, ed.), Academic Press, London.

Schulte, F. J., Lasson, U., Parl, U., Nolte, R., and Jürgens, U. 1969*b*, Brain and behavioural maturation in newborn infants of diabetic mothers, *Neuropaediatrie* **1**:36.

Schulte, F. J., Hinze, G., and Schrempf, G., 1971, Maternal toxemia, fetal malnutrition and bioelectric brain activity of the newborn, *Neuropaediatrie* **4**:439.

Schulte, F. J., Busse, C., and Eichhorn, W., 1979, REM sleep, motoneurone inhibition and apnoeic spells in preterm infants (in preparation).

Schwieler, G. H., 1968, Respiratory regulation during postnatal development in cats and rabbits and some of its morphological substrate, *Acta Physiol. Scand., Suppl.* 304.

Sears, T. A., 1965, The role of segmental reflex mechanisms in the regulation of breathing, in: *Studies in Physiology* (D. R. Curtis and A. K. McIntyre, eds.), Lange and Springer, Berlin.

Shannon, R., and Zechmann, F. W., 1972, The reflex and mechanical response of the inspiratory muscles to an increased airflow resistance, *Resp. Physiol.* **16:**61.

Skoglund, S., 1966, Muscle afferents and motor control in the kitten, in: *Muscular Afferents and Motor Control, Nobel Symposium I* (R. Granit, ed.), Almqvist & Wiksell, Stockholm.

5

Ontogenesis of Brain Bioelectrical Activity and Sleep Organization in Neonates and Infants

C. DREYFUS-BRISAC

1. Introduction

Interest in electrophysiological studies during development has increased over the last 25 years both in animals and humans. Comparison of fetuses and neonates of different species has allowed some generalizations, showing the existence of similar trends regardless of the degree of maturation at birth (Ellingson and Rose, 1970; Ellingson, 1972; Dreyfus-Brisac, 1975).

Descriptions of the EEG of premature (implied here and later, preterm) and full-term neonates and of infants were made between 1951 and 1965 (see references in Dreyfus-Brisac,1966). Some of them confirmed the pioneer descriptions made on newborn babies and infants as early as 1938–1942 and in the detailed study of Gibbs and Gibbs (1950). Other neurophysiological studies, including evoked potentials and peripheral nerve conduction, appeared a little later. After the description by Aserinsky and Kleitman (1955) of a cyclic organization of sleep in newborn infants (subsequently described also in human adults and in cats), polygraphic studies of sleep renewed the interest for further electrophysiological studies in animals and humans, including neonates and infants (see Minkowski, 1967; Robinson, 1969; Clemente *et al.,* 1972; Berenberg *et al.,* 1974; Lairy and Salzarulo, 1975; Passouant, 1975).

This review will not deal with all the electrophysiology of development which

---

C. DREYFUS-BRISAC • Centre de Recherches de Biologie du Développement Foetal et Neonatal, Hôpital Port Royal, Université René Descartes, Paris, France.

should, to be complete, include: (1) the development of the EEG up to 16 years of age as recently described by Eeg-Olofson (1970), Petersen and Eeg-Olofson (1971), and Petersen *et al.* (1975) (see Lairy, 1975); (2) the development of visual, auditory, and somatosensory evoked potentials and the maturation of nerve conduction; (3) the study of their variations in normal and abnormal infants and children; and (4) the results of computer frequency analysis such as power spectra and coherence.

We shall limit this review to the electrophysiological development of the EEG and sleep. The EEG, an important electrophysiological variable, can no longer be considered as the unique variable to study during sleep. Extraocular movements, chin myogram, motor behavior, and respiration are now recognized as important complementary data. Additionally, it is not possible to consider an EEG study complete before 3 years of age without both waking and sleeping records. It is important to determine the state of sleep at which the recording is obtained, i.e.: is it in active sleep with REMs, in quiet sleep without REMs, or in an intermediate, undetermined, or transitional state of sleep? Recording and analysis of other aspects of sleep, as previously mentioned, are necessary to answer this question.

The complexity of sleep studies in neonates and infants is partially explained by the great interindividual variability at a given age (Ellingson, 1975). Individual differences between infants may persist and be stable for as long as a 6-month period of life (Dittrichova and Paul, 1975). These interindividual differences may be secondary to mother–infant interaction (Sander *et al.*, 1970) or to some other factors. There is no general agreement on the influence of the infant's sex upon the sleep states. Korner (1969) feels that sex is responsible for some of the differences between infants, whereas some American authors have attributed such differences to the stress following circumcision, a procedure known to modify sleep states (Emde *et al.*, 1971; Anders and Chalemian, 1974; Richard *et al.*, 1976). The position of the infant during recording may modify sleep and wakefulness (Brackbill *et al.*, 1973; Casaer *et al.*, 1973) as does the thermal environment (Parmelee *et al.*, 1962) and phototherapy (Bernuth and Janssen, 1974).

The complexity of such studies is still more apparent when one studies the reactivity to sensory stimulation. The response will differ according to the state of the infant at the time of stimulation, and to a possible change in state following the stimulation (Eisenberg, 1969; Ashton, 1971*a*; Campos and Brackbill, 1973); to intensity, duration, and quality of stimuli (Ellingson, 1958; Dittrichova and Paul, 1974); and finally to the particular physiological variable studied (Monod and Garma, 1971). The existence of habituation in newborns, first suggested by Dreyfus-Brisac *et al.* (1957), is still debated. Schaeffer (1975) reports that habituation exists, whereas Hutt *et al.* (1968) found no evidence for such a phenomenon. Some forms of stimulation may increase the duration of quiet sleep (Brackbill, 1975) or even induce sleep. This empirical statement has been verified experimentally by Wolff (1966) and Murray and Campbell (1971). They state that a continuous white noise or a rhythmic noise (cardiac beat) induces sleep in neonates. Bertini *et al.* (1970) found augmentation of active sleep in infants of 1 month submitted to a heartbeat sound.

Despite the great variability and complexity of the cerebral electrical activity in infants, and the response of the brain to various stimuli, certain statements can be made (Samson-Dollfus, 1955; Dreyfus-Brisac *et al.*, 1957):

1. The bioelectrical development of the brain, as expressed by the EEG, is similar at 40 weeks of postmenstrual* age (i.e., 38 weeks postconception), regard-

---

*In my preceding publications conceptual age (CA) was used in place of postmenstrual age (PMA).

less of the gestational age (GA) at birth. For example, two EEGs, one of a premature baby born at 30 weeks attaining 40 weeks of postmenstrual age and the other of a full-term newborn, 2 or 3 days after birth, cannot be differentiated. This is true only if the premature infant has not suffered any central nervous system injury. Although Michaelis *et al.* (1973) have pointed out a difference between the duration of the interburst interval of the "tracé alternant" pattern (see below) in such a situation, clearly, EEG development is regulated independently of its external environment *in utero* or in the incubator. This also holds true of evoked potentials, nerve conduction, and neurologic development, although it is not the case for sleep cycle organization.

2. Birth weight (BW) does not influence EEG development, unless the central nervous system has been damaged during pregnancy or at birth, e.g., by rubella or neonatal hypoglycemia. Two babies born at 40 weeks, one with a normal BW and the other with a BW of 1000–1500 g will have similar EEGs (Dreyfus-Brisac *et al.*, 1962; Dreyfus-Brisac and Minkowski, 1968; Dreyfus-Brisac, 1972). Schulte *et al.* (1971), however, have pointed out that some abnormalities of maturation can be present in small-for-date infants born after maternal toxemia. In fact these anomalies, which appear only in quiet sleep, are more probably related to metabolic anomalies than to a delay in maturation (Larroche and Korn, 1977).

3. This strong relationship between the EEG and brain maturation, regardless of the gestational age and the birth weight, has allowed a description of EEG patterns according to postmenstrual age. Such a classification may rely solely upon wave patterns (Dreyfus-Brisac, 1962) or a code developed from the wave patterns by Parmelee *et al.* (1967). This progressive transformation of the EEG is now widely recognized and was recently confirmed by Katz *et al.* (1972) and Ellingson (1972; Ellingson *et al.*, 1974). Although some lag in maturation may occur following birth in premature infants, the EEG still allows a precise assessment of gestational age at birth and during the following weeks of life, as does the neurological examination (see bibliography in Casaer and Akiyama, 1971).

4. As previously discussed, an adequate interpretation of the EEG in the newborn infant requires a definition of the infant's state. The normality or abnormality of the cerebral electrical activity must be determined in relation to the particular state in which it is recorded. To determine this state, the EEG of newborns and infants must be recorded concomitantly with a recording of heart and respiratory rates and, if possible, the myographic activity of the chin and eye movements. A careful observation of body motility, eye movements, and behavior is necessary throughout the recording. In addition, biological artifacts transmitted to the scalp or the electrodes may be recognized with the aid of the extracerebral monitors. Prolonged disorders such as apnea and bradycardia may also be diagnosed by these comprehensive polygraphic recordings (Pajot, 1974).

2. EEG and Sleep of Full-Term Neonates

In full-term (FT) neonates, polygraphic recordings confirmed the statement, first made after observations by Denissova and Figurin (1926), and by Aserinsky and Kleitman (1955), that two states of sleep can be recognized (Roffwarg *et al.*, 1964, 1966; Wolff, 1966).

These two sleep states are commonly called quiet sleep (QS) and active sleep (AS). These terms are used by those who consider ocular and body motility the most reliable variables for the identification of these two states. They are sometimes called regular or irregular sleep if respiration appears to be the most prominent character (Wolff, 1966). QS (or regular sleep) is grossly similar to slow-wave sleep of adults, and AS (or irregular sleep) to the paradoxical sleep of adults, even if important differences exist between newborn and adult sleep (see below). Prechtl uses behavioral variables to classify sleep into state 1 which corresponds to QS, and state 2 which is similar to AS.

Quiet sleep is characterized by the absence of REMs and of localized body movements, regular respiration, and tonic activity of the chin muscles (which may be absent 20% of QS time) (Eliet-Flescher and Dreyfus-Brisac, 1966; Curzi-Dascalova and Plassart, 1976) as in adults. The mean cardiac rate is regular.

Active sleep is characterized by gross and localized body movements (including penile erection), REMs (isolated or in bursts), irregular respiratory rate, absence of tonic chin myographic activity interrupted by phasic activity which occurs in conjunction with facial movements, and irregular mean cardiac rate. Crying and fussing may also interrupt the polygraphic recording of AS, generally considered as a light sleep in newborn infants. Emde and Koenig (1969) have described different states in REM sleep as drowsy REM, crying REM, and so on. They also have paid special attention to smiling during sleep. A good description of body and facial motility has been recently detailed by Gambi *et al.* (1974).

Wakefulness may be difficult to identify and differentiate from periods of active sleep with gross body movement or short periods of crying. It may be also difficult to differentiate wakefulness and quiet sleep when newborn infants sleep with their eyes open. Although occasionally seen in normal babies, it is more commonly recorded in those with central nervous system pathology.

The presence of scanning eye movements and chin myographic activity characteristic of wakefulness may help to differentiate it from active sleep. Waking records are often difficult to interpret because of the abundant movement artifacts. Some periods of quiet wakefulness, however, will allow EEG interpretation of this important state.

Detailed descriptions of the EEG and sleep organization at this period of life have been given in a bibliography up to 1971 by Lairy (1975); see also Lenard (1970*a*), Evsyukova (1971*a*), Shepavolnikov (1971), Monod *et al.* (1972), Prechtl (1974), Engel (1975), and Anders (1976). A manual describing terminology, techniques, and criteria for scoring states of sleep and wakefulness in full-term newborn infants has been edited by Anders *et al.* (1971).

In wakefulness, during the first ten days of life, the main pattern is a diffuse low-voltage EEG, consisting of more or less well-defined theta activity (4–7 cps), of 25–50 μV. It has been called "activité moyenne" by Samson-Dollfus (1955). Central bursts of 7–8 cps can occur for 1–3 sec. Slower rhythms (1–4 cps) (25–100 μV) may also appear admixed with the low-voltage theta. This delta activity is often overlooked when its voltage is less than 25 μV. This activity is normally recorded in healthy babies. However, as in adults, the limits of normality in wakefulness are difficult to assess. In some normal newborn infants a very low-voltage EEG (<25 μV) may be recorded during the first days of life (Monod *et al.*, 1972) and, with increased gain, this activity will appear normal. The degree of rhythmicity may also vary considerably in this age group. A slight asymmetry is generally considered insignificant. However, Varner *et al.* (1976) consider an amplitude asymmetry exceeding 35% to be abnormal. Asymmetry may appear during photic driving, and for Crowell *et al.* (1973) such an asymmetry, normally present in most neonates, is considered as a sign of functional cerebral dominance. This photic driving, which is

161

*ONTOGENESIS OF
BRAIN BIOELECTRICAL
ACTIVITY AND SLEEP
ORGANIZATION IN
NEONATES AND
INFANTS*

not routinely performed in neonatal EEG, may be obtained with low-frequency flashes at 3–5 per sec.

In active sleep, the same EEG (Figure 1A) patterns already described in wakefulness exist. AS occurs either between two periods of QS or at the onset of sleep. In the former, the EEG is characterized by a low-voltage pattern which is somewhat more rhythmic than in the waking state; sleep onset REM, which is common in the newborn infant, is characterized by this low-voltage EEG pattern generally superimposed with slow waves (Figure 1B). Besides this difference in EEG pattern, the AS at the onset of sleep is also characterized by a lower density of ocular movements (Monod and Dreyfus-Brisac, 1965; Petre-Quadens, 1966, 1969; Evsyukova, 1971*b;* Cianchetti *et al.,* 1974). Progressive changes in cerebral activity may occur at the end of a period of AS, with the appearance of continuous slow waves (Monod and Pajot, 1965; Prechtl and Lenard, 1967; Dittrichova, 1969; Dittrichova *et al.,* 1972).

In quiet sleep, the typical EEG pattern is a "tracé alternant" (Figure 1D): bilaterally synchronous 3- to 5-sec bursts of slow waves (1–3 cps, 50–100 μV). The interburst activity is similar to the low-voltage pattern seen in wakefulness and AS and has the same duration as the bursts. At the beginning or end of QS, a continuous slow-wave EEG pattern (Figure 1C) (1–3 cps, 50–100 μV), is often seen in normal infants. The slow waves may also occur for several minutes during any portion of QS (Monod and Pajot, 1965; Prechtl *et al.,* 1968; Dittrichova, 1969; Dittrichova *et al.,* 1972).

Most of the abnormalities appearing in babies with pathology occur in QS. The limits of normality are particularly difficult to establish in this state in part due to the variety of transient or paroxysmal patterns which can appear in the record of normal neonates (Crawley and Kellaway, 1963; Kellaway and Crawley, 1964; Dreyfus-Brisac, 1966; Ellingson, 1967; Monod *et al.,* 1972; Engel, 1975). These include bilateral frontal sharp transients, monomorphic or polymorphic bifrontal slow waves, rapid low-voltage occipital rhythms (Figure 1D), uni- or bilateral rolandic sharp waves (humps) (Figure 2), low-voltage ill-defined spindles, and short bursts of rhythmic central alpha-like activity (Engel, 1974). A low-voltage EEG may also be seen for a few days.

The duration of periods of QS is fairly constant at about 20 min. The duration of periods of AS, which represent 50–60% of newborn sleep, varies from 10 to 45 or even 50 min. The first period of AS at onset of sleep is shorter (10–20 mins) (Ashton, 1971*b;* Fabiani *et al.,* 1975). The onset of sleep may also be in QS (Fabiani *et al.,* 1975).

Even during well-defined QS or AS periods, variations of the respiratory and cardiac rates and frequency of eye movements may occur with changes in the EEG pattern (Monod and Pajot, 1965; Prechtl *et al.,* 1968; Dittrichova, 1969; Paul *et al.,* 1973; Prechtl, 1974).

There are also progressive variations which are particularly striking during short periods of transition between two well-defined states (Prechtl *et al.,* 1968; Dittrichova *et al.,* 1972; Campbell and Raeburn, 1973). These periods have been carefully analyzed by Monod and Curzi-Dascalova (1973). The dynamic aspect of organization of sleep, and the complexity of the physiological mechanisms of sleep regulation appear during these periods which have a mean duration of 5 min. Besides these periods of "transitional sleep," there are other periods of sleep which do not fulfill all the criteria of AS or QS: They do not represent a high percentage of sleep time in normal neonates, but are often seen in babies with pathological states.

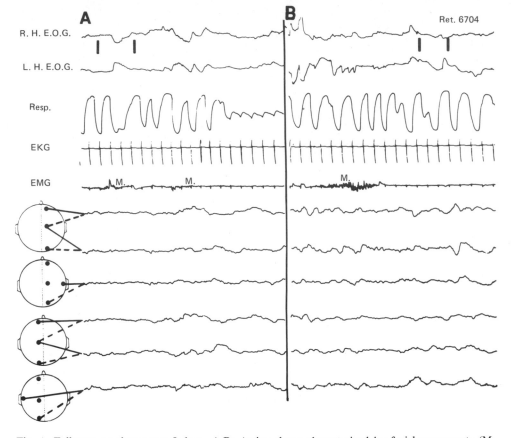

Fig. 1. Full-term newborn; age 5 days. A,B: Active sleep; characterized by facial movements (M: mouthing) and REMs (indicated by vertical bars), irregular respiration, absence of tonic myographic activity at chin level. Different EEG patterns in A and B. A: Low-voltage EEG. B: Low-voltage superimposed on slow waves. C,D: Quiet sleep; without REMs and body movements, regular respiration. At the beginning of QS, tonic myographic activity at the chin level is absent and the EEG pattern consists of continuous slow waves (C). Myographic activity and "tracé alternant" appear a little later (D) with occipital fast rhythm. R. H. E.O.G. = right horizontal eye movements; L. H. E.O.G. = left horizontal eye movements; Resp. = respiratory rate; EKG = electrocardiogram; EMG = myographic activity recorded at the chin level. Calibration: 50 μV and 1 sec.

Their duration varies according to the number of variables used for classification of sleep states (Ellingson, 1975).

Important differences between neonate and adult sleep exist. The newborn infant's EEG is characterized by the "tracé alternant" pattern, never seen in normal adult sleep, and lacks the well-developed sleep spindles of the adult. Curzi-Dascalova et al. (1973) have also shown that the variations in skin potentials (recorded from the extremities) predominate during AS with REMs in infants and during non-REM sleep in adults. This had previously been noted by Bell (1970) who attempted to stage sleep by variations in the skin potentials. The respiratory rhythm is also characteristic in neonates: Short periods of apnea (≤ 5 sec) occur more frequently in AS than in QS (Monod et al., 1976; Gabriel et al., 1976; Curzi et al., 1976). They are of central or obstructive type and differ from apnea seen in babies with pathological states by their shorter duration and lack of accompanying clinical

163

**ONTOGENESIS OF
BRAIN BIOELECTRICAL
ACTIVITY AND SLEEP
ORGANIZATION IN
NEONATES AND
INFANTS**

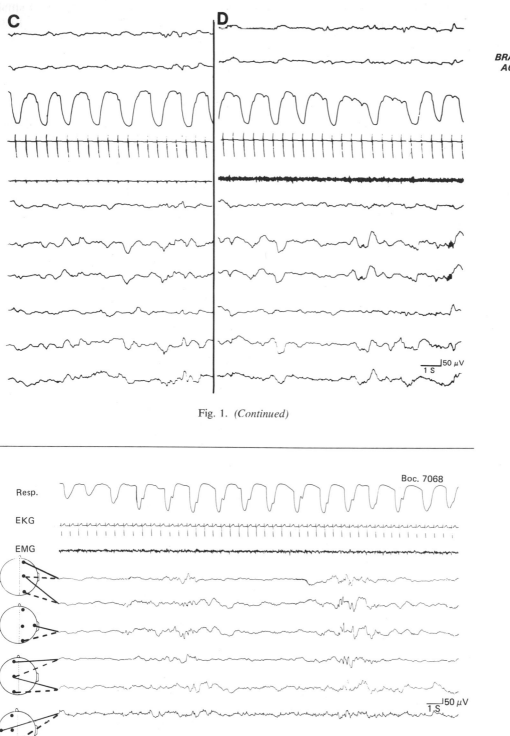

Fig. 1. *(Continued)*

Fig. 2. Full-term newborn; age 5 days. Quiet sleep; "tracé alternant". This EEG pattern is different from that in Figure 1D. Note the central spikes. Figures 1D and 2 illustrate the interindividual variability in full-term newborns. Legends and calibration: See Figure 1.

symptoms of seizures such as, for example, opening of the eyes, mouthing, or small localized clonic movements (Dreyfus-Brisac and Monod, 1964; Monod *et al.,* 1969; Schulte and Jurgens, 1969).

3. EEG and Sleep of Preterm Infants

The sleep of premature newborn infants has been studied less extensively than that of full-term infants.

3.1. Fetuses and Very Early Premature Newborn Infants

Electrical activity appears initially in subcortical regions. Bergström (1969) recorded continuous activity in the pons of fetuses at 70 days gestational age, a few minutes after separation from the maternal circulation. The amplitude of this activity increases until 120 days of age and subsequently remains stable. Intermittent wave complexes are also recorded in the rostral part of the brain stem and of the hippocampus from 17 weeks gestational age.

Electrical activity may be recorded from the scalp of 24- to 27-week-old fetuses surviving for a few hours or a few days. This EEG activity consists of a high-amplitude discontinuous polymorphic pattern (Dreyfus-Brisac, 1962, 1967, 1968). Occasional periods of more continuous EEG activity, up to 40 sec or even 1–2 min may occur. This more continuous activity may occur concomitantly with intense body motility and crying. Fetuses are never alert at this age, and a differentiation between wakefulness and sleep cannot be made. Body motility is continuous and consists of a variety of jerks and noncoordinated movements; limb tone is low (Saint-Anne Dargassies, 1966). Eyelids are closed and eye movements are rare. At the chin and tongue, clonic discharges of movements lasting 3–10 sec occur with regular frequency (2–3/sec). Cardiac and respiratory rates are regular. This behavior is similar to that described in chicken embryos (Hamburger, 1971) and rat fetuses (Narayan *et al.,* 1971). Polygraphic recording and observation of such human fetuses after birth did not give any support to the existence of a brain rest activity cycle (BRAC) at this early period of life. Such a BRAC has been detected by Sterman and Hoppenbrouwers (1971), recording movements of the fetus *in utero.* The differences found between *in utero* and *ex utero* recordings could be explained by the influence of the mother's own rest activity cycle upon the fetus recorded *in utero.*

3.2. After 28 Weeks Gestational Age

EEG and sleep studies have been carried out by Dreyfus-Brisac *et al.* (1957), Dreyfus-Brisac (1962, 1966, 1967, 1968, 1970), Polikanina (1966), Ellingson (1967), Parmelee *et al.* (1967, 1968*a,b,* 1969), Parmelee (1974, 1975), Graziani *et al.* (1968, 1974), Goldie *et al.* (1971), and the group headed by Watanabe (Watanabe and Iwase, 1972; Watanabe *et al.,* 1972, 1974).

Parmelee and his colleagues were able to demonstrate a periodicity in EEG (Parmelee *et al.,* 1969) and in sleep organization (Parmelee *et al.,* 1972).

Long-term survival is possible in premature infants after 28 weeks gestational age. The complex EEG pattern found before 28 weeks is replaced by a much

165

*ONTOGENESIS OF
BRAIN BIOELECTRICAL
ACTIVITY AND SLEEP
ORGANIZATION IN
NEONATES AND
INFANTS*

simpler one consisting of bursts of theta waves (4–6 cps), lasting 1 or 2 sec with a voltage between 25 and 100 μV, separated by long quiescent periods. During a short period of development (28–29 weeks gestational age) this theta activity appears synchronously in all areas of the same hemisphere. This hemispheric isosynchronism disappears after 29 weeks of gestation. Bursts of theta rhythms, however, remain the predominant activity until 30 or 31 weeks of age. During this latter period, a new EEG pattern, which will predominate between 32 and 36 weeks of life, appears. This typical pattern is composed of slow activity (1 cps, 25–100 μV) with superimposed rapid rhythms (10–14 cps, 10–20 μV). At 30–31 weeks gestational age the EEG is composed of 3- to 10-sec bursts of this characteristic pattern and the persistence of the theta waves previously described (Figure 3).

At this age (28–31 weeks) the EEG patterns are similar in wakefulness and sleep. Wakefulness is not yet clearly established, and states of sleep are not easily differentiated (Monod and Garma, 1971). The relationship between a continuous EEG pattern and AS on the one hand, and a discontinuous EEG pattern and QS on the other, which will characterize the sleep cycle organization between 32 and 38 weeks gestational age is still very poor. Body motility consisting of localized movements of face and limbs is relatively persistent, only being interrupted for short periods (Gesell and Amatruda, 1945). Behavior characteristics of AS and QS in older infants are often admixed at this period of life. Rapid eye movements may occur simultaneously with clonic discharge of chin movements. Relations between behaviorally assessed sleep states and EEG patterns are not clearly established. Irregular respiration (typical of AS in older infants) may appear with the discontinuous EEG pattern characteristic of QS. Most of the record is discontinuous (around 60%), as if the brain structures were still unable to generate a sustained activity. A large percentage of sleep time cannot be classified and has been described as "transitional" or "indeterminate."

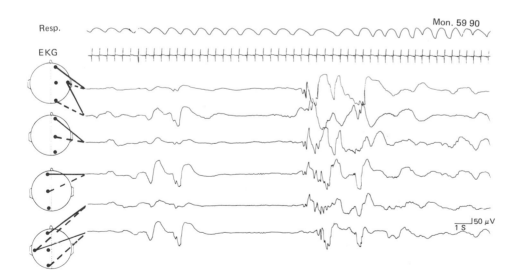

Fig. 3. Premature infant born at 28 weeks GA; age 4 days. Discontinuous EEG made of bursts of slow waves which predominate in the occipital areas and of diffuse bursts of theta waves. Between the bursts, the EEG is quite silent. The respiration is irregular. Legends and calibration: See Figure 1.

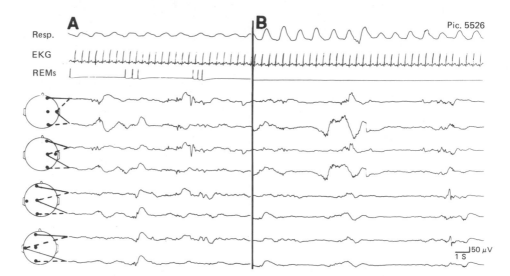

Fig. 4. Premature infant born at 31 weeks GA; PMA 35 weeks (1 month of life). A: Active sleep with REMs (indicated by a signal on the third channel) and a continuous slow-wave EEG pattern superimposed with fast rhythms. Note a sharp frontal transient. B: Quiet sleep; absence of REMs. The EEG pattern is discontinuous. Note the asynergy between both hemispheres, particularly in QS. Legends and calibration: See Figure 1.

3.3. From 32 Weeks of Age

Slow waves with superimposed rapid rhythms (Figure 4), similar to those already described at 30 weeks, are seen in the occipital, temporal, and central areas. Theta bursts are no longer present. The amplitude of the slow waves and rapid rhythms which appear in wakefulness and sleep varies from one infant to another. A continuous EEG pattern is present during the short periods of wakefulness which begin to appear at this age as well as during AS. On the contrary, EEG activity becomes discontinuous during QS with a delta activity having a similar morphology to that of AS, interrupted by silent periods of variable duration. The degree of discontinuity varies with depth of sleep, increasing in deep sleep. Some sharp waves may also be seen in deep quiet sleep.

The earlier characteristics of quiet sleep are a discontinuous record, absence of REMs, and presence of clonic movements of the chin. The amount and quality of body motility, as well as the chin myogram, are not well correlated with the two states of sleep at this age.

3.4. From 36 Weeks of Age

Three different EEG patterns are seen during wakefulness, active sleep, and quiet sleep: (1) wakefulness—a diffuse low-voltage EEG pattern similar to that of full-term newborn infants; (2) active sleep—a continuous slow-wave pattern with superimposed low-voltage rapid rhythms at 10–14 cps of maximum amplitude in the occipital areas; and (3) quiet sleep—a discontinuous EEG pattern consisting of bursts of variable morphology and duration.

Between 36 and 38 weeks the waking record is similar to the waking record of full-term newborn infants; the AS record is similar to the AS record found between 32 and 35 weeks, whereas the discontinuous pattern of QS is slightly different from the record seen earlier, being sharper and less discontinuous.

167

ONTOGENESIS OF
BRAIN BIOELECTRICAL
ACTIVITY AND SLEEP
ORGANIZATION IN
NEONATES AND
INFANTS

3.5. From 38 Weeks of Age

A low-voltage EEG pattern, often accompanied by superimposed slow-wave EEG, may be present in AS. In QS, a "tracé alternant" or a continuous slow activity may be seen.

3.6. Development of Some Variables from 28 to 40 Weeks of Age

Watanabe and Iwase (1972) paid special attention to the fast rhythms seen in the EEG of premature infants (Figures 4 and 5). They state that besides rapid rhythms superimposed on occipital slow waves, central and temporal fast rhythms exist even near term. Lombroso (1975) reported a higher number of rapid rhythms in QS than in AS after 35 weeks postmenstrual age.

The degree of synergy or synchrony between hemispheric bursts of activity in the two hemispheres has not been precisely studied. In a recent paper by Lombroso (1975), the percentage of synchrony between hemispheres in QS is reported to vary from 50–70% between 28 and 32 weeks to 100% at term.

Motor reactivity to auditory stimuli diminishes with increasing postmenstrual age in QS, whereas palpebral responses increase with postmenstrual age in AS. This difference in reactivity in the two sleep states begins at 32 weeks postmenstrual age (Monod and Garma, 1971). Vertex spikes provoked by auditory stimuli are present and easily recognized in QS before 36 weeks postmenstrual age (Ellingson and Rose, 1970; Uzemaki and Morrell, 1970).

Developmental studies of the respiratory rate have been made by Iwase and Watanabe (1971) and Parmelee *et al.* (1972). The relation between "periodic" respiration and states of sleep is complex in all age groups. Its percentage in total sleep time remains constant between 30 weeks and 37 weeks and decreases thereafter (Dreyfus-Brisac, 1970; Curzi-Dascalova *et al.*, 1976). Isolated periods of apnea are more frequent in AS.

The electrodermal activity appears for the first time at 28 weeks postmenstrual age with the onset of sweat-gland function at the plantar and palmar levels. As in full-term neonates, electrodermal responses predominate in active sleep. Their number increases between 28 and 32 weeks and thereafter remains stable until term (Curzi-Dascalova *et al.*, 1973).

From a behavioral point of view, it would be of interest to know the ontogenetic relationships between the rhythmic discharges of chin movements (present in fetuses and in QS of prematurely born children) and nonnutritive sucking, described in sleeping newborns (Wolff, 1968; Dreir and Wolff, 1972) and in premature babies (Dubignon *et al.*, 1969).

Sleep organization is not easy to describe in premature babies between 32 and 38 weeks of age. Periods of AS and QS are generally short and vary in duration (Stern *et al.*, 1973). The percentage of AS and QS at this early period is difficult to quantify accurately, and depends upon the variables used for classification. It appears that AS is well developed before QS, the latter being fully developed only after 35–36 weeks (Dreyfus-Brisac, 1967, 1970). QS, when defined by the following criteria—eyes closed, lack of body movements, and regular respiration—accounts

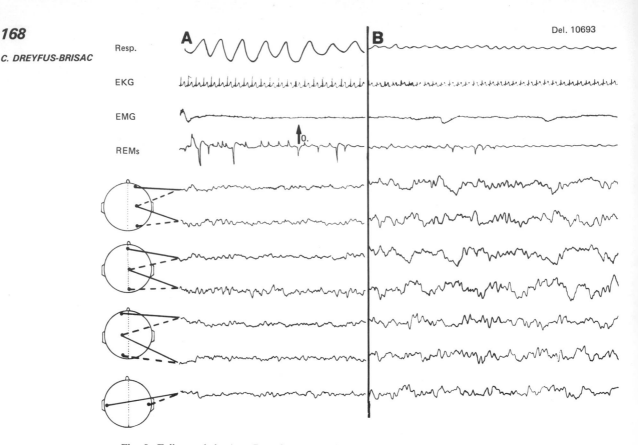

Fig. 5. Full-term baby; age 7 weeks. A: Awake and alert. Scanning movement of the eyes. When the infant opens the eyes (indicated by an arrow), no blocking reaction. B: Onset of sleep, in AS. Eye movements are present. C: AS, with body movements (M, mouthing) and eye movements. Saw-tooth waves are present on the median occipital area. D: QS with a regular respiration, a tonic activity at the chin level, and a slow-wave EEG pattern. Vertex spindle bursts appear. Eye movements are recorded by an accelerometer on the third channel. Legends and calibration: See Figure 1.

for 25–35% of sleep time at 38–40 weeks postmenstrual age (Parmelee *et al.,* 1972). Periods of transitional or indeterminate sleep (i.e., of a state which cannot be classified according to a definition based on ocular movements, body motility, and respiration rate), diminish after 34 weeks postmenstrual age. Sleep periodicity remains stable during the remainder of life after a rapid increase in duration of the periods between 32 and 36 weeks of age (Stern *et al.,* 1973).

During early phases of development, the rapid evolution of the brain (development of sulci, thickening of the cortex, etc.) is paralleled by a rapid transformation of EEG patterns and sleep organization.

Interspecies comparisons are of great interest. For example, in the kitten, Hoppenbrouwers and Sterman (1975) have demonstrated that, prior to 4 weeks of age, polygraphic patterns of sleep differ significantly from those of adult cats and that the most outstanding feature of polygraphic data between birth and 4 weeks of age was the gradual coalescence of initially discordant physiological measures of sleep states. We have noted that the same statement has been made for humans. However, an important difference exists between the kitten and the premature human baby: Between 32 and 35 weeks postmenstrual age, the EEG activity during

169

*ONTOGENESIS OF
BRAIN BIOELECTRICAL
ACTIVITY AND SLEEP
ORGANIZATION IN
NEONATES AND
INFANTS*

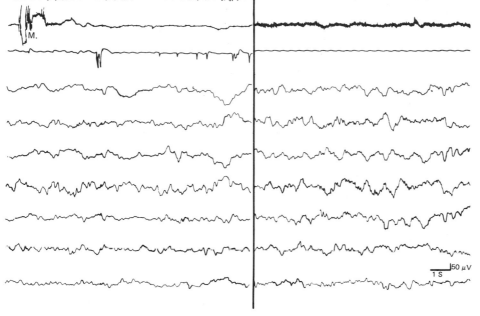

Fig. 5. *(Continued)*

AS and wakefulness is similar and differs significantly from that of QS. In cats AS and QS records are similar during the first days of life (Valatx *et al.,* 1964).

4. Development of EEG and Sleep Organization during the First Three Years of Life

The pioneer descriptions of EEG maturation in wakefulness were given by Lindsley (1939) and Henry (1944). Gibbs and Gibbs (1950) described the maturation of sleep patterns. Kellaway (1957) gave the first complete description of EEG development. The main publications in this field are those of Brandt and Brandt (1955), Dreyfus-Brisac and Blanc (1956), Dreyfus-Brisac *et al.* (1957, 1958), Drey-fus-Brisac and Curzi-Dascalova (1975), Dumermuth (1965), Ellingson (1967), Ohta-hara *et al.* (1967), Hagne (1968, 1972), Eeg-Olofson (1970), Pampiglione (1965, 1971), Petersen *et al.* (1975), Petersen and Eeg-Olofson (1971), Netchine and Lairy (1975). The preceding publications deal mainly with EEG patterns. Sleep organization has been studied primarily during the first year of life by Stern *et al.* (1973), Dittrichova *et al.* (1972), Paul *et al.* (1973), and Koslacz-Folga (1973). Recent review articles have been published by Anders (1976) and Passouant (1975).

Careful longitudinal studies have been carried out in Sweden (Hagne, 1972) and mainly in Czechoslovakia (Dittrichova *et al.,* 1972; Paul *et al.,* 1973) and Poland (Kolacz-Folga, 1973) on infants up to 6 months of age living in institutions for mothers and infants, and staying in such institutions for social reasons.

EEG and sleep cycle change more slowly during the first year of life compared to the rapid maturation that occurs in the premature baby. However, their modifications are striking.

The first of the important events in EEG maturation is the disappearance of the "tracé alternant," which was the typical pattern seen in QS of full-term neonates. By 3–4 weeks, the EEG of QS is made up of a continuous slow-wave pattern (Dittrichova, 1969; Kolacz-Folga, 1973) (Figure 5D).

The second important milestone is the appearance of well-developed central sleep spindles around 6 weeks of age. Ill-defined spindles have been described in younger babies by Metcalf (1970). The spindles appearing in the second month of life are grossly similar to those seen in adults although their frequency is often a little slower and more variable (10–16 cps), and their morphology is sharper. They appear immediately after the onset of QS and persist throughout QS, disappearing with the beginning of AS. Careful studies of sleep spindles in infancy have been made by Lenard (1970*b*), and Metcalf (1970). According to Lenard (1970*b*), low-voltage spindles of short duration appear during the second month of life and increase in amplitude and duration during the subsequent month with generally short interburst intervals persisting until 7 months of age. Hagne (1972) describes nearly continuous spindling between 3 and 6 months of life with bursts of at least 10-sec duration. These spindle bursts are sometimes asymmetric in amplitude and may appear asynchronously in the two hemispheres (Figure 6). Interhemispheric asynchrony of sleep spindles may occur in normal babies during the first year of life and even later (Figures 6 and 9B).

The third major change in the EEG occurs around 3 months of age and is characterized by the appearance of ill-defined rhythmic activity in the occipital area with a dominant frequency of 3–4 cps which blocks with eye opening. EEG changes little between 1 and 3 months of life and is characterized by diffuse irregular 25- to

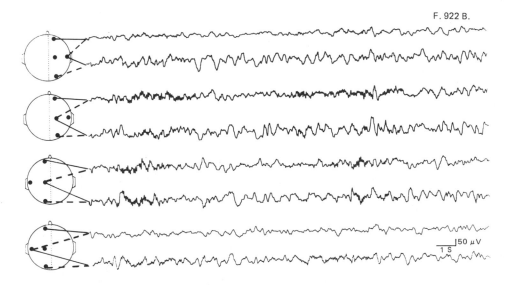

F. 922 B.

50 μV
1 S

Fig. 6. EEG pattern in quiet sleep at 3 months of age with long-duration burst of spindles appearing asynchronously on both hemispheres. Calibration: See Figure 1.

171

***ONTOGENESIS OF
BRAIN BIOELECTRICAL
ACTIVITY AND SLEEP
ORGANIZATION IN
NEONATES AND
INFANTS***

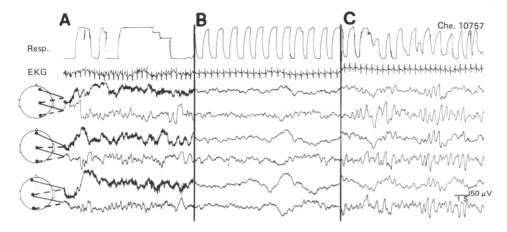

Fig. 7. EEG patterns; age 5 months. A: Wakefulness, eye closed (during periods of crying, with a myographic activity in the frontal leads). B: Wakefulness, eyes open. A blocking reaction exists. C: Drowsiness with a theta rhythm EEG, often present at this age, where the hypnagogic theta diffuse activity is not yet fully developed (see the following figure). Calibration: See Figure 1.

35-μV slow activity (2–6 cps) (Figures 5 and 8A). Occasionally one notes amplitudes higher than those of the newborn period and an inconstant rhythmicity in the occipital regions.

The age of appearance of a fairly well-developed occipital rhythm in the theta frequency band (Figure 7A,B) is not precisely defined in the literature and ranges from 4 months (Hagne, 1968) to 5 months (Dreyfus-Brisac and Blanc, 1956; Dreyfus-Brisac *et al.*, 1958; Pampiglione, 1965, 1971). Pampiglione (1971) has also described significant interindividual and interracial variations in this occipital activity. The frequency of this occipital rhythm accelerates after 4–5 months of age reaching 6–7 cps at 1 year when it is often intermingled with 4- to 5-cps waves. It is not surprising that there are differences of opinion as to the precise age of appearance of the occipital rhythm in the awake infant. This rhythm appears primarily during periods when the eyes are closed. Because voluntary eye closure is difficult to obtain in this age group, some authors have recorded awake infants with the eyes open (Hagne, 1968). Closure of the eyes can be obtained, however, either passively with a slight pressure on the eyelids (not always well accepted) or during periods of crying when infants always spontaneously close their eyes.

The fourth important change, which occurs at 6–8 months of age, is the appearance during drowsiness of a high-amplitude (75- to 200-μV) theta rhythm at a frequency of 4 cps (hypnagogic hypersynchrony, (Figure 8B). This pattern was first well described by Gibbs and Gibbs (1950). Prior to that age, a well-defined drowsy pattern cannot be recognized (Samson-Dollfus *et al.*, 1964) (Figures 7C and 8A).

Other maturational changes occur during sleep in the first year of life. A rhythmic 3- to 4-cps pattern may transiently appear in the central regions at the end of the first month of life in QS. Repetitive 75-μV sharp waves are noted in the occipital areas during AS (Figures 5C and 9A). These so-called "saw-tooth" waves have a frequency of 2 cps at 6 weeks of age and 2–4 cps at 12–16 weeks (Dittrichova *et al.*, 1972). Central humps appear around 4–5 months of age (Metcalf *et al.*, 1971) either as isolated sharp waves or in 1-sec bursts of 3–4 repetitive sharp waves. They often appear asynchronously in the two hemispheres. Central humps are present in

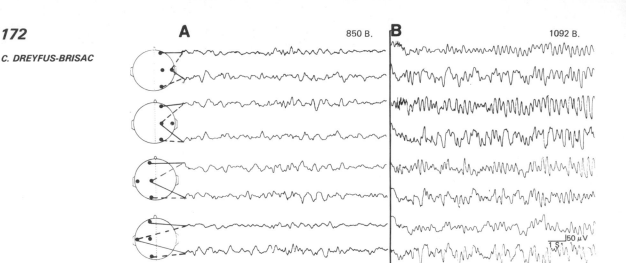

Fig. 8. EEG patterns in drowsiness. A at 3 months of age, B at 15 months of age. In A the EEG pattern does not differ clearly from that of wakefulness (see Figure 5A). In B the hypnagogic theta rhythm at 4 cps is fully developed. Calibration: See Figure 1.

the records of full-term infants, and the reason for their disappearance during the next 3–4 months of life is unknown. The relationship between the central sleep patterns seen in neonates and those of older infants is unclear. Central humps have similar morphological and topographical characteristics in these two age groups and appear only in QS. There is less similarity between the faster rhythms appearing during sleep, i.e., central spindles of infancy and the fast rhythms of premature infants [called ''spindle-like'' by Watanabe and Iwase (1972)]. The former occur in the central areas during QS, whereas the latter appear during all sleep stages and even in wakefulness, particularly in the occipital regions.

Reactivity in sleep has not been systematically investigated. Dreyfus-Brisac *et al*. (1958) have shown that reactivity differs according to depth of sleep and to intensity and repetition of stimuli. After stimulation, the EEG response consists of diffuse slow waves or diffuse sharp waves, generally of short duration but sometimes appearing as repetitive bursts lasting up to 10 sec. After an arousal provoked by stimulation, hypersynchronous EEG pattern appears, which is similar to the pattern seen in drowsiness (Figure 9C). When awakening follows this arousal reaction, the occipital theta rhythm of wakefulness is generally slower than the theta rhythm seen in wakefulness preceding the onset of sleep (Brandt and Brandt, 1955). The frequency of the theta rhythm is thus related to the quality of the waking state.

Photic driving is easily obtained at a flash frequency of 3–5 sec. High-intensity flashes are necessary to obtain driving at higher frequencies of stimulation (Vitova and Hrbek, 1970).

The distribution of skin potential responses (SPR) in sleeping infants after 3 months of age is similar to that of the adult; until this age, SPR are more frequent in AS than in QS. After 3 months of age, they are more frequent in QS than in AS. This modification of distribution from a neonatal type to an adult type is temporally related to the appearance of well-defined spindles and to the rapid increase of SPR in QS around 1–2 months of age (Curzi-Dascalova and Dreyfus-Brisac, 1976).

173

*ONTOGENESIS OF
BRAIN BIOELECTRICAL
ACTIVITY AND SLEEP
ORGANIZATION IN
NEONATES AND
INFANTS*

Sleep onset in neonates and during the first 2–3 months of life is usually by AS; after 3 months of age QS characterizes sleep onset. This change from AS-onset to QS-onset is not clearly related to age but is influenced by environmental conditions, e.g., the recording being in a home or laboratory setting. Even under similar recording conditions, the sleep onset pattern may change from one session to another (Bernstein *et al.*, 1973; Kligman *et al.*, 1975).

The classification of sleep states is not well established at this period of life. When is it possible to identify REM and AS, non-REM and QS, and to distinguish states 1, 2, 3, and 4 during QS? Further collaborative discussion will be needed in order to define the EEG criteria (amplitude and frequency of delta waves, number of spindles, etc.) for classifying QS during the first year of life into states 1, 2, 3, and 4. Some authors, for example Lenard (1972) and Metcalf and Jordan (1972), distinguish state 2 of non-REM sleep as early as 2–4 months. We agree with those who do not use the adult classification until 1 year of age.

The organization of the sleep cycle is part of the circadian rhythm. It is no longer acceptable to study only active and quiet sleep during the day; night recordings are also necessary. Paul *et al.* (1973) have shown that, in the course of a night, changes occur in some of the quiet sleep variables: The respiratory rate decreases gradually, the number of slow waves increases, and spindles become less frequent. During the night there is also a linear increase in the length of REM bursts during successive periods of nocturnal AS (Ornitz *et al.*, 1971). With increasing age, percentage of REM sleep decreases rapidly from around 50% in the full-term neonate to 40% at 3–5 months and 30% at 12 months (Roffwarg *et al.*, 1966; Stern *et al.*, 1969). The length of AS periods decreases significantly with age, from 30 min at 2 weeks of age to 15 min at 6 weeks, but the absolute number of REMs remains unchanged because the same number of eye movements occurs during a shorter period (Dittrichova *et al.*, 1972, 1976; Dittrichova and Paul, 1975).

Periods of quiet sleep are more prolonged than in newborn infants and periodic-

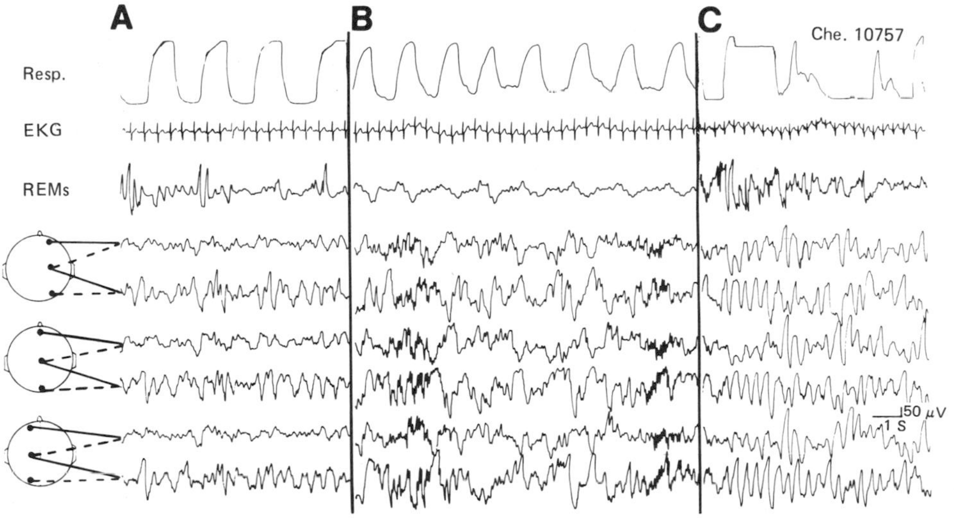

Fig. 9. EEG patterns at 5½ months of age. A: Active sleep with REMs, note the "saw-tooth" waves on the occipital areas. B: Quiet sleep with spindle bursts and a delta occipital rhythm. C: Arousal after an auditory stimulation. The EEG pattern at arousal is similar to the hypnagogic activity shown in Figure 8B. Legends and calibration: See Figure 1.

ity of both QS and AS remains stable and is the same at 3 and 8 months as that of full-term neonates (Stern *et al.,* 1973). The coalescence of variables of QS and AS may still be imperfect.

The transition between the two states of sleep has been carefully analyzed by Samson-Dollfus and Poussin (1973). It is similar to that of neonates in duration (4–5 min) and in the order of changes of the different variables: REMs disappear first when entering QS and reappear last when entering AS (Dittrichova *et al.,* 1972). The same pattern is found at 18 and 30 months, but the duration of the transitional periods between QS and AS is longer (8–10 min), whereas the duration of periods of transition from AS to QS has the same duration as in the full-term neonate (5 min).

In wakefulness and sleep, individual variations are important, and in the same infant unexpected variations from one record to another may appear (Hagne, 1972). However, the normal range of variation in the appearance of the occipital rhythm in wakefulness, the hypnagogic activity in drowsiness, and sleep spindles has now been well defined. It is now known that absence of spindles after 3 months of age, or an occipital rhythm after 5 months, is unequivocally abnormal, as is a hypnagogic rhythm less than 3 cps or an occipital rhythm slower than 4 cps at 12 months of age (Monod and Ducas, 1968). Interindividual differences have been also found for the different variables of active and quiet sleep (Dittrichova *et al.,* 1976).

4.2. From 1 to 3 Years of Age

Between 1 and 3 years of age, the maturation of the EEG patterns is progressive. The frequency of the occipital rhythm accelerates with occasional short periods of stabilization. Sleep patterns do not show important changes except in central humps (Figure 10). Torres (1970) found that central humps become more symmetrical than during the first year and are sharper, more repetitive, and intermingled with slow waves. The percentage of asynchronous spindle bursts,

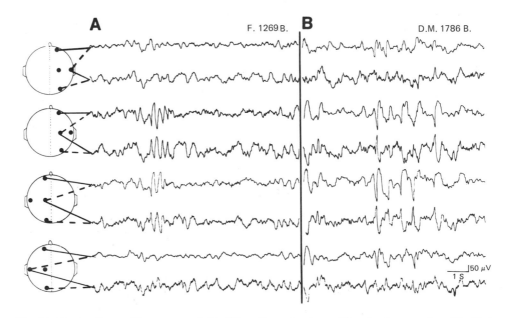

Fig. 10. Bursts of central humps recorded in QS at 18 months of age (A) and at 3 years of age (B). Their shape and duration differ clearly. At 3 years, central humps are sharper and admixed with slow waves.

175

*ONTOGENESIS OF
BRAIN BIOELECTRICAL
ACTIVITY AND SLEEP
ORGANIZATION IN
NEONATES AND
INFANTS*

which can reach 76% during the first year of life, falls to 10% at 2 years of age. The duration of central spindle bursts decreases, and frontal spindles of lower frequency than those in the central areas appear. At 2 years of age, spindles are no longer present in the lightest or very deepest stages of sleep. At 2 years a classification of slow-wave sleep into stages 1, 2, 3, and 4 is easily made. The amplitude of slow waves in sleep may exceed 200 μV. Around 3 years, the hypnagogic 4-cps rhythm of drowsiness, which has appeared around 7 months, disappears, but still persists at arousal. The hypnagogic rhythm is often discontinuous after 2 years of age. In some children, hypnagogic hypersynchrony of drowsiness is absent during infancy, a diffuse theta rhythm of 50 μV assuming its place.

Variability of the EEG is more marked between 1 and 3 years than during the first years of life, particularly in wakefulness. This is due to the variability of acceleration of the occipital frequency which can reach 11 cps at 3 years in some normal children. This dominant activity is sometimes intermingled with a 2.5- to 4.5-cps occipital activity (Eeg-Olofson, 1970), the latter appearing to be genetically inherited (Doose and Gerken, 1972).

Paroxysmal spikes are sometimes seen in records of children of 2–3 years of age free of clinical symptoms. A prolonged follow-up of such infants is necessary to confirm the benign significance of such paroxysmal activity.

Sleep cycling at night changes little during this period. Kohler *et al.* (1968) state that nocturnal sleep cycles are well organized with 7–8 periods of REM sleep having a mean length of 23 min. Most of the REM sleep time (48%) occurs at the end of the night; its mean percentage for the entire night is 29%, which is still higher than in adults. REM latency and burst REM organization increase with age (Tanguay *et al.*, 1974). Most of state 4 (66%) occurs during the first third of the night. The shifts of stages are smoother and less frequent (3.5/hr) than in young adults (5.3/hr). Similar observations have been made by Hirai *et al.* (1968). A high percentage of active sleep (50%) is still found by these authors until 3 years.

5. Conclusions

To conclude this review, it is of interest to point out some important facts:

1. Ontogenetic studies of EEG and sleep organization have demonstrated that the transformation which occurs during maturation is not regular and continuous. The different structures of the central nervous system mature at different rates. Histological and enzymatic studies of the brain have clearly shown that different areas of the central nervous system differ in maturation. Behavioral and bioelectrical changes would appear to reflect these maturational changes seen at the anatomical level.

2. Important electroencephalographic landmarks have been described in this maturational process; e.g., at 36 weeks gestational age a clear distinction appears between the EEG of wakefulness and that of sleep, and 3 months after term a theta rhythm appears which is characterized by its occipital predominance and its reactivity (blocking reaction). Despite significant interindividual variability, these landmarks are relatively predictable and represent major bioelectrical changes.

3. The fundamental biological rhythm of sleep periodicity is present very early in life and does not vary with maturation.

4. Electroencephalographic maturation does not appear to be influenced by the environment which surrounds the infant during the perinatal period (intra- vs extrauterine growth). However, Metcalf (1969) has described a slight advance in the

maturation of sleep spindles in babies born prematurely when compared with term infants. In contrast, sleep organization seems to be defective in prematurely born infants studied at 40 weeks postmenstrual age. Extrauterine life does not offer any advantage for such behavioral activity as visual pursuit (Parmelee and Sigman, 1976). The alterations in sleep maturation induced by extrauterine life need to be verified by controlled studies and, if confirmed, could provide indicators for modification of conditions in nurseries—such as increased sensory stimulation, as suggested by Rothschild (1967). During the first month after birth there is a constant interaction between congenital patterns and environmental influences. These external influences appear to have more effect on sleep organization, mode of onset of sleep, and motor activity (Campbell *et al.*, 1971), than on neurologic and electroencephalographic maturation.

5. Essential differences exist between the sleep of adults and infants (Feinberg, 1969). For example, the relationship between the release of growth hormone and the sleep cycle, well established in older children and adults, does not exist during the first week of life (Shaywitz *et al.*, 1971). Sleep deprivation is difficult to achieve in human neonates (Anders and Roffwarg, 1973); selective deprivation of stages of sleep by manual awakening has not been accomplished. Attempts at selective deprivation have led to reductions of both sleep states (non-REM, REM) and increased wakefulness. The interrupted sleep state demonstrated a considerable "tenacity," and during recovery, total sleep time was markedly increased, with a preferential recovery of non-REM sleep and an absence of REM rebound. Anders and Roffwarg (1973) suggest that this lack of REM rebound in the full-term newborn infant is due not only to the lack of adequate REM deprivation but also to the immaturity of the infant and the inadequacy of REM compensatory mechanisms. Differences between adult and infant sleep exist also in other species: Selective lesions of the serotonergic neurons of the anterior raphe nuclei in the brain stem of the cat, which in adult animals depresses (or eliminates) non-REM sleep, has no suppressing effect on the development of non-REM sleep in 3-week-old kittens (Adrien, 1976). Further experimental studies are needed in this field.

We hope that future sleep research in infants will include careful investigation of the differences between adult and infant sleep as well as of the role of environmental influences on sleep organization.

ACKNOWLEDGMENTS

I am grateful to Doctor Barry Tharp for reviewing this manuscript.

6. References

Adrien, J., 1976, Lesion of the anterior raphe nuclei in the newborn kitten and the effects on sleep, *Brain Res.* **103:**579–583.

Anders, T. F., 1976, Maturation of sleep patterns in the newborn infant, *Adv. Sleep Res.* **2:**43–66.

Anders, T. F., and Chalemian, R. J., 1974, The effects of circumcision on sleep-awake states in human neonates, *Psychosomat. Med.* **36:**174–179.

Anders, T. F., and Roffwarg, H. P., 1973, The effects of selective interruption and deprivation of sleep in the human newborn, *Dev. Psychobiol.* **6:**77–90.

Anders, T., Emde, R., and Parmelee, A., eds., 1971, A Manual of Standardized Terminology, Techniques and Criteria for Scoring of States of Sleep and Wakefulness in New-Born Infants, UCLA Brain Information Service, NINDS Neurological Information Network.

177

**ONTOGENESIS OF
BRAIN BIOELECTRICAL
ACTIVITY AND SLEEP
ORGANIZATION IN
NEONATES AND
INFANTS**

Aserinsky, E., and Kleitman, N., 1955, A motility cycle in sleeping infants as manifested by ocular and gross bodily activity, *J. Appl. Physiol.* **8:**11–18.

Ashton, R., 1971*a,* State and the auditory reactivity of the human neonate, *J. Exp. Child Psychol.* **12:**339–346.

Ashton, R., 1971*b,* Behavioral sleep cycles in the human newborn, *Child Dev.* **42:**2098–2100.

Bell, R. Q., 1970, Sleep cycles and skin potential in newborns studied with a simplified observation and recording system, *Psychophysiology* **6:**778–786.

Berenberg, S. R., Caniaris, M., and Masse, N. P. (eds.), 1974, Pre- and post-natal development of the human brain, in: *Modern Problem in Pediatrics,* Vol. 13, 359 pp., Karger, Basel.

Bergström, R. M., 1969, Electrical parameters of the brain during ontogeny, in: *Brain and Early Behavior* (R. J. Robinson, ed.), pp. 15–36, Academic Press, London.

Bernstein, P., Emde, R., and Campos, J., 1973, REM sleep in four month infants under home and laboratory conditions, *Psychosomat. Med.* **35:**322–329.

Bernuth, H. V., and Janssen, B., 1974, Behavioral changes in fototherapy. A polygraphic study, *Neuropädiatrie* **5:**369–375.

Bertini, M., Fornari, F., and Venturini, E., 1970, Observations on neonates' sleep under the influence of a heart-beat sound, in: *Psicofisiologica del sonno e del sogno* (M. Bertini, ed.), pp. 3–14, Vita e Pensiero, Milan.

Brackbill, Y., 1975, Continuous stimulation and arousal level in infancy: Effects of stimulus intensity and stress, *Child Dev.* **46:** 364–369.

Brackbill, Y., Douthitt, T. C., and West, H., 1973, Neonatal posture: Psychophysiological effects, *Neuropaediatrie* **4:**145–150.

Brandt, S., and Brandt, H., 1955, The electroencephalographic patterns in young healthy children from 0 to 5 years of age, *Acta Psychol. Scand.* **30:**77–89.

Campbell, D., Kuyek, J., Lang, E., and Partington, M. W., 1971, Motor activity in early life. II: Daily motor activity output in the neonate period, *Biol. Neonate* **18:**108–120.

Campbell, D., and Raeburn, J., 1973, Patterns of sleep in the newborn, in: *A Triune Concept of the Brain and Behavior* (T. J. Boag and D. Campbell, eds.), pp. 156–165, Toronto University Press, Toronto.

Campos, J. J., and Brackbill, Y., 1973, Infant state: Relationship to heart rate, behavioral response, and response decrement, *Dev. Psychobiol.* **6:**9–20.

Casaer, P., and Akiyama, Y., 1971, Neurological criteria for the estimation of the post-menstrual age of newborn infants, in: *Normal and Abnormal Development of Brain and Behavior* (G. B. A. Stoeling and J. J. Van der Werff Ten Bosch, eds.), pp. 148–178, Leiden University Press, Leiden.

Casaer, P., O'Brien, M. J., and Prechtl, H. F. R., 1973, Postural behavior in human newborns, 2nd Symposium International de Posturographie, *Agressologie* **14(b):**49–56.

Cianchetti, C., Fabiani, D., Favilla, A., Ferroni, A., and Massai-Piredda, E., 1974, Differentiation between two states in the active sleep of the new born, *IRCS* **2:**1583.

Clemente, C. D., Purpura, D. P., and Mayer, F. E. (eds.), *Ontogeny of Sleep Patterns,* 470 pp., Academic Press, New York.

Crawley, J., and Kellaway, P., 1963, The electroencephalogram in pediatrics, *Pediatr. Clin. North Am.* **10:**17–51.

Crowell, D. H., Jones, R. H., Kapuniai, L. E., and Nakagawa, J. K., 1973, Unilateral cortical activity in newborn humans: An early index of cerebral dominance? *Science* **180:**205–207.

Curzi-Dascalova, L., and Dreyfus-Brisac, C., 1976, Distribution of skin potential responses according to states of sleep during the first months of life in human babies, *Electroenceph. Clin. Neurophysiol.* **41:**399–407.

Curzi-Dascalova, L., Pajot, N., and Dreyfus-Brisac, C., 1973, Spontaneous skin potential response in sleeping infants between 24 to 41 weeks of conceptional age, *Psychophysiology* **10:**478–487.

Curzi-Dascalova, L., and Plassart, E., 1976, Activité tonique du muscle mentonnier pendant le sommeil du nouveau-né à terme et du nourrisson, *J. Physiol.* **72:**5A.

Curzi-Dascalova, L., Radvanyi, M. F., Couchard, M., Monod, N., and Dreyfus-Brisac, C., 1976, Some data on respiration in human ontogenesis, in: *Respiratory Centers and Afferent Systems* (B. Duron, ed.), pp. 287–297, INSERM Publ., Paris.

Denissova, M. P., and Figurin, N. L., 1926, [Phénomènes périodiques au cors du sommeil des enfants,] [*Nouv. Réflexol. Physiol. Syst. Nerv.*], **2:**338–345.

Dittrichova, J., 1969, Development of sleep in infancy, in: *Brain and Early Behavior* (R. Robinson, ed.), pp. 193–204, Academic Press, London.

Dittrichova, J., and Paul, K., 1974, Responsivity in newborns during sleep, *Act. Nerv. Super.* **16:**112.

Dittrichova, J., and Paul, K., 1975, Stability of individual differences during paradoxical sleep in infants, *Act. Nerv. Super.* **17:**49.

Dittrichova, J., Paul, K., and Pavlikova, E., 1972, Rapid eye movements in paradoxical sleep in infants, *Neuropädiatrie* **3**:248–257.

Dittrichova, J., Paul, K., and Vondracek, J., 1976, Individual differences in infants sleep, *Dev. Med. Child Neurol.* **18**:182–188.

Doose, H., and Gerken, H., 1972, The genetic of EEG anomalies, in: *Handbook of Electroencephalography and Clinical Neurophysiology* (A. Remond, ed.), Vol. 15B, pp. 5–13, Elsevier, Amsterdam.

Dreir, T., and Wolff, F. H., 1972, Sucking, state, and perinatal distress in newborns, *Biol. Neonate* **21**:16–24.

Dreyfus-Brisac, C., 1962, The electroencephalogram of the premature infant, *World Neurol.* **3**:5–15.

Dreyfus-Brisac, C., 1966, Development of the nervous system in early life: Part IV: The bioelectrical development of the central nervous system during early life, in: *Human Development* (F. Falkner, ed.), pp. 286–305, Saunders, Philadelphia.

Dreyfus-Brisac, C., 1967, Ontogénèse du sommeil chez le prématuré humain: Etudes polygraphiques à partir de 24 semaines d'âge conceptionnel, in: *Regional Maturation of the Brain in Early Life* (A. Minkowski, ed.), pp. 437–457, Blackwell, Oxford.

Dreyfus-Brisac, C., 1968, Sleep ontogenesis in early human prematurity from 24 to 27 weeks of conceptional age, *Dev. Psychobiol.* **1**:162–169.

Dreyfus-Brisac, C., 1970, Ontogenesis of sleep in human prematures after 32 weeks of conceptional age, *Dev. Psychobiol.* **3**:91–121.

Dreyfus-Brisac, C., 1972, Critères éléctrophysiologiques de maturation et retard de croissance intra utérin, XXIII Congr. Assoc. Pédiatr. Langue Fra. **1**:261–268.

Dreyfus-Brisac, C., 1975, Neurophysiological studies in human premature and full-term newborns, *Biol. Psychiatry* **10**:485–496.

Dreyfus-Brisac, D., and Blanc, C., 1956, Electroencéphalogramme et maturation cérébrale, *Encéphale* **45**:205–245.

Dreyfus-Brisac, C., and Curzi-Dascalova, L., 1975, The EEG during the first year of life, in: *Handbook of Electroencephalography and Clinical Neurophysiology* (A. Remond, ed.), Vol. 6B, pp. 20–24, Elsevier, Amsterdam.

Dreyfus-Brisac, C., and Minkowski, A., 1968, Electroencephalographic maturation and low birth weight, in: *Clinical Electroencephalography of Children* (P. Kellaway and I. Petersen, eds.), pp. 49–60, Almqvist and Wiksell, Stockholm.

Dreyfus-Brisac, C., and Monod, N., 1964, Electro-clinical studies of status epilepticus and convulsions in the newborn, in: *Neurological and Electroencephalographic Correlative Studies in Infancy* (P. Kellaway and I. Petersen, eds.), pp. 251–272, Grune & Stratton, New York.

Dreyfus-Brisac, C., Fischgold, H., Samson, D., Saint-Anne Dargassies, S., Ziegler, T., Monod, N., and Blanc, C., 1957, Veille, sommeil et réactivité sensorielle chez le prématuré et le nouveau-né, Activité électrique cérébrale du nourrisson, *Electroenceph. Clin. Neurophysiol. Suppl.* **6**:418–440.

Dreyfus-Brisac, C., Samson, D., Blanc, C., and Monod, N., 1958, L'électroencéphalogramme de l'enfant normal de moins de 3 ans. Aspect fonctionnel bioélectrique de la maturation nerveuse, *Etud. Neonat.* **7**:143–175.

Dreyfus-Brisac, C., Flescher, J., and Plassart, E., 1962, L'électroencèphalogramme: Critère d'âge conceptionnel du nouveau-né à terme et prématuré, *Biol. Neonat.* **4**:154–173.

Dubignon, J. M., Campbell, D., and Partington, M. W., 1969, The development of non-nutritive sucking in premature infants, *Biol. Neonat.* **14**:270–278.

Dumermuth, G., 1965, *Elektroencephalographie im Kindesalter,* 287 pp., Georg Thieme Verlag, Stuttgart.

Eeg-Olofsson, O., 1970, The development of the electroencephalogram in normal children and adolescents from the age of 1 to 21 years, *Acta Paediatr. Scand. Suppl.* **208**:1–46.

Eisenberg, R. B., 1969, Auditory behavior in the human neonate: Functional properties of sound and their ontogenesis implications, *Int. Audio.* **8**:34–45.

Eliet-Flescher, J., and Dreyfus-Brisac, C., 1966, Le sommeil du nouveau-né et du prématuré. II. Relations entre l'électroencéphalogramme et l'électromyogramme mentonnier au cours de la maturation, *Biol. Neonat.* **10**:316–339.

Ellingson, R. J., 1958, Electroencephalograms of normal, full-term newborns immediately after birth with observations on arousal and visual evoked responses, *Electroencephalogr. Clin. Neurophysiol.* **10**:31–50.

Ellingson, R. J., 1967, The study of brain electrical activity in infants, in: *Advances in Child Development and Behavior* (L. P. Lipsitt and C. C. Spiker, eds.), Vol. 3, pp. 53–97, Academic Press, New York.

Ellingson, R. J., 1972, Development of wakefulness sleep cycles and associated EEG patterns in mammals, in: *Sleep and Maturing Nervous System* (C. D. Clemente, D. C. Purpura, and F. E. Mayer, eds.), pp. 165–174, Academic Press, New York.

179

*ONTOGENESIS OF
BRAIN BIOELECTRICAL
ACTIVITY AND SLEEP
ORGANIZATION IN
NEONATES AND
INFANTS*

Ellingson, R. J., 1975, Ontogenesis of sleep in the human, in: *Experimental Study of Human Sleep. Methodological Problems* (G. C. Lairy and P. Salzarulo, eds.), pp. 129–149, Elsevier, Amsterdam.

Ellingson, R. J., and Rose, C. H., 1970, Ontogenesis of the electroencephalogram, in: *Developmental Neurobiology* (W. A. Himvich, ed.), pp. 441–474, Charles C. Thomas, Springfield, Illinois.

Ellingson, R. J., Dutch, S. J., and McInteir, M. S., 1974, EEG's of prematures: 3–8 year follow-up study, *Dev. Psychobiol.* **7**:529–538.

Emde, R. N., and Koenig, K. L., 1969, Neonatal smiling, frowning and rapid eye movement states. II Sleep cycle study, *J. Am. Acad. Child Psychiatry* **8**:637–656.

Emde, R. N., Harmon, R. J., Metcalf, D., Koenig, K. L, and Wagonfeld, S, 1971, Stress and neonatal sleep, *Psychosomat. Med.* **33**:491–497.

Engel, R., 1974, Alpha-like burst in the neonatal period. *Electroencephalogr. Clin. Neurophysiol.* **36:** 210.

Engel, R. C., 1975, *Abnormal Electroencephalograms in the Neonatal Period,* 128 pp., Charles C. Thomas, Springfield, Illinois.

Evsyukova, J. J., 1971*a*, [Formation of circadian rhythms of sleep in newly born babies], *Vopr. Okhr. Materin. Det.* **1**:3–8.

Evsyukova, J. J., 1971*b*, [Paradoxical sleep in newborn infants], *Zh. Vyssh. Nervn. Deyat. im. I. P. Pavlova* **21**:1230–1237.

Fabiani, D., Favilla, A., and Massai-Piredda, E., 1975, Aspetti del passagio veglia-sonno nel primo mese di vita, *Riv. Neurol.* **45**:128–130.

Falkner, F. (ed.), 1966, *Human Development,* 644 pp., Saunders, Philadelphia.

Feinberg, I., 1969, Effects of age on human sleep patterns, in: *Sleep: Physiology and Pathology* (A. Kales, ed.), pp. 39–52, J. B. Lippincott, Philadelphia.

Gabriel, M., Albani, M., and Schulte, F. J., 1976, Apneic spells and sleep states in preterm infants, *Pediatrics* **57**:142–147.

Gambi, D., Vacchini, F., Bertini, M., and Gagliardi, F., 1974, Observazioni comportamentali durante il sonno del neonato delle prime notti di vita, *Neuropsichiatr. Infant.* **160–161** (nuova serie): 1015–1029.

Gesell, A., and Amatruda, C. S., 1945, *The Embryology of Behavior: The Beginning of the Human Mind,* Harper, New York.

Gibbs, F. A., and Gibbs, E. L., 1950, *Atlas of Electroencephalography,* 324 pp., Addison-Wesley, Reading, Massachusetts.

Goldie, L., Svedens-Rhodes, J., and Roberton, E. N. R. C., 1971, The development of sleep rhythms in short gestation infants, *Dev. Med. Child Neurol.* **13**:40–56.

Graziani, L. J., Weitzman, E. D., and Velasco, M. S. A., 1968, Neurologic maturation and auditory evoked responses in low birth weight infants, *Pediatrics* **41**: 483–494.

Graziani, L. J., Katz, L., Cracco, R. Q., Cracco, J. B., and Weitzman, E. D., 1974, The maturation and interrelationship of EEG patterns and auditory evoked response in premature infants, *Electroenceph. Clin. Neurophysiol.* **36**:367–376.

Hagne, I., 1968, Development of the waking EEG in normal infants during the first year of life. A longitudinal study including automatic frequency analysis, in: *Clinical Electroencephalography of Children* (P. Kellaway and I. Petersen, eds.), pp. 97–118, Almqvist and Wiksell, Stockholm.

Hagne, I., 1972, Development of the EEG in normal infants during the first year of life. A longitudinal study, *Acta Paediatr. Scand. Suppl.* **232**:1–53.

Hamburger, V., 1971, Development of embryonic motility. in: *The Biopsychology of Development,* pp. 45–65, Academic Press, New York.

Henry, Ch. E., 1944, *Electroencephalograms of Normal Children,* 71 pp., Society for Research in Child Development, National Research Council, Washington, D.C.

Hirai, T., Takano, R., and Uchimma, Y., 1968, An electroencephalographic study on the development of nocturnal sleep, *Folia Psychiatr. Neurol. Jpn.* **22**:157–166.

Hoppenbrouwers, T., and Sterman, M. B., 1975, Development of sleep state patterns in the kitten, *Exp. Neurol.* **49**:822–839.

Hutt, C., von Bernuth, V., Lenard, H. G., Hutt, S. J., and Prechtl, H. F. R., 1968, Habituation in relation to state in the human neonate, *Nature* **220**:618–620.

Iwase, K., and Watanabe, K., 1971, Studies on the respiratory rate in premature babies in relation to conceptional and sleep stage, *Acta Neonatol. Jpn.* **7**:56–61.

Katz, C., Graziani, J., Cracco, R. G., and Weitzman, E., 1972, A method for analyzing EEG maturation in low birth weight premature infants, *Electroencephalogr. Clin. Neurophysiol.* **33**:452.

Kellaway, P., 1957, Ontogenic evolution of the electrical activity of the brain in man and animals, IVth Intern. Meeting of EEG and Clinical Neurophysiology, pp. 141–154, Acta Medica Belgica, Brussells.

Kellaway, P., and Crawley, J., 1964, *A Primer of Electroencephalography Infants, Methodology and Criteria of Normality,* Privately printed for the National Institute of Health, Bethesda, Maryland.

Kligman, D., Smyrl, R., and Emde, R., 1975, A "nonintrusive" longitudinal study of infant sleep, *Psychosomat. Med.* **37**:448–453.

Kohler, W. C., Caddington, R. D., and Agnew, H. W., 1968, Sleep patterns in 2-year-old children, *J. Pediatr.* **72**:228–233.

Korner, A. F., 1969, Neonatal startles, smiles, erections and reflex sucks as related to state, sex and individuality, *Child Dev.* **40**:1039–1053.

Koslacz-Folga, A., 1973, [Etude électrophysiologique de la maturation et réactivité du S.N.C. pendant la première année dans le sommeil d'enfants normaux], *Probl. Med. Wieki Rozwojowego,* **2a(13a)**:195–266 (summary and translation, 1974, **4**:215–227).

Lairy, G. C., 1975, The evolution of the EEG from birth to childhood, in: *Handbook of Electroencephalography and Clinical Neurophysiology* (A. Remond, ed.), Vol. 6B, 105 pp., Elsevier, Amsterdam.

Lairy, G. C., and Salzarulo, G. P. (eds.), 1975, *The Experimental Studies of Human Sleep. Methodological Problems,* 493 pp., Elsevier, Amsterdam.

Larroche, J. C., and Korn, G., 1977, Brain damage in intrauterine growth retardation, *Symposium on Intrauterine Asphyxia and the Developing Fetal Brain,* in: *Intrauterine Asphyxia and the Developing Fetal Brain,* pp. 25–35, Yearbook Med. Publ., Chicago.

Lenard, H. G., 1970a, Sleep studies in infancy, *Acta Paediatr. Scand.* **59**:572–581.

Lenard, H. G., 1970b, The development of sleep spindles in the EEG during the first two years of life, *Neuropädiatrie* **1**:264–276.

Lenard, H. G., 1972, The development of sleep behavior in babies and small children, *Electroencephalogr. Clin. Neurophysiol.* **32**:710.

Lindsley, D. B., 1939, A longitudinal study of the occipital alpha rhythm in normal children. Frequency and amplitude standards, *J. Gen. Psychol.* **55**:197–213.

Lombroso, C. T., 1975, Neurophysiological observations in diseased newborns, *Biol. Psychiatry* **10**:527–558.

Metcalf, D. R., 1969, The effect of extrauterine experience on the ontogenesis of EEG spindles, *Psychosomat. Med.* **31**:393–399.

Metcalf, D. R., 1970, EEG sleep spindle ontogenesis, *Neuropädiatrie* **1**:428–433.

Metcalf, D. R., and Jordan, K., 1972, EEG ontogenesis in normal children, in: *Drug, Development and Cerebral Function* (Th. Smith, ed.), pp. 125–144, Charles C. Thomas, Springfield, Illinois.

Metcalf, D. R., Mondale, J., and Butler, F. K., 1971, Ontogenesis of spontaneous K-complexes, *Psychophysiology* **8**:340–347.

Michaelis, R., Parmelee, A. H., Stern, E., and Haber, A., 1973, Activity states in premature and term infants, *Dev. Psychobiol.* **6**:209–215.

Minkowski, A. (ed.), 1967, *Regional Maturation of the Brain in Early Life,* 540 pp., Blackwell, Oxford.

Monod, N., and Curzi-Dascalova, L., 1973, Les états transitionnels de sommeil chez le nouveau-né à terme, *Rev. Neurophysiol.* **3**:87–96.

Monod, N., and Dreyfus-Brisac, C., 1965, Les premières étapes de l'organisation du sommeil chez le prématuré et le nouveau-né, in: *Le sommeil de nuit, normal et pathologique,* pp. 116–146, Masson, Paris.

Monod, N., and Ducas, P., 1968, The prognosis value of the electroencephalogram during the first two years of life, in: *Clinical Electroencephalography of Children* (P. Kellaway and I. Petersen, eds.), Almqvist and Wiksell, Stockholm, pp. 61–76.

Monod, N., and Garma, L., 1971, Auditory responsivity in the human premature, *Biol. Neonate* **17**:292–316.

Monod, N., and Pajot, N., 1965, Le sommeil du nouveau-né et du prématuré. I: Analyse des études polygraphiques (mouvements oculaires, respiratoires et EEG) chez le nouveau-né à terme, *Biol. Neonat.* **8**:281–307.

Monod, N., Dreyfus-Brisac, C. , and Sfaello, Z., 1969, Dépistage et pronostic de l'état de mal néonatal d'après l'étude électroclinique de 150 cas, *Arch. Fr. Pediatr.* **26**:1085–1102.

Monod, N., Pajot, N., and Guidasci, S., 1972, The neonatal EEG: Statistical studies and prognostic value in full-term and pre-term babies, *Electroencephalogr. Clin. Neurophysiol.* **32**:529–544.

Monod, N., Curzi-Dascalova, L., Guidasci, S., and Valenzuela, S., 1976, Pauses respiratoires et sommeil chez le nouveau-né et le nourrisson, *Rev. EEG Neurophysiol.* **6**:105–110.

Murray, B., and Campbell, D., 1971, Sleep states in the newborn: Influence of sound, *Neuropädiatrie* **2**:335–342.

Narayan, C. I., Fox, M. V., and Hamburger, V., 1971, Prenatal development of spontaneous and evoked activity in the rat, *Behavior* **40**:100–134.

181

**ONTOGENESIS OF
BRAIN BIOELECTRICAL
ACTIVITY AND SLEEP
ORGANIZATION IN
NEONATES AND
INFANTS**

Netchine, S., and Lairy, G. C., 1975, The EEG and psychology of the child, in: *Handbook of Electroencephalography and Clinical Neurophysiology* (A. Remond, ed.), Vol. 6B, pp. 69–104, Elsevier, Amsterdam.

Ohtahara, S., Kajitani, T., Shimo, M., Ishida, T., Takabataka, Y., and Mukai, Y., 1967, Studies on the electroencephalogram in normal children, XIVth Annual Meeting of the Japan EEG Society.

Ornitz, E. M., Wechter, V., Hartman, D., Tanguay, P. E., Lee, J. M. C., Ritvo, E. R., and Walter, R. D., 1971, The EEG and rapid eye movement during REM sleep in babies, *Electroencephalogr. Clin. Neurophysiol.* **30:**350–353.

Pajot, N., 1974, EEG recording technique in full term and premature newborn infant, *Am. J. Technol.* **14:**108–119.

Pampiglione, G., 1965, Brain development and the EEG of normal children of various ethnical groups, *Brit. Med. J.* **5461:**573–575.

Pampiglione, G., 1971, Some aspects of development of cerebral function in mammals, *Proc. R. Soc. Med.* **64:**429–435.

Parmelee, A. H., 1974, Ontogeny of sleep patterns and associated periodicities in infants, in: *Pre- and Postnatal Development of the Human Brain* (S. P. Berenberg, M. Caniaris, and N. P. Masse, eds.), pp. 298–311, Karger, Basel.

Parmelee, A. H., 1975, Neurophysiological and behavioral organization of premature infants in the first months of life, *Biol. Psychiatry* **10:**501–512.

Parmelee, A. H., and Sigman, M., 1976, Development of visual behavior and neurological organization in pre-term and full-term infants, *Minnesota Symposia on Child Development,* Vol. 10 (A. D. Prick, ed.), pp. 119–155, University of Minnesota Press.

Parmelee, A. H., Brück, K., and Brück, M., 1962, Activity and inactivity cycles during the sleep of premature infants, *Biol. Neonat.* **4:**317–339.

Parmelee, A. H., Wenner, W. H., Akiyama, Y., Stern, E., and Flescher, J. 1967, Electroencephalography and brain maturation, in: *Regional Maturation in Early Life* (A. Minkowski, ed.), pp. 459–476, Blackwell, Oxford.

Parmelee, A. H., Akiyama, Y., Schultz, M. A., Wenner, W., Schulte, F. J., and Stern, E., 1968*a*, The electroencephalogram in active and quiet sleep in infants, in: *Clinical Electroencephalography of Children* (P. Kellaway and 1. Petersen, eds.), pp. 77–88.

Parmelee, A. H., Schulte, F. J., Akiyama, Y., Wenner, W., Schultz, M. A., and Stern, E., 1968*b*, Maturation of EEG activity during sleep in premature infants, *Electroencephalogr. Clin. Neurophysiol.* **24:**319–329.

Parmelee, A. H., Akiyama, Y., Stern, E., and Harris, M. A., 1969, A periodic cerebral rhythm in newborn infants, *Exp. Neurol.* **25:**575–584.

Parmelee, A. H., Stern, E., and Harris, M. A., 1972, Maturation of respiration in prematures and young infants, *Neuropädiatrie* **3:**294–304.

Passouant, P. (ed.), 1975, EEG and sleep, in: *Handbook of Electroencephalography and Clinical Neurophysiology* (A. Remond, ed.), Vol. 7A, pp. 1–114, Elsevier, Amsterdam.

Paul, K., Dittrichova, J., and Pavlikova, E., 1973, The course of quiet sleep in infants, *Biol. Neonate* **23:**78–89.

Petersen, I., and Eeg-Olofsson, O., 1971, The development of the electroencephalogram in normal children from the age of 1 through 15 years, *Neuropädiatrie* **2–3:**247–304.

Petersen, I., Sellden, U., and Eeg-Olofsson, O., 1975, The evolution of the EEG in normal children and adolescents from 1 to 21 years, in: *Handbook of Electroencephalography and Clinical Neurophysiology* (A. Remond, ed.), Vol. 6B, pp. 31–68, Elsevier, Amsterdam.

Petre-Quadens, O., 1966, On the different phases of the sleep of the newborn with special reference to the activated phase or phase d, *J. Neurol. Sci.* **3:**151–161.

Petre-Quadens, O., 1969, Contribution à l'étude de la phase dite paradoxal du sommeil. Thesis, Université Libre, Brussels.

Polikanina, R. I., 1966, Development of the higher nervous activity in prematurely born babies during the early post natal period of life, *Meditsina,* 246 pp., Leningradskoe Otgenonie.

Prechtl, H. F. R., 1974, The behavioural status of the newborn infants (a review), *Brain Res.* **76:**185–212.

Prechtl, H. F. R., and Lenard, H. G., 1967, A study of eye movements in sleeping newborn infants, *Brain Res.* **5:**477–493.

Prechtl, H. F. R., Akiyama, Y., Zinkin, P., and Grant, D. K., 1968, Polygraphic studies of the full-term newborn. I: Technical aspects and quantitative analysis, in: *Studies in Infancy. Clinic in Developmental Medicine,* Vol. 27 (M. Bax and R. C. MacKeith, eds.), pp. 1–2, SIMP/Heinemann, London.

Richard, M. P. H., Bernal, J. F., and Brackbill, Y., 1976, Early behavioral differences: Gender or circumcision, *Dev. Psychobiol.* **9:**89–95.

Robinson, R. J., 1969, *Brain and Early Behavior,* 374 pp., Academic Press, London.

Roffwarg, H. P., Dement, W. C., and Fischer, C., 1964, Observation on the sleep dream pattern in neonates, infants, children and adults, *Child Psychiatry Monographs II* (E. Harms, ed.), pp. 60–71, Pergamon Press, New York.

Roffwarg, H. P., Muzio, J. N., and Dement, W. C., 1966, Ontogenic development of the human sleep-dream cycle, *Science* **152:**604–619.

Rothschild, B. F., 1967, Incubator isolation as a possible contributing factor to the high incidence of emotional disturbance among prematurely born persons, *J. Genet. Psychol.* **110:**287–304.

Saint-Anne Dargassies, S., 1966, Neurological maturation of the premature infants of 28 to 41 weeks' gestational age, in: *Human Development* (F. Falkner, ed.), pp. 306–325, W. B. Saunders, Philadelphia.

Samson-Dollfus, D., 1955, L'EEG du prématuré jusqu'à l'âge de 3 mois et du nouveau-né à terme. Thesis Méd., Foulon, Paris.

Samson-Dollfus, D., and Poussin, A., 1973, Les états transitionnels précédant et suivant les phases de sommeil rapide avec M.O. chez l'enfant normal de 8 à 30 mois, *Rev. Electroencephalogr. Neurophysiol. Clin.* **3:**97–103.

Samson-Dollfus, D., Forthomme, J., and Capron, E., 1964, EEG of the human infant during sleep and wakefulness during the first year of life, in: *Neurological and Electroencephalographic Studies in Infancy* (P. Kellaway and I. Petersen, eds.), pp. 208–229, Grune & Stratton, New York.

Sander, L. W., Stechler, G., Burns, P., and Julia, H., 1970, Early mother–infant interaction and 24 hour patterns of activity and sleep, *J. Am. Acad. Child Psychiatry* **9:**103–123.

Schaeffer, A. B., 1975, Newborn responses to nonsignal auditory stimuli: I: Electroencephalographic desynchronization, *Psychophysiology* **12:**359–366.

Schulte, F. J., and Jurgens, U., 1969, Apnoen bei reifen und unreifen Neugeborenen, *Monats. Kinderheild.* **117:**595–601.

Schulte, F. J., Hinze, G., and Schrempp, G., 1971, Maternal toxemia, fetal malnutrition and bioelectric brain activity of the new-born, *Neuropädiatrie* **2:**439–460.

Shaywitz, B. A., Finkelstein, J., Hellman, L., and Weitzman, E. O., 1971, Growth hormone in newborn infants during sleep-wake periods, *Pediatrics* **48:**103–110.

Shepavolnikov, A. N., 1971, *Activity of the Sleeping Brain. The Electropolygraphic Study of Physiological Sleep in Infants,* 182 pp., Nauka, Leningrad.

Sterman, M. B., and Hoppenbrouwers, T., 1971, The development of sleep–waking and rest–activity patterns from fetus to adult in man, in: *Brain Development and Behavior* (M. B. Sterman, D. J. McGinty, and A. M. Adinolfi, eds.), pp. 203–227, Academic Press, New York.

Stern, E., Parmelee, A. H., Akiyama, Y., Schultz, M. A., and Wenner, W. H., 1969, Sleep cycle characteristics in infants, *Pediatrics* **43:** 65–70.

Stern, E., Parmelee, A. H., and Harris, M. A., 1973, Sleep state periodicity in prematures and young infants, *Dev. Psychobiol.* **6:**357–366.

Tanguay, P. E., Ornitz, E. M., and Bozzo, E., 1974, Age related changes in rapid eye movement activity in childhood during REM sleep, *Electroencephalogr. Clin. Neurophysiol* **37:**207.

Torres, F., 1970, Electrographic development of sleep patterns in children: Longitudinal studies for normative criteria and EEG clinical correlations, *Electroencephalogr. Clin. Neurophysiol* **28:**421.

Uzemaki, H., and Morrell, F., 1970, Development study of photic evoked responses in premature infants, *Electroencephalogr. Clin. Neurophysiol.* **28:**55–63.

Valatx, J. L., Jouvet, D., and Jouvet, M., 1964, Evolution électroencéphalographique des différents états de sommeil chez le chaton, *Electroencephalogr. Clin. Neurophysiol.* **17:**218–233.

Varner, J. L., Ellingson, R. J., Danahy, T., and Nelson, B., 1976, Interhemispheric amplitude symmetry of the EEGs of normal full-term newborns, *Electroencephalogr. Clin. Neurophysiol.* **40:**215–216.

Vitova, Z., and Hrbek, A., 1970, Ontogeny of cerebral responses to flickering light in infants during wakefulness and sleep, *Electroencephalogr. Clin. Neurophysiol.* **28:**391–398.

Watanabe, K., and Iwase, K., 1972, Spindle like fast rhythms in the EEGs of low birth weight infants, *Dev. Med. Child Neurol.* **14:**373–381.

Watanabe, K., Iwase, K., and Hara, K., 1972, EEG of early premature infants. A polygraphic study, *Brain Dev.* **4:**34–43.

Watanabe, K., Iwase, K., and Hara, K., 1974, Development of slow-wave sleep in low-birth weight infants, *Dev. Med. Child Neurol.* **16:**23–31.

Wolff, P. H., 1966, The causes, control and organization of behavior in the neonate, Psychological Issues Monograph, International University Press, New York.

Wolff, P. H., 1968, Sucking patterns in infants mammals, *Brain Behav. Evol.* **1:**354–367.

6

Sexual Differentiation of the Brain

PAMELA C. B. MACKINNON

1. Introduction

It is abundantly clear that men and women are different, different in their genetic composition, their structure, and their behavior. Whether some of these differences can be attributed to differentiation of central nervous system mechanisms, brought about as a result of changes in the hormonal environment at a critical stage of brain development, is an important and intriguing question. But before considering this issue, it is as well to recall all the fundamental bases of the differences between sexes.

The genetic basis of sex determination and differentiation in mammals has been well documented, and experiments have shown that the development of a testis from the indifferent gonad depends on the presence of a gene or set of genes on the Y chromosome (Welshons and Russell, 1959). It has recently been suggested that the product of such a gene, or set of genes, may be a protein with antigenic properties (the histocompatibility antigen) which is present on the surface of the germ cells (Wachtel *et al.,* 1975; Bennett *et al.,* 1975).

The process of testicular differentiation is characterized by a rapid proliferation of medullary cords of cells which attract and surround the germ cells at the same time as a regression of cortical tissue is taking place. In the absence of the Y chromosome a prospective ovary is formed by a much slower proliferation of cortical elements and a regression of medullary tissue (Witschi, 1962). Once the developing testis begins to function, a local diffusion of androgens takes place which is responsible for differentiating the male (or Wolffian) reproductive tract (Siiteri and Wilson, 1974), while a further nonandrogenic testicular factor, possibly protein in nature (Josso, 1972, 1973), causes the presumptive female or Müllerian

PAMELA C. B. MACKINNON • University of Oxford, Oxford, England.

tract to regress. In contrast, the development of the female reproductive tract does not require a hormonal stimulus as experiments with antiandrogens have shown (Elger, 1966; Neumann *et al.,* 1966; see also Neumann *et al.,* 1970); thus, in the absence of either gonad a female type of reproductive tract will develop (Jost, 1970; Price, 1970; Jost *et al.,* 1973).

At a later stage of embryonic growth, when the circulatory system is becoming established, testicular hormones are carried to the region of the urogenital sinus where masculinization of the external genitalia takes place. Again, in the absence of testes or of ovarian tissue, the ultimate appearance will be female in type.

Of particular interest from the point of view of the present chapter has been the considerable body of evidence, much of it related to rodents, which has shown that the presence of testicular hormones at an early stage of neural development also leads to sexual differentiation of those mechanisms which are responsible for the control of both gonadotropin output and sexual behavior (Harris, 1970). The early presence of androgens ensures an almost constant or tonic output of gonadotropins in adult life and predisposes the animal toward certain male patterns of behavior. On the other hand the lack of such an early hormonal stimulus results in a superimposed cyclic or phasic output of gonadotropins, and in certain types of behavior which are characteristically female. Stated in a somewhat different way— the presence of a testis is required to prevent certain functions and activities which are typically female; while the presence of ovaries does not appear to be essential for the normal development of the reproductive tract, external genitalia, and brain of the female (Figure 1). However, it is worth noting that recent evidence has

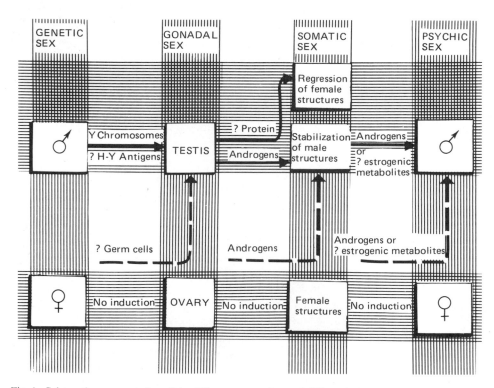

Fig. 1. Schematic representation of the different stages of sexual differentiation (after Dr. N. Neumann).

suggested a role for adrenal progesterone in the organization of the neonatal female brain (see Shapiro *et al.*, 1976a,b).

A fascinating aspect of sexual evolution and development is that despite the normal dependence of testicular formation and secretion on the presence of a Y chromosome, the masculinizing action of testosterone is, according to Ohno (1971), mediated by a single product of a gene on the X chromosome. Such a product (or products) may be represented by a protein which forms part of both the cytosol and nuclear androgen receptors and on which the manifestations of the male phenotype depend (Fang *et al.*, 1969; Mueller *et al.*, 1972; King and Mainwaring, 1974; Kato, 1975).

In general, the corpus of work in this field suggests that the overall pattern of sexual development is similar in both the primate and the nonprimate. Nevertheless, there is little evidence at present to indicate that the brain of higher species undergoes sexual differentiation during development; however, the primate brain, in particular that of man, is considerably more complex than that of lower animals, and it is possible that a greater subtlety is required in our experimental approach before the problem is resolved.

2. Evidence for Sexual Differentiation of the Brain in Nonprimates

2.1. Gonadal Steroids and the Brain

2.1.1. Effects of Neonatal Androgen

The most important contribution to the early literature on sexual differentiation of the brain was provided by Pfeiffer (1936). Principally he showed that if genetic male and female rats which had been castrated at birth were allowed to grow up, and were then transplanted with an ovary, corpora lutea (indicative of cyclic gonadotropin secretion and ovulation) would be formed. On the other hand if male rats were castrated at birth while at the same time testicular tissue was transplanted to the neck, when the animal reached maturity ovarian grafts failed to show true corpora lutea. Similarly, many females which received a testicular transplant at birth lost the capacity to form copora lutea when adult and, moreover, the vaginal epithelium of these animals was in an almost constant state of vaginal cornification, From this work Pfeiffer concluded that rats of either sex were born with undifferentiated pituitary tissue which was capable of secreting gonadotropins in a cyclic fashion; however, if, at an early stage of development, the pituitary gland was exposed to testicular secretion then the tissue became differentiated and unable to support cyclic ovarian and reproductive function. At the time that this concept was proposed the importance of the hypothalamohypophysial portal system had not been appreciated, still less the possibility of pituitary hormone-releasing factors.

Almost two decades later Harris and Jacobsohn (1952), in a classical experiment based on earlier work by Greep (1936), showed that male pituitary tissue when transplanted to the sella turcica of hypophysectomized females, gained a specific blood supply from the median eminence and was thereafter capable of sustaining normal female reproductive function. With confirmation of this work in mice (Martinez and Bittner, 1956) and the accumulation of further evidence derived from pituitary transplantation or stalk section on the one hand and the effects of brain stimulation or lesioning on the other, the importance of the central nervous system in the control of gonadotropin output and reproductive activity was substantiated (Harris, 1955). The tide of scientific thought then turned to the consideration of the brain rather than the anterior pituitary as a possible site of sexual differentiation. Subsequent work was greatly facilitated when Barraclough and Leathem (1954), using mice, and Barraclough (1961), using rats, showed that a single injection of testosterone propionate (TP) into neonates sufficed to "masculinize" the female and cause postpubertal anovulation and constant vaginal estrus. This simple technique enabled the presumptive period of neural plasticity during which sexual differentiation takes place to be defined: The maximal effects of a single TP (1.25 mg) injection were found to extend from the late stages of fetal life (Swanson and van der Werff ten Bosch, 1964, 1965) to the fifth postnatal day, while the effects became progressively less effective between 5 and 10 days of age, and after about 10 days of age only very heavy doses caused sterility (Brown-Grant, 1974a). However, if much smaller doses of TP (e.g., 0.125 mg) were given between 0 and 5 days of age, then the onset of sterility in the adult female rat was greatly delayed and was preceded by a period of apparently normal cyclicity and reproductive activity (Swanson and van der Werff ten Bosch, 1965). This effect, which does not occur if a testis is transplanted to a normal female neonate, has been termed the "delayed anestrous syndrome," and is thought to be due to an incomplete or partial differentiation of neurons by a "subthreshold" dose of androgen (Gorski, 1968).

The results of studies relating to the period of sexual differentiation of neural tissue in male rats were commensurate with those in neonatally androgenized females: Yazaki (1960) castrated groups of male rats of different ages from birth to 40 days of age and subsequently grafted ovarian and vaginal tissue beneath the abdominal skin. Daily vaginal smears and ovarian histology showed the occurrence of 4-day ovarian cycles in a majority of the animals which had been castrated before the third day of life, whereas in animals castrated later in life vaginal cornification and polyfollicular ovaries persisted.

Plasma testosterone concentrations in neonatal male rats as measured by radioimmunoassay have shown that they are low (0.5 ng/ml) in comparison with those recorded in young adult males (approx. 2.0 ng/ml). Nevertheless, since neonatal testicular transplants in neonatal females result in constant vaginal estrus and sterility at the onset of puberty (Gorski, 1968), these testosterone levels are presumably adequate to "masculinize" the brain.

2.1.2. Localization of Androgens in the Brain

To identify the possible site or sites at which androgens may act on brain mechanisms, Sar and Stumpf (1972) investigated the localization of high-specific-activity [^{3}H]testosterone in different areas of the brain using the technique of dry-mount autoradiography. These workers showed that in both immature and adult male rats radioactivity concentrated and was retained in cell nuclei of certain hypothalamic nuclei, in addition to the cortical and medical nuclei of the amygdala and the subiculum of the hippocampus (Figure 2). Although in this study the chemical nature of the radioactivity in the brain was not determined, nevertheless the administration of an antiandrogen, which reduced the retention and concentration of the label in the tissue, supported the possibility of its still being attached to testosterone, rather than a product of its metabolism.

Although testosterone may be located in specific brain nuclei, these sites may not necessarily be those at which the steroid exerts its effects. However, it is

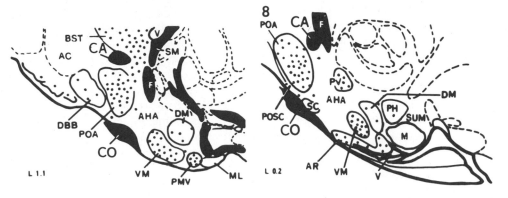

Fig. 2. Schematic drawings prepared after serial section autoradiograms from immature intact and mature castrated male rats 1 hr after the injection of 0.5–1.0 μg of [1,2-³H]testosterone, specific activity 50 Ci mM. Sagittal sections according to the atlas of DeGroot at the level of L 1.1 (Figure 7), and at the level of L 0.2. The black dots represent areas of concentration of neurons labeled with radioactivity after [³H]testosterone administration. Abbreviations: AC, nucleus (n.) accumbens septi; AHA, area anterior hypothalami; AR, n. arcuatus; BST, n. interstitialis striae terminalis (bed nucleus); CA, commissura anterior; CO, chiasma opticum; DBB, gyrus diagonalis (diagonal band of Broca); DM, n. dorsomedialis hypothalami; F, fornix; M, n. mammillaris medialis; ML, n. mammillaris lateralis; PH, n. posterior hypothalami; PMV, n. premammillaris ventralis; POA, area preoptica; POSC, n. preopticus suprachiasmaticus; PV, n. paraventricularis hypothalami; SC, n. suprachiasmaticus; SM, stria medullaris thalami; SUM, area supramammillaris; V, ventricle; VM, n. ventromedialis hypothalami. (From Sar and Stumpf, 1973.)

currently accepted that the action of steroids on their target tissues depends on acceptance of the hormone by receptor proteins in the cytosol and its subsequent transport to nuclear receptors; once the steroid has reached the nucleus, its action is expressed via the genome by induction of specific RNA and protein synthesis (King and Mainwaring, 1974; Gorski and Gannon, 1976). Therefore, if testosterone were to play a part in masculinizing central nervous mechanisms, it is reasonable to expect the presence of androgen receptor proteins in specific parts of the brain.

Although certain groups of workers have been unable to detect nuclear androgen binding in the neonatal brain (Barnea *et al.*, 1972), Kato (1975), using sucrose density gradient techniques, has reported the presence of cytosol receptors for dihydrotestosterone (DHT), a 5α reduced metabolite of testosterone in the brains of both neonatal males and females; furthermore he found that both cytosol and nuclear DHT receptors were concentrated in the middle rather than in the anterior-hypothalamic area, while a lesser but definite concentration existed in the amygdala. Barley *et al.* (1974, 1975), in an investigation of the whole brain of young adult rats by means of gel-filtration techniques, have also reported the presence of DHT receptors in the cytosol fraction.

In an attempt to localize the intracellular effects of neonatally administered androgens, Clayton *et al.* (1970) studied the incorporation of [³H]uridine into RNA in the brain of 2-day-old female rats which had been injected 4 hr before death with a low (nonmasculinizing) dose of testosterone. In autoradiographs of brain sections the uptake of uridine in most parts of the brain was found to be reduced in comparison with uninjected control females; in the preoptic and amygdaloid areas, however, the reduction of uridine uptake was significantly less than that observed in the rest of the brain. The interpretation of these data was that the administered androgens had affected RNA mechanisms at those sites which, in the male, would normally be expected to differentiate. The possibility that mechanisms of DNA, RNA, and probably also protein synthesis are involved at an early stage in the process of sexual differentiation was supported by Salaman and Birkett (1974), who were able to protect neonatal female rats from the effects of TP by the additional injection of α-amanitin, an inhibitor of nucleoplasmic or messenger–precursor RNA synthesis; hydroxyurea, an inhibitor of DNA synthesis with no effect on overall RNA synthesis, was also able to provide a high degree of protection. On the other hand, inhibition of protein synthesis was less effective while inhibition of ribosomal RNA synthesis alone was apparently completely ineffective.

2.1.3. Effect of Neonatal Estrogen

The effect of testosterone on neonatal females in causing postpubertal sterility can be elicited with equal facility by injecting instead its immediate precursor, androstenedione (Stern, 1969; Luttge and Whalen, 1970a; Edwards, 1971). With respect to the breakdown products of testosterone, however, the ring A reduced metabolite, dihydrotestosterone (DHT), is incapable of causing masculinization (Luttge and Whalen, 1970a; Brown-Grant *et al.*, 1971), while the ring A aromatized products, i.e., estrogens (Figure 3), are far more potent than is testosterone. A single injection of estradiol benzoate (EB) will cause anovulation and constant vaginal estrus in adulthood even when a dose equivalent to only $\frac{1}{10}$ that required for testosterone to have the same effect is given (Gorski, 1968; Gorski and Barraclough, 1963). A persuasive argument in favor of the concept that estrogen, rather than testosterone, is the ultimate effective agent in masculinizing the brain is the observation that pretreatment of neonatal female rats with antiestrogens blocks the effects of testosterone (MacDonald and Doughty, 1972). However, in one study, hypothalamic implants of an estrogen antagonist in the neonatal brain failed to block neonatal androgen sterilization (Hayashi, 1976); other evidence which runs counter to the "estrogen hypothesis" has been discussed by Brown-Grant (1972).

Apart from the masculinizing effects on adult female rats of a single small dose of EB in neonatal life, repeated small doses of EB during the first few days of life (Arai, 1971a), or a single large dose on the fourth postnatal day (Brown-Grant, 1974a), can cause a much more severe impairment of gonadotropin control mechanisms. The latter two treatments lead to a state of constant vaginal diestrus, a failure to ovulate following electrical stimulation of the preoptic area, and a markedly diminished gonadotropin rise in response to ovariectomy (Arai, 1971a). Since none of these features is characteristic of the female which has been masculinized either by TP or by a single small dose of EB, the possibility remains that high levels of estrogen may differentiate (or perhaps damage) neural or extraneural sites additional to those at which testosterone or small doses of estrogen may normally be expected to act.

It also seems likely that ovarian secretions might continue to have a potentially deleterious effect on gonadotropin throughout life. By judicious transplantation of ovarian tissue, Arai (1971a) showed that removal of ovaries from rats treated with small doses of TP in neonatal life either prevented or postponed the onset of the "delayed anestrous syndrome." Brown-Grant (1974a), using a different experimental approach, administered heavy doses of either TP or EB to normal young adult females of 60–70 days of age; from the subsequent histological appearance of the reproductive tract, and the gonadotropin response to estrogen and progesterone stimuli at various times after hormone treatment, it was clear that the normal

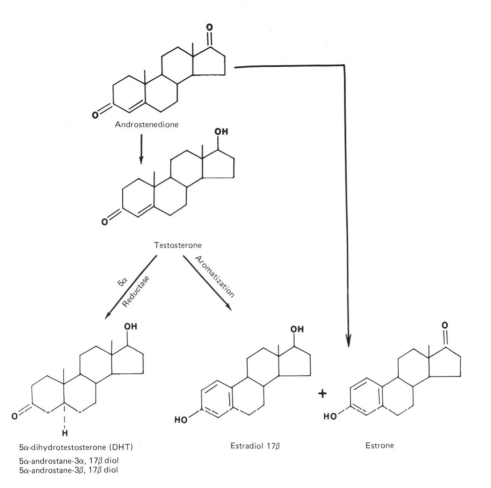

Fig. 3. Conversion of androstenedione to active metabolites.

functioning of the hypothalamopituitary system (but not the ovary) had been affected in some animals. The concept of a single circumscribed period in neonatal development during which steroid hormones can affect neural mechanisms concerned with ovulation may therefore need to be modified.

2.1.4. Localization of Estrogen in the Brain

Dry-mount autoradiographs of brain sections obtained from adult female rats which had been previously injected with [³H]estradiol-17β have shown that the isotope concentrates in the cell nucleus of several different hypothalamic nuclei (Stumpf, 1968), in the medial preoptic nucleus, and in nuclei of the corticomedial complex of the amygdala (Stumpf and Sar, 1971). Similar accumulations of [³H]estradiol-concentrating neurons have been seen in the mouse, tree shrew, and squirrel monkey (Stumpf *et al.,* 1974). A recent study of [³H]estradiol uptake in the ovariectomized rhesus monkey has shown concentrations of the steroid in the medial preoptic–anterior hypothalamic areas, the ventromedial n., arcuate n., medial n. of amygdala, bed n. of the stria terminalis, and both basophil and acidophil cells of the anterior pituitary (Pfaff *et al.,* 1976).

In 2-day-old female rats a similar pattern of uptake was observed following injection of the labeled steroid, while competition experiments showed that nuclear concentration of radioactivity was inhibited by either "cold" estradiol or testosterone (except in the medial preoptic area where it was only partly reduced) but not by DHT (Sheridan *et al.*, 1974). This characteristic distribution overlaps to some extent with that observed in autoradiographs obtained from [^{3}H]testosterone-injected male rats, particularly with respect to the medial preoptic area (Sar and Stumpf, 1972); but this might be explained by the androgen having at least in part metabolized to estradiol. Nevertheless, the possibility that certain cells may concentrate both hormones, and that separate binding sites for both androgens and estrogens might exist, cannot be ignored. It is worth noting that after subcutaneous injection of [^{3}H]testosterone the percentage of labeled cells in the hypothalamus and the density of grains per labeled cell was lower than after an injection of [^{3}H]estradiol (Tuohimaa, 1971), a finding which could be compatible with the idea that there may be fewer androgen than estrogen binding sites in the brain and/or that the macromolecular binding of androgens is weaker than that of estrogens. Also worth mentioning is the observation that uptake of labeled estradiol in the hypothalamus, preoptic area, and cortex (relative to that in the plasma) in gonadectomized male and female rats, which had been injected with the tritiated hormone and killed 2 or 24 hr later, was the same in both sexes (Whalen and Luttge, 1970). Likewise there were no sex differences in the regional distribution of estrogen metabolites in the brain after administration of [^{3}H]estradiol-17β (Luttge and Whalen 1970*b*). Although certain reports have suggested that androgenized females and immature males take up and retain less labeled estrogen than do normal females (McEwen and Pfaff, 1970; Clark *et al.*, 1972), other studies have failed to detect such a difference (Eisenfeld and Axelrod, 1966; Whalen and Luttge, 1970; Maurer and Woolley, 1975).

The presence of estrogen receptors in the cytosol fractions of brain homogenates is well documented (Kato, 1971; Kato *et al.*, 1974; see also Zigmond, 1975); however, with the use of a nuclear exchange assay, Westley and Salaman (1976; 1977) were able to demonstrate the presence of nuclear estrogen-receptors in the ventral portion of the hypothalamus and the amygdala of the neonatal rat. These receptors were shown to have a high affinity for estradiol, which was suppressible by synthetic estrogen (diethylstilbestrol). In the intact neonatal male or the androgen-injected neonatal female, the numbers of available (unoccupied) nuclear estrogen receptor sites in the hypothalamoamygdaloid region of the brain were significantly fewer than in 4-day-old females or 4-day-old males which had been castrated 24 hr earlier. According to Westley and Salaman (1977), these findings can be attributed to the occupation of nuclear binding sites by endogenous estrogen arising as a result of the aromatization of androgens in the "hypothalamoamygdaloid" area.

Strong support for a possible role of estrogen in masculinizing the brain has been provided by Naftolin *et al.* (1975). These workers homogenized various areas of the rat brain, and incubated the resultant homogenates with either [^{14}C]androstenedione or [^{3}H]testosterone; after various purification and separation procedures it was possible to identify small amounts of either estrone or estradiol, respectively, in the incubates. These results suggested that aromatization of the androgens had taken place, and subsequent work substantiated this view. The presence of aromatizing systems in the hypothalamus and the limbic system (i.e., the hippocampus and amygdala) have been demonstrated not only in the guinea pig and rabbit but also in the rhesus monkey and the human fetus. The data obtained from *in vitro* preparations was later supported by *in vivo* studies (Knapstein *et al.*, 1968; Weisz and Biggs, 1974). Of particular interest were those in which an isolated cranial vault containing the brain and pituitary gland of immature male rhesus monkeys were perfused. In this preparation both free and conjugated [³H]estrone and free [³H]estradiol were identified in the venous effluates after injection of [³H]androstenedione into the perfusate; moreover, dissection and monitoring of the tissues showed that these labeled steroids were present only in the hypothalamus and the limbic system (Flores *et al.*, 1973).

2.1.5. Plasma Estrogen Levels in the Neonatal and Infant Rat

If indeed estrogen is the effective hormone which masculinizes the brain, then it is curious that infant rats do not succeed in sterilizing themselves, since total plasma estrogen levels are very high in the infant female (e.g., Ojeda *et al.*, 1975). However, this state of affairs may in part be accounted for by the presence of a protein, namely α-fetoprotein (AFP), with a high affinity for estrogen, which is also present in the plasma of neonatal rats and which may help to sequester the hormone (Raynaud *et al.*, 1971; Nunez *et al.*, 1971; Puig-Duran *et al.*, 1978). An additional factor may be the presence of intracellular AFP which has been localized in presumed selective neurons of the hypothalamus and amygdala of the developing rat brain by means of the unlabeled antibody peroxidase–antiperoxidase technique (Benno and Williams, 1978). A further feature in developing rats is the high plasma FSH values (see, e. g., MacKinnon *et al.*, 1976) which decline at about the same time as AFP levels are diminishing. Although there is no rigorous evidence to suggest that estrogen binding to AFP accounts entirely for the failure of negative feedback inhibition, nevertheless it is reasonable to assume that as the estimated concentrations of the biologically active estrogen are low between 0 and 23 days of age (Puig-Duran *et al.*, 1978), while its affinity for AFP is very strong, the resultant activity of the hormone is too weak either to inhibit FSH levels or to masculinize the brain.

2.2. Gonadotropin Response to a Steroid Stimulus

In those spontaneously ovulating animals from which experimental evidence has been forthcoming, it appears that the central nervous system of the female differs from that of the male in that the control of gonadotropin output, which is responsible for ovarian function and ovulation, is of a cyclic nature; in contradistinction the secretory pattern of the male is relatively constant. The preovulatory LH surge which occurs in the evening of proestrus is characteristic of the cycling female rat (Brown-Grant *et al.*, 1970), and can be stimulated in the ovariectomized adult female by either two spaced injections of estradiol benzoate (EB) (Caligaris *et al.*, 1971), or by a single injection of progesterone after estrogen priming (Brown-Grant, 1974*b*). A simpler experimental model using either the adult ovariectomized female or the intact immature female (Figure 4) is that in which only a single dose of EB is given at noon on the first experimental day (day 1); in response to the steroid, LH concentrations are initially inhibited (negative feedback) until about 54 hr later, in the evening of day 2, when a surge of LH occurs (so-called "positive feedback") (Table I) which is repeated on successive evenings over the next few days (Burnet and MacKinnon, 1975).

Although the LH response to estrogen, or to progesterone after estrogen priming, can be readily elicited in castrate females and in males which have been

PAMELA C. B.
MACKINNON

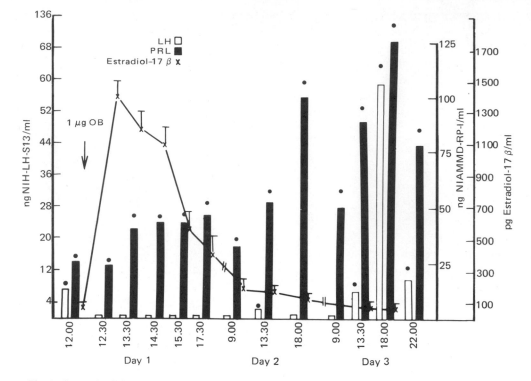

Fig. 4. Serum luteinizing hormone, prolactin, and 17β-estradiol concentrations (mean ± SEM) in 21-day-old female rats injected with 1 μg estradiol benzoate at 1200 hr on day 1 and killed at variable intervals of time thereafter. 4–8 animals per group. (From Puig-Duran and MacKinnon, 1978a.)

castrated at birth, no such effect can be obtained in either adult castrate males or neonatally androgenized females castrated as adults (Brown-Grant, 1972). In the immature intact female in which circulating free estrogen levels are apparently negligible (Puig-Duran *et al.*, 1978), the response seems to develop at about 14 days of age, since it cannot be provoked before that time either by progesterone in the estrogen-primed animal or by synthetic estrogens, for which plasma receptor proteins have little affinity (Puig-Duran and MacKinnon, 1978a). These results suggest that although mechanisms underlying the preovulatory LH surge may be sexually differentiated at an early age, they become overtly so between 12 and 15 days of age.

With respect to serum prolactin concentrations, Neill (1972) observed an increased evening output of the hormone 23 hr after a second daily noon injection of estrogen in ovariectomized adult females, but not in adult castrate males. A study of the response in intact immature rats of different sexual status (i.e., normal males and females exposed to androgens in neonatal life; normal females and males castrated within 24 hr of birth) (Puig-Duran and MacKinnon, 1978b) has shown that a single dose of estrogen injected at noon on day 1 will give rise to a peak of prolactin 54 hr later (day 3) in both 21-day-old females and males castrated within 24 hr of birth, but not in normal males. Somewhat unexpectedly, however, it was possible to elicit the response in androgen-sterilized females; this suggests either that the optimum time of development, during which androgen treatment is able to differentiate mechanisms concerned with PRL release, is different from that of which LH mechanisms are affected, or that PRL mechanisms are differentiated by substances other than testicular androgens or their metabolites.

It is of particular note that the LH response to an estrogen stimulus which is central to the question of sexual differentiation in spontaneous ovulators has not been elicited in voles (Milligan, 1978) or apparently in any other reflex ovulator.

2.3. Possible Sites of Sexual Differentiation: Mechanisms Controlling Gonadotropin Output

2.3.1. Preoptic–Anterior Hypothalamic Area

The technique devised by Halasz and Pupp (1965), in which a curved knife held in a stereotaxic instrument enabled the hypothalamus to be disconnected entirely or in part from other areas of the brain, has done much to support the concept that the integrity of the preoptic–anterior hypothalamic area is essential for the phasic output of gonadotropins. Although there would appear to be certain inconsistencies (see Butler and Donovan, 1971), in general it was found that cuts placed caudal to the preoptic–suprachiasmatic area led to acyclicity and constant estrus, while frontal cuts, which are designed to leave the preoptic hypothalamopituitary complex intact, were commensurate with cyclicity (albeit irregular) and ovulation. Complete deafferentation of the mediobasal hypothalamus, on the other hand, resulted in acyclic animals with leukocytic vaginal smears (Halasz and Gorski, 1967; Halasz, 1969; Köves and Halasz, 1970). Much further experimentation has supported this work (see Schwartz and McCormack, 1972), including that in which ovulation provoked by stimulation of the preoptic area was blocked by small bilateral cuts immediately caudal to the suprachiasmatic n. (Tejasen and Everett, 1967).

With respect to pituitary sensitivity, it is well accepted that a rise in plasma estrogen, especially if it is followed by an increase in progesterone levels, will facilitate the LH release which occurs on injection of luteinizing hormone-releasing hormone (LH-RH) (Debeljuk *et al.*, 1972; Aiyer and Fink, 1974). It is notable, therefore, that ovariectomized animals with long-term deafferentation of the mediobasal hypothalamus [Norman *et al.*, 1972 (in hamsters); Blake *et al.*, 1973;

Table I. Serum Luteinizing Hormone and Prolactin Levels (ng/ml) in 21-Day-Old Male, Female, Neonatally Androgenized Female, and 1-Day-Castrate Male Rats following a Single Injection of 1 μg Estradiol Benzoate or Oil at Noon (1200 hr) on Day 1[a]

Groups	Treatment	Basal values LH	Basal values PRL	Day 2 (1200 hr) 24 hr postinjection LH	Day 2 (1200 hr) 24 hr postinjection PRL	Day 3 (1800 hr) 54 hr postinjection LH	Day 3 (1800 hr) 54 hr postinjection PRL
♀	Oil	6.0 ± 4.4	16.5 ± 5.1 (6)	3.6 ± 2.9	18.4 ± 6.0 (6)	5.7 ± 3.8	17.1 ± 9.4 (8)
	EB			0.3 ± 0.05	60.5 ± 13.3 (6)	59.5 ± 9.1	116.3 ± 11.7 (8)
♂	Oil	2.5 ± 0.9	25.1 ± 4.9 (6)	1.9 ± 0.8	28.3 ± 5.6 (5)	$.9 \pm 0.8$	30.1 ± 4.2 (5)
	EB			0.3 ± 0.02	25.1 ± 2.1 (8)	$.4 \pm 0.03$	30.1 ± 2.6 (8)
♀ A	Oil	1.6 ± 0.1	42.5 ± 3.6 (6)	1.4 ± 0.6	40.2 ± 4.2 (5)	$.1 \pm 0.9$	38.9 ± 4.7 (5)
	EB			0.9 ± 0.02	69.8 ± 7.7 (8)	$.4 \pm 0.06$	115.6 ± 2.4 (8)
1-day ♂	Oil	24.4 ± 3.3	26.2 ± 2.7 (6)	19.7 ± 6.7	30.2 ± 4.1 (5)	20.3 ± 5.9	29.5 ± 11.2 (5)
	EB			2.3 ± 0.7	59.6 ± 3.6 (8)	28.8 ± 1.4	125.8 ± 12.0 (8)

[a] Values are mean ± SEM. Numbers of animals in parentheses (Puig-Duran and MacKinnon).

Fink and Henderson, 1977*b*] or pituitary stalk section (Greeley *et al.,* 1975) showed an estrogen-facilitated increase in LH output to exogenously administered LH-RH. Nevertheless, in ovariectomized rats with acute hypothalamic deafferentation (Fink and Henderson, 1977*b*), in which the hypothalamic content of LH-RH was unlikely to have been greatly affected (see Brownstein *et al.,* 1976), somewhat different results were observed; the facilitation of the LH response to exogenous LH-RH, which occurs after exposure to estrogen and progesterone, was significantly reduced in these animals when compared with results obtained in unlesioned but similarly treated rats. These findings were attributed to prevention (by virtue of the neural lesion) of the normal priming effect on pituitary tissue which results from the release of endogenous LH-RH (Fink *et al.,* 1976).

The combined findings suggest that neural mechanisms outside the mediobasal–hypothalamus–pituitary unit are also concerned with gonadotropin output and the elicitation of a steroid-stimulated "preovulatory-type" LH surge. However, although there is evidence to suggest that the midbrain may have a part to play in these mechanisms (Sawyer, 1957; Coen and MacKinnon, 1976), as a site of sexual differentiation the preoptic area appears to be the stronger candidate. Indeed, a number of workers have provided evidence of metabolic changes (Moguilevski *et al.,* 1969; Moguilevski and Rubinstein, 1967) and of changes in nuclear and nucleolar size of nerve cells in the anterior hypothalamic–preoptic area of normal male and female rats and in experimentally treated animals (Döcke and Kolocyek, 1966; Pfaff, 1966; Arai and Kusama, 1968; Dörner and Staudt, 1968). The nuclear size of nerve cells in this area appears to be less in male rats than in normal females and in males castrated on the first day of life. However, the animals were investigated at only one time of day, and possible changes in nuclear size due to sexual differences in circadian periodicities (see ter Haar *et al.,* 1974) were unaccounted for.

A high incidence of anovulatory sterility in adult females following TP implants in the dorsal preoptic area has been observed by Lobl and Gorski (1974) as late as 11 days postnatally. However, similar implants in the cerebral cortex were stated to be as successful in producing sterility as were those in the preoptic area, and although estrogen receptor sites have been reported in the cortex of young rats (Barley *et al.,* 1974), the possible effects of hormone diffusion in Lobl and Gorski's animals to extracortical areas should not be dismissed. This is especially so since stereotaxic implantation of micropellets of TP (insufficient in amount to cause masculinization if given systemically) in the mediobasal hypothalamus or the ventromedial arcuate region were reported to be more successful in causing constant vaginal estrus in adulthood than were implants placed in the anterior hypothalamus (Wagner *et al.,* 1966; Nadler, 1973; Döcke and Dörner, 1975).

Of much interest is the observation that electrical or chemical stimulation of the preoptic area will cause ovulation not only in female rats but also in neonatally androgenized females which have been kept under conditions of normal or continuous lighting (Terasawa *et al.,* 1969; Everett *et al.,* 1970). A similar effect can be obtained in adult male rats bearing ovarian grafts (Quinn, 1966), while correct parameters of electrical stimulation of the preoptic area in males can cause a "female-type" LH surge of preovulatory dimensions (Fink and Jamieson, 1976). This combination of results suggests that the mechanisms lying between the preoptic area and the anterior pituitary gland which are concerned with gonadotropin output are similar in both the male and female, and it has been argued that an afferent connection to the preoptic area rather than the preoptic neurons themselves might be sexually differentiated (Everett, 1969).

By cutting the projection pathway from the amygdala to the preoptic area (POA) and allowing orthograde degeneration to occur, Raisman and Field (1971, 1973) demonstrated that fibers of the stria terminalis, in rats of both sex, establish synaptic contacts in the neuropil of the dorsal part of the POA and of a region between the ventromedial n. and the arcuate n. of the hypothalamus. The majority of nondegenerated terminals, which were nonstrial and therefore not of amygdaloid origin, were found to contact dendritic shafts (approx. 50%) while the remainder contacted dendritic spines (approx. 3–5%). It was of particular interest, however, that a significantly greater number of fibers making synaptic contacts with dendritic spines in the POA in cyclic animals (i.e., in normal females, females which had been treated with androgens at 16 days of age, and in males which had been castrated within 24 hr of birth) than in noncyclic animals (i.e., normal males, males which had been castrated 7 days after birth, and in females which had been androgenized at 4 days of age). No such differences were found in the ventromedial area (Table II).

Further work showed that the higher incidence of nonamygdaloid synapses on dendritic spines in the POA could be correlated not only with the ability to initiate a preovulatory surge of gonadotropins, but also with the ability of progesterone to facilitate a behavioral response, i.e., lordosis. The significance of this potentially important finding is not yet clear since Brown-Grant and Raisman (1972) found that lesions of the "sexually differentiated" part of the preoptic area, although causing an acute blockade of ovulation and a high incidence of both immediate and delayed pseudopregnancies, are nevertheless commensurate with ultimate cyclicity and reproductive function.

Investigations of a quite different nature have recently been reported in which single unit recordings were made from neurons in the preoptic and anterior hypothalamic areas of rats of different sexual status (Dyer et al., 1976). The neurons were classified according to their responses to single-shock stimulation of the mediobasal hypothalamus, and subsequently according to their response following stimulation of the corticomedial amygdala. The cells were either antidromically invaded, orthodromically activated or inhibited, or unaffected by the stimulus, and all

Table II.   Incidences per Grid Square (Mean ± Standard Error) of the 4 Types of Synapse in the Preoptic Area (POA) and Ventromedial Nucleus (VMH) in the 6 Groups of Animals[a]

		Endocrine status					
		Cyclic			Noncyclic		
		F (16)	F16 (7)	MO (9)	M (11)	M7 (7)	F4 (14)
POA							
Nonstrial	Shaft	50.0 ± 2.0	53.1 ± 3.0	54.3 ± 3.0	55.2 ± 1.4	52.6 ± 1.3	48.4 ± 2.4
	Spine	5.3 ± 0.3	5.4 ± 0.5	5.0 ± 0.3	3.3 ± 0.2	3.9 ± 0.4	3.5 ± 0.2
	Shaft	1.2 ± 0.1	1.0 ± 0.1	1.2 ± 0.1	1.2 ± 0.1	0.7 ± 0.1	0.9 ± 0.1
	Spine	1.6 ± 0.2	1.4 ± 0.3	1.8 ± 0.2	1.7 ± 0.1	1.3 ± 0.1	1.6 ± 0.2
VMH							
Nonstrial	Shaft	46.3	44.4 ± 1.6	43.6 ± 1.6	42.4 ± 1.5	42.3 ± 1.4	42.8 ± 1.8
	Spine	12.9	12.6 ± 0.8	13.2 ± 0.9	12.0 ± 0.7	13.9 ± 0.9	13.1 ± 0.9
Strial	Shaft	2.3	1.6 ± 0.2	1.9 ± 0.3	1.9 ± 0.3	1.8 ± 0.2	1.9 ± 0.2
	Spine	6.2	6.2 ± 0.6	6.6 ± 0.5	7.0 ± 0.6	6.7 ± 0.4	6.5 ± 0.5

[a]The "cyclic" group consists of F (normal females), F16 (females treated with androgen on day 16), and MO (males castrated within 12 hr of birth); the noncyclic group consists of M (normal males), M7 (males castrated on day 7), and F4 (females androgenized on day 4). Number of rats is given in parentheses. Statistical comparisons between the mean incidence of spine synapses of nonstrial fibers in the different groups of cyclic and noncyclic animals were, in almost all cases, significant (Raisman and Field, 1973).

combinations were observed. The results showed that in males or in androgenized females the antidromically identified cells were more likely to receive a synaptic connection from the amygdala than the same cells in females and neonatally castrated males. From these and other findings the authors suggested that exposure to androgens in neonatal life may facilitate the formation of both excitatory and inhibitory synaptic contacts in the preoptic–anterior hypothalamic area. In addition, it was found that neurons which were not influenced by stimulation of the mediobasal hypothalamus fired twice as fast in normal females and neonatally castrated males as the same cell type in normal males and androgenized females.

Although a sexual dimorphism in the preoptic area has apparently been shown by both neuroanatomical and neurophysiological studies, the two sets of data do not appear to be directly compatible. Dyer *et al.* (1976) have found a sexually differentiated neural input from the amygdala, whereas Raisman and Field's sexual dimorphism was with respect to fibers of nonamygdaloid origin. But as the former authors point out, "it is possible that a fixed number of amygdaloid afferents could be more effective in altering the firing pattern of a cell if that cell receives less spine synapses from another, non-amygdaloid, pathway." Yet again, Dyer and his colleagues found that only a small percentage of cells which were orthodromically influenced by the amygdala was located in the dorsomedial preoptic area described by Raisman and Field (1973) as being sexually differentiated. But it is possible that the sexually differentiated input shown in the neurophysiological study may have been from efferent fibers of the stria terminalis which terminate outside Raisman and Field's area. This and other possibilities which could explain the discrepancies between the two studies are fully discussed by Dyer *et al.* (1976).

2.3.2. Amygdala

In addition to the work previously cited by Clayton *et al.* (see p. 187) in relation to the possible intracellular effects of androgen on the neonatal brain, and by Dyer and his colleagues (see p. 196) with respect to sexual dimorphism of amygdaloid neurons, there is certain further, although less direct, support for the involvement of the amygdala in sexual differentiation. Staudt and Dörner (1976) have reported that neuronal nuclear size in the medial and central parts of the amygdala was significantly larger in adult male rats which had been castrated within 24 hr of birth than in adult intact males (with presumably normal levels of circulating androgens) or males castrated after neonatal life. The nuclear size of rats castrated at one day of age approximated to that of the female; furthermore the administration of neonatal androgens prevented the difference.

Electrical stimulation of the corticomedial complex of the amygdala has been shown to increase plasma LH levels and to provoke ovulation in "constant-estrus" females exposed to constant illumination (Velasco and Taleisnik, 1969); furthermore it has been reported that stimulation in the same area can cause a rise in plasma LH levels in estrogen-primed ovariectomized animals (Kawakami and Terasawa, 1973). It is worth noting therefore that LH responses to electrical stimulation were unobtainable in either males transplanted with ovarian tissue (Arai, 1971) or neonatally androgenized females which had been ovariectomized and primed with estrogen as adults (Kawakami and Terasawa, 1973).

2.3.3. Suprachiasmatic Nucleus

Remarkably scant attention has been paid to the role that the suprachiasmatic nucleus (Sch.n.), which lies within the preoptic–anterior hypothalamic area, may play in the control of the preovulatory surge in cyclic rats, more especially since electrical stimulation of the nucleus apparently causes as great an increase in the concentration of luteinizing hormone releasing factor in pituitary portal blood as does stimulation of the medial preoptic area (Chiappa, Fink, and Sherwood, 1977).

Ever since 1950 when Everett and Sawyer first demonstrated a critical period of the proestrous day during which prevention of a neural event by a central nervous depressant (sodium pentobarbital; Nembutal) led to a 24-hr delay in ovulation, the importance of a daily neural signal for the maintenance of regular estrous cycles has been recognized (see Schwartz and McCormack, 1972), and it is notable that lesions of the Sch.n., which receives fibers directly from the retina (Moore, 1974), prevent circadian rhythms of both neural and hormonal origin (Moore and Eichler, 1972; Coen and MacKinnon, 1976; Raisman and Brown-Grant, 1977). Furthermore the integrity of the Sch.n. appears to be essential for the estrogen-stimulated preovulatory LH surge (Coen and MacKinnon, 1976). It is tempting, therefore, to think that this nucleus, or mechanisms controlling its daily signal, might in some way be sexually differentiated. There is evidence of a biochemical nature, albeit indirect, for this hypothesis. A study of the incorporation of ^{35}S from methionine into TCA-precipitable proteins of the brain at intervals throughout a 24-hr period has indicated the presence of a circadian rhythm of incorporation, the phase of which is sexually differentiated (Haar *et al.,* 1974) (Figure 5). In adult intact females and in males castrated within 24 hr of birth, peak levels of incorporation were observed in the latter part of the day while a nadir occurred in the morning. In contradistinction, the peaks and nadirs of ^{35}S incorporation in intact males and androgenized females were the reverse of those seen in the cyclic animals, being about 120° out of phase. Further work showed that lesions of the Sch.n. prevented not only the recurrent evening increases in ^{35}S incorporation in brain tissue but also the evening LH surges in ovariectomized estrogen-stimulated rats (Coen and MacKinnon, 1977). These findings appear to gain support from recent work in which stereological techniques were used to assess the area of synaptic contact in electron micrographs of the Sch.n. A significant difference was apparently found between data obtained from intact female rats and that from intact males; the area was significantly greater in males studied at intervals from birth to adulthood (Dr. Patrick Thomas, personal communication).

2.3.4. Anterior Pituitary

The conclusion drawn from Pfeiffer's (1936) original studies was that the anterior pituitary was implicated in the process of sexual differentiation. However, after the demonstration that male pituitary tissue transplanted to the sella turcica of hypophysectomized females was able to sustain female reproductive activity (Harris and Jacobsohn, 1952; Martinez and Bittner, 1956; Segal and Johnson, 1959), and so too was pituitary tissue obtained from androgenized females (Adams-Smith and Peng, 1966), the focus of attention was directed to the brain. Recently, however, there has been renewed interest in the possibility of sexual dimorphism of anterior pituitary tissue. Pituitaries of neonatally androgenized rats are capable of releasing an ovulatory quota of LH in response both to exogenous gonadotropin-releasing hormone (GnRH) (Borvendeg *et al.,* 1972) and to stimulation of the medial preoptic area which causes a release of endogenous releasing hormone (Gorski and Barraclough, 1963; Terasawa *et al.,* 1969; Jamieson and Fink, 1976). However, the pituitary response to LH-RH is dependent on its hormonal environment (Debeljuk *et al.,* 1972; Aiyer and Fink, 1974), and a study of its responsiveness after injection of the releasing hormone in ovariectomized androgen-sterilized rats which had been stimulated with estrogen alone, or with progesterone after estrogen priming, showed a significant reduction in the increment of LH compared with that found in either similarly treated castrate males or nonsterilized females (Barraclough and Turgeon, 1974). These results were supported by those of Fink and Henderson (1977a), who showed that the LH response to LH-RH in ovariectomized rats could be restored to that observed on the evening of proestrus provided that an appropriate estrogen–progesterone regime was instituted; however, in comparison, the steroid facilitated LH response to LH-RH in castrated androgenized animals, although present, was distinctly less than that observed in ovariectomized rats, while that in castrate males was significantly reduced. In addition, in castrate

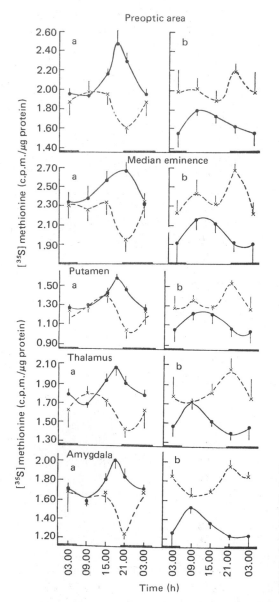

Fig. 5. Circadian rhythms in the incorporation of ^{35}S from methionine into proteins (cpm/μg protein) of the preoptic area, medial eminence, putamen, thalamus, and amygdala of (a) normal adult female ($\bullet$) and normal adult male ($\times$) rats, and (b) neonatally castrated adult male ($\times$) and neonatally androgenized female ($\bullet$) rats over a 24-hr period. The black horizontal bars indicate hours of darkness. The vertical lines are $\pm$ SEM. (From ter Haar *et al.*, 1974.)

androgenized females the FSH responses to LH-RH which followed various steroid treatments were significantly different from those observed in either gonadectomized male or female animals. Yet Mennin *et al.* (1974) administered LH-RH in varying dosage to groups of proestrous rats just prior to the critical period, and to androgenized females, all of which had received an ovulation-blocking dose of sodium pentobarbital, which presumably prevented pituitary priming due to central nervous activity and liberation of endogenous LH-RH (see Aiyer *et al.,* 1974). The pituitary response to LH-RH as measured by the magnitude and time-course of LH release in these two groups of animals was largely identical (with the exception of the response to a heavy dose of LH-RH), as was the reduction in pituitary LH concentration.

The combined results of these studies suggest that there are differences between the pituitary responses to LH-RH in rats of different sexual status under different steroid regimes, but whether these are due to direct effects on neural mechanisms concerned with endogenous LH-RH release (and therefore pituitary sensitivity) and/or on pituitary tissue is unclear.

In an investigation of cytoplasmic estrogen receptors in the hypothalamus and anterior pituitary of adult and immature rats of both sexes and of neonatally androgen-treated females, no differences in receptor levels were found among any of the groups belonging to the same strain (Cidlowski and Muldoon, 1976). In contrast, when receptor depletion and later receptor replenishment were studied in response to an estrogen stimulus, marked differences were observed between the sexes. In respect to the anterior pituitary, the rate of replenishment of the receptor was much faster in the androgenized female than in the untreated female, while the level 15 hr after estrogen injection was appreciably greater. The pattern of receptor depletion and replenishment in the androgenized female was similar to that in the male which had also received androgens in neonatal life. In the hypothalamus on the other hand, receptor depletion was far greater in the untreated than in the androgenized animals. In the course of the same investigation Cidlowski and Muldoon could not find evidence of a defective synthesis of estrogen receptor. These data, therefore, fail to support an earlier hypothesis which suggested that nuclear receptor activity and the ability to synthesize new estrogen receptor was impaired by androgenization (Vertes *et al.,* 1973).

There is now good evidence that the pituitary gland is responsible for the maintenance of sex differences in hepatic steroid metabolism (Colby *et al.,* 1973; Denef, 1974; Gustafsson and Stenberg, 1974; Lax *et al.,* 1974), and that this difference occurs at about 30 days of age (Stenberg, 1976). If pituitary tissue obtained from female rats of 28 days and older is placed beneath the kidney capsule of hypophysectomized adult males, it causes a change in hepatic steroid metabolism to that of female type; if donor tissue is taken from females below 28 days of age no such change is seen. The transitional period during which a "feminizing factor" begins to be secreted by anterior pituitary tissue appears to be between 28 and 35 days of age. Recent evidence, based on results of electrothermic lesions of the hypothalamus, has indicated that the release of feminizing factor from the pituitary is controlled by a release-inhibiting factor from the hypothalamus of male but not female rats (Gustafsson *et al.,* 1976). The combined results suggested that, with respect to hepatic steroid metabolism, the hypothalamic pituitary axis in rats (of the Sprague-Dawley strain) matures between 28 and 35 days of age.

It may also be noteworthy that in both rats and monkeys characterization studies of pituitary LH and FSH have shown that after gonadectomy their properties are modified (Bogdanove *et al.*, 1974; Peckham and Knobil, 1976*a*). Furthermore the molecular size of LH and FSH from male pituitaries is apparently smaller, and their disappearance from the circulation faster, than that of the corresponding hormones in female monkeys (Peckham and Knobil, 1976*b*). In contrast the molecular weight of FSH in the female rat is apparently smaller than that in the male, while the clearance of endogenous FSH in the androgen-treated orchidectomized rat takes longer than that in the androgen-deprived orchidectomized rat (Bogdanove *et al.*, 1974).

2.4. Sexual Behavior: Sexual Differentiation

A further manifestation of the sexually differentiated brain to which a good deal of attention has been paid is that of sexual behavior. However, sexual differentiation is not as well defined in terms of this parameter as it is in respect of patterns of gonadotropin output. The reason for this difference is that changes in behavior, far from being clear-cut, tend to occur as a result of alterations in the *thresholds* of response to certain hormonal stimuli.

2.4.1. Female-type Behavior in the Rat

Females. Although the characteristics of mammalian female behavior may best be considered in terms of sexual attractiveness, proceptivity, and receptivity (Beach, 1976), the behavioral feature to which most attention has been paid in the rat is that of lordosis. The lordosis quotient (LQ), defined as the ratio between the number of lordoses to the number of mounts in response to an experienced male, can be elicited with maximal ease around the time of ovulation. After ovariectomy the response is depleted, but it can be rapidly restored by estrogen either alone or in combination with progesterone (Whalen and Edwards, 1967).

Early and even some recent reports on the effects of prenatal androgens or of estrogen on subsequent adult behavior indicated almost universally that lordosis is markedly depressed by such treatment (see, e.g., Hendricks and Welton, 1976), but the results of a recent extensive and careful study have suggested otherwise (Brown-Grant, 1975; see also Mullins and Levine, 1968). Adult females treated either with varying doses of TP or EB neonatally showed a marked and progressive capacity for lordosis on exposure to concentrated periods of mounting activity (see Table III). According to Brown-Grant (1975), the discrepancies between his data and the earlier reports were probably due to the latter having been based on tests consisting of a low number of mounts, or short test periods with only a single male. The failure of perinatal androgen treatment to disrupt female receptive behavior in rats has also been reported in ferrets (Baum, 1976) and in rhesus monkeys (Goy, 1970) and is supported by the results of a study in women with the adrenogenital syndrome (Money and Ehrhardt, 1972). An interesting suggestion put forward to account for the progressive increase of the LQ in neonatally androgenized animals during mating tests was the possibility of a mating-induced release of GnRH into the central nervous system (Brown-Grant, 1975). This hypothesis was based on previous observations of facilitated LQs in ovariectomized estrogen-primed rats injected systemically with GnRH (Moss and McCann, 1973; Pfaff, 1973) and is supported by recent work which has shown that lordosis is facilitated by hypophysectomy (Crawley *et al.*, 1976), which increases the hypothalamic content of LH-RH.

Table III.   Incidence of Persistent Vaginal Cornification, Failure of Ovulation, Mating in Anovulatory Rats at 160–180 Days of Age and Ovulation after Mating in Female Rats Treated with Testosterone Propionate (TP) on Day 4 of Life

Dose of TP (μg)	No.	Persistently cornified (%)		Anovulatory (%)		Mating (%)	Ovulating after mating (%)	LQ[a]
		Day 70	Day 160	Day 70	Day 160			
3	25	0	28	0	20	100	40	
10	38	18	63	8	68	93	12	92 ± 2 (18,0)
30	32	41	94	34	94	93	18	82 ± 6 (18,0)
100	44	27	95	23	95	98	10	89 ± 4 (27,1)
250	36	39	83	36	83	97	3+	93 ± 2 (14,0)
1250	46	59	98	65	98	100	0	82 ± 3 (31,0)
2500	8	88	100	88	100	100	0	81 ± 4 (8,0)

[a] Group means ± SEM. First figure in parentheses is the number of animals tested, and second figure is number with LQ = 0 (Brown-Grant, 1975).

The effect of estrogen on adult neonatally androgenized female rats was observed to increase the LQ still further, although if the steroid was followed by an additional dose of progesterone a few hours before testing then the LQ was significantly depressed, as it is in normal male rats (Brown-Grant, 1975). Södersten (1976) has also examined the effects of progesterone on the LQ in estrogen-primed animals and in general confirmed Brown-Grant's findings, but although progesterone in different doses consistently facilitated the LQ in female rats (see also Clemens *et al.*, 1969; Brown-Grant, 1975), the comparative effects in castrate males and androgenized females did not appear to be so clearly different as those obtained in the earlier study; the discrepancies may have been due to a species difference.

Males. The normal adult male rat is fully capable of exhibiting lordosis when mounted by vigorous male partners, but this is greatly enhanced if the animals are treated with estrogen and more especially if the animals have been previously castrated (Davidson, 1969; Davidson and Levine, 1969; Brown-Grant, 1975).

In adult genetic male rats which had been treated with antiandrogens both before and after birth, elicitation of lordosis was reported to be achieved as frequently in response to EB, followed by an injection of progesterone, as it was in normal females (Nadler, 1969). This effect has also been observed in adult males which had been castrated within a few hours of birth (Harris, 1964; Harris and Levine, 1965; Grady *et al.*, 1965). In adult one-day castrate males which had been primed with estrogen, injected with progesterone, and tested 4–6 hr later, the lordotic response was enhanced, although according to Gerall *et al.* (1967) it did not compare in quantity and quality with that observed in ovariectomized females which had been similarly treated.

2.4.2. Male-type Behavior in the Rat

Male patterns of sexual behavior are analyzed in terms of the latency and frequency of mounting, thrusting, intromission, and ejaculation when the animal is placed with a receptive female. But some of these male parameters are as difficult to assess as female behavior patterns since mounting is quite commonly observed in normal females of various species. A further complication in the appraisal of male sexual behavior patterns is the fact that intromission and ejaculation are dependent on sensory feedback information derived from the penis. Ossification of the penile bone and its sensitization are both heavily dependent on exposure of the organ to androgens in neonatal life (Beach *et al.*, 1969), and the degree of development of the phallus is of considerable importance in the interpretation of data derived from behavioral studies, more particularly in animals which have been exposed to manipulations of the hormonal environment in neonatal life.

Males. Most normally developing postpubertal males will attempt to mount other animals of the same species and a high frequency of mounting can be obtained if large doses of testosterone (or estrogen) are administered beforehand (Davidson, 1969). If mating tests are undertaken after the testes have been removed in adulthood, then mating will eventually cease although it can be restored with the appropriate dose of testosterone. Adult male rats which have been castrated at birth will exhibit mounting behavior in response to estrous females, an effect which can be increased by injections of either testosterone or EB (Gerall *et al.*, 1967; Beach and Holz, 1946; Beach, 1971), but they are less likely than are testosterone-treated adult castrates to achieve intromission and ejaculation (see Whalen, 1968; Davidson and Levine, 1972). However, a high ejaculation score can be achieved if neonatally castrated males are injected with either testosterone or DHT in addition to RU 2858 (a synthetic estrogen) during the first few days of life (Booth, 1977); DHT or RU 2858 alone do not have this effect. These results give full support to the concept that neonatal estrogen is of vital importance in the neural organization of male behavior patterns, provided that DHT is also present to complete the differentiation of the male external genitalia.

Females. Mounting of estrous females by other females is commonplace and in every mammalian species that has been studied there would seem to be a pre- or perinatal period during which exposure to androgens enhanced some aspect of male-type behavior, but here again criticism of the data with regard to phallic development and sensory feedback obtains. Nevertheless, studies in the female ferret have been useful since differentiation of the phallus in response to perinatal androgen treatment occurs in this species at an earlier time of development than that at which the potential for masculine responses is heightened (Baum, 1976). In postnatally androgenized female ferrets that lacked phallic sheaths, estrogen administration in adulthood significantly stimulated male-type behavior patterns in the absence of any clitoral stimulation (Baum, 1976). Thus, in this species at least, there is an effect of perinatal androgens on adult male-type behavior which is divorced from any effects on the penis.

Recently the concept that female behavior is basic in type and that its development is not dependent on differentiation of neural mechanisms by steroid hormones has been brought to question. Apparently, the androgen-insensitive genetically male rat (testicular feminization) fails to display either masculine or feminine sexual behavior when primed with the appropriate sex hormones. The implication of these results is that "female differentiation" of the brain in respect to mechanisms controlling behavior may require active imprinting by perinatal hormones (Shapiro *et al.*, 1976).

In summary, although measurements of sexual behavior between male and female rats fail to show absolute differences, nevertheless from the collective data a useful test of sexual differentiation has emerged. The experimental model is that in which the adult castrate animal which has been exposed to estrogen is subsequently injected with a single dose of progesterone. In those animals that have been exposed either naturally or experimentally to androgens or estrogen in neonatal life the LQ is diminished by progesterone after estrogen priming, whereas in those which have lacked the differentiating effects of androgenic hormones at a critical period of development the LQ is increased by such hormonal stimuli (Figures 6 and 7).

Evidence for a sexual dimorphism of sexual behavior in the rat is supported by work derived from three other species of rodent, namely the guinea pig (Phoenix *et al.,* 1959), golden hamster (Swanson and Crossley, 1970), and the mouse (Edwards and Burge, 1971). Moreover, there are other forms of behavior patterns which show sexual dimorphism, such as open-field behavior in the rat (Gray *et al.,* 1969) and hamster (Swanson, 1967; Carter *et al.,* 1972), aggression in the mouse (Bronson and Desjardins, 1969; Edwards, 1971), saccharin preference in drinking water in the rat (Valenstein *et al.,* 1967; Wade and Zucker, 1969), urination with hind-leg raising in the dog (Beach, 1975), and urination with squatting in the ewe (Short, 1974; Clark *et al.,* 1976).

2.5. Possible Sites of Sexual Differentiation of Mechanisms Controlling Sexual Behavior

The existence of a site or sites in the rat brain which respond to androgen by permitting a manifestation of male sexual behavior was originally demonstrated by stereotaxic implantation of the steroid in the preoptic–anterior hypothalamic area (Davidson, 1966; Lisk, 1967) and by the elimination of mating behavior by lesions of

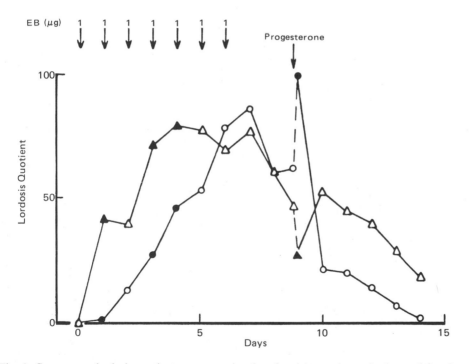

Fig. 6. Group mean lordosis quotients on successive days for eight ovariectomized normal female rats (O——O) and seven ovariectomized female rats that had received 1.25 mg TP on day 4 of postnatal life (Δ——Δ). The arrows indicate the injection of estradiol benzoate (EB) at the dose indicated. Progesterone indicates the injection of 1.25 mg of progesterone to each rat. The solid symbols (● and ▲) indicate that the LQ values for the two groups of animals were significantly different from one another at that stage of the experiment. (From Brown-Grant, 1975.)

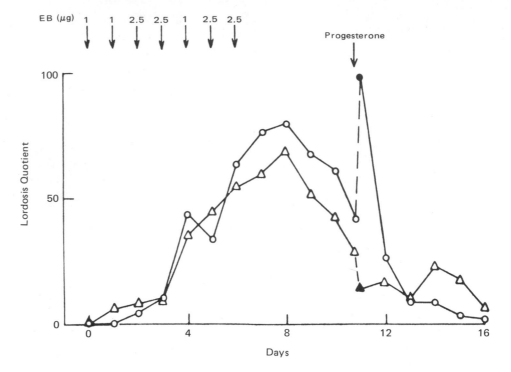

Fig. 7. Group mean lordosis quotients on successive days for six ovariectomized normal rats (O——O) and six ovariectomized female rats that had received 250 μg estradiol benzoate (EB) on day 4 of postnatal life (Δ——Δ). The arrows indicate the injection of EB at the dose indicated. Progesterone indicates the injection of 1.25 mg progesterone to each rat. The solid symbols (● and ▲) indicate that the LQ values for the two groups of animals were significantly different from one another at that stage of the experiment. (From Brown-Grant, 1975.)

the medial preoptic nucleus (Larsson and Heimer, 1964). Whether this result was a direct effect of androgen or a product of its metabolism has been questioned, although systemic administration of the 5α-reduced metabolite DHT is as ineffective in provoking male behavior (Davidson and Bloch, 1972) as it is in causing subsequent masculinization when injected into the female neonate (Luttge and Whalen, 1970*b*; Brown-Grant *et al.*, 1971). In contrast, however, DHT–estrogen combinations are highly effective (Feder *et al.*, 1974). That aromatization of androgen to estrogen may be essential for eliciting changes in behavior is further supported by the results of studies in which antiestrogens were found to block androgen-induced sex behavior in both rats (Whalen *et al.*, 1972; Södersten and Larsson, 1974) and rabbits (Beyer and Vidal, 1971).

Whether or not estrogen is the effective agent at the intracellular level that triggers the complex pattern of male sexual behavior, it is unlikely that only the preoptic area (POA) is concerned. Early work showed that male rats, following destruction of the neocortex, fail to respond to, and will not mate with, receptive females (Beach, 1940). Moreover, parasaggital cuts which sever the connections between the POA and the medial forebrain bundle markedly inhibit male sexual behavior (Paxinos and Bindra, 1973). Yet further studies employing lesions or electrical stimulation have implicated the amygdala and, perhaps importantly, the midbrain (for a review see Gorski, 1974).

With respect to female behavior in the rat, early work showed that although the cortex is not essential for mating, almost complete removal of the neocortex interferes with the integration of normal lordotic behavior (Beach, 1974), and this has been supported by more recent studies (Merari *et al.*, 1975). It is generally accepted that the POA, which contains estrogen receptors, is important in the elicitation of lordosis; although there have been contradictory reports, it appears that lesions of the POA abolish this behavior while estrogen implants will restore it (see Gorski, 1974).

Other neural areas are also implicated in the control of female receptive behavior. Yanasi and Gorski (1976) implanted estrogen in indwelling cannulae in the preoptic–diagonal band area in ovariectomized rats while progesterone was subsequently implanted (3 days later) in either the midbrain or the preoptic area. The results of the study demonstrated that progesterone in either area facilitates the estrogen-primed lordotic response while control cholesterol implants had no such effect. The results of further experiments using implants of antiestrogen in the midbrain suggested that although exposure of either the POA or the midbrain to estrogen facilitated the response to systemically injected progesterone, prior exposure of the midbrain to estrogen may be required for the action of locally implanted progesterone. Gorski (1976) has recently reviewed the possible sites of steroid facilitation of lordosis in the rat and proposed a model for regulation of lordosis behavior. Since the midbrain appears to be an important site of progesterone action, it may also prove to be a site of sexual differentiation.

The interesting possibility also arises as to whether a single neural mechanism or set of mechanisms exists which is capable of controlling either masculine or feminine behavior or whether two separate sets of mechanisms control the two different patterns of behavior; it is outside the scope of this chapter to discuss this aspect of the work and readers are referred to articles by Goy (1970), Beach (1971, 1975), Goy and Goldfoot (1973), and Baum (1976).

3. Evidence for Sexual Differentiation of the Brain in Primates

3.1. Gonadotropin Response to a Steroid Stimulus

Using the rhesus monkey, which has a similar pattern of hormone output during its menstrual cycle to that observed in the human, Knobil (1974) and his co-workers have attempted to delineate those areas of the brain which are essential for the control of phasic gonadotropin output. "Hypothalamic islands," which ensure complete disconnection of the mediobasal hypothalamus and its attached pituitary complex from the preoptic–Sch.n. area and all other extrahypothalamic structures were made in female monkeys (Krey *et al.*, 1973) with the use of a Halasz knife (Halasz, 1969). Radioimmunoassay measurements of plasma LH and FSH levels in blood-samples obtained before and after the operation showed that basal levels were unaffected by the cerebral trauma and that in some animals normal menstrual cycles and ovulation continued unabated. Furthermore, in ovariectomized animals in which circhoral discharges of LH are characteristically found (Dierschke *et al.*, 1970), a similar operative procedure was entirely consistent with the retention of this tonic rhythm. In respect to the simulated preovulatory LH surge, the administration of EB in the early part of the follicular phase, in dosages designed to imitate the rise of estrogen at midcycle, occasioned an initial inhibition of LH levels followed closely by a facilitatory release. The conclusions which were drawn from this work were that cerebral sites responsible for the control of both inhibitory and facilitatory influences on LH release were likely to be situated within the hypothalamohypophysial unit. These findings were substantiated in rhesus monkeys in which all brain structures anterior and dorsal to the optic chiasma, including the preoptic area, had been removed, thus leaving the mediobasal hypothalamus and connections with the brain stem intact (Hess *et al.*, 1977).

Norman *et al.* (1976), also using rhesus monkeys but a different experimental design from that used by Knobil, placed bilateral radio-frequency lesions in the ventral preoptic area which significantly damaged the Sch.n. Monkeys which were later found to have had effective lesions showed a tonic secretion of gonadotropins which was maintained at a level commensurate with their ovarian follicular development; however, these animals failed to show spontaneous LH surges, and all attempts to induce a gonadotropin response to estrogen likewise failed. On the other hand, animals with unilateral destruction or lesions which lay outside the target site showed spontaneous LH surges.

The reason for the discrepancies between the two sets of experiments is unclear, but conceivably they may be explained by differential damage to neural pathways arising in the brain stem that have either stimulatory or inhibitory influences on gonadotropin release. Whatever the answers to these problems may be, the possibility that spontaneous LH surges in the monkey may be dependent, as in the rodent, on mechanisms associated with the preoptic area and the Sch.n. and its connections appears at present to be unlikely. A further question then arises: If these areas which are important in the rat in terms of the LH surge are sexually differentiated, is their apparent unimportance in the primate associated with a lack of sexual differentiation in the primate brain and therefore an ability to elicit an LH surge in both sexes?

Karsch *et al.* (1973) castrated groups of adult male and female rhesus monkeys and suppressed the resulting high serum LH levels with crystalline estradiol-17β contained in silastic capsules placed beneath the skin. A subsequent subcutaneous injection of EB led to unambiguous discharges of LH 24 hr later, which were similar in all respects in the two sexes (Figure 8). These results were supported by similar findings in the human. Stearns *et al.* (1973) demonstrated a stimulatory effect on LH output of progestin in estrogen-primed castrate adult men, while other authors (e.g., Yen *et al.*, 1972) have observed a similar rise in gonadotropins after administration of a progestin to estrogen-pretreated castrate and postmenopausal women. Collectively, these results suggest that mechanisms controlling the LH surge in primates may not be sexually differentiated, on the other hand more subtle stimuli may be required to expose such a potential difference.

In line with the above data are the results of studies on the offspring of rhesus monkeys which had been treated with androgen repeatedly from day 24 of pregnancy to just prior to birth (average gestation 168 days) (Goy, 1970; Goy and Resko, 1972). Although the onset of puberty in these animals was delayed, their subsequent menstrual cycles were normal as regards hormonal output and were no more irregular than those observed in control animals.

Although there are no substantial indications that mechanisms subserving positive feedback responses of LH to estrogen are sexually differentiated in primates, it is possible that mechanisms underlying negative feedback responses of LH to estrogen may be. Recent evidence has shown that in groups of gonadectomized male, female, and "androgenized" rhesus monkeys in which silastic capsules

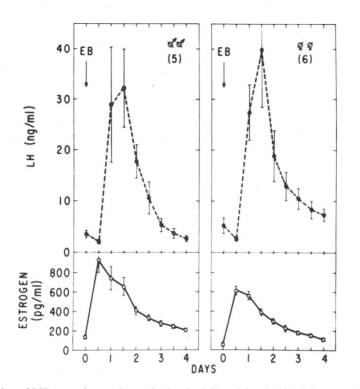

Fig. 8. Induction of LH surges in gonadectomized male (left) and female (right) rhesus monkeys by the administration of estradiol benzoate (EB). Before EB was injected plasma LH concentrations were suppressed by the subcutaneous implantation of silastic capsules containing crystalline 17β-estradiol. Plasma concentrations of estrogen effected by these treatments are shown in the lower panels. The mean ± standard error is shown; the number of observations is in parentheses. (From Karsch *et al.*, 1973.)

containing estradiol had been implanted beneath the skin one week earlier, LH levels in the female were significantly lower than in the other two groups (Steiner *et al.*, 1976). Negative feedback mechanisms in the female would therefore appear to be more sensitive to the effects of long-term feedback inhibition of estrogen than are those in males or androgenized females.

3.2. Sexual Behavior; Sexual Differentiation

3.2.1. Rhesus monkeys

In an attempt to alter the hormonal environment of the fetus, van Wagenen and Hamilton (1943) and later Wells and van Wagenen (1954) administered heavy doses of testosterone to pregnant rhesus monkeys. When the offspring were born, the genetic females were clearly hermaphroditic with morphological changes of the external genitalia. Using a similar technique, Goy (1970) and other workers (Goy and Resko, 1972; Phoenix, 1974) likewise produced a small number of hermaphroditic rhesus monkeys which were reared from 3 months of age to 4 years of age with other hormone-treated or untreated monkeys of similar age. At frequent intervals during development various behavior patterns were carefully assessed; the outcome was that rough-and-tumble play, threats, play-initiation, and chasing play were observed significantly more often in males and in hermaphrodites than in females.

On the other hand such forms of behavior as grooming, huddling, or fear-grimacing did not show signs of being sexually differentiated. Castration of males at birth or at 90 days of age did not affect the frequencies with which these behavior patterns were exhibited when a comparison was made with scores obtained from intact male peers. Likewise, the behavior patterns of ovariectomized monkeys were similar to those observed in intact control females.

Mounting behavior too was studied, and again this behavior was more frequently observed in hermaphrodites than in normal females. There was also a qualitative difference in this behavior; when normal male rhesus monkeys mount a female, they clasp the partner's shanks with their hind legs while holding her back with the fore paws. This characteristic "shank-clasp" position was more often displayed by the small number of hermaphrodites which had been reared with mothers and their peer group than by the normal control females (Goy and Goldfoot, 1973).

Tests of male-type behavior in adult hermaphrodites after castration showed that the hermaphrodites were significantly more aggressive and likely to sit next to an EB-stimulated female than were control females, and this difference was not altered by administration of TP (Eaton et al., 1973).

The combined evidence suggests that prenatal exposure to androgens, or to the products of aromatization (see Naftolin et al., 1975) modifies social interaction patterns independently of the hormonal environment during the late stages of development. Circulating androgen levels in the fetal monkey are therefore of considerable interest, and radioimmunoassay measurements of umbilical artery samples have shown that testosterone levels in the fetal male rhesus are higher on average than in the female from day 59 to just before birth (Resko et al., 1973).

3.2.2. Man

There are two available sources of human data from which an attempt can be made to ascertain whether certain components of human sexuality (see Whalen, 1966) are determined by pre- or postnatal hormonal influences on central neural development or depend entirely on the current hormonal status. The first is a variety of prenatal abnormalities which are associated with pseudohermaphroditism, and the second is the condition of transsexualism which is not associated with any physical or genetic anomaly (Money and Ehrhardt, 1972; Bancroft, 1977).

Of the prenatal states of hormonal dysfunction, the most common is the adrenogenital syndrome in which there exists a genetic deficiency of the hydroxylating enzyme at the 21 position of the steroid molecule. This deficiency leads to a failure of cortisol production with a subsequent increase in adrenocorticotrophin (ACTH) output and, in turn, an excessive production of androgens. Should the dysfunction be sufficiently severe, then the external genitalia of the female fetus becomes masculinized *in utero* and the child might be assigned at birth to the genetically incorrect sex. Some of these children have therefore been reared as boys and others as girls, with the consequent development of a self image or gender identity which is relevant to the sex of assignment. Of particular interest are those in whom the endocrine state and the external genitalia are of male type, in spite of which they have been reared as girls. Two such groups of children have been studied (Ehrhardt et al., 1968b). In one group the diagnosis was made at birth and cortisol treatment was instituted immediately. In the other, recognition of the condition was delayed and treatment was started in adolescence or in adulthood

(Ehrhardt *et al.*, 1968*a*). Behavior patterns in these two groups of children were studied and the results were reminiscent of those observed in masculinized infant monkeys, in that they exhibited more "tomboy" behavior than did a matched control group. These observations are further supported by data obtained from a group of children who had inadvertently been exposed to excess progestins during intrauterine development because their mothers had been treated with these drugs in the early weeks of pregnancy in order to prevent abortion (Ehrhardt and Money, 1967). Again, an analysis of play patterns indicated distinct tomboyish qualities.

In the "late-treatment" group of children with adrenogenital dysfunction, there appeared to be differences in regard to sexual preference—as pertaining to characteristics of a desirable sexual partner or erotic stimulus. A greater degree of bisexual fantasy and bisexual experience was shown by this group than by subjects in the "early-treatment" group (Ehrhardt *et al.*, 1968). In another study, Lev-Ran (1974) reported that in women born with the adrenogenital syndrome in whom treatment designed to lower androgen levels was started (i.e., women similar to the late-treatment group), sexual dreams virtually ceased.

A naturally occurring experimental model in genetic males which is of considerable interest is the state of testicular feminization in which there is a deficiency of androgenic binding proteins and thus a failure of hormonal expression. The phenotype of these testes-carrying subjects is typically female, and most of these subjects are reared from birth as girls and thus develop a female gender indentity. In adulthood their behavior is very close to the female stereotype, for among other parameters they are said to make good mothers to their adopted children. On the criteria of sexual preference they also appear to be strongly feminine, with little evidence of homosexuality or of bisexual imagery (Money and Ehrhardt, 1972).

Equivalent perhaps to the girls born to progestin-treated mothers are males born to diabetic mothers who, in order to reduce the high fetal mortality rate associated with the disorder, had been treated with high doses of estrogen and low doses of progestins during pregnancy. It is notable that these boys were found to be less aggressive and less athletic than were a control group of children born to untreated mothers (Yalom *et al.*, 1973).

From these observations of genetic males and females who have been exposed to a variety of naturally occurring or drug-induced endocrine abnormalities, it appears that gender identity is of paramount importance in the development of human sexuality; moreover, it is strongly dependent on parental attitudes and social learning. Yet there are, in addition, indications that prenatal influences may also play a part in human sexuality, in that exposure to androgens during development appears to be conducive to the formation of a masculine gender identity. Of relevance to such a consideration, however, is a report on a group of genetic males with a 5α-reductase deficiency (Imperato-McGinley *et al.*, 1974). Such an enzyme deficiency prevents the metabolism of testosterone to DHT and thus leads to a failure of the external genitalia to masculinize. As a consequence of the appearance of female-type genitalia at birth, these 5α-reductase-deficient children were reared as girls. But, unexpectedly, hormonal changes occurred at puberty which led to a partial masculinization of the genitalia, and subsequent alterations toward a masculine gender identity; in addition interest increased with respect to the opposite genetic sex.

The second source of human data from which evidence might be gleaned in support of a pre- or postnatal hormonal influence on gender identity is that of transsexualism. This condition is one in which the individual believes himself to be, or has a very strong desire to be, the gender opposite to that of his genetic and anatomical sex. The characteristic behavior pattern is often evolved from fetishistic transvestism in early childhood and becomes more marked at puberty (Bancroft, 1972). There is, however, little evidence to link the condition with hormonal changes during development either in the genetic male (Migeon *et al.,* 1969) or the genetic female (Jones and Saminy, 1973; Fulmer, 1973). But in the knowledge that these subjects are prone to taking exogenous hormones, a carefully controlled hormonal investigation would be of considerable interest and importance.

In respect to a further aspect of human sexuality, namely sexual preference, several groups of clinical research workers have sought to account for homosexuality on the basis of a pre- or postnatal hormonal imbalance, but so far these attempts have been negative, with one possible exception in which the estrogen-stimulated LH response was investigated in groups of heterosexual, homosexual, and bisexual men (Dörner *et al.,* 1975). Of 21 homosexuals, 13 appeared to show a facilitatory LH response to a single injection of estrogen in comparison with 2 of the 20 heterosexuals and none of the bisexuals. The facilitatory LH response of all the male groups was said to be less marked than that observed in normal women, but this was accounted for by androgen levels which are higher in the male than the female and which may alter the pituitary response to estrogen. In the light of previous results which have shown that a facilitatory LH response can be elicited by injection of gonadal steroids in castrate male and female monkeys or castrate men and women (see p. 206), the above report requires confirmation.

4. Differentiation of the Female Brain

The possibility that a hormonally directed neural differentiation may occur in the female brain at a critical period of development, similar in a sense to that observed in the male, has recently been raised. The question was highlighted by the observation that sexual behavior patterns of testicular-feminized rats which had been gonadectomized and subsequently injected with steroids were neither male nor female in type when compared with those of similarly treated littermate males and nonlittermate females (Shapiro *et al.,* 1976). Since both serum and adrenal progesterone concentrations are significantly higher in 3-day-old female rats than in males of the same age, it has been suggested that the gonadotropin-adrenal axis may be required for adequate feminization of the brain.

5. Summary and Conclusions

There is considerable evidence for sexual differentiation of the brain in the rat, some of which is supported by data obtained in other nonprimate species: Cow (Jost *et al.,* 1963), sheep (Short, 1974; Karsch and Foster, 1975), dog (Beach, 1970, 1975), guinea pig (Brown-Grant and Sherwood, 1971), hamster (Swanson, 1967; Alleva *et al.,* 1969), gerbil (Turner 1976), and mouse (Barraclough and Leathem, 1954).

It appears that testosterone, perhaps by virtue of its conversion to estrogen, can, at an early stage of development, organize and differentiate basal female-type mechanisms into those of male type. The presence of testes in male rats or the administration of testosterone to neonatal females prevents the possibility of a preovulatory LH surge and therefore of regular estrous cycles after puberty. On the other hand, if the male is castrated within 24 hr of birth, it retains the potential to exhibit a phasic output of LH.

The maintenance of phasic gonadotropin output in the rat depends not only on the presence of neurons which synthesize LH-RH but on neural mechanisms which appear to signal the time of day and control its release. The site or sites of sexual differentiation within the brain with respect to phasic gonadotropin output are still unclear. However, stimulation of the preoptic area in male rats leads to an LH surge and ovulation if the animals have been implanted with ovarian tissue; it may be expected therefore that an important site of differentiation may lie outside the preoptic–median eminence–anterior pituitary unit.

The results of a single-unit recording study have suggested that two categories of neuron in the preoptic–anterior hypothalamic area may be sexually differentiated, while sexual dimorphism of synaptic connections in a circumscribed area of the preoptic region has also been reported. But neither of these studies was directly connected with mechanisms known to control gonadotropin output or specific behavioral responses. A further biochemical study has indicated that a circadian rhythm of incorporation of ^{35}S into brain proteins exists, the phase of which is sexually differentiated; peak values in the adult female occur in the evening and are closely associated with the time at which LH surges can be elicited. Both the circadian rhythm of incorporation and estrogen-stimulated LH surges are prevented by discrete lesion of the Sch.n. It is possible, therefore, that mechanisms associated with this nucleus, which is responsible for the control of circadian rhythmicity and presumably time of day signals, may be sexually differentiated.

In the monkey, menstrual cyclicity does not appear to depend on the time of day or on the presence of the preoptic–Sch.n. area. Menstrual cycles continue unabated in this species even if the mediobasal hypothalamus–anterior pituitary unit is surgically separated from the preoptic–Sch.n. area and the rest of the brain. Furthermore the LH response to an estrogen (or estrogen/progesterone) stimulus, which is so clearly sexually differentiated in rodents and certain other nonprimates, can be elicited with ease in castrate male and female monkeys or in castrate humans. In line with these findings are those in which normal menstrual cycles or LH surges were observed in female monkeys which had been exposed to increased androgen levels during intrauterine development.

With respect to sexual behavior, castration of the male rat on the day of birth or administration of testosterone or estrogen to neonatal females tends to change thresholds of sexual behavior patterns toward those of the opposite sex. The most useful test of sexual differentiation in this field is perhaps the lordotic response to a progesterone stimulus in estrogen-primed gonadectomized animals; the response is clearly facilitated in females and one-day castrate males but not in acyclic animals. The site or sites of differentiation in the brain, which are likely to be different to and more diverse than those responsible for the control of gonadotropin output, are still unclear. In the rhesus monkey there are apparently significant differences in social-interaction patterns and mounting displays between, on the one hand, males and females treated with androgens *in utero,* and on the other, untreated females. These patterns of behavior are considered to have a counterpart in the tomboy activities and attitudes displayed by female children who have, as a result of hormonal dysfunction or inadvertent drug treatment, also been exposed to the effects of androgen in uterine life.

In conclusion, there is at present a paucity of evidence for sexual differentiation of the brain not only in primates but in nonprimate reflex ovulators. Nevertheless, if such a fundamental neural difference exists in certain nonprimate spontaneous ovulators, then it might be expected to show itself in one form or another throughout the mammalian class. Perhaps we have yet to ask those scientific questions, the answers to which will provide convincing evidence for sexual differentiation of the brain of man.

ACKNOWLEDGMENTS

My cordial and very grateful thanks are due to Professor B. T. Donovan who read this manuscript and made helpful criticisms.

6. References

Adams-Smith, W. N., and Peng, M. T., 1966, Influence of testosterone upon sexual maturation in the rat, *J. Physiol.* **185:**655–666.
Aiyer, M. S., and Fink, G., 1974, The role of sex steroid hormones in modulating the responsiveness of the anterior pituitary gland to luteinizing hormone releasing factor in the female rat, *J. Endocrinol.* **62:**553–572.
Aiyer, M. S., Chiappa, S. A., and Fink, G., 1974, A priming effect of luteinizing hormone releasing factor on the anterior pituitary gland in the female rat, *J. Endocrinol.* **62:**573–588.
Alleva, F. R., Alleva, J. J., and Umberger, E. J., 1969, Effect of a single prepubertal injection of testosterone propionate on later reproductive functions of the female golden hamster, *Endocrinology* **85:**312–318.
Arai, Y., 1971, Some aspects of the mechanisms involved in steroid induced sterility, in: *Steroid Hormones and Brain Function* (C. H. Sawyer and R. A. Gorski, eds.), pp. 185–191, Univ. of California Press, Los Angeles.
Arai, Y., and Kusama, T., 1968, Effect of neonatal treatment with estrone on hypothalamic neurons and regulation of gonadotrophin secretion, *Neuroendocrinology* **3:**107–114.
Bancroft, J. H. J., 1972, The relationship between gender identity and sexual behaviour: some clinical aspects, in: *Gender Differences; Their Ontogeny and Significance* (C. Ounsted and D. C. Taylor, eds.,) pp. 57–72, Churchill Livingstone, Edinburgh.
Bancroft, J. H. J., 1977, Biological determinants of sexual behaviour, in: *Hormones and Sexual Behaviour in the Human* (J. Hutchinson, ed.), pp. 493–519, Wiley, New York.
Barley, J., Ginsburg, M., Greenstein, B. D., MacLusky, N. J., and Thomas, P. J., 1974, A receptor mediating sexual differentiation?, *Nature* **252:**259–260.
Barley, J., Ginsburg, M., Greenstein, B. D., MacLusky, N. J., and Thomas, P. J., 1975, An androgen receptor in rat brain and pituitary, *Brain Res.* **100:**383–393.
Barnea, A., Weinstein, A., and Lindner, H. R., 1972, Uptake of androgens by the brain of the neonatal female rat, *Brain Res.* **46:**391–402.
Barraclough, C. A., 1961, Production of anovulatory, sterile rats by single injections of testosterone propionate, *Endocrinology* **68:**62–67.
Barraclough, C. A., and Leathem, J. H., 1954, Infertility induced in mice by a single injection of testosterone propionate, *Proc. Soc. Exp. Biol. Med.* **85:**673–674.
Barraclough, C. A., and Turgeon, J. L., 1974, Further studies of the hypothalamo-hypophysea-gonadal axis of the androgen-sterilized rat, Intern. Symp. on Sexual Endocrinology of the Perinatal Period, *INSERM* **32:**339–356.
Baum, M. J., 1976, Effects of testosterone propionate administered perinatally on sexual behaviour of female ferrets, *J. Comp. Physiol. Psychol.* **90:**399–410.
Beach, F. A., 1940, Effects of cortical lesions upon the copulatory behavior of male rats, *J. Comp. Physiol. Psychol.* **29:**193–244.
Beach, F. A., 1944, Effects of injury to the cerebral cortex upon sexuality—receptive behavior in the female rat, *Psychosomat. Med.* **6:**40–45.
Beach, F. A., 1970, Hormonal effects of socio-sexual behaviour in dogs, in: *Mammalian Reproduction* (H. Gibian and E. J. Plotz, eds.,) pp. 437–466, Springer-Verlag, Berlin–Heidelberg–New York.

Beach, F. A., 1971, Hormonal factors controlling the differentiation development and display of copulatory behaviour in the ramstergig and related species, in: *Biopsychology of Development* (E. Tobach, L. R. Aronson, and E. Shaw, eds.), pp. 247–296, Academic Press, New York.

Beach, F. A., 1975, Hormonal modification of sexually dimorphic behaviour, *Psychoneuro-endocrinology* 1:3–23.

Beach, F. A., 1976, Sexual attractivity, proceptivity and receptivity in female mammals, *Horm. Behav.* 7:105–138.

Beach, F. A., and Holz, A. M., 1946, Mating behavior in male rats castrated at various ages and injected with androgen, *J. Exp. Zool.* 101:91–142.

Beach, F. A., Noble, R. G., and Arndorf, R. K., 1969, Effects of perinatal androgen treatment on responses of male rats to gonadal hormones in adulthood, *J. Comp. Physiol. Psychol.* 68:490–497.

Bennett, D., Boyse, E. A., Lyon, M. F., Mathieson, B. J., Scheid, M., and Yanagisawa, K., 1975, Expression of H-Y (male) antigen in phenotypically female Tfm/Y mice, *Nature* 257:236–238.

Benno, R. H., and Williams, T. H., 1978, Evidence for intracellular localization of alpha-fetoprotein in the developing rat brain, *Brain Res.* 142:182–186.

Beyer, C., and Vidal, N., 1971, Inhibitory action of MER-25 on androgen-induced oestrous behaviour in the ovariectomised rabbit, *J. Endocrinol.* 51:401–402.

Blake, C. A., Norman, R. L., and Sawyer, C. G., 1973, Effects of hypothalamic deafferentation and ovarian steroids on pituitary responsiveness to LH-RH in female rats, in: *Hypothalamic Hypophysiotropic Hormones* (C. Gual and E. Rosemberg, eds.), pp. 33–38, Excerpta Medica, Amsterdam.

Bogdanove, E. M., Campbell, G. T., Blair, E. D., Mula, M. E., Miller, A. E., and Grossman, G. M., 1974, Gonad–pituitary feedback involves qualitative change: Androgens alter the type of FSH secreted by the rat pituitary, *Endocrinology* 95:219–228.

Booth, J. E., 1977, Sexual behaviour of neonatally castrated rats injected during infancy with oestrogen and dihydrotestosterone, *J. Endocrinol.* 72:135–141.

Borvendeg, J., Hermann, H., and Bajusz, S., 1972, Ovulation induced by synthetic luteinizing hormone releasing factor in androgen-sterilized female rats, *J. Endocrinol.* 55:207–208.

Bronson, F. H., and Desjardins, C., 1969, Aggressive behavior and seminar vesicle function in mice: Differential sensitivity to androgen given neonatally, *Endocrinology* 85:971–974.

Brown-Grant, K., 1972, Recent studies on the sexual differentiation of the brain, in: *Foetal and Neonatal Physiology* (K. S. Comline, K. W. Cross, G. S. Dawes, and P. W. Nathanielsz, eds.), pp. 527–545, Cambridge Univ. Press, Cambridge.

Brown-Grant, K., 1974a, On "critical periods" during the post-natal development of the rat, in: *Les Colloques de l'Institut National de la Santé et de la Recherche Médicale, INSERM* 32:357–376.

Brown-Grant, K., 1974b, Steroid hormone administration and gonadotrophin secretion in the gonadectomised rat, *J. Endocrinol.* 62:319–332.

Brown-Grant, K., 1975, A re-examination of the lordosis response in female rats given high doses of testosterone propionate or estradiol benzoate in the neonatal period, *Horm. Behav.* 6:351–378.

Brown-Grant, K., Exley, D., and Naftolin, F., 1970, Peripheral plasma oestradiol and luteinizing hormone concentrations during the oestrous cycle of the rat, *J. Endocrinol.* 48:295–296.

Brown-Grant, K., Munck, A., Naftolin, F., and Sherwood, M. R., 1971, The effects of the administration of testosterone propionate alone or with phenobarbitone and of testosterone metabolites to neonatal female rats, *Horm. Behav.* 2:173–182.

Brown-Grant, K., and Raisman, G., 1972, Reproductive function in the rat following selective destruction of afferent fibres to the hypothalamus from the limbic system, *Brain Res.* 46:23–42.

Brown-Grant, K., and Sherwood, M. R., 1971, The "early androgen syndrome" in the guinea-pig, *J. Endocrinol.* 49:277–291.

Brownstein, M. J., Arimura, A., Schally, A. V., Palkovits, M., and Kizer, J. S., 1976, The effect of surgical isolation of the hypothalamus on its luteinizing hormone-releasing hormone content, *Endocrinology* 98:662–665.

Burnet, F. R. B., and MacKinnon, P. C. B., 1975, Restoration by oestradiol benzoate of a neural and hormonal rhythm in the ovariectomized rat, *J. Endocrinol.* 64:27–35.

Butler, J. E. M., and Donovan, B. T., 1971, The effect of surgical isolation of the hypothalamus on reproductive function in the female rat, *J. Endocrinol.* 49:293–304.

Caligaris, L., Astrada, J. J., and Taleisnik, S., 1971, Release of luteinizing hormone induced by estrogen injection into ovariectomised rat, *Endocrinology* 88:810–815.

Carter, C. S., Clemens, L. G., and Hockemeyer, D. J., 1972, Neonatal androgen and adult sexual behavior in the golden hamster, *Physiol. Behav.* 9:89–95.

Chiappa, S. A., Fink, G., and Sherwood, N. M., 1977, Immunoreactive luteinizing hormone releasing factor (LRF) in pituitary stalk plasma from female rats: Effects of stimulating diencephalon, hippocampus and amygdala, *J. Physiol.* 267:625–640.

Cidlowski, J. A., and Muldoon, T. G., 1976, Sex related differences in the regulation of cytoplasmic estrogen receptor levels in responsive tissues of the rat, *Endocrinology* **98**:833–841.

Clark, J. M., Campbell, P. S., and Peck, E. J., Jr., 1972, Receptor–oestrogen complex in the nuclear fraction of the pituitary and hypothalamus of male and female immature rats, *Neuroendocrinology* **10**:218–228.

Clarke, J., Scaramuzzi, R. J., and Short, R. V., 1976, Effects of testosterone implants in pregnant ewes on their female offspring, *J. Embryol. Exp. Morphol.* **36**:87–99.

Clayton, R. B., Kogura, J., and Kraemer, H. C., 1970, Sexual differentiation of the brain: Effects of testosterone on brain RNA metabolism in newborn female rats, *Nature, London* **226**:810–812

Clemens, L. G., Hiroi, M., and Gorski, R. A., 1969, Induction and facilitation of female mating behavior in rats treated neonatally with low doses of testosterone propionate, *Endocrinology* **84**:1430–1438.

Coen, C. W., and MacKinnon, P. C. B., 1976, Serotonin involvement in oestrogen induced LH release in ovariectomised rats, *J. Endocrinol.* **71**:49P–50P.

Coen, C., and MacKinnon, P. C. B., 1977, Evidence for a critical period and a possible site of action for serotonin involvement in oestrogen-induced luteinizing hormone release in ovariectomized rats, *J. Endocrinol.* **75**:43P–44P.

Colby, H. D., Gaskin, J. H., and Kitay, J. I., 1973, Effect of anterior pituitary hormones on hepatic corticosterone metabolism in rats, *Steroids* **24**:679–686.

Crawley, W. R., Rodriguez-Sierra, J. P., and Komisaruk, B. R., 1976, Hypophysectomy facilitates sexual behavior on female rats, *Neuroendocrinology* **20**:328–338.

Davidson, J. M., 1966, Activation of the male rat's sexual behavior by intracerebral implantation of androgen, *Endocrinology* **79**:783–794.

Davidson, J. M., 1969, Effects of oestrogen on the sexual behavior of male rats, *Endocrinology* **84**:1365–1372.

Davidson, J. M., 1974, Gonadotrophin feedback and sexual behavior: Functions differentiated by perinatal androgen, in: *Les Colloques de L'Institut National de la Sante et de la Recherche Medicale, INSERM* **32**:277–394.

Davidson, J. M., and Bloch, G. J., 1972, Neuroendocrine aspects of male reproduction, *Biol. Reprod.* **1** **(Suppl. 1)**:67–92.

Davidson, J. M., and Levine, S., 1969, Progesterone and heterotypical sexual behavior of male rats, *J. Endocrinol.* **44**:129–130.

Davidson, J. M., and Levine, S., 1972, Endocrine regulation of behavior, *Annu. Rev. Phys.* **34**:375–408.

Debeljuk, L., Arimura, A., and Schally, A. V., 1972, Effect of testosterone and estradiol on the LH and FSH release induced by LH-releasing hormone (LH-RH) in intact male rats, *Endocrinology* **90**:1578–1581.

Denef, C., 1974, Effect of hypophysectomy and pituitary implants at puberty on the sexual differentiation of testosterone metabolism in rat liver, *Endocrinology* **94**:1577–1582.

Dierschke, D. J., Bhattacharya, A. N., Atkinson, L. E., and Knobil, E., 1970, Circhoral oscillations of plasma LH levels in the ovariectomised rhesus monkey, *Endocrinology* **87**:850–853.

Döcke, F., and Dörner, G., 1975, Anovulation in adult female rats after neonatal intracerebral implantation of oestrogen, *Endokrinologie* **65**:375–377.

Döcke, F., and Kolocyek, G., 1966, Einfluss einer postnatalen Androgen-behandlung auf den Nucleus Hypothalamicus anterior, *Endokrinologie* **50**:225–230.

Dörner, G., and Staudt, J., 1968, Structural changes in the preoptic anterior hypothalamic area of the male rat, following neonatal castration and androgen substitution, *Neuroendocrinology* **3**:136–140.

Dörner, G., Rohde, W., Stahl, F., Krell, L., and Masius, W. G., 1975, A neuroendocrine predisposition for homosexuality in men, *Arch. Sex. Behav.* **4**:1–8.

Dyer, R. G., MacLeod, N. K., and Ellendorff, F., 1976, Electrophysiological evidence for sexual dimorphism and synaptic convergence in the preoptic and anterior hypothalamic areas of the rat, *Proc. R. Soc. London* **193**:421–440.

Eaton, G. G., Goy, R. W., and Phoenix, C. H., 1973, Effects of testosterone treatment in adulthood on sexual behaviour of female pseudohermaphrodite rhesus monkeys, *Nature (London) New Biol.* **242**:119–120.

Edwards, D. A., 1971, Neonatal administration of androstenedione, testosterone or testosterone propionate. Effects on ovulation, sexual receptivity and aggressive behavior in female mice, *Physiol. Behav.* **6**:223–228.

Edwards, D. A., and Burge, K. G., 1971, Early androgen treatment and male and female sexual behavior in mice, *Horm. Behav.* **2**:49–58.

Ehrhardt, A. A., and Money, J., 1967, Progestin-induced hermaphroditism I.Q. and psychosexual identity in a study of ten girls, *J. Sex Res.* **3**:83–100.

Ehrhardt, A. A., Evers, K., and Money, J., 1968a, Influence of androgen and some aspects of sexually dimorphic behaviour in women with late-treated adreno-genital syndrome, *Johns Hopkins Med. J.* **123:**115–122.

Ehrhardt, A. A., Epstein, R., and Money, J., 1968b, Fetal androgens and female gender identity in the early treated adrenogenital syndrome, *Johns Hopkins Med. J.* **122:**160–167.

Eisenfeld, A. J., and Axelrod, J., 1966, Effect of steroid hormones, ovariectomy, estrogen pretreatment, sex and immaturity on the distribution of ³H-estradiol, *Endocrinology* **79:**38–42.

Elger, W., 1966, Die Rolle der fetalen Androgene in der Sexual differenzierung des Kaninchens und ihre Abgrenzung gegen andere hormonale und somatische Faktoren durch Anwendung eines starken Antiandrogens, *Arch. Anat. Microsc. Morphol. Exp.* **55:**657–743.

Everett, J. W., 1969, Neuroendocrine aspects of mammalian reproduction, *Annu. Rev. Physiol.* **31:**383–416.

Everett, J. W., and Sawyer, C. H., 1950, A 24-hour periodicity in the LH-release apparatus of female rats disclosed by barbiturate sedation, *Endocrinology* **47:**198–218.

Everett, J. W., Holsinger, J. W., Zeilmaker, G. H., Redmond, W. C., and Quinn, D. L., 1970, Strain differences for preoptic stimulation of ovulation in cyclic, spontaneously persistent-oestrus, and androgen-sterilized rats, *Neuroendocrinology* **6:**98–108.

Fang, S., Anderson, K. M., and Liao, S., 1969, Receptor proteins for androgens. On the role of specific proteins in selective retention of 17β-hydroxy-5α-andiostan-3-one by rat ventral prostate *in vivo* and *in vitro*, *J. Biol. Chem.* **244:**6584–6595.

Feder, H. H., Naftolin, F., and Ryan, K. J., 1974, Male and female sexual responses in male rats given estradiol benzoate and 5α-androstan-17β-ol-3-one propionate, *Endocrinology* **94:**136–141.

Fink, G., and Jamieson, M. G., 1976, Immunoreactive luteinizing hormone releasing factor in rat pituitary stalk blood: Effects of electrical stimulation of the medial preoptic area, *J. Endocrinol.* **68:**71–87.

Fink, G., and Henderson, S. R., 1977a, Steroids and pituitary responsiveness in female, androgenized female and male rats, *J. Endocrinol.* **73:**157–164.

Fink, G., and Henderson, S. R., 1977b, Site of modulatory action of oestrogen and progesterone on gonadotrophin response to luteinizing hormone releasing factor, *J. Endocrinol.* **73:**165–170.

Fink, G., Chiappa, S. A., and Aiyer, M. S., 1976, Priming effect of luteinizing hormone releasing factor elicited by preoptic stimulation and by intravenous infusion or multiple injection of the synthetic decapeptide, *J. Endocrinol.* **69:**359–372.

Flores, F., Naftolin, F., Ryan, K. J., and White, R. J., 1973, Estrogen formation by the isolated perfused rhesus monkey brain, *Science* **180:**1074–1075.

Fulmer, G. P., 1973, Testosterone levels and female-to-male transsexualism, *Arch. Sex. Behav.* **2:**399–400.

Gerall, A. A., Hendricks, S. E., Johnson, L. L., and Bounds, T. W., 1967, Effects of early castration in male rats on adult sexual behavior, *J. Comp. Physiol. Psychol.* **64:**206–212.

Gorski, R. A., 1968, Influence of age on the response to paranatal administration of a low dose of androgen, *Endocrinology* **82:**1001–1004.

Gorski, R. A., 1974, The neuroendocrine regulation of sexual behavior, in: *Advances in Psychobiology,* Vol. 2 (G. Newton and A. G. Riesen, eds.), pp. 1–58, Wiley, New York.

Gorski, R. A., 1976, The possible neural sites of hormonal facilitation of sexual behavior in the female rat, *Psychoneuroendocrinology* **1:**371–387.

Gorski, R. A., and Barraclough, C. A., 1963, Effects of low dosages of androgen on the differentiation of hypothalamic regulatory control of ovulation in the rat, *Endocrinology* **73:**210–216.

Gorski, J., and Gannon, F., 1976, Current models of steroid hormone action: A critique, *Annu. Rev. Physiol.* **38:**425–450.

Goy, R. W., 1970, Experimental control of psychosexuality, *Phil. Trans. R. Soc. London, Ser. B* **259:**149–162.

Goy, R. W., and Goldfoot, D. A., 1973, Hormonal influences on sexually dimorphic behavior, in: *Handbook of Physiology, Endocrinology,* Section 7, Vol. 2, Part 1 (R. O. Greep and E. B. Astwood, eds.), pp. 169–186, Williams and Wilkins, Baltimore.

Goy, R. W., and Resko, J., 1972, Gonadal hormones and behaviour of normal and pseudohermaphrodite non human, female primates, *Rec. Prog. Horm. Res.* **28:**707–733.

Grady, K. L., Phoenix, C. H., and Young, W. C., 1965, Role of the developing rat testis in differentiation of the neural tissues mediating mating behavior, *J. Comp. Physiol. Psychol.* **59:**179–182.

Gray, J. A., Lean, J., and Keynes, A., 1969, Infant androgen treatment and adult openfield behavior: Direct effects and effects of injections to siblings, *Physiol. Behav.* **4:**177–181.

Greenley, G. H., Jr., Allen, M. B., Jr., and Mahesh, V. B., 1975, Potentiation of luteinizing hormone release by estradiol at the level of the pituitary, *Neuroendocrinology* **18:**233–241.

Greep, R. O., 1936, Functional pituitary grafts in rats, *Proc. Soc. Exp. Biol. N.Y.* **34:**754–755.

Gustafsson, J.-A., and Stenberg, A., 1974, Masculinization of rat liver enzyme activities following hypophysectomy, *Endocrinology* **95:**891–896.

Gustafsson, J.-A., Ingelman-Sundberg, M., Stenberg, A., and Hokfelt, T., 1976, Feminization of hepatic steroid metabolism in male rats following electrothermic lesion of the hypothalamus, *Endocrinology* **98:**922–926.

Halasz, B., 1969, The endocrine effects of isolation of the hypothalamus from the rest of the brain, in: *Frontiers in Neuroendocrinology* (W. F. Ganong and L. Martini, eds.), pp. 307–342, Oxford University Press, New York.

Halasz, B., and Pupp, L., 1965, Hormone secretion of the anterior pituitary gland after physical interruption of all nervous pathways to the hypophysiotropic area, *Endocrinology* **77:**553–562.

Harris, G. W., 1955, *Neural control of the pituitary gland,* Edward Arnold, London.

Harris, G. W., and Jacobsohn, D., 1952, Functional grafts of the anterior pituitary gland, *Proc. R. Soc. London B* **139:**263–276.

Harris, G. W., 1970, Hormonal differentiation of the developing central nervous system with respect to patterns of endocrine function, *Phil. Trans. Roy. Soc. Lond. B* **259:**165–177.

Hayashi, Y., 1976, Failure of intrahypothalamic implants of an estrogen antagonist, ethamoxytriphetol (MER-25), to block neonatal androgen-sterilization, *Proc. Soc. Exp. Biol. Med.* **152:**389–392.

Hendricks, S. E., and Welton, M., 1976, Effect of estrogen given during various periods of prepubertal life on the sexual behaviour of rats, *Physiol. Psychol.* **4(1):**105–110.

Hess, D. L., Wilkins, R. H., Moossy, J., Chang, J. L., Plant, T. M., McCormack, J. T., Nakai, Y., and Knobil, E., 1977, Estrogen-induced gonadotropin surges in decerebrated female rhesus monkeys with medial basal hypothalamic (MBH) peninsulae, *Endocrinology* **101:**1264–1271.

Imperato-McGinley, J., Gerrero, L., Gautier, T., and Peterson, R. E., 1974, Steroid 5α-reductase deficiency in man: An inherited form of male pseudohermaphroditism. *Science* **186:**1213–1215.

Jamieson, M., and Fink, G., 1976, Immunoreactive luteinizing hormone releasing factor in rat pituitary stalk blood: Effects of electrical stimulation of the medial preoptic area, *J. Endocrinol.* **68:**71–87.

Jones, J. R., and Saminy, J., 1973, Plasma testosterone levels and female transsexualism, *Arch. Sex. Behav.* **2:**251–256.

Josso, N., 1972, Permeability of membranes to the Mullerian inhibiting substance synthetized by the human fetal testis *in vitro:* A clue to its biochemical nature, *J. Clin. Endocrinol. Metab.* **34:**265–270.

Josso, N., 1973, *In vitro* synthesis of Mullerian inhibiting hormone by seminiferous tubules isolated from the calf fetal testis, *Endocrinology* **93:**829–834.

Jost, A., 1970, Hormonal factors in the sex differentiation of the mammalian foetus, *Phil. Trans. R. Soc. London B* **259:**119–130.

Jost, A., Chodkiewicz, H., and Mauleon, P., 1963, Intersexualité du foetus de veau produite par des androgènes. Comparison entre l'hormone foetale responsable du free-martinisme et l'hormone testiculaire adulte, *C.R. Seanc. Acad. Sci. Paris* **256:**274–276.

Jost, A., Vigier, B., Prepin, J., and Perchellet, J. P., 1973, Studies on sex differentiation in mammals, *Rec. Prog. Horm. Res.* **29:**1–35.

Karsch, F. J., and Foster, D. L., 1975, Sexual differentiation of the mechanism controlling the preovulatory discharge of luteinising hormone in sheep, *Endocrinology* **97:**373–379.

Karsch, F. J., Dierschke, D. J., and Knobil, E., 1973, Sexual differentiation of pituitary function: Apparent difference between primates and rodents, *Science* **179:**848.

Kato, J., 1971, Estrogen receptors in the hypothalamus and hypophysis in relation to reproduction, in: *Proceedings of the Third International Congress on Hormonal Steroids* (V. H. T. James and L. Martini, eds.), pp. 764–773, Excerpta Medica, Amsterdam.

Kato, J., 1975, The role of hypothalamic and hypophyseal 5α-dehydrotestosterone, estradiol and progesterone receptors in the mechanism of feedback action, *J. Steroid Biochem.* **6:**979–987.

Kato, J., Atsumi, Y., and Inaba, M., 1974, Estradiol receptors in female rat hypothalamus in the developmental stages and during pubescence, *Endocrinology* **94:**309–317.

Kawakami, M., and Terasawa, E., 1973, Further studies on sexual differentiation of the brain: Response to electrical stimulation in gonadectomized and estrogen primed rats, *Endocrinol. Jpn.* **20:**595–607.

King, R. J. B., and Mainwaring, W. I. P., 1974, *Steroid-Cell Interactions,* 245 pp., Butterworths, London.

Knapstein, P., David, A., Wu, C. H., Archer, D. F., Flickinger, G. L., and Touchstone, J. C., 1968, Metabolism of free and sulfoconjugated Dhea in brain tissue *in vivo* and *in vitro, Steroids* **11:**885–896.

Knobil, E., 1974, On the control of gonadotrophin secretion in the rhesus monkey, *Rec. Prog. Horm. Res.* **30**:1–36.

Köves, K., and Halasz, B., 1970, Location of the neural structures triggering ovulation in the rat, *Neuroendocrinology* **6**:180–193.

Krey, L. C., Butler, W. R., Weiss, G., Weick, R. F., Dierschke, D. J., and Knobil, E., 1973, Influences of endogenous and exogenous gonadal steroids on the actions of synthetic LRF in the rhesus monkey, in: *Hypothalamic Hypophysiotropic Hormones* (C. Gaul and E. Rosemberg, eds.), pp. 39–47, Int. Congr. Ser. No. 263, Excerpta Medica, Amsterdam.

Larsson, K., and Heimer, L., 1964, Mating behaviour of male rats after lesions in the preoptic area, *Nature* **202**:413–414.

Lax, E. R., Hoff, H. G., Ghraf, R., Schroder, E., and Schreifers, H., 1974, The role of the hypophysis in the regulation of sex differences in the activities of enzymes involved in hepatic steroid hormone metabolism, *Hoppe-Seyler's Z. Physiol. Chem.* **355**:1325–1331.

Lev-Ran, A., 1974, Sexuality and educational level of women with the late-treated adreno-genital syndrome, *Arch. Sex. Behav.* **3**:27–32.

Lisk, R. D., 1967, Neural localization for androgen activation of copulatory behavior in the male rat, *Endocrinology* **80**:754–761.

Lobl, R. T., and Gorski, R. A., 1974, Neonatal intrahypothalamic androgen administration: The influence of dose and age in androgenization of female rats, *Endocrinology* **94**:1325–1330.

Luttge, W. G., and Whalen, R. E., 1970*a*, Dihydrotestosterone, androstenedione, testosterone; comparative effectiveness in masculinizing and defeminizing reproductive systems in male and female rats, *Horm. Behav.* **1**:265–281.

Luttge, W. G., and Whalen, R. E., 1970*b*, Regional localization of estrogenic metabolites in the brain of male and female rats, *Steroids* **15**:605–612.

MacDonald, P. G., and Doughty, C., 1972, Inhibition of androgen-sterilization in the female rat by administration of an anti-oestrogen, *J. Endocrinol.* **55**:455–456.

MacKinnon, P. C. B., Mattock, J. M., and ter Haar, M. B., 1976, Serum gonadotrophin levels during development in male, female and androgenized female rats and the effect of general disturbance on high luteinizing hormone levels, *J. Endocrinol.* **70**:361–371.

Martinez, C., and Bittner, J. J., 1956, A non-hypophysial sex difference in oestrous behaviour of mice bearing pituitary grafts, *Proc. Soc. Exp. Biol. Med.* **91**:506–509.

Maurer, R. A., and Woolley, D. E., 1975, ³H-Estradiol distribution in female, androgenized female and male rats at 100 and 200 days of age, *Endocrinology* **96**:755–765.

McEwen, B. S., and Pfaff, D. W., 1970, Factors influencing sex hormone uptake by rat brain regions. 1. Effects of neonatal treatment, hypophysectomy and competing steroid on estradiol uptake, *Brain Res.* **21**:1–16.

Mennin, S. P., Kubo, K., and Gorski, R. A., 1974, Pituitary responsiveness to luteinizing hormone-releasing factor in normal and androgenized female rats, *Endocrinology* **95**:412–416.

Merari, A., Frenk, C., Hirwig, M., and Ginton, A., 1975, Female sexual behavior in the male rat: Facilitation by cortical application of potassium chloride, *Horm. Behav.* **6**:159–164.

Migeon, C. J., Rivarola, M. A., and Forest, M. G., 1969, Studies of androgens in male transsexual subjects; effects of oestrogen therapy, in: *Transsexualism and Sex Re-assignment* (R. Green and J. Money, eds.) pp. 203–211, Johns Hopkins Press, Baltimore.

Milligan, S. R., 1978, The feedback of exogenous steroids on LH release and ovulation in the intact female vole *(Microtus agrestis)*, *J. Reprod. Fert.* (in press).

Moguilevski, J. A., and Rubinstein, L., 1967, Glycolytic and oxidative metabolism of hypothalamic areas in prepubertal androgenized rats, *Neuroendocrinology* **2**:213–221.

Moguilevski, J. A., Libertun, C., Schiaffini, O., and Scacchi, P., 1969, Metabolic evidence of the sexual differentiation of the hypothalamus, *Neuroendocrinology* **4**:264–269.

Money, J., and Ehrhardt, A. A., 1972, *Man and Woman; Boy and Girl. The Differentiation and Dimorphism of Gender Identity from Conception to Maturity,* Johns Hopkins Univ. Press, Baltimore.

Moore, R. Y., 1974, Visual pathways and the central neural control of diurnal rhythms, in: *The Neurosciences. Third Study Program* (F. O. Schmitt and F. G. Worden, eds.), pp. 537–542, The MIT Press, Cambridge, Massachusetts.

Moore, R. Y., and Eichler, V. B., 1972, Loss of a circadian adrenal corticosterone rhythm following suprachiasmatic lesions in the rat, *Brain Res.* **42**:201–206.

Moss, R. L., and McCann, S. M., 1973, Induction of mating behaviour in rats by luteinizing hormone-releasing factor, *Science* **181**:177–179.

Mueller, G. C., Vanderhaar, B., Kim, U. H., and Le Mahieu, M., 1972, Estrogen Action: An inroad to cell biology, *Rec. Prog. Horm. Res.* **281**:1–45.

Mullins, R. F., and Levine, S., 1968, Hormonal determinants during infancy of adult sexual behaviour in the female rat, *Physiol. Behav.* **3:**333–343.

Nadler, R. D., 1969, Differentiation of the capacity for male sexual behavior in the rat, *Horm. Behav.* **1:**53–63.

Nadler, R. D., 1973, Further evidence on the intrahypothalamic locus for androgenization of female rats, *Neuroendocrinology* **12:**110–119.

Naftolin, F., Ryan, K. H., Davies, I. J., Reddy, V. V., Flores, F., Petro, Z., Kuhn, M., White, R. J., Takaoka, Y., and Wolin, L., 1975, The formation of oestrogens by central neuroendocrine tissues, *Rec. Prog. Horm. Res.* **31:**295–315.

Neill, J. D., 1972, Sexual differences in the hypothalamic regulation of prolactin secretion, *Endocrinology* **90:**1154–1159.

Neumann, F., Elger, W., and Kramer, M., 1966, The development of a vagina in male rats by inhibiting the androgen receptors with an anti-androgen during the critical phase of organogenesis, *Endocrinology* **78:**628–632.

Neumann, F., von Berswordt-Wallrabe, R., Elger, W., Steinbeck, M., Hahn, J. D., and Kramer, M., 1970, Aspects of androgen-dependent events studied by antiandrogens, *Rec. Prog. Horm. Res.* **26:**337–405.

Norman, R. L., and Speis, H. G., 1974, Neural control of the estrogen-dependent twenty-four-hour periodicity of LH release in the golden hamster, *Endocrinology* **95:**1367–1372.

Norman, R. L., Blake, C. A., and Sawyer, C. H., 1972, Effects of hypothalamic deafferentation on LH secretion and the estrous cycle in the hamster, *Endocrinology* **91:**95–100.

Norman, R. L., Resko, J. A., and Spies, H. G., 1976, The anterior hypothalamus: How it affects gonadotropin secretion in the rhesus monkey, *Endocrinology* **99:**59–71.

Nunez, E., Engelmann, F., Benessayag, C., Savu, L., Crepy, O., and Jayle, M.-F., 1971, Identification et purification préliminaire de la foeto-proteine liant les oestrogènes dans le serum des rats nouveau-nés, *C.R. Acad. Sci. D* **273:**831–834.

Ohno, S., 1971, Simplicity of mammalian regulatory systems inferred by single gene determination of sex phenotypes, *Nature* **234:**134–137.

Ojeda, S. R., Kalra, P. S., and McCann, S. M., 1975, Further studies on the estrogen negative feedback on gonadotropin release in the female rat, *Neuroendocrinology* **18:**242–255.

Paxinos, G., and Bindra, D., 1973, Hypothalamic and midbrain neural pathways involved in eating, drinking, irritability, aggression and copulation in rats, *J. Comp. Physiol. Psychol.* **82:**1–14.

Peckham, W. D., and Knobil, E., 1976*a*, Qualitative changes in the pituitary gonadotropins of the male rhesus monkey following castration, *Endocrinology* **98:**1061–1064.

Peckham, W. D., and Knobil, E., 1976*b*, The effects of ovariectomy, estrogen replacement and neuraminidase treatment on the properties of the adenohypophysial glycoprotein hormone of the rhesus monkey, *Endocrinology* **98:**1054–1060.

Pfaff, D. W., 1966, Morphological changes in the brains of adult male rats after neonatal castration, *J. Endocrinol.* **36:**415–416.

Pfaff, D. W., 1973, Luteinising hormone-releasing factor potentiation lordosis behavior in hypophysectomized ovariectomized female rats, *Science* **182:**1148–1149.

Pfaff, D. W., Gerlach, J. L., McEwen, B. S., Ferin, M., Carmel, P., and Zimmerman, E. A., 1976, Autoradiographic localization of hormone and concentrating cells in the brain of the female rhesus monkey, *J. Comp. Neurol.* **170:**279–291.

Pfeiffer, C. A., 1936, Sexual differences of the hypophyses and their determination by the gonads, *Am. J. Anat.* **58:**195–225.

Phoenix, C. H., 1974, Prenatal testosterone in the non human primate and its consequences for behavior, in: *Sex Differences in Behavior* (R. C. Friedman, R. M. Richart, and R. L. Vande Wiele, eds.), pp. 19–32, Wiley, New York.

Phoenix, C. H., Goy, R. W., Gerall, A. A., and Young, W. C., 1959, Organizing action of prenatally administered testosterone propionate on the tissue mediating mating behavior in the female guinea-pig, *Endocrinology* **65:**369–382.

Price, D., 1970, *In vitro* studies on differentiation of the reproductive tract, *Phil. Trans. R. Soc. London B* **259:**133–139.

Puig-Duran, E., and MacKinnon, P. C. B., 1978*a*, The ontogeny of luteinizing and prolactin hormone responses to oestrogen stimuli and to a progesterone stimulus after priming with either a natural or synthetic oestrogen in immature female rats, *J. Endocrinol.* **76:**311–320.

Puig-Duran, E., and MacKinnon, P. C. B., 1978*b*, The effect of oestrogen and progesterone given at various times of day on serum luteinizing hormone and prolactin concentration in 21 day old Wistar rats of different sexual status, *J. Endocrinol.* **76:**321–331.

Puig-Duran, E., Greenstein, B. D., and MacKinnon, P. C. B., 1978, Serum oestrogen-binding components in developing female rats and their effects on the unbound oestradiol-17β fraction, serum FSH concentrations and uterine uptake of ³H-oestradiol-17β, *J. Reprod. Fert.* (in press.)

Quinn, D. L., 1966, Luteinising hormone release following preoptic stimulation in the male rat, *Nature* **209**:891–892.

Raisman, G., 1969, Neuronal plasticity in the septal nuclei of the adult rat, *Brain Res.* **14**:25–48.

Raisman, G., and Brown-Grant, K., 1977, Endocrine and behavioral abnormalities following lesions of the suprachiasmatic nuclei in the female rat, personal communications.

Raisman, G., and Field, P. M., 1971, Sexual dimorphism in the preoptic area of the rat, *Science* **173**:731–733.

Raisman, G., and Field, P. M., 1973, Sexual dimorphism in the neuropil of the preoptic area of the rat and its dependence on neonatal androgen, *Brain Res.* **54**:1–29.

Raynaud, T. P., Mercier-Bodard, C., and Baulieu, E. E., 1971, Rat estradiol binding plasma protein (EBP), *Steroids* **18**:767–788.

Resko, J. A., Feder, H. H., and Goy, R. W., 1968, Androgen concentrations in plasma and testis of developing rats, *J. Endocrinol.* **40**:485–491.

Resko, J. A., Malley, A., Begley, D. E., and Hess, D. L., 1973, Radioimmunoassay of testosterone during fetal development of the rhesus monkey, *Endocrinology* **93**:156–161.

Salaman, D. F., and Birkett, S., 1974, Androgen-induced sexual differentiation of the brain is blocked by inhibitors of DNA and RNA synthesis, *Nature, London* **247**:109–112.

Sar, M., and Stumpf, W. E., 1972, Cellular localization of androgens in the brain and pituitary after the injection of tritiated testosterone, *Experientia* **28**:1364–1366.

Sawyer, C. H., 1957, Activation and blockade of the release of pituitary gonadotropin as influenced by the reticular formation, in: *Reticular Formation of the Brain* (H. H. Jasper, L. D. Proctor, R. S. Knighton, W. C. Noshay, and R. T. Costello, eds.), pp. 221–223, J. and A. Churchill, London.

Schwartz, N. B., and McCormack, C. E., 1972, Reproduction: Gonadal function and its regulation, *Annu. Rev. Physiol.* **34**:425–472.

Segal, S. J., and Johnson, D. C., 1959, Inductive influence of steroid hormones on the neural system: Ovulation controlling mechanisms, *Arch. Anat. Microsc. Morphol. Exp.* **48**:261–273.

Shapiro, B. H., Goldman, A. S., Steinbeck, H. F., and Neumann, F., 1976*a*, Is feminine differentiation of the brain hormonally determined? *Experientia* **32**:650–651.

Shapiro, B. H., Goldman, A. S., Bongiovanni, A. M., and Marino, J. M., 1976*b*, Neonatal progesterone and feminine sexual development, *Nature* **264**:795–796.

Sheridan, P. J., Sar, M., and Stumpf, W. E., 1974, Autoradiographic localization of ³H-estradiol or its metabolites in the central nervous system of the developing rat, *Endocrinology* **94**:1386–1390.

Short, R. V., 1974, Sexual differentiation of the brain of sheep, in: *Les Colloques de l'Institut National de la Santé et de la Recherche Médicale, INSERM* **32**:121–142.

Siiteri, P. K., and Wilson, J. D., 1974, Testosterone formation and metabolism during male sexual differentiation in the human embryo, *J. Clin. Endocrinol. Metab.* **38**:113–129.

Södersten, P., 1976, Lordosis behaviour in male, female and androgenized female rats, *J. Endocrinol.* **70**:409–420.

Södersten, P., and Larsson, K., 1974, Lordosis behavior in castrated male rats treated with estradiol benzoate or testosterone propionate in combination with an estrogen antagonist, MER-23, and in intact male rats, *Horm. Behav.* **5**:13–18.

Staudt, J., and Dörner, G., 1976, Structural changes in the medial and central amygdala of the male rat, following neonatal castration and androgen treatment, *Endokrinologie* **67**:296–300.

Stearns, E. L., Winter, J. S. D., and Faiman, C., 1973, Positive feedback effect of progestin upon serum gonadotropins in estrogen primed castrate men, *J. Clin. Endocrinol.* **37**:635–648.

Steiner, R. A., Clifton, D. K., Spies, H. G., and Resko, J. A., 1976, Sexual differentiation and feedback control of luteinizing hormone secretion in the rhesus monkey, *Biol. Reprod.* **15**:206–212.

Stenberg, A., 1976, Developmental, diurnal and oestrous cycle dependent changes in the activity of liver enzymes, *J. Endocrinol.* **68**:265–272.

Stern, J. J., 1969, Neonatal castration, androstenedione and the mating behavior of the male rat, *J. Comp. Physiol. Psychol.* **69**:608–612.

Stumpf, W. E., 1968, Estradiol concentrating neurons. Topography in the hypothalamus by dry-mount autoradiography, *Science* **162**:1001–1003.

Stumpf, W. E., and Sar, M., 1971, Estradiol concentrating neurons in the amygdala, *Proc. Soc. Exp. Biol. Med.* **136**:102–114.

Stumpf, W. E., Sar, M., and Kepper, D A., 1974, Anatomical distribution of estrogen in the central nervous system of mouse, rat, tree shrew and squirrel monkey, *Adv. Biosci.* **15**:77–84.

Swanson, H. H., 1967, Alteration of sex typical behavior of hamsters in open field and emergence tests by neonatal administration of androgen or oestrogen, *Anim. Behav.* **15:**209–216.

Swanson, H. E., and van der Werff ten Bosch, J. J., 1964, The "early androgen" syndrome: Differences in response to pre-natal and post-natal administration of various doses of testosterone propionate in female and male rats, *Acta Endocrinol., Copenhagen* **47:**37–50.

Swanson, H. E., and van der Werff ten Bosch, J. J., 1965, The 'early androgen' syndrome; effects of pre-natal testosterone propionate. Acta Endocrinol., Copenhagen, **50:**379–390.

Swanson, H. H., and Crossley, D. A., 1971, Sexual behaviour in the golden hamster and its modification by neonatal administration of testosterone propionate, in: *Proceedings of the International Conference on Hormones in Development* (M. Hamburgh and E. J. W. Berrington, eds.), pp. 677–687, Appleton–Century–Crofts, New York.

Tejasen, T., and Everett, J. W., 1967, Surgical analysis of the preopticotuberal pathway controlling ovulatory release of gonadotropins in the rat, *Endocrinology* **81:**1387–1396.

Terasawa, E., Kawakami, M., and Sawyer, C. H., 1969, Induction of ovulation by electrochemical stimulation in androgenized and spontaneously constant oestrous rats, *Proc. Soc. Exp. Biol. N.Y.* **132:**497–501.

ter Haar, M. B., MacKinnon, P. C. B., and Bulmer, H. G., 1974, Sexual differentiation in the phase of the circadian rhythm of ^{35}S methionine incorporation into cerebral proteins, and of serum gonadotrophin levels, *J. Endocrinol.* **62:**257–265.

Tuohimaa, P., 1971, The radioautographic localization of exogenous tritiated dihydrotestosterone, testosterone, and oestradiol in the target organs of female and male rats, in: *Basic Actions of Sex Steroids on Target Organs* (P. O. Hubinot, F. Leroy, and P. Galand, eds.), pp. 208–214, Karger, Basel.

Turner, J. W., 1976, Influence of neonatal androgen on the display of territorial marking behavior in the gerbil, *Physiol. Behav.* **15:**265–276.

Valenstein, E. S., Kakolewski, J. W., and Cox, V. C., 1967, Sex differences in taste. Preference for glucose and saccharine solutions, *Science* **146:**942–943.

van Wagenen, G., and Hamilton, J. B., 1943, Experimental production of pseudohermaphroditism, in: *Essays in Biology* (T. Cowles, ed.), pp. 581–607, University of California Press, Berkeley.

Velasco, M. E., and Taleisnik, S., 1969, Release of gonadotropins induced by amygdaloid stimulation in the rat, *Endocrinology* **84:**132–139.

Vertes, M., Barnea, A., Lindner, H. R., and King, R. J. B., 1973, Studies on androgen and estrogen uptake by rat hypothalamus, in: *Receptors for Reproductive Hormones* (B. W. O'Malley and A. R. Means, eds.), pp. 137–173, Plenum, New York.

Wachtel, S. S., Ohno, S., Koo, G. C., and Boyse, E. A., 1975, Possible role for H-Y antigen in the primary determination of sex, *Nature* **257:**235–236.

Wade, G. H., and Zucker, I., 1969, Taste preferences of female rats modification by neonatal hormones, food deprivation and prior experience, *Physiol. Behav.* **4:**935–943.

Wagner, J. W., Erwin, W., and Critchlow, V., 1966, Androgen sterilization produced by intracerebral implants of testosterone in neonatal female rats, *Endocrinology* **79:**1135–1142.

Weisz, J., and Gibbs, C., 1974, Metabolites of testosterone in the brain of the newborn female rat after an injection of tritiated testosterone, *Neuroendocrinology* **14:**72–86.

Wells, L. J., and van Wagenen, H. G., 1954, Androgen induced female pseudohermaphroditism in the monkey *(Macaca mulatta)* anatomy of the reproductive organs, *Contrib. Embryol.* **35:**93–106.

Welshons, W. J., and Russell, L. B., 1959, The Y-chromosome as the bearer of male determining factors in the mouse, *Proc. Natl. Acad. Sci. U.S.A.* **45:**560–566.

Westley, B. R., and Salaman, D. F., 1976, Role of oestrogen receptor in androgen-induced sexual differentiation of the brain, *Nature* **262:**407–408.

Westley, B. R., and Salaman, D. F., 1977, Nuclear binding of the oestrogen receptor of neonatal rat brain after injection of oestrogens and androgens: Localization and sex differences, *Brain Res.* **119:**375–388.

Whalen, R. E., 1966, Sexual motivation, *Psychol. Rev.* **73:**151–163.

Whalen, R. E., 1968, Differentiation of the neural mechanisms which control gonadotropin secretion and sexual behavior, in: *Perspectives in Reproduction and Sexual Behavior* (M. Diamond, ed.), pp. 303–340, Indiana University Press, Bloomington.

Whalen, R. E., and Edwards, D. A., 1967, Hormonal determinants of the development of masculine and feminine behavior in male and female rats, *Anat. Rec.* **157:**173–180.

Whalen, R. E., and Luttge, W. G., 1970, Long-term retention of tritiated estradiol in brain and peripheral tissues of male and female rats, *Neuroendocrinology* **6:**255–263.

Whalen, R. E., Battie, C., and Luttge, W. G., 1972, Testosterone, androstenedione and dihydrotestosterone: Effects on mating behavior of male rats, *Behav. Biol.* **2:**117–125.

Witschi, E., 1962, Embryology of the ovary, in: *The Ovary* (H. G. Grady and D. E. Smith, eds.), pp. 1–10, Int. Acad. Path. Monograph No. 3, Williams and Wilkins, Baltimore.

Yalom, I., Green, R., and Fisk, H., 1973, Prenatal exposure to female hormones—effect on psychosexual development in boys, *Arch. Gen. Psychiatry* **28**:554–561.

Yanasi, M., and Gorski, R. A., 1976, Sites of oestrogen and progesterone facilitation of lordosis behavior in the spayed rat, *Biol. Reprod.* **15**:536–543.

Yazaki, I., 1960, Further studies on endocrine activity of subcutaneous ovarian grafts in male rats by daily examination of smears from vaginal graft, *Annot. Zool. Jpn.* **33**:217–225.

Yen, S. S. C., Tsai, C. C., Vandenberg, G., and Rebar, R., 1972, Gonadotropin dynamics in patients with gonadal dysgenesis: A model for the study of gonadotropin regulation, *J. Clin. Endocrinol. Metab.* **35**:897–904.

Zigmond, R. E., 1975, Binding, metabolism and action of steroid hormones in the central nervous system, in: *Handbook of Psychopharmacology,* Vol. 5 (L. L. Iversen, S. D. Iversen, and S. H. Snyder, eds.), pp. 239–328, Plenum Press, New York.

7

Critical Periods in Organizational Processes

J. P. SCOTT

1. Introduction

Of all the known phenomena of development, that of critical periods lends itself most readily to practical applications, both for preventive mental hygiene and for positive control of behavioral development. Once a critical period has been identified, there is little difficulty in putting the information to use. Until recently, however, our understanding of this phenomenon had progressed very little further than its description and the recognition of its generality. It is the principal purpose of this chapter to develop and illustrate its underlying theoretical principles.

2. Physical Growth and Behavioral Development—A Comparison

2.1. The Concept of Systems

A system may be defined as a group of interacting entities. As such, it is one of our most general scientific concepts, applying to both living and nonliving phenomena and ranging in size from an atom to the solar system and perhaps to the universe beyond. I shall limit my discussion to living systems and point out only their most basic characteristics: reciprocal stimulation between the units of the system; a tendency for organization to change toward increasing complexity and, concurrently, toward stability; the differentiation of function between units; finally, a tendency for the interactions between the units to favor the survival of the system or, in biological terms, to be adaptive.

There are various dimensions along which the organization of a system can be measured, but two of the most important are the predictability of the interactions

---

J. P. SCOTT • Center for Research on Social Behavior, Bowling Green State University, Bowling Green, Ohio.

and the differentiation of function between the individuals. Another basic characteristic of systems which is implicit in the concept of interaction is communication, which can be measured in various ways, the simplest of which is mere quantity. Without some form of mutual stimulation, there can be no system.

2.2. The Concept of Development

Within the framework of the systems concept, development can be defined as change in the organization of systems that persists for only one generation. The qualification differentiates development from evolutionary change and ecological change, both of which persist over more than the lifetime of an individual. Generally, the organizational changes that we call development are in the direction of increasing complexity of organization and, as such, involve greater adaptivity and more efficient functioning. They are wholly or partially irreversible, which is related to the general systems characteristic of stability. One cannot take an adult human and reprocess him back to an embryo, attractive as this idea might be to a student of science fiction.

2.3. Growth and Behavioral Development

Both of these are organizational processes, but they take place on different levels of systems organization. Growth is largely a process of organization taking place within subsystems of an organism, whereas behavioral development, although it involves internal physiological processes and systems, is primarily a process of organizing the organismic system as a whole with respect to the external world and, in particular, to other living systems. Both of these organizational processes are still incompletely understood except on a descriptive basis. Growth involves changes in an enormously complex set of physiological systems and subsystems. We have determined some of the methods through which units of these systems interact, but we are still very far from being able to understand the process as a whole. From a descriptive viewpoint we can say that physical growth is a cumulative process, an integrative process in that the units become increasingly related to each other and react in a more and more predictable fashion, and finally, that it is a process of specialization of function. We also know in general that the last process—specialization or differentiation—is opposed to the cumulative part of the process, or growth in size.

Actually, behavioral development could be described in much the same terms. It is cumulative in that more and more behaviors are added, and as these are added, they are integrated into systems and subsystems of behavior. Finally, the development of specialized sorts of behavior may interfere with the acquisitions of new behaviors.

Besides the differences with respect to levels of organization, behavioral development and physical development are different with respect to time relationships, since behavior is defined as the activity of an organism as a whole which can only be expressed when the organismic system is sufficiently well organized to be capable of activity, whereas growth takes place from the very outset of development. Although there is a considerable period during which these two processes overlap, the processes of physical growth must be initiated earlier and generally achieve complete stability and so cease development before the process of behavioral development reaches a similar stability.

Seen within the concept of developmental change in systems, an infant is born with its internal physiological organization already far advanced, although not complete, but with no organization with respect to the social systems of which it will become a part. This suggests that the important problems are the time relationships between behavioral and physiological organizational processes and their effects on each other, for it is not only the entities within a system that interact with each other but also systems at higher and lower levels.

These considerations place the old "innate–acquired" problem in a modern focus. The central problem now becomes one of empirically determining the time relationships between the organizing processes of growth and behavioral development, and the nature of the interactions between the systems produced by each. This replaces the older dichotomy, which was at best merely descriptive.

3. Theory of Organizational Change

Since organizational processes on different levels have similar general characteristics, it should be possible to formulate general theories of organization as well as descriptive laws.

3.1. The Concept of Processes

In its most general definition, a process is something that proceeds or goes forward. In a system the thing that proceeds is organizational change. Therefore, to call something a process is simply to give it a name; we must discover the nature of the change or changes. An initial step in this direction is to divide processes into classes. A mature system must process outside materials and alter their organization in order to maintain the system. We can, therefore, make a distinction between *maintenance processes* that do not alter the organization of the system, and *developmental processes* that result in the organization of the system itself.

We know that maintenance processes are not unitary but consist of many different ongoing processes. Likewise, it is highly unlikely that developmental processes are simple and unitary even within a relatively simple system. Developmental change results in a single overall system, the organism, but we know that this consists of many subsystems. Therefore, any theory of organizational change must include the possibility of multiple processes.

3.2. The Concept of Time

A fundamental dimension of organizational change is time, and no study of development can omit it. It is important both in a relative and an absolute sense. However, anyone who studies the development of more than one organism is immediately confronted with the problem of variation.

3.3. The Concept of Variation

Differences in the state of development of individuals can result from a variety of causes, the most obvious of which are differences in starting points. In addition, various environmental accidents can speed up or slow down developmental rates. Similarly, genetic variation can affect developmental change. Variation in organizational change is, therefore, inevitable; and this brings up the question of the role of genetics relative to such changes.

3.4. The Role of Genetics

The genes are the fundamental units of organizational change. Without going into the details of gene structure and gene action, concerning which so many fundamental discoveries have been made in recent years, we can say that genes are large organic molecules. The genes not only reproduce themselves but produce enzymes, substances that alter the speed of chemical reactions. Each gene produces one primary enzyme (Beadle and Tatum, 1941). Therefore, each gene represents not only a structure but a different organizing process. This confirms the conclusion that multiple organizing processes must be involved in development. To some extent genetic processes go forward independently, resulting in variation in degrees of organization in different parts of an organism. Since each gene is subject to mutational change, variation in rates of development are inevitable, not only within an individual but between individuals.

In a higher organism there may be 100,000 or more genes, each controlling a different organizational process but interacting with others to produce an integrated whole. How integration is accomplished is still a mystery. We are only beginning to make progress in understanding gene action in bacteria, which are little more than collections of a few genes with very little higher organization. Thus, while the genes are the fundamental units of organizational change, progress in understanding developmental change must involve the study of organizing processes at higher levels, both in order to achieve more rapid progress and in order to understand the activities of the genes themselves. Furthermore, genes can affect both developmental processes and maintenance processes, and these two functions must be considered separately.

4. Critical Periods

The phenomenon of critical periods in development was first discovered by Stockard (1907, 1921) in the early part of this century. He was experimenting with the results of altering the chemical composition of sea water in which eggs of the fish *Fundulus* were developing. These eggs developed into various kinds of monsters. Stockard at first thought that the deformed embryos were produced by specific ions, but his more detailed experiments showed that it was not the nature of the chemical but the time at which it was applied that was important. Earlier or later application produced different effects, and in general, the effects were much smaller as the embryo grew older.

In the same era Child (1921) established the principle that the most rapidly growing, or as he put it, the most rapidly metabolizing, parts of an organism are the most vulnerable to toxic agents of any sort, thus suggesting that organizing processes are vulnerable when they are proceeding rapidly. He did not, however, make a clear distinction between maintenance processes and developmental processes.

These findings and subsequent confirmation represent an important contribution to the understanding of teratology. Thus, when I studied the development of Wright's genetically determined polydactylous monster in the guinea pig (Scott, 1937) it was readily apparent that the condition of extra toes was produced by the activity of some genetically determined agent at the time when limb buds were developing most rapidly. In this case, the agent appeared to enhance growth rather than inhibit it, but it was possible to interpret this effect in terms of critical periods. By now, it is possible to predict the effects of various noxious agents and developmental accidents on human development in relation to time and thus to make specific recommendations regarding the use of drugs and the avoidance of other hazards (Moore, 1974).

4.1. Critical Periods in Behavioral Development

In the 1930s, Lorenz drew attention to the importance of the phenomenon of imprinting in birds (Lorenz, 1935). The tendency for hand-reared birds to become attached to the experimenter had been noticed by others, such as Heinroth (1910) and Spalding (1873), but they had not appreciated its theoretical importance. A brief contact with a developing bird during its early life will transform it from a wild animal that cannot even be approached by a human into one that is not only unafraid of people but strongly attached to them. Similar periods of contact in later development are ineffective, although it is possible to reduce the fear responses of wild birds as adults. In addition, the adult bird may become tame, but show no sign of attachment.

A large volume of research on imprinting confirmed the existence of the phenomenon and demonstrated its generality in bird species, although the time of the critical period may be later in altricial birds than in the precocial ones (Hess, 1973). This research can now be criticized as devoting too much attention to side issues and as considering the process of attachment as a phenomenon affecting only the young bird, whereas it is really a process of the formation of a social system involving the mother bird and other birds as well.

Nevertheless, certain generalizations were established. The process of attachment is largely an internal one and is little affected by external rewards or punishments. It is also a perceptual one in that the bird must perceive the objects to which it becomes attached and discriminate between these and others to which it is not attached.

Later, I discovered a similar phenomenon in mammals, first in sheep and later in dogs (Scott, 1945; Scott and Marston, 1950). I called this process "primary socialization" in order to emphasize the fact that the basic phenomenon is the formation of the animal's first and presumably most important social relationships. At first I did not sufficiently emphasize the fact that the process is a mutual one, but this soon became apparent in research with sheep.

If a young lamb is taken from its mother at birth and raised on a bottle by human beings for the first 10 days of life before being offered contact with other sheep, it forms a strong attachment to human beings and does not follow the flock. In one case, a female lamb trained in this fashion was still feeding and walking independently three years later, even though she had been confined with the flock in the same small field and had been mated and produced lambs of her own. Collias (1956) showed that the ability of the young lamb to form attachments was limited not only by developmental change but by the activity of other sheep. The mother will accept and become attached to any lamb, including her own, only during a short period varying from 2 to 4 hr after giving birth. Afterwards she will reject any lamb that approaches and attempts to nurse. Thus there is in the mother a very short critical period for attachment, the physiological basis having never been determined but presumably being related to hormonal changes. The period may be prolonged for a few hours if the mother is kept isolated from other sheep (Moore, 1960). The lamb, on the other hand, has a somewhat longer critical period that permits it to become attached not only to its mother but also to other young lambs which, unlike the adults, do not reject it. A similar phenomenon is present in all other ungulate species that have been studied.

In dogs, which are nonprecocious animals, behavioral development can be readily divided into periods characterized by different processes. The *neonatal period,* chiefly devoted to the establishment of the process of neonatal nutrition, or nursing, lasts for approximately 2 weeks. During the neonatal period all behavior is adaptive to the environment in which the mother takes complete care of the puppy. There follows a *transition period* that includes a rapid transformation of behavior into forms that are adaptive for more independent living. Following this, and beginning at approximately 3 weeks of age, there is a period which is primarily characterized by the process of socialization, or the formation of attachments, both with the physical environment and with any living things that are present, normally the littermates and mother.

Recent evidence indicates that the process of attachment first appears during the early transition period but does not reach its full capacity until approximately 4 weeks of age (Gurski *et al.,* 1974). The process is maintained at a peak level for several weeks but declines to a relatively low level by 12 weeks, partially influenced by the development of a fear response to strange individuals that prevents prolonged contact (Figure 1). Thus, in the dog, behavioral development has evolved in two kinds of social environments, one in which complete care is provided by the mother and the other in which the puppies are providing more and more care for themselves and interacting with other individuals. It is significant that during the *period of socialization* the mother spends less and less time with her offspring, resulting in the strongest attachments being formed with littermates. This is obviously related to the normal pack organization of dogs and wolves (Scott, 1963).

The following conclusions were reached as a result of a long series of observations and experiments with developing puppies.

1. The process of attachment is based on the ability to discriminate between familiar and unfamiliar individuals. Evidence of attachment does not appear until the senses of sight and hearing begin to be well developed.

2. Attachment is dependent on the ability to subsequently perceive individuals as familiar or unfamiliar, implying the development of memory. Research on the development of learning in the dog is still incomplete but indicates that memory develops gradually during the time when the first indications of attachment are appearing.

3. The best index that attachment has occurred is the emotional reaction to separation in the form of intense and persistent vocalization. This does not appear in its most intense form until approximately 4 weeks of age.

4. Separation distress is most effectively eliminated by active social interaction either with another dog or with a person. The same kind of interaction speeds up the process of attachment.

5. As with birds, the process of attachment is not dependent upon external rewards and punishments. It must therefore be largely an internal process.

6. The process does not completely disappear with age but is slowed to a much lower rate by various interfering factors, including a developing fear response to the strange, and previous attachments that limit contact with new individuals.

7. The process of attachment to places and inanimate objects is similar to that of social attachment, both with respect to time and the response to separation. The combination of separation from familiar individuals and a familiar site produces a

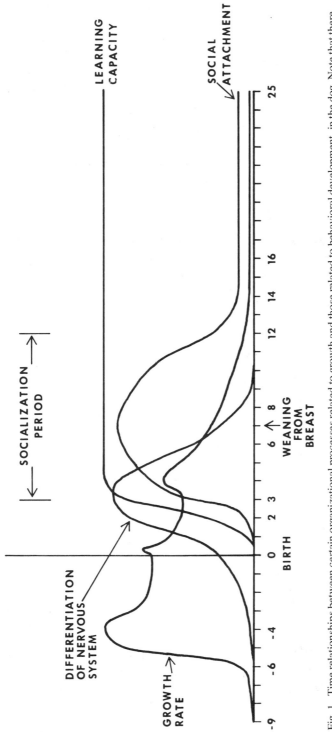

Fig. 1. Time relationships between certain organizational processes related to growth and those related to behavioral development, in the dog. Note that there is considerable overlap between the two, permitting possible interaction. (From Scott *et al.*, 1974, by permission of John Wiley and Sons.)

much more intense response. Further, the process of attachment to people appears to be identical with that of attachment to other dogs.

The similarity of the attachment process in animals as widely different as mammals and birds suggests that the process of attachment originally evolved in connection with site attachment, a phenomenon that can be demonstrated in almost all vertebrate species. The process is, therefore, available for evolution into one with a social function, depending on the species involved, but without being fundamentally changed. These facts and conclusions cry out for a fundamental neurological and biochemical analysis of the process.

With respect to primates other than man, little has been done to define critical periods of social attachment. In the case of the rhesus monkey, the infants are relatively precocious and cling to the mother from the day of birth; thus, the infants are physically as well as psychologically attached to their mothers. Mason and Kenney (1974) have showed that rhesus infants will readily clasp tolerant dogs and appear to be emotionally disturbed when separated from them. If a monkey is taken off one dog and given to another, it clasps the new one within a short time and appears to become readily attached. This kind of behavior takes place over many months, indicating that the infant does not have a short critical period for making such attachment. On the other hand, if a monkey mother carrying an infant is removed from a normal monkey group, the infant becomes greatly disturbed, first becoming very active and excited and eventually depressed. The infant does not seek out another monkey female as it would a dog foster mother.

Some light is thrown on these facts by the work of Kaufman and Rosenblum (1969) on other species of macaques. Adult female pigtail monkeys generally remain apart from other members of a group, retaining their infants and guarding them. When a mother is removed, the infant goes through a period of depression similar to the rhesus. Bonnet monkeys, however, characteristically huddle in groups; when a mother is removed, the infant simply associates with another female and shows little distress. This suggests that the critical period for social attachment in the rhesus and pigtail monkeys may be similar to that in the sheep; i.e., there may be only a short period during which a female will accept an infant. At any rate, adoptions are almost unknown in these species. The limitation of attachment may be the result of maternal behavior of two types: possessive behavior by the true mother and rejection by others. The substitute dog mother behaves more like the bonnet monkey.

Bowlby (1951) called attention to the possibility of a critical period of attachment in humans and its effects on adoption. In general, babies adopted before the age of 6 months later show very few psychological problems compared with those adopted as older infants.

From a descriptive viewpoint, human infants are born in a more advanced state than puppies (Scott, 1963) in that their eyes and ears are functional, but they are less advanced in motor behavior because of the inability to crawl. As with puppies, human infants show a neonatal period (lasting approximately 5 or 6 weeks) in which their behavior is almost totally adapted to complete care by the mother. During this period they are unresponsive to changes in places and caretakers, and if warm, dry, and well-fed, will be contented under almost any circumstances.

The first change in behavior is the appearance of the social smile in response to a human face. From this point until approximately 6 or 7 months, a human infant will give positive responses to anyone and will form attachments to strangers after very short periods of contact (Fleener, 1967); this, then, is a period of socialization.

At approximately 7 months, the baby begins to make a transition to adult forms of behavior such as crawling (progressing eventually to walking) and eating semisolid food with the aid of newly acquired teeth. Thus, the period of socialization precedes the transition period. An associated phenomenon is the appearance, during the transition period, of a fear of strangers which would, of course, have the effect of preventing contact and attachment. This occurs after the baby has formed his primary attachment and before he is capable of independent motion and can approach or move away from strangers at will.

The human infant also has a second transition period, found in no other species. Beginning when he is approximately 2 years old, the human infant becomes gradually capable of transmitting and receiving information through verbal language in addition to the nonverbal signals that have been previously exchanged.

In general, critical periods for social attachment are characterized by very rapid organizational processes. Attachments can be made within a few minutes in chicks (Hess, 1973), in as little as 2 hr in dogs (Cairns and Werboff, 1967), and in as little as 7½ hr in humans (Fleener, 1967).

The phenomenon of critical periods for social attachment and site attachment has been reviewed chiefly because it has been so thoroughly documented. There are numerous other developmental processes that include critical periods. In fact, any such process may have a critical period provided certain conditions are met.

5. General Theory of Critical Periods

5.1. The Nature of Critical Periods

The term critical is derived from the Greek word *krisis,* and its basic meaning is that it involves a decision. A critical period is therefore one in which a decision is made, but in its more restricted meaning it involves a decision that cannot be repeated or reversed at a later time. For example, contact with a newly hatched turkey may produce an adult bird that will be attached to humans and will even attempt to mate with them (Schein, 1963). Similar contact with an adult bird will not produce such an effect. The irreversibility of decisions is even more strikingly illustrated by embryological examples. There is only a brief period in embryological development when an individual can be influenced to develop more than the normal number of digits. No interference in later periods can modify the number so produced except by amputation. Waddington (1962) has used the metaphor of canalization to describe such effects. Early in development organizational changes may proceed in many directions but as they proceed it is as if the process is following a deeper and deeper canal whose boundaries limit further choices.

From an analytical viewpoint, developmental organizational processes tend to be irreversible simply because such processes produce more and more stable systems; thus, the phenomenon of critical periods is inherent in the concept of systems and therefore it becomes important to determine the stability of any system that is analyzed.

It is obvious that species may vary with respect to stability of organization. For example, a *Planaria* worm can be cut into small pieces, each of which will regenerate into an entire worm. In the higher vertebrates, regeneration is largely limited to healing of minor wounds. Any drastic disorganization such as produced by a major wound usually results in the destruction of some vital organ and death. This is one of the penalties of complex organization. It produces greater stability and greater efficiency of function but inhibits the power of regeneration, i.e., reorganization. Similar species differences can be seen in behavior. Most birds, which have evolved a relatively rapid course of development in connection with acquiring the ability for flight, show much more stability of behavioral organization than do mammals, whose behavior can be extensively modified even as adults. We must conclude that the stability of a system can be modified by genetics, at least between species and, by implication, within species.

5.2. The Theory of Critical Periods

This theory, which has been developed in detail elsewhere (Scott *et al.,* 1974) is largely based on the empirical finding that organizational processes are modified most easily at the time that they are proceeding most rapidly. On this basis we can reach the following conclusions:

An organizational process cannot have a critical period if it proceeds at a uniform rate. It can be modified just as easily at one time as another. Since maintenance processes generally proceed at fairly uniform rates over the life of a system, they should not manifest critical periods.

On the other hand, developmental processes do not and cannot proceed at uniform rates. We can, therefore, state that all developmental processes must exhibit critical periods whose duration and importance will depend on the rates of change in these processes. The courses of their changes in rates may take several forms. In the classical case of the critical period, the process may proceed rapidly for a short time and then cease completely. This is true of a large number of embryological processes. A second possibility is that the process may proceed rapidly then fall off to a low level but never entirely cease. This seems to be characteristic of the social attachment process in the higher mammals including man.

Still another possibility is that the organizing process may proceed intermittently at high rates with a consequent possibility of multiple critical periods. Such is the process of learning in the higher animals.

Because of the complexity of organization in living systems, we must also consider the consequences of multiple processes interacting with each other. This can be best illustrated with the simplest case, that of two processes. Their interaction will chiefly depend on the relative times of most rapid activity. If they coincide, the interaction of two processes could produce one brief critical period. A prolonged period may occur if they immediately follow each other, and two separate critical periods may be manifested if the phases of rapid organization take place at separate times. The addition of other processes simply multiplies the possible number of separate critical periods.

In order to delimit the critical periods involved in any developmental process, it becomes necessary to determine the nature of the process and its relative timing. In the case of the general process of social attachment in the dog, we can identify at least three necessary processes:

1. development of sensory capacities that permit discrimination
2. development of learning capacities that permit memory
3. development of the capacity for separation distress

Behavior is defined as the activity of the entire organism and, therefore, is dependent on all those organizing processes that have produced a functioning organism. Of particular relevance to behavior is the development of the capacity for movement which is, of course, dependent upon the organization of the muscular and skeletal systems. These systems are particularly responsive to functional differentiation. Muscles not only increase in size but also develop additional circulation in response to exercise, and there is an increasing body of evidence that indicates a critical period for this effect in young rats (Rakusan and Poupa, 1970). Another system that is concerned with behavior is the endocrine system. This system also can be organized in response to function. The critical period for the effects of handling in young rats appears to involve this phenomenon (Denenberg and Zarrow, 1971). The most essential physiological system involved in behavior integration is, of course, the nervous system. As with other systems, it becomes more and more stable in its organization with maturity and has quite limited powers of regeneration or reorganization in the adult mammal.

In addition to the above, the process of learning has major relevance for behavioral organization. In spite of its importance, we still do not know its precise physiological basis. We know that it must be biochemical in nature and we know that it is related to the organization of nerve tracts produced by other organizing processes, but its mode of function is still known largely through description.

The major characteristics of the learning process are, first, that it may occur very rapidly, in fact, within a fraction of a second in one trial learning. It can be organized in a more stable fashion, however, by repetition.

Second, the evidence from higher organisms indicates that learning is permanent and that forgetting is not a matter of erasure but of inability to retrieve information. This implies that learning is a cumulative process.

Considered in more detail, the process consists of: (1) attempts at adaptation, (2) varying these attempts, (3) the choice of one of these attempts as a solution, and (4) the fixing of this solution by habit formation. This implies that there is a critical period involving the choice or decision between adaptive behaviors.

Still another characteristic of learning is that it tends to be self-limiting. That is, the formation of certain habits inhibits the formation of conflicting new ones. In this way, the learning process shares the general characteristic of systems organization of producing stability. Finally, in contrast to most growth processes, considerable but not complete reorganization may be possible, even in adulthood.

Learning is, therefore, an example of an organizational process that involves repeated or intermittent critical periods. Each new problem that is solved involves a new critical period for that particular organization of behavior. On a higher level, the early period of mutual adjustment seen in the formation of a social relationship is a critical period for determining the nature of that relationship.

From the nature of the process in higher vertebrates, decisions made in the learning process are not as critical as those that are involved in growth. Nevertheless, the same general principle applies; there can be *no reorganization without disorganization.*

The flexibility of behavior organized by the process of learning is dependent upon the fact that each system tends to be organized around a particular environmental situation. Thus, new behavioral organization is created in each new situation. Behavioral organization becomes less flexible when many subsystems of behavior are organized around a particular event. Human language lends itself to generalization based on verbally stated principles around which very extensive organization of behavior can occur. If such a principle is found to be false or ineffective, the resulting organization is faulty and presents great difficulty if it is to be modified. One suspects that there may be a period shortly after verbal capacities are fully developed that may be critical for the establishment of such generalizations.

5.4. Established Cases of Critical Periods of Development

Critical period phenomena have been described in a wide range of behaviors (Scott, 1962). A partial list includes: early infantile attachment (Cairns, 1972), neurohumeral organization (Denenberg and Zarrow, 1971; Bronson and Desjardins, 1968), hormonal modification of secondary sex organs in birds and mammals (Young *et al.,* 1964; Reinisch, 1974), sex role determination in humans (Money and Ehrhardt, 1972), language learning (Lenneberg, 1967), song learning in birds (Thorpe, 1965; Marler and Mundinger, 1971), and development of normal heterosexual behavior in rhesus monkeys (Mason, 1963; Harlow and Harlow, 1965). On theoretical grounds it can be predicted that a critical period can be found for any organizing process or processes that operate for a limited period and result in stable organization of a system or subsystem.

6. Optimal Periods and Critical Periods

The time of most rapid organization in a particular process is not only a critical period for decisions but also is an optimal period for producing desirable (or at least desired) changes in organization. Change cannot be produced before the organizational process begins nor after it has ceased.

6.1. The Optimal Period for Adoption in Dogs

In this case, the usual desired effect is for a dog to be a well-adjusted member of both canine and human societies. If a puppy is removed from its canine companions before the critical period begins and no later than 4 weeks and is given no further contact with other dogs, it will form all of its social relationships with humans but none with other dogs and consequently be maladjusted in its relationships with canines. A male puppy adopted at 4 weeks into a human family may as an adult be antagonistic toward all dogs, even females in estrus. On the other hand, if a puppy is kept with other dogs and has no contact with humans, it develops strong relationships with other dogs but never develops them with people. If such a puppy is given no human contact for 14 weeks, it becomes acutely maladjusted in human relations although it can be laboriously tamed like a wild animal. The situation is somewhat modified when there is minimal contact with humans, such as dogs would receive in an ordinary kennel. The dogs will show signs of social attachment to humans during the critical period even if they are only able to see them from a distance (Scott and Fuller, 1965). On the basis of these empirical facts, we have placed the optimal time for adoption by humans at 6–8 weeks, with a week or two leeway thereafter. The best results are obtained if the puppy is adopted at approximately 8 weeks. As happens in wolf packs, the age of approximately 2 months is also the time when the adults will move a litter from the den and relocate it at a somewhat distant spot known as a rendezvous site (Theberge and Pimlott, 1969). This optimal period is, therefore, associated with a change in site attachment which has obvious practical implications in a hunting animal. If wolf puppies were kept permanently in a den and became attached only to that locality, they would have difficulty in making the kind of adjustments that are necessary for a wide-ranging hunting animal.

6.2. Adoption in Humans

Here the desired result is to have a child that is well adjusted and strongly attached to close social relatives. Attachment to the personnel of an orphanage or to the natural parents is, in this case, undesirable. Since the period of most rapid social attachment in human beings occurs from approximately 5–6 weeks to 6–7 months, the optimal period for adoption should be as early as possible within this period and preferably soon after birth when no complications could occur. The later that the adoption procedure is put off beyond 6 months, the more difficulties should develop. Since such adoptions are often necessary, however, it is possible to minimize difficulties through an understanding of the attachment process itself.

Two things interfere with new attachments after the critical period. First, is the development of fear responses toward strangers, and every precaution should be taken not to frighten such a child. Also, young infants in an orphanage could be brought into frequent contact with strangers so that being adopted by a stranger would not be regarded as frightening. Second, and even more important, is the emotional disturbance resulting from breaking previous social and site attachments. The old method of adoption was to remove the infant immediately and completely from all early contact, automatically ensuring maximum emotional disturbance. A better procedure is to proceed more gradually in order to reduce emotional disturbance. In the beginning, the adoptive parent should visit the infant several times in its own familiar setting and attempt to establish pleasant relations by active social interaction. After this, the infant could be taken away to visit the new home for several short periods of a few hours before finally moving there permanently. Also it would be well to visit the old environment occasionally so the infant could maintain some continuity between its earlier and later environments (Scott, 1971).

Xenophobia, the fear of strangers, is one of the major obstacles to human understanding and world peace. It would be desirable if young infants became attached not only to their own social relatives but to the entire human race. It is not possible for a young infant to have contact with large and widespread samples of human populations, but it should be helpful to expose the young infant to as many kinds of people as possible during early infancy and childhood. The racial integration of the public schools in the USA, for example, is a desirable step in this direction, but the age of six may already be beyond the optimal period. Empirical studies could be made of the behavior of children who are being brought up in restricted or isolated conditions in order to determine the reality of these effects.

6.3. Optimal Periods for Site Change

In the dog, this coincides with the optimal period for adoption. The period is also one in which young puppies are reasonably mature in their motor capacities and are beginning to move out and investigate new parts of the adjacent environment. Human infants, on the other hand, are not capable of leaving their parents for even short distances until long after the period of socialization. Infants begin to walk on the average of about 14 months and do not become really active until somewhat later. Rheingold and Eckerman (1973) have studied this phenomenon of the child leaving his mother. On the basis of their work and some observations of the times at which young children get lost by straying too far from their parents, I would estimate that the optimal time for changing a child's environment is around two years of age. This could, of course, be either in the form of a vacation trip or a permanent move. Since moving is such a common practice in our society, it would be desirable to get some empirical data on the effects of moving at different ages. It may turn out that there are other optimal periods as well.

6.4. Optimal Period for Motor Learning

In spite of its obvious practical implications, very few studies of this have been done. McGraw's (1935) early study on two identical twins that were given differential training was marred by the fact that her twins later turned out not to be identical, but the study nevertheless produced certain interesting suggestions. One twin could be taught to do such things as swimming and roller skating at an astonishingly early age, before he had even learned to walk well, but showed no benefit over the other twin when retaught at a later age. In fact, if anything, roller skating was more difficult because it had been learned at an age when the proportions of leg length and body length were different from later, and these habits had to be relearned. One would predict that motor learning involving whole-body coordination could be acquired most rapidly at the time when the nervous system, sensory capacities, and motor organs were close to the adult form. Judging from the development of the EEG, brain development is not complete until 7 or 8 years although there is a great deal of individual variation. It is at approximately this time that most children first begin to be well coordinated physically. In the absence of better information, we could predict that this age would be a good one in which to initiate the development of physical skills.

McGraw herself stated that there was a different optimal time for learning each kind of coordinated physical activity. These processes are complicated by many different factors which might determine that one age is good for one skill, another for another. For example, learning to ride a tricycle at an early age should not interfere with learning to ride a bicycle later since the two skills are quite different and the first might contribute to later confidence. Another complicating factor in any learning process is that of motivation, and this may be a major factor in determining an optimal period. The child must have at least some interest in learning, and his motivation will be strengthened or diminished by his success or failure. Therefore, it is important not to try to attempt learning any activity before the age when success is possible.

6.5. Optimal Periods for Intellectual Learning

Many of these same considerations apply also to the learning of intellectual skills. Of the various factors, motivation is perhaps the most important. In the dog, young puppies achieve a mature EEG at about 8 weeks. For the next 4 weeks or so, puppies are interested in almost any new activity and are quite easy to teach, provided one does not require skilled motor performance or long-continued persistence. The usual rule is to achieve one or perhaps two successes and then quit for the day.

Similarly, in human children there are ages at which there is active intellectual curiosity but poor staying power. This is particularly noticeable in children around the age of 8 and during the next few years prior to puberty, but it may also exist much earlier with respect to certain activities. In humans the basic intellectual skill is language. One of the things that children begin to learn and actively try to learn at a very early age is language itself.

Indeed the acquisition and organization of language forms an almost classical case of a critical period (Lenneberg, 1967). The period is a long one extending between the ages of approximately 2 and 12 years. During the early part of this period, the organization of language may be easily modified in almost any direction, i.e., any language may become the native language of an individual who is adopted from one culture into another at an early age. Toward the end of the period, it becomes more and more difficult (although not impossible) to learn a second language. By the age of 13 or so, it can only be learned with a foreign accent.

Should one try to enlarge a child's vocabulary when he is just beginning to learn words? The answer is possibly that undue emphasis on individual words may interfere with sentence learning and the acquiring of skills in rapid, uninterrupted speech. Emphasis might better be placed on learning organized language rather than its isolated parts.

It is obvious that the optimal time for learning a second language is during the time the child is learning a primary one, but it has never been satisfactorily demonstrated exactly when this should take place because a second language may interfere with mastery of the first.

Similarly, the optimal period for acquiring secondary language skills such as reading and writing has never been determined. The obvious guideline would be the time at which the child can do these things easily and is also well motivated to do them. One would expect that reading would come before writing and, in fact, when writing skills are first acquired at an early age, they tend to persist in an immature and somewhat unskilled fashion because of the early formation of habits. There is also a tendency to try to teach skills at the same age in all children, in spite of obvious differences in developmental rates. The age of six may be too early for one child to learn to read and later than the optimal period for another.

The critical period for language acquisition can be more precisely defined by the evidence from children who have suffered brain lesions from various accidents. Before 20 months, language development may be delayed in some cases, but otherwise there is no deficit. This is prior to the age when language appears, and obviously the organization process has not yet begun. From 21 to 36 months, a lesion has the effect of destroying all language abilities, which then reappear as they did in the beginning. From 3 to 10 years, lesions may produce aphasia; but eventually there is complete recovery except for some effects on reading and writing. At the same time, the child will continue to acquire new language skills in addition to the old. By 11–14 years, some changes produced by lesions are irreversible; and after 14 years, the symptoms that remain 5 months after the injury tend to be irreversible.

The evidence concerning the organization of language acquisition is similar in form to that of socialization, namely, that there is an early period of very rapid organization which then falls off to a lower rate but never goes down to zero. These facts concerning language organization then bring up the problem of the relationship between the optimal periods and periods that are vulnerable to injury.

7. Vulnerable and Sensitive Periods

In contrast to the concept of optimal periods, which implies the modification of organization in a favorable direction either by speeding it up or modifying it, the concept of vulnerability implies that the organizing process is unfavorably affected by either slowing it down or distorting it. The concept of sensitivity similarly implies damage to an organizing process although of a less serious degree. For example, a bird becoming attached to a human being instead of to its own species has obviously been damaged but not to the extent that its existence is threatened. The period in which such events occur easily is called a sensitive period by some authors (Hinde, 1970; Hess, 1973).

If an injurious agent or event is applied to an organizational process and affects only that process, it should act as follows (Scott *et al.,* 1974):

Prior to the critical period, defined as the time during which the organizational process is taking place, the agent should produce no effect. During the critical period, it should produce effects proportional to the speed of organization and inversely proportional to time. That is, if the agent is applied early in the period when organization is still proceeding slowly, it should produce relatively little effect and, because reorganization is possible, there should be almost complete recovery. Later, the effect should be more serious because the process is proceeding more rapidly and also because there is less time for reorganization and recovery. Finally, once the organizational process is complete, the agent should produce no effect.

On the other hand, if the injurious agent affects not the organizing process but the organized system itself, the results should be minimal early in the critical period, when reorganization is still possible, and increase as the organization becomes stable and fixed. Once the organizational process ceases, the system can be destroyed and never recover. Such effects as these can be observed with lesions of the nervous system.

The general theory of critical periods states that an organizational process is most easily modified when it is proceeding most rapidly. Such modifications can produce either favorable or unfavorable effects. Thus, vulnerable periods and optimal periods should be essentially the same with respect to the time periods involved, the distinction between these concepts being the nature of their effects on organization.

With respect to critical periods, a further distinction can be made on the basis of basic and developmental research. The concept of critical periods is concerned with basic research in the description and identification of organizational processes. The concepts of optimality and vulnerability, on the other hand, lead to applied and developmental research and practical uses of this information.

8. Conclusion

As presented in this paper, the theory of critical periods is a general one applying to developmental change in the organization of any living system. It applies equally well to growth processes, behavioral development, and even organizational change in social systems. The application of the critical period concept to both embryology and behavioral development is no mere analogy but is based on the fundamental nature of all living systems.

At the same time, organizational processes differ at different levels of system organization. The critical nature of any given decision made during organizational change depends on three variable factors:

1. its importance for survival, or adaptive value
2. the degree of irreversibility, i.e., the degree of system stability
3. the ease with which the organizational processes can be modified

Comparing growth processes with those of behavioral development, it can be generally stated that the decisions made in behavioral development are somewhat less important for survival, are somewhat more reversible, and are considerably easier to modify in optimal directions. These generalities may not hold in every pair of processes that may be compared, as certain behavioral systems may achieve extraordinary stability while certain physiological systems are relatively labile.

A major new theoretical contribution of this chapter is the distinction made between maintenance processes, which organize external materials to be included in a living system, and developmental processes, which produce organizational change of the system itself. Being more or less continuous, the former should not show critical periods, whereas the latter must exhibit critical periods as they produce stable systems or subsystems. Further, the critical nature of a given developmental decision is variable, depending on the degree of stability of organization of the resulting system. Finally, the concepts of optimality, vulnerability, and sensitivity are related to that of critical periods.

The usefulness of a theory depends on its ability to provide a causal explanation of a given set of facts and one that will give predictable results when controlling factors or processes are manipulated. The theory of critical periods meets this test. In addition, a theory is scientifically valuable if it leads to further research. As presented here, the theory of critical periods demands in each particular case the identification of organizing processes and the measurement of time relationships between them.

9. References

Beadle, G. W., and Tatum, E. L., 1941, Genetic control of biochemical reactions in Neurospora, *Proc. Natl. Acad. Sci. U.S.A.* **27**:499.

Bowlby, J., 1951, *Maternal Care and Mental Health,* World Health Organization Monograph, Geneva.

Bronson, F. H., and Desjardins, C., 1968, Aggression in adult mice: Modification by neonatal injections of gonadal hormones, *Science* **161**:705.

Cairns, R. B., 1972, Attachment and dependency: A psychobiological and social learning synthesis, in: *Attachment and Dependency* (J. L. Gewirtz, ed.), pp. 29–80, V. H. Winston, New York.

Cairns, R. B., and Werboff, J. A., 1967, Behavior development in the dog: An interspecific analysis, *Science* **158**:1070.

Child, C. M., 1921, *The Origin and Development of the Nervous System,* University of Chicago Press, Chicago.

Collias, N. E., 1956, The analysis of socialization in sheep and goats, *Ecology* **37**:228.

Denenberg, V. H., and Zarrow, M. X., 1971, Effects of handling in infancy upon adult behavior and adrenal activity: Suggestions for a neuroendocrine mechanism, in: *Early Childhood: The Develop-*

ment of Self-regulatory Mechanisms (D. N. Walcher and D. L. Peters, eds.), pp. 39–71, Academic Press, New York.

Fleener, D. E., 1967, Attachment Formation in Human Infants, Doctoral dissertation, Indiana University, Ann Arbor, Michigan, University Microfilms, No. 6872-12.

Gurski, J. C., Davis, K. L., and Scott, J. P., 1974, Onset of separation distress in the dog: The delaying effect of comfortable stimuli, Paper presented at the International Society for Developmental Psychobiology, St. Louis, Missouri.

Harlow, H. F., and Harlow, M. K., 1965, The affectional systems, in: *Behavior of Nonhuman Primates* (A. M. Schrier, H. F. Harlow, and S. F. Stollnitz, eds.), pp. 287–334, Academic Press, New York.

Heinroth, O., 1910, Beitrage zur Biologie, namentlich Ethologie und Psychologie der Anatiden, *Verh. Ver. Int. Ornithol. Kongr.* **5:**589.

Hess, E. H., 1973, *Imprinting,* D. Van Nostrand, New York.

Hinde, R. A., 1970, *Animal Behavior: A Synthesis of Ethology and Comparative Psychology,* 2nd ed., McGraw-Hill, New York.

Kaufman, I. C., and Rosenblum, L. A., 1969, Effects of separation from mother on the emotional behavior of infant monkeys, *Ann. N.Y. Acad. Sci.* **159:**681.

Klinghammer, E., 1967, Factors influencing choice of mate in altricial birds, in: *Early Behavior* (H. W. Stevenson, E. H. Hess, and H. L. Rheingold, eds.), pp. 5–42, Wiley, New York.

Lenneberg, E. H., 1967, *The Biological Foundations of Language,* Wiley, New York.

Lorenz, K., 1935, Der Kumpan in der Umwelt des Vogels, *J. Ornithol.* **83:**137.

Marler, P., and Mundinger, P., 1971, Vocal learning in birds, in: *Ontogeny of Vertebrate Behavior* (H. Moltz, ed.), pp. 389–450, Academic Press, New York.

Mason, W. A., 1963, The effects of environmental restriction on the social development of rhesus monkeys, in: *Primate Social Behavior* (C. H. Southwick, ed.), pp. 161–173, Van Nostrand, New York.

Mason, W. A., and Kenney, M. D., 1974, Redirection of filial attachments in rhesus monkeys: Dogs as mother surrogates, *Science* **183:**1209.

McGraw, M. B., 1935, *Growth: A Study of Johnny and Jimmy,* Appleton-Century, New York.

Money, J., and Ehrhardt, A. A., 1972, *Man and Woman, Boy and Girl,* Johns Hopkins Press, Baltimore.

Moore, A. U., 1960, Studies on the formation of the mother–neonate bond in sheep and goats, *Am. Psychol.* **15:**413.

Moore, K. L., 1974, *The Developing Human: Clinically Oriented Embryology,* Saunders, Philadelphia.

Rakusan, K., and Poupa, O., 1970, Ecology and critical periods of the developing heart, in: *The Post Natal Development of Phenotype* (S. Kazda and V. H. Denenberg, eds.), pp. 359–368, Academia, Prague.

Reinisch, J. M., 1974, Fetal hormones, the brain, and human sex differences: A heuristic, integrative review of the recent literature, *Arch. Sex. Behav.* **3:**51.

Rheingold, H. L., and Eckerman, C. O., 1973, The infant separates himself from his mother, in: *Child Development and Behavior,* 2nd ed. (F. Rebelsky and L. Dormon, eds.), Alfred A. Knopf, New York.

Schein, M. W., 1963, On the irreversibility of imprinting, *Z. Tierpsychol.* **20:**462.

Scott, J. P., 1937, The embryology of the guinea pig. III. Development of the polydactylous monster. A case of growth accelerated at a particular period by a semidominant lethal gene, *J. Exp. Zool.* **77:**123.

Scott, J. P., 1945, Social behavior, organization and leadership in a small flock of domestic sheep, *Comp. Psychol. Monogr. No. 96* **18**(4):1.

Scott, J. P., 1962, Critical periods in behavioral development, *Science* **138:**949.

Scott, J. P., 1963, The process of primary socialization in canine and human infants, *Monogr. Soc. Res. Child Dev.* **28**(1):1.

Scott, J. P., 1971, Attachment and separation in early development, in: *The Origins of Human Social Relations* (H. R. Schaffer, ed.), pp. 227–243, Academic Press, London.

Scott, J. P., and Fuller, J. L., 1965, *Genetics and the Social Behavior of the Dog,* University of Chicago Press, Chicago.

Scott, J. P., and Marston, M. V., 1950, Critical periods affecting normal and maladjustive behavior in puppies, *J. Genet. Psychol.* **77:**25.

Scott, J. P., Stewart, J. M., and DeGhett, V. J., 1974, Critical periods in the organization of systems, *Dev. Psychobiol.* **7:**489.

Spalding, D. A., 1873, Instinct, with original observations on young animals, *McMillan's Mag.* **27:**282.

Stockard, C. R., 1907, The artificial production of a single median cyclopian eye in the fish embryo by means of sea water solutions of magnesium chloride, *Arch. Entwicklungsmech.* **23:**249.

Stockard, C. R., 1921, Developmental rate and structural expression, *Am. J. Anat.* **28:**115.

Theberge, J. B., and Pimlott, D. H., 1969, Observations of wolves at a rendezvous site in Algonquin Park, *Can. Field Nat.* **83:**122.

Thorpe, W. H., 1965, The ontogeny of behavior, in: *Ideas in Modern Biology* (J. A. Moore, ed.), pp. 483–518, Natural History Press, Garden City, New York.

Waddington, C. H., 1962, *New Patterns in Genetics and Development,* Columbia University Press, New York.

Young, C., Goy, W., and Phoenix, H., 1964, Hormones and sexual behavior, *Science* **143:**212.

8

Patterns of Early Neurological Development

INGEBORG BRANDT

1. Introduction

1.1. Relevance of Early Neurological Development

The pattern of early neurological development is used for maturational assessment in preterm infants since there are phases of some reactions and reflexes that appear at clear-cut times before term. Repeated neurologic examination in the perinatal period allows the achievement of successive neuromuscular "milestones" to be followed. The procedures may easily be performed, and no instruments or highly specialized techniques are necessary. In contrast external characteristics, used in the assessment of gestational age (Farr *et al.*, 1966 *a,b*), need to be scored within a time limit from 12 to 36 hr postnatal age because the "results apply only to babies examined at this age."

This chapter will only deal with neurological examinations that are clearly defined and easy to perform. The more complex techniques for neuromaturational assessment, such as motor nerve conduction velocity (Schulte, 1968; Schulte *et al.*, 1968*a*, 1969), and electromyographic evaluation of the reactions and reflexes (Schulte *et al.*, 1968*b*; Schulte and Schwenzel, 1965), which require special equipment, will not be discussed.

Knowledge of the postmenstrual age at birth is a prerequisite for the proper evaluation of an infant and for differentiating between small-for-gestational-age (SGA) infants and appropriate-for-gestational-age (AGA) infants. Infants of the same weight but of different gestational age behave quite differently as regards risks in the newborn period (Battaglia and Lubchenco, 1967), neurological development (Robinson, 1966), and prognosis (Brandt, 1975; Schröder, 1977). Defining prematur-

---

INGEBORG BRANDT • Universitäts-Kinderklinik und Poliklinik, Bonn, Germany.

ity by low birth weight only, i.e., 2500 g and below, is misleading. By this method, SGA full-term infants would be included and many preterm infants excluded.

Also, for later developmental assessment of preterm infants, at least in the first and second year of life, knowledge of postmenstrual age at birth is essential since the amount in time of prematurity must be subtracted from chronological age, i.e., corrected age (Gesell and Amtruda, 1967; Parmelee and Schulte, 1970). According to the results of the Bonn longitudinal study (Brandt and Schröder, 1977; Schröder, 1977), the developmental and intelligence quotients (Griffiths, 1964; and Stanford-Binet of Terman and Merrill, 1965) differ significantly from 3 to 54 months if chronological and corrected age are compared.

1.2. Definitions

1.2.1. Postmenstrual Age, Conceptional Age, Synonyms

Postmenstrual, fetal, or gestational age is the time from the first day of the mother's last menstrual period until birth. In the preterm infant, postmenstrual age should be used at all examinations up to, and often after, 40 postmenstrual weeks, irrespective of birth. The definitions of postmenstrual age and postconceptional age (time from conception to birth) should be clearly distinguished. The confusing use of ''conceptional age'' for the time from the first day of the mother's last menstrual period until birth, plus the postnatal period of life (e.g., by Saint-Anne Dargassies, 1966; Graziani *et al.*, 1968; Schulte, 1974; Prechtl and Beintema, 1976) is to be avoided.

André-Thomas and Saint-Anne Dargassies (1952) use the term conceptional age, but they clearly define that they mean time from conception. They distinguish exactly the different ''ages'' of a preterm infant, e.g., (1) intrauterine life (vie utérine), (2) life outside the uterus (vie aérienne), and (3) total age (vie total). An infant, born at six months intrauterine life, and examined at three months outside the uterus, is considered as nine months old (total age) and may be compared with a full-term infant. So long as these clear definitions are observed, conceptional age is considered by André-Thomas to be of great importance.

If, in the present literature, the term conceptional age is used as postmenstrual age plus postnatal age combined, this confusion adds to the complication of newborn neurology. This chapter will show there are considerable differences between infants of, for example, 34 postmenstrual weeks and 36 postmenstrual weeks; i.e., the difference between postmenstrual age and conceptional age of 14 days, provided that conception is calculated as occurring 14 days after the first day of the last menstrual period.

For differentiation between a newly born infant of 35 postmenstrual weeks and an infant of the same age born at 28 postmenstrual weeks and with an age of 7 weeks postnatally, it is suggested that two ages be given, the postmenstrual age at birth and the postnatal age.

1.2.2. Perinatal Period

The perinatal period is defined as the time from 28 postmenstrual weeks to the 7th day after term. In the preterm infant, the perinatal period corresponds to the third trimester of pregnancy of the full-term infant.

1.2.3. Differentiation between Appropriate-for-Gestational-Age and Small-for-Gestational-Age Infants

Preterm infants are defined as those born with a postmenstrual age of less than 38 weeks (i.e., 37 weeks + 6 days or less), irrespective of birth weight (McKeown and Gibson, 1951; Battaglia and Lubchenco, 1967; Brandt, 1977).

To quantitate normal fetal growth and to diagnose fetal growth retardation, intrauterine growth standards, based on postnatal measurements of infants born at different postmenstrual ages, are necessary. A further prerequisite is the knowledge of postmenstrual age. The differentiation between AGA and SGA preterm infants is necessary since infants of the same weight behave neurologically quite differently if they are not of the same postmenstrual age. Neurologic development—with few exceptions—is a function of age and not of weight.

Preterm infants with normal intrauterine development and with birth weights between the 10th and 90th percentiles of Lubchenco *et al.* (1963) are defined here as appropriate for gestational age. Small-for-gestational-age infants are defined as those with a birth weight below the 10th percentile of Lubchenco *et al.* (1963) or the 10th percentile of Hosemann (1949).

1.2.4. Early Neurological Development

"Early" may be defined differently, depending on the starting point, i.e., conception, premature birth, or birth at expected date of delivery.

Three periods of early development may be differentiated: (1) intrauterine, during pregnancy; (2) after premature birth from 28 postmenstrual weeks to extrauterine the 7th day after term—the perinatal period; and (3) from birth at term to the 7th day—the newborn period. In the present chapter the emphasis will be on (2), the perinatal period.

2. Survey of Methods

2.1. Assessment of Gestational Age before Birth: Maternal History; Ultrasound Examinations

Knowledge of the gestational age of the fetus is important in assessment of the intrauterine development. Primarily this is determined by the maternal history, i.e., last menstrual period. A basal temperature curve may be used to show the date of conception.

Recently more exact determination of fetal age has become possible by ultrasound measurements, which offer today the best opportunities for serial studies of fetal development. The earliest assessment is made by measurement of the diameter of the gestational sac (Hackelöer and Hansmann, 1976). Measurements of fetal crown–rump length in the first trimester of pregnancy, introduced by Robinson (1973; Robinson and Fleming, 1975), prove to be the most reliable method for gestational age assessment. Between the 9th and 15th week the fetus grows in length with a high velocity, i.e., 1.6 mm/day on average (Hackelöer and Hansmann, 1976). During this time, estimation of gestational age, based on only one measurement of crown–rump length, is possible to ±4.7 days, and by three independent measurements even to ±2.7 days with a reliability of 95% (Robinson and Fleming, 1975).

These results are even more precise than assessment of the expected date of delivery from the last menstrual period of the mother after the rule of Naegele, i.e., to subtract 3 months from the first day of the mother's last menstrual period, and to add 7 days.

In the second trimester ultrasound measurements of the biparietal diameter—before it exceeds 78 mm, corresponding to the 29th week of gestation—enable the obstetrician to estimate fetal age with confidence limits of ±10 days in 96% of cases (Hansmann, 1974).

Determination of fetal age in the third trimester by a biparietal diameter greater than 90 mm is clinically no longer relevant (range of two standard deviations = ±24 days) (Hansmann, 1976).

2.2. Observation of Intrauterine Fetal Movements by Ultrasonography

In recent years, with the perfection of ultrasound techniques, intrauterine fetal movements can be directly observed. Real-time scanning (''live'' sonograms) offer the opportunity for continuous observation of fetal movements *in utero*. With this method neurophysiological examinations are possible in the normal physiological environment (Reinold, 1971, 1976), in contrast to earlier studies on the fetus, made outside the uterus after premature termination of pregnancy (Humphrey, 1964).

Further, the measurement of fetal breathing movements *in utero* is used in the clinical management of problem cases, especially in SGA fetuses (Dawes, 1976). These measures are considered a useful diagnostic tool for the early detection of fetal distress. A scan ultrasound method is used, but the technique is not easy and requires experience and relatively expensive apparatus.

2.3. Tests Other Than Neurological Signs Used as Maturation Criteria after Birth

The examination of joint mobility in the newborn is considered helpful for maturational assessment (Dubowitz *et al.*, 1970; Dubowitz and Dubowitz, 1977; Saint-Anne Dargassies, 1966, 1974). Extensibility of the joints in the full-term newborn infant may be demonstrated, e.g., in the dorsiflection angle of the foot, and in the volarflection angle of the hand. This is related to the maturational age reached *in utero*. With increasing gestational age, near the expected date of delivery, joint mobility is increased due to synergistic influences of placental hormones. This hormonal factor is absent in preterm infants.

2.4. Neurological Examination of Reflexes and Reactions in the Perinatal Period

The great contribution of the Paris School (André-Thomas and Saint-Anne Dargassies, 1952; Saint-Anne Dargassies, 1955, 1966, 1974) is to have demonstrated age-dependent changes in the neurologic behavior of preterm infants before their expected date of delivery. Periodic testing of the preterm infant offers the opportunity to observe extrauterine neurologic maturation, which normally occurs *in utero* in the third trimester of pregnancy, since many responses develop similarly under both conditions. The close relationship of gestational age with many developmental patterns allows maturational assessment based on clinical–neurological examination.

The neurological examination elaborated for full-term newborn infants by Prechtl and Beintema (1964) is well defined and quantifiable. Moreover, Beintema (1968) has demonstrated the inter- and intraindividual variability of these neurologic responses in normal newborn infants in the first 9 postnatal days.

In many studies of neurological development of preterm infants (Robinson, 1966; Graziani *et al.,* 1968; Meitinger *et al.,* 1969; Finnström, 1971; Michaelis *et al.,* 1975), the examination methods are based on those of Prechtl and Beintema (1964). The design of a neurologic examination must take into account the age-specific properties of the developing nervous system, and this postulate of Prechtl (1970) is especially valid for preterm infants. Further, the neurological tests must be centered on sufficiently complex functions of the nervous system and not on artificially fragmented responses (Prechtl, 1970).

2.5. Consideration of Infant State

Prechtl (1970) emphasizes that it is necessary to standardize the examination technique with respect to the infant's behavioral state, because of the wide but systematic fluctuations in the readiness of particular nervous system functions to act. Particular functions should only be assessed in particular states. Since the behavioral state of an infant often is influenced by the examination procedure, Prechtl (1970; Prechtl and Beintema, 1964) recommends strict adherence to a rigid sequence of tests.

The concept of state is a prerequisite for a quantitative neurological assessment (Prechtl, 1974). The criteria and definitions of states vary between investigators. Prechtl and Beintema (1964) distinguish five different behavioral states in newborn infants.

State 1: Eyes closed, regular respiration, no movements.
State 2: Eyes closed, irregular respiration, no gross movements.
State 3: Eyes open, no gross movements.
State 4: Eyes open, gross movements, no crying.
State 5: Eyes open or closed, crying.

Beintema (1968) defines the term "state of an infant" according to Ashby (1956): "By a state of a system is meant any well-defined condition or property that can be recognised if it occurs again." In *Neurological Study of Newborn Infants,* Beintema (1968) uses the states described by him and Prechtl (1964). In addition, he uses the following scores for alterability of a state by handling:

0: Almost impossible to alter the state of the infant at any time in the examination.
1: Usually difficult to obtain the optimal state at any time.
2: The examiner sometimes succeeds in altering the state.
3: Fairly easy to obtain the optimal state.
4: Very easy to obtain the optimal state.

The relationship of 14 reflexes and reactions in the newborn to behavioral state has been studied by Lenard *et al.* (1968). The reflexes were elicited in three states only: state I, regular sleep; state II, irregular sleep; and state III, quiet wakefulness.

According to the relationship with state, three groups of reflexes are distinguished: (1) Reflexes equally strong during regular sleep and wakefulness, but weak or absent during irregular sleep. To this group belong, among others, the knee jerk and the Moro reflex. (2) Responses mostly absent during regular sleep, weak during irregular sleep, and strongest during wakefulness; all these are exteroceptive reflexes. To this group belong the palmar grasp and the glabella reflex. (3) Reflexes which do not show any alterations, but are easily obtained in all three states. To this group belongs, among others, the Babinski reflex.

Robinson (1966) uses "state of arousal" for each response following the system of Prechtl and Beintema (1964) for preterm infants. If there existed a "required state" for a response, the babies were only tested by him in this state. Also Finnström (1971) applies the states, as defined by Prechtl and Beintema (1964), for the preterm infants in his study of maturity in newborn infants. Graziani *et al.* (1968) did not consider different groups of states, but described the general state during testing: e.g., "awake state; gross body movements struggling and stretching, but in general not crying, interfered with reliable scores." Amiel-Tison (1968) and Babson and McKinnon (1967) do not mention the states in their studies. Also Dubowitz *et al.* (1970; Dubowitz and Dubowitz 1977) do not consider states in the neurologic assessment of newborn infants.

The 30 preterm infants in the study of Meitinger *et al.* (1969) were mostly in state 2 according to Prechtl and Beintema (1964), except for four infants who later reached state 3.

Michaelis, in his studies of SGA infants (Michaelis, 1970; Michaelis *et al.*, 1970), uses the states of Prechtl and Beintema (1964).

Saint-Anne Dargassies (1966, 1974) describes, instead of states, the quality of vigilance of an infant, i.e., 0 = no vigilance during the examination; 1 = good and lasting; 2 = good but of short duration; 3 = bad but lasting; 4 = bad and disappearing "par éclipse."

Since there exists no adequate classification of states for preterm infants, those described by Prechtl (1974; Prechtl and Beintema, 1964) for full-term infants are used in our Bonn study. The state is recorded as that demonstrated by the infant for most of the duration of examination, i.e., more than 50%. Since with every neurological examination anthropometric measurements are also made, sometimes it is possible to alter the state to the optimal one for the corresponding neurological response. Throughout every examination 6–8 photographs are made so that the states can be checked in evaluating the results.

2.6. Drawbacks of Early Neurological Examination

The neurologic examination of a preterm infant during the first two days of life is of little value for the determination of gestational age, because factors greatly influencing the general condition of the infant will prejudice the results. This is a serious shortcoming, since the early determination of gestational age—especially in the SGA infant—is important because of the need for special care immediately after birth. Clinical external characteristics and nerve conduction velocity assessment are more valuable in these conditions.

If the examination is performed after the third day, short-acting factors that are directly related with birth are usually eliminated (Escardó and deCoriat, 1960). Further, in infants with a normal Apgar (1953) score at birth, the examination results from the third to fifth postnatal day correspond to the true gestational age.

In infants in a poor general state of health, e.g., with hyperbilirubinemia, general hypotonia, marked edema (especially at the extremities), infections, or electrolyte disturbances, only unreliable results are to be expected from the neurological examination. Hypotonia may lead one erroneously to consider the infant as younger than it is. As soon as the general state improves, the infant exhibits the reactions and reflexes corresponding to his postmenstrual age. This "catch-up" should not be misinterpreted as "accelerated" neurological development.

Some neurologic responses are impaired in infants born in breech position (Prechtl and Knol, 1958). Depressant drugs, such as barbiturates or intravenous narcotics, given the mother immediately before or during birth, accentuate and prolong the disorganizing effects which seem to occur as a normal result of the birth process (Brazelton, 1961). The extent of relative central nervous system disorganization seems positively correlated with the type, amount, and timing of the medication given to the mother, whereas inhalant anesthesia has a more transient effect than premedication, for example, with a barbiturate.

According to Schulte (1970a), on the basis of neurological criteria, in about 70% of normal newborn infants the gestational age may be assessed with an accuracy of two weeks; for 97% of infants this is possible to 3–4 weeks. In abnormal infants the responses are abnormal too, and an age assessment becomes impossible because the abnormal responses may imitate immature behavioral patterns. This pertains for motor development and also for the electroencephalogram. Even nerve conduction velocity assessment does not allow a very exact age determination. Schulte (1970a) further states that neurological criteria are reliable only in completely normal infants; but in abnormal infants, where knowledge of postmenstrual age is of clinical significance, they are useless. Abnormal results from an examination, then, may either signify that the infant is younger than reported by the mother or that the infant is abnormal.

The critical objections of Schulte (1970a) hold when neurologic development phases are used for the determination of gestational age in ill infants. In cases where the age of the newborn infant is known with some certainty, the neurologic behavior is a valuable diagnostic tool and allows one to give a prognosis (Schulte, 1973). Moreover, with increasing use of ultrasound measurements in early pregnancy, especially in high-risk cases, there will be an increasing number of newborn infants with an exact age assessment, and neurologic development will regain its significance for longitudinal observations in the perinatal period.

Comparison of some of the neurologic responses described in the literature is made difficult by differences in examination methods, in methods of reporting results, and in definition of abnormal responses.

2.7. Longitudinal Neurological Examinations of Low-Risk Preterm Infants

Out of the Bonn longitudinal study (Brandt, 1975, 1976, 1978a), a group of 29 AGA low-risk preterm infants (16 girls and 13 boys), born between 28 and 32 postmenstrual weeks, were studied. The selection was made according to the following criteria:

1. The menstrual cycles of the mother were normal. No oral contraceptives were used in the last 6 months before conception.

2. Reliable data on the last menstrual period of the mother; regular obstetrical examinations during pregnancy. In some cases the date of conception was known by measurement of the basal temperature; in others ultrasound follow-up examinations in early pregnancy were performed.

3. The mother had a normal prenatal course, and no severe complications of labor and delivery.

4. The recognition of fetal movements was within the normal range, i.e., for primipara between 18 and 22 postmenstrual weeks, and for multipara between 16 and 20 weeks (Hansmann, 1977).

5. The neonatal course was uncomplicated.

6. All the infants had a birth weight between the 10th and 90th percentiles of Lubchenco *et al.* (1963) at their gestational age.

7. The long-term outcome of the infants was normal, i.e., their neuromotor and psychological development, followed longitudinally at least until the age of 3 years, was normal when compared with full-term control infants.

These preterm infants represented a relatively homogeneous group in respect to postmenstrual age and low risk. Also, with advancing postmenstrual age there were no abnormal conditions observed which might have influenced neurological behavior.

After birth, the external maturational criteria were assessed according to von Harnack and Oster (1958) and Farr *et al.* (1966*a,b*).

The first neurological examination was carried out on the 4th to 10th postnatal day, 1½–2 hr after the last feed if possible. Until discharged, the infants were examined once a week or every 2–3 weeks. All examinations were made by the author, a total of 153, performed on the 29 infants until the expected date of delivery. As in the study of Robinson (1966), the responses were tested as described by Prechtl and Beintema (1964) for the full-term infant, with slight modifications according to the developmental particularities of preterm infants of low postmenstrual age. An attempt was made to test the babies in the state required for a response to be elicited.

The state of the infants during examination was recorded as defined by Prechtl and Beintema (1964). Even if the judgment of states is considered problematic in preterm infants of low postmenstrual age (i.e., it is difficult to discriminate between sleeping and being awake because in both states the eyes may be open), one tried to describe the state during an examination and to get the infant in the optimal state for a specific response. The sequence of the examination was the same for all infants.

Out of the 55 reflexes and reactions examined in the Bonn study, only 20 will be reported and discussed. Measurements of head circumference, supine length, and body weight were made at the time of each neurological examination.

The Bonn results are based not only on the 153 longitudinal examinations in 29 AGA preterm infants which already have been analyzed statistically, but also on experiences from about 500 longitudinal observations on a further 35 AGA and 44 SGA preterm infants and 80 full-term control infants in the perinatal period.

3. Development

3.1. Fetal Movements—First and Second Trimester of Pregnancy

Many observers agree that the nervous or neural phase of early fetal movements is preceded by a purely muscular or aneural phase, where the muscles contract because of their own excitability (M. Minkowski, 1924, p. 242). This early fetal aneural phase of movement is combined with nervous influences early on (end of second month), and the integrative effect of the nervous system and a greater variability of reactions appear.

In the past it has not been possible to study fetal movements *in utero*. In earlier examinations of very young fetuses in a physiologic salt solution outside the uterus and without a continuous oxygen supply, anoxia may have influenced fetal motricity. As anoxia progresses, the reflexes of the fetus are suppressed (Humphrey, 1964). Nowadays ultrasonic real-time scanning is the method of choice for studying early fetal behavior.

Echoes generated by the fetal body within the gestational sac become measurable around the 6th week of gestation (Reinold, 1976). Around the 8th week, slight changes in fetal position can often be seen. Only at the 10th or 11th week do these fetal motions become increasingly extensive and prominent.

Since the fetal body floats in the amniotic fluid—its specific weight being only slightly above that of the fluid—very little effort is needed for the fetus to move. On the basis of his observations, Reinold (1976, pp. 116–117) has introduced a classification of spontaneous fetal movements for clinical use: (1) strong and brisk movements, forceful initial motor impulse, movement involves entire body, change in location and posture; and (2) slow and sluggish movements, no initial motor impulse, movements confined to fetal parts, no change in location and posture. Reinold (1976) distinguishes further between "spontaneous movements" and "passive fetal movements" elicitable in the absence of spontaneous movements, i.e., in the resting fetus.

Passive fetal movements may be elicited by palpation of the maternal abdomen, by lifting the portio of the uterus with the finger from the vagina, or by the mother's vigorous coughing. The resulting forceful passive movements may change fetal position and posture. The fetus rises from its resting position and will then drift downward again (Reinhold, 1976). The fetus can turn somersaults and fling about arms and legs, as has been demonstrated by Hoffbauer (1977). Normally, the number of spontaneous fetal movements tends to increase gradually from 10 to 20 weeks during early pregnancy. In the resting state the fetus is positioned with legs flexed or else squatting (Hoffbauer, 1977).

Reinold (1976) has studied fetal behavior by the "motor provocation test." After passive movement of the fetus, there may occur brisk or sluggish movements, or no movements may be seen. There are several reasons for each motor response: The fetus may be asleep or in a condition that precludes movements. According to Reinold (1976), the absence of any response of the fetus to the motor provocation test appears to be a sign of being at risk. Haller *et al.* (1973) have shown that in ultrasound examinations, between the 9th and 21st week of gestation the frequency of movements increases up to the 21st week in normal pregnancies. It is difficult to describe fetal movements quantitatively (Henner *et al.*, 1975). A new method of quantifying and analyzing fetal movements has been described by Henner *et al.* (1975) in which two-dimensional fetal movements, seen on a TV screen, are transferred to a geometrical diagram which is then quantified.

The frequency and intensity of fetal movements in a group of abnormal pregnancies shows significant differences from the norm as regards frequency, type, and intensity.

At about 20 postmenstrual weeks, fetal movements become more vigorous and are felt by the mother as quickening (Hoffbauer, 1977); the fetus often changes position. From 24 weeks the amount of somersaults and loopings decreases, and the fetus, in turning, exhibits marked motility of the extremities (Hoffbauer, 1977).

Timor-Tritsch *et al.* (1976) have studied fetal movements from 26 postmenstrual weeks to term by continuous recording with a tocodynamometer. The authors distinguish four types of movement: (1) rolling movement, associated with motion of the whole body; (2) simple movement, i.e., relatively short and easily palpable movement, possibly originating from an extremity; (3) high-frequency movement—either single or repetitive—i.e., short, easily palpable and sometimes visible movement, felt by the mother through the entire abdomen; (4) respiratory movement, due to "breathing movements" of the fetal chest wall.

In the third trimester, fetal movements become more and more limited with decreasing space in the uterus. The extent is dependent upon the amount of amniotic fluid; the findings are variable and are not appropriate diagnostic criteria (Hansmann, 1977).

3.2. Extrauterine Neurological Development in the Perinatal Period after Premature Birth

3.2.1. Posture in Supine Position

Spontaneous Body Posture. In the same way as the intrauterine fetus, the extrauterine preterm infant shows spontaneous movements and changes of position. No age-typical body posture occurs before tone of the flexor muscles has increased to such a degree that the extremities remain in a more or less flexed position. Meitinger *et al.* (1969) studied 26 preterm infants between 30 and 38 weeks gestation who showed no typical age-dependent posture. In some infants they observed the asymmetric tonic neck reflex (see pp. 298–299) or a paradoxical pattern of the reflex, i.e., flection of the arm that the face is turned toward.

A similar age-independency has been shown by Prechtl *et al.* (1975) in a pilot study of motor behavior and posture of 12 preterm infants born between the 28th and 36th postmenstrual weeks. Before 35 postmenstrual weeks there was hardly any particular posture prevailing within the 120 min of each observation. After the postmenstrual age of 37 weeks, they found no infant in an arms-extended posture. Others (Saint-Anne Dargassies, 1955, 1966, 1974; Koenigsberger, 1966; Amiel-Tison, 1968; Graziani *et al.,* 1968; Dubowitz *et al.,* 1970) have reported typical age-related changes in posture in preterm infants: At 28 postmenstrual weeks the completely hypotonic infant lies with arms and legs extended and with the head in a markedly lateral position. The less mature the infant, the more pronounced is the lateral position of the head. This weakness of the neck muscles leads to the dolichocephalus in preterm infants (Brandt, 1976).

At 30 postmenstrual weeks there is a beginning of flection of the thigh at the hip; this flection becomes stronger at 32 weeks (Amiel-Tison, 1968). At 32 postmenstrual weeks Saint-Anne Dargassies (1974) describes a slight flection of the legs of the infant. According to Koenigsberger (1966), there is "total extension" at 28–32 weeks.

At 34–35 postmenstrual weeks the froglike (batrachian) attitude is prevalent with flection of the lower limbs and an abduction or outward rotation of the upper thighs. This flection contrasts with the extension of the upper limbs. At 36–37 postmenstrual weeks the attitude of flection of the four extremities occurs. This posture is also prevalent at the expected date of delivery.

At term in 74% of the 150 full-term infants—all born in vertex position—in the study of Saint-Anne Dargassies (1974, p. 22), the four extremities were flexed; in 22% only the arms were flexed, with the legs extended; 1% showed a froglike position; and 1% an extension of the upper, and flection of the lower, extremities.

Posture after Passive Extension of the Extremities: Imposed Posture. Although a typical age-dependent posture is rejected by Prechtl *et al.* (1975), there is a close age relationship with *imposed posture* in the preterm infant, i.e., posture after a slow passive extension of the extremities or after putting the head in the midline of the trunk. The posture the baby assumes after this manipulation for at least 2–3 min—before being changed by spontaneous movements—can be recorded. The typical changes in imposed posture with increasing postmenstrual age are due to increasing resistance against gradual stretching of the extremities, caused by tonic myotatic reflexes (Schulte, 1974). Imposed posture is closely related to the recoil of the lower and upper extremities. These responses are also ascribed to the tonic myotatic reflex activity of the infant (Schulte, 1974).

Imposed posture of the lower and upper extremities in the Bonn study was evaluated longitudinally in 29 AGA preterm infants of 32 weeks and under, until the expected date of delivery. An attempt was made to get the infant in Prechtl states 3, 4, or 5 for this examination. We excluded evaluation of the posture of the lower extremities for the first 2 postnatal weeks of the 6 infants born in breech position because the responses of the lower extremities are different from those of infants born in the vertex position. Out of 19 newborn infants born in complete breech position, Prechtl and Knol (1958) have shown that in 18 cases there was a resting posture with the lower limbs in an extended, or nearly extended, position.

Photographs at each observation are correlated with the records made (Figure 1).

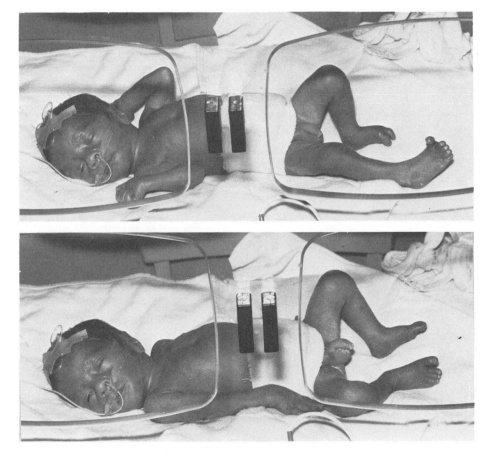

Fig. 1. Above: Spontaneous posture with the arms flexed; the left leg is flexed due to the asymmetric tonic neck reflex. Below: Imposed posture, i.e., after passive extension of the extremities, the arms remain in the extended position, and the legs return to flection. Preterm infant born at 32 postmenstrual weeks, four days after birth.

Table I. Imposed Posture, Lower Extremities—Results of the Bonn Study

| | Postmenstrual age in weeks | | | |
Response	28–33	34	35–38	39–40
Flection	5	6	31	18
Semiflection	28	8	12	18
Extension	4	0	1	0
No. of observations	37	14	44	36
No. of infants	23	14	29	29

1. Imposed posture, lower extremities: results are shown in Table I. These findings contrast with reports in the literature where at 35–36 weeks a froglike attitude prevails, and at the expected date of delivery there is flection of the legs. No clear-cut changes of the imposed posture of the lower extremities in relation to gestational age were observed. The frequency of flexed position from 35 to 38 weeks of 70% was not significantly different from 50% at 39–40 weeks. Posture of the lower extremities is of little value for gestational age assessment.

2. Imposed posture, upper extremities: results are shown in Table II. "Imposed posture" of the upper extremities may be used to differentiate infants of 33 weeks and under, and 37 weeks and over. At 34–36 weeks the results are intermediate, and no conclusions as to gestational age may be drawn. Obviously, the results of slowly extending the arms by gentle manipulation are different from those following a quick extension and release to elicit recoil.

3.2.2. Recoil

The quick recoil of the extremities into the flexed position of the normal full-term infant after extension depends on the tonic myotatic reflex activity. The motoneuron activity demonstrated electromyographically by Schulte (1974) during a quick recoil maneuver of the forearms "provides clear evidence that the phenomenon is a spinal reflex rather than the consequence of the elastic properties of muscles and ligaments."

Lower Extremities. The recoil of the lower extremities after a brief passive extension has been used for the assessment of neurologic maturation by Saint-Anne Dargassies (1955, 1966, 1970, 1974, p. 260), Koenigsberger (1966), Graziani *et al.* (1968), and Dubowitz *et al.* (1970; Dubowitz and Dubowitz, 1977). Prechtl and Beintema (1964) and Beintema (1968) do not mention this response. Koenigsberger (1966) observed a "slight" recoil of the lower extremities at 32 postmenstrual weeks

Table II. Imposed Posture, Upper Extremities—Results of the Bonn Study

| | Postmenstrual age in weeks | | |
Response	28–33	34–36	37–40
Flection	0	13	53
Semiflection	1	8	2
Extension	40	18	0
No. of observations	41	39	55
No. of infants	29	29	29

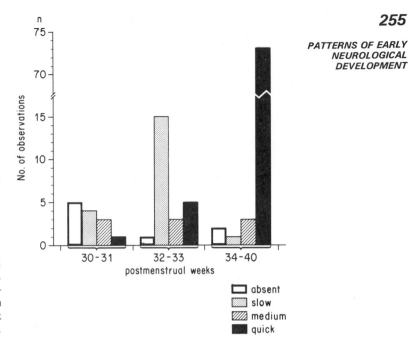

Fig. 2. Recoil of legs. From 30 to 31 weeks all kinds of responses are seen. From 32 to 33 weeks a slow recoil prevails in most of the observations; from 34 to 40 weeks, there is seen a quick recoil in most of the cases (73 out of 79).

and a good recoil from 34 weeks. Saint-Anne Dargassies (1966) reported that for the gestational age of 35 weeks, extension of the lower extremities is difficult, and an infant returns to its customary flexed position like a spring. Graziani *et al.* (1968), on 16 AGA preterm infants and 15 SGA infants, showed the midpoint of age range of rapid recoil of lower extremities was 34 postmenstrual weeks, whereas a slow response was observed from 30 weeks with a midpoint age of 31 weeks. There was no difference between AGA and SGA infants. Dubowitz *et al.* (1970; Dubowitz and Dubowitz, 1977) used "leg recoil" in their scoring system for neurologic criteria, but did not give the age range involved.

Saint-Anne Dargassies (1974, p. 260) reported an excellent return into flection of the lower extremities at 35 postmenstrual weeks. In her quantitative study on the neurologic behavior of full-term infants, the response is not mentioned (Saint-Anne Dargassies, 1974).

Data on the recoil of the legs in the supine position in the Bonn study (Brandt, 1978*b*) agree with the findings of Graziani *et al.* (1968) and with the data in the table of Michaelis, published by Schulte (1974). Infants born in the breech position were excluded. For the reaction, four scores are given: slow, medium, or quick recoil, and response absent (Figure 2). The results are shown in Table III.

Table III.   *Recoil of Legs—Results of the Bonn Study* [a]

Response	Postmenstrual age in weeks		
	30–31	32–33	34–40
Quick	1	5	73
Medium	3	3	3
Slow	4	15	1
Absent	5	1	2
No. of observations	13	24	79

[a] $\chi^2 = 13.39$, $df = 4$, $P < 0.01$.

The differences among the three gestational age groups are significant ($P <$ 0.001, Table III) (chi-square test, see Siegel, 1956). This response can seemingly be used for the assessment of neuromaturational age; it distinguishes between infants of 33 postmenstrual weeks or under, and of 34 weeks or over.

In spite of the significant differences of this response in the different gestational age groups, a disadvantage of the test is that it is not valid for infants delivered in the breech position (Prechtl and Knol, 1958); furthermore there has to be a differentiation between "slow" and "quick," implying a certain experience of the examiner. Therefore this response can only be recommended as of limited value in gestational age assessment.

Upper Extremities. Recoil of the forearms depends upon the tonic myotatic reflex activity of the infant which is similar to many other "primitive reflexes" and reactions (Schulte, 1974). In a wakeful full-term infant, a brief extension of the forearms at the elbow is followed by a quick recoil into flection, which is due to biceps muscle activity as has been demonstrated electromyographically by Schulte (1974). Recoil of the forearms at the elbow is listed among the recognized neuromaturational criteria by Koenigsberger (1966), Amiel-Tison (1968), Graziani *et al.* (1968), Dubowitz *et al.* (1970), Finnström (1971), and Saint-Anne Dargassies (1955, 1970, 1974). Koenigsberger (1966) reports a negative response at 34 postmenstrual weeks, a "slow recoil" of the upper extremities at 37 weeks, and a "good recoil" at 41 weeks. According to Graziani *et al.* (1968) the "midpoint of age range" for slow recoil of the upper extremities is at 31 postmenstrual weeks, and of rapid recoil at 35 weeks.

In the table on "neurological evaluation of the maturity" of Amiel-Tison (1968) at 34 weeks, a "flexion of forearms begins to appear, but very weak." A "strong return to flexion" is reported at 36 postmenstrual weeks; "the flexion tone is inhibited if forearm maintained 30 seconds in extension." From 38 to 40 weeks the forearms return very promptly to flection after being extended for 30 sec.

Saint-Anne Dargassies (1970, 1974, p. 254) regards a quick recoil of the upper extremities as among the six most reliable neuromaturational signs. She found that at 37 postmenstrual weeks recoil of the upper extremities could occur, but was inhibited by a slow and prolonged passive extension. At 41 weeks there is a quick recoil of the upper extremities; a strong resistance against extension is exhibited.

Finnström (1971) studied preterm and full-term infants divided into gestational age groups of 2-week intervals, six of which had a gestational age of 32 weeks and below. The recoil of the upper extremities was among the 6 responses out of 26 neurologic tests with a good correlation with postmenstrual age. According to his frequency tables, 33% of 6 infants under 32 postmenstrual weeks showed no recoil and 67% a slow one. From 32 to 34 weeks, in 80% of 10 infants there was a slow return of the forearms to flection, and in 20%, a quick one. From 34 to 36 weeks, 44% of 18 infants exhibited a slow, and 56% a quick, recoil. From 36 to 40 weeks, about 20% of the infants showed a slow, and 80% a quick, recoil; from 40 weeks in 4% the upper extremities returned slowly to flection and 96% quickly. There were only small differences between AGA and SGA infants. In all full-term infants (n = 49) between one and nine days in the study of Beintema (1968), recoil of the forearms at the elbow—mostly a marked, quick recoil in both arms—was present. Out of the 150 full-term infants in the study of Saint-Anne Dargassies (1974, pp. 31, 134), 136 showed a positive response, and in 14 infants the reaction was missing on both sides.

In the Bonn study, the recoil of the forearms at the elbow in the supine position was tested according to Prechtl and Beintema (1964): both forearms were simultaneously and briefly passively extended at the elbows and then released (Figure 3). The required state was 3, 4, or 5. The results for this reaction were similar to those reported by Amiel-Tison (1968) and by Saint-Anne Dargassies (1970, 1974). They agreed also with the figures in the developmental table of Michaelis cited by Schulte (1974).

Because of increase in muscle tone (i.e., resistance against passive movements) in a caudocephalic direction, one would expect that recoil of the upper extremities after a brief passive extension would occur at a later age than that of the lower extremities. Reactions are classified in terms of recoil: quick, medium, slow, or absent. The results are shown in Table IV and Figure 4.

The differences among the three gestational age groups, 28–33 weeks, 34–36 weeks, and 37–40 weeks, are highly significant, $P < 0.001$ (Figure 4, Table IV). Recoil of the forearms has a relatively clear-cut time of appearance and is a useful indicator of gestational age in infants in states 3–5. The test allows one to distinguish

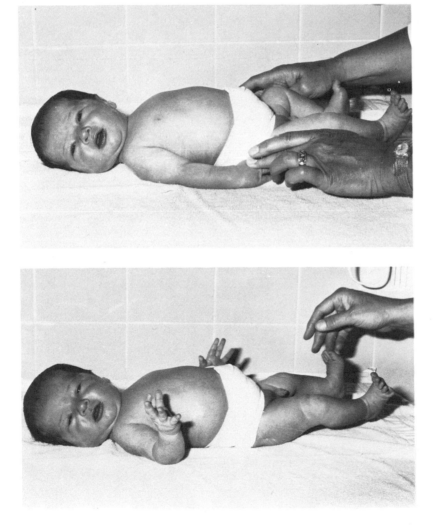

Fig. 3. Recoil of the forearms at the elbows. Full-term infant at the fourth postnatal day.

Table IV. Recoil of the Forearms at the Elbow—Results of the Bonn Study[a]

| | Postmenstrual age in weeks | | |
Response	28–33	34–36	37–40
Quick recoil	0	4	27
Medium recoil	0	2	8
Slow recoil	0	15	6
No recoil	41	18	0
No. of observations	41	39	41

[a]$\chi^2 = 34.33$, $df = 3$, $P < 0.001$.

between infants of 33 weeks and under, where the reaction is absent completely, and infants of 37 weeks and over, where the reaction is quick, or medium, in 85% of infants.

3.2.3. Resistance against Passive Movements

The term "resistance against passive movements" is used to describe muscle tone. Prechtl, in his manual for *The Neurological Examination of the Full-Term Newborn Infant* (Prechtl and Beintema 1964), avoided the words "tone" and "tonus" because there has arisen so much confusion about them. Beintema (1968) has given scores separately for the resistance against passive movements of shoulders, elbows, wrists, hips, knees, and ankles. A scoring system for "general tone" has been elaborated by Brazelton (1973). Muscle tone comprises both cellular and contractile components (Schulte, 1974), and posture of the infant reflects tone to a

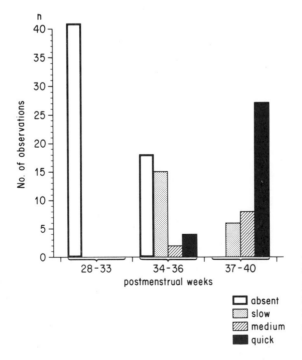

Fig. 4. Recoil of forearms at the elbow. This reaction distinguishes between infants at 36 weeks and under, where the response is absent or slow (85%), and of 37 weeks and over, where a quick (66%) or medium (19%) recoil is seen in most of the cases.

large extent (Saint-Anne Dargassies, 1966, 1974; Brazelton, 1973).

In a study of muscle tone in preterm infants compared to full-term infants at the gestational ages of 38–41 weeks, Michaelis *et al.* (1975) showed significant differences between the groups. Passive tone, examined at shoulder, elbow, trunk, hip, and knee, was significantly less in preterm infants than in full-term control infants. Also, the active tone, evaluated by the flection angle of elbows and knees after traction, was decreased. The duration of extrauterine life had no influence on muscle tone between 38 and 41 postmenstrual weeks.

Saint-Anne Dargassies (1955, 1966, 1970, 1974), Koenigsberger (1966), Amiel-Tison (1968), and Finnström (1971) have used increasing muscle tone in a caudocephalic direction as maturational criteria. André-Thomas *et al.* (1960; André-Thomas and Saint-Anne Dargassies, 1952) and Saint-Anne Dargassies (1955) have elaborated some responses for evaluating muscle tone.

Lower Extremities. The amount of resistance against passive movements in the lower extremities is usually tested by the popliteal angle, the heel-to-ear maneuver, and the abduction angle of the hips.

The popliteal angle decreases between 28 and 32 postmenstrual weeks from 180° to 150°, between 33 and 34 weeks to 120°, between 35 and 37 weeks to 90°, and remains so until the expected date of delivery (Saint-Anne Dargassies, 1966, 1974; Koenigsberger, 1966). Saint-Anne Dargassies (1974, p. 254) counts this response among the "particularly unreliable signs" of maturity. The figures of Amiel-Tison (1968) show an earlier closing of the popliteal angle than reported by Saint-Anne Dargassies: at 32 weeks, 110°; at 34 weeks, 100°. Dubowitz *et al.* (1970; Dubowitz and Dubowitz, 1977) have included the changes of the popliteal angle in their scoring system, but absolute values are not given. In her study of full-term infants, Saint-Anne Dargassies (1974) reports a considerable variability: only 63% have a popliteal angle of 90° or less, whereas in 37%, the angle is between 100° and 180°.

The maturational changes of the popliteal angle, i.e., decrease from 150° to 90°, are not valid for the infant born in the breech position, as an extended spontaneous position of the knees is often found (Prechtl and Knol, 1958); this may persist for some weeks after birth.

According to the Bonn results, the popliteal angle decreases in the last weeks before term; but the relationship of the angle to postmenstrual age is not clear-cut before 37 postmenstrual weeks (Figure 5). In evaluation, we excluded the 6 infants born in breech position in the first 2–3 weeks. The results are shown in Table V.

The popliteal angle distinguishes between infants of 33 weeks and under and 37 weeks and over. But even if this response shows clear-cut changes from 37 postmenstrual weeks in this homogeneous group of preterm infants, it may behave more variably in infants of the same postmenstrual age but of different postnatal age. Moreover, the interval of 4 weeks for a clear-cut change is insufficient for maturational assessment. The chi-square test yielded significant differences of the popliteal angle among the three gestational age groups, 28–33 weeks, 34–36 weeks, and 37–40 weeks ($P < 0.001$), but these are of little practical use.

The heel-to-ear maneuver has been considered useful for gestational age assessment by Koenigsberger (1966), Amiel-Tison (1968), Dubowitz *et al.* (1970), and Saint-Anne Dargassies (1966, 1974). At 28–32 postmenstrual weeks there is no resistance against drawing the infant's foot as near to the head as it will go without forcing it. With increasing postmenstrual age this maneuver becomes more and more difficult and is impossible in the normal infant at term. Among the 150 full-term infants in the study of Saint-Anne Dargassies (1974, pp. 33, 128), strong

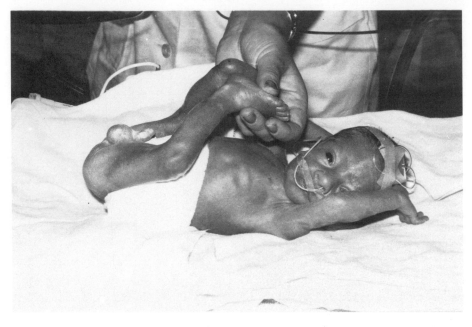

Fig. 5. Popliteal angle 130° to 140° at 32 postmenstrual weeks in a preterm infant of four days. The arms—in a reflex movement—are also moved upwards.

resistance, a normal response, was observed in 40%. Forty-three percent of the infants showed only slight resistance. Since this response is closely related to the popliteal angle, and its relation to postmenstrual age is not very clear-cut, it cannot be recommended as a reliable maturational criterion.

The dorsiflection angle of the foot also has been used for maturational assessment (Amiel-Tison, 1968; Dubowitz *et al.*, 1970; Dubowitz and Dubowitz, 1977; Saint-Anne Dargassies, 1974) and is not a neurologic criterion (see p. 291). Saint-Anne Dargassies (1974, p. 254) counts this test as among the six most reliable maturational criteria.

Upper Extremities. Resistance against passive movements in the upper extremities and in the shoulder girdle is usually tested with an approximation of the "hand-to-the-opposite-shoulder," the well-known *scarf maneuver* (mouvement du foulard), as described by André-Thomas *et al.* (1960), Saint-Anne Dargassies (1966, 1974), Amiel-Tison (1968), and Dubowitz *et al.* (1970; Dubowitz and Dubowitz,

Table V. Popliteal Angle—Results of the Bonn Study[a]

Response (degrees)	Postmenstrual age in weeks		
	28–33	34–36	37–40
Less than 90	0	0	1
90–110	1	12	51
120–140	10	20	2
150 and greater	24	7	1
No. of observations	35	39	55

[a] $\chi^2 = 100.79$, $df = 6$, $P < 0.001$.

Table VI.   Scarf Maneuver—Results of the Bonn Study[a]

Response	Postmenstrual age in weeks			
	28–33	34–36	37–39	40
Strong resistance	0	0	9	25
Moderate resistance	2	10	16	3
Slight resistance	9	22	2	0
No resistance	30	7	0	0
No. of observations	41	39	27	28

[a] $\chi^2 = 75.29$, $df = 9$, $P < 0.001$.

1977). One observes in this response how far the elbow will move across when one draws the hands to the opposite shoulder, and then attempts to put them around the neck (like a scarf). Up to a postmenstrual age of 32 weeks, no resistance is found against the scarf maneuver (Amiel-Tison, 1968). According to Koenigsberger (1966), the maneuver is possible until 33 weeks without resistance and minimal resistance against approximation of the hands to the opposite shoulder occurs at 34 weeks; Graziani *et al.* (1968) report a "midpoint of age range" of 33 weeks for slight to moderate resistance. From 36 to 38 weeks, a moderate to marked resistance is demonstrated (Graziani *et al.,* 1968), with the elbows just passing the midline (Amiel-Tison, 1968). At term, the scarf maneuver is limited (Amiel-Tison, 1968), and the elbows will not reach the midline. Because of the changes with increasing postmenstrual age, this response is recommended for assessment of maturity by Koenigsberger (1966), Amiel-Tison (1968), Graziani *et al.* (1968), and Dubowitz *et al.* (1970). Contrary to earlier publications, Saint-Anne Dargassies recently counts the scarf sign (1974, p. 254) among the six particularly unreliable criteria of maturity.

According to the Bonn results, the scarf sign showed typical changes with increasing postmenstrual age. The attitude of the elbow was scored, i.e., how far the elbow will cross the midline: (1) no resistance, elbow nearly reaches opposite axillary line; (2) elbow between midline and opposite axillary line (slight resistance); (3) elbow reaches midline (moderate resistance); (4) elbow only reaches axillary line (strong resistance). The results are shown in Table VI.

The Bonn results agree with the reports of Koenigsberger (1966), Amiel-Tison (1968), and Graziani *et al.* (1968). The scarf sign may be used for the assessment of gestational age when the attitude of the elbows is recorded and the infant is in states 3–5. The scoring presupposes that the examiner has some experience. The differences among the four gestational age groups are highly significant (Table VI). The scarf maneuver permits one to differentiate among infants of 33 weeks and under (mostly no resistance), 34–39 weeks (mostly slight or moderate resistance), and 40 weeks and over (mostly strong resistance).

3.2.4. Spontaneous Motor Activity

Unlike the fetus *in utero,* the preterm infant is exposed to the force of gravity, and this influences motility. The degree of spontaneous movements, i.e., speed, intensity, and amount (Prechtl and Beintema, 1964) is used as a maturational criterion by Koenigsberger (1966) and Finnström (1971); Saint-Anne Dargassies (1955, 1966, 1974) counts it among the six most reliable signs.

According to Prechtl *et al.* (1975), a big change in general movements of preterm infants occurs after 35 postmenstrual weeks, probably due to central inhibition. The authors show a significant decrease in general movements, twitches, stretches, cloni, and startles after 35 postmenstrual weeks ($P = 0.025$ to $P < 0.005$). A similar change in general movements is observed *in utero* by ultrasound examinations (Hoffbauer, 1977).

Saint-Anne Dargassies (1955, 1966) describes as characteristic of the stage of development at 28 postmenstrual weeks "repetitive movements of the upper limbs that have a catatonic appearance," which are later incorporated into a generalized movement and further bursts of movements. The movements are more frequent in the lower extremities where alternating pedaling movements are often seen. At 32 weeks the spontaneous motility has become generalized and is "dominated by vigorous movements of the whole body," especially the trunk with lateral incurvations. Typical is the raising of an extended leg which is held up by the infant at right angles to the pelvis, and the pelvis may be raised. At 35 weeks, spontaneous motility has increased and continues. The preterm infant is able to change his position completely in the incubator and can raise the pelvis on heels and occiput. In the following weeks until the expected date of delivery, motility of the extremities and the trunk becomes more limited in time and space, caused by increasing muscle tone, whereas good lateral rotation of the head can be observed. At term, motility is more marked in a preterm infant than in a recently born full-term infant (Saint-Anne Dargassies, 1966, 1974).

Finnström (1971) reported that below 32 postmenstrual weeks there are no spontaneous movements in 100% of the infants ($n = 6$) in his study; from 32 to 36 weeks, about 50% of the infants exhibited no spontaneous movements and 50% a medium amount. From 36 to 40 weeks, in most of the infants (79–86%), the spontaneous motor activity was medium, as is considered normal for a full-term infant. Beintema (1968) in his study of full-term newborn infants has shown an increase in spontaneous motor activity during the early postnatal days.

In the Bonn study we observed no easily quantifiable age-dependent changes in spontaneous motor activity during the examinations evaluated to date. Only the amount of spontaneous head rotation showed clear-cut changes, as detailed in Table VII.

Exact scoring of spontaneous motor activity necessitates continuous monitoring. A reliable and objective method for this activity has been introduced recently by Bratteby and Andersson (1976) with two instruments (Animex I and II). All movements observed are classified according to amplitude, velocity, and also in relation to the size of the moving object (i.e., hand, arm). The advantage of the

Table VII. Spontaneous Head Rotations—Results of the Bonn Study[a]

Response	Postmenstrual age in weeks		
	28–33	34–39	40
Absent	30	3	0
Moderate	11	58	8
Good but not against resistance	0	3	13
Against slight resistance	0	2	7
No. of observations	41	66	28

[a] $\chi^2 = 88$, $df = 6$, $P < 0.001$.

instruments is that they provide an integrated record of all movements without any attachment of cables or electrodes to the infant. To date, no longitudinal observations on preterm infants have been reported.

3.2.5. Reflexes and Reactions Dependent on Postmenstrual Age

The designation "reflex" is not quite appropriate for these specific reactions or automatims which have been known for a long time. The pathway of the reflex arc is not known in detail for most of the reflexes studied (Robinson, 1966).

To this group belong primary reflexes and reactions with a relatively clear-cut time of appearance in the perinatal period which is the same in appropriate and small-for-gestational-age infants. Thus, they can be used as an indicator of postmenstrual age. These reactions and reflexes allow one to follow neurological maturation in preterm infants until 40 weeks, the expected date of delivery, when comparison with full-term infants may be made. Each examination improves the value of the preceding one. The value of the different reactions for maturity assessment is rated differently by different authors.

Moro Reflex. The Moro response (Moro, 1918; McGraw, 1937; Saint-Anne Dargassies, 1955; Parmelee, 1964; Schulte *et al.,* 1968*b*) is one of the most frequently examined. There is no general agreement about the different components of the reflex. The classification of Prechtl and Beintema (1964, 1976) with its four phases for the pattern of arm movements is most used: (1) abduction at the shoulder; (2) extension at the elbow; (3) adduction of the arms at the shoulder; (4) flection at the elbow. Phases (3) and (4) are described as bowing by McGraw (1937). In phase (2) an observation of the hands is useful, the fingers may be all fanned, fanned with flection of the distal phalanges of the thumb and index finger, all semiflexed, or all flexed. A cry as a further component of the reflex has been described by McGraw (1937) and Saint-Anne Dargassies (1955, 1966).

The Moro response is usually elicited by a rapid but gentle head drop of a few centimeters with the baby suspended horizontally and the head in the midline (Prechtl and Beintema, 1964). In order to observe all phases of the test, it should be carried out at least three times. Another method of elicitation is pulling the infant up by the wrists so that the shoulders and head are lifted a few centimeters off the table or bed; then the hands are released so that the head falls back to the examination table. There is no agreement about the main afferent pathway of the Moro reflex. According to André-Thomas and Saint-Anne Dargassies (1952) and Saint-Anne Dargassies (1954), it is a response to proprioceptive stimulation from the neck by a sudden change of the position of the head in relation to the trunk. Peiper (1964) considers the Moro response a labyrinthine reaction. Further details are given by Parmelee (1964) in a critical evaluation.

There are characteristic developmental changes of the Moro response with increasing postmenstrual age. At 28 postmenstrual weeks only extension of the arms and spreading of the fingers occur (Saint-Anne Dargassies, 1966, 1974). Amiel-Tison (1968) describes the Moro response at 28–30 weeks as "weak, obtained just once and not elicited every time."

At 32 weeks Saint-Anne Dargassies (1966) and Amiel-Tison (1968) report that the Moro reflex "has become complete" whereby abduction and extension of the arms with spreading of the fingers and a cry are described; the phases of adduction and flection are not mentioned. From 32 to 34 postmenstrual weeks Finnström (1971), using the scoring system of Saint-Anne Dargassies (1966), reports that in

100% of the infants in his study an extension of forearms at the elbow with extension of the fingers and an abduction of the arms is observed and in no case adduction. Schulte *et al.* (1968*b*) have shown electromyographically that at 32 postmenstrual weeks the Moro response consists of the extension and abduction of the upper extremities.

From 34 to 36 weeks Finnström (1971) reports an adduction (embrace) in 56% of the infants tested. Graziani *et al.* (1968) have observed a partial embrace at a mean age of 35 weeks.

From 36 to 38 weeks 80% of the infants in Finnström's study showed an adduction, and 98–100% in the weeks thereafter. Graziani *et al.* (1968) report a midpoint of age range for "complete embrace" at 39 weeks. Amiel-Tison (1968) does not mention the adduction phase. Saint-Anne Dargassies (1970, 1974, p. 249) reports "complete embrace" at 41 weeks but does not consider this phase in her quantitative neurologic study on full-term infants.

Robinson (1966) reports the adduction phase of the Moro response as "more constantly present after the period 32 to 36 weeks than before" but considers the time of appearance too unpredictable to be an accurate measure of postmenstrual age.

Babson and McKinnon (1967) observed the embrace component of the Moro reflex at 35 ± 1 weeks of postmenstrual age in 82% of the infants tested.

In summary, it may be said that at 28 weeks there is only an extension of the arms and fingers; from 32 to 35 postmenstrual weeks the Moro response consists of an extension and abduction of the arms with extension of the fingers and occasional crying. From 36 to 40 weeks the phases of arm adduction and flection are observed more and more distinctly (complete embrace) with increasing postmenstrual age, whereas the phases of extension and abduction become less prominent. These results agree with the descriptions in the table on the developmental sequence of various reflexes (Michaelis, cited by Schulte, 1974).

The age-dependent changes of the Moro reaction of AGA preterm and full-term infants and of SGA full-term infants have been evaluated by Schulte *et al.* (1968*b*) in an electromyographic study. The duration and amount of biceps and triceps activity in at least four Moro responses were tested. For calculation of the overall reaction pattern, a triceps/biceps Moro response activity quotient is used. With increasing postmenstrual age, duration and amount of biceps activity increased significantly, whereas triceps activity showed no significant alteration. The triceps/biceps activity quotients were <1 in full-term newborn infants and >1 in preterm newborn infants of a mean postmenstrual age of 36.2 weeks—a highly significant difference ($P < 0.001$)—but the variance in each group is very high. The SGA full-

Table VIII. *Moro Response; Adduction of the Arms and Flection at the Elbow—Results of the Bonn Study* [a]

	Postmenstrual age in weeks	
Response	28–35	37–40
No adduction and flection	66	9
Adduction and flection	2	39
No. of observations	68	48

[a] $\chi^2 = 72.12$ (Yates' correction), $P < 0.001$.

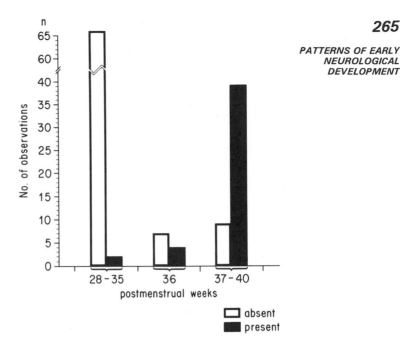

Fig. 6. Moro response, adduction phase. The response distinguishes between infants of 35 postmenstrual weeks or under and 37 weeks or over.

term infants with a mean weight of 2118 g exhibited a mean triceps/biceps activity quotient intermediate between full-term infants of the same gestational age and preterm infants of the same weight. In spite of the significant increase in biceps activity, the Moro response is considered unsuitable for an exact estimation of postmenstrual age because of "considerable variability in all parameters" in infants of the same postmenstrual age. Schulte *et al.* (1968*b*) suggest that this might be due, in part, to the fact that even minimal pathological influences in the newborn period affect the Moro response more than age differences. Schulte's electromyographic findings have confirmed the results of behavioral studies of preterm infants.

According to Lenard *et al.* (1968), the Moro reflex is equally strong during regular sleep and wakefulness, but weak or absent during irregular sleep. The infants of the Bonn study were tested only when awake.

The Bonn results agree with the developmental trend of the Moro response as reported in the literature. The adduction phase of the Moro appears at a relatively clear-cut time; one reason might be that the preterm infants who were included represented a relatively homogeneous group with respect to gestational age at birth, i.e., 32 weeks and under, and of low risk. In evaluation of the results each phase was scored separately. The results are shown in Table VIII and in Figure 6.

From 29 to 35 postmenstrual weeks the Moro response consists of extension of the arms with abduction, and with spreading and extension of the fingers (68 examinations). At 28 weeks only extension of the arms and fingers was observed.

At 36 weeks the response was intermediate; in 4 of 11 cases an adduction was observed and in 7 cases extension of the arms with abduction, and extension and spreading of the fingers only (Figure 6). From 37 to 40 weeks adduction of the arms with flection was observed in 39 out of 48 examinations, whereas at 37 weeks, the flection component is not complete in all cases (Figure 6; Table VIII).

The position of the fingers—scored separately—yields the following results: up to 34 postmenstrual weeks extension and fanning was observed; from 35 to 38 weeks in 20% of the infants a flection of the thumb and index finger occurred; and from 39 to 40 weeks this occurred in 50% (as described by McGraw, 1937, and Saint-Anne Dargassies, 1974).

Crying during the Moro response has not been recorded up to 32 postmenstrual weeks; at 33 weeks, 50% of the infants cried; and from 34 weeks to term, a cry was registered in 80–90% of the examinations. Saint-Anne Dargassies (1974) reports crying in 87% of normal newborn full-term infants.

The timing of the adduction phase of the Moro response allows one to distinguish between infants of 28–35 postmenstrual weeks and of 37 weeks and over, the difference between the two gestational age groups being highly significant, $P < 0.001$ (Table VIII). The adduction/flection phase of the Moro reflex may be recommended as a maturational criterion according to the results of Babson *et al.* (1967), Graziani *et al.* (1968), Finnström (1971), and Michaelis (cited by Schulte, 1974). This is contrary to the findings of Robinson (1966), Schulte *et al.* (1968*b*), and recently, Saint-Anne Dargassies (1974, p. 254), who counts the Moro reflex among the six particularly unreliable maturational signs.

Neck-Righting Reflex. With this reflex Schaltenbrand (1925) introduced into the neurologic examination of human infants the results of the animal experiments of Magnus and de Kleijn (1912) and Magnus (1924) concerning neck and righting reflexes. The neck-righting reflex is elicited with the infant in a supine position and the head being rotated by 90° to the left or right side. This lateral head movement is followed by a reflex rotation of the spine, i.e., trunk and legs, in the same direction (Figure 7). In the full-term newborn infant, the marked neck-righting reflex leads to a "vehement roll over of the body in line with head rotation" (Schaltenbrand, 1925). The neck-righting reflex has not been tested in the quantitative neurologic studies of full-term newborn infants of Beintema (1968) and Saint-Anne Dargassies (1974).

Robinson (1966) first published data on the maturational changes of the neck-righting reflex in AGA and SGA preterm infants. According to his results, this reflex usually appears between 34 and 37 postmenstrual weeks in both AGA and SGA infants. Between the postmenstrual age periods less than 37 weeks and more than 37 weeks, there is a significant difference ($P < 0.01$).

Graziani *et al.* (1968) scored the movement of the opposite shoulder, and of the trunk and pelvis, after head rotation toward the left and then to the right. At 37 postmenstrual weeks a "sustained shoulder elevation of >1.5 in., trunk and pelvis may turn" response is described. The midpoint of age range for "sustained shoulder, trunk and pelvis response" is 39 weeks.

Meitinger *et al.* (1969) tested the neck-righting reflex in 26 preterm and 4 full-term infants and reported a good turning of the trunk in 5 infants (36, 37, 39, and 2 of 40 postmenstrual weeks), and a suggestion of turning of the trunk in 2 infants (33 and 36 weeks). In 23 infants (the age is not mentioned), the reflex was negatively scored.

Dubowitz *et al.* (1970; Dubowitz and Dubowitz, 1977) consider the neck-righting reflex "inconsistent because of the difficulty in defining a positive response."

In the Bonn preterm infants the pattern of the neck-righting reflex in relation to gestational age was similar to the results of Robinson (1966). The results are shown in Table IX and Figure 8.

The difference between the gestational age groups 34–35 weeks and 37–40 weeks is highly significant, $P < 0.001$ (Table IX). The neck-righting reflex allows one to differentiate between infants of 35 postmenstrual weeks or under and 37

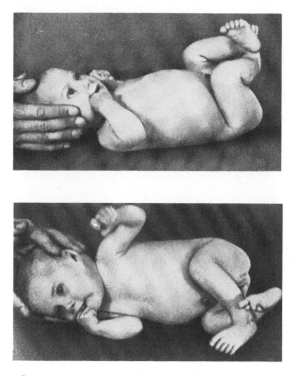

Fig. 7. Neck-righting reflex. Above: Supine position. Below: After turning the head of the baby to the right side, the whole body is turned to the right. (Original photograph from Schaltenbrand, 1925.)

weeks or over. Therefore, it is recommended as being useful for maturational assessment.

The neck-righting reflex requires state 3, 4, or 5, and it is not elicitable in states of hypotonia of the neck muscles.

Traction Response—Head Control in Sitting Position. Head control in the sitting position is listed among the labyrinthine righting reactions. Based on the animal experiments of Magnus (1924) and Rademaker (1926), Schaltenbrand (1925) first introduced this response in child neurology. His method of elicitation was slightly different from that used now: The infants were blindfolded in order to eliminate optical righting reactions, and held by the examiner at the pelvis in an upright position. Of the 28 full-term infants tested in the first week by Schaltenbrand (1925), in 12 cases a moderate head control was observed, 13 infants were not able to lift the head up, and in 3 infants the response was questionable.

Head control in the sitting position represents the second part of the traction response as described by Prechtl and Beintema (1964). In the first part the infant is slowly pulled up by the wrists to a sitting position from being supine with the head in the midline. The resistance to extension of the arms at the elbow is observed.

For testing the head control in the sitting position in small preterm infants before the expected date of delivery, the method described by Saint-Anne Dargassies (1954) and Beintema (1968, p. 96) is more practical and is used in the Bonn study: The infant, after being brought to the sitting position, is supported at the shoulders and trunk with both hands of the examiner (Figure 9). From shortly

Table IX.   Neck-Righting Reflex—Results of the Bonn Study[a]

Response	Postmenstrual age in weeks	
	34–35	37–40
No following of trunk and legs	24	3
Trunk and legs follow	4	38
No. of observations	28	41

[a] $\chi^2 = 39.71$ (Yates' correction), $P < 0.001$

before term and thereafter, head control in the sitting position is recorded with the infant held at the wrists, as described by Prechtl and Beintema (1964).

The recording of the response is as recommended by Prechtl and Beintema (1964). The four scores are: (1) head hangs passively down; (2) head is raised once or twice momentarily, or head is lifted up and drops backward without remaining in the upright position; (3) head is raised and remains in the upright position for at least 3 sec, oscillations may occur; (4) head is maintained in the upright position for more than 5 sec with no, or only little, oscillation (Figure 10).

In preterm infants head control in the sitting position develops with increasing postmenstrual age. Results from other studies are hardly comparable because methods of elicitation of the reflex and scoring were different.

Robinson (1966) scored the arm flection and head-raising parts of the traction response together, because both parts tend to appear between 33 and 36 weeks gestation and are present in all cases of 36 weeks and over. The results are similar in AGA and SGA preterm infants. The difference in the three gestational age groups

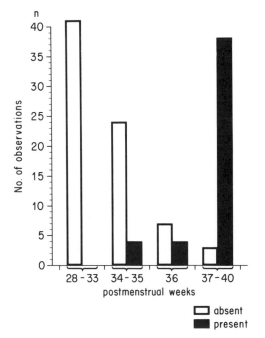

Fig. 8. Neck-righting reflex. The response distinguishes between infants of 35 postmenstrual weeks or under and 37 weeks or over.

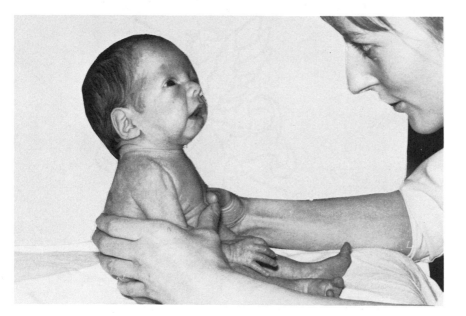

Fig. 9. Traction response, head control in sitting position. The infant, after being brought to the sitting position, is supported at the trunk. AGA preterm boy at 39 postmenstrual weeks with a postnatal age of 8 weeks.

"under 33 weeks," "33 to 36 weeks," and "36 weeks and over" are highly significant ($P < 0.001$).

Saint-Anne Dargassies (1966) describes head control "fleetingly" at 35 postmenstrual weeks and states that the head can be kept in line with the trunk in the sitting position at 41 weeks. Amiel-Tison (1968) reports a good righting of the head, but inability of the infant to hold it, at 36 postmenstrual weeks. At 38 weeks the infant begins to maintain the head's position for a few seconds, and at 40 weeks keeps the head in line with the trunk for more than a few seconds. Finnström (1971) uses "strong" or "moderate" stimulation for the observation of "head extension when sitting." Up to 34 postmenstrual weeks no infant could extend the head without stimulation. A "slow extension" of the head without stimulation is observed from 34.1 to 36 weeks in 33%, and from 38.1 to 40 weeks in 48% of the infants. Babson and McKinnon (1967), who scored the "maintenance of head nearly parallel to trunk" during traction, report that "neck flexion on traction developed at 36 ± 1 weeks of postmenstrual age in 68 percent" of the infants.

In their comparative neurologic study of 97 preterm infants, 32 of whom were of a gestational age of 32 weeks and below, and 97 full-term infants at term, Howard *et al.* (1976) report head control in the sitting position (posterior neck muscles) to be among the neurologic items with "significantly greater incidence of weak responses" in the preterm than in the full-term infants. In this study "anterior and posterior neck muscle control" of the infant, brought to a sitting position and supported at the shoulders, was scored. A similar observation is reported by Saint-Anne Dargassies (1966) that in preterm infants who have reached the expected date of delivery, the "straightening responses will be less firm and lasting."

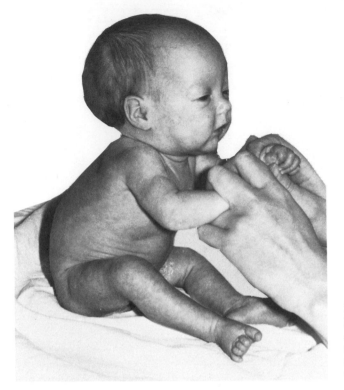

Fig. 10. Traction response, good head control in sitting position, AGA preterm infant at expected date of delivery, born at 32 post-menstrual weeks with a birth weight of 1370 g.

The full-term infants in the study of Beintema (1968) have lower scores for head control in the sitting position at one and two postnatal days than thereafter. From three days in about 50% of the infants, the head remains in the upright position for some seconds.

Good head control in the sitting position is reported by Saint-Anne Dargassies (1974, pp. 104–105) in 53% of the full-term infants in her study and a medium reaction in 25%.

In the Bonn study we observed clear-cut changes of the head-raising part of the traction response in relation to postmenstrual age. The results are shown in Table X and Figure 11. The required state of the infants was 3, 4, or 5. The arm flection part of the response behaves less predictably and was omitted from this report.

The difference between the gestational age periods 34 weeks and under and 36–40 weeks is highly significant ($P < 0.001$, Table X). Because of these maturational changes, head control in the sitting position is considered useful for the assessment of postmenstrual age by Robinson (1966) and Amiel-Tison (1968). The response differentiates between infants of 34 weeks or under and of 36–40 weeks.

Head Lifting in the Prone Position. Head lifting in the prone position is also listed among the labyrinthine righting reactions. Schaltenbrand (1925) observed momentary head lifting in 13 out of 30 full-term infants in the first postnatal week; in 13 cases the response was absent; and in four, doubtful.

Neck extension in the prone position occurred at 37 ± 1 weeks in 84% of the

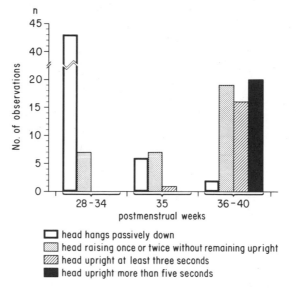

Fig. 11. Traction response, head control in sitting position. The response distinguishes between infants of 34 weeks or under and 36 weeks or over.

□ head hangs passively down
▨ head raising once or twice without remaining upright
▨ head upright at least three seconds
■ head upright more than five seconds

preterm infants in the study of Babson and McKinnon (1967), who scored "raising of head off the bed." According to the table on "developmental sequence of various reflexes" of Michaelis (cited by Schulte, 1974), head raising is present from 38 postmenstrual weeks; from 35 to 37 weeks the response is "not yet present in all cases."

Head raising in the prone position was also tested in the comparative study on preterm and full-term infants at term by Howard *et al.* (1976), but further details about this reflex were not given. Head raising in the prone position is not counted among those responses "with significantly greater incidence of weak responses in the preterm than in the term infant."

Beintema (1968) observed significantly lower scores for head lifting in the prone position of full-term infants in the first 5 days than at 9 days. From 6 days, about 50% of the infants are able to lift the head for at least 2 sec. Between 6 and 9 postnatal days only 12–18% of the infants are not able to lift the head off the bed, as against 69 and 54% percent at the age of 1 and 2 days, respectively.

Table X. *Traction Response—Head Control in Sitting Position—Results of the Bonn Study*[a]

Response	Postmenstrual age in weeks	
	28–34	36–40
Head hangs passively down	47	2
Head raising once or twice without remaining upright	7	19
Head upright at least 3 sec	0	16
Head upright more than 5 sec	0	20
No. of observations	54	57

[a] $\chi^2 = 68.40$ *df* = 3, P < 0.001

Table XI.   Head Lifting in the Prone Position—Results of the Bonn Study[a]

Response	Postmenstrual age in weeks	
	28–38	39–40
No head lifting	73	6
A short lift of the head once or twice	7	15
Momentary head lifting at least three times	3	9
Repeated head lifting for 3–5 sec or more	0	6
No. of observations	83	36
No. of infants	29	29

[a] $\chi^2 = 86.3$, $df = 3$, $P < 0.001$.

In the Bonn study, head lifting in the prone position was scored according to Prechtl and Beintema (1964) and Beintema (1968), with little modification for the preterm infant (Table XI; Figure 12). The infants were in states 3, 4, or 5.

For this reaction, photographs of the infants, made at each examination, serve as controls in statistical evaluation of the results. Figure 13 shows head lifting in the prone position in a preterm girl at the expected date of delivery.

Up to 38 postmenstrual weeks no head lifting was observed in 73 out of 83 examinations. Seven infants lifted the head shortly once or twice, and three infants did so at least 3 times. From 39 to 40 weeks, no head lifting was observed in 6 out of 36 examinations. Fifteen infants lifted the head shortly once or twice, nine infants at least 3 times, and six infants exhibited repeated head lifting for 3–5 sec (Table XI; Figures 12 and 13).

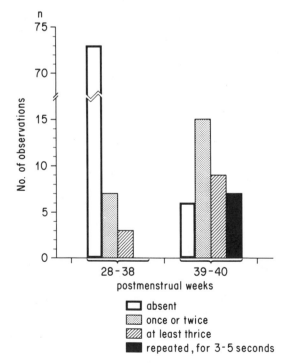

Fig. 12. Head lifting in the prone position. The response distinguishes between infants of 38 postmenstrual weeks or under and 39 weeks or over.

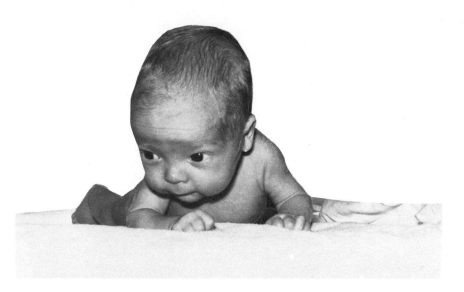

Fig. 13. Head lifting in the prone position for 3–5 sec. AGA preterm girl at expected date of delivery, born at 32 postmenstrual weeks.

The difference between the gestational age periods 38 weeks and under and 39–40 weeks is highly significant ($P < 0.001$, Table XI). A differentiation is possible between infants of 38 postmenstrual weeks and under, and of 39 weeks or over, using this response.

In the prone position a further lateral rotation of the head, when placed in mid-position, can be seen. This response may be considered as a protective movement in order to free the nose and mouth for unhindered respiration.

From 31 to 33 postmenstrual weeks, such lateral rotation of the head in the prone position was observed in about 60% of the Bonn cases (24 examination) and from 34 to 40 weeks, it was present in 80–100% of 72 examinations.

Ventral Suspension—Head Lifting. Ventral suspension is used as a maturational criterion by Dubowitz *et al.* (1970) in their scoring system. The pattern in relation to postmenstrual age was not given. In the quantitative study of Beintema (1968), most of the full-term infants in prone suspension were able to lift their heads to the horizontal plane for at least some seconds. Never did the majority of the infants maintain the head in the horizontal plane for at least 5 sec. Ventral suspension is also used by Lubchenco (1976, 1977) in the estimation of gestational age. At term she observes a straight back and the head held in line with the body. The postterm infant holds the head above body level.

For ventral suspension in the Bonn study, the relationship of the head to the trunk was scored, i.e., whether the head lifted or not. The following scores have been used: (1) no head lifting; (2) head lifting to the horizontal plane sustained for 1–3 sec; (3) head lifting to the horizontal plane sustained for more than 3 sec; and (4) sustained head lifting with retroflection of the head above the horizontal plane for 1–3 sec. The results are shown in Table XII.

According to the Bonn results, head lifting in ventral suspension may be used as a reliable maturational criterion. The differences between the gestational age

Table XII.   Ventral Suspension—Head Lifting—Results of the Bonn
Study[a]

Response	Postmenstrual age in weeks		
	28–37	38	39–40
No head lifting	46	7	9
Head lifting for 1–3 sec	4	5	14
Head lifting > 3 sec	2	2	13
No. of observations	52	14	36

[a] $\chi^2 = 38.15$, $df = 4$, $P < 0.001$.

groups 28–37 weeks and 39–40 weeks are significant ($P < 0.001$). This reaction allows one to distinguish between infants of 37 weeks and under and of 39–40 weeks. At 38 weeks the results are intermediate.

Righting Reactions—Lower Extremities, Trunk, Head, Arms. These reactions demonstrate a righting of (1) the legs, (2) the trunk, and (3) the head; in the Bonn study, a righting of the arms (4) was also scored. A full-term newborn infant is able to right himself when his feet are placed on a table and to maintain an upright posture; therefore, this response has also been called "primary standing" by André-Thomas *et al.* (1960). This "standing response" is produced by a tonic myotatic stretch reflex (Schulte, 1970*b*).

The righting reactions were introduced by André-Thomas and Saint-Anne Dargassies (1952; André-Thomas *et al.*, 1960) and Saint-Anne Dargassies (1954, 1955, 1966) to newborn neurology. On the basis of maturational changes of these reactions with increasing postmenstrual age, they have been used for age assessment (Brett, 1965; Koenigsberger, 1966; Amiel-Tison, 1968; Saint-Anne Dargassies, 1966, 1970).

For an infant of 28 postmenstrual weeks, Saint-Anne Dargassies (1966) states that "the active tone is already well enough developed to permit extension of the lower limbs (although weak and supporting little weight . . .) when the infant is held upright." "The beginning of extension of the lower leg on the thigh upon stimulation of the soles in a lying position" at 30 weeks is reported by Amiel-Tison (1968).

At 32 weeks, according to Saint-Anne Dargassies (1966), straightening of the lower limbs when the infant is held in the upright position happens more quickly and extension of the legs on the pelvis is gradually firmer. Straightening of the trunk appears for the first time. Amiel-Tison (1968) reports "good support" of body weight with the baby in the standing position, but only very briefly, at 32 weeks. She did not describe trunk righting.

At 35 weeks "straightening of the lower limbs in the upright position becomes stronger and begins to include the muscles of the trunk" (Saint-Anne Dargassies, 1966). At 34 weeks, Amiel-Tison (1968) observes excellent righting reaction of the leg and a transitory righting of the trunk; at 36 weeks there is good righting of the trunk with the infant held in vertical suspension.

According to Koenigsberger (1966), who has scored "trunk elevation," the reaction is "slight" at 34 weeks, and from 37 weeks there is good trunk elevation on the hips.

At 37 weeks Saint-Anne Dargassies (1966) reports that "all straightening responses are . . . excellent. Straightening of the lower limbs leads to straightening of the trunk," which leads to "a similar response of the whole body axis." Amiel-Tison (1968) observes at 38 weeks "good righting of trunk with the infant held in a walking position." At 40 weeks she reports a "straightening of the head and trunk together," as does Saint-Anne Dargassies (1970, 1974, p. 109), who observes a solid righting reaction or holding of the head in alignment with the trunk. In her quantitative study of full-term newborn infants, a righting reaction of the trunk and head was observed in 68% of 150 full-term infants; in 18% the reaction is incomplete; in 13% absent; and in 1% "abnormal."

The righting reactions were mostly tested in the Bonn study with the infant in the standing position as described by Amiel-Tison (1968). Then, after exerting a short light pressure on the soles of the feet, we observed an extension of the legs, a plantar support of body weight, and a righting of the trunk, head, and arms. Near the expected date of delivery, the righting reaction was also tested by having the infant on a firm surface, then placed squatting on the haunches with the feet flat against the surface (André-Thomas *et al.,* 1960; André-Thomas and Saint-Anne Dargassies, 1952). Intermittent pressure of the feet against the resting plane resulted in the righting reactions. Figure 14 demonstrates the righting reaction in an AGA preterm infant at the expected date of delivery, elicited according to André-Thomas and Saint-Anne Dargassies (1952, pp. 24, 43).

In evaluation of the Bonn results, in most of the cases the type of reaction and the scoring could be checked with the help of photographs made at each examination in a standardized position. A righting of the lower extremities, trunk, head, and arms is scored separately on the examination forms.

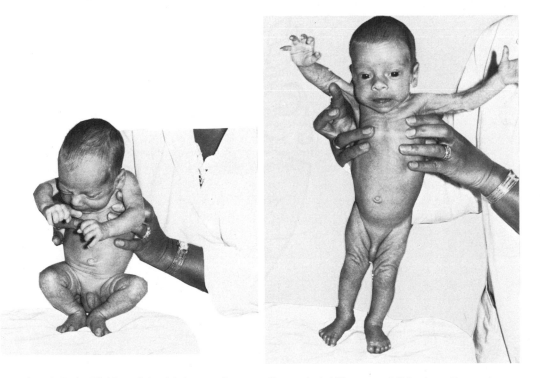

Fig. 14. Left: Eliciting of the righting reaction according to André-Thomas and Saint-Anne Dargassies (1952). Right: Righting of lower extremities, trunk, head, arms. AGA preterm infant at expected date of delivery, born at 32 postmenstrual weeks.

1. From 30 postmenstrual weeks a righting reaction of the legs on the thighs upon stimulation of soles was observed in all the infants (105 observations). With the infant lying in an incubator, at least an extension of the lower extremities could be tested. Near term, a righting of the lower extremities has become less pronounced in some preterm infants, possibly because of increasing flexor tone. Sometimes there is a marked resistance against full extension of the legs in full-term newborn infants.

2. The righting reaction of the trunk in the infant held in a standing position was seen from 34 postmenstrual weeks (73 examinations). Up to 33 weeks the reaction was negative in most of the cases (20 out of 22 examinations). This is in accordance with the results of the Paris school (Saint-Anne Dargassies, 1966; Amiel-Tison, 1968).

3. The righting reaction of the head in the infant held in a standing position was scored as absent if there was no true head righting and present if there was head righting for a few seconds (Table XIII). Up to 35 weeks head righting is absent in most of the cases (38 out of 43 examinations). At 36 weeks the response is intermediate, a head righting being observed in half the cases. From 37 to 40 weeks there is a head righting in most of the infants (in 50 out of 53 examinations, Figure 14). The difference in head righting between the two gestational age groups of 35 weeks and under and 37 weeks and over is highly significant ($P < 0.001$). The righting reaction of the head allows one to differentiate between infants of 35 weeks or under and 37 weeks or over.

4. The righting reaction of the arms was scored as present when they are lifted above the horizontal (Figure 14; Table XIV). The difference between the two gestational age groups 35 weeks and under and 36 weeks and over is highly significant ($P < 0.001$). With the righting reaction of the arms, one may distinguish between infants of 35 weeks and under and 36 weeks and over.

In spite of the clear-cut changes demonstrated in relation to postmenstrual age, the righting reactions can be recommended for maturational assessment only with reservation since they are too easily influenced by any impairment of the infant's general state. Elicitation needs a healthy infant who may be examined outside an incubator in an upright position. All states that reduce muscle tone have to be rejected; states 4 or 5 are required.

Head Turning to Diffuse Light—Phototropism. This reaction was described as long ago as 1859 by Kussmaul (1859) and was observed by him in an infant of "seven months gestation." Also, Stirnimann (1940) observed a tropism to soft light in newborn infants who often turn their heads to the window. With a bright light—as coming from the snow-covered Alps—the infants turn their heads away as in a protective reaction. Peiper (1964) reports that "subdued light is generally so attrac-

Table XIII. Righting Reaction of the Head—Results of the Bonn Study [a]

Response	Postmenstrual age in weeks	
	28–35	37–40
Absent	38	3
Present	5	50
No. of observations	43	53

[a] $\chi^2 = 63.03$ (Yates' correction), $P < 0.001$.

Table XIV. *Righting Reaction of the Arms—Results of the Bonn Study*[a]

Response	Postmenstrual age in weeks	
	28–35	36–40
Absent	39	4
Present	3	54
No. of observations	42	58

[a] $\chi^2 = 69.96$ (Yates' correction), $P < 0.001$

tive for infants that, already at an early age, not only the eyes but also the head is turned toward it.''

According to Saint-Anne Dargassies (1954) and Robinson (1966), to elicit the response, the infant, in a state of quiet alertness, is held with the back against the upper part of the body of the examiner in a comfortable position with the head supported. Then by slow turning, the infant is brought near to a diffuse light (e.g., a window) so that diffuse light falls on the lateral side of his face. In a positive reaction the infant slowly turns the eyes and head toward the light. While turning slowly away with the baby from the window to the other side, the infant maintains the direction of gaze by turning his head until the light of the window disappears. If the light then comes from the other side, the baby orientates again to it. The turning toward light of the eyes, and of the eyes and the head, is scored separately on both sides. It is difficult to test this reflex in an incubator, so it cannot be evaluated in very immature infants.

Robinson (1966) tested the head turning to light reflex longitudinally, and reports that in 4 babies born at less than 32 weeks gestation, in whom the response was initially absent, the response first appears at 32, 33, 34, and 36 weeks. He states that ''the response usually appears between 32 and 36 weeks'' in AGA and probably also in SGA preterm infants. According to his data, under 32 weeks the reflex is absent in all observations; from 32 to 36 weeks the reflex is present in most (11 AGA and 8 SGA preterm infants); and at 36 weeks and over, head turning to light is present in nearly all observations (10 AGA and 15 SGA). Because of significant differences between gestational age periods this reflex is considered of value in assessing gestational age (Robinson, 1966).

Head turning to light is not mentioned by Saint-Anne Dargassies (1974) among the six reliable, or the six unreliable, maturational signs. She describes turning of the eyes and then of the head toward a soft light source (phototropism) to be present first at 37 postmenstrual weeks as a new acquisition (Saint-Anne Dargassies, 1966, 1974, p. 238). The pattern of this response in the earlier weeks was not given.

Studying full-term infants, she (1974, pp. 147–149) observed in 90 out of 150 infants an ''active and complete'' head turning to light; in 30 infants the reaction was incomplete, i.e., turning only of the eyes to light or only to one side; it was absent in 30. She states phototropism also exists in blind infants.

Dubowitz and Dubowitz (1977) consider it impossible to obtain any consistent response to diffuse light, even in full-term infants.

In the Bonn study, the head-turning-to-light reaction (positive phototropism) is elicited according to Saint-Anne Dargassies (1954) (Figure 15). The response is scored for each side separately. Turning of the eyes (1) and turning of the eyes and the head (2) are scored. The reflex has rarely been tested in very young infants because they could not be moved from the incubator without risk. The observed response was bilaterally symmetric in all 29 infants of the study group. The results are shown in Table XV.

There are clear-cut differences between the gestational groups 28–33 weeks and 35–40 weeks (Figure 16). At 34 weeks the results are intermediate. We agree with Robinson (1966) that phototropism may be counted among the reflexes of value in assessing gestational age.

In some other infants of the Bonn study, a unilateral response was observed (as reported by Saint-Anne Dargassies, 1974), i.e., the head-turning response appeared first on one side and one or two weeks later also on the other. In longitudinal observations the response is consistent.

Some preterm infants of more than 34 weeks gestation who are still in the incubator turn their heads constantly to the window, i.e., to one side under the influence of phototropism. With this asymmetric position the pattern of the asymmetric tonic neck reflex may be induced with a marked flection of the leg on the side toward which the occiput is turned. If this position is maintained for several days, or even some weeks, there exists a risk of hip dysplasia in the constantly flexed and adducted leg (Brandt, 1978*b*).

The head turning to light belongs to the primitive reactions. In normal infants, it becomes negative at the corrected age of 2–4 months, although in cases of brain damage it may persist longer. Perhaps the appearance of this primitive reaction in the perinatal period in preterm infants is more related to the ability of spontaneous head rotation (present from 34 weeks in most of the Bonn infants, see p. 262) than to the phototropism itself (Brandt, 1978*b*).

3.2.6. Reflexes and Reactions Almost Constantly Present in the Perinatal Period Independent of Postmenstrual Age

Some of these reactions are also called "fetal reflexes" (M. Minkowski, 1921, 1924; Humphrey, 1964) because they are elicitable in immature fetuses and become negative within the first three months. Langreder (1949) discussed an intrauterine function of such "reflexes," e.g., active adaptation to attitude, stabilization of attitude, and active participation during birth. These "reflexes" are already present in infants from 28 postmenstrual weeks and do not essentially change until the expected date of delivery.

To this group belongs the well-known lateral curving of the trunk, or Galant's response, which is elicited with the infant in the prone position by softly scratching the paravertebral region on the right or left side. This results in a curving of the trunk to the stimulated side. The response was elicited according to Galant (1917) in all but one of 164 infants from birth to 6 months; and with increasing age, the response became feeble. According to Paine (1960) the lateral trunk curving is seen as early as 26 postmenstrual weeks and the response remains positive until 2 months. Robinson (1966) observed a Galant response in 87% of preterm infants, even in babies of 26 weeks gestation. Therefore it is of little value in estimating fetal age.

The withdrawal response on stimulation of the sole of the foot is present in 98% of preterm infants between 26 and 40 weeks (Robinson, 1966). A fetal precursor of the Babinski (1896) reflex has been elicited by M. Minkowski (1921) in a fetus at the fifth month by stimulation of the sole of the foot, resulting in flection of the toes. The Babinski reflex, with extension of the big toe and fanning of the other toes, is present in all infants in the early postnatal days (Galant, 1917).

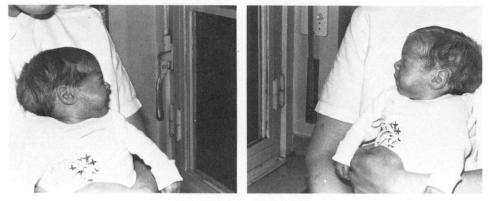

Fig. 15. Head turning to light (positive phototropism) in a preterm boy at expected date of delivery, born at 31 postmenstrual weeks. The response is present to both sides.

Table XV.   Head Turning to Light—Results of the Bonn Study

Response	Postmenstrual age in weeks		
	28–33	34	35–40
Absent	28	6	7
Turning of eyes	0	4	11
Turning of eyes and head	0	2	57
No. of observations	28	12	75
No. of infants	28	12	29

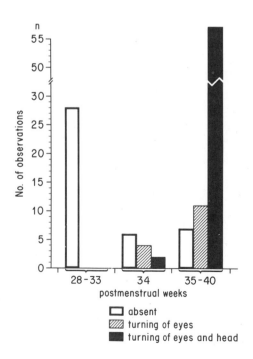

Fig. 16. Head turning to light (positive phototropism). The response distinguishes between infants of 35 weeks or over and 33 weeks or under.

The methods used to elicit the plantar response and the descriptions of resulting movements vary greatly. In the Bonn study the Babinski reflex was elicited according to Wolff (1930) by passing the smooth handle of a reflex hammer softly over the lateral part of the soles of the feet. The toe responses to the plantar stimulus were scored as: (1) dorsal extension of all toes with or without fanning; (2) plantar flection of the toes; and (3) isolated slow hyperextension of the big toe. Dorsal extension of all toes with or without fanning was the most frequent response from 28 to 40 postmenstrual weeks. It was observed in 125 out of 135 examinations. Eight infants showed plantar flection of the toes, and in two cases an isolated slow hyperextension of the big toe was seen.

According to Lenard *et al.* (1968) the Babinski reflex of the infant may be regarded as part of a withdrawal reaction to a nociceptive stimulus; it is easily obtainable during regular sleep, irregular sleep, and quiet wakefulness.

The knee jerk belongs to the group of reflexes which may be elicited even in very immature infants. According to Schulte (1974) this tendon reflex is usually seen in newborn infants beyond 30 weeks gestation. He has elicited the patellar reflex also in a preterm infant of 25 postmenstrual weeks. Robinson (1966) reports that the correlation of the knee jerks with gestational age is a loose one and therefore cannot be recommended for age assessment. Beintema (1968) never found an absent response in newborn infants. Only Finnström (1971) reports a correlation coefficient of 0.51 with gestational age for the patellar reflex. The response is scored absent in 50% of the infants below 32 weeks, and in 10% of the gestational age group of 32.1–36 weeks. In the Bonn study, knee jerks could be elicited in all infants from 28 postmenstrual weeks. The knee jerk belongs to the reflexes "which are equally strong during regular sleep and wakefulness, but are weak or absent during irregular sleep" (Lenard *et al.*, 1968).

We will discuss in greater detail a few examples out of the numerous reactions and reflexes which are almost constantly present from birth and also in preterm infants. They are selected because their value for maturational assessment is disputed in the literature.

Glabella Tap Reflex. This reflex is elicited by tapping sharply on the glabella (Prechtl and Beintema, 1964). The response is a tight closure of the eyes of short duration. The glabella reflex is mostly absent during regular sleep, weak during irregular sleep, and strongest during wakefulness (Lenard *et al.*, 1968).

Robinson (1966) scores a "blink of the eyelids in response to a tap by a finger on the glabella." A grimace of the whole face, which is sometimes seen in very immature infants, is not considered a positive response. In his study of AGA and SGA preterm infants, the glabella reflex was absent under 32 gestational weeks in 17 out of 21 infants. At 32–34 weeks the result was intermediate, in half of the cases the reflex being present. Serial examinations on 5 preterm infants, in whom the reflex was initially absent, showed that it first appeared between 32 and 34 postmenstrual weeks. From 34 weeks the reflex was present in most of the observations, with no difference between AGA and SGA infants. According to Joppich and Schulte (1968), the glabella reflex appears at 32–34 weeks gestational age.

Koenigsberger (1966), who uses the expressions supraciliary tap or MacCarthy reflex, elicits it on each side by tapping the supraciliary region above the eyebrow, which produces a homolateral and sometimes bilateral blink. He suggests that this is "probably the same reflex arc as the corneal reflex (afferent cranial nerve V, efferent VII)." At 28 postmenstrual weeks the reflex is inconstant, and from 32 weeks is present according to the table of findings on neurologic examination at various fetal ages.

According to Schulte (1970*b*), this response probably consists of two components: the short blink, a monosynaptically mediated propioceptive reflex of the musculus orbicularis oculi (cranial nerve VII) and the following tonic contraction, maintained by polysynaptic activation of the same motoneurons.

Graziani *et al.* (1968) also elicit the reflex—which they call "supraorbital blink response"—on both sides by tapping the supraorbital area lightly when the infant's eyes are open. Presence of unilateral or bilateral blink is observed and the symmetry of the contralateral response. The midpoint of age range for unilateral blink, or asymmetrical responses, is 30 weeks.

Dubowitz *et al.* (1970) have seen the glabellar tap consistently present in all infants over 30 weeks gestation and thus regard it as of little value for age assessment. Neither Beintema (1968), Peiper (1964), or Saint-Anne Dargassies (1974) mention this reaction.

In the Bonn study the glabella tap reflex was elicited as described by Prechtl and Beintema (1964). Before 31 postmenstrual weeks the reflex has not been tested regularly in all infants. From 29 to 30 weeks a positive response was observed in 5 of 6 infants. From 31 to 32 weeks the reflex was seen in 18 of 19 cases. From 33 to 40 weeks there was a positive response in all examinations ($n = 95$).

A similar blink of the eyelids as a response to a bright light (flashlight for the photographs) was observed in the Bonn study in all preterm infants from 28 to 40 postmenstrual weeks (153 examinations). This blink reflex in its response is very similar to the glabella tap reflex. It is constantly present from birth, even at an early age, and shows no changes in the perinatal period. Even during sleep, the eyes are closed more tightly and it is sometimes accompanied by grimacing in infants of very low gestational age. This reflex is considered a defense reaction by Peiper (1964).

Crossed Extensor Reflex. The crossed extensor reflex is counted among the "crossed reflexes," i.e., where the stimulus of the stimulated extremity overlaps the other one, and there elicits an effect. In a fetus of four months Minkowski (1924) reports distinct signs of antagonistic innervation in the lower extremities, where pinching of one leg produces a flection of it, and an extension of the crossed one. Peiper (1964) mentions only a crossed flection reflex. The reflex is weak or absent after breech delivery with extended legs (Prechtl and Beintema, 1964).

To elicit the response one leg is extended and kept in this position by pressing the knee down. After stimulating the lateral part of the sole of the foot of the passively extended leg, the unstimulated (free) leg is first flexed, then extended and adducted slowly to the stimulated one with fanning and dorsiflection of the toes.

Saint-Anne Dargassies (1954, 1966, 1974) attaches great importance to the crossed extensor reflex. She distinguishes three phases: flection, extension, and then adduction of the free leg which vigorously approaches the stimulated one; when the stimulation is continued, the free leg is placed on the stimulated one with fanning of the toes. This phase of adduction is considered the most important phase of the reflex. She has demonstrated its gradual appearance in the course of the last month of gestation, and it is not complete before near term (41 weeks). This phase is characteristic of this fetal age. A prerequisite for eliciting the reflex is that the infant does not cry, as the reaction is inhibited by the cry, by pain, and by hypertonia (Saint-Anne Dargassies, 1974, p. 218).

The crossed extensor reflex is counted by Saint-Anne Dargassies (1974, p. 254) among the six maturational signs which are reliable in a high percentage of cases. At 28 postmenstrual weeks the real reflex does not exist and instead only a gesture of defense is observed. At 32 weeks the unstimulated free leg is flexed and then abducted. At 35 weeks the reflex is said to be excellent in its two first phases, flection and extension, but without adduction. The abduction, observed at 32 weeks, has become less. At 37 weeks the crossed extensor reflex remains imperfect in its adduction movement. At 41 weeks the adduction phase of the reflex is perfect.

In her quantitative study with 150 full-term infants, examined between the fifth and seventh day, Saint-Anne Dargassies (1974, pp. 91–93) reports a perfect adduction in 90% of the cases (135), and the reaction was absent in only 2 infants. In 7% of the infants the crossed extensor reflex is incomplete and weak.

Koenigsberger (1966) recommends the crossed extension for differentiating fetal ages 34 through 40 weeks. At 28 weeks there is slight withdrawal, from 32 to 34 weeks withdrawal, at 37 weeks withdrawal and extension, and at 41 weeks withdrawal, extension, and adduction.

Robinson (1966) states the crossed extensor reflex proved an unreliable index of maturity. As a positive response he scored "extension of the opposite leg"—the second reflex phase according to Saint-Anne Dargassies (1974). He found no significant differences in the gestational periods "under 32 weeks," "32 to 36 weeks," and "36 weeks and over." In most groups the results are intermediate. There is no difference between AGA and SGA infants. He suggests "that the reflex is more commonly present after 32 weeks gestation than before." In longitudinal studies the extension of the opposite leg did not appear at a particular postmenstrual age and remain so, but varied in its presence or absence.

Amiel-Tison (1968) described the following three components of the crossed extensor reflex: (1) extension of the free leg after a rapid flection; (2) adduction of the free leg with the foot going toward the stimulated one; (3) fanning of the toes. At 28–30 weeks she reports flection and extension in a random pattern. At 32 weeks there is extension but no adduction. At 34 weeks the reflex remains still incomplete. Only from 36 weeks is the response good with extension, adduction, and fanning of the toes.

Graziani *et al.* (1968) score the reflex differently. They distinguish: (1) flection of contralateral extremity; (2) flection, then delayed extension; and (3) flection, then rapid extension and occasional adduction. The midpoints of age range for these three reflex phases are 30 weeks for (1), 34 weeks for (2), and 39 weeks for (3).

Finnström (1971), who gives a correlation coefficient of 0.56 with gestational age, uses the following scores: (1) no response, i.e., no flection of the contralateral leg; (2) flection and extension; (3) flection, extension, and adduction of the free leg. Below 32.1 weeks he reports in 50% an absent response and in 50% flection and extension. From 32 to 34 weeks score (2) is given in 90%; from 36 to 40 weeks score (3) is prevailing in 76–88% of the infants.

According to Dubowitz and Dubowitz (1977) most infants object to the elicitation of the crossed extension reflex and start crying and resisting. This makes interpretation of the reaction difficult. The authors report that the reflex also fatigues easily and thus cannot be elicited consecutively.

Beintema (1968) scores the extension phase of the crossed extensor reflex as recommended by Prechtl and Beintema (1964) in his study of full-term newborn infants. On the first and second day the reflex is absent or doubtful in 12–25% of the observations, and thereafter it is present in 95–100%.

In the Bonn study the crossed extensor reflex was elicited as described by Prechtl and Beintema (1964) and the following phases were scored: (1) flection; (2) extension; (3) abduction; (4) adduction. The flection phase (1) was present in most of the observations from 29 postmenstrual weeks (89 of 92 examinations). The extension phase (2) was present in most of the cases from 29 postmenstrual weeks (90 of 92 examinations). From 31 to 36 weeks the abduction phase (3) was observed in about half of the examinations (21 of 45), and from 37 to 40 weeks in most of the cases (38 of 42). Up to 35 weeks the adduction phase (4) was absent in most of the examinations (38 of 41). From 36 to 37 weeks the results were intermediate, the response being positive in half of the cases (7 of 13), and from 38 to 40 weeks this fourth phase was present in most of the cases (35 of 38 examinations).

This reflex is of limited value because the infants object to the elicitation and cry and resist. For many preterm infants a fixation of the leg and stimulation of the sole of the foot represents a nociceptive stimulus because of frequent extractions of blood mostly from the heel. Moreover, in infants with a low postmenstrual age (i.e., 31 weeks and under), who are fussing about this unpleasant stimulation, the degree of excitement is easy to recognize, a monitor showing a remarkable rise of heart rate and respiration irregularities. In these infants the reaction cannot be considered as normal for the reflex, but as a reaction to pain. Therefore, in recent years the reflex has no longer been performed.

Since different authors have scored the reflex components differently, the varying results in their reports are easy to understand and comparison is difficult. In summary, it may be said this very often used and cited reflex is too dependent on the state of the infant. Moreover, it has proven to be inconsistent in longitudinal studies, so that this—for the infant an unpleasant reflex—cannot be recommended for age assessment.

Palmar Grasp. A palmar grasp has been shown to occur already in early fetal life. Humphrey (1964, p. 95) reports a quick partial closure of all fingers on stimulation of the palm of the hand, first elicited at 10–10.5 postmenstrual weeks; at 13–14 weeks, finger closure is complete but momentary; and at 27 postmenstrual weeks, Humphrey (1964, p. 117) describes a "maintained grasp with ability to support almost the entire body weight momentarily."

The grasp reflex of the hands is among the responses that are constantly present in preterm infants from 28 postmenstrual weeks (Robinson, 1966). It is mostly absent during regular sleep, weak during irregular sleep, and strongest during wakefulness (Lenard *et al.,* 1968). With increasing postmenstrual age, the grasp reflex of the hands becomes firmer and the reaction extends to the shoulder and neck muscles allowing the infant to be lifted completely whereby the head is also raised. No clear-cut changes which allow an age assessment are observed.

Saint-Anne Dargassies (1966, 1970) describes developmental stages of palmar grasp: at 28 postmenstrual weeks an isolated flection of the fingers; at 32 weeks an involvement of the muscles of the forearms; at 37 weeks a steady and firm grasping; and at 41 weeks a firm grasp reflex by which the infant may be lifted. More recently Saint-Anne Dargassies (1974, p. 254) counts the palmar grasp among the six particularly unreliable neurological signs.

Koenigsberger (1966) recommends the palmar grasp as "a fairly characteristic reflex for given age" with the scoring: feeble at 24 weeks; fair at 28 weeks; solid at 32 weeks; and "may pick infant up" from 37 to 41 weeks—similar to Amiel-Tison (1968).

Graziani *et al.* (1968) give three different scores for the grasp reflex, and report a "slight brief" grasp at 29 postmenstrual weeks (midpoint of age range) and a "firm, sustained" grasp and "brief support of partial weight" at 34 weeks. A "sustained" grasp and "brief support of entire weight" is observed at 39 weeks (midpoint of age range). In addition a reaction called "grasp reinforcement" is scored.

In the study of Meitinger *et al.* (1969) on neurological behavior of preterm infants, a palmar grasp can always be elicited, with a weaker response only in one infant of 31 postmenstrual weeks.

The grasp reflex was easily elicited in all preterm infants in the study of Dubowitz and Dubowitz (1977). Because the most marked variation in the response seems to be in relation to the infant's state of wakefulness, it was thought unlikely to be a useful sign for scoring maturity.

In Beintema's (1968) quantitative study of full-term infants, an absent response for palmar grasp was never found, in accordance with most previous reports, but the intensity of the palmar grasp increased with postnatal age. Saint-Anne Dargassies (1974, p. 26) reports an absent palmar grasp in 2 of 150 full-term infants in her study.

Palmar grasp is observed in all preterm infants of the Bonn study from 28 postmenstrual weeks. A scoring of the firmness of grasp did not yield additional information for maturational assessment.

Rooting Response. The rooting reflex is counted by Minkowski, who observed a rooting response already in fetuses of 2 and 3 months gestational age, among the earliest and most constant fetal reflexes (cited by Prechtl, 1958).

According to Humphrey (1964) the earliest response to an exteroceptive stimulation was observed in a human fetus of 7.5 weeks after stroking the perioral region with a hair. The stimulation caused a contralateral flection in the neck region, i.e., away from the stimulus. At 9.5 postmenstrual weeks, she describes mouth opening by lowering of the mandible after stimulation of the edge of the lower lip, and at 10 weeks, after stimulation of the lower lip area and over the mandible, a ventral flection, i.e., toward the stimulus. At 11–11.5 weeks a perioral stimulation is followed by an ipsilateral rotation of face.

Prechtl (1958) gives a comprehensive review of this response. In preterm infants of 28 postmenstrual weeks he observed weak side-to-side movements of the head with or without perioral stimulation. The "physiological conditions of sleep" (acting as a brake) "and hunger" (facilitating the reaction) "are the main factors influencing" the rooting response.

According to the method of Prechtl and Beintema (1964) the rooting response is elicited by stimulation of the perioral region by fingertip at the angles of the mouth, the upper lip, and the lower lip. Stimulation at the edges elicits a head turning to the stimulated side. Stimulation of the upper lip results in opening of the mouth and retroflection of the head. After stimulation of the lower lip, the mouth is opened and the infant tries to grasp the stimulating finger with the lips and to flex the neck. In preterm infants of low postmenstrual age, (i.e., 31 weeks and under), who cannot keep the head in the midline, the head has to be supported a little.

Koenigsberger (1966) considers rooting a "fairly characteristic reflex for given age." The reflex is elicited by "gentle stimulation around the mouth," the response a "turning of the head in the direction of the stimulus." He recommends that one "reinforce this reflex at 24 and 28 weeks by holding the neck of the infant up slightly." He states that rooting at 28 postmenstrual weeks is "good (with reinforcement)," and from 32 weeks "good (no reinforcement)."

Amiel-Tison (1968) considers the rooting reflex dependent on postmenstrual age. From 28 to 30 weeks the observed response is slow and imperfect with a long latency period. At 32 weeks rooting is "complete and more rapid." From 34 to 40 weeks rooting is "brisk, complete and durable."

Finnström (1971) elicits the response according to Prechtl and Beintema (1964) and uses as reinforcement lifting the infant's head a little by putting the hand behind the neck. Of the six infants 32.1 postmenstrual weeks and under in his study, in two the reflex is lacking in spite of reinforcement; in three there was a "minimal response after reinforcement"; and in one a "complete response." From 32.1 to 34 weeks one of 10 infants showed a moderate response, and nine infants a "complete response." From 34.1 to 40 weeks the response is complete in most of the cases.

The rooting reflex is eliminated for maturational assessment by Dubowitz and Dubowitz (1977) because it is considered too dependent on the time of feeding.

Stirnimann (1940) reports that of 100 newborn infants on the first and second day only eight show no response to stimulation at the upper lip, lower lip, and at the angles of the mouth; the reaction is absent in 73 infants after stimulation at one, two, or three sites. Head movements are often supplemented by movements of the mouth. After stimulation at the lower lip, 43 of 100 infants replace head flection by simpler lowering of the lower lip, whereby the stimulating finger also enters the mouth; the object of the movement is thus reached with less effort.

Beintema's (1968) quantitative study of full-term newborn infants found that on the first day 10% of the infants had an absent response and 12% showed a lip movement only. On the first 6 days the rooting response was significantly weaker than at 9 days, where 88% of the infants showed "full turn towards the stimulated side, sustained for at least three seconds while the finger maintained contact with the angle of the mouth, and lip-grasping."

Saint-Anne Dargassies (1974) differentiates three phases of the rooting response—lateral rotation of the head, extension, and flection—and counts this response among the six maturational criteria, which are reliable in a high percentage of cases. At 28 postmenstrual weeks, where support of the head is indispensable for eliciting the reflex, she reports lateral rotation of the head and limited extension, and a stimulation at the midpoint of the lower lip produces no response (i.e., no flection). Yawning is still found as a primary response to perioral excitation (Saint-Anne Dargassies, 1966). Yawning instead of rooting as a response to stimulation of the perioral region has also been described in full-term infants (Peiper, 1964, p. 410).

According to Saint-Anne Dargassies (1974, p. 225) at 32 weeks the rooting response is quick and complete "but extension of the head is difficult and its flexion is still impossible because of weak neck muscle tone." At 35 weeks the rooting response is "perfect in its course" and the fourth phase (flection of the head) is only suggested (Saint-Anne Dargassies, 1974, pp. 229, 231). At 37 weeks the rooting response is "perfect to the four directions" and at 41 weeks "immediate" and "inexhaustible" (Saint-Anne Dargassies, 1970) and "perfect even in the flexion of the head" (Saint-Anne Dargassies, 1974, p. 249). The rooting response is "completely present" in 142 of 150 full-term infants; in six, incomplete or slow; and in two infants, absent.

In the Bonn study, a rooting response was observed in all examinations ($n = 153$) from 28 postmenstrual weeks.

Besides postmenstrual age, there are other factors that might influence a rooting response being weaker in infants of a postmenstrual age of 30 weeks and under. Since there is a high correlation for rooting with the level of alertness (Prechtl and Beintema, 1964), a weaker response is to be expected in states 1 (eyes closed, regular respiration, no movements) and 2 (eyes closed, irregular respiration, rarely gross movements) prevailing in these babies. Another factor may be that the infants are not hungry because they are fed every hour and often get intravenous glucose in addition.

Since marked intraindividual variations in the rooting response have been attributed to factors like hunger and state (Beintema, 1968), low postmenstrual age alone cannot be identified with certainty as being responsible for a slow and imperfect response. Therefore the rooting response, as the earliest response to an exteroceptive stimulation in fetal life, cannot be recommended for maturational assessment of the preterm infant. The reported changes in this reaction during the perinatal period cannot be ascribed to the rooting reflex itself.

Stepping Movement. The stepping movements are elicited, according to Prechtl and Beintema (1964), by holding the infant upright with the soles of the feet touching the surface of a table; the infant is then moved forward to accompany the stepping. The response requires state 4 (eyes open, gross movements, no crying) or state 5 (eyes open or closed, crying). Stepping movements are "absent in infants born after breech delivery, who either extend or flex the legs" (Prechtl and Beintema, 1964).

Automatic walking belongs among the primitive patterns of motility. Peiper (1964, p. 225) describes the newborn infant making stepping movements in every body position, even with the head downward. Automatic stepping movements have been observed in decerebrated animals that walk until they collapse (Peiper, 1964, p. 226). This automatic function of the central nervous system is inhibited with increasing maturity of the brain.

According to Langreder (1958), stepping movements of the fetus can already be demonstrated *in utero* by impulses on the abdominal wall and by experiences with women in labor. It is suggested that stepping movements, like other newborn reflexes such as spontaneous crawling, may serve as an active intrauterine "reflex participation (or aid) in birth" of the fetus.

The stepping movements are recommended for maturational assessment by Brett (1965), Koenigsberger (1966), Amiel-Tison (1968), and Graziani *et al.* (1968). Finnström (1971) reports a low correlation coefficient with gestational age, and Saint-Anne Dargassies (1974, p. 254) counts the stepping movement among the six particularly unreliable maturational signs.

Brett (1965) reports that an evolution is seen in automatic walking reactions with increasing maturity: walking on the toes, walking with the soles flat, and, finally, walking with the heels coming down first.

Koenigsberger (1966) describes automatic walking at 34 postmenstrual weeks as "minimal," at 37 weeks as "fair on toes," and at 41 weeks as "good on heels." This response is considered a "fairly characteristic reflex for given age."

According to Amiel-Tison (1968), at 32 postmenstrual weeks the infant "begins tip-toeing with good support on the sole and a righting reaction of the legs for a few seconds." From 34 to 36 weeks automatic walking is "pretty good" with "very fast tip-toeing." From 38 to 40 weeks the preterm infant "walks in a toe–heel progression or tip-toes," contrary to a full-term newborn, who "walks in a heel–toe progression on the whole sole of the foot."

Graziani *et al.* (1968) report a midpoint of age range of "few steps" at 35 postmenstrual weeks, for "more than three steps on toes" at 37 weeks, and for "sustained stepping on plantar surface" at 39 weeks, where premature infants continue to walk on the toes.

Automatic walking proved difficult to elicit in infants who need high oxygen and constant temperature (Dubowitz and Dubowitz, 1977); therefore they consider this reaction unsuitable as the aim of their study was "that each of the tests should be completely comparable over the whole gestational range."

Stirnimann (1938, 1940) has studied stepping movements in 75 newborn infants in the first 24 hr after birth and at 9–14 days. On the first day only 16% of the infants make stepping movements. Some infants lift only one leg or both without a tendency to move forward; 40% of the infants show no reaction.

At the age of 9–14 days, 35% of the infants "stepped" adequately, 18% stepped with crossed legs, and some infants made only one step. Altogether stepping movements were seen in 58%, contrary to 16% within the first 24 hr. According to Stirnimann (1938), in stepping movements, neither the type of delivery nor the maturational state is of great significance. Only dyspnea seems to inhibit the reaction.

Beintema (1968) observed stepping movements in 66.5–91.5% of full-term infants 4–9 days after birth. On the first day, stepping movements were seen in only 36.5%, these mostly consisting of one or two steps.

In the study on 150 full-term infants, examined at the fifth to seventh postnatal day, Saint-Anne Dargassies (1974) reports regular stepping movements in 58%, incomplete or suggestive in 15%, abnormal or with crossing over of the feet in 9%, and absent in 18%.

Following the examination of the righting reactions, the walking reflex is tested by leaning the infant a little forward. Then the infant makes alternating flection and extension movements of the legs with good plantar support. By these movements the infant is brought forward when the trunk is supported. The full-term infant "walks" with the knees and hips slightly flexed and more on the heels. In the Bonn study the response is scored present when at least two steps are made. Further, the position of the legs is scored.

In the Bonn study stepping movements are rarely tested before 32 postmenstrual weeks in order to avoid unnecessary disturbance for the infants; they can be elicited from 30 postmenstrual weeks. From 32 to 40 weeks stepping movements were present in most observations (in 65 of 69). No preterm infant exhibited "tiptoeing." In 65 observations they walked on the sole of the foot (Figure 17), different

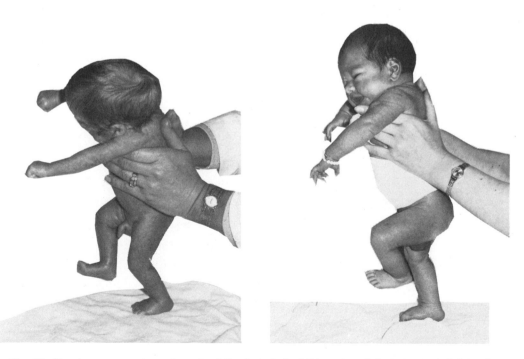

Fig. 17. Stepping movements on the sole of the foot. Left: AGA preterm infant at expected date of delivery, born at 31 postmenstrual weeks, birth weight 1100 g. Right: Full-term infant at the postnatal age of 7 days.

from the full-term newborn who starts stepping movements on the heel (Figure 17). Only at 36–37 weeks do a few infants ($n = 4$) walk more on the anterior part of the foot than on the sole.

Contrary to reports of others (e.g., Illingworth, 1967; Saint-Anne Dargassies, 1974) are found the preterm infants who "walk" mostly on the soles of the feet and not on tip-toes. The greater dorsiflection angle of the foot of 20–40°, on average, in preterm infants compared to 0° in normal full-term infants, is no hindrance to normal walking movements, since the ankle dorsiflection in normal infants during the early years is also at least 30°.

3.2.7. Asymmetric Tonic Neck Reflex

This reflex belongs to those reactions which may be present in the perinatal period but are not demonstrable in all infants. By turning the head to one side, the asymmetric tonic neck reflex may be elicited, consisting of an extension of the extremities on the side toward which the face is turned and of a flection on the contralateral side, i.e., of the occiput. The flection pattern may be present only in the arm, only in the leg, or in both. While turning the head to one side, an elicitation of the neck-righting reflex pattern has to be avoided, i.e., the trunk has to be kept in a supine position.

Magnus and de Kleijn (1912), who examined the influence of head position on tone and posture of the extremities (p. 456) in decerebrate cats and dogs, showed that after head turning to one side, the extremities toward which the face is turned have an increased extensor tone, whereas on the contralateral side (of the occiput) extensor tone is decreased. These reactions, called neck reflexes, persist as long as the head is kept in its specific position. If the head is turned to one side, the extensor tone in the leg where the face is turned to increases, and in the leg the occiput is turned to it decreases. These tonic neck reflexes have been observed by Minkowski (1921, p. 1115) in fetuses of low postmenstrual age. Schaltenbrand, who introduced the results of animal experiments of Magnus and de Kleijn (1912) and Magnus (1924) in infant neurology, reported that in young infants "asymmetric neck reflexes are often to be seen"; 4 of 21 newborns in the first week show an asymmetric tonic neck reflex, and in 5 the pattern is doubtful.

Gesell (1938) has found the asymmetric tonic neck response "an ubiquitous, indeed, a dominating characteristic of normal infancy in the first three months of life," this being contrary to Magnus (1924), who was unable to elicit the reflex in normal infants from the newborn period to 16 postnatal weeks and considered it a pathologic phenomenon. Gesell documented his findings in a series of photographs of normal infants. The reflex pattern is similar to a fencing stance. Gesell (1938) has induced the asymmetric tonic neck reflex indirectly by slowly moving a stimulus across the field of vision, thereby inducing head rotation. In longitudinal observations during the first three months, Gesell (1938) found that at least one third of the infants consistently assumed a right tonic neck reflex, a third the left, and the remainder either a right or a left reflex in an ambivalent manner. The asymmetric tonic neck posture is considered by Gesell (1952) "a reflex attitude so basic that it is already present in the fetal infant, born two months prematurely."

According to Schulte (1971) this reflex is, in general, not seen in normal full-term newborn infants; but an increase of extensor muscle activity in the extremities the face is turned to can be substantiated electromyographically. The receptors for this reflex are located in the neck muscles and neck articulations. Robinson (1966) reports only a weak correlation with gestational age of the asymmetric tonic neck reflex in his study of preterm infants.

Saint-Anne Dargassies (1974, pp. 165, 166) distinguishes two kinds of reflex patterns: (1) spontaneous, constant, quick, inexhaustible; and (2) slow, inconstant, incomplete, needing a long time of latency. The first is considered a highly pathological sign, whereas the second may be observed in newborn babies. Contrary to Gesell (1938), André-Thomas (cited by Saint-Anne Dargassies, 1974) has reported the reflex to be inconstant in normal infants.

In full-term infants, Saint-Anne Dargassies (1974) rarely observed the asymmetric tonic neck reflex. The first reaction pattern, described as slow or asymmetric, was absent in 118 of 150 infants. Eighteen infants showed the reflex only in the legs, nine infants only in the arms, and five infants in arms and legs. The second pattern of reaction, called rapid, was absent in 135 of 150 infants, 12 infants showed the reflex in the upper and lower extremities, and three only in the upper ones.

In the Bonn study, the asymmetric tonic neck reflex was examined to the right and to the left side and was evaluated for each side separately. It was scored present when either in the arms, the legs, or both, a positive response was observed. On the examination form, arms and legs are scored separately.

Further, a differentiation is made between "spontaneously assumed" and "reflex can be elicited." "Spontaneously assumed" means that the infant at the beginning of examination lies there in the pattern of the reflex or assumes this position spontaneously during the observation without being moved (Figure 18). The two photographs in Figure 18 show an AGA preterm infant of 32 weeks, assuming spontaneously the reflex posture. "Reflex can be elicited" is scored when the reflex is present after isolated turning of the head to one side as described by Gesell (1938). It is also scored if the reflex pattern is surmounted easily, or with difficulty, and if it is a prevailing posture during the examination. For evaluation of the results, the photographs, taken at each examination, again serve as a control. The reflex needs a few seconds for elicitation.

We found the asymmetric tonic neck reflex showed no clear-cut changes in relation to postmenstrual age, but there was a trend toward more positive reactions with increasing age. The reflex was scored present when one extremity showed a

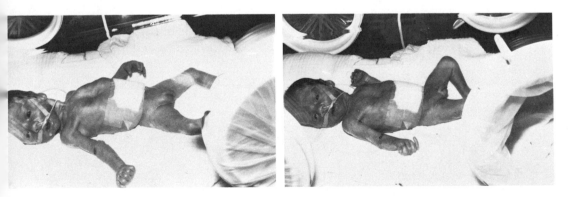

Fig. 18. AGA preterm infant of 7 days born at 31 postmenstrual weeks. The asymmetric tonic neck reflex is assumed spontaneously. The face arm and face leg are in extension, the contralateral extremities in flection. Between the two photographs a time of 10–20 sec has elapsed.

Table XVI.   Asymmetric Tonic Neck Reflex—Results of the Bonn Study[a]

Response	Postmenstrual age in weeks		
	28–32	33–36	37–40
Absent	21	21	17
Spontaneously assumed	3	13	18
Elicitable	5	14	20
No. of observations	29	48	55

[a]$\chi^2 = 13.39$, $df = 4$, $P < 0.01$.

reaction. The results are shown in Table XVI. The reflex pattern was scored "easy to surmount" in most of the infants at all examinations (69 of 73).

The response was examined to the right and to the left. A comparison of both sides yielded no difference in the group of 29 AGA preterm infants, followed longitudinally, i.e., bilateral, except at 39 weeks where the response was unilateral (only to the right) in one infant.

Even though a comparison of the three gestational age groups (28–32 weeks, 33–36 weeks, and 37–40 weeks) yields a statistically significant difference ($P <$ 0.01), this reflex is too unpredictable and cannot be recommended for maturational assessment. This is in agreement with Robinson (1966). As demonstrated in Figure 19, this reaction has no clear-cut time of appearance and, by a negative response, no gestational age can be excluded with certainty.

In most of the infants (two thirds) with a positive reaction, the reflex is consistent, i.e., the response remains positive at every examination; in one third, the reflex is inconsistently present, i.e., present in one examination, absent in the next, and present or absent in the next. In most of the infants with a negative reaction the finding is consistent in the perinatal period. Later we observed no difference between the infants of each group, and all were developing normally.

3.2.8. Nonneurological Tests as Maturational Criteria

Joint mobility in the hand and foot—the degrees of flection of the ankle and wrist—are used as maturational criteria but they are not neurologically based since they are not dependent upon the development of resistance to passive movements or of muscle tone. On the contrary, they behave quite the opposite. These angles are related more to the general flexibility of the joints, which are relatively stiff early in gestation and become relaxed closer to term, allowing the infant to mold himself to the uterine space (Lubchenco, 1976). Possibly the placental hormones responsible for relaxation of maternal joints late in pregnancy also influence fetal joints (Lubchenco, 1976, 1977).

Purely mechanical factors—as supposed in the past (Saint-Anne Dargassies, 1974)—do not seem to cause the decrease of these joint angles in the last weeks of gestation, from 20–40° up to 0° at term. These tests of joint mobility are valuable for age assessment only within the first postnatal days since the intrauterine development is different from the extrauterine development in the corresponding time. They are not applicable for longitudinal observations before term; at the expected date of delivery there is a difference between preterm and full-term infants.

Volarflection Angle of the Hand—Square Window. The angle between the hypothenar eminence and the ventral aspect of the forearm or volarflection angle of the hand, called "square window" by Dubowitz *et al.* (1970; Dubowitz and Dubowitz, 1977), is tested by flexing the hand on the forearm between the thumb and index finger of the examiner. In the scoring system for the neurologic criteria of Dubowitz *et al.* (1970) scores 0 to 4 are given for angles of 90° to 0°. The angles in relation to postmenstrual age are not given.

The volarflection angle of the hand is considered by Saint-Anne Dargassies (1974, p. 254) to be particularly unreliable for maturational assessment because of its great individual variability. At term she observes oscillations between 0° and 45°. At 28 postmenstrual weeks, Saint-Anne Dargassies (1974, p. 220) reports a volarflection angle of the hand of 25–30°. In the full-term infants of her study this angle is zero in 40% of cases, and 10–30° in 50%, i.e., in 90% less than 30°. In 9% of the infants she observes an angle between 30° and 45° and in 1% an angle between 46 and 70°.

Dorsiflection Angle of the Foot. For this reaction the foot is dorsiflexed onto the anterior aspect of the leg. The dorsiflection angle between the dorsum of the foot and the anterior aspect of the leg (ankle dorsiflection) is measured according to Dubowitz *et al.* (1970; Dubowitz and Dubowitz, 1977). The authors use the ankle dorsiflection as a maturational criterion; for a decrease from 90 to 0°, the scores 0–4 are given. There are no figures in relation to postmenstrual age.

The dorsiflexion angle of the foot is counted by Saint-Anne Dargassies (1974, pp. 125, 254) among the six maturational criteria which are reliable in a high percentage of cases. In her study of 150 full-term infants, she observes in 82% an angle of 0°; 18% exhibited an angle of 10–20°. The dorsiflection angle of the foot is considered characteristic for a full-term newborn infant at term. According to Saint-Anne Dargassies (1974, pp. 131–132), the dorsiflection angle of the foot is greater the more premature the infant is at birth. At 28 postmenstrual weeks she observes a dorsiflection angle of the foot of 35–40° which remains so at 32 and 37 weeks. At 41 weeks she reports an angle of 40–60° in preterm infants contrary to 0° in full-term infants, and she attributes this difference to mechanical reasons, i.e., compression in the uterus.

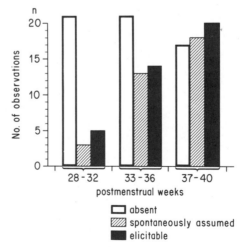

Fig. 19. Asymmetric tonic neck reflex. This reaction is inconsistent and has no clear-cut time of appearance; no gestational age can be excluded with certainty by a negative response.

3.3. Comparison between SGA and AGA Infants

Robinson (1966), comparing AGA and SGA preterm infants, showed a lack of influence of intrauterine growth retardation on neurologic maturation, confirming the results of Saint-Anne Dargassies (1955, 1974).

Michaelis *et al.* (1970), however, report differences in motor behavior of SGA infants, born at term, and AGA full-term infants as: a weak adduction and flection phase of the Moro reflex, a marked asymmetric tonic neck reflex of the legs, and poor or absent head lifting in the prone position.

Also Schulte *et al.* (1971), who have compared 21 SGA infants of toxemic mothers with a mean gestational age of 39 weeks, examined between the third and eight postnatal day, with 21 control infants, have demonstrated differences between the groups. In the SGA "skeletal muscle tone and general excitability were significantly reduced and the motor behavior was usually not consistent with the infant's conceptional age."

Finnström (1971) showed the mean difference in maturity scores between SGA and AGA infants corresponds to a gestational age difference of ten days. Since the infants are not homogeneous, it cannot be stated whether the observed reduction in neurological score is characteristic of only certain types of SGA infants. SGA infants are more often underestimated in development than AGA infants.

Saint-Anne Dargassies (1974, p. 293) states that nutrition and fetal growth do not play a role in the characteristic quality of muscle tone (righting reactions, dorsiflection angle of the foot) and do not alter the neurological maturational criteria.

Since SGA preterm infants represent a heterogeneous group in respect to pathogenesis and prognosis (Brandt, 1975, 1977), *it is also to be expected that their neurologic behavior will be diverse.* This may explain the different findings cited in the literature.

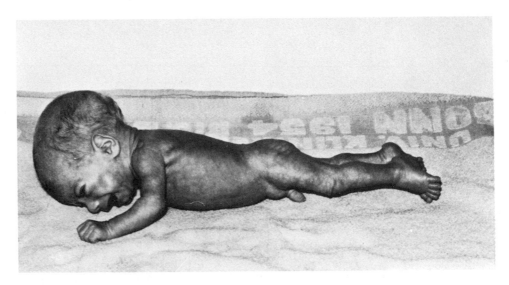

Fig. 20. Head lifting in the prone position. SGA preterm boy at expected date of delivery, born at 34 postmenstrual weeks with a birth weight of 850 g.

According to the Bonn results, the neurologic maturation of SGA preterm infants with normal development at later follow-up is similar to that of the AGA group. SGA infants born at term are not part of the study. Figure 20 shows head lifting in the prone position in an SGA preterm boy, born at 34 postmenstrual weeks with a birth weight of 850 g, and at the expected date of delivery with weight of 1620 g. He belongs to a group of SGA infants with catch-up growth of head circumference after early and high-caloric feeding (Brandt, 1975; Weber *et al.*, 1976).

4. Value for Determination of Postmenstrual Age: A Critical Evaluation

Which reflexes and reactions show clear-cut changes ·in their response in relation to increasing postmenstrual age and may serve as indices of maturity?

To begin with the ·prerequisites: (1) extrauterine development in preterm infants until the expected date of delivery has to be similar to the intrauterine development in full-term infants, so that the test will be completely comparable over the whole gestational range; (2) the developmental course is not affected by premature environmental influences; (3) the reaction is clearly defined and easily elicitable; and (4) the component of the response considered the mature phase has also to be constantly present in normal full-term infants, at least from the third postnatal day. If a reaction varies greatly in a group of normal full-term newborn infants, it cannot be recommended for developmental assessment of preterm infants.

A. Minkowski (1966) states that an infant born prematurely at 28 weeks will, at the postnatal age of six weeks, have the same neurological development as an infant born at 34 weeks.

The results of Robinson (1966) virtually support the concept of Saint-Anne Dargassies (1955, 1974) that neurological maturation in the third trimester of pregnancy is neither accelerated nor retarded by premature extrauterine environmental influences.

The classification of neurologic reactions and reflex phases, the methods used to elicit the response, and the descriptions of resulting movements vary greatly in the literature. This makes an evaluation of the results of different studies difficult. Moreover, often missing are the age range of appearance of reactions and information about the constancy of the response and about its inter- and intraindividual variability.

Tables XVII and XVIII present a critical evaluation of the neurological signs, found to be useful by earlier observers, based on our own experience from some hundred examinations in the perinatal period. An attempt is made to quantify the signs which have been proven reliable by statistical evaluation of, so far, 153 examinations of 29 AGA preterm infants born at 32 postmenstrual weeks and below (Table XIX).

Table XVII shows a review of different reflexes and reactions in the perinatal period, which are considered *not useful* for determination of gestational age according to the Bonn results. Most of these reactions belong to "fetal reflexes" (see p. 278). A "plus" signifies a reported good correlation, and a "minus" signifies a weaker or lower correlation with gestational age. "No statement" means that the reflex or reaction considered is not referred to in the cited reports.

Table XVII. Synopsis of Reflexes and Reactions in the Perinatal Period Considered Not Useful for Determination of Gestational Age According to the Bonn Results—View of Different Authors[a]

	Koenigsberger (1966)	Robinson (1966)	Babson and McKinnon (1967)	Amiel-Tison (1968)	Graziani et al. (1968)	Dubowitz and Dubowitz (1977)	Finnström (1971)	Saint-Anne Dargassies (1974)
Degree of spontaneous movements	+	n.st.	n.st.	n.st.	n.st.	n.st.	+	+
Heel-to-ear maneuver	+	n.st.	n.st.	+	n.st.	+	–	n.st.
Popliteal angle	+	n.st.	n.st.	+	n.st.	+	n.st.	–
Moro reflex abduction, extension	+	–	n.st.	+	+	–	+	–
Glabella tap reflex	+	+	n.st.	n.st.	+	–	–	n.st.
Crossed extensor reflex	+	–	n.st.	+	+	–	+	+
Knee jerk	n.st.	–	n.st.	n.st.	n.st.	–	+	n.st.
Palmar grasp	+	–	n.st.	+	+	–	+	–
Rooting response	+	n.st.	n.st.	+	n.st.	–	–	+
Stepping movements	+	n.st.	n.st.	+	+	–	–	–

[a] (+) = useful for determination of gestational age; (−) = not useful for determination of gestational age; n.st. = no statement.

Table XVIII. Synopsis of Reflexes and Reactions in the Perinatal Period Considered Useful for Determination of Gestational Age According to the Bonn Results—View of Different Authors[a]

	Koenigsberger (1966)	Robinson (1966)	Babson and McKinnon (1967)	Amiel-Tison (1968)	Graziani et al. (1968)	Dubowitz and Dubowitz (1977)	Finnström (1971)	Saint-Anne Dargassies (1974)
Posture[b]	+	n.st.	n.st.	+	−	+	−	+
Recoil lower extremities	+	n.st.	n.st.	n.st.	+	+	n.st.	n.st.
Recoil upper extremities	+	n.st.	n.st.	+	+	+	+	+
Scarf maneuver	+	n.st.	n.st.	+	+	+	n.st.	−
Moro reflex adduction, flection	+	−	+	n.st.	+	−	+	n.st.
Neck-righting reflex	−	+	n.st.	n.st.	+	−	n.st.	n.st.
Traction response,[c] head control	+	+	+	+	n.st.	+	+	+
Head lifting in prone position	n.st.	n.st.	+	n.st.	n.st.	n.st.	n.st.	n.st.
Righting reactions legs, trunk, head	+	n.st.	n.st.	+	n.st.	n.st.	−	+
Ventral suspension, head lifting	n.st.	n.st.	n.st.	n.st.	n.st.	+	n.st.	n.st.
Head turning to light, positive phototropism	n.st	+	n.st.	n.st.	n.st.	−	−	+

[a] (+) = useful for determination of gestational age; (−) = not useful for determination of gestational age; n.st. = no statement.
[b] Imposed posture in the Bonn study.
[c] Head control during traction or when sitting.

Table XIX. Reflexes and Reactions Useful for Determination of Gestational Age—Results of the Bonn Study[a]

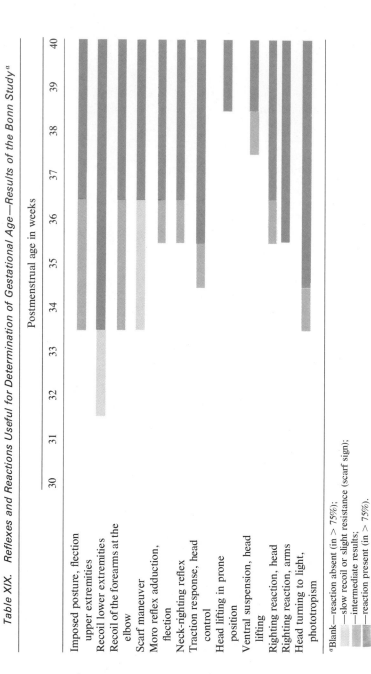

[a]Blank—reaction absent (in > 75%);
—slow recoil or slight resistance (scarf sign);
—intermediate results;
—reaction present (in > 75%).

In agreement with the Bonn results, most of the reactions are shown to be useful only by a minority of the authors cited, with the exception of the crossed extensor reflex, which is considered valuable in five out of eight reports, and the rooting response in half the studies. The stepping movements, for instance, that are considered useful for age assessment by three out of eight authors in Table XVII cannot be recommended for determination of gestational age because they may be absent in normal full-term infants and can be observed even in young preterm infants (see Section 3.2.6., p. 286).

Table XVIII gives a review of different reflexes and reactions in the perinatal period which are considered *useful* for determination of gestational age according to the Bonn results compared with the view of different authors. The key for the signs is the same as for Table XVII. Recoil of the upper extremities is considered useful by six out of eight authors. The traction response—head control in sitting position—is evaluated positively in seven out of eight reports. The adduction phase of the Moro reflex is evaluated contradictorily: four authors report a close correlation with gestational age, two authors consider this reaction useless, and two give no statement.

Head lifting in ventral suspension in Table XVIII is considered only by Dubowitz *et al.* (1970; Dubowitz and Dubowitz, 1977). This reaction has proved to have a clear-cut appearance according to the Bonn results and is considered useful for maturational assessment also by Lubchenco (1977).

Table XIX includes the reflexes and reactions considered useful for determination of gestational age according to the Bonn results. Reactions which may be present at birth and from 28 postmenstrual weeks, the so-called fetal reflexes (see p. 278) (e.g., palmar grasp, rooting response, crossed extensor reflex, and stepping movements), are omitted, because they are not predictive of maturational development. If a positive reaction has been graded for scoring, the positive results are combined for the table, i.e., for head lifting in prone position scores two to four are combined (see p. 272); the reaction is considered present if the head is lifted at least once.

In the following, the time of appearance of the responses in Table XIX is compared with other reports:

The timing for recoil of the lower extremities agrees well with the findings of Koenigsberger (1966), Saint-Anne Dargassies (1966), Graziani *et al.* (1968), and Michaelis (cited by Schulte, 1974).

Recoil of the upper extremities appears at a similar time in the studies of Koenigsberger (1966), Finnström (1971), and Saint-Anne Dargassies (1974); according to Amiel-Tison, the response is present at 36 weeks and Graziani *et al.* report a rapid recoil at 35 weeks.

The scarf maneuver shows similar responses in relation to age in the studies of Koenigsberger (1966) and Graziani *et al.* (1968).

The appearance of the adduction phase of the Moro reflex is similar as reported by Finnström (1971); in the Babson study it is present at 35 ± 1 weeks, where a "partial embrace" is reported by Graziani *et al.* (1968).

The timing of the neck-righting reflex of 37 weeks is the same in the London study of Robinson (1966). Graziani *et al.* (1968) locate the response at 39 weeks.

The head control phase of the traction response agrees with Robinson (1966), Babson and McKinnon (1967), and Finnström (1971).

Head lifting in the prone position occurs earlier in the Babson study (37 ± 1 weeks) and according to the table of Michaelis (cited by Schulte, 1974).

Head lifting in ventral suspension is not considered in relation to postmenstrual age in other studies.

Righting reaction of the head occurs at a similar time in the study of Amiel-Tison (1968).

Righting of the arms is scored for the first time in the Bonn study and cannot be compared with other studies.

For head turning to light, the Bonn results agree with the findings of Robinson (1966).

As demonstrated in Table XIX, most of the clear-cut age-dependent changes in the phases of the cited reflexes and reactions occur only late in the perinatal period. This could suggest that some of them serve as preparation and facilitation for birth at the expected date of delivery.

On the basis of their study, Prechtl *et al.* (1975) conclude that ''there is no indication that preterm infants benefit from the prolonged exposure to extrauterine life.'' ''Environmental conditions, provided they are not adverse, seem to have very little influence on the development of various neural functions'' before 40 postmenstrual weeks.

5. Prognostic Value

In their foreword to the book of André-Thomas *et al.*, Polani and MacKeith (1960) state that the lack of correlation between subsequent development and states considered abnormal in the early days of life is at times leveled as a criticism against the value of neurological examination of the neonate. ''We are, it appears, still at the stage of finding out what is normal and what the normal variation is.'' These remarks made in 1960 still seem valid. The prognostic value is further obscured by the plasticity and adaptability of the developing brain—as often stressed by Prechtl (1973).

Prechtl (1965) demands three criteria for a neurological examination: (1) It must be reliable; that is, it should give the same results when carried out by different examiners. (2) It must be valid; that is, it must be sensitive enough not only to detect gross disturbances, but minor abnormalities as well. (3) It should predict the risk of permanent damage. Criteria (2) and (3) cannot be fulfilled completely in the newborn period because most of the reflexes and reactions studied belong to the ''primitive'' ones, some of which are observed even in decerebrate animals. In any case, it may be assumed that the influence of the cerebrum on the reactions in the neonatal period is small. Inversely, a pathological ''primitive'' reaction may signify that damage to the inferior brain centers may have occurred, and also to the cerebrum.

Long-term prognostic value can be evaluated only by a follow-up study over many years. Prechtl (1965) reports a high correlation between the neonatal findings of hyperexcitability syndrome, apathy syndrome, and hemisyndrome, and the results of the follow-up examination at 2–4 years. In 102 neonates without abnormal neurological signs in the newborn period, 14 were found to be abnormal and 88 normal at the follow-up. In 150 neonates with abnormal neurological signs, 110 were still abnormal and 40 normal between 2 and 4 years. Prechtl (1965) has shown ''that neurological abnormalities diagnosed by a rigorously standardized and reliable examination at birth are highly prognostic signs.''

According to Prechtl (1968), there exists a strongly positive correlation between nonoptimal obstetrical conditions and neurological abnormality as measured by his standardized procedure (Prechtl and Beintema, 1964). Using this technique, he considers neonatal abnormalities as predictive of continuing abnormality and future behavioral disturbance.

On the basis of a detailed neurologic study of 192 mature babies as part of a group of 2100 babies in the Collaborative Project of the National Institute of Neurological and Communicative Disorders and Stroke, it was concluded that the conventional neurologic signs and postural automatisms in the newborn infant seem of only very limited predictive value with regard to abnormalities at one year. Among the few responses that may be more ominous than others are: persistent overall depression of the Moro reflex, universal muscular hypotonia, persistently shrill cry, and asymmetric automatisms.

Illingworth (1967) also counts the persistent absence of the Moro reflex to the neurological responses as suggesting a poor prognosis. Further, the absence of oral reflexes and blink reflexes, and the absence of the Galant response are considered ominous.

Prechtl (1972) states, regarding the problem of early assessment of neural dysfunction, that people often want to identify neurological conditions, such as spasticity or mental deficiency, at a moment when these signs or conditions are not yet present as such. Therefore, it is necessary to know signs that do indicate, with high prognostic value, the risk of the development of these handicapping conditions.

Schulte (1973) recommends the use of the term "pathological" only rarely in the newborn period. All pathological findings are abnormal, but these may be seen also in normal healthy infants. Prognostic value has nothing in common with the definition "abnormal" or "pathological." Schulte (1973) states that the more exact the examination in the newborn period, the safer a prognosis on later development will be. An infant judged completely normal after a neurologic examination will not develop cerebral palsy caused by natal encephalopathy, even if the birth history is indicative of potential risk (Schulte, 1973).

The prognosis on the basis of neurological examination of preterm infants is often considered too favorably, since the chronic encephalopathies do not manifest themselves, or only by small deviations, in motor behavior (Schulte, 1973). This is true, for example, in the preterm infant (SGA) with intrauterine malnutrition in combination with postnatal undernutrition (Brandt, 1975; Brandt and Schröder, 1976).

According to Saint-Anne Dargassies (1977), a "symptomatic gap" after the newborn period may occur before definite signs of damage appear later. This causes much difficulty in establishing an early prognosis. The particularly severe sequelae (i.e., the "bedridden vegetative children") do not manifest such a gap. An extensive symptomatic gap is reported in cases with cerebral palsy: 80% of the infants first appeared to be normal or were improving, and 20% had unspecific symptoms of sequelae.

The value for maturational assessment and the prognostic value are increased by repeated neurologic examinations, although the purpose of newborn neurology is not only the assessment of gestational age. In infants where the age is known from reliable dates of the mother, from a curve of basal temperature, from ultrasound examinations in early pregnancy, or from age assessment at birth using the external maturational criteria, then the assessment of neurological behavior (appropriate for age or not) can lead to a conclusion of benefit to the infant.

Before the infant shows neurologic responses corresponding to his postmenstrual age, the timing, extent, and duration of abnormal neurologic signs in the perinatal period can lead to conclusions concerning residual neurologic deficit. This is an important contribution neonatal neurology can make.

According to the fundamental research of Prechtl and Beintema (1964) and Beintema (1968) in full-term infants, a better standardization of the neurologic reactions and their phases in preterm infants, together with a quantification of the results—which is more than simple recording of the presence or absence of a response—in relation to postmenstrual age, is a major task for the future.

ACKNOWLEDGMENTS

This research was carried out with grants from the Bundesminister für Jugend, Familie und Gesundheit, from the Minister für Wissenschaft und Forschung des Landes Nordrhein-Westfalen, and from the Fritz Thyssen Stiftung which are gratefully acknowledged.

For support of this study at the Universitäts-Frauenklinik Bonn (Director: Professor Dr. Ernst Jürgen Plotz) and at the Universitäts-Kinderklinik Bonn (Director: Professor Dr. Heinz Hungerland until 1974, and Professor Dr. Walther Burmeister from 1975) my thanks are due to my colleagues and assistants.

I particularly wish to thank all the mothers and fathers in the study for their faithful cooperation.

My thanks are due to Dr.rer.nat. Herbert Kurz of the Gesellschaft für Mathematik and Datenverarbeitung Bonn for his help in writing the computer programs. The computations were performed on the IBM/370-168 of the Regionales Hochschul-Rechenzentrum der Universität Bonn.

6. References

Amiel-Tison, C., 1968, Neurological evaluation of the maturity of newborn infants, *Arch. Dis. Child.* **43:**89–93.

André-Thomas, and Saint-Anne Dargassies, S., 1952, *Etudes Neurologiques sur le Nouveau-Né et le Jeune Nourrisson,* Masson, Paris.

André-Thomas, Chesni, Y., and Saint-Anne Dargassies, S., 1960, The neurological examination of the infant, Little Club Clinics Develop. Med., No. 1.

Apgar, V., 1953, A proposal for a new method of evaluation of the newborn infant, *Curr. Res. Anesth.* **32:**260–267.

Ashby, W. R., 1956, *An Introduction to Cybernetics,* Chapman and Hall, London.

Babinski, M. J., 1896, Sur le réflexe cutané plantaire dans certaines affections organiques du système nerveux central, *C.R. Soc. Biol.* **3:**207–208.

Babson, S. G., and McKinnon, C. M., 1967, Determination of gestational age in premature infants, *J.-Lancet* **87:**174–177.

Battaglia, F., and Lubchenco, L. O., 1967, A practical classification of newborn infants by weight and gestational age, *J. Pediatr.* **71:**159–163.

Beintema, D. J., 1968, A neurological study of newborn infants, *Clin. Dev. Med.* **28:** 178 pp.

Brandt, I., 1975, Postnatale Entwicklung von Früh-Mangelgeborenen, *Gynaekologe* **8:**219–233.

Brandt, I., 1976, Dynamics of head circumference growth before and after term, in: *The Biology of Human Fetal Growth* (D. F. Roberts and A. M. Thomson, eds.), pp. 109–136, Taylor & Francis, London.

Brandt, I., 1978*a,* Growth dynamics of low birth weight infants with emphasis on the perinatal period, in: *Human Growth. A Comprehensive Treatise,* Vol. 2, (F. Falkner and J. M. Tanner, eds.), pp. 557–617, Plenum, New York.

Brandt, I., 1978*b,* Neurological development of preterm and full term infants with emphasis on the perinatal period (in preparation).

Brandt, I., and Schröder, R., 1976, Postnatales Aufholwachstum des Kopfumfanges nach intrauteriner Mangelernährung, *Monatsschr. Kinderheilkd.* **124:**475–477.

Brandt, I., and Schröder, R., 1977, Neuromotor and psychological development of preterm and full term infants from birth to the sixth year (in preparation).

Bratteby, L.-E., and Andersson, L., 1976, Continuous monitoring of motor activity in the newborn, *5th European Congress of Perinatal Medicine,* pp. 141–149, Uppsala.

Brazelton, T. B., 1961, Psychophysiologic reactions in the neonate. II. Effect of maternal medication in the neonate and his behavior, *J. Pediatr.* **58:**513–518.

Brazelton, T. B., 1973, Neonatal behavioral assessment scale, *Clin. Dev. Med.* **50:** 66 pp.

Brett, E. M., 1965, The estimation of foetal maturity by the neurological examination of the neonate, in: *Gestational Age, Size and Maturity* (M. Dawkins and B. MacGregor, eds.), *Clinics in Developmental Medicine,* No. 19, pp. 105–115, Heinemann Medical, London.

Dawes, G. S., 1976, The potential of fetal breathing measurements in clinical research, *5th European Congress of Perinatal Medicine,* pp. 175–176, Uppsala.

Donovan, D. E., Coues, P., and Paine, R. S., 1962, Prognostic implications of neurological abnormalities in the neonatal period, *Neurology* **12:**910–914.

Dubowitz, L. M. S., and Dubowitz, V., 1977, *Gestational Age of the Newborn,* Addison-Wesley, London.

Dubowitz, L. M. S., Dubowitz, V., and Goldberg, C., 1970, Clinical assessment of gestational age in the newborn infant, *J. Pediatr.* **77:**1–10.

Escardó, F., and de Coriat, L. F., 1960, Development of postural and tonic patterns in the newborn infant, *Pediatr. Clin. North Am.* **7:**511–525.

Farr, V., Mitchell, R. G., Neligan, G. A., and Parkin, J. M., 1966*a,* The definition of some external characteristics used in the assessment of gestational age in the newborn infant, *Dev. Med. Child Neurol.* **8:**507–511.

Farr, V., Kerridge, D. F., and Mitchell, R. G. (1966*b*), The value of some external characteristics in the assessment of gestational age at birth, *Dev. Med. Child Neurol.* **8:**657–660.

Finnström, O., 1971, Studies on maturity in newborn infants. III. Neurological examination, *Neuropaediatrie* **3:**72–96.

Galant, S., 1917, *Der Rückgratreflex. (Ein neuer Reflex im Säuglingsalter). Mit besonderer Berücksichtigung der anderen Reflexvorgänge bei den Säuglingen,* Diss, Basel.

Gesell, A., 1938, The tonic neck reflex in the human infant, *J. Pediatr.* **13:**455–464.

Gesell, A., 1952, *Infant Development. The Embryology of Early Human Behavior,* Harper, New York.

Gesell, A., and Amatruda, C. S., 1967, *Developmental Diagnosis. Normal and Abnormal Child Development,* 2nd ed., Harper & Row, New York.

Graziani, L. J., Weitzman, E. D., and Velasco, M. S. A., 1968, Neurologic maturation and auditory evoked responses in low birth weight infants, *Pediatrics* **41:**483–494.

Griffiths, R., 1964, *The Abilities of Babies. A Study in Mental Measurement,* 3rd ed., University of London Press, London.

Hackelöer, B.-J., and Hansmann, M., 1976, Ultraschalldiagnostik in der Frühschwangerschaft, *Gynaekologe* **9:**108–122.

Haller, U., Rüttgers, H., Wille, F., Heinrich, D., Müller, P., and Kubli, F., 1973, Aktive Kindsbewegungen im schnellen Ultraschall-B-Bild, *Gynaekol. Rdsch.* **13**(Suppl. 1):118–119.

Hansmann, M., 1974, Kritische Bewertung der Leistungsfähigkeit der Ultraschalldiagnostik in der Geburtshilfe heute, *Gynaekologe* **7:**26–35.

Hansmann, M., 1976, Ultraschall-Biometrie im II. und III. Trimester der Schwangerschaft, *Gynaekologe* **9:**133–155.

Hansmann, M., 1977, personal communication.

Henner, H., Haller, U., Wolf-Zimper, O., Lorenz, W. J., Bader, R., Müller, B., and Kubli, F., 1975, Quantification of fetal movement in normal and pathologic pregnancy, Excerpta Medica International Congress Series, No. 363, pp. 316–319, Excerpta Medica, Amsterdam.

Hoffbauer, H., 1977, Personal communication.

Hosemann, H., 1949, Schwangerschaftsdauer und Neugeborenengewicht, *Arch. Gynaekol.* **176:**109–123.

Howard, J., Parmelee, A. H., Kopp, C. B., and Littman, B., 1976, A neurologic comparison of pre-term and full-term infants at term conceptional age, *J. Pediatr.* **88:**995–1002.

Humphrey, T., 1964, Some correlations between the appearance of human fetal reflexes and the development of the nervous system, *Prog. Brain Res.* **4:**93–135.

Illingworth, R. S., 1967, *The Development of the Infant and Young Child, Normal and Abnormal,* 3rd ed., E. & S. Livingstone, Edinburgh.

Joppich, G., and Schulte, F. J., 1968, *Neurologie des Neugeborenen,* Springer-Verlag, Heidelberg.

Koenigsberger, R. M., 1966, Judgement of fetal age. I. Neurologic evaluation, *Pediatr. Clin. North Am.* **13:**823–833.

Kussmaul, A., 1859, *Untersuchungen über das Seelenleben des neugeborenen Menschen,* Verlag Franz Pietzcker, Tübingen.

Langreder, W., 1949, Welche Fötalreflexe haben eine intrauterine Aufgabe?, *Dtsch. Med. Wochenschr.* **1949**(1):661–667.

Langreder, W., 1958, Die geburtshilflich bedeutsame Motorik des Neugeborenen, *Kinderärztl. Prax.* **26:**559–560.

Lenard, H.-G., Bernuth, H. von, and Prechtl, H. F. R., 1968, Reflexes and their relationship to behavioral state in the newborn, *Acta Paediatr. Scand.* **57:**177–185.

Lubchenco, L. O., 1976, *The High Risk Infant,* W. B. Saunders, Philadelphia.

Lubchenco, L. O., 1977, Personal communication.

Lubchenco, L. O., Hansman, C., Dressler, M., and Boyd, E., 1963, Intrauterine growth as estimated from live born birth-weight data at 24 to 42 weeks of gestation, *Pediatrics* **32:**793–800.

Magnus, R., 1924, *Körperstellung,* Springer-Verlag, Berlin.

Magnus, R., and de Kleijn, A., 1912, Die Abhängigkeit des Tonus der Extremitätenmuskulatur von der Kopfstellung, *Pfluegers Arch. Ges. Physiol.* **145:**455–548.

McGraw, M. B., 1937, The Moro reflex, *Am. J. Dis. Child.* **54:**240–251.

McKeown, T., and Gibson, J. R., 1951, Observations on all births (23,970) in Birmingham 1947. IV. Premature birth, *Br. Med. J.* **2:**513–517.

Meitinger, C., Vlasch, V., and Weinmann, H. M., 1969, Neurologische Untersuchungen bei Frühgeborenen, *Münch. Med. Wochenschr.* **20:**1158–1168.

Michaelis, R., 1970, Risikofaktoren, neonatale Adaption und motorische Automatismen bei untergewichtigen Neugeborenen, *Fortschr. Med.* **88:**525–528.

Michaelis, R., Schulte, R. J., and Nolte, R., 1970, Motor behavior of small for gestational age newborn infants, *J. Pediatr.* **76:**208–213.

Michaelis, R., Nolte, K., and Sonntag, I., 1975, Der Muskeltonus Frühgeborener im Vergleich zum Muskeltonus reifer Neugeborener, Presented at the Congress meeting of South German pediatricians in Ulm.

Minkowski, A., 1966, Development of the nervous system in early life. I. Introduction, in: *Human Development* (F. Falkner, ed.), pp. 254–257, W. B. Saunders, Philadelphia.

Minkowski, M., 1921, Sur les mouvements, les réflexes et les réactions musculaires du foetus humain de 2 à 5 mois et leurs rélations avec le système nerveux foetal, *Rev. Neurol.* **37:**1105–1118; 1235–1250.

Minkowski, M., 1924, Zum gegenwärtigen Stand der Lehre von den Reflexen in entwicklungsgeschichtlicher und anatomisch-physiologischer Beziehung, *Schweiz. Arch. Neurol. Psychiatr.* **15:**239–259.

Moro, E., 1918, Das erste Trimenon, *Münch. Med. Wochenschr.* **65:**1147–1150.

Paine, R. S., 1960, Neurologic examination of infants and children, *Pediatr. Clin. North Am.* **7:**471–510.

Parmelee, A. H., 1964, A critical evaluation of the Moro reflex, *Pediatric* **33:**773–788.

Parmelee, A. H., and Schulte, F. J., 1970, Developmental testing of pre-term and small-for-date infants, *Pediatrics* **45:**21–28.

Peiper, A., 1964, *Die Eigenart der kindlichen Hirntätigkeit,* 3rd ed., VEB Georg Thieme, Leipzig.

Polani, P. E., and MacKeith, D. M., 1960, Foreword, in: *The Neurological Examination of the Infant* (André-Thomas, Y. Chesni, and S. Saint-Anne Dargassies, eds.), Little Club Clinics Develop. Med., No. 1, pp. 2–8.

Prechtl, H. F. R., 1958, The directed head turning response and allied movements of the human baby, *Behaviour* **13:**212–242.

Prechtl, H. F. R., 1965, Prognostic value of neurological signs in the newborn infant, *Proc. R. Soc. Med.* **58:**1–4.

Prechtl, H. F. R., 1968, Neurological findings in newborn infants after pre- and paranatal complications, in: *Aspects of Praematurity and Dysmaturity* (J. H. P. Jonxis, H. K. A. Visser, and J. A. Troelstra, eds.), pp. 303–321, Nutricia Symposium, Groninger, 1967, H. E. Stenfert Kroese N.V., Leiden.

Prechtl, H. F. R., 1970, Hazards of oversimplification, *Dev. Med. Child Neurol.* **12:**522–524.

Prechtl, H. F. R., 1972, Strategy and validity of early detection of neurological dysfunction, in: *Mental Retardation. Prenatal Diagnosis and Infant Assessment* (C. P. Douglas and K. S. Holt, eds.), pp. 41–46, Butterworths, London.

Prechtl, H. F. R., 1973, *Das leicht hirngeschädigte Kind. Theoretische Überlegungen zu einem praktischen Problem,* pp. 282–305, Van Loghum Slaterus Deventer.

Prechtl, H. F. R., 1974, The behavioural states of the newborn infant (A review), *Brain Res.* **76:**185–212.

Prechtl, H. F. R., and Beintema, D. J., 1964, *The Neurological Examination of the Full-Term Newborn Infant,* Little Club Clinics Develop. Med., No. 12, Heinemann Medical, London.

Prechtl, H. F. R., and Beintema, D. J., 1976, Die neurologische Untersuchung des reifen Neugeborenen, 74 pp., 2nd ed., Georg Thieme Verlag, Stuttgart.

Prechtl, H. F. R., and Knol, A. R., 1958, Der Einfluss der Beckenendlage auf die Fusssohlenreflex beim neugeborenen Kind, *Arch. Psychiatr. Nervenkr.* **196**:542–553.

Prechtl, H. F. R., Fargel, J. W., Weinmann, H. M., and Bakker, H. H., 1975, Development of motor function and body posture in pre-term infants, in: *Aspects of Neural Plasticity/Plasticité nerveuse* (F. Vital-Durand and M. Jeannerod, eds.), Vol. 43, pp. 55–66, INSERM, Paris.

Rademaker, G. G. J., 1926, Die Bedeutung der Roten Kerne und des übrigen Mittelhirns für Muskeltonus, Körperstellung und Labyrinthreflexe, *Monogr. Gesamtgeb. Neurol. Psychiatr.* **44**: 344 pp.

Reinold, E., 1971, Beobachtung fetaler Aktivität in der ersten Hälfte der Gravidität mit dem Ultraschall, *Paediatr. Paedol.* **6**:274–279.

Reinold, E., 1976, Ultrasonics in early pregnancy. Diagnostic scanning and fetal motor activity, *Contrib. Gynecol. Obstet.* **1**: 148 pp.

Robinson, H. P., 1973, Sonar measurement of fetal crown–rump length as means of assessing maturity in first trimester of pregnancy, *Br. Med. J., ***4**:28–31.

Robinson, H. P., and Fleming, J. E. E., 1975, A critical evaluation of sonar "crown–rump length" measurements, *Br. J. Obstet. Gynaecol.* **82**:702–710.

Robinson, R. J., 1966, Assessment of gestational age by neurological examination, *Arch. Dis. Child.* **41**:437–447.

Saint-Anne Dargassies, S., 1954, Méthode d'examen neurologique du nouveau-né, *Etud. Néo-Natal.* **3**:101–123.

Saint-Anne Dargassies, S., 1955, La maturation neurologique des prématurés, *Etud. Néo-natal.* **4**:71–116.

Saint-Anne Dargassies, S., 1966, Neurological maturation of the premature infant of 28 to 41 weeks' gestational age, in: *Human Development* (F. Falkner, ed.), pp. 306–325, W. B. Saunders, Philadelphia.

Saint-Anne Dargassies, S., 1970, Détermination neurologique de l'âge foetal néonatal (tableaux—schéma évolutif—grille d'examen maturatif), *J. Paris. Pediatr.* **1**:310–326.

Saint-Anne Dargassies, S., 1974, *Le Développement Neurologique du Nouveau-Né à Terme et Prématuré,* Masson, Paris. English translation: 1977, *Neurological Development in the Full-term and Premature Neonate,* Elsevier/North-Holland, Amsterdam and New York.

Saint-Anne Dargassies, S., 1977, Long-term neurological follow-up study of 286 truly premature infants. I. Neurological sequelae, *Dev. Med. Child Neurol.* **19**:462–478.

Schaltenbrand, G., 1925, Normale Bewegungs- und Lagereaktionen bei Kindern, *Dtsch. Z. Nervenheilkd.* **87**:23–59.

Schröder, R., 1977, Longitudinalstudie über Wachstum und Entwicklung von früh- und reifgeborenen Kindern von der Geburt bis zum 6. Lebensjahr. Ergebnisse der bisherigen Auswertung der psychologischen Untersuchungsbefunde, *Forschungsbericht des Landes Nordrhein-Westfalen,* Westdeutscher Verlag GmbH, Opladen.

Schulte, F. J., 1968, Gestation, Wachstum und Hirnentwicklung, in: *Fortschritte der Pädologie.*, Vol. II, pp. 46–64 (F. Linneweh, ed.), Springer-Verlag, Heidelberg.

Schulte, F. J., 1970a, Sinn und Unsinn der Altersschätzung von Neugeborenen aufgrund neurologischer Kriterien, *Perinat. Med.* **2**:261–262.

Schulte, F. J., 1970b, Neonatal brain mechanisms and the development of motor behavior, in: *Physiology of the Perinatal Period,* Vol. 2 (U. Stave, ed.), pp. 797–841, Appleton-Century-Crofts, New York.

Schulte, F. J., 1971, Das motorische Verhalten von Früh- und Neugeborenen, in: *Handbuch der Kinderheilkunde,* I/1 (H. Opitz and F. Schmid, eds.), pp. 108–124, Springer Verlag, Berlin.

Schulte, F. J., 1973, Besonderheiten bei der neurologischen Untersuchung und Beurteilung von Neugeborenen und Säuglingen, in: *Neuropädiatrie* (A. Matthes and R. Kruse, eds.), pp. 169–191, Georg Thieme Verlag, Stuttgart.

Schulte, F. J., 1974, The neurological development of the neonate, in: *Scientific Foundations of Paediatrics* (J. A. Davis and J. Dobbing, eds.), pp. 587–615, Heinemann Medical Books, London.

Schulte, F. J., and Schwenzel, W., 1965, Motor control and muscle tone in the newborn period. Electromyographic studies, *Biol. Neonat.* **8**:198–215.

Schulte, F. J., Michaelis, R., Linke, I., and Nolte, R., 1968a, Motor nerve conduction velocity in term, preterm and small-for-dates newborn infants, *Pediatrics* **42**:17–26.

Schulte, F. J., Linke, I., and Nolte, R., 1968b, Electromyographic evaluation of the Moro reflex in preterm, term and small-for-dates newborn infants, *Dev. Psychobiol.* **1**:41–47.

Schulte, F. J., Albert, G., and Michaelis, R., 1969, Gestationsalter und Nervenleitgeschwindigkeit bei normalen und abnormen Neugeborenen, *Dtsch. Med. Wochenschr.* **94**:599–601.

Schulte, F. J., Schrempf, G., and Hinze, G., 1971, Maternal toxemia, fetal malnutrition, and motor behavior of the newborn, *Pediatrics* **48:**871–882.

Siegel, S., 1956, *Nonparametric Statistics for the Behavioral Sciences,* International Student Edition, McGraw-Hill, Tokyo.

Stirnimann, F., 1938, Das Kriech- und Schreit-phaenomen der Neugeborenen, *Schweiz. Med. Wochenschr.* **19:**1374–1376.

Stirnimann, F., 1940, *Psychologie des neugeborenen Kindes,* Kindler Verlag GmbH, Munich.

Terman, M. L., and Merrill, M. A., 1965, *Stanford–Binet Intelligenz-Test, Deutsche Bearbeitung von H. R. Lückert,* Verlag für Psychologie C. J. Hogrefe, Göttingen.

Timor-Trisch, I., Zador, I., Hertz, R. H., and Rosen, M. G., 1976, Classification of human fetal movement, *Am. J. Obstet. Gynecol.* **126:**70–77.

von Harnack, G.-A., and Oster, H., 1958, Quantitative Reifebestimmung von Frühgeborenen, *Monatsschr. Kinderheilkd.* **106:**324–328.

Weber, H. P., Kowalewski, S., Gilje, A., Mollering, M., Schnaufer, I., and Fink, H., 1976, Unterschiedliche Calorienzufuhr bei 75 ''low birth weights'': Einfluss auf Gewichtszunahme, Serumeiweiss, Blutzucker und Serumbilirubin, *Eur. J. Pediatr.* **122:**207–216.

Wolff, L. V., 1930, The response to plantar stimulation in infancy, *Am. J. Dis. Child.* **39:**1176–1185.

9

Early Development of Neonatal and Infant Behavior

E. TRONICK, H. ALS, and
T. B. BRAZELTON

1. Introduction

To understand the behavior of the neonate, one must see it as adaptive. Not only is the newborn well-organized for survival, but for eliciting the nurturing necessary for his development. His dependency led, in the past, to the view that he was undifferentiated and poorly organized. An alternate view is that this very dependency requires that he is able to elicit nurturing from his environment. This demands that he have a set of organized and complex behaviors. Thus, if we are to look for the infant's primary behavioral capabilities, we must look for them in his exchanges with the social environment. In that context he emerges as organized in the motor, perceptual, and emotional spheres, able to shape his environment and ready to be shaped by it (Als, 1975).

The infant's behavior is built up from a genetic base which has evolved through adaptation to the environments in which our species evolved. Much of the reflexive behavior seen in neonates can be related to similar behavior in other mammals. The Moro reflex, for instance, is still adaptive for some monkeys; the placing reflex is a residuum of the adaptation to the tree-living primate's need to grab for a branch as it brushes the plantar aspect of the hand or foot as it falls. This reflex behavior unfolds with maturation of the central nervous system in the uterus, and during the period *in utero* "external" events influence these developments. Maternal nutrition, infection, drugs, neuroendocrine factors, for example, all affect the development of fetal behavior. We have every reason to believe that the emotional state of the mother may play an important role in shaping the fetus's behavioral reactions. Hence, the

E. TRONICK • Department of Psychology, University of Massachusetts, Amherst, Massachusetts. *H. ALS and T. B. BRAZELTON* • Child Development Unit, Harvard University Medical School, Boston, Massachusetts.

neonate's behavior is phenotypic, and his individuality at birth is the result of an interaction between his genotype, maturational processes, and the intrauterine environment.

The neonate is in active transaction with his environment, shaping his caretakers as much as they shape him. Since the newborn is able to select input from the environment by the mechanisms of habituation and orientation, he is able to show behavioral preferences among these external stimuli. However, while self-regulation is gradually achieved, the neonate has important behaviors that provide information to the caretaker so that she will provide the appropriate nurturing. The newborn is a sophisticated elicitor.

In this chapter we will review the growing evidence on newborn behavioral capabilities, sensory capacities, ability to learn from social and nonsocial situations. Much of the data have been collected from nonsocial situations which do not afford the input which produces the best behavioral responses. Most of his behavioral organization is richer than a simple stimulus–response model will represent, and we must begin to look at a more integrated, process-oriented model of behavior that identifies the real depth of adaptive organization available *at birth* to the human neonate.

2. Intrauterine Psychological Capabilities

Sontag (1938) first suggested that conditioning of fetal behavioral responses represents ''learning'' *in utero* which contributes to the striking individual differences seen in the neonate's behavior. When the heart rate of the fetus was monitored by a cardiotachometer placed on the mother's abdomen, the fetal heart responded to three kinds of stimuli: (1) auditory, (2) cigarette smoking, and (3) emotional shocks administered to the mother. As these stimuli were repeated, the fetal heart rate response became diminished and its latency prolonged. The infant was ''learning'' not to respond to such stimuli. In another study, Ando and Hattori (1970) demonstrated that infants near an airfield in Okinawa who were exposed to high noise levels during the first four months of gestation were significantly less reactive to loud sounds after birth. Thus, one can expect that all kinds of information and stimulation—both those directly received by the fetus and those received via the neurological and chemical responses of the mother—shape his behavior. But the effects may not always be the same. For example, the fetus may respond to anxiety in the mother by becoming more active and reactive, or it may ''learn'' to cope with the stress induced by her anxiety, and become quieter, having learned to shut down on its own responses to her signals. At birth, the infant may be intensely driving, overreactive, or he may be able to quietly handle stimulation; he may demonstrate a mixture of both these mechanisms.

3. Behavior of the Premature Infant

Prenatal behavioral organization has been well outlined by Hooker (1969), Gesell (1945), and Windle (1950). The sequential maturation of neurologic responses gives us a matrix for the developmental schedule of reflex behavior. It is from this base that the scaling of reflex responses expected at different maturational ages of premature infants has been documented by Dubowitz and Dubowitz (1970).

However, there has been little documentation so far of sensorimotor behavior in the premature infant. We can now see that even premature infants of 34 weeks do fix on, and track, a visual object, however briefly. They can be quiet, attend to, and even turn their heads toward a human voice or soft rattle, although their response often seems more obligatory and rigid than equivalent responses appear to be in full-term infants (Als, 1976). There seem to be mechanisms for control of behavioral states available, although such states in premature infants are less well defined than those in full-term infants. Their states are more fragile, and the impinging demands of physiological stimuli become major sources of competition for organization of behavioral states. Thus, eliciting positive responses from the premature infant requires more skill and energy from the examiner because of the greater physiological demands the premature has to overcome in order to respond.

4. Behavior of the Newborn

4.1. Behavioral State

The concept of behavioral state as a matrix for organization in the neonate has gained recognition since Prechtl's neurological assessment of the newborn (Prechtl and Beintema, 1964). Within the context of optimal state of alertness, he demonstrated that the newborn's midbrain reflexive behavior improved and could be seen as correlated with his behavioral states.

Sleep states have been recognized and defined by Wagner (1937). She described some of the behaviors seen in deep sleep—jerky startles and relative unresponsiveness to external stimuli—and more regulated smooth movements accompanied by responsiveness to stimuli in lighter sleep. Wolff (1959) added his observations of regular, deep respirations with sudden spontaneous motor patterns such as sobs, mouthing and sucking, and erections, which occurred in deep sleep at fairly regular intervals in an otherwise inactive baby. He observed that babies were more responsive to stimuli in light sleep. Aserinsky and Kleitman (1960) described cycles of quiet, regular sleep followed by active periods of body movements and rapid eye movements (REMs) under closed lids. Hence, light sleep has come to be designated as REM sleep. More recently, electrophysiologists have differentiated quiet and active sleep into four stages as designated by electroencephalographic (EEG) activity (Anders and Weinstein, 1972). By behavioral observational techniques, two main sleep states—quiet and active—must suffice in the full-term neonate. In prematures and abnormal infants, and to some extent in term infants, there is a third, indeterminate, sleep state observable (Anders and Weinstein, 1972). At term, active sleep (REM) occupies 45–50% of total sleep time, indeterminate sleep occupies 10%, and quiet sleep 35–45%. The predominance of active sleep has led to the hypothesis that REM sleep mechanisms stimulate the growth of the neural systems by cyclic excitation, and it is in REM sleep that much of the differentiation of neuronal structures and neurophysiological discharge patterns occurs (Anders and Roffwarg, 1973). Quiet sleep seems to serve the purpose of inhibiting CNS activity and is truly an habituated state of rest.

The length of sleep cycles (REM active and quiet sleep) changes with age. At term they occur in a periodicity of 45–50 min, but immature babies have even shorter, less well-defined cycles. Newborn infants have as much active REM in the first half of the deep sleep period as in the second half. Initial brief, individualized sleep and wake patterns coalesce as the environment presses the neonate to develop diurnal patterns of daytime wakefulness and night sleep. Appropriate feeding patterns, diet, absence of excessive anxiety, sufficient nurturing stimulation, a fussing period prior to a long sleep have all been implicated as reinforcing to the CNS maturation necessary to the development of diurnal cycling of sleep and wakefulness. Sleep polygrams, consisting of both EEG and activity monitoring, are sensitive indicators of neurological maturation and integrity in the neonatal period. Steinschneider (1973) has analyzed the occurrence of apnea episodes during sleep as part of a study of sudden infant death syndrome (SIDS). He found that these episodes are more likely to occur during REM sleep and has suggested that prolonged apnea, a concomitant of sleep, is part of a final physiological pathway culminating in SIDS.

In our behavioral assessment of neonates, we have utilized the two states of deep, regular and active REM sleep described by Wolff (1959) and Prechtl (1963) and four states of wakefulness. These states can be reliably determined by observation and without any instrumentation.

State 1. Deep sleep with regular breathing, eyes closed, spontaneous activity confined to startles and jerky movements at quite regular intervals. Responses to external stimuli are partially inhibited and any response is likely to be delayed. No eye movements. State changes are less likely after stimuli or startles than in other states.

State 2. Active, REM sleep with irregular breathing, sucking movements, eyes closed but rapid eye movements underneath the closed lids, low activity level, irregular smooth organized movements of extremities and trunk. Startles or startle equivalents as response to external stimuli often with change of state.

Alert states have been separated into four states for our behavioral assessment:

State 3. Drowsy, a transitional state with semidozing, eyes may be open or closed, eyelids often fluttering, activity level variable, with interspersed mild startles, possibly fussy vocalizations, and slow smoothly monitored movements of extremities at periodic intervals; reactive to sensory stimuli but with some delay, state change frequently after stimulation.

State 4. Quiet alert with bright look; focused attention on sources of auditory or visual stimuli; motor activity suppressed in order to attend to stimuli. Impinging stimuli break through with a delayed response.

State 5. Active awake with eyes open, considerable motor activity, thrusting movements of extremities, occasional startles set off by activity and with fussing. Reactive to external stimulation with an increase in startles or motor activity, discrete reactions difficult to distinguish because of general high activity level.

State 6. Intense crying with jerky motor movements, difficult to break through with stimulation.

The waking states are easily influenced by fatigue, hunger, or other organic needs, and may last for variable amounts of time. However, in the neonatal period, they are at the mercy of the sleep cycles, and surround them in a fairly regular fashion. Waking states are infrequently observed in noisy, overlit neonatal nurseries, but in a rooming-in situation or at home, they become a large part of each cycle, and the neonate lies in his crib looking around for as much as 20–30 min at a time. Appropriate stimulation can bring him up to a responsive state 4. Rocking, gently jiggling, crooning, stroking, setting off vestibular responses by bringing him upright or by rotating him all serve to open his eyes. Then his interest in visual and auditory stimuli helps him to maintain a quiet alert state. In the alert state his respirations are regular at a rate of 50–60 a min, his cardiac rate too is regular and fairly slow (around 100–120/min), his eyes are wide, shiny, and capable of conjugate movements to scan and to follow with head turning to appropriate objects; his limbs, trunk, and face are relaxed and inactive; the skin is uniform in color.

Alert inactive states occur in the first 30–60 min after delivery but then are likely to decrease in duration and occurrence over the next 48 hr as the infant recovers. They return after the first 2 days, and make up as much as 8–16% of total observation time in the first month (Wolff, 1959).

Kleitman (1963) speaks of wakefulness of ''necessity'' and wakefulness of ''choice.'' Wakefulness of necessity is brought about by stimuli such as hunger, cold, bowel movements, or external stimuli which disturb sleep cycles. Sleep recurs as soon as the response to the disturbing stimuli is completed. Wakefulness of choice is related to neocortical activity and is a late acquisition coinciding with the emergence of voluntary motor actions and a mature capacity to achieve and maintain full consciousness. Absence of hunger and fatigue, bowel and bladder activity, as well as of gross motor activity are necessary to this state. After a few weeks this state can accompany gross motor activity as long as the infant is not too active. Pursuit movements of his extremities accompany visual fixation and following as early as 2–3 weeks (Brazelton *et al.*, 1966). The occurrence and duration of these quiet inactive states may be highly correlated with intactness and mature organization of the neonate's central nervous system.

4.2. Crying

Crying serves many purposes in the neonate—not the least of which is to shut out painful or disturbing stimuli. Hunger and pain are also responded to with crying which brings the caretaker to him. And there is fussy crying which occurs periodically throughout the day, usually in a cyclic fashion. It seems to act as a discharge of energy, and an organizer of the states which ensue. After a period of such fussy crying, the neonate may be more alert, or he may sleep more deeply.

As a behavior for organizing his day and for reducing disturbance within his central nervous system (CNS), crying seems to be of real importance in the neonatal period. Most parents can distinguish cries of pain, hunger, and fussiness by 2–3 weeks, and learn quickly to respond appropriately (Wolff, 1968). So the cry is of ethological significance, with the function of eliciting appropriate caretaking for the infant.

The neonate's cry has been analyzed by Lind *et al.* (1966), Wolff (1969), and others for diagnostic purposes. It is a complex serially organized acoustical pattern that is directly regulated by the CNS. Wolff (1969) found that there were three distinct acoustical patterns of cries in normal infants, signifying hunger, anger, and pain. Lind *et al.* (1966) found four patterns: birth, hunger, pain, and pleasure. The fundamental frequency of the normal infant's cries are between 250 and 600 Hz, predominantly between 350 and 400 Hz. Lind *et al.* (1966) found that experienced researchers can discriminate pain cries in normal infants from the cries of infants who had asphyxia or CNS afflictions. The discriminating feature appears to be an increase in the pitch of the infant's cry with a fundamental frequency around 650–800 Hz in brain-damaged infants (Wolff, 1969; Wasz-Hockert *et al.*, 1968). Moreover, brain-damaged infants require more stimuli to elicit a cry, show a larger latency to their cry, and have a less sustained and more arhythmical cry.

4.3. Sensory Capacities

4.3.1. Visual Performance

The newborn is equipped with the capacity for processing complex visual stimulation and showing organized visuomotor behavior. He demonstrates the capacity to alert to, turn his eyes and head to, and follow and fix upon a stimulus which appeals to him. Fantz (1961) first pointed out neonatal preference for complex visual stimuli. More recently, Goron (1975) showed that, immediately after delivery, a human neonate would not only fix upon a drawing which resembled a human face, but would follow it for 180° arcs with eyes and head turning to follow. A scrambled face did not produce the same kind of attention, nor did the infant follow the distorted face with his eyes and head for the same degree of lateral following. We found that the capacity of neonates to fix upon and follow a red ball was a good predictive sign of neurological integrity. However, its absence might not be a serious predictor since it is so dependent on the infant's being in an alert state. Moreover many conditions interfere with the neonate's capacity to come to an alert state—such as CNS depression following anoxia or similar stress, medication effects, recovery from delivery, transient effects of metabolic derangement, or illness. Nonetheless, most newborn infants should be capable of brief alert periods and of fixing and following during alert periods.

Visual acuity of the newborn is still undetermined. Gorman et al. (1957) used the neonate's opticokinetic responses to a moving drum lined with stripes and found that 93 of 100 infants responded to stripes subtending a visual angle of only 33.5 minutes of arc. We found that premature infants were less reliable but could also fix upon and follow the same lined drum. Dayton et al. (1964) found at least 20/150 vision in the newborn by this same technique.

Tronick and Clanton (1971) have experimentally demonstrated coordination of head and eye movements. Using electro-oculargram recordings of eye movements and transducer-recorded head movements they examined the pattern of head and eye movements in 3-week-old infants. They found that when infants were in an upright position both head and eyes became aimed at the target. When the infant moved his line of sight from one target to another the eyes typically moved first, with a rapid saccadic shift, followed by a slower head movement. They found four basic patterns. First, a rapidly shifting pattern in which a fairly rapid head movement was integrated with a brief saccadic movement of the eyes to one side. Second, a search pattern involving a slow movement of the head with a series of fixations and saccades of the eyes. Third, a focal pattern in which the head remained stable while a series of fixations scanned a portion of the visual field. Last, a compensatory pattern in which the line of sight could be maintained by the eyes even when the head moved in an opposite direction. These complex patterns suggested that the infant has a cortically controlled visual system at birth that coordinates head and eye for the extraction of information from the environment.

Adamson (1977) has demonstrated the importance of vision to the neonate by covering an alert baby's eyes with both an opaque and a clear plastic shield. He swipes at and attempts vigorously to remove the opaque shield, building up to frantic activity to do so, and quieting suddenly when it is removed. When the clear shield is substituted, he stays calm and looks interestedly through it.

Bower *et al.* (1970) point out that visual attention to an object brings about reaching behaviors, such as directed swipes at the object. These occur in the first few weeks when a baby is propped up and alert and an attractive object is brought into his "reach space" (about 9–12 in. in front of him in the midline).

Bower *et al.* (1970) and Ball and Tronick (1971) have demonstrated that neonates will actively defend themselves from looming visual targets with a defensive reaction to an approaching object. This defense includes head turning and directed arm movements.

Several observers (Hershenson, 1964; Stechler and Latz, 1966) have demonstrated that neonates prefer moving and complex visual patterns. Further, the duration and degree of attention may be correlated with both a middle range of complexity and the similarity of the target to the ovoid shape and the structures involved in the human face. Salapatek and Kessen (1966) and Haith *et al.* (1969) have found that the highest points of concentration of fixation in the neonatal period are on the contrasting edges of an object. In the neonate, the eyes or sides of the head seem to be the most compelling features of the face. Thus, the neonate seems to be highly programmed for visual learning from birth. The visual stimuli which are "appropriate" or appealing, such as the human face and eyes, or a moving object, seem to capture his attention very early. This allows for very early learning about his human caretakers and the world around him. If his physiological systems are not overwhelmed by too much information, or by too prolonged a period of attention, he will attend for quite long alert periods.

4.3.2. Auditory Performance

The neonate's auditory responses are specific and well-organized. Preyer (1888) had doubts about the neonate's capacity to hear, but few observers have questioned it since. With a rattle as a stimulus we have described the following sequence. First, the infant moves from sleep states to alert states. His breathing becomes irregular, his face brightens, eyes open, and if completely alert his eyes and head turn toward the sound. In the case of well-organized neonates, this head turning is followed by a searching scanning of eyes and face to find the source of the auditory stimulus. Hence a routine test for hearing should include several stimuli—animate as well as inanimate, with careful pairing of the stimulus so that it will break through the neonate's present state of consciousness, e.g., a rattle in light sleep, a voice in awake states, and a clap in deep sleep. As well as the more obvious behavioral responses to auditory stimuli, respirations and eyeblinks should be monitored for responses.

Evoked cortical responses demonstrate the neonate's auditory integrity, and increasing latency of evoked responses with repeated stimuli demonstrates the intactness of the reticular activating system. Eisenberg (1965) has determined the differential responses which are available to the infant. In the range of human speech (500–900 cycles per second), the neonate will inhibit behavior, and he will demonstrate cardiac deceleration and evidences of attention such as alerting and head turning to the source of sound will follow. Outside this range, there is a less complex behavioral response. With a high-pitched or too-loud sound, the infant will startle initially, turn his head away from the sound, his heart rate and breathing accelerating as he does so, and his color will redden. He will attempt to shut out the sound by habituation, and, if unsuccessful, he will start to cry in order to control the disturbing effects of such auditory input. The strikingly narrow range of stimuli for these positive attending responses can be demonstrated by linking devices that record sucking in response to auditory input (Lipsitt, 1967). Within a narrow range, sucking will cease as an initial response to an interesting stimulus, to be followed by a burst–pause pattern of sucking, as if the infant were pausing off and on to receive more of the interesting stimulation.

Lower and higher frequencies have different functional properties. Signals above 4000 Hz are more effective in producing a response, even in crying or sleep states, but they are likely to produce distress. Lower frequencies, at 35–40 db, are effective inhibitors of distress, especially as continuous "white noise" (Lipton *et al.*, 1966). White noise at these levels will most often induce a sleep state, even in a crying neonate (Wolff, 1969). Kearsley (1973) has demonstrated the importance of auditory rise time on the neonate's behavioral responses. Sounds with prolonged onset times and low frequencies produced eye opening and cardiac deceleraton, while sounds with rapid onset and high frequency produced eye closing, cardiac acceleration, and increased head movements. Thus different quality sounds produced an alerting or a defensive reaction.

In assessing neonatal responses to sound one must be aware of and account for what Lacey (1956; Lacey and Lacey, 1962) called the law of initial values. He pointed out that a response to an external stimulus, such as an auditory one, must be seen as pressing the neonate toward a homeostatic resting level. Using heart rate responses as his measure, he demonstrated that when the cardiac rate was high, an appropriate response brought the heart rate down, and when the cardiac rate was low, the response brought the rate up. This same paradigm can be seen in all behavioral systems, i.e., when the baby is active or crying, an auditory stimulus will produce quieting in the baby as he attends to it. When he is quiet or asleep, he will become more activated in order to attend to the stimulus. Thus, an appropriate response becomes one that leads him behaviorally or autonomically in the direction of his normal, alert but resting level. In a sinusoidal fashion the neonate builds up when down, and goes down when already up, on receiving appropriate stimuli. This enables the neonate to overcome his ambient "state" to attend to and interact with an auditory stimulus.

In summary then, one sees a series of regular steps in a neonate's behavioral response to an appropriate sound. As the sound is localized, cardiac rate first increases and may be accompanied by a mild startle. If the auditory stimulus is attractive to the infant his face will brighten, his heart rate will decelerate, his breathing will slow and he will alert and search with his eyes until the source of the sound is localized and in the *en face* midline of the baby. This train of behavior which occurs as a response to an attractive auditory stimulus, such as the human voice, becomes a measure of the neonate's capacity to organize his central nervous system.

4.3.3. Olfactory Performance

Engen *et al.* (1963) have demonstrated differentiated responses to acetic acid, anise, asafetida, and alcohol in the neonatal period. More recently, MacFarlane (1975) has shown that 7-day-old neonates can reliably distinguish their own mothers' breast pads from those of other mothers. They turn their heads toward their own mothers' breast pads with 80% reliability, after controls for laterality are imposed. We have seen that breastfeeding infants at 3 weeks may refuse to accept a formula from their mothers. This refusal seems to be related to the infants' ability to choose the available breast by smell. Yet fathers are successful in giving a bottle to these same babies. Thus it seems that the neonate does have the capacity to make choices from olfactory stimuli and that this information too can be used as part of the attachment process.

4.3.4. Taste Performance

The newborn has fine differential responses to taste. Stuart and Stevenson (1959) observed differentiated sucking responses to sugar, salt, quinine water, and citric acid solutions, with increased sucking to sugar and decreased sucking to the others. More recently, it has been reported (MacFarlane, 1975) that a newborn's differentiation is expressed in an even more complex fashion. An infant is fed different fluids through a monitored nipple, and his sucking pattern is recorded. With a cow's milk formula he will suck in a rather continuous fashion, pausing at irregular intervals. If breast milk is fed to him in this paradigm, he will register his recognition of the change after a short latency, then feed in bursts of sucks with frequent pauses at regular intervals. The pauses seem to be directly related to breast milk as if he were programmed for other kinds of stimuli, such as social communication, in the pauses.

4.3.5. Tactile Performance

The sensitivity of the infant to handling and to touch is quite apparent. A mother's first response to an upset baby is to contain him—to diminish his disturbing motor activity by touching or holding him. Swaddling has been used in many cultures to replace the important constraints offered first by the uterus, and then by mothers and caretakers. As a restraining influence on the overreactions of hyper-reactive neonates, the supportive control which is offered by a steady hand on a baby's abdomen or by holding his arms so that he cannot startle reproduces the swaddling effects of holding or wrapping a neonate. This added control of disturbing motor responses allows the neonate to attend and interact with his environment.

First used for containment, tactile stimuli can be used to rouse or sooth the neonate. A patting motion of 3 times per second seems to be soothing, whereas at 5–6 times a second he becomes more excited and alerted for other kinds of stimuli. As with auditory stimuli, the law of initial values seems to be of primary importance. When a baby is quiet, a rapid intrusive tactile stimulus serves to alert and bring him up to an alert state. When he is upset, a slow, modulated tactile stimulus seems to serve to reduce his activity. When an infant cannot use tactile stimuli to help him to adapt his behavioral state, one must consider a diagnosis of CNS irritability, for a baby with CNS irritation from bleeding or from infection demonstrates constantly increasing irritability in response to most external stimuli, especially tactile ones.

4.4. Sucking

The awake, hungry newborn exhibits rapid searching movements in response to tactile stimulation in the region around the mouth, and even as far out on the face as the cheek and sides of the jaw and head (Peiper, 1963; Blauvelt and McKenna, 1961). This is called the rooting reflex and is present in the premature infant even before sucking itself is effective (Prechtl, 1963). Peiper describes three sets of oral pads in the cheeks and mouths, which help maintain and establish negative pressure. Sucking is facilitated by the thorax in inspiration (Jensen, 1932) and by fixing the jaw to maintain it between respirations. A second mechanism, "expressing," is made by the tongue as it moves up against the hard palate and from the front to the back of the mouth. Swallowing and respiration must be coordinated, and the depth and rate of respiration is handled differentially in nutritive and nonnutritive sucking. Peiper (1963) argues for a hierarchical control of swallowing, sucking, and breathing in which swallowing controls sucking, and sucking controls breathing. An absence of coordination among these three systems in a neonate is indicative of discoordination within the central nervous system, and it may be seen in brain-damaged infants and in the very immature infant.

Gryboski (1965) describes a technique of monitoring the three components of sucking with three transducers: one at the front of the tongue, another at the base of the tongue, and a third in the upper esophagus. The timing among these three components is a measure of the immaturity of the CNS of a premature infant, as well as a measure of the disruption of the central processes which control these mechanisms. Since one can set off and feel the three components by inserting a finger into the newborn's mouth when the infant is sucking, an examiner can determine for himself the presence and coordination of the three components. This, then, is an easily available and valuable measure of the infant's stage of relative maturity and CNS functioning.

The infant sucks in a more or less regular pattern of bursts and pauses. During nonnutritive sucking, his rate varies around two sucks a second (Wolff, 1968). Bursts seem to occur in packages of 5–24 sucks per burst (Kaye, 1967). The pause between bursts has been considered to be a rest and recovery period as well as a period during which cognitive information is being processed by the neonate (Halverson, 1944; Bruner, 1969). Kaye and Brazelton (1971) found that the pauses were important ethologically since they are used by mothers as signals to stimulate the infant to return to sucking. However, the mother's jiggling actually prolongs the pause as the infant responds to the information given to him by his mother.

Factors affecting sucking, such as age, hunger, fatigue, rate of milk flow, and state of arousal, affect but do not change the individual burst–pause pattern which seems to be a relatively stable inborn pattern (Kron *et al.*, 1963).

Sucking, because of this stability produced through central control, is used by infant researchers to measure sensory discrimination, conditioning and learning (Kessen *et al.*, 1970), attention, and orienting (Haith *et al.*, 1969; Kaye, 1967; Sameroff, 1968).

Finger sucking is common in the neonatal period, and there is evidence that the insertion of parts of the hand in the mouth occur commonly in the uterus. The importance of sucking as a regulatory response can be seen in a newborn as he begins to build up from a quiet state to crying. His own attempts to achieve hand to mouth contact in order to keep his activity under control are fulfilled when he is able to insert a finger into his mouth, suck on it, and quiet himself. The sense of satisfaction and gratification at having achieved this self-regulation are apparent. His face softens and alerts as he begins to concentrate on maintaining this kind of self-regulation. A pacifier can achieve this same kind of quieting in an upset baby, but may not serve the self-regulatory feedback system as richly as the baby's own maneuver.

4.5. Habituation

The neonate's capacity to attend to external stimuli is coupled with the capacity to habituate, or tune them out. When a bright light is flashed into a neonate's eyes, not only do his pupils constrict, but he blinks, his eyelids and whole face contract, he withdraws his head by arching his whole body, often setting off a complete startle as he withdraws, his heart rate and respirations increase, and there is an evoked response registered on his visual occipital EEG. Repeated stimulation of this nature will induce diminishing responses. For example, in a series of 20 bright-light stimuli presented at 1-min intervals, we found that the infant rapidly "habituated" the behavioral responses, and by the tenth stimulus had decreased not only his observable motor responses, but his cardiac and respiratory responses had begun to decrease markedly also. The latency to evoked responses as measured by EEG traces was increasing, and by the fifteenth stimulus his EEG reflected the induction of a quiet, unresponsive state similar to that seen in sleep (Brazelton, 1975). His capacity to shut out repetitive disturbing visual stimuli protects him from having to respond to visual stimulation and at the same time forces him to save his energy to meet his physiological demands.

This capacity on the part of the neonate to suppress his responses to visual stimuli has been designated as a kind of neurological habituation (Bartoshuck and Tennant, 1964; Bridger, 1961) and has been found to be active in neonates with intact central nervous systems (Ellingson, 1960). The capacity to habituate to visual stimuli is decreased, although it is present, in immature infants (Hrbek and Mares, 1964) and is depressed by drugs such as barbiturates given to mothers as premedication at the time of delivery. This has led Brazier (1959) to postulate that the primary focus for this mechanism is in the reticular formation and midbrain. However, the cortical control over this mechanism seems apparent as one observes a neonate, who is initially in irregular, light sleep, become drowsier with repeated stimulation. He then becomes deeply asleep, with tightly flexed extremities, little movement except jerky startles; no eyeblinks, deep, regular respirations; and rapid, regular heart rate. This state seems to resemble a defensive one against the assaults of the environment and upon cessation of the stimulation the infant almost immediately goes back to his initial state or even a more alert state. One can often see neonates in noisy, brightly lit neonatal nurseries in this "defensive" sleep state.

Hutt *et al.* (1968) have argued, in fact, that habituation results from a change from one state to another and that it may not occur within a single state. They examined habituation in the neonate to three different stimuli (125-cps square-wave tone, a female voice saying "baby," and a 125-cps sine-wave tone). They found no evidence of response decrement or of dishabituation in the EMG unless it was accompanied by a state change, except occasionally when a weak stimulus was followed by a strong stimulus. However, Tronick and Scanlon (1975) have demonstrated habituation in newborns to an air-puff stimulus administered to the sole of their feet. The response decreased from full body startles to a tonic response of the stimulated foot alone. Dishabituation and recovery of the full startle occurred when the other foot was stimulated. The shutdown was found to occur in sleep states and quiet awake states.

Heart rate has been used as a primary measure of habituation. But there are certain limitations on any interpretation of heart rate as a response system. The response of heart rate closely parallels the infant's activity level and the physiological demands of any stress under which he may be. Hence, rate which is poorly responsive to stimulation becomes a clear signal of distress in the neonatal period. The kind of stimulus may determine the kind of behavioral reaction of the infant. An intrusive or painful stimulus is more likely to bring about a startle or an active response with an accompanying increase in heart rate, whereas an attractive or interesting stimulus will bring the infant to a quiet alert state in order to attend to and learn about the stimulus. This quiet alertness brings with it a cardiac deceleration which can be used as a measure of attention (Clifton, 1974; Porges, 1974; Graham and Clifton, 1966). Porges (1974) has defined a biphasic curve of heart-rate response to novel stimuli with neonates in an alert period. This consists of two components—first a reflexive component in which the eliciting properties of the stimulus determine whether the infant startles and his heart rate accelerates or whether he produces an initial orienting response with cardiac deceleration. Porges thinks this is subcortically initiated. Secondly the response has a tonic, instrumental component in which the infant begins to attend to and learn about the stimulus and in which a period of cardiac deceleration can be seen. This phase is indicative of stimulus complexity, of the infant's capacity to gather information at a cortical level, and of the physiological demands on his cardiorespiratory system. Thus, this measure becomes a significant indicator of the infant's capacity to overcome physiological demands in order to attend to his environment.

The law of initial values plays an important role in any assessment of cardiac responses. Bridger and Reiser (1959) and Lipton *et al.* (1966) have demonstrated that the prestimulus heart rate is likely to be inversely correlated with the response, as well as the magnitude of heart rate toward which his poststimulus rate will tend in a cyclic homeostatic curve. If his prestimulus rate is high, a response will bring about relative deceleration; if his prestimulus rate is low, his response will be toward acceleration. The magnitude of response is correlated with the distance of the prestimulus rate from the resting rate. Thus, theoretically, at one particular rate, a stimulus might cause no cardiac change. But in reality, the initial response is that of acceleration as part of a defensive or startle reaction to be followed by deceleration which is correlated with the properties of the stimulus. Negative or disturbing stimuli produce little or no deceleration, soothing stimuli produce a mild decrease in base rate and a stable low-grade homeostatic response, whereas an attractive or interesting stimulus may bring about marked brief acceleration followed by a period of observable deceleration.

Habituation to repeated stimuli can be monitored by this technique (Bartoshuck and Tennant, 1964; Bridger, 1961). Since habituation is generally thought to reflect cortical storage of information about a stimulus, the decrease in heart-rate response over repeated trials with a stimulus might reflect the infant's cortical functioning. With an intrusive stimulus, one sees a gradually diminishing shutdown on the degree of acceleratory cardiac responses to the repetitious stimulus. With an interesting stimulus, the acceleratory component may decrease slightly, but the deceleratory phase is markedly affected with repeated presentations. This, then, becomes a measure of cortical function. Dishabituation, or recovery of the total response, when the stimulus is changed, can be seen in the neonatal period and can be used as a further measure of cortical and subcortical function in the neonatal period (Brackbill, 1970; Eisenberg, 1965).

Since these measures are dependent on the integrity of the central nervous system as well as upon the demands upon the cardiac system, it is not surprising that stressed infants do not demonstrate this complex behavior. Lester (1975) demonstrated a substantial orienting response followed by rapid habituation to auditory signals with dishabituation to changes in tonal frequency in well-nourished infants. But infants who had been undernourished showed attenuation or complete absence of the orienting response and dishabituation to changes in tonal frequency. The decrease in these responses seemed highly correlated with the degree of malnutrition to which the infants had been exposed. The correlation of this lack of capacity of attentional mechanisms suggests itself as a precursor for disabilities in learning later on.

5. Modal Newborn Behavioral Development during the Neonatal Period

An integrated picture of the newborn's behavioral repertoire, and its initial organization and development is provided by a study (Tronick *et al.,* 1976*d*) using the Neonatal Behavioral Assessment Scale (Brazelton, 1973*a,b*). The behavioral assessment represents an attempt to incorporate the above information on newborn behavior into a scale that can be used to record the behavioral capacities of the newborn and his individual style in expressing those capacities. Through a set of maneuvers, it uses 26 behavioral items to assess the infant's capacity to interact with his social and inanimate environment, his capacity to integrate and organize his motor behavior, his ability to regulate his states of consciousness from sleep to waking, and to record indicators of his physiological capacity to cope with stress. Its conceptual underpinning is that the infant is a highly integrated and complexly organized organism, that his behavioral capacities reflect the organization of his central nervous system and that these in turn have a major effect on shaping his caretakers.

The behavioral assessment tests for neurological adequacy with 20 reflex measures scored on a 4-point scale and for interactive capacities with 26 behavioral responses to environmental stimuli, including the kind of interpersonal stimuli which mothers use in their handling of the infant as they attempt to help him adapt to the world. The 26 behavioral items are each scored on a 9-point scale. In the exam, there is a graded series of procedures—talking, hand on belly, restraint, holding, and rocking—designed to soothe and alert the infant. His responsiveness to animate stimuli, e.g., voice and face, and to inanimate stimuli, e.g., rattle, bell, red ball, white light, and temperature change are assessed. Estimates of vigor and attentional excitement are measured as well as motor activity and tone, and autonomic responsiveness as he changes state. With this examination, given on successive days we have been able to outline (1) the initial period of alertness immediately after delivery—presumably the result of stimulation of labor and the new environmental stimuli after delivery; (2) the period of depression and disorganization which follows and lasts for 24–48 hr in infants with uncomplicated deliveries and no medication effects, but for longer periods of 3–4 days if they have been compromised by medication given their mothers during labor; and (3) the curve of recovery to "optimal" function after several days. This third period may be the best single predictor of individual potential function, and it seems to correlate well with the neonate's retest ability at 30 days. The shape of the curve made by several examinations may be the most important assessment of the basic CNS intactness of the neonate's ability to integrate CNS and other physiological recovery mechanisms and the strength of his compensatory capacities when there have been compromising insults to him during labor and delivery. Interscorer reliability of 90% can be achieved with training and maintained for at least a year. The exam takes 30 min to perform and 10 min to score reliably.

This neonatal behavioral exam has been used in cross-cultural studies to outline genetic differences (Brazelton *et al.*, 1975), with premature infants to predict their outcome successfully (Als *et al.*, 1976), to document behavioral correlates of intrauterine protein depletion (Als *et al.*, 1976; Brazelton *et al.*, 1977), to determine the effects of uteri affected by rapidly repeated pregnancies (Brazelton *et al.*, 1975), and to assess the influence of heavy medication given the mother during labor, as well as maternal addiction to heroin and methadone (Strauss *et al.*, 1976; Tronick *et al.*, 1976).

The behavioral items of the neonatal assessment are

> Response decrement to repeated visual stimuli
> Response decrement to repeated auditory stimuli
> Response decrement to repeated tactile stimuli
> Orienting responses to inanimate visual and auditory stimuli
> Orienting responses to the examiner's face and voice
> Quality and duration of alert periods
> General muscle tone—in resting and in response to being handled (passive and active)
> Motor maturity
> Traction responses as the neonate is pulled to sit
> Responses to being cuddled by the examiner
> Defensive reactions to a cloth over his face
> Consolability with intervention by examiner
> Attempts to console self and control state behavior
> Rapidity of buildup to crying state
> Peak of excitement and his capacity to control himself
> Irritability during the exam
> General assessment of kind and degree of activity
> Tremulousness
> Amount of startling
> Lability of skin color (measuring autonomic lability)
> Lability of states during entire exam
> Hand-to-mouth activity

We feel that the behavioral items give important evidences of cortical control and responsiveness in the neonatal period. The neonate's capacity to manage and overcome the physiological demands of this adjustment period in order to attend to, to differentiate, and to habituate to the complex stimuli of an examiner's maneuvers are an important predictor of his future central nervous system organization. Certainly, the curve of recovery of these responses over the first neonatal week must be of more significance than the midbrain responses detectable in routine neurological exams.

Our study (Tronick *et al.*, 1976) to assess the course of neonatal behavioral development was carried out on a group of 54 white newborn infants who were carefully selected for optimal *in utero* and obstetrical histories in order to minimize confounding stress factors. The mothers of these newborn babies received minimal amounts of medication during their deliveries. The infants were examined with the neonatal assessment on days 1, 2, 3, 4, 5, 7, and 10. The data on these infants provide a picture of their behavioral capacities over the first 10 days of life.

Table I presents the median score and interquartile range on the items of the neonatal assessment that describe the organization of behavioral state in the infant. For these infants their state of organization remains relatively constant. However, on day 1 the infant's predominant state is crying, and the second most frequent state is alertness or sleep during the examination; on day 10 the predominant state during the exam is alertness. During the 30-min examination the infant shows 6–8 state shifts from either sleep to awake, awake to crying, crying to sleep, or vice versa. His build-up to a crying state from sleep to awake is gradual, and then only after several disturbing stimuli have been administered. His control is evidenced in the overall intensity of his reactions. The modal performance of these infants was to reach a crying state in response to stimulation but to remain predominantly in a less aroused state. Their irritability was moderate with fussing/crying being produced by only four out of seven definitely aversive stimuli. More importantly, these infants were able to bring themselves from a crying state back into an alert state with no assistance from the examiner. They accomplished this with a hand-to-mouth maneuver or by locking onto a visual or auditory stimulus.

This modulated pattern of state change and moderate irritability reflects the capacity of neonates for self-organization. This capacity is further reflected in their ability to decrease their responsiveness to disturbing stimuli. For the repeated bell, light, and rattle stimuli the infants were able to shut down startles and gross body movements and diminish initial respiratory changes after only 3–4 applications of each stimulus. To the repeated pinprick, they could shut down generalized body movement and respond with a simple withdrawal of the stimulated leg after 5 trials. The overall stability over the first 10 days of these capacities to modulate reactivity stands in contrast to the improvement in other areas of functioning and clearly reflects their relatively stress-free condition at birth.

The motor organization of these infants was moderate at birth and improved over the 10 days (Table II). The movements of these infants when alert were smooth, with only sporadic jerkiness and the arcs of their movements were often as wide as 60°. By day 10, there was a significant improvement, their movements being smoother and wider. Few of the infants' movements on day 10 were jerky, and many of their movements were as wide as 90°. Their tone improved over the first 10 days. Initially their tone was good when handled, but they were relatively flaccid when left alone. By day 10, when handled their tone had increased and their tone when left alone was more moderate.

The neonates' capacity for integrated motor performance was evidenced at birth and also improved during this period. In the pull-to-sit maneuver, their active participation increased. At first they made attempts to control their heads but could sustain them in the upright position for less than 1–2 sec. Later in the first week, their shoulder tone had improved and their heads stayed balanced for as long as 10 sec. Defensive movements in response to a cloth over the face improved from rooting and some lateral head turning to active neck stretching, arching backwards, and directed swipes at the cloth to free themselves of it.

The ability of the infant to bring his hand to his mouth was apparent at birth. On day 1, the infants swiped at the mouth area several times, their hands approaching the mouth; by day 10 they were able to achieve repeated insertion of parts of their hands and began to suck actively on their fists.

Finally, their activity level increased over these first few days. This can be described in a combination of how much the infant responds to stimulation and how much activity is spontaneous. At first, the infants showed a slight amount of

Table I. *Items Reflecting the Newborn's Organization in Respect to State Control*

Items	Day 1					Day 5					Day 10				
	Mean	SD[b]	Median	Mode	Range	Mean	SD[b]	Median	Mode	Range	Mean	SD[b]	Median	Mode	Range
Habituation															
Light[a]	6.30	1.66	6.00	5.00	2–9	7.16	1.57	8.00	8.00	4–9	7.00	1.28	7.00	8.00	4–9
Rattle	6.87	2.10	8.00	8.00	1–9	6.96	1.86	7.00	9.00	4–9	6.00	2.06	7.00	8.00	1–7
Bell[a]	7.00	2.09	8.00	9.00	1–9	7.59	1.56	8.50	9.00	4–9	6.90	2.09	7.50	9.00	1–8
Pinprick	4.37	1.51	4.00	3.00	3–8	3.92	1.32	4.00	3 + 4	2–7	4.42	1.60	4.00	3.00	3–8
Peak of excitement	5.50	1.29	5.50	7.00	4–8	5.89	1.54	7.00	7.00	2–8	5.70	1.47	6.00	7.00	3–8
Lability of state	7.24	1.23	3.00	3.00	1–6	7.30	1.23	3.00	3.00	1–5	7.35	1.35	3.00	3.00	1–6
Rapidity of buildup	6.41	2.25	7.00	9.00	2–9	6.29	2.20	6.50	9.00	2–9	6.45	2.14	7.00	9.00	2–9
Irritability	5.92	1.55	4.00	4.00	1–8	5.57	2.05	4.50	6.00	1–8	5.75	1.78	5.00	5.00	1–7
Self-quieting	5.22	1.66	6.00	6.00	1–9	4.96	1.54	5.00	5.00	2–9	5.42	2.06	6.00	6.00	1–9

[a] $P < 0.05$ significant change over days.
[b] Standard deviation.

Table II. *Items Reflecting the Newborn's Motoric Organization*

Items	Day 1					Day 5					Day 10				
	Mean	SD[b]	Median	Mode	Range	Mean	SD[b]	Median	Mode	Range	Mean	SD[b]	Median	Mode	Range
Tonus[a]	5.40	0.95	6.00	6.00	3–7	5.56	0.90	6.00	6.00	2–8	5.85	0.93	6.00	6.00	2–8
Pull-to-sit	5.60	1.25	5.00	5.00	3–8	5.74	1.56	6.00	6.00	1–9	6.07	1.39	6.00	6.00	1–8
Motor maturity[a]	4.20	0.59	4.00	4.00	3–6	4.89	0.96	5.00	5.00	3–7	5.28	1.05	5.00	5.00	4–8
Defensive movement[a]	4.88	2.37	5.00	2.00	1–8	6.96	1.13	7.00	8.00	3–8	6.84	1.24	7.00	7.00	1–8
Hand-to-mouth[a]	4.96	2.18	5.00	5 + 7	1–9	5.65	2.00	6.00	6.00	1–9	5.80	2.15	6.00	5.00	1–9
Activity[a]	3.98	0.85	4.00	4.00	2–6	4.76	1.15	5.00	5.00	2–7	4.85	1.22	5.00	6.00	1–7

[a] $P < 0.05$ significant change over days.
[b] Standard deviation.

spontaneous activity (25% of the time spent in spontaneous activity). Stimulation produced a slight response but it did not perpetuate itself and was soon over. Over the next 10 days the infants' activity level built up, and one cycle of activity tended to produce a second cycle or a continuation of the movement. Few infants were continuously in motion or were dominated by random, uncontrolled, and unstoppable movement.

The physiological stability of the organism shows improvement over the first week also (Table III). On the scale items, by day 10, tremulousness, which was seen in fussy and crying states, had largely disappeared and generally occurred only as part of Moro responses or startles. Startles, too, had decreased by day 10. Initially, four or more were seen during the exam, but on day 10 only one spontaneous startle generally occurred. The improvement in the infant's autonomic system was further indicated by the greater stability of the infant's skin color. Color changes, mottling, and reddening still occurred with crying or uncovering, but these became much less evident and recovery of good color was much faster.

The infant's capacities for focused attention are perhaps best captured in his abilities to orient toward positive environmental events (Table IV). These abilities, however, are possible only because of the infant's abilities to control his states of consciousness, to shut out disturbing and distracting stimuli, and to moderate his motoric and physiological processes as he attends. Often we see an infant's motor activity decrease, his startles or jerky movements disappear, his color return as he comes alert, with eyes and face brightening to an auditory or visual stimulus.

With an appropriate visual stimulus he is not only able to fix on the object but he is able to follow the object with smooth eye movements and head movements for 30–60° as well. With an inanimate sound he becomes quiet and brightens and shows eye movements, then head and eye movements, in the direction of the sound. When the human face is presented, his performance is even better than with an inanimate object. Although he can follow the object through 60° of the visual field with smoothly coordinated head and eye movements, he follows the face over a 90° track. The infant's performance improves over 10 days. In the neonatal period he is able to turn his head 90° to a voice and fixate on the face and then to follow it in complete circles. The most compelling stimulus is the face and voice combined. On day 1 the infant brightens and fixes on the face and then is able to track it through 60° of the visual field with smooth head and eye movements. On day 10 this increases to 90° of the visual field and when he loses track of the face and voice he is able to locate it again without assistance.

The infant's span of attention increased over these initial days. At first he could stay alert for a moderate amount of time for two or three stimuli with little delay in his responsiveness. By day 10 he responded with long periods of alertness to stimuli as they were presented.

The infant also became more cuddly over this period. He started out by molding his body to the examiner's and actively moving in order to participate. On day 10, he was more active in this participation. He nestled, leaned into the examiner and turned toward the examiner's face. On the other hand, the infant is less consolable on day 10 than he was on day 1. At first gentle talking and holding of the infant's arms was enough to quiet him, but on day 10 it became necessary to pick him up and rock him. This increase in the need for outside constraint actually reflects the increased vigor of these infants. By day 10 they were more able to console themselves; but when this failed, they demanded more input.

Table III. *Items Reflecting the Newborn's Physiological Stability in Response to Stress*

Items	Day 1					Day 5					Day 10				
	Mean	SD[b]	Median	Mode	Range	Mean	SD[b]	Median	Mode	Range	Mean	SD[b]	Median	Mode	Range
Tremor[a]	4.92	1.76	5.00	5.00	1–8	6.00	1.97	5.00	5.00	1–7	6.30	2.10	4.00	1.00	1–7
Color change[a]	5.61	1.42	5.00	5.00	2–7	5.88	1.55	3.50	4.00	1–8	6.26	1.67	4.00	4.00	1–9
Startle[a]	5.50	1.79	6.00	6.00	2–7	6.18	1.50	4.00	3.00	2–7	6.68	1.49	3.00	2.00	1–7

[a]$P < 0.05$ significant change over days.
[b]Standard deviation.

Table IV. *Items Reflecting the Newborn's Interactive Capacities*

Items	Day 1					Day 5					Day 10				
	Mean	SD[b]	Median	Mode	Range	Mean	SD[b]	Median	Mode	Range	Mean	SD[b]	Median	Mode	Range
Orient to inanimate visual	5.77	1.42	6.00	4.00	3–8	5.67	1.55	5.50	7.00	2–8	6.08	1.52	7.00	7.00	1–8
Orient to inanimate auditory	5.60	1.14	6.00	6.00	2–8	5.80	1.13	6.50	6.00	3–8	6.16	1.16	6.00	6.00	4–8
Orient to animate visual (face)[a]	6.30	1.25	7.00	7.00	4–8	6.43	1.12	7.00	7.00	2–8	7.06	0.93	7.00	7.00	4–9
Orient to animate auditory (voice)[a]	5.56	1.35	6.00	6.00	1–8	6.02	1.33	6.00	7.00	1–8	6.27	1.07	6.00	7.00	4–8
Orient to animate visual and auditory (face and voice)[a]	6.67	1.17	7.00	7.00	4–8	6.72	1.10	7.50	7.00	2–8	7.35	1.08	8.00	8.00	4–9
Alertness[a]	4.85	2.28	5.50	7.00	1–8	6.00	2.07	6.00	8.00	1–9	6.17	1.85	6.00	8.00	1–9
Cuddliness[a]	5.63	1.04	6.00	6.00	3–8	5.94	1.09	6.00	6.00	3–9	6.59	1.35	7.00	6 + 7	3–9
Consolability[a]	6.50	1.73	7.50	8.00	1–8	5.70	1.55	5.00	8.00	2–8	5.98	1.95	5.00	5.00	1–9

[a]$P < 0.05$ significant change over days.
[b]Standard deviation.

Observational studies of mother–infant interactions clearly demonstrate that a complex communication system exists during the first months of the infant's life (Bowlby, 1969; Goldberg, 1971; Richards and Bernal, 1972; Brazelton *et al.*, 1973*a*; Richards, 1971). Most of the existing studies have analyzed the interactions to show the one-to-one contingencies between maternal and infant behavioral displays. These displays were at least several seconds and even minutes long and were typically observed while the pair was engaged in functional or task-oriented situations (Lytton, 1971). Impressive reciprocity exists between the infant and the mother under normal conditions. For example, mothers respond differentially, and their latency of response varies according to the type of infant cry (Wolff, 1969); infant rooting produces characteristic adjustments by the mother to aid the infant's access to the breast (Blauvelt and McKenna, 1961), and the infant's cessation of sucking produces a jiggling of the infant by the mother (Kaye and Braxelton, 1971). Furthermore, although the relative frequencies of specific behaviors vary widely, there is nonetheless a similar relationship between infant and mother behaviors over many cultures and social classes (Rebelsky, 1971; Korner and Thoman, 1972; Freedman, 1968; Brazelton *et al.*, 1966).

Nonfunctional tasks, play, and face-to-face situations have also been observed. Such situations lack the supporting constraints that exist in functional situations and force complete reliance on the communicative skills of the participants. This demand to communicate without supports makes these situations more revealing of the cognitive and affective skills of the participants. In these situations, it has been shown that simple contingencies exist between adult and infant displays of short durations: smiles, eye-to-eye contact, and vocalizations (Gewirtz and Gewirtz, 1969; Bullowa, 1978; Ambrose, 1961; Moss, 1965, 1967), and that the contingent responding can modify their rate of occurrence (Rheingold *et al.*, 1959).

There is a large variation in the expressive qualities of individual normal infants. Moreover, infants who are small for gestational age (Als *et al.*, 1976), malnourished (Brazelton *et al.*, 1975), premature (Als *et al.*, 1976), drug addicted (Strauss *et al.*, 1976), and those with high levels of drugs received from the mother during delivery (Standley *et al.*, 1974; Tronick and Scanlon, 1975; Scanlon, 1974) have markedly modified behaviors and appearances. There are also variations in infants of different cultural groups (Brazelton *et al.*, 1976, 1977; Freedman and Freedman, 1969).

These variations in infant behavioral characteristics have been shown to have significant effects on caretakers. In normal hospital settings, Als (1975) found that mothers fit their reactions to their newborn infants' visual or auditory responsiveness and to their feeding patterns. Similarly, cuddly infants produced more cuddling by adults (Schaffer and Emerson, 1964). State of arousal, activity level, and irritability markedly effect parental responses (Lewis, 1972; Brazelton, 1975*a*; Korner and Thoman, 1972; Escalona, 1968; Als and Lewis, 1975). Adults become disturbed by gaze aversion (Hutt and Ounsted, 1966; Robson, 1967), and also by the irritability of brain-damaged infants (Prechtl, 1963), and mothers soon give up breast feeding with infants who fight the breast (Gunther, 1961). In institutions, difficult-to-cuddle infants are passed around among caretakers more than the more responsive ones who are successfully nurtured by one caretaker (David and Appell, 1963). In families, hyperkinetic, hypersensitive infants produce parental desperation and eventually an overreactive hostile environment (Heider, 1966).

As another kind of example, Chavez's (Chavez *et al.,* 1974) studies with disadvantaged Mexican families demonstrate that the effect of neonatal behavior in shaping families' reactions to them is seen most powerfully in the striking difference in child-rearing practices of parents of well-nourished families, as contrasted to those of malnourished mothers and infants. The former were more active, reactive, and expressive, and the parents were able to modify their reactions to fit their babies, whereas stressed babies in this environment created a stressed and a less than optimally responsive environment around them. Thus difficult or too-quiet infants born into an already stressed environment can expect less than optimal reinforcement for their behavior (Brazelton *et al.,* 1977).

Bowlby (1969), Bell (1971), Goldberg (1971), and others have theoretical positions that describe how the behavioral signals emitted by the infant produce effects on the caretaker. For the most part, they still view the infant as a passive emitter of signals. Bowlby does not describe the infant as making goal-corrected actions until he is more mobile at 8 months of age. Prior to that time, although the infant has potent signaling capacities, he does not modify them significantly to serve his own purposes. Bowlby's view of interactive behavior can be compared to Lorenz's view of the infant's physical appearance. The infant's ''Kewpie'' doll physiognomy and size elicit and inhibit particular responses from adults. The infant has this appearance but has no control over it. Similarly for Bowlby, the infant emits signals, physiognomic and behavioral, but does not modify them. These signals affect the adult, but the infant is not aware of their effect, nor did he intend them (Tronick *et al.,* 1977).

Our view, in contrast, is that the infant adjusts and changes his actions in relation to the ongoing actions of his partner. He actively regulates these evolved behavioral adaptations to regulate the adult–infant communication system. Als (1975) has demonstrated this for the first 3 days of life of newborns when with their mothers. We have found that infants as young as 3 weeks of age showed significantly different attentional cycles and behavior while interacting with objects as contrasted with people (Brazelton *et al.,* 1973*a,b*). When interacting with people during face-to-face exchanges, there are smooth, rhythmical accelerations and decelerations of behavioral responses as they greet, then interact with, a caretaker. During these interchanges the hands are used gesturally and are linked rhythmically to the infant's vocalizations. In contrast, with objects the infant evidences prolonged attentional cycles broken by abrupt turning away. His body is still but his movements are jerky, and the movements of his extremities come in bursts accompanied by swipes at the object. With the object the infant's goal seems to be to reach for it, while with people the goal is to reciprocate in an affective interchange.

This latter goal of the infant is evidenced in studies where the normal behavior of the adult in relation to the infant is distorted. For example, if the mother remains still-faced and immobile in front of an infant, the infant attempts to get the interaction back on track (Tronick *et al.,* 1977, 1978*a,b*). Initially the infant orients to the adult and smiles, but when the adult fails to respond, the infant sobers. His facial expression is serious and he becomes still. He stares at the adult and smiles again, but briefly. Then he looks away. He repeatedly looks toward and away from the adult, smiling briefly in conjunction with the look toward, and sobering with the look away. Eventually he slumps in his seat with his chin tucked, his head and eyes oriented away from the adult, looking hopeless and helpless in the face of an unresponsive adult. This pattern of behavior occurs reliably within 2–3 min of a distorted interaction.

In another situation, in which the adult sits sideways to the infant, the infant uses a different set of behaviors, but again his goal is to get the interaction back on track (Tronick *et al.*, 1977, 1978*a,b*). The infant looks at the adult and makes cooing, calling types of vocalizations and eventually leans toward and reaches out to the adult. Sometimes he makes cries which are interposed with the periods of intense looking. These infant behaviors are indicative of the infant's capacity to regulate his expressive behaviors in a goal-directed fashion as early as the first month of life.

7. Summary

The newborn's system is adapted to operate within a social matrix. His abilities for self-regulation, intake of information, and responsiveness are organized and develop rapidly during the neonatal period. A primary function of these abilities is to elicit regulatory input from adults. It appears that the infant is able to modify intentionally his responses in following his own goals which are appropriately adjusted to the input from the environment.

8. References

Adamson, L., 1977, Infants' Response to Visual and Tactile Occlusions, PhD dissertation, University of California, Berkeley.

Als, H., 1975, The Human Newborn and His Mother: An Ethological Study of Their Interaction, PhD dissertation, University of Pennsylvania.

Als, H., 1976, unpublished data.

Als, H., and Lewis, M., 1975, The contribution of the infant to the interaction with his mother, paper presented at Society for Research in Child Development, Denver, Colorado.

Als, H., Tronick, E., Adamson, L., and Brazelton, T. B., 1976, The behavior of the fullterm but underweight newborn infant, *Dev. Med. Child Neurol.* **18**:590–602.

Als, H., Tronick, E., and Brazelton, T. B., 1976, Manual for the Behavioral Assessment of the Premature and At-Risk Newborn (An Extension of the Brazelton Neonatal Behavioral Assessment Scale), mimeo (in preparation).

Ambrose, J. A., 1978, The development of the smiling response in early infancy, in: *Determinants of Infant Behavior,* Vol. 1 (B. M. Foss, ed.), pp. 179–201, Methuen, London.

Anders, T. F., and Roffwarg, H., 1973, The effects of selective interruption and total sleep deprivation in the human newborn, *Dev. Psychobiol.* **6**:79–81.

Anders, T. F., and Weinstein, P., 1972, Sleep and its disorders in infants and children: A review, *Pediatrics* **50**:312.

Ando, Y., and Hattori, M., 1970, Effects of intense noise during fetal life upon postnatal adaptability, *J. Acoust. Soc. Am.* **47**(4):1128

Aserinsky, E., and Kleitman, N., 1960, A motility cycle in sleeping infants, *Curr. Res. Anesthesiol. Analgesiol.* **32**:260.

Ball, W., and Tronick, E., 1971, Infant response to impending collision: Optical and real, *Science* **171**:818–820.

Bartoshuck, A. K., and Tennant, J. M., 1964, Human neonatal correlates of sleep wakefulness and neural maturation, *J. Psychiatr. Res.* **2**:73.

Bell, R. Q., 1971, Stimulus control of parent or caretaker behavior by offspring, *Dev. Psychol.* **4**:63.

Blauvelt, H., and McKenna, J., 1961, Mother neonate interaction: Capacity of the human newborn for orientation, in: *Determinants of Infant Behavior* (B. M. Foss, ed.), pp. 3–29, Methuen, London.

Bower, T. G. R., Broughton, J. M., and Moore, M. K., 1970, The coordination of visual and tactual input in infants, *Percept. Psychophys.* **8**(1):51–53.

Bowlby, J., 1969, *Attachment and Loss, Vol. I: Attachment*, Basic Books, New York.

Brackbill, Y., 1970, Continuous stimulation and arousal level in infants: Additive effects, *Proc. 78th Annu. Conv. Am. Psychol. Assn.* **5:**271; and 1973, in: *The Competent Infant* (J. V. Stone, H. T. Smith, and L. B. Murphy, eds.), p. 300, Basic Books, New York.

Brazelton, T. B., 1973a, Assessment of the infant at risk, *Clin. Obstet. Gynecol.* **16:**361–375.

Brazelton, T. B., 1973b, Neonatal Behavioral Assessment Scale Clinics in Developmental Medicine, No. 50, Spastics International Medical Publications, William Heinemann Medical Books, Ltd., London.

Brazelton, T. B., 1975, Newborn behavior, in: *Scientific Foundations of Obstetrics and Gynecology* (E. Philips, J. Barnes, and M. Newton, eds.), pp. 961–991, William Heinemann Medical Books, London.

Brazelton, T. B., Scholl, M. L., and Robey, J., 1966, Visual behavior in the neonate, *Pediatrics* **37:**284.

Brazelton, T. B., Kowlowski, B., and Main, M., 1973a, Early mother–infant reciprocity, in: *The Origins of Behavior* (M. Lewis, and L. Rosenblum, eds.), pp. 49–77, John Wiley and Sons, New York.

Brazelton, T. B., Robey, J. S., and Scholl, M. L., 1973b, Infant development in the Zincanteco Indians of Southern Mexico, in: *The Competent Infant* (J. Stone, H. Smith, and L. Murphy, eds.), p. 529, Basic Books, New York.

Brazelton, T. B., Tronick, E., Lechtig, A., Lasky, R., and Klein, R., 1977, The behavior of nutritionally deprived Guatemalan infants, *Dev. Med. Child Neurol.* **19:**364–372.

Brazelton, T. B., Koslowski, B., and Tronick, E., 1976, Neonatal behavior among urban Zambians and Americans, *J. Acad. Child Psychiatry* **15:**97.

Brazier, M. A. B. (ed.), 1959, *The Central Nervous System and Behavior*, Trans. 2nd Conf., Josiah Macy Foundation.

Bridger, W. H., 1961, Sensory habituation and discrimination in the human neonate, *Am. J. Psychiatry* **117:**991.

Bridger, W. H., and Reiser, M. F., 1959, Psychophysiologic studies of the neonate: An approach toward the methodological and theoretical problems involved, *Psychosomat. Med.* **21:**265–276.

Bruner, J. S., 1969, Eye, hand and mind, in: *Studies in Cognitive Development* (D. Elkind and J. H. Flavell, eds.), p. 223, Oxford University Press, New York.

Bullowa, M. (ed.), 1978, *Before Speech: The Beginnings of Communication*, Cambridge University Press, London (in press).

Chavez, A., Martinez, C., and Yaschine, T., 1974, Nutrition, mother–child relations and behavioral development in the young child from a rural community (unpublished manuscript).

Clifton, R. K., 1974, Cardiac conditioning and orienting in the human infants, in: *Cardiovascular Physiology* (P. Obrist, A. H. Black, J. Brener, and L. DiCara, eds.), p. 479, Aldine Press, Chicago.

David, M., and Appell, G., 1963, A study of nursing care and nurse–infant interaction, in: *Determinants of Infant Behavior*, Vol. 2 (B. M. Foss, ed.), pp. 121–136, Methuen, London.

Dayton, G. O., Jr., Jones, M. H., Ain, P., Rawson, R. A., Steele, B., and Rose, M., 1964, Developmental study of coordinated eye movements in the human infant, *Arch. Ophthalmol.* **71:**865.

Dubowitz, L., and Dubowitz, V., 1970, Clinical assessment of gestational age in the newborn infant, *J. Pediat.* **77:**1.

Eisenberg, R. B., 1965, Auditory behavior in the human neonate: Methodologic problems, *J. Audiol. Res.* **5:**159.

Ellingson, R. V., 1960, Cortical electrical responses to visual stimulation in the human infant, *Electroencephalogr. Clin. Neurophysiol.* **12:**663.

Engen, T., Lipsitt, L. P., and Kaye, H., 1963, Olfactory responses and adaptation in the human neonate, *J. Comp. Physiol. Psychol.* **56:**73.

Escalona, S. K., 1968, *Roots of Individuality*, Aldine, Chicago.

Fantz, R., 1961, The origin of form perception, *Sci. Amer.* **204:**459.

Freedman, D. G., 1968, An evolutionary framework for behavioral research, in: *Progress in Human Behavior Genetics* (S. G. Vandenberg, ed.), pp. 1–6, Johns Hopkins Press, Baltimore.

Gesell, A., 1945, *The Embryology of Behavior*, Harper, New York.

Gewirtz, H. B., and Gewirtz, J. L., 1969, Caretaking settings, background events, and behavior differences in four Israeli child-rearing environments: some preliminary trends, in: *Determinants of Infant Behavior*, Vol. III (B. M. Foss, ed.), pp. 229–253, Methuen, London.

Goldberg, S., 1971, Some stimulus properties of the human infant (unpublished manuscript).

Gorman, J. J., Cogan, D. G., and Gellis, S. S., 1957, An apparatus for grading the visual acuity of infants on the basis of opticokinetic nystagmus, *Pediatrics* **19:**1088.

Goron, C., 1975, Form perception, innate form preference and visually-mediated head-turning in the human neonate, paper presented at Society for Research in Child Development, Denver, Colorado.

Graham, F. K., and Clifton, R. K., 1966, Heart rate change as a component of the orienting response, *Psychol. Bull.* **65:**305–320.

Gryboski, J. D., 1965, The swallowing mechanism of the neonate: Esophageal and gastric motility, *Pediatrics* **35**:445.

Gunther, M., 1961, Infant behavior at the breast, in: *Determinants of Infant Behavior,* Vol. I (B. M. Foss, ed.), pp. 37–40, Methuen, London.

Haith, M. M., Kessen, W., and Collens, D., 1969, Response of the human infant to level of complexity of intermittent visual movement, *J. Exp. Child Psychol.* **7**:52.

Halverson, H. M., 1944, Mechanisms of early feeding, *J. Genet. Psychol.* **64**:185.

Heider, G. M., 1966, Vulnerability in infants and young children, *Genet. Psychol. Monogr.* **73**:1.

Hershenson, M., 1964, Visual discrimination in the human newborn, *J. Comp. Physiol. Psychol.* **58**:270.

Hooker, D., 1969, *The Prenatal Origin of Behavior,* 3rd ed., Hafner, New York.

Hrbek, A., and Mares, P., 1964, Cortical evoked responses to visual stimulation in full term and premature infants, *EEG Clin. Neurophysiol.* **16**:575.

Hutt, C., and Ounsted, C., 1966, The biological significance of gaze aversion with particular reference to the syndrome of infantile autism, *Behav. Sci.* **11**:346.

Hutt, C., Von Bernuth, H., Lenard, H. G., Hutt, S. J., and Prechtl, H. F. R., 1968, Habituation in relation to state in the human neonate, *Nature* **220**:618.

Jensen, K., 1932, Differential reactions to taste and temperature stimuli in newborn infants, *Genet. Psychol. Monogr.* **12**:361.

Karelitz, S., and Fisichelli, V., 1962, The cry thresholds of normal infants and those with brain damage, *J. Pediatr.* **61**:679.

Kaye, K., 1967, Infant sucking and its modification, in: *Advances in Child Development and Behavior,* Vol. III (L. P. Lipsitt and C. C. Spiker, eds.), pp. 1–52, Academic Press, New York.

Kaye, K., and Brazelton, T. B., 1971, Mother–infant interaction in the organization of sucking, presented at Society for Research in Child Development, Minneapolis.

Kearsley, R. B., 1973, The newborn's response to auditory stimulation: A demonstration of orienting and defensive behavior, *Child Dev.* **44**:582.

Kessen, W., Haith, M. M., and Salapatek, P. H., 1970, Human infancy: A bibliography and guide, in: *Carmichael's Manual of Child Psychology,* Vol. 1 (P. Mussen, ed.), p. 287, Wiley, New York.

Kleitman, N., 1963, *Sleep and Wakefulness,* 2nd ed., Chicago University Press, Chicago.

Korner, A. F., and Thoman, E. B., 1972, Relative efficacy of contact and vestibular-proprioceptive stimulation in soothing neonates, *Child Dev.* **43**:443.

Kron, R. E., Stain, M., and Goddard, K. E., 1963, A method of measuring sucking behavior of newborn infants, *Psychosomat. Med.* **25**:181.

Lacey, J. I., 1956, The evaluation of autonomic response. Toward a general solution, *Ann. N.Y. Acad. Sci.* **67**:123–164.

Lacey, J. I., and Lacey, B. C., 1962, The law of initial value in the longitudinal study of autonomic constitution: Reproducibility of autonomic responses and response patterns over a four-year interval, *Ann. N.Y. Acad. Sci.* **98**:1257–1290 (article 4); 1322–1326.

Lester, B. M., 1975, Cardiac habituation of the orienting response to an auditory signal in infants of varying nutritional status, *Dev. Psychol.* **11**:432.

Lewis, M., 1972, State as an infant–environment interaction: An analysis of mother–infant behavior as a function of sex, *Merrill-Palmer Q.* **18**:95.

Lind, J., Wasz-Hockert, O., Vuorenkoski, F., Partanen, T., Theorell, K., and Valanne, E., 1966, Vocal responses to painful stimuli in newborn and young infants, *Ann. Paediatr. Fenn.* **12**:55–63.

Lipsitt, L. P., 1967, Learning in the human infant, in: *Early Behavior, Comparative and Behavioral Approaches* (H. N. Stevenson, H. L. Rheingold, and E. Hess, eds.), pp. 225–247, Wiley, New York.

Lipton, E. L., Steinschneider, A., and Richmond, J., 1966, Auditory sensitivity in the infant: effect of intensity on cardiac and motor responsivity, *Child Dev.* **37**:233.

Lytton, H., 1971, Observational studies of parent–child interaction: A methodological review, *Child Dev.* **42**:3.

MacFarlane, J. A., 1975, Olfaction in the development of social preferences in the human neonate, in: *Parent–Infant Interaction,* Ciba Foundation Symposium 33, pp. 103–113, Elsevier, New York.

Moss, H., 1967, Sex, age, and state as determinants of mother–infant interaction, *Merrill-Palmer Q.* **13**:19.

Peiper, A., 1963, *Cerebral Function in Infancy and Childhood,* Consultants Bureau, New York.

Porges, S. W., 1974, Indices of newborn attentional responsivity, *Merrill Palmer Q.* **20**:231.

Prechtl, H. F. R., 1963, The mother–child interaction in babies with minimal brain damage, in: *Determinants of Infant Behavior,* Vol. II (B. M. Foss, ed.), pp. 53–59, Methuen, London.

Prechtl, H., and Beintema, D., 1964, *The Neurological Examination of the Full-term Newborn Infant,* Little Club Clinics In Developmental Medicine No. 12, William Heinemann Medical Books, London.

Preyer, W., 1888, *The Mind of the Child,* Appleton, New York.

Rebelsky, F. G., 1971, Infancy in two cultures, in: *Child Development and Behavior* (F. G. Rebelsky and E. Dorman, eds.), Knopf, New York.

Rheingold, H. L., Gewirtz, J. L., and Ross, A. W., 1959, Social conditioning of vocalizations in the infant, *J. Comp. Physiol. Psychol.* **52**:68.

Richards, M. P. M., 1971, Social interaction in the first two weeks of human life, *Psychiatria, Neurologia, Neurochirurgia* **14**:35.

Richards, M. P. M., and Bernal, J. F., 1972, An observational study of mother–infant interaction, in: *Ethological Studies of Child Behavior* (N. J. Blurton-Jones, ed.), pp. 175–197, Cambridge University Press, London.

Robson, K. S., 1967, The role of eye-to-eye contact in maternal–infant attachment, *J. Child Psychol. Psychiatry* **8**:13.

Salapatek, P. H., and Kessen, W., 1966, Visual scanning of triangles by the human newborn, *J. Exp. Child Psychol.* **6**:607.

Sameroff, A. J., 1968, The components of sucking in the human newborn, *J. Exp. Child Psychol.* **6**:607.

Scanlon, J. W., 1974, Obstetric anesthesia as a neonatal risk factor in normal labor and delivery, *Clinics Perinatal.* **1**:465.

Schaffer, H. R., and Emerson, P. E., 1964, Patterns of response to physical contact in early development, *J. Child Psychol. Psychiatry* **5**:1.

Sontag, L. W., and Richardo, T. W., 1938, On intrauterine conditioning, studies in fetal behavior: Fetal heart rate as a behavioral indicator, Monograph, Society for Research in Child Development 3, No. 4.

Stechler, G., and Latz, E., 1966, Some observations on attention and arousal in the human infant, *J. Acad. Child Psychiatry* **5**:517.

Steinschneider, A., 1973, Prolonged apnea and the sudden infant death syndrome: Clinical and laboratory observations, *Pediatrics* **50**:646.

Strauss, M. E., Lessen-Fireston, J., Starr, R., and Ostrea, E., 1976, Behavior of narcotics addicted newborns, *Child Dev.* (in press).

Stuart, H. C., and Stevenson, S. S., 1959, Physical Growth and Development in: *Textbook of Pediatrics* (W. E. Nelson, ed.), p. 32, W. B. Saunders, Philadelphia.

Tronick, E., and Clanton, C., 1971, Infant looking patterns, *Vision Res.* **2**:1479.

Tronick, E., and Scanlon, J., 1978, Habituation to a localized somatosensory stimulus in the neonate, *Infant Behavior and Development* (submitted).

Tronick, E., Wise, S., Als, H., Adamson, L., Scanlon, J., and Brazelton, T. B., 1976, Regional obstetric anesthesia and newborn behavior: Effect over the first ten days of life, *Pediatrics* **58**:94.

Tronick, E., Als, H., and Brazelton, T. B., 1977, Mutuality in mother–infant interaction, *J. Communication* **27**(2):74.

Tronick, E., Als, H., and Adamson, L., 1978a, Structure of early face-to-face communicative interactions, in: *Before Speech: The Beginnings of Communication* (M. Bullowa, ed.), Cambridge University Press, London (in press).

Tronick, E., Als, H., Adamson, L., Wise, S., and Brazelton, T. B., 1978b, The infant's response to entrapment between contradictory messages in face-to-face interaction, *J. Amer. Acad. Child Psychiatry* **17**(1):1.

Wagner, I. F., 1937, The establishment of a criterion of depth of sleep in the newborn infant, *J. Genet. Psychol.* **51**:17.

Wasz-Hockert, O., Lind, J., Vuorenkoski, V., Partanen, T., and Valanne, E., 1968, The infant cry, *Clin. Dev. Med.* **29**.

Weitzman, E. D., Fishbein, W., and Graziani, L., 1965, Auditory evoked responses obtained from the scalp EEG of the full term infant during sleep, *Pediatrics* **35**:458.

Windle, W., 1950, Reflexes of mammalian embryos and fetuses, in: *Genetic Neurology* (P. Weiss, ed.), pp. 214–222, University of Chicago Press, Chicago.

Wolff, P. H., 1969, The natural history of crying and other vocalizations in early infancy, in: *Determinants of Infant Behavior,* Vol. 4 (B. M. Foss, ed.), p. 81, Methuen, London.

Wolff, P. H., 1968, The serial organization of sucking in the young infant, *Pediatrics* **42**:943.

Wolff, P. H., 1959, Observations on newborn infants, *Psychosomat. Med.* **21**:110.

Wolff, P. H., Matsumiya, Y., Abroms, I. F., Van Velzer, C., and Lombroso, C. T., 1974, The effect of white noise on the somatosensory evoked response in sleeping newborn infants, *Electroencephalogr. Clin. Neurophysiol.* **37**:269.

VII
Nutrition

10

Nutrition and Growth in Infancy

RENATE L. BERGMANN and KARL E. BERGMANN

1. Introduction

Whereas no direct information on intrauterine growth and nutrition can be obtained, postnatal growth and nutrition and their interactions are accessible to study by exact measurements.

There seems to be little question about the causative factor in this relationship; growth appears to be affected by nutrient intake: In the presence of dequate nutrition, a healthy infant will grow normally; in malnutrition and overnutrition, failure to thrive and excessive weight gain appear to be caused by inappropriate food availability.

It seems that an infant will voluntarily take the amount of food that will allow him to grow adequately. Are actual body size or target size or even a target body composition and its changes during maturation the driving forces of food intake? We have to treat both food intake and growth as independent entities until we know more about the causative factors in that relationship.

We are on safe ground if we consider the growth of normal infants of a well-defined population; first, breast-fed infants and then infants fed various formulas *ad libitum*. Model considerations on the interrelationship of growth and nutritional requirements follow. Finally, we shall comment on the impact of some single nutritional factors on growth disorders.

2. Food Intake and Growth of Normal Infants under Controlled Conditions

The considerations to follow will refer mainly to studies by Fomon and associates between 1966 and 1975. Normal full-term Caucasian infants with birth

---

RENATE L. BERGMANN and KARL E. BERGMANN • Department of Pediatrics, J. W. Goethe University, Frankfurt/Main, West Germany.

weights of 2500 g or more were enrolled between ages 6 and 9 days in prospective studies of food intake and growth. Nearly all the subjects were children of students or younger staff members of the University of Iowa, living continuously at home with their parents. The infants were either breast-fed or received ready-to-feed infant formulas *ad libitum,* and in both instances adequate intake of vitamins and minerals was provided. Specified strained foods in limited amounts were introduced from age 28 days. Daily amounts of formula ingested and consumption of strained food were calculated from the weight of bottles, cans, or jars before and after use.

From careful determinations of length and weight (Fomon, 1974*b*) at regular intervals (ages 8, 14, 28, 42, 56, 84, 112, and in one study also 140 and 168 days), daily gains in weight and length could be computed.

2.1. Growth of Breast-Fed Infants

This study of Fomon *et al.* (1970) comprised 149 infants, 83 males and 66 females, of whom 70% completed a 4-month observation period. Except for one formula feeding per day (Similac®, 67 kcal/100 ml), which not all parents elected to give, the infants were breast-fed.

Figure 1 indicates the 10th, 50th, and 90th percentiles of weight and length of the 58 male and 46 female infants who completed the 4-month study. With one exception (10th weight percentile at age 28 days) all percentile values for weight and length were higher in male than in female infants during the whole observation period. There appears to be an unexpectedly broad interval between the 50th and 10th weight percentile in male infants lasting from age 8 to 56 days, and in female infants from age 42 to 112 days, which will be discussed later (Section 2.2).

Table I provides the 10th, 50th, and 90th percentile values of increments in weight and length for specified age intervals. Rates of growth may change significantly in any individual between short-term observation periods (Figure 2), but there is a tendency to equilibrate for preceding high or low gains during successive growth periods. This is demonstrated by the following phenomenon: summation of all the short-term gains at the 10th percentile results in a lower value than the 10th percentile gain for the total (8 through 112 days) intervals, and vice versa: Summation of the 90th percentile gains during successive short-term interval gives a higher result than the 90th percentile gain for the total observation period.

Gain in weight was larger and increase in length slightly larger in boys than in girls. Weight gain per unit of gain in length was greater in males than in females between 8 and 112 days of age.

For two major reasons nutrient intake of these breast-fed infants was not evaluated: The natural environment could have been disturbed by weighing infants precisely before and after nursing and by collecting samples of breast milk. Also composition of breast milk is known to change between feedings and even during nursing (Tarján *et al.,* 1965; Hall, 1975), prohibiting any accurate determination of nutrient consumption at the breast.

For an insight into the interplay between growth and nutrient intake, it is necessary to rely on observation of bottle-fed infants, whose food consumption and food composition were carefully recorded during the corresponding growth intervals.

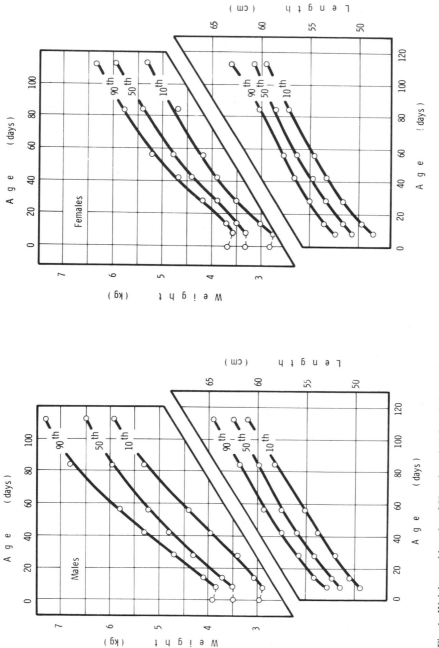

Fig. 1. Weights and lengths of 58 male and 46 feamle breast-fed infants during the first 112 days of life; 10th, 50th and 90th percentiles. (From Fomon *et al.*, 1970.)

RENATE L. BERGMANN
and KARL E. BERGMANN

Table Ia.   Daily Changes in Weight (g/day) of Male Breast-Fed Infants (Fomon et al., 1970)

Age (days)	Percentile	Day 14	28	42	56	84	112
8	10	13	24	28	28	24	23
	50	40	39	39	37	33	32
	90	59	54	50	49	44	37
14	10		24	27	28	25	22
	50		40	39	37	33	31
	90		57	52	49	42	36
28	10			26	23	22	21
	50			40	36	31	29
	90			52	48	42	36
42	10				17	20	18
	50				35	29	26
	90				48	40	36
56	10					16	18
	50					26	25
	90					39	35
84	10						14
	50						22
	90						34

Table Ib.   Daily Changes in Weight (g/day) of Female Breast-Fed Infants (Fomon et al., 1970)

Age (days)	Percentile	Day 14	28	42	56	84	112
8	10	12	22	24	23	21	20
	50	32	33	33	31	28	26
	90	50	48	45	41	35	32
14	10		20	25	23	20	20
	50		36	33	31	28	26
	90		48	42	40	35	31
28	10			17	18	18	17
	50			33	30	27	25
	90			43	38	33	30
42	10				15	16	16
	50				24	25	23
	90				37	31	28
56	10					14	15
	50					24	22
	90					31	28
84	10						11
	50						20
	90						29

2.2. Growth, Body Size, and Energy Consumption of Infants Fed Milk-Based Formulas

The considerations that follow are again evolved from data of Fomon and associates (1971*b*) obtained from 142 infants: 65 males and 77 females. Each infant was fed one of nine milk-based formulas *ad libitum* between age 8 and 112 days. All of the formulas provided 67 kcal/100 ml, containing 7–13% of calories as protein, 41–48% as carbohydrate, and 38–52% as fat.

Table Ic.   Daily Changes in Length (mm/day) of Male Breast-Fed Infants
(Fomon et al., 1970)

Age (days)	Percentile	Day			
		42	56	84	112
8	10	1.0	1.0	1.0	1.0
	50	1.3	1.2	1.1	1.1
	90	1.6	1.4	1.3	1.2
14	10		0.9	0.9	1.0
	50		1.2	1.1	1.1
	90		1.4	1.2	1.2
18	10			0.9	0.9
	50			1.1	1.0
	90			1.3	1.2
42	10			0.8	0.9
	50			1.0	1.0
	90			1.2	1.1
56	10				0.8
	50				1.0
	90				1.1

Table Id.   Daily Changes in Length (mm/day) of Female Breast-Fed Infants
(Fomon et al., 1970)

Age (days)	Percentile	Day			
		42	56	84	112
8	10	1.1	1.0	0.9	0.9
	50	1.3	1.2	1.1	1.0
	90	1.5	1.4	1.3	1.2
14	10		0.9	0.9	0.8
	50		1.1	1.0	1.0
	90		1.3	1.2	1.2
28	10			0.8	0.8
	50			1.0	0.9
	90			1.2	1.1
42	10			0.7	0.7
	50			1.0	0.9
	90			1.1	1.1
56	10				0.7
	50				0.9
	90				1.1

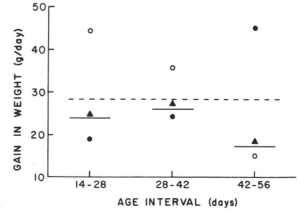

Fig. 2. Patterns of weight gain of three individual infants (●,○,▲) in relation to the 10th percentiles of weight gain in 58 male breast-fed infants; solid lines represent three consecutive 14-day intervals, dashed line indicates entire 42-day interval. (From Fomon *et al.*, 1970.)

Table II. Percentile Values for Energy Intake of Infants Fed Milk-Based Formulas (Fomon et al., 1971b)

Percentile	Age interval (days)					
	8–13	14–27	28–41	42–55	56–83	84–111
Males						
Calorie intake (kcal/day)						
10th	292	362	408	456	455	503
50th	383	466	520	531	568	617
90th	496	580	665	701	700	706
Calorie intake (kcal/kg/day)						
10th	82	95	91	91	83	81
50th	111	121	116	108	100	96
90th	142	143	140	133	119	106
Females						
Calorie intake (kcal/day)						
10th	279	342	382	386	437	463
50th	374	440	470	489	500	545
90th	500	508	573	577	612	654
Calorie intake (kcal/kg/day)						
10th	82	86	90	83	87	82
50th	113	117	111	108	97	94
90th	143	136	136	125	114	109

2.2.1. Energy Consumption and Body Weight

Rather large variations were observed in the intakes of the infants fed *ad libitum*, both if expressed on an absolute basis (kcal/day) or if related to body mass (kcal/kg/day). Table II provides the 10th, 50th, and 90th percentile values of energy intake during successive age intervals: Energy intake increased with age. Per unit of body weight, intake was greatest during the interval 14 through 27 days of age,

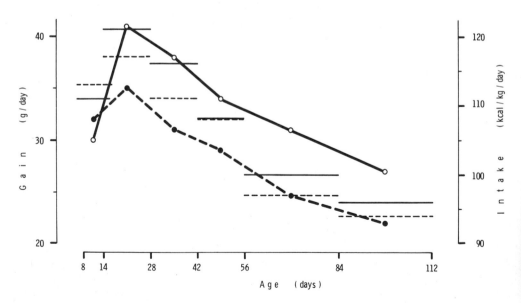

Fig. 3. Weight gain and energy intake by 65 male and 77 female infants fed milk-based formulas (Fomon *et al.,* 1971b). The horizontal lines indicate the 50th percentile values of energy intake during consecutive age intervals, the graph is connecting the corresponding 50th percentile values of weight gain. Solid lines pertain to males, dashed lines to females.

Table III. Relation of Calorie Intake to Body Weight during Consecutive Age
Intervals (Fomon et al., 1971b)

Age interval (days)	Regression equation[a]	Correlation coefficient	Residual variance (s_e^2)
Males			
8–13	$Y = 86\,x + 90$	0.33	88
14–27	$Y = 108\,x + 52$	0.42	89
28–41	$Y = 107\,x + 48$	0.42	97
42–55	$Y = 89\,x + 114$	0.44	84
56–83	$Y = 103\,x - 14$	0.60	74
84–111	$Y = 75\,x + 128$	0.61	63
Females			
8–13	$Y = 25\,x + 291$	0.13	80
14–27	$Y = 24\,x + 337$	0.14	65
28–41	$Y = 42\,x + 292$	0.21	77
42–55	$Y = 70\,x + 151$	0.37	73
56–83	$Y = 69\,x + 147$	0.48	55
84–111	$Y = 66\,x + 158$	0.44	67

[a] Y is calorie intake (kcal/day) and x is body weight (kg).

decreasing thereafter (Figure 3). Mean daily intakes (kcal/day) were significantly greater in males than in females. Energy consumption, expressed in kcal/kg/day, was also significantly greater in boys than in girls if the comparison was restricted to groups receiving the same type of formula. Caloric intake was therefore related to age, sex, and body weight of the infant (Table III).

2.2.2. Energy Consumption and Growth

The correlation between gain in weight and caloric intake was statistically significant for both sexes during each age interval (Table IV). Per unit of energy consumption, mean daily gain in weight and length was significantly greater for age

Table IV. Relation of Weight Gain to Calorie Intake during Consecutive Age Intervals
(Fomon et al., 1971b)

Age interval (days)	Regression equation[a]	Correlation coefficient	Residual variance (s_e^2)
Males			
8–13	$Y = 0.058\,x + 8$	0.36	190
14–27	$Y = 0.067\,x + 10$	0.66	55
28–41	$Y = 0.046\,x + 15$	0.51	70
42–55	$Y = 0.079\,x - 9$	0.61	91
56–83	$Y = 0.062\,x - 6$	0.70	35
84–111	$Y = 0.044\,x - 1$	0.49	39
Females			
8–13	$Y = 0.008\,x + 28$	0.06	142
14–27	$Y = 0.065\,x + 8$	0.48	59
28–41	$Y = 0.051\,x + 6$	0.49	51
42–55	$Y = 0.062\,x - 0.7$	0.60	43
56–83	$Y = 0.056\,x - 3$	0.57	25
84–111	$Y = 0.039\,x + 0.7$	0.39	48

[a] Y is gain in weight (g) and x is calorie intake (kcal).

RENATE L. BERGMANN
and KARL E. BERGMANN

Table Va. Size of Breast-Fed Infants and Infants Fed Milk-Based Formulas: Percentile Values
(Fomon et al., 1971b)

Age (days)	Percentile	Weight (g)			
		Males		Females	
		Fed milk-based formulas (65)[a]	Breast-fed (58)	Fed milk-based formulas (77)	Breast-fed (46)
Birth	10	2870	2947	2816	2798
	25	3125	3216	3135	2994
	50	3410	3470	3350	3188
	75	3705	3642	3615	3520
	90	3804	3916	3964	3730
8	10	2878	2881	2903	2746
	25	3248	3216	3274	2978
	50	3502	3459	3401	3260
	75	3688	3642	3698	3465
	90	3906	3839	4031	3555
14	10	3132	3089	3124	2990
	25	3399	3375	3426	3199
	50	3665	3681	3575	3500
	75	3876	3881	3894	3630
	90	4089	4065	4127	3698
28	10	3636	3406	3605	3511
	25	3970	3960	3824	3763
	50	4235	4271	4134	3935
	75	4470	4507	4440	4148
	90	4761	4692	4593	4230
42	10	4250	3962	4028	3930
	25	4506	4466	4275	4136
	50	4828	4808	4579	4360
	75	5080	5040	4915	4600
	90	5437	5312	5060	4692
56	10	4685	4387	4441	4243
	25	4967	4933	4603	4498
	50	5255	5250	4953	4758
	75	5599	5550	5308	4959
	90	5918	5791	5500	5161
84	10	5292	5320	5000	4707
	25	5740	5589	5388	5161
	50	5973	5955	5675	5424
	75	6501	6236	6027	5654
	90	7015	6773	6357	5800
112	10	5949	5933	5568	5282
	25	6305	6186	5976	5582
	50	6697	6482	6288	5987
	75	7253	7051	6669	6204
	90	7993	7306	7018	6428

[a]Values in parentheses are number of subjects.

interval 8–55 days than for interval 56–111 days. For the total interval (8–111 days) mean gain in weight per unit of energy intake was significantly greater in boys than in girls. Mean gain in length per unit of calorie intake was nearly identical for males and females.

Similar relationships have been described by Ashworth *et al.* (1968) and

*Table Vb. Size of Breast-Fed Infants and Infants Fed Milk-Based Formulas: Percentile Values
(Fomon et al., 1971b)*

Age (days)	Percentile	Length (cm)			
		Males		Females	
		Fed milk-based formulas (65)[a]	Breast-fed (58)	Fed milk-based formulas (77)	Breast-fed (46)
Birth	10				
	25				
	50				
	75				
	90				
8	10	48.9	49.5	48.8	48.5
	25	50.1	50.4	49.9	49.5
	50	51.7	51.2	51.1	50.7
	75	52.6	52.3	52.0	51.4
	90	53.4	52.8	53.2	52.2
14	10	49.8	50.6	49.4	49.7
	25	50.9	51.5	50.7	50.4
	50	52.6	52.4	52.0	51.6
	75	53.4	53.3	53.0	52.2
	90	54.1	54.2	54.1	53.4
28	10	51.6	52.0	51.5	51.6
	25	52.8	53.3	52.4	52.0
	50	54.6	54.2	53.8	53.3
	75	55.3	55.1	55.1	53.9
	90	55.9	55.7	56.0	55.0
42	10	53.0	53.8	53.0	53.2
	25	54.8	54.8	54.0	53.7
	50	56.2	56.0	55.2	54.7
	75	57.3	56.8	56.5	55.5
	90	57.7	57.5	57.6	56.6
56	10	54.3	55.2	54.6	54.6
	25	56.3	56.3	55.3	55.1
	50	57.6	57.6	56.8	56.3
	75	58.7	58.4	58.1	56.9
	90	59.3	59.4	59.2	57.7
84	10	57.8	58.3	57.0	57.2
	25	59.3	59.0	58.2	57.7
	50	60.8	59.9	59.5	58.6
	75	61.6	61.0	60.8	59.7
	90	62.6	61.9	61.7	60.2
112	10	60.1	61.1	59.6	59.7
	25	61.4	61.8	60.3	60.0
	50	63.0	62.6	61.9	60.9
	75	64.5	63.8	63.2	62.2
	90	65.4	64.2	64.0	63.1

[a]Values in parentheses are number of subjects.

Rutishauser and McCance (1968), who demonstrated positive correlations between rate of weight gain and energy intake in children recovering from severe malnutrition. Chávez *et al.* (1973) concluded from a study on milk consumption and growth in breast-fed infants from a poor rural community, that male infants utilized nutrients better than did female infants.

2.2.3. Comparison of Size and Gain between Breast-Fed and Formula-Fed Infants

Percentile values for weight and length of breast-fed and bottle-fed infants are presented in Table V and Figure 4. Although breast-fed boys had a tendency toward greater birth weight if compared to bottle-fed ones, at age 112 days their 50th and 90th percentile values for weight were smaller by 215 g and 687 g, respectively, the 10th percentile values being similar. Except for a larger dispersion of values in bottle-fed infants, lengths were not different in the two groups. In female infants, differences in weight between the formula-fed and breast-fed series at 112 days were larger than in males, and even differences in length became apparent. The 10th, 50th, and 90th percentile values for gain in weight and length of formula-fed infants are provided in Table VI. There was a slight tendency toward greater gains in the formula-fed group if compared to the breast-fed group (Table I). This is in accordance with the findings of Mellander *et al.* (1959), who described higher weight gains in Sweden from age 3 months if the formula-fed infants were compared to breast-fed infants. Differences in weight gain have also been reported to be significantly greater in bottle-fed than in breast-fed infants for the first 6 months of life by Steward and Westropp (1953), and by Belton *et al.* (1977) in English infants. The study of Dugdale (1971) in Malayan infants did not reveal similar differences before age 20 weeks, and Nitzan and Schönfeld (1976) did not observe differences in weight gain between breast- and bottle-fed infants in Israel between the ages 4 and 8 weeks. Although weight gain between breast- and bottle-fed infants in England during the first 6 weeks of life was not different in a study by Oakley (1977), increase of skinfold thickness was significantly higher in male as well as in female breast-fed infants compared to bottle-fed infants. This may be at variance with the calculations of Fomon (1974*b*), revealing higher gains in weight for length in bottle-fed versus breast-fed infants, although it should be noted that Fomon is only referring to the first 4 months of life.

2.3. Considerations on the Interplay of Food Intake and Growth of Breast-Fed and Formula-Fed Infants

There may be substantial differences between breast- and bottle-feeding (Fomon, 1971; Jelliffe and Jelliffe, 1976). These could possibly explain why the 10th weight percentile of breast-fed infants appears so unexpectedly low (Section 2.1): Breast-feeding techniques prevailing in developed countries, e.g., daily nursing rhythms, may leave some of the children unsatisfied (Blurton Jones, 1972; Richards, 1975). On the other hand, formula-fed infants may be encouraged to drain the last drop from the bottle (Fomon, 1971), which would explain the weight distribution skewed toward heavier infants.

Despite these differences, similarity of growth performance in breast- and formula-fed infants by far outranks the deviations: Provided that infants are offered nutritionally adequate food upon demand, they will be able to choose an amount that covers their requirement for maintenance and growth. On the basis of the data on breast-fed infants (Fomon *et al.*, 1970) and of the data on infants fed milk-based formulas (Fomon *et al.*, 1971*b*), the sex-specific relationships between pairs of variables have been analyzed, including body weight at age 8 days, weight at age 112 days, length at age 8 days, length at age 112 days, and weight gain and length gain during the age interval 8 through 112 days.

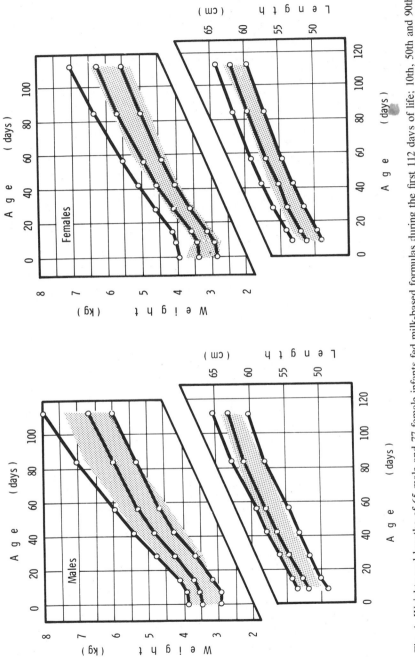

Fig. 4. Weights and lengths of 65 male and 77 female infants fed milk-based formulas during the first 112 days of life; 10th, 50th and 90th percentiles. The shaded areas include the 10th to 90th percentile values of weight and length in breast-fed infants. (From Fomon *et al.*, 1971*b*.)

Table VIa. Daily Changes in Weight (g/day) of Male Infants Fed Milk-Based Formulas (Fomon et al., 1971b)

Age (days)	Percentile	Day 14	28	42	56	84	112
8	10	12	29	30	29	26	25
	50	30	38	38	36	34	32
	90	54	51	49	48	46	43
14	10		30	33	30	26	25
	50		41	40	38	35	32
	90		54	51	50	48	44
28	10			29	26	25	23
	50			38	36	33	30
	90			56	52	46	43
42	10				22	22	21
	50				34	32	30
	90				53	46	41
56	10					20	19
	50					31	28
	90					42	38
84	10						15
	50						27
	90						35

Table VIb. Daily Changes in Weight (g/day) of Female Infants Fed Milk-Based Formulas (Fomon et al., 1971b)

Age (days)	Percentile	Day 14	28	42	56	84	112
8	10	15	24	24	23	21	20
	50	32	34	34	33	30	27
	90	47	46	42	39	37	34
14	10		25	24	23	21	20
	50		35	34	32	30	28
	90		49	42	39	37	34
28	10			20	21	20	20
	50			31	30	28	26
	90			41	39	37	34
42	10				20	19	19
	50				29	27	25
	90				49	35	32
56	10					18	17
	50					25	23
	90					34	32
84	10						14
	50						22
	90						31

There is a strong relationship between body size at age 8 days and size at age 112 days reflected by the correlation coefficients computed in Table VII. On the other hand, when analyzing the relationship between weight at age 8 days and weight gain during the age interval 8 through 112 days and between length at age 8 days and length gain during this interval, a tendency toward counterbalance could be recognized (Table VIII). Whether this tendency is emerging from the self-regulating power of the organism or from a bounded food supply to the infant

*Table VIc.   Daily Changes in Length (mm/day) of Male Infants Fed Milk-
Based Formulas (Fomon et al., 1971b)*

Age (days)	Percentile	Day 42	Day 56	Day 84	Day 112
8	10	1.1	1.1	1.0	1.0
	50	1.4	1.3	1.2	1.1
	90	1.6	1.5	1.4	1.3
14	10		1.0	1.0	1.0
	50		1.2	1.2	1.1
	90		1.4	1.3	1.2
28	10			1.0	0.9
	50			1.2	1.1
	90			1.3	1.3
42	10			0.9	0.8
	50			1.1	1.0
	90			1.3	1.2
56	10				0.8
	50				1.0
	90				1.2

*Table VId.   Daily Changes in Length (mm/day) of Female Infants Fed Milk-
Based Formulas (Fomon et al., 1971b)*

Age (days)	Percentile	Day 42	Day 56	Day 84	Day 112
8	10	1.0	1.0	1.0	0.9
	50	1.3	1.2	1.1	1.0
	90	1.5	1.4	1.3	1.2
14	10		1.0	0.9	0.9
	50		1.2	1.1	1.0
	90		1.4	1.2	1.1
28	10			0.9	0.8
	50			1.0	1.0
	90			1.2	1.1
42	10			0.8	0.8
	50			1.0	1.0
	90			1.2	1.1
56	10				0.7
	50				0.9
	90				1.1

*Table VII.   Relations between Body Weight at Age 8 and at Age 112 Days and between Body
Length at Age 8 and at Age 112 Days in Breast-Fed and Formula-Fed Males and Females
(Calculations on the Basis of Data by Fomon et al., 1970, 1971b)*

	Breast-fed		Formula-fed	
	Males	Females	Males	Females
Weight 8 day/112 day				
Number of subjects	58	46	65	78
Correlation coefficient	0.478	0.357	0.388	0.374
Significance level	$P < 0.001$	$P < 0.05$	$P < 0.01$	$P < 0.001$
Length 8 day/112 day				
Number of subjects	58	46	65	78
Correlation coefficient	0.799	0.625	0.815	0.808
Significance level	$P < 0.001$	$P < 0.001$	$P < 0.001$	$P < 0.001$

Table VIII. Relations between Body Weight or Length at Age 8 Days and Increment in Body Weight or Length, Respectively, During the Age Interval 8 through 112 Days in Breast-Fed and Formula-Fed Males and Females (Calculations on the Basis of Data by Fomon et al., 1970, 1971b)

| | Breast-fed | | Formula-fed | |
	Males	Females	Males	Females
Weight 8 day/weight gain 8–112 day				
Number of subjects	58	46	65	78
Correlation coefficient	−0.058	−0.261	−0.242	−0.378
Significance level	N.S.	N.S.	N.S.	$P < 0.001$
Length 8 day/length gain 8–112 day				
Number of subjects	58	46	65	78
Correlation coefficient	−0.206	−0.446	−0.125	−0.228
Significance level	N.S.	$P < 0.01$	N.S.	$P < 0.05$

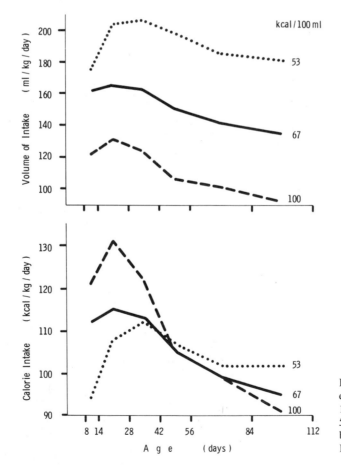

Fig. 5. Mean volume of intake and energy intake of female infants fed 100 kcal/100 ml, 67 kcal/100 ml, or 53 kcal/100 ml milk-based formulas between age 8 and 112 days. (From Fomon, 1974c.)

cannot be recognized by the analysis of this relatively short life-span. Because both groups of infants, breast- and bottle-fed, received their food *ad libitum* and the range of the voluntary intake was rather wide (see above), it can be speculated that this trend is a feature of growth itself. This is consistent with the observation on equilibrating trends in weight and length gain during successive periods (see above) and is in accordance with the findings of Ounsted and Sleigh (1975), who described larger weight gains in small-for-gestational-age infants, and smaller weight gains in large-for-gestational-age infants during the first 2 months of life.

2.4. Food Intake and Growth of Infants Fed Milk-Based Formulas of Different Caloric Concentrations

Further insight into causative factors of the growth–nutrient interplay may be offered by another study of Fomon and associates (1975): 29 female infants received one of two milk-based formulas, which were nearly identical in nutrient content except for carbohydrate and fat. The concentration of these constituents was about twice as high in one formula as in the other, resulting in caloric densities of 100 and 54 kcal/100 ml, respectively. During the first 6 weeks of life, although mean quantity of food consumed by infants fed the low-calorie formula was greater, the mean energy intake was less (Figure 5; Table IX). Thereafter, infants receiving the 54 kcal/100 ml formula consumed nearly twice the volume taken by the other group. Related to body mass, energy intake was 103 ($\pm$12) and 97 ($\pm$9) kcal/kg/day in the low- and high-calorie concentration groups, respectively. For comparison, Figure 5 and Table IX also present the food intake of 77 female infants receiving conventional milk-based formulas of 67 kcal/100 ml energy density (Fomon *et al.*, 1971*b*).

Table IX. Food Intake and Growth by Normal Female Infants Fed 100 kcal/100 ml, 53 kcal/100 ml, or Various Milk-Based Formulas (67 kcal/100 ml) (Fomon et al., 1971b; Fomon, 1974c)

		Formula concentration		
		100 kcal/100 ml (14)[a]	53 kcal/100 ml (15)	67 kcal/100 ml (77)
Volume of intake (ml/kg/day)				
8–41 days	Mean	126	201	164
	SD	17	31	24
42–111 days	Mean	99	186	141
	SD	10	21	14
Calorie intake (kcal/kg/day)				
8–41 days	Mean	126	107	110
	SD	18	16	16
42–111 days	Mean	97	103	95
	SD	9	12	10
Gain in weight (g/day)				
8–41 days	Mean	41.0	29.8	32.9
	SD	10.4	4.9	6.8
42–111 days	Mean	24.0	24.6	25.0
	SD	5.3	4.6	5.4
Gain in length (mm/day)				
8–41 days	Mean	1.23	1.26	1.30
	SD	0.17	0.17	0.20
42–111 days	Mean	0.92	0.88	0.95
	SD	0.15	0.13	0.13

[a]Number of subjects.

RENATE L. BERGMANN
and KARL E. BERGMANN

Gain in weight between age 8 and 42 days was significantly larger in infants fed the high-calorie formula than in infants receiving the low-calorie feeding (Table IX). But corresponding gains thereafter up to age 112 days were nearly identical and similar to the weight gain of bottle-fed girls receiving the 67 kcal/100 ml formulas. Length gains were similar in all feeding groups during both intervals.

These observations appear to indicate that if infants older than 6 weeks of age are fed, *ad libitum,* a rather wide range of energy concentrations, they are able to adjust the energy intake to match their requirements for growth and maintenance.

2.5. Influence of Fat-to-Carbohydrate Ratio in Formula on Energy Intake and Growth

Infants fed *ad libitum* obviously control food intake according to their caloric needs. For further differentiation of the contribution of the major calorific constituents, a study was undertaken by Fomon and associates (1976), where two isocaloric formulas similar in all respects, except for one having a high fat (60% of calories) and a low carbohydrate content (31% of calories) and the other having a high carbohydrate (60% of calories) and a low fat content (31% of energy), were fed to 30 male infants. There was no significant difference in intake of energy and gain in weight and length between the two groups during the various age intervals. If the ratio of dietary energy contributed by carbohydrate and fat may differ to such an extent without affecting food intake and growth, the long-term control of energy intake could well be stronger than the short-term satiety depending on the composition of the single meal (van Itallie *et al.,* 1977).

2.6. Influence of Protein Concentration in Formula on Food Intake and Growth

Results related to the following considerations are taken from investigations by Fomon and associates (1971*a,c*) on food intake, growth, and serum chemical values of 32 male infants, fed one of three formulas similar in composition, except for protein concentration being 1.15 and 1.30 g of cow milk protein per 100 ml and 0.99 g of casein per 100 ml, respectively.

Table X provides data on food consumption, growth, and serum albumin concentration, as an additional variable of protein nutritional status (Waterlow and Alleyne, 1971), for the interval 8–56 days. Data from the study of breast-fed infants (Fomon *et al.,* 1970) are included for comparison. Protein content of samples of breast milk for the ages 14, 56, and 112 days, determined by Jensen *et al.* (1972), were 1.54 ($\pm$ 0.18), 1.09 ($\pm$ 0.18), and 0.87 ($\pm$ 0.26) g/100 ml, respectively (Fomon and Filer, 1974).

There was no apparent trend to compensate for low protein content by an increase of food intake, energy consumption being similar with all regimens; consequently protein intake corresponded to protein concentration in formula. Mean gain in weight by infants receiving the formula with the lowest protein content was not significantly different from any of the other groups. However, between ages 8 and 55 days, their gain in length was significantly lower than in any other group, and serum concentrations of albumin were also significantly less in these infants than in breast-fed males or infants receiving the 1.30 g/100 ml protein formula.

Because all of the formulas were similar except for protein, it seems reasonable

Table X. Growth, Intake of Energy and Protein of Male Infants Fed Formulas with Different Concentrations of Protein, 8 through 56 Days of Age (Fomon et al., 1971b)

	Breast-fed	Formula protein concentration (g/100 ml)		
		1.15	1.30	0.99
Number of subjects	58	10	11	11
Calorie intake		470	539	497
(kcal/day)		(55)[a]	(93)	(72)
Intake of protein		8.7	10.6	7.5
(g/day)		(1.0)	(1.8)	(1.1)
Gain in weight	37.4	35.8	41.1	38.8
(g/day)	(8.4)	(6.7)	(6.1)	(8.1)
Gain in length	1.25	1.24	1.33	1.16
(mm/day)	(0.17)	(0.13)	(0.09)	(0.12)
Number of subjects	36	8	8	5
Albumin	4.14	3.90	4.08	3.58
concentration in serum at 52–58 days (g/100 ml)	(0.34)	(0.29)	(0.41)	(0.47)

[a]Values in parentheses are standard deviations.

to conclude that quantity or quality of protein or both factors were marginally insufficient in the formula with 0.99 g protein per 100 ml for optimal protein synthesis during this period (8–56 days) of most rapid growth, assuming that gain in length and serum albumin concentration are sensitive indicators of protein sufficiency.

During the successive age interval (56–112 days), no significant differences of food intake and growth among the three experimental groups could be observed, indicating that protein intake was adequate to allow for normal growth during that period of slower growth velocity. Serum albumin concentrations were available from only some of the infants and were still lowest in the low-protein-formula group.

Gross protein deficiency in growing animals will result in reduction of appetite, food intake, and severe growth failure with a tendency for the animals to become relatively fat (McCance, 1975; Geist et al., 1972).

To examine the influence on growth of a generous protein concentration in the formula, the published data of two groups of female infants, receiving milk-based formulas of similar caloric density (Fomon et al., 1971b) have been further analyzed. One formula, fed to 14 girls, had a relatively low protein concentration, i.e., 1.3 g/100 ml; 14 girls in the other group received a formula with 2.1 g/100 ml protein. Whereas during the interval 8 through 56 days calorie consumption did not differ in both groups, protein intake was significantly different. No difference could be observed in weight and length gain during this interval between infants fed the formula with low protein content and infants receiving the formula with a high protein concentration (Table XI).

These results are in accordance with studies on premature infants (Goldman et al., 1969; Davidson, 1967; Räihä et al., 1976) and on infants recovering from protein–energy malnutrition (Graham et al., 1963; Ashworth et al., 1968; Rutishau-

RENATE L. BERGMANN
and KARL E. BERGMANN

Table XI. Growth, Intake of Energy and Protein of Female Infants Fed Formulas with Different Concentrations of Protein, 8 through 56 Days of Age (Fomon et al., 1971b)

	Protein concentration in formula (g/100 ml)		
	1.30	2.10	P^a
Number of subjects	15	15	
Calorie intake (kcal/day)	479 (63)[b]	449 (59)	N.S.
Intake of protein (g/day)	9.0 (1.1)	13.9 (1.8)	0.001
Gain in weight (g/day)	32.7 (7.3)	30.0 (5.8)	N.S.
Gain in length (mm/day)	1.17 (0.17)	1.27 (0.15)	N.S.

[a]Statistical analysis by two-sample t test.
[b]Values in parentheses indicate standard deviation.

ser and McCance, 1968), which did not find significant improvement in growth rate by increasing protein concentration in formula above an optimal value. The surplus of protein has to be metabolized and may even be harmful to the infant (see Sections 3.2, 3.3, and 4.2).

2.7. Food Intake and Growth of Infants Fed Skim Milk

In recent years an increasing number of infants in the United States are being fed skim milk from 4–6 months of age onwards, because their parents or physicians believe that prevention of obesity and of atherosclerosis should be instituted early in life. Two investigations were therefore designed by Fomon *et al.* (1977) to study the influence of this type of infant nutrition on food consumption and growth during the age interval 112–167 days. We are referring here to the results of the second study comprising 58 male and female infants, who received either Similac® (Ross Laboratories, Ohio) or a modified skim milk (36 kcal/100 ml) containing 0.23 g of safflower oil per 100 ml to assure an adequate intake of linoleic acid with the addition of fat-soluble vitamins.

Table XII. Food Consumption and Growth of Infants Fed Skim Milk or Similac®, Age Interval 112–167 Days (Fomon et al., 1977)

	Food consumption		Weight gain (g/day)	Length gain (mm/day)
	g/day	kcal/day		
Males				
Skim milk ($n = 14$)	1353	565	11.4	0.68
	(175)[a]	(86)	(6.5)	(0.17)
Similac® ($n = 14$)	1021	681	19.3	0.76
	(85)	(56)	(5.3)	(0.19)
Females				
Skim milk ($n = 14$)	1250	519	11.3	0.69
	(117)	(48)	(3.5)	(0.15)
Similac® ($n = 16$)	907	609	16.1	0.72
	(97)	(64)	(4.2)	(0.14)

[a]Values in parentheses indicate standard deviation.

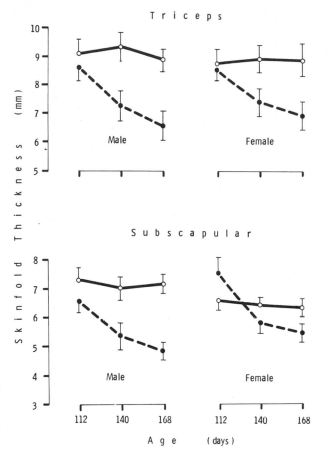

Fig. 6. Triceps and subscapular skinfold thicknesses at various ages of infants fed Similac (solid lines) or modified skim milk (dashed lines). (From Fomon *et al.*, 1977.)

Table XII summarizes data on food consumption and growth of these infants. In spite of the large volumes of formula consumed and the greater intake of strained food by the infants fed skim milk, their total energy intake was significantly less than in those fed Similac. This resulted in a significantly lower gain in weight in the skim-milk group whereas gain in length was not significantly different.

Subscapular and triceps skinfold measurements were taken at 112, 140, and 168 days of age. At the time of enrollment in the study (112 days), there were no significant differences in skinfold thickness for either site or sex between feeding groups. Figure 6 shows little change of skinfold thickness occurred during the time 112–168 days in infants fed Similac. By contrast a statistically significant decrease at both sites was observed in skim-milk-fed infants, amounting in males to 24% shrinkage of triceps and 27% of subscapular skinfold thickness after 8 weeks. Corresponding decrease in female infants was 19% and 27%, respectively. Obviously, in skim-milk feeding the infants had to mobilize energy from body fat stores to permit growth of fat-free body mass. Children receiving the conventional formula (67 kcal/100 ml) consumed the amount of food that allowed them to maintain their subcutaneous adipose tissue. This goal was not attainable in skim-milk feeding because presumably maximum volumes of intake could not compensate for low caloric density of the formula.

RENATE L. BERGMANN
and KARL E. BERGMANN

3. Interrelationship of Growth and Nutritional Requirements: Model Considerations

3.1. Partitioning of Nutritional Requirements between Growth and Nongrowth

Maintenance of body functions requires energy in relation to body mass; for growth, energy is needed relative to amount, extent, and composition of accreted tissues. For the sake of simplicity, energy for maintenance and energy lost in feces and urine (less than 10% of energy intake) will be assumed to be energy for nongrowth.

An estimate of the partitioning between energy for growth and nongrowth can be based upon the following considerations: Presuming that the vast majority of energy is required for synthesis of fat and protein in tissue growth (Kirchgessner and Müller, 1974), we shall concentrate on computations regarding these major constituents. Taking the body composition of the male reference infant and the reference fetus as a basis (Fomon, 1967; Fomon *et al.,* 1974), protein and lipid content of the daily accreted tissue has been estimated for the male reference infant at ages 3 weeks, 4 months, 12 months, 2½ years, and for a 3-week-old premature infant (assuming that composition of his gain is similar to that *in utero*) (Table XIII).

Energy cost for synthesis of protein and fat can be derived from animal studies: 1 g of body fat in piglets has an energy content of 9 kcal (Kirchgessner and Kellner, 1972). With a high fat concentration in the diet, as in human milk and milk-based formulas, we can expect a human infant to incorporate the dietary fatty acids into tissue lipids without much extra energy (van Es, 1977). The efficiency for this process can be as high as 85, 90, or 95% (van Es, 1977). Utilizing the median value, deposition of 1 g tissue lipid will therefore require 10 kcal. With this presumption, the energy requirement for the synthesis of 1 g protein in growing piglets, fed a diet with an optimal concentration of milk protein, has been calculated to be 10.7–12.0 kcal (Kirchgessner and Müller, 1974). In other experiments with piglets (Kielan-owski and Kotarbinska, 1970; Sharma and Young, 1970), pigs (Thorbek, 1970; Oslage *et al.,* 1971), and growing rats (Schiemann *et al.,* 1969), calculated values of energy cost for synthesis of 1 g protein ranged from 10.0 to 14.6 kcal.

Because the human infant is growing more slowly than the experimental animals (McCance and Widdowson, 1964), which implies a less rapid protein turnover rate (Millward *et al.,* 1976), the lowest calculated value for synthesis of protein appears to be the most applicable one. With this figure, 10 kcal for synthesis of either 1 g fat or 1 g protein, energy for the synthesis of protein and lipid in the male reference infant has been calculated (Table VIII).

Daily energy intake for the infant at the specified ages (see above) are assumed to be 475, 675, 1050, and 1400 kcal, respectively (Fomon *et al.,* 1971*b;* Beal, 1970) and for the premature infant at age 3 weeks to be 240 kcal, a value slightly above the energy cost of 200 kcal/day for intrauterine growth (van Es, 1977).

The 3-week-old full-term infant, accordingly, will utilize 44% of his energy intake for growth. Because growth, and especially lipid deposition, are decreasing with age, less of the energy intake is required for growth and more for maintenance (Figure 7).

Similar calculations may be applied to constituents of the body, e.g., protein. Provided that intake of a high-quality protein is just at the requirement level,

Table XIII.   Energy Requirement for Growth

Age	Body weight (kg)	Gain (g/day) Total	Protein	Lipid	Energy intake (kcal/day)	% intake for growth
3 weeks, premature	1.7	30	3.78	4.20	240 (200)[a]	33 (40)[a]
3 weeks, male reference infant	4.0	40	4.56	16.56	475	44
4 months	7.0	25	3.50	8.00	675	17
12 months	10.5	10	2.07	1.43	1050	3
2½ years	14.0	5	1.05	0.17	1400	1
Adult	70.0	0	0	0	2800	0

[a]Values in parentheses indicate intrauterine growth.

percentage of protein intake utilized for growth has been calculated by Fomon (1974*a*) for the male reference infant (Figure 8). As with the energy requirement, increment in body protein of a rapidly growing infant may account for a high percentage of the protein requirement, whereas, with increasing age and decelerating growth, less of the dietary protein is needed for growth.

The calculations of gain and nutrient intake are interpolated from observations of two weeks' duration or more. Irregularities of intake and growth from day to day are not apparent, because the organism has been able to level them off during successive periods. Partially responsible for this homeostasis in infants is their ability to store energy and dietary constituents in tissues. Depending on availability of the respective dietary components, body pool size, turnover rate, loss or

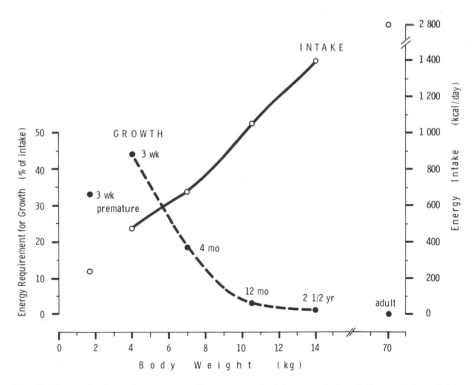

Fig. 7. Energy intake and percentage of energy required for growth, in relation to body weight.

RENATE L. BERGMANN
and KARL E. BERGMANN

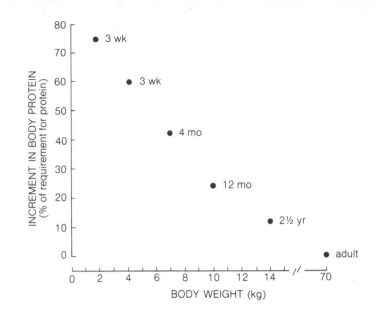

Fig. 8. Increment in body protein, expressed as percentage protein requirement, in relation to body weight. (From Fomon, 1974a.)

excretion, and growth of body mass, their concentration at the reaction site will either be high enough or be insufficient to maintain optimal body function.

3.2. Dilution of Body Stores by Growth

While nutritional components are needed for growth, expansion of the body mass, on the other hand, may be diluting their concentration to values below the threshold for normal function, and nutritional disorders may develop. Using iron as an example, this dualistic feature of growth can be illustrated.

For the considerations to follow, it is assumed that the total body iron of two full-term newborn infants A and B were 293 and 129 mg, respectively, both of the values being taken from two previous studies (Sturgeon, 1956; Widdowson, 1969). No available iron was present in the diet, inevitable losses were 57 mg/year (estimated after Green *et al.*, 1968; Bergmann, 1974), and at age 1 year both infants had body weights of 10.5 kg and blood volumes of 75 ml/kg. If at this time all of the body iron could be assumed to be in hemoglobin, both of the infants A and B would

Table XIV. Estimate of Hemoglobin Concentration in Two Hypothetical Infants, A and B, at 1 Year of Age[a]

Infant	A	B
Total body iron at birth (mg)	293	129
Iron loss during first year of life (mg)	57	57
Total body iron at 1 year of age (mg)	236	72
Hemoglobin concentration at 1 year of age (g/100 ml)	8.8	2.7

[a]For further details, see text.

be anemic (Table XIV), and furthermore the condition of infant B might not be compatible with life.

Similar "diluting" effects of growth may also occur with other trace elements. In a sample of 36 male and 40 female infants and toddlers (Bergmann *et al.*, 1978), zinc concentration in hair correlated inversely with preceding growth rate (with yearly length gain in girls and weight gain in boys, both, $r = -0.36$; $P < 0.025$). Lowest hair zinc concentrations of a sample of 337 children were observed following the most rapid rate of growth in infancy (Figure 9) (Bergmann *et al.*, 1978).

3.3. Easing of Strains on Renal Function by Growth

On one hand, we have seen that growth may dilute the concentration of essential substances; on the other hand, while using up dietary components, it may relieve the organism of the task of metabolizing and excreting them. This "homeostasis by growth" (McCance and Widdowson, 1961) may be illustrated by the example of renal solute load (Bergmann *et al.*, 1974). Renal solute load consists of metabolic endproducts, especially nitrogenous compounds and electrolytes, that must be excreted by the kidney. These solutes arise from endogenous sources and from diet. An estimate of the potential renal solute load originating from the diet may be obtained by adding up the molar intake of the major electrolytes and the contribution of protein, i.e., 5.7 mosmol of urea per g protein. Part of this nutritional intake by the infant will be utilized for growth. It has been estimated that 1 g of weight gain, during the first weeks of life uses up 0.98 mosmol of renal solute load; the corresponding value for the age interval 5 through 12 months would be 1.6 mosmol/g of weight gain, corresponding roughly to the removal of potential renal solute load arising from consumption of 300 ml/day of humanized infant formula.

Consideration of the dietary acid burden and net acid intake (Kildeberg *et al.*, 1969), reveals that "homeostasis by growth" (McCance and Widdowson, 1961) may also contribute to normalizing acid–base equilibrium. The late metabolic acidosis of premature infants has been explained by higher rates of endogenous acid production in nongrowing infants receiving milk-based formula together with functional immaturity of the kidney (Kildeberg, 1973). As acidosis is corrected, growth reappears and aids the prevention of further acid–base derangements.

4. Adverse Effects of Deficient or Excessive Intake of Some Nutrients on Infant Growth

Availability of energy and nutrients is one of the basic conditions for growth. Maintenance metabolism and deposition of new tissue may become competitive in the case of an energy and nutrient shortage. Nutritional deprivation is most likely to interfere with growth during periods when a high growth velocity would normally occur, especially in infancy. Fortunately the organism is able to recover and catch up (Prader *et al.*, 1963), depending upon the extent and duration of the deficiency state, when adequate nutrition is reinstituted during a developmental period of high growth potential. Catch-up growth may even serve as a diagnostic tool retrospectively. On the other hand, when the total amount and constituents of the diet forced into an infant are in excess of his needs, overnutrition and adverse effects of food may result. "Catch-down" growth (Fomon, 1974*b*) may follow overnutrition, when food intake is reduced again.

RENATE L. BERGMANN
and KARL E. BERGMANN

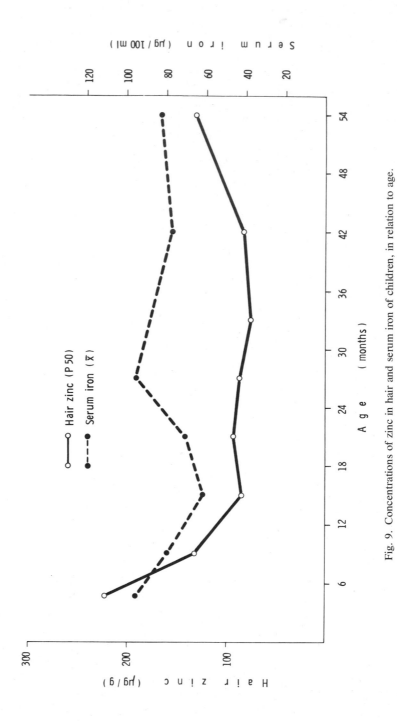

Fig. 9. Concentrations of zinc in hair and serum iron of children, in relation to age.

Carbohydrates, fat, and protein can substitute for each other, to a certain extent, as energy sources. Multinutritional deficiencies may accompany an energy deprivation, and vary the clinical picture of failure to thrive, "wasting," and "stunting" (Waterlow and Rutishauser, 1974).

Most frequently, inadequate amount of food offered to an infant and/or too low caloric density of food are the primary causes for growth failure: In developing countries this results from food shortage or maldistribution within the family (McLaren, 1976). In affluent societies, objectionable psychosocial circumstances are the most frequent cause for this symptom of child neglect (Barbero and Shaneen, 1967).

Infants to whom adequate food is offered may not be able to consume it (because of a physical handicap), may refuse it (because of an underlying acute or chronic disease), or may regurgitate and vomit it. Anorexia, and therefore food refusal, is one of the causative factors for growth failure in renal disease (Holliday, 1972). Offering additional calories to such children may improve their growth rate (Holliday, 1972) or their activity level (Betts *et al.*, 1977).

Malabsorption is frequently a contributing factor in energy deprivation. Although infants are able to balance intestinal losses to a certain extent by greater energy intakes (Fomon, 1974*c*), in cystic fibrosis of the pancreas energy losses in stool fat cannot be compensated for, in spite of the insatiable appetites of these infants: Growth in weight and to a lesser extent in height are impaired (Kreissl *et al.*, 1972). Increased rates of weight and length gain (Allan *et al.*, 1973), and rapid head growth (Bray and Herbst, 1973) have been observed after treatment has been commenced. Infants suffering from celiac disease show impressive catch-up growth in weight and later in height after institution of a gluten-free diet (Barr *et al.*, 1972).

Higher metabolic rates (Stocker *et al.*, 1972) together with reduced food intake (Krieger, 1970) and gastrointestinal absorptive defects (Pittman and Cohen, 1964) are contributing factors in the growth retardation of infants with cardiac disease. Catch-up growth after surgery has been described to be most evident in the youngest age group (Suoninen, 1972). Growth acceleration could be achieved by forced feeding of infants with cardiac disease (Krieger, 1970).

Although infants are able to adjust their food intake according to requirements, abnormally high weight gains have been observed due to excessive energy intake (Fomon *et al.*, 1975; Sveger *et al.*, 1975). Such excess may be caused by parental intervention, e.g., forcing an infant to drain the last drop of a bottle (Fomon, 1971), overconcentrating infant formula (Fomon *et al.*, 1969; Taitz, 1976), and introducing cereal at an early age (Shukla *et al.*, 1972; Oakley, 1977). On the other hand, the infant may demand more volume of an overconcentrated formula because he is thirsty (Taitz, 1976), he may perceive short-term satiety signals at a higher threshold (van Itallie *et al.*, 1977), the long-term regulation may be set to encourage the intake of a small excess of energy every day (Widdowson and McCance, 1975), and, furthermore, some of the infants may be more prone to overeating on a hereditary basis (Brook *et al.*, 1975; Börjeson, 1976).

A surplus of energy may be stored in body fat as well as in lean body mass (Frisancho *et al.*, 1977; van Es, 1977) accelerating, to a certain extent, tissue maturation and growth. In experimental animals, food intake, growth rate in early infancy, and accelerated development have been shown to be negatively correlated with life-span (Ross, 1972; Widdowson and McCance, 1975). One is tempted to

RENATE L. BERGMANN
and KARL E. BERGMANN

speculate that this might also be true for the human, but there is a lack of adequate data to analyze this hypothesis.

Adult overweight is known to be a risk factor in morbidity and mortality rates in the developed countries. However, relationship of excessive weight and fat gain of the human infant to juvenile and adult obesity is still a controversial issue (Hernesniemi *et al.*, 1974; Charney *et al.*, 1976; Mellbin and Vuille, 1976*a,b;* Taitz, 1976; Dörner *et al.*, 1977; Poskitt and Cole, 1977).

4.2. Protein

While infants are able to adjust their food consumption according to the energy density of formula, no such adaptive process could be observed with regard to different protein concentrations in formula (Fomon *et al.*, 1971*a,c*). Infant rhesus monkeys consumed less of a low-protein diet than did control animals on a high-protein food (Geist *et al.*, 1972). Piglets, with a shortage of protein in their diet, had no appetite and remained small, but tended to become quite fat (McCance, 1975).

As human milk is balanced in regard to protein and energy content, protein deficiency, as a predominant nutritional disorder, typically occurs after weaning to a low-protein food, whose energy is usually derived from flour or other starchy staples (see Chapter 11). Protein deficiency, with accompanying growth failure, may also develop in infants with amino acid disorders receiving a formula low in protein content (Smith and Francis, 1976).

Synthesis of protein for replacement and growth of tissue requires a proper composition of amino acids to be present in food. A diet deficient in one essential amino acid (Snyderman, 1973) or with an excess of one or more amino acids may lead to metabolic derangements, "amino acid toxicity," "amino acid imbalance" (Harper *et al.*, 1970), with cessation of growth.

A surplus of protein in the diet may also impose a strain on renal regulating ability (Section 3.3). Premature infants are especially sensitive to excessive protein intakes or to a suboptimal composition of its amino acids (Goldman *et al.*, 1969, 1974; Svenningsen and Lindquist, 1973; Räihä *et al.*, 1976; Rassin *et al.*, 1977).

4.3. Trace Elements

Essential fatty acid deficiency, vitamin deficiencies, or trace element deficiencies cause either specific clinical syndromes, where growth retardation may only be a less important feature of the disease itself, or they are associated with other nutritional deficiencies, e.g., protein–energy malnutrition. Under these circumstances, growth failure is seen primarily as a consequence of the gross nutritional disorders.

The relationship between single trace element deficiency and growth has been studied more extensively in humans for iron and zinc: Iron deficiency in the United States has been shown to be more frequent in infants whose weight and height were below the 25th percentile than in those with higher weight and height (Owen *et al.*, 1971). But this association could also be explained by the fact that other nutritional disorders are more frequent in low socioeconomic groups (Owen *et al.*, 1971). Judisch and associates (1966) demonstrated catch-up growth in weight of infants and young children with iron-deficiency anemia after treatment with iron.

In six of eight infants and children under 4 years of age, in a study of Hambidge and associates (1972), exceedingly low hair zinc concentrations were associated with height and/or weight values below the 10th percentile. In a double-blind trial, male infants receiving a zinc-supplemented formula had significantly higher mean increments in weight and height during the first 6 months of life than did male infants on the same formula with no supplement (Walravens and Hambidge, 1976).

With the increasing consumption of highly refined foods in developed countries, incidence of deficiencies of minor nutrients may be expected to increase, and more attention must be paid to trace-element deficiency in growth disorders.

5. Summary and Conclusions

From existing information on the interplay of food intake and growth of normal infants fed *ad libitum,* it is concluded that energy needs for maintenance and growth determine food intake. Requirements for nutritional constituents like water, protein, and other essential substances appear to be less influential as driving forces of food intake. When energy availability is deficient, growth and maintenance become competitive, and growth disorders, as well as lower activity levels occur. Growth retardation can also result from a deficiency of single nutritional factors in calorically adequate feedings given *ad libitum,* because infants eat mainly for energy.

ACKNOWLEDGMENTS

We wish to thank Miss A. Hendry for her invaluable help in revising the English text, Mr. W. Herborn for his untiring efforts in preparing the manuscript, and Mr. W. Rosenberger for kindly sketching the figures.

6. References

Allan, J. D., Mason, A., and Moss, A. D., 1973, Nutritional supplementation in treatment of cystic fibrosis of the pancreas, *Am. J. Dis. Child.* **126:**22.

Ashworth, A., Bell, R., James, W. P. T., and Waterlow, J. C., 1968, Calorie requirements of children recovering from protein-calorie malnutrition, *Lancet* **2:**600.

Barbero, G. J., and Shaheen, E., 1967, Environmental failure to thrive: A clinical view, *J. Pediatr.* **71:**639.

Barr, D. G. D., Shmerling, D. H., and Prader, A., 1972, Catch-up growth in malnutrition, studied in celiac disease after institution of gluten-free diet, *Pediatr. Res.* **6:**521.

Beal, V. A., 1970, Nutritional intake, in: *Human Growth and Development* (R. W. McCammon, ed.), p. 63, Charles C. Thomas, Springfield, Illinois.

Belton, N. R., Cockburn, F., Forfar, J. O., Gles, M. M., Kirkwood, J., Smith, J., Thistlethwaite, D., Turner, T. L., and Wilkinson, E. M., 1977, Clinical and biochemical assessment of a modified evaporated milk for infant feeding, *Arch. Dis. Child.* **52:**167.

Bergmann, K. E., 1974, Spurenelemente in der Säuglingsernährung, *Monatsschr. Kinderheilkd.* **122:**285.

Bergmann, K. E., Stegner, W., Klarl, A., Makosch, G., Tews, K. H., and Bergmann, R. L., 1978, Zinkmangel im Kindesalter, in: *Spurenelemente* (E. Gladke, G. Heimann, and I. Eckert, eds.) Georg Thieme Stuttgart.

Bergmann, K. E., Ziegler, E. E., and Fomon, S. J., 1974, Water and renal solute load, in: *Infant Nutrition* (S. J. Fomon), p. 245, W. B. Saunders, Philadelphia.

Betts, P. R., Magrath, G., and White, R. H. R., 1977, Role of dietary energy supplementation in growth of children with chronic renal insufficiency, *Br. Med. J.* **1977**(1):416.

Blurton Jones, N., 1972, Comparative aspects of mother–child contact, in: *Ethological Studies of Child Behaviour* (N. Blurton Jones, ed.), p. 305, University Press, Cambridge.

Börjeson, M., 1976, The aetiology of obesity in children, *Acta Paediatr. Scand.* **65:**279.

Bray, P. F., and Herbst, J. J., 1973, Pseudotumor ceribri as a sign of "catch-up" growth in cystic fibrosis, *Am. J. Dis. Child.* **126:**78.

Brook, C. G. D., Huntley, R. M. C., and Slack, J., 1975, Influence of heredity and environment in determination of skinfold thickness in children, *Br. Med. J.* **1975**(2):719.

Charney, E., Chamblee Goodman, H., McBride, M., Lyon, B., and Pratt, R., 1976, Childhood antecedents of adult obesity, *N. Engl. J. Med.* **295**:6.

Chávez, A., Martinez, C., Ramos Galván, R., Coronado, M., Bourges, H., Diaz, D., Basta, S., and Olinda y Garcidueñas, N., 1973, Nutrition and development of infants from poor rural areas. IV Differences attributable to sex in the utilization of mother's milk, *Nutr. Rep. Int.* **7**:603.

Davidson, M., Levine, S. Z., Bauer, C. H., and Dann, M., 1967, Feeding studies in low-birth-weight infants, *J. Pediatr.* **70**:695.

Dörner, G., Grychtolik, H., and Julitz, M., 1977, Überernährung in den ersten drei Lebensmonaten als entscheidender Risikofaktor für die Entwicklung von Fettsucht und ihrer Folgeerkrankungen, *Dtsch. Gesundh.-Wes.* **32**:6.

Dugdale, A. E., 1971, The effect of the type of feeding on weight gain and illnesses in infants, *Br. J. Nutr.* **26**:423.

Fomon, S. J., 1967, Body composition of the male reference infant during the first year of life, *Pediatrics* **40**:863.

Fomon, S. J., 1971, A pediatrician looks at early nutrition, *Bull. N.Y. Acad. Med.* **47**:569.

Fomon, S. J., 1974a, Nutritional requirements in relation to growth, *Monatsschr. Kinderheilkd.* **122**:236.

Fomon, S. J., 1974b, Normal growth, failure to thrive and obesity, in: *Infant Nutrition* (S. J. Fomon), p. 34, W. B. Saunders, Philadelphia.

Fomon, S. J., 1974c, Voluntary food intake and its regulation, in: *Infant Nutrition* (S. J. Fomon), p. 20, W. B. Saunders, Philadelphia.

Fomon, S. J., and Filer, L. J., Jr., 1974, Milks and formulas, in: *Infant Nutrition* (S. J. Fomon), p. 359, W. B. Saunders, Philadelphia.

Fomon, S. J., Filer, L. J., Thomas, L. N., Rogers, R. R., and Proksch, A. M., 1969, Relationship between formula concentration and rate of growth of normal infants, *J. Nutr.* **98**:241.

Fomon, S. J., Filer, L. J., Jr., Thomas, L. N., and Rogers, R. R., 1970, Growth and serum chemical values of normal breastfed infants, *Acta Paediatr. Scand.* Suppl. 202.

Fomon, S. J., Filer, L. J., Jr., Ziegler, E. E., and Thomas, L. N., 1971a, Protein requirement of normal infants between 56 and 112 days of age, in: *Proc. 13th Int. Congr. Pediatrics,* Vol. II, *Nutrition and Gastroenterology,* p. 217, Verlag der Wiener Medizinischen Akademie, Vienna.

Fomon, S. J., Thomas, L. N., Filer, L. J., Jr., Ziegler, E. E., and Leonard, M. T., 1971b, Food consumption and growth of normal infants fed milk-based formulas, *Acta Paediatr. Scand.* Suppl. 223.

Fomon, S. J., Ziegler, E. E., Thomas, L. N., and Filer, L. J., Jr., 1971c, Protein requirement of normal infants between 8 and 56 days of age, in: *Nutritia Symposium: Metabolic Processes in the Foetus and Newborn Infant* (J. H. P. Jonxis, H. K. A. Visser, and J. A. Troelstra, eds.), p. 144, Stenfert Kroese, Leiden.

Fomon, S. J., Ziegler, E. E., and O'Donnell, A. M., 1974, Infant feeding in health and disease, in: *Infant Nutrition* (S. J. Fomon), p. 472, W. B. Saunders, Philadelphia.

Fomon, S. J., Filer, L. J., Jr., Ziegler, E. E., Bergmann, K. E., and Bergmann, R. L., 1977, Skim milk in formula concentration on caloric intake and growth of normal infants, *Acta Paediatr. Scand.* **64**:172.

Fomon, S. J., Thomas, L. N., Filer, L. J., Jr., Anderson, T. A., and Nelson, S. E., 1976, Influence of fat and carbohydrate content of diet on food intake and growth of male infants, *Acta Paediatr. Scand.* **65**:136.

Fomon, S. J., Filer, L. J., Jr., Ziegler, E. E., Bermann, K. E., and Bergmann, R. L., 1977, Skim milk in infant feeding, *Acta Paediatr. Scand.* **66**:17.

Frisancho, A. R., Klayman, J. E., and Matos, J., 1977, Newborn body composition and its relationship to linear growth, *Am. J. Clin. Nutr.* **30**:704.

Geist, C. R., Zimmerman, R. R., and Strobel, D. A., 1972, Effect of protein-calorie malnutrition on food consumption, weight gain, serum proteins, and activity in the developing rhesus monkey (*Macaca mulatta*), *Lab. Anim. Sci.* **22**:369.

Goldman, H. I., Freudenthal, R., Holland, B., and Karelitz, S., 1969, Clinical effects of two different levels of protein intake on low-birth-weight infants, *J. Pediatr.* **74**:881.

Goldman, H. I., Goldman, J. S., Kaufman, I., and Liebman, O. B., 1974, Late effects of early dietary protein intake on low-birth-weight infants, *J. Pediatr.* **85**:764.

Graham, G. G., Cordano, A., and Baertl, J. M., 1963, Studies in infantile malnutrition. II. Effect of protein and calorie intake on weight gain, *J. Nutr.* **81**:249.

Green, R., Charlton, R., Seftel, H., Bothwell, T., Mayet, F., Adams, B., Finch, C., and Layrisse, M., 1968, Body iron excretion in man: A collaborative study, *Am. J. Med.* **45**:336.

Hall, B., 1975, Changing composition of human milk and early development of an appetite control, *Lancet* **1**:779.

Hambidge, K. M., Hambidge, C., Jacobs, M., and Baum, J. D., 1972, Low levels of zinc in hair, anorexia, poor growth, and hypogeusia in children, *Pediatr. Res.* **6**:868.

Harper, A. E., Benevenga, N. J., and Wahlhueter, R. M., 1970, Effects of ingestion of disproportionate amounts of amino acids, *Physiol. Rev.* **50**:428.

Hernesniemi, I., Zachmann, M., and Prader, A., 1974, Skinfold thickness in infancy and adolescence, *Helv. Paediatr. Acta* **29**:523.

Holliday, M. A., 1972, Calorie deficiency in children with uremia: Effect upon growth, *Pediatrics* **50**:590.

Jelliffe, D. B., and Jelliffe, E. F. P., 1976, Nutrition and human milk, *Postgrad. Med.* **60**(1):153.

Jensen, R. L., Thomas, L. N., Bergmann, K. E., Filer, L. J., Jr., and Fomon, S. J., 1972, Composition of milk of Iowa City women (unpublished data).

Judisch, J. M., Naiman, J. L., and Oski, F. A., 1966, The fallacy of the fat iron-deficient child, *Pediatrics* **37**:987.

Kielanowski, J., and Kotarbinska, M., 1970, Further studies on energy metabolism in the pig, in: *Proc. 5th Symp. Energy Metabolism, Vitznau 1970* (EAAP Publ. No. 13), p. 145, Juris Druck, Zürich.

Kildeberg, P., 1973, Late metabolic acidosis of premature infants, in: *The Body Fluids in Pediatrics* (R. W. Winters, ed.), p. 338, Little, Brown and Company, Boston.

Kildeberg, P., Engel, K., and Winters, R. W., 1969, Balance of net acid in growing infants, *Acta Paediatr. Scand.* **58**:321.

Kirchgessner, M., and Kellner, B. B., 1972, Stickstoff- und Energieansatz früh abgesetzter Ferkel bei Diäten auf Kasein-Trockenmilch-Basis, *Arch. Tierernähr.* **22**:249.

Kirchgessner, M., and Müller, H. L., 1974, Energieaufwand für Wachstum und Proteinansatz frühentwöhnter Ferkel bei Ernährung mit verschiedenen Proteinen, *Arch. Tierernähr.* **24**:215.

Kreissl, T., Bender, S. W., Mörchen, R., and Hövels, O., 1972, The physical development of children with cystic fibrosis, *Z. Kinderheilkd.* **113**:93.

Krieger, I., 1970, Growth failure and congenital heart disease, *Am. J. Dis. Child.* **120**:497.

McCance, R. A., 1975, The determinants of growth and form, in: *Modern Problems in Paediatrics, Vol. 14, Nutrition, Growth, and Development* (F. Falkner, N. Kretchmer, and E. Rossi, eds.), p. 167, S. Karger, Basel.

McCance, R. A., and Widdowson, E. M., 1961, Mineral metabolism of the foetus and newborn, *Br. Med. Bull.* **17**:132.

McCance, R. A., and Widdowson, E. M., 1964, Protein metabolism and requirements in the newborn, in: *Mammalian Protein Metabolism* (M. Allison, ed.), p. 225, Academic Press, New York.

McLaren, D. S., 1976, Protein energy malnutrition, in: *Textbook of Paediatric Nutrition* (D. S. McLaren and D. Burman, ed.), p. 105, Churchill Livingstone, Edinburgh.

Mellander, O., Vahlquist, B., Mellbin, T., and Collaborators, 1959, Breast feeding and artificial feeding, *Acta Paediatr. Scand., Suppl.* **116**:Chapter VII.

Mellbin, T., and Vuille, J.-C., 1976a, Relationship of weight gain in infancy to subcutaneous fat and relative weight at 10½ years of age, *Br. Prev. Soc. Med.* **30**:239.

Mellbin, T., and Vuille, J.-C., 1976b, Weight gain in infancy and physical development between 7 and 10½ years of age, *Brit. Prev. Soc. Med.* **30**:233.

Millward, D. C., Garlick, P. J., James, W. P. T., Sender, P. M., and Waterlow, J. C., 1976, Protein turnover in: *Protein Metabolism and Nutrition, Proc. 1st Int. Symp. on Protein Metabolism and Nutrition* (D. Lewis and D. J. Cole, eds.), p. 49, Butterworths, London.

Nitzan, M., and Schönfeld, T. M., 1976, Excessive weight gain during early infancy as related to breast feeding versus bottle feeding, in: *Pediatric and Adolescent Endocrinology, Vol. I, The Adipose Child* (Z. Laron, ed.) p. 66, S. Karger, Basel.

Oakley, J. R., 1977, Differences in subcutaneous fat in breast- and formula-fed infants, *Arch. Dis. Child.* **52**:79.

Oslage, H. J., Gädeken, D., and Fliegel, H., 1971, Über den energetischen Wirkungsgrad der Protein- und Fettsynthese bei wachsenden Schweinen, *Landwirtsch. Forsch.* **24**:34.

Ounsted, M., and Sleigh, G., 1975, The infant's self-regulation of food intake and weight gain, *Lancet* **2**:1393.

Owen, G. M., Lubin, A. H., and Garry, P. J., 1971, Preschool children in the United States: Who has iron deficiency? *J. Pediatr.* **79**:563.

Pittman, J. G., and Cohen, P., 1964, The pathogenesis of cardiac cachexia, *N. Engl. J. Med.* **271**:403.

Poskitt, E. M. E., and Cole, T. J., 1977, Do fat babies stay fat? *Br. Med. J.* **1977**(1):7.

Prader, A., Tanner, J. M., and von Harnack, G. A., 1963, Catch-up growth following illness or starvation, *J. Pediatr.* **62**:646.

Räihä, N. C. R., Heinonen, K., Rassin, D. K., and Gaull, G. E., 1976, Milk protein quantity and quality in low-brithweight infants: I. Metabolic responses and effects on growth, *Pediatrics* **57**:659.

Rassin, D. K., Gaull, G. E., Heinonen, K., and Räihä, N. C. R., 1977, Milk protein quantity and quality in low-birth-weight infants: II. Effects on selected aliphatic amino acid in plasma and urine, *Pediatrics* **59**:407.

Richards, M. P. M., 1975, Feeding and the early growth of the mother–child relationship, in: *Modern Problems in Paediatrics, Vol. 15, Milk and Lactation* (F. Falkner, N. Kretchmer, and E. Rossi, eds.), p. 143, S. Karger, Basel.

Ross, M. H., 1972, Length of life and caloric intake, *Am. J. Clin. Nutr.* **25**:834.

Rutishauser, I. H. E., and McCance, R. A., 1968, Caloric requirements for growth after severe undernutrition, *Arch. Dis. Child.* **43**:252.

Schiemann, R., Chudy, A., and Herceg, O., 1969, Der Energieaufwand für die Bildung von Körperprotein beim Wachstum, nach Modellversuchen an Ratten, *Arch. Tierernähr.* **19**:395.

Sharma, V. D., and Young, L. G., 1970, Energy requirement for maintenance and production in Lacombe and Yorkshire pigs, *J. Anim. Sci.* **31**:210.

Shukla, A., Forsyth, H. A., Anderson, C. M., and Marwah, S. M., 1972, Infantile overnutrition in the first year of life: A field study in Dudley, Worcestershire, *Br. Med. J.* **1972**(4):507.

Smith, J., and Francis, D. E. M., 1976, Disorders of amino acid metabolism, in: *Textbook of Pediatric Nutrition* (D. S. McLaren and D. Burman, eds.), p. 263, Churchill Livingstone, Edinburgh.

Snyderman, S. E., 1973, The protein and amino acid requirements of the infant, in: *Therapie lebensbedrohlicher Zustände bei Säuglingen und Kleinkindern* (K. Lang, R. Frey, and M. Halmágyi, eds.), p. 21, Springer-Verlag, Berlin.

Steward, A., and Westropp, C., 1953, Breastfeeding in the Oxford child health survey, Part II, Comparison of bottle- and breast-fed babies, *Br. Med. J.* **1953**(2):305.

Stocker, F. P., Wilkoff, W., Miettinen, O. S., and Nadas, A. S., 1972, Oxygen consumption in infants with heart disease, *J. Pediatr.* **80**:43.

Sturgeon, P., 1956, Iron metabolism. A review with special consideration of iron requirements during normal infancy, *Pediatrics* **18**:267.

Suoninen, P., 1972, Physical growth of children with congenital heart disease, *Acta Paediatr. Scand.* Suppl. 225.

Sveger, T., Lindberg, T., Weibull, B., and Olsson, U. L., 1975, Nutrition, overnutrition and obesity in the first year of life in Malmö, Sweden, *Acta Paediatr. Scand.* **64**:635.

Svenningsen, N. W., and Lindquist, B., 1973, Incidence of metabolic acidosis in term, preterm and small-for-gestational age infants in relation to dietary protein intake, *Acta Paediatr. Scand.* **62**:1.

Taitz, L. S., 1976, Relationship of infant feeding patterns to weight gain in the first weeks of life, in: *Pediatric and Adolescent Endocrinology, Vol. I, The Adipose Child* (Z. Laron, ed.), p. 60, S. Karger, Basel.

Tarján, R., Krámer, M., Szöke, K., Lindner, K., Szarvas, T., and Dworschák, E., 1965, The effect of different factors on the composition of human milk. II. The composition of human milk during lactation, *Nutr. Diet.* **7**:136.

Thorbek, G., 1970, The utilization of metabolizable energy for protein and fat gain in growing pigs, in: *Proc. 5th Symp. Energy Metabolism, Vitznau* (EAAP Publ. No. 13), p. 129, Juris Druck, Zürich.

van Es, A. J. H., 1977, The energetics of fat deposition during growth, in: *Second European Nutrition Conference* (N. Zöllner, G. Wolfram, and C. Keller, eds.), p. 88, S. Karger, Basel.

van Itallie, T. B., Schupf Smith, N., and Quartermain, D., 1977, Short-term and long-term components in the regulation of food intake: Evidence for a modulatory role of carbohydrate status, *Am. J. Clin. Nutr.* **30**:742.

Walravens, P. A., and Hambidge, K. M., 1976, Growth of infants fed a zinc supplemented formula, *Am. J. Clin. Nutr.* **29**:1114.

Waterlow, J. C., and Alleyne, G. A. O., 1971, Protein malnutrition in children: Advances in knowledge in the last ten years, in: *Advances in Protein Chemistry,* Vol. 25 (C. B. Anfinsen, Jr., J. T. Edsall, and F. M. Richards, eds.), p. 117, Academic Press, New York.

Waterlow, J. C., and Rutishauser, J. H. E., 1974, Malnutrition in man, in: *Early Malnutrition and Mental Development* (J. Cravioto, L. Hambraeus, and B. Valhquist, eds.), p. 13, Almqvist and Wiksell. Sweden.

Widdowson, E. M., 1969, Trace elements in human development, in: *Mineral Metabolism in Pediatrics. A Glaxo Symposium* (D. Barltop and W. L. Burland, eds.), p. 85, Davis, Philadelphia.

Widdowson, E. M., and Dickerson, J. W. T., 1964, Chemical composition of the body, in: *Mineral Metabolism,* Vol. II, Part A (C. L. Comar and F. Bronner, eds.), p. 1, Academic Press, New York.

Widdowson, E. M., and McCance, R. A., 1975, A review: New thoughts on growth, *Pediatr. Res.* **9**:154.

11

Protein–Energy Malnutrition and Growth

LAURENCE MALCOLM

1. Introduction

Growth is a dynamic process which begins with the conjunction of two sets of genes in the fertilized ovum. The ensuing growth and development on the pathway toward the mature adult is at any stage the result of a number of competing, complementary, and interacting influences. Of these influences no other single environmental factor affects the tempo and patterns of human growth to a greater extent than malnutrition. This is hardly unexpected. Growth is manifestly dependent upon the supply of sufficient nutrients of adequate quantity and quality, and the failure of such supply imposes stresses upon the growing child which may compel a major adaptation of the genetically programmed sequence of events and pathways laid down at conception.

While malnutrition in its broadest sense may influence growth through such factors as mineral and vitamin deficiencies, e.g., iodine, these effects are small in proportion to the general problem of protein–energy malnutrition. This chapter will therefore focus upon ways in which this form of malnutrition plays an overriding role in determining patterns of human growth in different ecosystems.

Three stages of growth are examined—prenatal, preschool, and school age through to maturity. The effect of malnutrition may be observed not only upon such physical factors as height and weight, but also upon secondary sex characteristics, dental and skeletal development, and body proportions.

2. Prenatal Growth

Although birth is the earliest stage at which growth can usually be examined, significant influences have by then led to a wide variation in birth size as indicated

---

LAURENCE MALCOLM • Health Planning and Research Unit, Christchurch, New Zealand.

by the range of birth weights observed in different societies (Meredith, 1970). Although such variation is undoubtedly due to a wide variety of factors (Bergner and Susser, 1970), including both the intrauterine environment, manifested in, for example, maternal weight and parity, and the wider maternal environment, such as tobacco smoking, infection, and socioeconomic variables, the evidence that malnutrition is the major determinant of the differences between populations in birth size at full-term is overwhelming.

Studies of the outcome of war-time starvation (Antonov, 1947; Stein *et al.,* 1975) point to the effect of gross but short-lived malnutrition during pregnancy. There is a close association between birth size and socioeconomic status which also relates to maternal size and weight gained during pregnancy. In Papua New Guinea, mean birth weights among genetically similar populations range from 2.32 to 3.25 kg (Malcolm, 1974)—a variation which is closely correlated with the differing nutritional circumstances of these populations. Confirmation of the place of malnutrition in prenatal growth is found in animal experiments and in the positive results of supplementation of diets of nutritionally deprived women during pregnancy (Iyengar, 1967).

The effect of such malnutrition is most pronounced upon weight and body fat and is least upon head circumference and length. This indicates the capacity of the growing organism to preserve and assign priorities to the more vital somatic components.

The mechanism through which such effects are mediated is far from clear. While the fetus may be regarded as an efficient maternal parasite, in that mild maternal malnutrition has little or no effect on fetal growth, greater degrees of malnutrition lead to decompensation and a direct effect on fetal and placental growth and on postnatal mortality (Stein *et al.,* 1975). However, there is a possibility that intervening variables such as intra- and extrauterine infection in deprived socioeconomic circumstances could play an important part (Malcolm, 1974), especially in view of the established cyclical relationship between malnutrition and infection in the young child.

The extent to which severe malnutrition in the early and rapid growth period of life can permanently affect growth patterns has not yet been determined in human growth. There is clear evidence for this in rats and other animals (McCance, 1967), but no such effect has been satisfactorily demonstrated, for example, in such studies as the Dutch famine (Stein *et al.,* 1975) or the follow-up of malnourished children (Chase and Martin, 1970; Garrow and Pike, 1967).

3. Early Postnatal Growth

Protein–energy malnutrition produces its most serious and profound effects upon growth patterns during the first three years of life. Two broad patterns of growth, especially in weight, may be described. The first is that of the well-nourished, typically bottle-fed child of industrial societies and advanced sectors of developing countries where there is a steady increase in weight with doubling of birth weight at 5 months and tripling at 12. There is some propensity toward obesity, fostered by competitiveness and cultural attitudes which have tended to place a premium upon size.

On the other hand, the typical child of the preindustrial society is breast fed on demand for a prolonged period but the quantity, quality, and timing of supplementary foods is inadequate, with the result that, while weight gain in the first few months of life is rapid, there tends to be a break in the continuity of the curve and a faltering in growth. This effect is most pronounced in nutritionally deprived societies. Figure 1 illustrates this in a series of curves as recorded in Papua New Guinean societies compared with Harvard Standards (Malcolm, 1974). The most rapid growth rates were achieved in urban children, particularly those from a more privileged army community, while the slowest rates were found among a primitive little-contacted group subsisting on tuberous sweet potato as the basic staple with only 3% of the energy derived from protein. This is the slowest rate of growth in weight reported from any world population, with the average child of 18 months having the same weight as the Harvard Standard of 5 months. The curves between these extremes are from populations whose environmental characteristics exhibit a corresponding spectrum of variation.

The reasons for this pattern have been extensively reviewed (Jelliffe, 1968; Malcolm, 1974). There is clear evidence that it is determined almost entirely by the varying level of nutritional intake, with energy being the predominant deficiency,

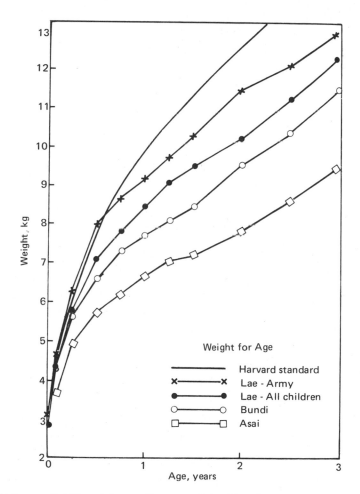

Fig. 1. Weight-for-age of children 0–3 years from some Papua New Guinean populations compared with Harvard Standards.

but with many intervening variables such as maternal nutrition, infection, cultural attitudes toward feeding, the addition of supplements, and food availability determining the actual food intake.

Variation in the patterns of skinfold thickness at this age is even more pronounced than that of weight, with severe depression of values in deprived populations (Neumann *et al.*, 1969; Malcolm, 1974) while height and head circumference are affected to a lesser extent. These differential effects on anthropometric growth patterns indicate the varying influence of protein and energy deficiencies upon growth. Height is affected most in a protein deficiency where energy intake is adequate. Weight and arm circumference, which are comprised of both fat and protein components, are affected by both energy and protein intake, whereas skinfold thickness responds more specifically to total energy intake (Malcolm, 1974).

4. School Age to Maturity

Growth curves beyond the preschool stage display a range of variation which, as for earlier ages, is closely correlated with the level of nutritional deprivation experienced in different societies. Figure 2 shows this variation in height from birth to maturity observed in Papua New Guinean socieites (Malcolm, 1974) as compared with the curve of British children (Tanner *et al.*, 1966). These curves are chosen as an example of the range of height-for-age patterns as they are, with the exception of Lae urban children, from traditional village populations unaffected by significant environmental changes which could distort this pattern, while the range of variation is wider than any other set of curves from developing countries. All curves are based upon known age with the exception of the Asai where dental eruption was used to locate critical points on the growth curve, eruption times being only slightly and predictably delayed by slower growth (Malcolm and Bue, 1970). Furthermore, they are of village populations unaffected by the selection process which can seriously distort growth curves derived from the more readily available, but anthropometrically biased, school populations used in many growth surveys. Where such selection bias is avoided, most studies of nutritionally deprived populations show a pattern similar to that of Figure 2.

The following characteristic features may be noted. The different rates of growth are maintained throughout the growth span with the result that there is a progressive divergence from the British pattern, a divergence which persists to the mature adult. As a consequence a varying height deficit remains, this being directly proportional to the rate of growth and to height at any age. In other words those children who grow faster are both taller at any age during growth and become taller adults when growth eventually ceases. As described for the preschool stage, the rate of growth is directly correlated with the degree of nutritional deprivation, being most pronounced in the Asai population. The most rapidly growing Papua New Guinean population reported is found in the urban area of Lae where children, genetically similar to village populations and who in almost all cases are first-generation urban dwellers, display a rate of growth only marginally slower than British children but widely divergent from that of village children.

This appears to be unquestionably related to the better urban environment, in particular the higher protein intake than is available to most village children. However, the adult height reached is shorter in urban residents. This may be due to either of two factors. First, these graphs are of cross-sectional and not longitudinal

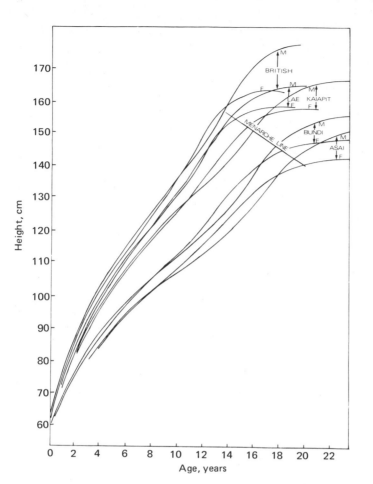

Fig. 2. Height-for-age from birth to maturity from some Papua New Guinean populations compared with British children.

data, so it is possible that those measured at the terminal stages of growth were shorter throughout the growth process than those of younger age groups, i.e., that this is an artifact of a rapid secular trend in growth rates in urban areas, this being the result of improving nutrition. On the other hand it could be that ultimate height responds more slowly to improved nutrition than does growth, this being a conse-quence of the differential rate of skeletal growth and maturation. While both epiphyseal growth and closure are delayed by malnutrition, it is possible that better nutrition may promote earlier closure than expected thus terminating growth in height earlier than anticipated from bone length (see below). A further speculative alternative is that ultimate adult height could be influenced by prenatal or early childhood malnutrition, although there would appear to be little good evidence for this possibility (Stein *et al.,* 1975).

Despite the wide variation in growth rates there is little variation in the patterns of growth (Figure 2). Male/female differences are similar in all curves, although the cross-over accompanying the female adolescent spurt seems to be earlier in slower growth, while the adolescent spurt is, if anything, slightly earlier in onset but more prolonged than in the European child. Maximum growth velocity is slower and peaks at a later age with slower growth.

There is little evidence from these curves that malnutrition has a more adverse effect on the growth of males in comparison to females as has been shown in large medical studies of children subjected to stresses such as atomic radiation (literature in Tanner, 1972). Sex differences in height are proportionally preserved throughout the growth span and the adult male and female height ratios in the most deprived populations are similar to those in well-nourished ones, a conclusion reached by Eveleth (1975) in a review of data on sexual dimorphism from a number of different populations. Genetic factors of course play an important part, especially in the later stages of growth and in determining adult height.

Further evidence of the depressing effect on growth of malnutrition comes from animal experiments (Widdowson, 1974) and from supplementary feeding studies in children (Subrahmanyan *et al.*, 1957; Tanner, 1972; Malcolm, 1972). In a longitudinal growth study in a boarding school at Bundi in the New Guinean highlands, in a nutritionally deprived population with low protein intake and comparatively slow growth of children (Figure 2), those children in the school environment were 4 cm shorter and 2.5 kg lighter than their village counterparts after 4 years of residence (Malcolm, 1970). This was attributable to the even poorer quality of the school diet in comparison with that of the village. The same children, when fed supplementary protein in the form of skim-milk powder, displayed an initially rapid catch-up phase following which they continued to grow at rates equal to that of European children of the same age, with the result that they were 7.3 cm taller and 3.1 kg heavier than the equivalent village child after a 4-year period.

Bundi children reared in the village displayed a small but significant variation in growth rates according to their position in the family. While the first child is nearly 1 cm taller than the mean, with increasing position in the family there is a decline in height and weight so that the third and fourth children are nearly 1 cm shorter than the mean. Thereafter, size increases again so that the sixth and subsequent children are larger even than the first child. A possible explanation, of what is obviously an environmental variation, is that the first child is born to a healthy and usually well-nourished mother. Subsequent children have a lesser advantage in this respect and successively compete for available food. Children born later, however, have the advantage of care, including feeding, by their older siblings, who may also contribute to the gathering of the family food and may forage, on behalf of their younger siblings, for the tidbits of animal protein such as insects, frogs, and rodents.

All the evidence suggests that the only environmental factor of any real significance at this stage, particularly in determining the variation in growth and height, is the level of protein intake. Supplementary feeding experiments in Bundi children show that height gains were highly significant when protein was added to the diet but no change was observed when only calories were added in the form of margarine (Malcolm, 1970). Height gains, from a wide range of anthropometric variables, gave the best measure of discrimination among different levels of protein intake (Malcolm, 1972).

In summary, therefore, it may be stated that while genetic influences play a major part in determining the patterns of growth, such as male/female differences, cross-over in curves, and adolescent spurt, malnutrition, especially of protein, is the major factor in determining the tempo of growth, size at any age, and ultimate adult height. Protein deprivation throughout the growth span results in a shortening of ultimate adult stature to an extent proportional to the degree and duration of the deprivation, growth in height being the best anthropometric indicator of discrimination among different levels of protein inadequacy.

It is now widely recognized that, contrary to earlier folklore, children in developing countries mature sexually at a later age than those in more advanced countries. Furthermore, children in urban areas of both developed, as well as developing, countries mature earlier than their rural counterparts.

All observations which have been reported indicate a high degree of correlation between physical and sexual development in different societies. Although the development of all secondary sex characteristics, such as breast formation, pubic and axillary hair, and a deepening of the voice in males, are correlated with physical growth, the traditional measure of sexual development used is the median age of onset of the menarche.

The close links between height for age, adult height, and the age of menarche is shown for Papua New Guinean societies by the "menarche line" drawn across Figure 2 (Malcolm, 1974). This is based upon cross-sectional data drawn from a number of societies and the median age determined by probit analysis (Malcolm, 1970). This is expressed as a regression equation of median age of menarche on the mean adult female height as follows:

Age of menarche (years) = 62.71 − [0.303 × adult female height (cm)] ($r = 0.85$)

The point of intersection with the female growth curve indicates the median age at which menarche occurs in that society. This ranges from 13.8 years in urban Lae to 18.0 years in Bundi, this latter figure being the highest age reported for any population. No data are available for the age of menarche in Asai girls who, on this basis, would be expected to reach menarche at just under 20 years.

A similar delay occurs in the onset of pubescence in both males and females, in the deepening of the voice in boys, while puberty, in keeping with the slower and more prolonged adolescent spurt, is extended over a longer period.

Studies from other countries show a similar pattern with earlier menarche occurring in the more advantaged sectors of the population, in particular urban as compared with rural children (Tanner, 1972; Burrell *et al.,* 1961; Wilson and Sutherland, 1953). Observations on the growth and development of Japanese children from 1900 to 1971 (Oiso, 1975) showed dramatic changes in height and weight for age associated with a reduction of nearly 4 years in the age of menarche from around 16 years in 1900, these changes being associated with improved nutrition, especially of protein intake.

A similar secular trend in the age of menarche has been reported from most countries (Tanner, 1972), although the evidence that nutrition has been an important factor is largely circumstantial. However, the close links with physical growth and the dramatic changes which have occurred in some countries, as well as the evidence of animal experiments (Widdowson, 1974), clearly lead to the conclusion that sexual development is markedly delayed by malnutrition. As for physical growth, this delay is proportional to the extent and duration of nutritional deprivation.

6. Skeletal Development

It is self-evident that the variation in growth already referred to, especially in height, is essentially an expression of the variation in rates of bone growth. This has two aspects—the rate of epiphyseal growth and the timing of the closure of epiphyses is a factor which determines the time at which growth in height ceases, and secondly, the rate of skeletal maturation, as determined by the time of appearance and the development of ossification centers. These two components of skeletal development are closely related, and both are affected by protein–energy malnutrition.

Malnutrition leads to slower growth at epiphyseal centers as evidenced by the effect on growth in height. In slower growing populations, growth continues for a longer period and some compensation is achieved through the delay in epiphyseal closure. For example, in Asai and Bundi males growth in height continues well into the 20s compared with cessation in European males at around 18 years.

Bone maturation as measured by hand/wrist radiography is delayed to a varying extent by malnutrition, this being dependent, as for other aspects of growth, upon its extent and duration. Correlations between physical and skeletal growth are high (Tanner, 1962), although apparently not so high as for sexual development. Bundi children showed a delay in skeletal age increasing from about 1 year at around 6–7 years of age, to in excess of 3 years around puberty with a correlation between sexual and skeletal development of approximately 0.5. Similar observations have been recorded for other populations (Roche *et al.,* 1975). Greulich (1957), in a study of Japanese children in the past, showed that skeletal development lagged by up to 2 years in comparison with children of Japanese ancestry living in California. Frisancho *et al.* (1970) reported a delay in skeletal maturation in Central American children and noted that this delay was most marked in childhood rather than adolescence. They noted a similar pattern in other studies and postulated that the lesser delay in skeletal maturation in adolescence, as compared with growth in size, could explain the smaller adult stature of these and other malnourished populations.

The significance of malnutrition as a factor in this delay is confirmed by supplementary feeding experiments. Spies *et al.* (1959) demonstrated a significant increase in height, weight, and skeletal development in malnourished, retarded American children with protein supplementation. Malcolm (unpublished data) showed a significant acceleration in skeletal age with protein supplementation in Bundi school children over an 8-month period as compared with the expected figure for these children and with matched controls. Similar findings have been reported with animal experiments (Roche *et al.,* 1975).

In conclusion, therefore, it may be stated that the available evidence suggests that both skeletal growth and skeletal maturation are delayed by protein–energy malnutrition, that this effect is correlated with physical growth but not to the same degree as sexual development, and that a possible differential between bone growth and more advanced bone maturation in adolescents may be the principal factor in the short stature of adults in situations of chronic malnutrition.

7. Dental Development

The eruption times in deciduous teeth show comparatively little variation between different world populations despite the wide variations in growth rates. In Papua New Guinean populations (Malcolm, 1973) there is no significant difference in eruption times of deciduous teeth between the most rapidly growing children of the urban area as compared to the slow-growing Bundi child. However, growth in height variation within populations is correlated with the number of teeth erupted, suggesting that this characteristic is under strong genetic influence and is therefore almost completely resistant to environmental factors.

On the other hand, eruption times of permanent teeth (Malcolm and Bue, 1970) do vary, although to a limited extent, with rates of growth. Bundi children show a mean delay in overall eruption times of 0.66 years in females and 1.01 years in males as compared with children of urban Lae. This delay is most marked in the later-erupting teeth, especially the molars, but eruption times are earlier than for European populations. Correlations between height variation and teeth erupted within populations agree very closely with between-population comparisons, suggesting that this variation can to a large extent be explained by environmental rather than genetic influences. However, eruption times between races are clearly genetically determined.

The relatively small influence of the environment, presumably nutritional, supports Tanner's conclusion (1972) that dental development is under strong genetic control and that environmental factors exert only a minor influence on the development of the head end of the body. As a consequence, dental eruption times, most conveniently assessed by counting the number of erupted teeth, provide a useful basis for the determination of age of groups of children in both the deciduous and permanent eruption phase where birth records of a population are not available.

8. Body Proportions

The various developmental patterns already discussed relate to age and are, hence, age dependent. Malnutrition, however, affects these various patterns differentially, depending largely upon the balance between, and total intakes of, energy and protein. Reference has already been made to the more pronounced effect of energy, as compared with protein, deficiency in children under 3 years old and hence the more marked effect upon weight and skinfold thickness as compared with height for age.

The most commonly reported relationship between body proportions is that of weight for height, on the assumption that a low weight-for-height ratio is an indicator of malnutrition. While this may be true of energy deficiency, it is almost certainly an erroneous assumption for the effect of general protein–energy malnutrition and is possibly complicated by genetic influences upon growth.

Hiernaux (1964) showed that a most useful way of comparing populations for weight/height ratio was to plot log mean weight against mean height which resulted in a linear function for all except the terminal stages of growth. Comparison of a number of racial groups (Hiernaux, 1964) revealed a strongly parallel relationship which seemed to reflect the contrasting nutritional backgrounds of the populations studied, the more deprived populations showing a low weight-for-height ratio throughout the growth span. However, not all populations conformed to the pattern described by Hiernaux. Growth curves of the Papua New Guinean populations studied (Malcolm, 1970) show both a wide variation in slope and of intercept, especially with highland populations where protein deficiency tends to exceed energy deficiency. As a consequence there is a high weight-for-height ratio which substantially exceeds that of European children. This suggests that where protein is inadequate but energy relatively satisfactory, the ratio will be high, whereas in the reverse situation, for example the Tutsi reported by Hiernaux, where protein intake is high in comparison with energy, children show a low weight-for-height ratio.

What is quite clear about this ratio is that it is highly susceptible to environmental influences but should be used with caution as an indicator of nutritional status.

On the other hand, the ratio of linear body proportions appears to be highly stable under a wide variety of ecological conditions. The relationship of sitting/standing height in different populations is almost linear and parallel throughout most of the growth span and shows almost no variation within different racial groups, but varies widely among them. Papua New Guinean populations show a remarkable constancy in this relationship despite a wide range in height for age. Greulich (1957) showed a similar stability in a comparison of native and American-born Japanese children, despite the more rapid growth and taller adult stature of the latter group. This suggests that nutritional and other environmental influences are relatively slight in influencing what must be strongly acting genetic factors in determining this linear body proportion.

9. Conclusion and Summary

The foregoing evidence clearly implicates protein–energy malnutrition as the major environmental determinant of the outcome of the growth process from the early stages of prenatal development through to the size and physique of the mature adult.

Protein–energy malnutrition, however, is not just a simple matter of food availability, nor is it, in itself, a simple concept. There are many intervening variables between the macroenvironmental factors of agricultural production, distribution, and availability, and the actual food intake of the child, including cultural attitudes, meal frequency, motivational factors, infection and illness, and interpersonal relationships between parents and siblings.

Adaptation to adverse environmental conditions must occur to some extent during the growth period. Growth is the evolution of the individual, and throughout the process individual characteristics are shaped into the environmental mold. Slow growth may be considered as an adaptational response to nutritional deprivation, and it is perhaps inappropriate to place a value judgment upon this outcome by the use of words such as "delayed" or "retarded." While the price of adaptation may be a high early childhood mortality, such adaptation is to some extent a successful one in that the older child, in nutritionally deprived societies, shows few signs of physical disability or illness. In Papua New Guinea highland adults are active, relatively free of sickness, show reasonable mental alertness, and are capable of hard physical work (Hipsley and Kirk, 1965). Efforts to remedy the problem should be directed at not merely improving the rate of growth, which of itself is not a problem, but at the clear and definite adverse effects which are higher morbidity and mortality, susceptibility to infection, and perhaps mental handicap.

The criteria for age estimation based upon size in populations exhibiting slow growth need to be carefully determined, especially when no good birth data are available. This is required especially for such purposes as admission to school and health and legal matters.

Recent work has gone a long way in disentangling the contribution of the various environmental and genetic factors to human growth patterns. While it is now clear that smaller size for age and smaller adult stature are primarily adaptive responses to protein malnutrition, there is still much that needs further study in elucidating the mechanisms involved, and their interplay in this most fascinating of human biological investigations.

10. References

Antonov, A. N., 1947, Children born during seige of Leningrad in 1942, *J. Pediatr.* **30**:250.

Bergner, L., and Susser, M., 1970, Low birth weight and pre-natal nutrition: An interpretative review, *Pediatrics* **46**:946.

Burrell, R. J., Healy, M. J., and Tanner, J. M., 1961, Age at menarche in South African Bantu schoolgirls living in the Transkei Reserve, *Hum. Biol.* **33**:250.

Chase, H. P., and Martin, H. P., 1970, Undernutrition and child development, *N. Engl. J. Med.* **282**:933.

Eveleth, P. B., 1975, Differences between ethnic groups in sex dimorphism of adult height, *Ann. Hum. Biol.* **1**:35.

Frisancho, A. R., Garn, S. M., and Ascott, W., 1970, Childhood retardation resulting in reduction of adult size due to a lesser adolescent skeletal delay, *Am. J. Phys. Anthropol.* **33**:325.

Garrow, J., and Pike, M. D., 1967, The long term prognosis of severe infantile malnutrition, *Lancet* **1**:1.

Greulich, W. W., 1957, A comparison of the physical growth and development of American-born and native Japanese children, *Am. J. Phys. Anthropol.* **15**:489.

Hiernaux, J., 1964, Weight–height relationship during growth in Africans and Europeans, *Hum. Biol.* **36**:273.

Hipsley, E. H., and Kirk, N. E., 1965, Studies of dietary intake and the expenditure of energy by New Guineans, South Pacific Commission Tech. Paper No. 147, Noumea.

Iyengar, Leela, 1967, Effects of dietary supplements late in pregnancy on the expectant mother and her newborn, *Ind. J. Med. Res.* **55**:85.

Jelliffe, D. B., 1968, *Infant Nutrition in the Sub-Tropics and the Tropics*, W.H.O. Monograph, No. 29, Geneva.

Malcolm, L. A., 1970, Growth and development in New Guinea: A study of the Bundi people of the Madang district, Institute of Human Biology, Madang, Monograph Series No. 1.

Malcolm, L. A., 1972, Anthropometric, biochemical and immunological effects of protein supplements in a New Guinean highland boarding school, IXth International Congress of Nutrition, Mexico.

Malcolm, L. A., 1973, Deciduous dental development and age assessment of New Guinea children, *Environ. Child Health* June (special issue).

Malcolm, L. A., 1974, Ecological factors relating to child growth and nutritional status, in: *Nutrition and Malnutrition* (A. F. Roche and F. Falkner, eds.), pp. 329–352, Plenum, New York.

Malcolm, L. A., and Bue, B., 1970, Eruption times of permanent teeth and the determination of age in New Guinean children, *Trop. Geogr. Med.* **22**:307.

McCance, R. A., 1967, The effect of calorie deficiencies and protein deficiencies on final weight and stature, in: *Calorie Deficiencies and Protein Deficiencies* (R. A. McCance and E. Widdowson, eds.), pp. 319–328, Churchill, London.

Meredith, Howard V., 1970, Body weight at birth of viable human infants: A world-wide comparative treatise, *Hum. Biol.* **42**:217.

Neumann, C. G., Stankar, H., and Uberoi, I. S., 1969, Nutritional and anthropometric profile of young rural Punjabi children, *Ind. J. Med. Res.* **57**:1.

Oiso, T., 1975, A historical review of nutritional improvement in Japan after World War II, in: *Physiological Adaptability and Nutritional Status of the Japanese* (K. Asahina and R. Shigiza, eds.), Human Adaptability, Vol. 3, JIBP Synthesis, Vol. 4, pp. 171–191, University of Tokyo, Tokyo.

Roche, A. F., Roberts, Jean, and Hamill, Peter V. V., 1975, Skeletal maturity of children 6–11 years: Racial, geographic area, and socioeconomic differentials, National Center for Health Statistics. Data from National Health Survey, Series 11, No. 149, U.S. Govt. Printing Office, Washington, D.C.

Spies, T. D., Dreizen, S., Snodgrasse, R. M., Arnett, C. M., and Webb-Peploe, H., 1959, Effect of dietary supplement of non-fat milk on human growth failure. Comparative response in undernourished children and undernourished adolescents, *Am. J. Dis. Child.* **98**:187.

Stein, Z., Susser, M., Saenger, G., and Marolla, F., 1975, *Famine and Human Development: The Dutch Hunger Winter of 1944–1945*, Oxford, London.

Subrahmanyan, V., Joseph, K., Doraiswamy, T. R., Narayannarae, M., Sankaran, A. N., and Swaminathan, M., 1957, The effect of a supplementary multipurpose food on the growth and nutritional states of schoolchildren, *Br. J. Nutr.* **11:**382.

Tanner, J. M., 1972, *Growth at Adolescence,* 2nd ed., Blackwell, Oxford.

Tanner, J. M., Whitehouse, R. H., and Takaishi, M., 1966, Standards from birth to maturity for height, weight, height velocity and weight velocity for British children, 1965, *Arch. Dis. Child.* **41:**454.

Widdowson, E. M., 1974, Changes in pigs due to undernutrition before birth and for one, two, and three years afterwards, and the effects of rehabilitation, in: *Nutrition and Malnutrition* (A. F. Roche and F. Falkner, eds.), pp. 165–181, Plenum, New York.

Wilson, D. C., and Sutherland, I., 1953, The age of the menarche in the tropics, *Br. Med. J.* **2:**607.

12

Population Differences in Growth: Environmental and Genetic Factors

PHYLLIS B. EVELETH

1. Introduction

There are differences among populations in body size and shape, and in rate of growth in children. These differences give rise to the population differences seen in adults. Even the casual observer will note that Japanese or Indian or Filipino children are smaller than their age peers in Europe or the United States. The more astute observer also perceives that the body shape of Japanese or Burmese children, for example, is different from the shape of West African or European children. School teachers and clinic workers may remark that American black adolescents seem to mature faster than their white peers.

Among the more important environmental influences on growth are nutrition, disease, socioeconomic status, urbanization, physical activity, psychological stress, season of year, and climate. Most of these, when analyzed, are seen to be directly or indirectly related to nutrition. It is generally in developing countries with borderline nutrition that children are small, and thus, it is quite rightly concluded that the slower growth and smaller size of these children is the result of poor nutrition. Nutrition does play a most important part in growth; however, not all differences among populations are the result of poor nutrition. If such were the case, by supplying all peoples with adequate nutrition, we could eliminate population differences. This thinking ignores the very real effect of the genes.

PHYLLIS B. EVELETH • Division of Research Grants, National Institutes of Health, Bethesda, Maryland.

Population differences are the result of both environmental and genetic factors, not acting singly, but interacting so that genetic factors may predispose an individual or group to have greater sensitivity to some environmental factors that may not affect other children in the same manner. During a period of famine the growth of all children will be retarded, but not all will be affected to the same degree because of their genetic differences.

We may endeavor to study the relative contributions of genetic and environmental factors by carefully selecting the populations used for comparison, although we can never be certain of completely separating all the factors. In an effort to identify existing genetic differences in growth, I have drawn from studies of economically well-off populations that differ physically. In discussions of environmental influences, I have selected groups within a population that vary in socioeconomic status or degree of urbanization. Throughout, one assumption has been made: that families in the high and middle socioeconomic classes provide their offspring with better nutrition, child care, and take better advantage of health and medical services than families at the lower levels; and that urban families are better off than rural when it comes to food supply, medical services, and sanitation facilities.

2. Population Differences in Body Size

2.1. Birth Weight

European populations and those of European ancestry have male birth weight means ranging from 3.3 to 3.5 kg and female weight means from 3.0 to 3.4 kg. Other populations have greater ranges and frequently lower means. The lowest means recorded are from the Lumi of New Guinea (2.4 kg, both sexes; Wark and Malcolm, 1969), Guatemalan Maya (2.56 kg, both sexes; Mata *et al.*, 1971), and the Nagaya pygmies (2.61 kg, both sexes; Vincent *et al.*, 1962). The highest means are among Europeans, North and South American upper and middle classes, North American Indians, New Zealand Maori, and well-off Near Easterners (see Eveleth and Tanner, 1976; Meredith, 1970).

One cause of lower weight at birth is believed to be malnutrition during pregnancy (Brasel and Winick, 1972; Naeye *et al.*, 1971; Smith, 1947). The placenta was considered to protect the fetus from maternal malnutrition, but recently it has been shown in the rat that fetal brain and placenta are reduced in cell number (DNA) and increased in cytoplasmic mass (RNA/DNA) with food restriction (Brasel and Winick, 1972). Similar results were obtained from analyses of human placentas from Chilean and Guatemalan infants with intrauterine growth retardation (Brasel and Winick, 1972). Supplementing the diets of pregnant mothers in Guatemalan villages with a high-calorie drink raised the incidence of babies over 3.0 kg and reduced the number of those under 2.5 kg (Read *et al.*, 1975).

2.1.1. Socioeconomic Status

In most countries upper- and middle-class families produce offspring with higher mean birth weight and a smaller incidence of infants under 2.5 kg than lower-class families (Table I). Some of these differences reflect the size of the mothers, who frequently are smaller in the lower social classes. In Britain when birth weight

375

POPULATION
DIFFERENCES IN
GROWTH:
ENVIRONMENTAL AND
GENETIC FACTORS

Table I.   Male Birth Weight Means (kg) Among Different Socioeconomic Groups

Place	Well-off	Poor	Source
Teheran, Iran	3.43	3.27	Hedayat *et al.*, 1971
Shiraz, Iran	3.18	3.02	Sarram and Saadatnejadi, 1967
Lebanon	3.50	3.40	Hasan *et al.*, 1969
Delhi, India	3.16	2.74	Banik *et al.*, 1969
Campinas, Brazil	3.41[a]	3.18[a]	Martins Filho *et al.*, 1974
Baltimore, USA, black	2.97[a]	2.91[a]	Penchaszadeh *et al.*, 1972
Baltimore, USA, white	3.27[a]	3.13[a]	Penchaszadeh *et al.*, 1972

[a]Both sexes.

differences were corrected for maternal height, social class differences turned out to be negligible (Butler and Alberman, 1969).

In the Maternity Hospital of Campinas, Brazil, private patients had infants with greater mean birth weights when data were grouped for gestational age than indigent patients or patients with social insurance (Martins Filho *et al.*, 1974). From various cities in India there are reports of higher mean birth weights in the higher socioeconomic groups (Achar and Yankauer, 1962; Banik *et al.*, 1969; Chandra, 1971), but in Delhi even the means of the upper class (3.16 kg, boys; 2.92 kg, girls; Banik *et al.*, 1969) were far below those of Europeans (Netherlands 3.5 kg, boys; 3.4 kg, girls; van Wieringen *et al.*, 1971). In Hyderabad, of those infants born into low middle-income families, 32.7% were below 2.5 kg (Chandra, 1971). In Baltimore, gestational age and birth weight increased with higher socioeconomic index in Americans, both of European and African ancestry (Penchaszadeh *et al.*, 1972). At comparable socioeconomic levels, however, the black infants had lower birth weights and lengths than the whites. Other studies in the United States likewise have reported a gap between black and white Americans in birth weight (Garn and Clark, 1976a; Robson *et al.*, 1975). Although these racial differences may be genetic (Naylor and Myrianthopoulous, 1967), they may also reflect maternal childhood nutrition or possibly different habits in the black subculture of the United States.

Population differences in size at birth may reflect not only the environment of the infants and their mothers, but also the nutrition of the mother while she was growing up. Results from the British Perinatal Mortality Survey showed that mothers under 152 cm had a higher incidence of babies under 2.5 kg and a much lower incidence of babies over 4.0 kg than taller mothers (Butler and Alberman, 1969). Moreover tall women are more likely to produce large babies (Ounsted and Ounsted, 1973; Thomson *et al.*, 1968). It is assumed that taller mothers have enjoyed better conditions during growth and have more successfully reached their growth potential than shorter mothers.

2.1.2. Genetic Factors

It is difficult to extract genetic components from world birth weight data, although birth weight has been shown to be to a large degree inherited (Morton, 1955; Robson, 1955; and see Chapter 10, Volume 1). It is likely that some characteristic of maternal constitution or maternal environment is more important than, or as important as, fetal genotype although more evidence along these lines is needed. Racial differences are consistent, as Meredith (1970) in his compilation of world birth weights has pointed out. He has estimated for various populations the frequency of viable infants under 2.5 kg as follows: 6% for northwest European ancestry, 11% for Mexican and Chinese, 13% for Africans and Afro-Americans, and 25% for Indian and Burmese newborn infants.

2.2. Childhood and Adolescence

After birth, size differences between black and white American infants are soon reversed. By one or two years of age the blacks are similar to or greater than white Americans of similar income groups in height (Garn and Clark, 1976*a;* Robson *et al.,* 1975) and weight (Eveleth and Tanner, 1976). This is not true of blacks in Africa unless they are from well-off families. In Africa the more common pattern is a gradual falling off in growth in weight and length after 6 months of age when compared to European reference standards (Eveleth and Tanner, 1976). This unquestionably is a reflection of a poor environment where malnutrition and infection interact in a deadly partnership, retarding growth and increasing morbidity and mortality.

2.2.1. Socioeconomic Status

Two longitudinal studies on well-off and slum children in Ibadan and Lagos, Nigeria (Rea, 1971; Janes, 1974, 1975) illustrate the overwhelming effect of the environment. Means of height and weight for the well-off are fully comparable to European children, at least for the period studied (1–10 years of age). In the slums, however, the average 1-year-old child is comparable to the 6-month-old well-off child in weight and to the 9-month-old child in length. Since genetically the groups are said to be similar, we may assume these differences to reflect the unfavorable environment in the slums.

Well-off preschool children in Shiraz, Baghdad, Tunis, and Beirut are larger than their lower-class peers and are also similar to Europeans in weight and height (Eveleth and Tanner, 1976). In Delhi the well-off are larger than the lower-class Indians but are still somewhat smaller than Europeans (Banik *et al.,* 1972). Figure 1 shows the socioeconomic differences that exist in school-age children in Port au Prince, Ibadan, Tunis, Hong Kong, Britain, and the United States. In Britain, the National Child Development Study revealed that upper- and middle-class 7-year-old children were on average 3.3 cm taller than children from unskilled working-class

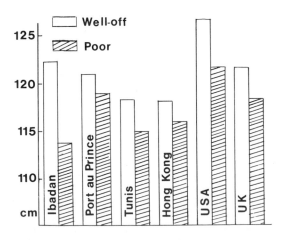

Fig. 1. Mean heights of poor and well-off 7-year-old boys in developing and industrial countries. Sources are Ibadan, Janes 1974, 1975; Port au Prince, King *et al.,* 1963; Tunis, H. B. Boutourline Young, unpublished; Hong Kong, Chang, 1969; U.S.A., Hamill *et al.,* 1972; U.K., Goldstein, 1971.

377

*POPULATION
DIFFERENCES IN
GROWTH:
ENVIRONMENTAL AND
GENETIC FACTORS*

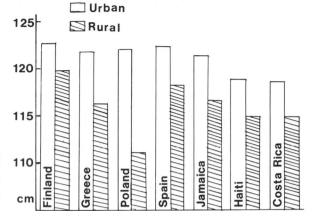

Fig. 2. Mean heights of urban and rural 7-year-old boys. Sources are Finland, Backström-Järvinen, 1964; Greece, Valaoras and Laros, 1969; Poland, Wolański and Lasota, 1964; Spain, Palacios and Vivanco, 1965; Jamaica, Ashcroft *et al.,* 1965, and Ashcroft and Lovell, 1966; Haiti, King *et al.,* 1963; Costa Rica, Villarejos *et al.,* 1971.

families (Goldstein, 1971). The U.S. Health Examination Survey and the Ten-State Nutrition Survey both showed that as family income and/or education increased so did the size of the children (Hamill *et al.,* 1972; Garn and Clark, 1975). A recent report reveals that father's income or occupational status are no longer important factors influencing a child's height and weight-for-age in Sweden where differences in living conditions between different socioeconomic groups are diminishing (Lindgren, 1976). A socioeconomic difference was observed, however, by a greater weight-for-height among the lower socioeconomic groups, especially for girls.

2.2.2. Urbanization

Life in urban areas is also apparently more favorable for child growth than that in rural zones. Children in cities are usually larger than comparable children in rural areas. Urbanization means a high population density where residents are provided with a regular supply of goods and services, health and sanitation facilities, large and specialized medical institutions, and education, recreational, and welfare facilities. Air and noise pollution in big cities have more recently come into consideration, and their effects on growth, if any, are only beginning to be investigated. Some areas of high population density, as the shanty slums of South America and Africa, have only a minimum of urban services and institutions and do not really fit the above definition. A European or North American city is considerably different from an urbanized area in Africa or even a city in India or Japan.

Investigators in Poland (Wolański and Lasota, 1964), Greece (Valaoras and Laros, 1969), Finland (Backström-Järvinen, 1964), Rumania (Cristesçu, 1969), Spain (Palacios and Vivanco, 1965), and Korea (Kwon, 1978) have reported that children are taller and weigh more in the cities than in the rural zones, although they frequently have more weight for height (Figure 2). Children of migrants to a new industrial town, Nowa Huta, in Poland were seen to be on average 4.3 cm taller and 1.5 kg heavier at ages 4–14 years than the children who grew up in the farm villages from whence the migrants came (Panek and Piasecki, 1971). These results suggest better conditions in the new city, but we cannot rule out some selection of migrants as well.

In the Netherlands (van Wieringen *et al.,* 1971), Australia (Jones *et al.,* 1973), and the United States (Hamill *et al.,* 1972), no significant differences were found in height and weight between urban and rural children, although socioeconomic differences in height and weight were present. Presumably in these countries conditions in the rural areas (in Australia, an inland town) are more favorable than in European countries where the rural villages date from medieval times.

In many developing countries, where the cities are newer and growing rapidly, urban slums spring up to house the new migrants. It is questionable whether life in such urban slums is an improvement over that in impoverished rural villages. In Nigeria (Janes, 1974; Morley *et al.,* 1968), Costa Rica (Villarejos *et al.,* 1971), South Africa (Richardson, 1973), and India (Indian Council of Medical Research, 1972) there was little difference in heights and weights between the urban slum and the rural village children. However, children living in better parts of the cities in Costa Rica (Villarejos *et al.,* 1971) and Jamaica (Ashcroft *et al.,* 1965; Ashcroft and Lovell, 1966), were considerably larger than rural children.

2.2.3. Genetic Influences

It is becoming increasingly apparent that in a good environment black children of African ancestry will be taller, and possibly heavier, than children of European ancestry. Well-nourished Ibadan boys and girls are above the British standards in height and height velocity but are comparable in weight (Janes, 1975, and unpublished). Numerous investigations from the United States, across the country from Pennsylvania to California, are in agreement in their conclusion that Afro-Americans are taller than Americans of European ancestry living in the same cities (Barr *et al.,* 1972; Garn and Clark, 1976*a;* Garn *et al.,* 1973; Hamill *et al.,* 1970; Krogman, 1970; Rauh *et al.,* 1967; Robson *et al.,* 1975). In some samples they were also heavier (Barr *et al.,* 1972; Robson *et al.,* 1975) or girls only were (Krogman, 1970; Rauh *et al.,* 1967). Thus we see that the deficit in birth weight discussed earlier is not only made up, it is sometimes reversed.

In contrast Asiatic children, according to data reported so far, are similar to Europeans only during the first 6 months of life, and many are smaller at birth. They then fall progressively behind. Even well-off children in Japan (Terada and Hoshi, 1965), Bangkok (Khanjanasthiti *et al.,* no date), Hong Kong (Chang, 1969), and Formosa (Kimura and Tsai, 1967) are below the European averages in height at 3 or 4 years (Formosa, not until 7 years) and after 1 year of age are lower in weight. Smaller size is also observed among Asiatics living outside Asia: Chinese in Jamaica (Ashcroft and Lovell, 1964; Ashcroft *et al.,* 1966), Japanese in California (Greulich, 1957; Kondo and Eto, 1972), Japanese in Brazil (Guaraciaba, 1967), and Orientals in San Francisco (Barr *et al.,* 1972), all of whom are experiencing an improvement in their environment.

An interesting comparison may be made between the children of Japanese immigrants to California studied in 1956 (Greulich, 1957) and of the children of the succeeding generation studied in 1971. Greulich found that Japanese children who grew up in California were larger than those who were living in Japan at that time and concluded that this was the result of a better environment. In view of Greulich's results and also because the secular increase in height in Japan itself has been considerable, some people suggested that the size differential between Japanese and European-descended children would eventually disappear. However, there does appear to be a limit to the increase in size, and this limit is probably imposed by the genes, for we see that those Japanese in California measured in 1971 (Kondo and Eto, 1972) were not perceptibly taller than Greulich's sample 15 years earlier (except 15- to 17-year-old girls) (Greulich, 1976; Eveleth and Tanner, 1976) (Figure 3).

379

**POPULATION
DIFFERENCES IN
GROWTH:
ENVIRONMENTAL AND
GENETIC FACTORS**

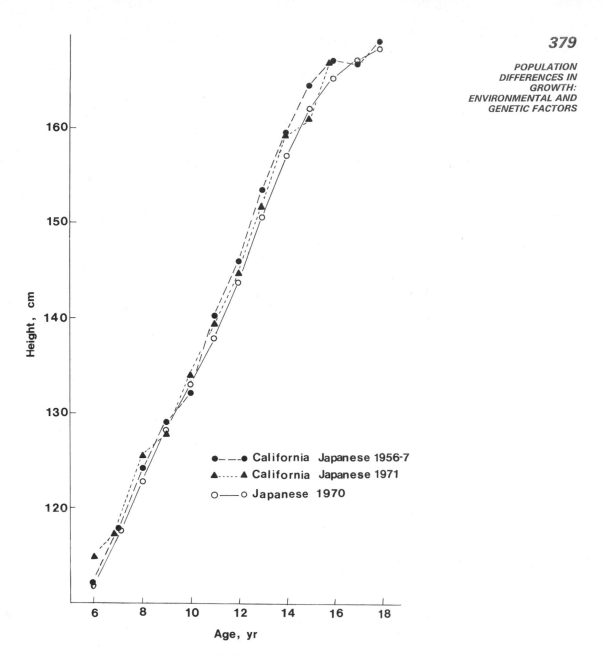

Fig. 3. Mean heights of Japanese boys in Japan, 1970 (Tokyo Dept. Maternal and Child Health, 1970), and in California, 1956–1957 (Greulich, 1957) and 1971 (Kondo and Eto, 1972). Although Japanese living in California are somewhat taller than those in Japan, the differences between the two California samples examined 15 years apart are negligible.

3. Population Differences in Body Shape

The characteristic adult shape differences among the major racial groups may be recognized during the growth period. For the most part shape had been considered to be genetically determined, although at least in one instance it has been shown to be influenced by the environment. The children of Japanese ancestry in

California, discussed above, were seen to have longer legs relative to sitting height than Japanese in Japan (Eveleth and Tanner, 1976; Greulich, 1957). This difference did not continue into adulthood (Greulich, 1976), and thus would be considered as an effect on growth, not a lasting change in adult physique. It should not be implied from this that populations with longer legs relative to sitting height are better nourished, because we know that is not the case. Changes in weight for height would be considered also as shape changes and may occur with improved nutritional circumstances or, more likely, with overnutrition, because when nutrition is satisfactory height and weight both are greater.

The population with the longest legs relative to sitting height is the Australian Aborigines who far exceed the African and African-descended populations. The Africans in turn have relatively longer legs than the Europeans (Meredith and Spurgeon, 1976) or Asiatics. A surprising observation has been made in reference to the latter population. The characteristic short legs of the Chinese and Japanese apparently arise at midchildhood, for before that time they have leg-to-trunk proportions that are greater than, or similar to, Europeans (Eveleth, 1978; Eveleth and Tanner, 1976). Figure 4 illustrates this relationship by regressions of sitting height on leg length for Hong Kong Chinese, Londoners, Ibadan Nigerians, and Australian Aborigines.

People of African ancestry not only have longer legs to sitting height, they also have longer arms, and this is seen during growth. Europeans and Asiatics are quite comparable in relative length of arm to trunk but the Chinese, at least from 6 to 17 years, have relatively longer arms to legs than either Africans or Europeans (Eveleth and Tanner, 1976).

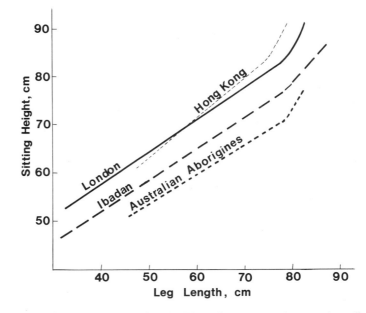

Fig. 4. Regression of sitting height on leg length of boys from one to 18 years. Australian Aborigines have longest legs to sitting height, while young and adolescent boys in Hong Kong have the shortest. The lack of parallelism between Hong Kong and other populations points to a different growth pattern in Chinese: growth in leg length slows down relative to trunk length beginning in midchildhood. Sources are: Australian Aborigines, Abbie, 1967; Hong Kong, Chang, 1969; Ibadan Nigerians, M. D. Janes, unpublished; London, J. M. Tanner, unpublished.

The relation of shoulder width to hip width portrays marked racial differences. Children of African origin, both males and females, have narrower hips in relation to shoulders than Asiatics or Europeans, while Chinese have broader hips to shoulders than the other groups (Eveleth and Tanner, 1976).

4. Differences in Amount of Subcutaneous Fat

Although population differences in subcutaneous fat deposition (measured by skinfold thickness) closely depend upon level of nutrition, there may be some genetic factors concerned, both in amount and site at which fat is deposited. In Africans, triceps skinfolds were smaller than in Europeans (see Eveleth and Tanner, 1976, for references); in Afro-Americans they were smaller than in Euro-Americans living in the same cities even though subscapular skinfolds were comparable (Garn and Clark, 1976b; Johnston et al., 1972, 1974; Malina, 1966; Rauh and Schumsky, 1968). Apparently less fat is deposited on lower limbs of African descendants as well, according to the results of the U.S. Health Examination Survey (Johnston et al., 1974). Medial calf skinfold was measured and reported to be considerably smaller in Afro-Americans than Euro-Americans. Blacks were also smaller in suprailiac and midaxillary skinfolds but similar in subscapular skinfold.

Asiatic and Amerindian populations, even those under nutritional stress, had subscapular skinfolds comparable to, or greater than, Europeans when, at the same time, their triceps skinfolds were smaller (Eveleth and Tanner, 1976). Malina et al. (1974) have noted among Guatemalan Ladinos that loss of fat in the first year after birth was greater in the limbs than trunk with the result that Ladinos had average subscapular values between the British 25th and 50th centiles, but triceps skinfold averaged between the British 3rd and 10th centiles. Subscapular skinfolds were greater than the British median in Quechua Indian and Alaskan Eskimo girls, but not boys (Frisancho and Baker, 1970; Jamison and Zegura, 1970). Eskimo adults do not have particularly large skinfolds as possible insulation against cold. In fact, it is uncertain how much body fat is situated subcutaneously for any group of people (Durnin and Womersley, 1974), but percentage body fat predicted from the European regression of total body water was no higher in Igloolik Eskimos than in many American college students (Shephard et al., 1973).

European infants gain fat rapidly after birth and lose it gradually beginning anywhere from 9 to 12 months until 7 or 8 years of age. Non-European infants in unfavorable environments seem to display a different pattern: There is a small postnatal increase with subsequent decrease in the first year. This is followed at two years of age by a considerable increase which lasts until approximately 3 years of age (Eveleth, 1975; Janes, 1974; Malina et al., 1974). Well-off infants in Nigeria exhibit this pattern as well, which Janes (1974) suggests may be due to the lack of suitable weaning foods to use between 4 and 12 months of age. Inadequate supply of breast milk along with little food supplementation may be involved in Guatemala, as Malina and his colleagues (1974) have suggested. Moreover age at weaning among Guatemala Ladinos and Nigerians, and the dip in subcutaneous fat, correspond well in time.

At adolescence, girls increase fat, and some boys do so as well. Adults may continue to gain fat until about 50 years of age (Garn and Clark, 1976b), and some individuals gain an extraordinary amount. In the United States sex and racial differences in obesity based on triceps skinfold have been observed: At 14 years of age 18.8% boys and 15.3% girls of European ancestry were considered obese (from triceps skinfold) as were 12.5% boys and 18.6% girls of African ancestry (Schaefer, 1975). By 30 years of age more black females than any other group (36.4%) were described as obese. Among Euro-American women obesity was not quite as frequent (25.9%), and men of European and African ancestry also had smaller incidents (21% and 12%, respectively). Obesity seems to be more prevalent at the lower socioeconomic levels in highly industrialized countries such as the United States and the United Kingdom (Garn and Clark, 1976b, women only; Goldblatt *et al.*, 1965; Silverstone *et al.*, 1969; Stunkard *et al.*, 1972; Whitelaw, 1971), but this does not indicate that the lower-class diet is nutritionally better (Cook *et al.*, 1973). More likely it reflects the increasing financial means to consume more food without social restrictions to maintain a lean body form or even with social encouragement to be overweight as a sign of affluence. Unfortunately as developing countries progress, the prevalence of obesity increases among their populations (Fernandez *et al.*, 1969).

5. Skeletal Maturation

5.1. Environmental Factors

An inadequate level of nutrition, severe disease, or period of famine will slow down the rate of skeletal maturation (Acheson, 1960). In an impoverished rural area of Guatemala where the average diet is protein-deficient, children were delayed after 1 year of age in the time of first appearance of ossification centers in the hand and wrist compared with American children in Ohio (Blanco *et al.*, 1972). During the first year of life, however, the rural Guatemalans were similar to Americans: The bones that begin ossification earliest did so in both populations at similar ages. Guatemalan children who were identified as having even mild protein-calorie malnutrition showed a deficiency in cortical bone growth as well as reduced body size (Himes *et al.*, 1975). In Delhi, Indian children from 1 to 5 years suffering from poor nutrition were behind those who were considered to be well nourished, although the latter were not quite up to American norms (Banik *et al.*, 1970). Among Hong Kong Chinese, Indians, and Baghdad Arabs, well-off children were more advanced in skeletal age than those from the lower socioeconomic groups (Low *et al.*, 1964; Maniar *et al.*, 1974; Shakir and Zaini, 1974). The first two samples were rated by the Greulich–Pyle system and the latter by the Tanner–Whitehouse I method.

In addition to nutrition, high altitude appears to have a considerable effect on skeletal maturation. The slowest maturing populations recorded are the Quechua Indians of Nuñoa, Peru, living at an altitude of 4000–5500 m (Frisancho, 1969), the Sherpas at 3475–4050 m, and the Tibetans at 1400 m (Pawson, 1977). At high altitudes these people suffer from hypoxia and low temperatures and, in some instances, unsatisfactory nutritional levels.

5.2. Genetic Factors

In the United States low-income Afro-Americans were advanced in timing of ossification centers over Euro-Americans of somewhat higher incomes, suggesting that genetic factors also have an important influence on skeletal maturation (Garn *et al.*, 1972a; Garn and Clark, 1975). There is some evidence pointing to the more

383

*POPULATION
DIFFERENCES IN
GROWTH:
ENVIRONMENTAL AND
GENETIC FACTORS*

rapid skeletal maturation of children of African ancestry, but in Africa and Jamaica this advancement is not maintained throughout the growth period (Malina, 1970; Marshall *et al.,* 1970; Massé and Hunt, 1963; Michaut *et al.,* 1972) (Figure 5). Although adverse circumstances may underlie the change in skeletal maturation in blacks, Wingerd *et al.* (1974) have pointed out that blacks apparently have a different maturational pattern from whites as far as timing and duration of some stages is concerned.

Asiatic populations show different timing when compared with European standards, but the differences are not in the same direction at all ages. A substantial number of investigations have shown either a delay or comparability among Japanese and Chinese during childhood when related to British or American standards; this then reverses to an advancement during adolescence (Ashizawa, 1970; Greulich, 1957; Kimura, 1972*a,b,* 1976*a,b,* 1977; Kondo and Eto, 1972; Low *et al.,* 1964) (Figure 5). Japanese–American hybrids in Tokyo studied longitudinally tended to be between the Japanese and British standards, although closer to the former than the latter in that they became quite advanced after 8 years of age as did the Japanese (Kimura, 1976*a*). Adolescence then appears to begin earlier and occupies a shorter period among Asiatics than in children of European ancestry.

More rapid skeletal maturation at adolescence may go a long way toward

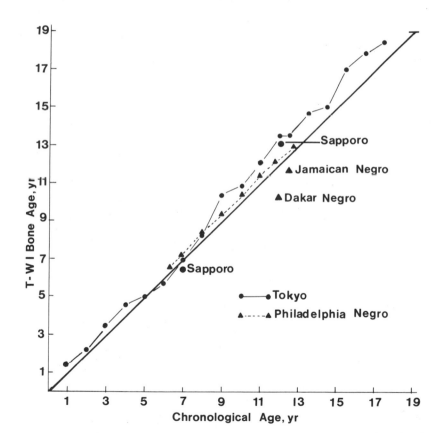

Fig. 5. Tanner–Whitehouse I bone age of some Negro and Asiatic populations in reference to the British standard: boys. Adolescent Asiatics mature considerably faster than British. Although Afro-American adolescents are advanced over British, Africans living in adverse conditions in Africa are not. Data from Tokyo, Ashizawa, 1970; Philadelphia, Malina, 1970; Dakar, Michaut *et al.,* 1972; Jamaica, Marshall *et al.,* 1970; Sapporo, Kimura, 1972*b.*

explaining the shorter stature and shorter legs relative to height of adult Asiatics. Growth in height occurs at the cartilaginous plates in the long bones which at maturity are completely ossified. The mechanism of skeletal maturation results in the eventual closure or ossification of these plates, but as long as they remain open growth in height occurs. Japanese and Chinese appear to go through adolescence faster than Europeans: In three chronological years they gain four bone age "years" (Eveleth and Tanner, 1976). This more rapid ossification, combined with the fact that they are shorter at the initiation of the spurt, may result in smaller adult height. Only longitudinal studies on Asiatic populations will give firm evidence to support this hypothesis, however, and these are still lacking.

6. Sexual Maturation

6.1. Menarche

6.1.1. Genetic Factors

Heredity influences the timing of menarche. Identical twin sisters growing up together in a good environment differ in age of menarche only by one or two months, while sisters differ by nearly a year on average (Tanner, 1962). There are also notable population differences in age at menarche (Table II).

Afro-American girls in the United States are among the earliest maturing in the world (median age 12.5; MacMahon, 1973). In Africa, however, even the well-off are later than Afro-Americans: Baganda upper class, 13.4 (Burgess and Burgess, 1964); Transkei Bantu, 15.0 (Burrell *et al.*, 1961); Ibadan upper class, 13.3 (Oduntan *et al.*, 1976); Nigerian Ibo, 14.1 (Tanner and O'Keefe, 1962). In the Middle East, most populations reach menarche relatively late except for well-off girls in Istanbul

Table II. Median Age at Menarche

Place	Median age ± S.E.	Year	Source
United States (African descent)	12.5	1960–1970	MacMahon, 1973
Ibadan, Nigeria	13.7 ± 0.03	1973–1974	Oduntan *et al.*, 1976
Rwanda, Tutsi	16.5 ± 0.16	1957–1958	Heintz, 1963
Rwanda, Hutu	17.0 ± 0.30	1957–1958	Heintz, 1963
United States (European descent)	12.8	1960–1970	MacMahon, 1973
Montreal, Canada	13.1 ± 0.04	1969–1970	Jeniček and Demirjian, 1974
Sydney, Australia	13.0	1970	Jones *et al.*, 1973
New Zealand (European descent)	13.0 ± 0.02[a]	1970	N.Z. Dept. Health, 1971
India, all (urban)	13.7	1956–1965	Indian Council of Medical Research, 1972
Japan (urban)	12.9 ± 0.01	1966–1967	Yanagisawa and Kondo, 1973
Brazil (Japanese descent)	12.9 ± 0.12	1965	Eveleth and Freitas, 1969
California (Japanese descent)	13.2 ± 0.04[a]	1971	Kondo and Eto, 1972
Oslo, Norway	13.1 ± 0.02	1970	Brundtland and Walløe, 1973
London, England	13.0 ± 0.03	1966	Tanner, 1973
Moscow, USSR	13.0 ± 0.08	1970	Miklashevskaya *et al.*, 1972
Carrara, Italy	12.6 ± 0.04	1968	Marubini and Barghini, 1969
Madrid (well-off)	12.8 ± 0.12	1968	Marin, 1971
New Guinea, Bundi	18.0 ± 0.19	1967	Malcolm, 1970
New Guinea, Chimbu	17.5 ± 0.35	1965	Malcolm, 1970
New Zealand, Maori	12.7 ± 0.07[a]	1970	N.Z. Dept. Health, 1971

[a]Recalculated by probits.

385

*POPULATION
DIFFERENCES IN
GROWTH:
ENVIRONMENTAL AND
GENETIC FACTORS*

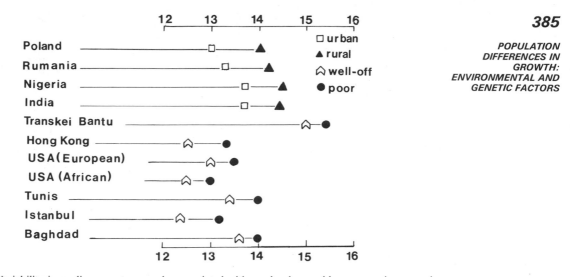

Fig. 6. Variability in median age at menarche associated with rural–urban residence or socioeconomic group. Data from Poland, Laska-Mierzejewska, 1970; Milicer and Szczotka, 1966; Rumania, Cristesçu, 1969; Nigeria, Oduntan *et al.*, 1976; India, Indian Council of Medical Research, 1972; Bantu, Burrell *et al.*, 1961; Hong Kong, Chang, 1969; U.S.A., MacMahon, 1973; Tunis, H. B. Boutourline-Young, unpublished; Istanbul, Neyzi *et al.*, 1975; Baghdad, Shakir, 1971.

(12.3; Neyzi *et al.*, 1975). The populations with the latest menarche are the New Guinean tribes having medians ranging from 15.5 to 18.4 years (Malcolm, 1970).

The above remarks on earlier skeletal maturation in Asiatics are supported by the results on sexual maturation. Japanese girls in 17 Japanese cities (Yanagisawa and Kondo, 1973) and in Bauru, Brazil (Eveleth and Freitas, 1969), Chinese in Singapore (Aw and Tye, 1970) and in Hong Kong (Chang, 1969) have been reported to reach menarche earlier than European girls (probit medians ranging between 12.5 and 12.9 years). Japanese girls in California (13.2; Kondo and Eto, 1972); and Burmese and Assamese (13.2; Foll, 1961) were later and comparable to Europeans.

European girls, except for those in the Mediterranean countries of Italy (12.6) and Spain (12.8), appear to be rather late-maturing in relation to other world populations (13.0–13.4) (see Eveleth and Tanner, 1976). The trend toward earlier maturation which has been occurring for the past century appears now to be stopping or slowing down in Europe (Bruntland and Walløe, 1973; Tanner, 1973): European girls may never mature as early as Asiatics. The trend is also stopping among girls of European ancestry in the U.S.A., but at a younger age (12.8; Zacharias *et al.*, 1976).

6.1.2. Environmental Factors

Evidence from studies on poorly nourished populations shows that poor environment will delay sexual maturation. Undernutrition slows down growth and delays the onset of maturity. In most populations (exceptions are Santiago: Rona and Pereira, 1974; and Montreal: Jeniček and Demirjian, 1974), where the comparison has been made, poor girls reached menarche later than the well-off, and the rural later than the urban (Eveleth and Tanner, 1976) (Figure 6). In a Swedish suburban town girls of all socioeconomic groups, as classified by father's occupation, had comparable mean menarcheal ages, presumably because of good social conditions (Furu, 1976). Roberts and his colleagues (1975) showed that in Newcastle, England, at least, family size rather than social class significantly affected menarcheal age.

In the U.S.A., Euro-American girls from high-income families had a mean age of 13.0 years at menarche; low-income girls 13.5. High-income Afro-Americans had a mean age of 12.6 versus 13.0 in the low income (retrospective data, MacMahon, 1973). Well-off Chinese in Hong Kong averaged 12.5; middle income, 12.8; and lower class, 13.3 years (Chang, 1969). Similar results come from Ibadan, Nigeria (13.3 for girls whose fathers were university-educated and 13.8 for girls whose fathers were illiterate, Oduntan *et al.,* 1976), Baghdad (13.6 well-off versus 14.0 poor, Shakir, 1971), and Tunis (13.5 well-off versus 14.0 poor, Boutourline-Young, unpublished data).

As pointed out above, urban residence is considered to provide better conditions for growth than rural. Thus we see that rural Indian girls have a median menarcheal age of 14.4 years compared with 13.7 in the cities (Indian Council of Medical Research, 1972), rural Rumanians 14.2 against 13.3 in urban areas (Cristesçu, 1969), rural Nigerians 14.5 versus 13.7 in Ibadan (Oduntan *et al.,* 1976) (see Tanner and Eveleth, 1975; Eveleth and Tanner, 1976, for further discussions). Undoubtedly nutritional level is involved here. Girls suffering from chronic malnutrition in a poor rural area of the United States were delayed in menarche and in skeletal maturation compared to controls. While they were shorter than the controls during growth, their mature height was not significantly different (Dreizen *et al.,* 1967).

6.2. Stages of Puberty

While, to date, investigations into the stages of pubertal development are scant, what is available shows population differences in median age of reaching each stage and possibly in the time it takes for a group to pass through a designated stage. Stages are numbered 1–5 for breast, pubic hair, and penis and have been described by Tanner (1962) and Tanner and Whitehouse (1975). Most data are from cumulative frequency distributions although probit analysis is recommended.

Hong Kong Chinese boys are later than English or Dutch boys in the first appearance of pubic hair although, as we have noted above, they are skeletally in advance of the English at adolescence. On the other hand, Chinese girls are earlier in the beginning of breast development than English, Dutch, Parisian, or Indian girls (see Eveleth and Tanner, 1976, for references). New Zealand Maori girls reach menarche earlier than non-Maori, yet they are apparently not earlier in pubic hair or breast development (New Zealand Department of Health, 1971). The different measures of pubertal development are, to a considerable degree, independent of each other, so that generalizations as to early or late development between or even within populations cannot be made (Marshall and Tanner, 1969, 1970).

7. Dental Maturation

7.1. Environmental Factors

The effect of the environment on the eruption of either the deciduous or permanent dentition is neither very well-established nor apparently very strong compared to the marked environmental influence on body size and skeletal maturation. Even in undernourished populations deciduous dentition is not delayed, while height, weight, and bone growth are (Truswell and Hansen, 1973; Robinow, 1973). Reasons for this may be that tooth germs are laid down prenatally and that the survival value of a full set of functioning teeth must be high.

387

POPULATION
DIFFERENCES IN
GROWTH:
ENVIRONMENTAL AND
GENETIC FACTORS

Results from studies dealing with dental eruption and effects of malnutrition are uneven. Severely malnourished Guatemalan infants did not differ in number of teeth present from those who were well-nourished up to 18 months of age. After 24 months the malnourished children had more, not fewer, teeth erupted than the well-nourished (Cifuentes and Alvarado, 1973). Infants in Caribbean countries and in Egypt showed no relation between weight-for-age and number of teeth erupted except in cases of severe malnutrition, but in Lagos, Nigeria, eruption was delayed even in mild malnutrition (Neill *et al.*, 1973), and rural Nigerians were considerably delayed compared to the well-off in Ibadan (Enwonwu, 1973).

Results are similarly inconclusive for environmental influences on permanent tooth eruption, although theoretically the adverse environment has had more time to have an effect. Well-off Hong Kong Chinese were advanced over poor Chinese only for some teeth: first molars, incisors, and canines (Lee *et al.*, 1965). Children of Japanese immigrants living under improved conditions in Brazil were advanced over Japanese in Japan in eruption of incisors, canines, and upper first premolars (Eveleth and Freitas, 1969). No social class differences were observed among French-Canadians (Sapoka and Demirjian, 1971), and consistent but nonsignificant social class differences were observed in British children (Clements *et al.*, 1953).

7.2. Genetic Factors

The genetic influence on the timing of eruption of permanent dentition does appear to be strong, and marked population differences are encountered. Some populations that have been cited above as being delayed in skeletal or sexual maturation or in height and weight growth are, on the contrary, advanced over, or comparable to, Europeans. New Guineans and Africans, both in Africa and the United States, are generally most advanced. As examples of eruption times of permanent teeth, we see that Ghanaian (Houpt *et al.*, 1967), Afro-American (Garn *et al.*, 1972*b*), and urban Polish boys (Charzewski, 1963) erupt their upper central incisors at a median age of 6.8 years; New Guinean boys in Bougainville at 6.4 years (Friedlaender and Bailit, 1969); Kaiapit, 6.5 years; Bundi, 7.2 years (Malcolm and Bue, 1970); and Euro-American boys in Michigan, 7.3 years (Garn *et al.*, 1972*b*). New Guinean boys have on average completed their second dentition by 11.5 years (Bundi), 11.2 years (Kaiapit) (Malcolm and Bue, 1970), 11.3 years (Bougainville) (Friedlaender and Bailit, 1969), Ghanaian boys 11.4 years (Houpt *et al.*, 1967) and Michigan boys not until 12.5 years (Garn *et al.*, 1972*b*).

In the eruption of the deciduous dentition, population differences are not very obvious. Chinese infants are comparable to British in Newcastle (Billewicz *et al.*, 1973) or to New Guineans (Bailey, 1964), although Australian Aborigines, Bengalis, Hindus, rural Nigerians, and Gambians may be somewhat delayed (see Eveleth and Tanner, 1976).

8. Summary and Generalization

Several generalizations may be made in summary. Asiatics reach puberty on average earlier than Europeans and pass through adolescence more rapidly. Evidence for this comes from timing of menarche, skeletal maturation, and the pubertal spurt. At the same time they are smaller in size than Europeans from preschool age on.

Children of African ancestry are usually taller and sometimes heavier than their peers of European ancestry when matched for income level. The principal reason for this appears to be that they are more mature throughout growth, since means of adult stature are similar in blacks and whites. They do not appear to go through adolescence faster than Europeans. Well-off Africans in Africa are also taller than Europeans according to the scant records. Even poorly off Africans are skeletally advanced early in life, but later become delayed. One can presume that under good conditions Africans in Africa would be developmentally more mature and taller than Europeans throughout the entire growth period.

European children appear to be less mature than Asiatics or Africans, taller than Asiatics, and shorter on average than Africans.

Even the best genetic programming may go awry in a poor environment. Because of the overwhelming effect of adverse conditions on growth, it is frequently difficult to see the genetic target. A poor environment is most likely not limited to one generation in its effects. Only recently have we been seeing that the effects may be witnessed in succeeding generations and thus may still be masking genetic factors.

Evidence of the effects from growing up under unfavorable circumstances comes from comparisons of well-off and poor groups, and urban and rural groups within a population. The well-off are always more mature and have a larger body size than the poor. The urban children are almost always more mature and larger than the rural, except in countries where marked socioeconomic differences between town and country have disappeared.

In assessing the growth of an individual child, or a group, one needs to be aware of population differences in growth. Standards and reference values in the past have been based on children in Europe or the United States because those countries are the ones where the standards were first made up. We know now that race-specific standards need to be developed. The African child who appears to be average by European standards may actually be retarded (Garn and Clark, 1976a; Robson et al., 1975). The Asiatic child who is judged to be undersized may actually be quite normal for his group (Barr et al., 1972) and should not be expected to be larger. It has only been recently, however, that enough data have been amassed from many world populations to show possible genetic patterns in growth, and many lacunae still exist to be filled by future research.

9. References

Abbie, A. A., 1967, Skinfold thickness in Australian Aborigines, *Archaeol. Phys. Anthropol. Oceania* **2**:207.

Achar, B. T., and Yankauer, A., 1962, Studies on the birth weight of South Indian infants, *Ind. J. Child Health* **11**:157.

Acheson, R. M., 1960, Effects of nutrition and disease on human growth, in: *Human Growth. Symposium of the Society for the Study of Human Biology,* Vol. 3 (J. M. Tanner, ed.), pp. 73–92, Pergamon Press, Oxford.

Adams, M. S., and Niswander, J. D., 1968, Birth weight of North American Indians, *Hum. Biol.* **40**:226.

Adams, M. S., and Niswander, J. D., 1973, Birth weight of North American Indians: A correction and amplification, *Hum. Biol.* **45**:351.

Amirhakimi, G. H., 1974, Growth from birth to two years of rich urban and poor rural Iranian children compared with Western norms, *Ann. Hum. Biol.* **1**:427.

Ashcroft, M. T., and Lovell, H. A., 1964, Heights and weights of Jamaican children of various racial origins, *Trop. Geogr. Med.* **4**:346.

389

*POPULATION
DIFFERENCES IN
GROWTH:
ENVIRONMENTAL AND
GENETIC FACTORS*

Ashcroft, M. T., and Lovell, H. G., 1966, Heights and weights of Jamaican primary school children, *J. Trop. Pediatr.* **11:**56.

Ashcroft, M. T., Lovell, H. G., George, M., and Williams, A., 1965, Heights and weights of infants and children in a rural community of Jamaica, *J. Trop. Pediatr.* **11:**56.

Ashcroft, M. T., Heneage, P., and Lovell, H. A., 1966, Heights and weights of Jamaican schoolchildren of various ethnic groups, *Am. J. Phys. Anthropol.* **24:**35.

Ashizawa, K., 1970, Maturation osseuse des enfants japonais de 6 à 18 ans, estimée par la methode de Tanner-Whitehouse, *Bull. Mem. Soc. Anthropol. Paris* **6:**265.

Aw, E., and Tye, C. Y., 1970, Age of menarche of a group of Singapore girls, *Hum. Biol.* **42:**329.

Backström-Järvinen, L., 1964, Heights and weights of Finnish children and young adults, *Ann. Paediatr. Suppl.* **23:**116 pp.

Bailey, K. V., 1964, Dental development in New Guinean infants, *J. Pediatr.* **64:**97.

Banik, N. D. Datta, Krishna, R., Mane, S. I. S., Raj, L., and Taskar, A. D., 1969, The influence of maternal factors on birth weight of the newborn, *Indian J. Pediatr.* **36:**278.

Banik, N. D. Datta, Nayar, S., Krishna, P., Raj, L., and Gadeker, N. G., 1970, Skeletal maturation of Indian children, *Indian J. Pediatr.* **37:**249.

Banik, N. D. Datta, Nayar, S., Krishna, P., and Raj, L., 1972, The effect of nutrition on growth of pre-school children in different communities in Delhi, *Indian Pediatr.* **9:**460.

Barr, G. D., Allen, C. M., and Shinefield, H. R., 1972, Height and weight of 7,500 children of three skin colours, *Am. J. Dis. Child.* **124:**866.

Billewicz, W. Z., Thomson, A. M., Baber, F. M., and Field, C. E., 1973, The development of primary teeth in Chinese (Hong Kong) children, *Hum. Biol.* **45:**229.

Blanco, R. A., Acheson, R. M., Canosa, C., and Salomon, J. B., 1972, Retardation in appearance of ossification centres in deprived Guatemalan children, *Hum. Biol.* **44:**525.

Boutourline-Young, H. B., Hamza, B., Louyot, P., ElAmouri, T., Rebjet, H., Boutourline, E., and Tesi, G., 1973, Social and environmental factors accompanying malnutrition, paper presented at Symposium of the International Association of Behavioural Sciences, Ann Arbor, Michigan, 1973 (mimeographed).

Brasel, J. A., and Winick, M., 1972, Maternal nutrition and prenatal growth, *Arch. Dis. Child.* **47:**479.

Brundtland, G. H., and Walløe, L., 1973, Menarcheal age in Norway, *Nature (London)* **241:**478.

Burgess, A. P., and Burgess, J. L., 1964, The growth pattern of East African schoolgirls, *Hum. Biol.* **36:**177.

Burrell, R. J. W., Tanner, J. M., and Healy, M. J. R., 1961, Age at menarche in South African Bantu girls living in the Transkei reserve, *Hum. Biol.* **33:**250.

Butler, N. R., and Alberman, E. D., 1969, *Perinatal Problems. The Second Report of the 1958 British Perinatal Mortality Survey,* E. & S. Livingstone, Edinburgh.

Chandra, H., 1971, Birth weights of infants of different economic groups in Hyderabad, *Proc. Nutr. Soc. India* **10:**99.

Chang, K. S. F., 1969, *Growth and Development of Chinese Children and Youth in Hong Kong,* University of Hong Kong, Hong Kong.

Charzewski, J., 1963, Some problems of the cutting of permanent teeth in children and youth in urban and rural environments, *Pr. Mater. Nauk. IMD* **1:**65 (in Polish with English summary).

Cifuentes, E., and Alvarado, J., 1973, Assessment of deciduous dentition in Guatemalan children, *Environ. Child Health* **19**(2A):211.

Clements, E. M. B., Davies-Thomas, E., and Pickett, K. G., 1953, Time of eruption of permanent teeth in British children in 1947–48, *Br. Med. J.* **1**(1953):1421.

Cook, J., Altman, D. G., Moore, D. M. C., Topp, S. G., Holland, W. W., and Elliott, A., 1973, A survey of the nutritional status of schoolchildren. Relation between nutrient intake and socio-economic factors, *Br. J. Prev. Soc. Med.* **27:**91.

Cristesçu, M., 1969, *Aspecte ale Cresterii si Dezvoltării Adolescentilor din Republica Socialistă România,* Editura Academiei, Republicii Socialiste România, Bucharest.

Dreizen, S., Spirakis, C. N., and Stone, R. E., 1967, A comparison of skeletal growth and maturation in undernourished and well-nourished girls before and after menarche, *J. Pediatr.* **70:**256.

Durnin, J. V. G. A., and Womersley, J., 1974, Body fat assessed from total body density and its estimation from skinfold thickness: measurements on 481 men and women aged from 16 to 72 years, *Br. J. Nutr.* **32:**77.

Enwonwu, C. O., 1973, Influence of socio-economic conditions on dental development in Nigerian children, *Arch. Oral Biol.* **18:**95.

Eveleth, P. B., 1975, Worldwide trends in growth: Infancy and adolescence with notes on recommended methodology, unpublished manuscript.

Eveleth, P. B., 1978, Differences between populations in body shape of children and adolescents, *Amer. J. Phys. Anthropol.* (in press).

Eveleth, P. B., and Freitas, J. A. de Souza, 1969, Tooth eruption and menarche of Brazilian-born children of Japanese ancestry, *Hum. Biol.* **41**:176.

Eveleth, P. B., and Tanner, J. M., 1976, *Worldwide Variation in Human Growth,* Cambridge University Press, Cambridge.

Fernandez, N. A., Burgos, J. C., and Asenjo, C. F., 1969, Obesity in Puerto Rican children and adults, *Bol. Asoc. Med. P.R.* **61**:153.

Foll, C. V., 1961, The age of menarche in Assam and Burma, *Arch. Dis. Child.* **36**:302.

Friedlaender, J. S., and Bailit, H. L., 1969, Eruption times of the deciduous and permanent teeth of natives on Bougainville Island, Territory of New Guinea: A study of racial variation, *Hum. Biol.* **41**:51.

Frisancho, A. R., 1969, Growth, physique and pulmonary function at high altitude: A field study of a Peruvian Quechua population. Ph.D. dissertation, Pennsylvania State University.

Frisancho, A. R., and Baker, P. T., 1970, Altitude and growth: A study of the patterns of physical growth of a high altitude Peruvian population. *Am. J. Phys. Anthropol.* **32**:279.

Furu, M., 1976, Menarcheal age in Stockholm girls, 1967, *Ann. Hum. Biol.* **3**:587.

Garcia Almansa, A., Fernańdez, M. D., and Palacios Mateos, J. M., 1969, Patrones de crecimiento de los niños españoles normales, *Rev. Clin. Espan.* **113**:45.

Garn, S. M., and Clark, D. C., 1975, Nutrition, growth, development, and maturation: Findings from the Ten-State Nutrition Survey of 1968–1970, *Pediatrics* **56**:306.

Garn, S. M., and Clark, D. C., 1976a, Problems in the nutritional assessment of black individuals, *Am. J. Phys. Anthropol.* **66**:262.

Garn, S. M., and Clark, D. C., 1976b, Trends in fatness and the origins of obesity, *Pediatrics* **57**:443.

Garn, S. M., Sandusky, S. T., Nagy, J. M., and McCann, M. B., 1972a, Advanced skeletal development in low income Negro children, *J. Pediatr.* **80**:965.

Garn, S. M., Wertheimer, F., Sandusky, S. T., and McCann, M. B., 1972b, Advanced tooth emergence in Negro individuals, *J. Dent. Res.* **51**:1506.

Garn, S. M., Clark, D. C., and Trowbridge, F. L., 1973, Tendency toward greater stature in American black children, *Am. J. Dis. Child.* **126**:164.

Goldblatt, P. B., Moore, M. E., and Stunkard, A. J., 1965, Social factors in obesity, *J. Am. Med. Assoc.* **192**:1039.

Goldstein, H., 1971, Factors influencing the height of seven-year-old children. Results from the National Child Development Study, *Hum. Biol.* **43**:92.

Greulich, W. W., 1957, A comparison of the physical growth and development of American-born and native Japanese children, *Am. J. Phys. Anthropol.* **15**:489.

Greulich, W. W., 1976, Some secular changes in the growth of American-born and native Japanese children, *Am. J. Phys. Anthropol.* **45**:553.

Guaraciaba, H. L. B., 1967, Physical growth of Japanese–Brazilian children, *Zinruigaku Zassi* **75**:1.

Hamill, P. V. V., Johnston, F. E., and Grams, W., 1970, Height and Weight of Children: United States, Vital Health Statistics Series 11, No. 104, U.S. Govt. Printing Office, Washington, D.C.

Hamill, P. V. V., Johnston, F. E., and Lemeshow, S., 1972, Height and Weight of Children: Socioeconomic Status, Dept. Health, Education and Welfare Publication No. (HSM) 73-1601, Vital Health Statistics Series 11, No. 119, U.S. Govt. Printing Office, Washington, D.C.

Hamill, P. V. V., Johnston, F. E., and Lemeshow, S., 1973, Body Weight, Stature and Sitting Height: White and Negro Youths 12–17 Years, United states, Dept. Health, Education and Welfare Publication, No. (HRA) 74-1608, Vital Health Statistics Series 11, No. 126, U.S. Govt. Printing Office, Washington, D.C.

Hasan, F., Najjar, S. S., and Asfour, R. Y., 1969, Growth of Lebanese infants in the first year of life, *Arch. Dis. Child.* **44**:131.

Hedayat, S., Koohestani, P. A., Ghassemi, H., and Kamili, P., 1971, Birth weight in relation to economic status and certain maternal factors, based on an Iranian sample, *Trop. Geogr. Med.* **23**:355.

Heintz, N. Petit-Maire, 1963, Croissance et puberté feminines au Rwanda, *Mem. Acad. R. Sci. Nat. Med.* **12**:1.

Himes, J. H., Martorell, R., Habicht, J.-P., Yarbrough, C., Malina, R. M., and Klein, R. E., 1975, Patterns of cortical bone growth in moderatley malnourished preschool children, *Hum. Biol.* **47**:337.

Houpt, M. J., Adu-Aryee, S., Ghana, K., and Grainer, R. M., 1967, Eruption times of permanent teeth in the Brong Ahafo region of Ghana, *Am. J. Orthodont.* **53**:95.

Indian Council of Medical Research, 1972, Growth and Physical Development of Indian Infants and Children, Technical Report Series No. 18, ICMR, New Delhi.

Jamison, P. L., and Zegura, S. L., 1970, An anthropometric study of the Eskimos of Wainwright, Alaska, *Arctic Anthropol.* **7**:125.

Janes, M. D., 1974, Physical growth of Nigerian Yoruba children, *Trop. Geogr. Med.* **26**:389.

391

*POPULATION
DIFFERENCES IN
GROWTH:
ENVIRONMENTAL AND
GENETIC FACTORS*

Janes, M. D., 1975, Physical and psychological growth and development, *Environ. Child Health* (special issue) **21**:26.

Jeniček, M., and Demirjian, A., 1974, Age at menarche in French–Canadian urban girls, *Ann. Hum. Biol.* **1**:339.

Johnston, F. E., Hamill, P. V. V., and Lemeshow, S., 1972, Skinfold Thickness of Children 6–11 Years, United States, Dept. Health, Education and Welfare Publication No. (HSM) 73-1602, Vital and Health Statistics Series 11, No. 120, U.S. Govt. Printing Office, Washington, D.C.

Johnston, F. E., Hamill, P. V. V., and Lemeshow, S., 1974, Skinfold Thickness of Youths 12–17 Years, United States. Dept. Health, Education and Welfare Publication No. (HRA) 74-1614, Vital and Health Statistics Series 11, No. 132, U.S. Govt. Printing Office, Washington, D.C.

Jones, D. L., Hemphill, W., and Meyers, E. S. A., 1973, Height, Weight, and Other Physical Characteristics of New South Wales Children, Part 1. Children Aged Five Years and Over, New South Wales Dept. Health.

Khanjanasthiti, P., Supchaturas, P., Mekanandha, P., Srimusikpohd, V., Leeuwan, V., and Choopanya, K., no date, The anthropometry of the growth of infant and preschool children, unpublished manuscript.

Kimura, K., 1972*a*, Skeletal maturation in Japanese according to the Oxford and Tanner–Whitehouse methods, *Acta Anat. Nippon* **47**:358.

Kimura, K., 1972*b*, Skeletal maturation in Japanese. A new analytical method, *J. Anthropol. Soc. Nippon* **80**:319.

Kimura, K., 1976*a*, On the skeletal maturation of Japanese–American white hybrids, *Am. J. Phys. Anthropol.* **44**:83.

Kimura, K., 1976*b*, Skeletal maturation of children in Okinawa, *Ann. Hum. Biol.* **3**:149.

Kimura, K., 1977, Skeletal maturity of the hand and wrist in Japanese children in Sapporo by the TW2 method, *Ann. Hum. Biol.* **4**:449.

Kimura, K., and Tsai, C. M., 1967, Comparative studies of the physical growth in Formosans. I. Height and weight, *J. Anthropol. Soc. Nippon* **75**:11 (in Japanese with English summary).

King, K. W., Foucauld, J., Fougere, W., and Severinghaus, 1963, Height and weight of Haitian children, *Am. J. Clin. Nutr.* **13**:106.

Kondo, S., and Eto, M., 1972, Physical growth studies on Japanese–American children in comparison with native Japanese, in: *Proceedings of Meeting for Review and Seminar of the US Japan Cooperative Research on Human Adaptabilities,* Kyoto, May 1972, Japan Society for the Promotion of Science and National Science Foundation, Kyoto.

Krogman, W. M., 1970, Growth of the head, face, trunk, and limbs in Philadelphia white and Negro children of elementary and high school age, *Monogr. Soc. Res. Child. Dev.* **35**:1.

Kurnjewicz-Witczakowa, R., Mięsowicz, I., Mazurczak, T., and Jarmolińska-Eska, H., 1972, Indices of somatic development of Warsaw children aged from 0 to 36 months, *Probl. Med. Wieku Rozwojowego* **2**(13):45 (in Polish with English summary).

Kwon, Y., 1978, A Comparison of the Physical Growth and Development of Korean Children Born and Raised in a Rural Area with the Descendants of Rural to Urban Migrants in Seoul, Korea, Ph.D. thesis, University of Pennsylvania, Philadelphia.

Laska-Mierzwjewska, T., 1970, Effect of ecological and socio-economic factors on the age at menarche, body height, and weight of rural girls in Poland, *Hum. Biol.* **42**:284.

Lee, M. M. C., Low, W. D., and Chang, K. S. F., 1965, Eruption of the permanent dentition of Southern Chinese children in Hong Kong, *Arch. Oral Biol.* **10**:849.

Lindgren, G., 1976, Height, weight and menarche in Swedish urban school children in relation to socio-economic and regional factors. *Ann. Hum. Biol.* **3**:501.

Low, W. D., Chan, S. T., Chang, K. S. F., and Lee, M. M. C., 1964, Skeletal maturation of Southern Chinese children in Hong Kong, *Child Dev.* **35**:1313.

MacMahon, B., 1973, Age at Menarche: United States, Dept. Health, Education and Welfare Publication No. (HRA) 74-1615, NHS Series 11, No. 133, National Center for Health Statistics, Rockville, Maryland.

Malcolm, L. A., 1970, Growth and development of the Bundi child of the New Guinea highlands, *Hum. Biol.* **42**:293.

Malcolm, L. A., and Bue, A., 1970, Eruption times of permanent teeth and the determination of age in New Guinea children, *Trop. Geogr. Med.* **22**:307.

Malina, R. M., 1966, Patterns of development in skinfolds of Negro and white Philadelphia children, *Hum. Biol.* **38**:89.

Malina, R. M., 1970, Skeletal maturation studied longitudinally over one year in American whites and Negroes six through thirteen years of age, *Hum. Biol.* **42**:377.

Malina, R. M., 1971, Skinfolds in American Negros and white children, *J. Am. Diet. Assoc.* **59**:34.

Malina, R. M., Habicht, J.-P., Yarbrough, C., Martorell, R., and Klein, R. E., 1974, Skinfold thickness at seven sites in rural Guatemalan Ladino children birth through seven years of age, *Hum. Biol.* **46**:453.

Maniar, B. M., Seervai, M. H., and Kapur, P. L., 1974, A study of ossification centres in the hand and wrist of Indian children, *Indian Pediatr.* **11**:203.

Marin, B., 1971, Edad actual de la menarquia en escolares españolas, *Arch. Fac. Med.* **5**:355.

Marshall, W. A., and Tanner, J. M., 1969, Variation in the pattern of pubertal changes in girls, *Arch. Dis. Child.* **44**:291.

Marshall, W. A., and Tanner, J. M., 1970, Variation in the pattern of pubertal changes in boys, *Arch. Dis. Child.* **45**:13.

Marshall, W. A., Ashcroft, M. T., and Bryan, G., 1970, Skeletal maturation of the hand and wrist of Jamaican children, *Hum. Biol.* **42**:419.

Martins Filho, J., Pinotti, J. A., Carvalho, J. F., Bueno, R. D., Paes de Freitas, N. A., Carvalho, M. B., and Moraes, L. P., 1974, Dénutrition intra-uterine variation du poids à la naissance en fonction de la classe socio-économique dans une maternité de la ville de Campinas, S. P., Brésil, *Courrier* **24**:122.

Marubini, E., and Barghini, G., 1969, Richerche sull'eta media di comparse della puberta nella populazione scolare femminilie di Carrara, *Minerva Pediatr.* **21**:281.

Massé, G., and Hunt, E. E., 1963, Skeletal maturation of the hand and wrist in West African children, *Hum. Biol.* **35**:3.

Mata, L. J., Urrutia, J. J., and Lechtig, A., 1971, Infection and nutrition of children of a low socio-economic rural community, *Am. J. Clin. Nutr.* **24**:249.

Meredith, H. V., 1970, Body weight at birth of viable human infants: A worldwide comparative treatise, *Hum. Biol.* **42**:217.

Meredith, H. V., and Spurgeon, J. H., 1976, Body size and form of black and white female youth measured during 1974–75 at Columbia, South Carolina, *Child Dev.* **47**:360.

Michaut, E., Niang, J., and Dan, V., 1972, La maturation osseuse pendant la pérode pubertaire, *Ann. Radiol.* **15**:767.

Miklashevskaya, N. N., Solovyeva, V. S., Godina, E. Z., and Kondik, V. M., 1972, Growth processes in man under conditions of the high mountains, *Trans. Moscow Soc. Nat.* **43**:181 (in Russian with English summary).

Milicer, H., and Szczotka, F., 1966, Age at menarche in Warsaw girls in 1965, *Hum. Biol.* **38**:199.

Miller, F. J. W., Billewicz, W. Z., and Thomson, A. M., 1972, Growth from birth to adult life of 440 Newcastle-upon-Tyne children, *Br. J. Prev. Soc. Med.* **26**:224.

Morley, D. C., Woodland, M., Martin, W. J., and Allen, J., 1968, Heights and weights of West African village children from birth to the age of five, *W. Afr. Med. J.* **17**:8.

Morton, N. E., 1955, The inheritance of human birth weight, *Ann. Hum. Genet.* **20**:125.

Naeye, R. L., Diener, M. M., Harcke, H. T., and Blanc, W. A., 1971, Relation of poverty and race to birth weight and organ and cell structure, *Pediatr. Res.* **5**:17.

Naylor, A. F., and Myrianthopoulos, N. C., 1967, The relation of ethnic and selected socio-economic factors to human birth weight, *Ann. Hum. Genet.* **31**:71.

Neill, J. J., Gurney, J. M., Kuti, O. R., Doherty-Akinkugbe, D., Hanafy, M. M., Kassem, S. A., El Lozy, M., Field, E. E., Mendoza, H. R., and McDowell, M. F., 1973, Deciduous dental eruption time and protein–calorie malnutriton from different parts of the world, *Environ. Child Health* **19**(2A):217.

New Zealand Department of Health, 1971, Physical Development of New Zealand Schoolchildren, 1969, Special Report No. 38, Health Services Research Unit, Department of Health, Wellington.

Neyzi, O., Alp, H., and Orhon, A., 1975, Sexual maturation in Turkish girls, *Ann. Hum. Biol.* **2**:49.

Oduntan, S. O., Ayeni, A., and Kale, O. O., 1976, The age of menarche in Nigerian girls, *Ann. Hum. Biol.* **3**:269.

Ounsted, M., and Ounsted, C., 1973, *On Fetal Growth Rate,* W. Heinemann Medical Books, London.

Palacios, J. Mateos, and Vivanco, F., 1965, Datos de talla y peso de 128,000 niños españoles, *Rev. Clin. Espan.* **99**:230.

Panek, S., and Piasecki, E., 1971, Nowa Huta: Integration of the population in the light of anthropological data, *Mater. Pr. Anthropol.* **80**:1.

Pawson, I. G., 1977, Growth characteristics of populations of Tibetan orgin in Nepal, *Amer. J. Phys. Anthropol.* **47**:473.

Penchaszadeh, V. B., Hardy, J. B., Mellits, E. D., Cohen, B. H., and McKusick, V. A., 1972, Growth and development in an "inner city" population: An assessment of possible biological and environmental influences. II. The effect of certain maternal characteristics on birthweight, gestational age and intra-uterine growth, *Johns Hopkins Med. J.* **131**:11.

Rauh, J. L., and Schumsky, D. A., 1968, An evaluation of triceps skinfold measures from urban school children, *Hum. Biol.* **40**:363.

Rauh, J. L., Schumsky, D. A., and Witt, M. T., 1967, Heights, weights and obesity in urban schoolchildren, *Child Dev.* **38**:515.

Rea, J. N., 1971, Social and economic influence on the growth of preschool children in Lagos, *Hum. Biol.* **43**:46.

Read, M. S., Habicht, J.-P., Lechtig, A., and Klein, R. E., 1975, Maternal malnutrition, birth weight and child development, in: *Nutrition, Growth and Development* (C. Canosa, ed.), pp. 203–215, S. Karger, Basel.

Richardson, B. D., 1973, Studies on the nutritonal status and health of Transvaal Bantu and white preschool children, *S. Afr. Med. J.* **47**:688.

Roberts, D. F., Danskin, M. J., and Chinn, S., 1975, Menarcheal age in Northumberland, *Acta Paediatr. Scand.* **64**:845.

Robinow, M., 1973, The eruption of the deciduous teeth (factors involved in timing), *Environ. Child Health* **19**(2A):200.

Robson, E. B., 1955, Birth weight in cousins, *Ann. Hum. Genet.* **19**:262.

Robson, J. R. K., Larkin, F. A., Bursick, J. H., and Perri, K. P., 1975, Growth standards for infants and children: A cross sectional study, *Pediatrics* **56**:1014.

Rona, R., and Pereira, G., 1974, Factors that influence age of menarche in girls in Santiago, Chile, *Hum. Biol.* **46**:33.

Sapoka, A. A., and Demirjian, A., 1971, Dental development of the French Canadian child, *J. Can. Dent. Assoc.* **37**:100.

Sarram, M., and Saadatnejadi, M., 1967, Birth weight in Shiraz (Iran) in relation to maternal social economic status, *Obstet. Gynecol.* **30**:367.

Schaefer, A. E., 1975, Epidemiology of pre- and postnatal malnutrition—USA, in: *Nutrition, Growth and Development* (C. A. Canosa, ed.), pp. 9–19, S. Karger, Basel.

Shakir, A., 1971, The age of menarche in girls attending school in Baghdad, *Hum. Biol.* **43**:265.

Shakir, A., and Zaini, S., 1974, Skeletal maturation of the hand and wrist of young children in Baghdad, *Ann. Hum. Biol.* **1**:189.

Shephard, R. J., Hatcher, J., and Rode, A., 1973, On the body composition of Eskimos, *Eur. J. Appl. Physiol.* **32**:3.

Silverstone, J. T., Gordon, R. P., and Stunkard, A. J., 1969, Social factors in obesity in London, *Practitioner* **202**:682.

Smith, C. A., 1947, The effect of wartime starvation in Holland upon pregnancy and its product, *Am. J. Obstet. Gynecol.* **53**:599.

Stunkard, A. J., d'Aquili, E., Fox, S., and Filion, R. D. L., 1972, Influence of social class on obesity in children, *J. Am. Med. Assoc.* **221**:579.

Tanner, J. M., 1962, *Growth at Adolescence,* 2nd ed., Blackwell Scientific, Oxford.

Tanner, J. M., 1973, Trend toward earlier menarche in London, Oslo, Copenhagen, the Netherlands and Hungary, *Nature (London)* **243**:95.

Tanner, J. M., and Eveleth, P. B., 1975, Variability between populations in growth and development at puberty, in: *Puberty, Biologic and Psychological Components* (S. R. Berenberg, ed.), pp. 256–273, Stenfert Kroese, Leiden.

Tanner, J. M., and O'Keefe, B., 1962, Age at menarche in Nigerian schoolgirls with a note on their heights and weights from age 12 to 19, *Hum. Biol.* **34**:187.

Tanner, J. M., and Whitehouse, R. H., 1975, Clinical longitudinal standards for height, weight, height velocity, weight velocity, and the stages of puberty, *Arch. Dis. Child.* **51**:170.

Terada, H., and Hoshi, H., 1965, Longitudinal study on the physical growth in Japanese. 2. Growth in stature and body weight during the first three years of life, *Acta Anat. Nippon* **40**:166.

Thomson, A. M., Billewicz, W. Z., and Hytten, F. E., 1968, The assessment of foetal growth, *J. Obstet. Gynaecol. Br. Commonw.* **75**:903.

Tokyo Department of Maternal and Child Health, 1970, Physical Status of Japanese Children in 1970, Department of Maternal and Child Health, Institute of Public Health.

Truswell, A. S., and Hansen, J. D. L., 1973, Eruption of deciduous teeth in protein-calorie malnutrition, *Environ. Child Health* **19**(2A):214.

Valaoras, V., and Laros, K., 1969, Biometric characteristics of Greek pupils in elementary schools, *IATPIKH* **15**:266 (in Greek with English summary).

van Wieringen, J. C., Wafelbakker, F., Verbrugge, H. P., and Haas, J. H., 1971, *Growth Diagrams 1965, Netherlands,* Walters-Noordhoff, Groningen.

Villarejos, V. M., Osborne, J. A., Payne, F. J., and Arguedes, J. A., 1971, Heights and weights of children in urban and rural Costa Rica, *Environ. Child Health* **17**:31.

Vincent, M., Jans, C., and Ghesquière, J., 1962, The newborn pygmy and his mother, *Am. J. Phys. Anthropol.* **20**:237.

Wark, M. L., and Malcolm, L. A., 1969, Growth and development of the Lumi child of the Sepik district of New Guinea, *Med. J. Aust.* **2**(1969):129.

Whitelaw, A. G. L., 1971, The association of social class and sibling number with skinfold thickness in London schoolboys, *Hum. Biol.* **43**:414.

Wingerd, J., Peritz, E., and Sproul, A., 1974, Race and stature differences in the skeletal maturation of the hand and wrist, *Ann. Hum. Biol.* **1**:201.

Wolański, N., and Lasota, A., 1964, Physical development of countryside children and youth aged 2 to 20 years as compared with the development of town youth of the same age, *Z. Morphol. Anthropol.* **54**:272.

Yanagisawa, S., and Kondo, S., 1973, Modernization of physical features of Japanese with special reference to leg length and head form, *J. Hum. Ergol.* **2**:97.

Zacharias, L., Rand, W. M., and Wurtman, R. J., 1976, A prospective study of sexual development and growth in American girls: The statistics of menarche, *Obstet. Gynecol. Surv.* **31**:325.

13

Epidemiological Considerations

DERRICK B. JELLIFFE and
E. F. PATRICE JELLIFFE

1. Introduction: Interrelationships of Growth and Epidemiology

The epidemiological approach assists in many public health activities concerned with the prevention and control of nutritional diseases, as Gordon (1976) has elaborated. These range through the monitoring of "good" nutrition (and the early detection of malnutrition, "plus" as well as "minus"), to the evaluation of the effectiveness of programs, and the management of famines.

Running through all the examples we shall give are two key interrelationships between growth and epidemiology.

1.1. Growth Data in Epidemiological Analysis

The field assessment of some forms of malnutrition has no particular dependency on growth data; iron deficiency and iodine lack, for example, are detected, respectively, by hemoglobin (or hematocrit) estimation and clinical inspection for goiter. However, early detection of two of the world's main community nutrition problems—protein—calorie malnutrition (PCM or PEM) and obesity—assessment of growth failure, growth excess, or growth disproportion is irreplaceable (Gordon, 1976).

As is well known, growth failure is characteristic of mild-to-moderate PCM. This can be shown serially in children attending clinics by flattening weight curves, as on Morley's (1973) "road to health" charts, and by diminution of velocity of other variables, including height. In communities, the prevalence of PCM in the children can be assessed by discovering what percentage have growth failure as indicated by body measurements, particularly weight and height, when compared with "levels of reference" for age (Jelliffe, 1966). When, as in most of the world,

DERRICK B. JELLIFFE and E. F. PATRICE JELLIFFE ● Division of Population, Family and International Health, School of Public Health, University of California, Los Angeles.

exact age is not known, so-called "age-independent anthropometry" may be used based on measurement of a body dimension unrelated to precise age, such as arm circumference in young children (Jelliffe and Jelliffe, 1969), or on a ratio based on the comparison of a nutritionally labile with a nutritionally less-affected tissue, such as weight-for-height in preschool children (Waterlow, 1974).

The usefulness of monitoring growth failure in public health nutrition can be illustrated by Haiti. In 1958, a republic-wide survey of preschool children based on simple anthropometry and clinical examination indicated widespread PCM, with the area of Fonds Parisien particularly affected (Jelliffe and Jelliffe, 1960). This study, based for the most part on uncomplicated information on growth failure, was in part responsible for the initiation of a network of "Mothercraft Centers" in this country, commencing at Fonds Parisien. Also, the evaluation of these centers, which have been quite successful, has been based on improved growth, especially indicated in weight curves, in affected children and their siblings (King *et al.*, 1977).

1.2. Epidemiological Factors and the Interpretation of Growth Data

Conversely, those engaged in analyzing the meaning of growth data have to appreciate the epidemiological factors that can cause it. A wide range of ecological factors can influence nutritional status and growth directly, as via restrictive food habits or poverty, or indirectly, as with malabsorption following diarrhea, or as a consequence of poor road communications.

In addition, experts in growth and its measurements can play a major role for the field worker in trying to discover the most appropriate "levels of reference" (Falkner and Roche, 1972) and to devise simplified, low-cost apparatus which will make the monitoring of growth less complicated, cheaper, and more accurate. The Zerfas insertion tape may be given as an example (Zerfas, 1975).

In sum, anthropometrists need epidemiological insight to appreciate the many, interacting forces influencing growth. The epidemiologist and field worker need to appreciate that abnormalities in growth patterns, including failure, excess, or disproportion, are irreplaceable methods to assess the commonness and distribution of two of the world's major nutritional problems—obesity and protein–calorie malnutrition of early childhood.

The nutritional status of a community is the end result of the interaction of a whole range of different physical, biological, and cultural factors in the ecology (Table I). The practical question is to identify which are the main ones in a particular community, paying due attention to the practical possibility of improving the situation at reasonable cost and rapidly, as part of realistic preventive programs. The use of the investigative techniques and methodologies of epidemiology is absolutely necessary if understanding of the complicated and varying nexus of causation is to be unraveled in any community.

As an oversimplified guide, the level of nutrition in a community, especially of its vulnerable young child population, may be considered as being due to the numerous interacting ecological forces shown in the following expression (Jelliffe, 1968a):

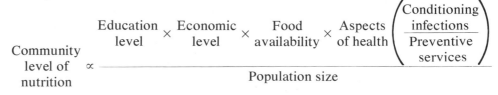

$$\text{Community level of nutrition} \propto \frac{\text{Education level} \times \text{Economic level} \times \text{Food availability} \times \text{Aspects of health} \left(\dfrac{\text{Conditioning infections}}{\text{Preventive services}} \right)}{\text{Population size}}$$

Table I. Miscellaneous Factors in the Etiology of Malnutrition, to Indicate the Range of
Information Collectable in Epidemiological Investigation (World Health Organization, 1963)

Geographic–climatic	Unproductive soil
	Climate (high temperature, extremes of rainfall)
Educational	Too few schools (illiteracy)
Social	Illegitimacy; family instability
	Absence of family planning (children too closely spaced; population pressure)
	Poor communications (food distribution)
	Alcoholism
Economic	National poverty (low gross national product)
	Family poverty (low per capita income)
	Low level of industrialization
Agronomic	Old-fashioned methods of agriculture
	Inadequate protein production (animal and vegetable)
	Concentration on inedible cash crops
	Poor food, storage, preservation and marketing
Medical	High prevalence of conditioning infections (measles, diarrhea. tuberculosis, whooping cough, malaria, intestinal parasites)
Sanitational	Unclean, inadequate water supply
	Defective disposal of excreta and rubbish
Cultural	Faulty feeding habits of young children
	Recent urbanization (changing habits)
	Limited culinary facilities
	Inequable intrafamilial food distribution
	Overwork by women (limited time for food preparation for children)
	Sudden weaning (psychological trauma)

Each of the components of the expression affects the community nutrition level directly, and via the other factors indicated. Each is, of course, shorthand for a complex situation, needing definition and clarification in the particular circumstances. Epidemiological analysis attempts to assess these different factors and judge their interrelationships and their relative causative significance.

For example, the education level includes not only the level of formal schooling and the availability of relevant technical personnel in a country, such as those in nutrition and in extension services, but also literacy in general. The education level affects nutritional status in various ways, such as by the number, quality, and relevance of technical training establishments, and the existence of schools and their use (or otherwise) for nutrition education. End results include the supply of technically trained personnel and the numbers of literates likely to be exposed to modern knowledge concerning food, nutrition, and related matters, as well as written misinformation, both from commercial sources and from exploitive faddist literature.

The economic level operates on a country-wide basis, when an increased national income can, at least theoretically, make more governmental funds available for social programs, such as schools and health services, as well as for specific nutrition programs directed at major problems. At the family level, improved earning capacity means that there will be more money which can be used for the purchase of a greater range of foods, especially more nutritious items, such as higher-cost protein-rich foods. It does not, of course, imply that improved nutrition will always occur. For example, the relatively wealthy usually have higher intakes of calories and fat in their diets, with increased risks of obesity, diabetes, and atheroma. In poorer groups in developing countries, working mothers in urban shanty towns may increase the families' cash income, yet infant nutrition may deteriorate because breast milk is denied to the baby. Instead inadequate amounts of formula are purchased and then given as contaminated bottled feeds, leading to the still further economic drain on the parents and community of prolonged medical treatment of the marasmus—diarrhea syndrome, which is very likely to ensue.

Food availability is shorthand both for the foods present in the country as a whole (and their distribution and relative costs) and for the foods within the economic reach of a family. It also includes consideration of their intrafamilial distribution, especially to young children, to pregnant and lactating women, and to the elderly. Frequently, the more nutritious items tend to deviate to adult males.

The term food availiability includes a composite of many activities, such as agricultural production, marketing methods, and losses during storage and processing. It includes the relative emphasis given to cash crops versus food production in the nation's economic planning, to the cultural significance of local foods in relation to physiologically vulnerable members of the family, and to dietary patterns in the community in general, but especially during infancy, the transitional (or weaning) period, pregnancy and lactation, and old age.

Certain aspects of health particularly influence the community nutritional level by the commonness of nutritionally "conditioning infections" (including, for example, diarrheal disease, tuberculosis, measles, malaria, and intestinal worms); by health services concerned with the prevention or early rehabilitation of child malnutrition by nutrition education and growth supervision at child health clinics; by the wise (or excessive) issue of food supplements; by measures designed to control relevant infectious diseases as with immunization; by child spacing programs; and by activities to improve environmental hygiene, especially water supply and excreta disposal.

It is becoming increasingly apparent that the level of nutrition in a community is related to the number of mouths to be fed. The fact, population size (including its rate of increase, degree of urban migration, family size, and the closeness of child spacing) is a universal denominator for other components of the expression. The present serious world food situation is due in considerable measure to the disproportionate, currently geometrical, increase in population size, constantly outstripping not only economic education and social development, but also per capita availability of food. As is well known, this trend particularly affects preindustrial, largely tropical, resource-poor countries where food supplies are often already inadequate. The situation has recently become still further worsened in some countries, especially those lacking petroleum, minerals, and key foods, which are dependent on costly imported food and/or oil products to continue modernizing their agriculture via the "green revolution" and the use of intermediate technology adapted to the needs of the area.

In addition, a tremendous shift from the rural to the urban areas is now taking place in many developing countries. This is producing a different epidemiological situation, and causing new problems altogether, which are related to the rapid change from rural to urban life styles—from a subsistence, food-growing economy, with the support of a closely knit pattern of relatives and kinship groups, to the often hostile world of the town, where all food has to be purchased, for which the "new townsmen" have little preparation, and in which the foods most available are often those which are cheapest and most easily transported and stored. The pattern of infant feeding has been particularly affected, with change from breast-feeding to bottle-feeding and a consequent increased incidence of the marasmus–diarrhea syndrome (Jelliffe and Jelliffe, 1977).

Lastly, the culture pattern of the community must be considered. The system of values, attitudes, beliefs, and customs is interwoven into every aspect of life and affects each component of the expression ranging from the dietary pattern and methods of treating illness to the sociopolitical priorities given to health and nutrition, particularly of mothers and young children, compared with other governmental activities, including defense.

Of particular importance are customs and practices which act as "cultural blocks" to particularly nutritious items, especially for the physiologically vulnerable groups (Jelliffe, 1957). This is most important when the range of available foods is limited (Jelliffe, 1974). For example, rural Malay mothers traditionally undertake a *pantang* for 40 days after delivery, when the diet is severely restricted and even fluid limited. They do not give their newborn colostrum, nor their secotrants (children in the second year of life) fish, which is the main animal protein food available, as this is believed to produce intestinal worms (Jelliffe, 1968*b*). Cumulatively, these cultural blocks considerably increase the infants' risks of protein deficiency and associated diarrheal disease.

1.3. Reasons for Epidemiological Approach

Rational public health plans to assist in dealing with undernutrition can only be based on epidemiological knowledge—that is, a continuing "community diagnosis" of local forms of malnutrition, including their nature, dimensions, and geographical location, the age groups involved, and the complex and interacting causes in the particular ecology. This has to be based on a similar process to that of clinical diagnosis in classical medicine, using corroborative evidence from different sources, but ideally should also include an assessment of the resources actually or potentially available for mounting preventive programs, including personnel, facilities, equipment, and funds for relevant activities in health, agriculture, and education (Gordon, 1976).

Information on the prevalence and causation of different forms of undernutrition is needed to convince planners and administrators of the magnitude of the problem in human and in economic terms. It is needed to ensure the mobilization of limited resources for rational programs geared (1) to at-risk areas (see below) (with, for example, low-protein tuber staple foods), (2) to major problems (for example, rickets in Ethiopia, kwashiorkor in southern Uganda, or iron deficiency in children in the U.S.A.), and (3) to vulnerable age groups (whether the first or second year of life). It also provides baseline data from which the effects of future developments, planned and accidental, may be measured, particularly by nutritional surveillance (p. 404).

Likewise, an assessment of the actual and potential resources for dealing with problems of community nutrition needs to be made, particularly including the type, location, function, and coverage of existing health, agricultural extension, and other relevant social services, and the duties, interrelationships, and training of available cadres of staff.

From a functional point of view, recent developments with regard to epidemiological investigation can be considered under two overlapping headings: (1) community diagnosis and practical programs and (2) nutritional surveillance. The at-risk concept will not be considered in this chapter (see *American Journal of Clinical Nutrition*, 1977).

DERRICK B. JELLIFFE
and E. F. PATRICE
JELLIFFE

2. Community Diagnosis and Practical Programs

2.1. Methods of Data Collection: Surveys

The nutritional status of a community and the prevalence of many forms of malnutrition often need to be measured by screening surveys (Jelliffe, 1966), which can be of greatly varying degrees of complexity and can be either longitudinal or, more usually, cross-sectional. Both methods have advantages and disadvantages (Jelliffe, 1974).

2.1.1. Longitudinal Data Collection

This approach is rarely possible because of the high cost of staff, logistics, and analysis, because of the long time involved (usually several years), and because of difficulties in maintaining continuity of interest by staff and population, and because of unpredictable ecological alterations that occur following unrelated events or as a result of the effects of the study itself.

Conversely, longitudinal surveys have obvious advantages. They reveal much more dynamic information concerning the interacting epidemiological factors involved, including seasonal variation in the diet, economics, customs, and conditioning infections, throughout the year and throughout the length of time of the investigation.

2.1.2. Cross-Sectional Data Collection

Information is most usually obtained by cross-sectional survey, as this is relatively cheap and quick. At the same time, this method has the disadvantage of giving a picture at one time of the year only, with emphasis on more chronic conditions, and with limited opportunity to gain insight into the interplay of factors responsible for the nutritional situation.

Direct Assessment. Cross-sectional surveys are often based on the examination of statistically representative sections of the population for body measurements indicative of growth failure (anthropometry), for selected clinical signs suggestive of nutrition deficiency, and for indications of various forms of malnutrition shown by biochemical tests (Table II). Major difficulties here include problems of defining normal limits for such tests, as well as the selection of appropriate standards (Falkner and Roche, 1972, 1974).

Such cross-sectional or point-prevalence surveys need to employ carefully defined, objective criteria for identification of malnutrition, whether anthropometric, clinical, or biochemical. Careful pretesting of methods is needed to ensure uniformity by observers, and statistically guided sampling, although always difficult in actual communities, is essential if an unbiased picture is to be obtained.

Direct assessments of human groups can be undertaken by examination of the particular population gathered at agreed collecting points or by home visiting. Both methods have their advantages and problems. Home visiting is very time-consuming, but gives insight into living conditions. Examination at collecting points can deal with much larger numbers, but depends on the reliability and representativeness of the selected samples which actually attend.

Such cross-sectional surveys have the disadvantages of identifying more chronic forms of malnutrition, rather than those that are relatively acute—for

Table II. Selected List of Direct Indicators of Nutritional Status Suitable for Less-Complicated Epidemiological Studies (Zerfas et al., 1977)

Direct indicator	Method	Target group	Nutritional deficiency	Objectivity	Health statistics	Hospital	Clinic	School	Surveys	Village
I. Existing										
1. Proportion low birth weights	Anth.	Fetus (mother)	PCM	Good	+/-	+	-	-	-	
2. Growth charts (Wt/Age)	Anth.	0–5	PCM	Good	-	+/-	+	-	+	
3. Edema rate	Clin.	0–5	PCM (restricted)	Fair	-	+/-	+	-	+	
4. Marasmus rate	Clin.	0–5	PCM	Poor	-	+/-	+	-	+	
5. Hemoglobin	Bio.	Any	Iron/Folic	Good	-	+	+/-	-	+	
6. Serum albumin	Bio.	Any	PCM (restricted)	Good	-	+	-	-	+	
II. Recommended										
1. Height 7-year-old	Anth.	0–7	PCM (past)	Good	-	-	-	+	-	
2. Arm circ. 7-year-old	Anth.	7	PCM (present)	Good	-	-	-	+	-	
3. Arm circ. 1–5 years	Anth.	1–5	PCM	Good	-	+	+	-	+	+
4. Arm circ. adults (1 restrict to females)	Anth.	Adults	PCM	Good	-	-	+/-	-	+	+
5. Fat fold (simplified)	Anth.	Any	Caloric reserve	Fair	-	-	+/-	+/-	+	
6. Weight for height	Anth.	Any	PCM	Good	-	-	+/-	-	+	-
7. Blindness	Clin.	1–3	Vit A (past)	Fair	+	-	-	-	+	+
8. Night blindness	Clin.	2–5	Vit A	Poor–Fair	-	-	+	-	+	+
9. Goiter (visible)	Clin.	Adults	Iodine	Fair	-	-	-	-	+	+
10. Severe pallor	Clin.	Any	Iron/Folic	Poor	-	-	+	+	+	

example, marasmus rather than kwashiorkor. In addition, they give a picture of the situation in the community at one time of the year, and, ideally, need to be repeated at representative seasons. This can be particularly important if there is marked seasonal variation—for example, with an annually changing pattern of infections, such as diarrhea or malaria, or with a yearly season for hunger, usually before the crops are ripe or harvested.

Health Service Data. Statistics concerning the occurrence of the more easily identified forms of malnutrition should also be sought from health service records, such as hospital admissions and health center attendances. Plainly, these results are biased, but still helpful if interpreted with caution, as they can give an approximate insight into year-round incidence and varying prevalence at different seasons. The collection of such information from hospitals and health centers is almost always in need of improvement, especially by recognizing that diagnoses in the ward (and at autopsy) usually only label the terminal illness (often an infection such as pneumonia or measles), rather than indicating the underlying background of malnutrition. This was, for example, clearly revealed by the recent detailed study in eight Latin American countries by the Pan-American Health Organization (Puffer and Serrano, 1973). Improvement of such data can easily be achieved by always recording the weight (and preferably the height) of all children admitted to hospital and by the use of weight charts in outpatients and clinics, on which different levels below the reference standards are indicated. Commonly, 80% (or 75%) and 60% levels are indicated to differentiate moderate and severe protein–calorie malnutrition. More complex charts may be used with advantage which enable changes in both weight and height to be revealed (see Waterlow *et al.,* 1978).

Detailed analysis of the histories and investigation of cases admitted to the ward can be helpful in suggesting the causative factors responsible in the particular community. It can, for example, suggest the main age group at risk and the locally important conditioning infections.

In addition, information of a more continuous nature may often be available from defined areas around health centers. This can be valuable but is atypical as it reflects the adjacent influence of the center's activities and staff.

Selected Statistics. In some circumstances, statistics on the mortality within age ranges known to be especially susceptible to malnutrition may be·available, such as the second year of life for kwashiorkor, or 2–5 months for infantile beriberi. In fact, in early childhood the breakdown of mortality statistics into smaller age segments than infancy (0–12 months) and the so-called preschool child (1–4 years) is usually revealing. However, with the limited and inaccurate statistical data often available, this is usually impossible. Indeed, even basic general data on infant mortality may be sparse, inaccurate, lacking, or suspect in most developing countries.

Ecological Diagnosis. In addition to information on the commonness of various forms of malnutrition in the community, epidemiological studies should also seek to analyze ecological factors which may be associated and/or causative. Such "ecological diagnoses" should include relevant information on socioeconomic status, conditioning infections, food availability, diet (food consumption) and cultural factors (Table III).

While the community diagnosis should ideally include assessment of a wide range of nonmedical factors, plainly, it can become a highly complex enterprise, taking considerable time, funds, trained staff, and equipment. Also, such studies can yield such a huge, complex mass of data, the analysis of which may take so long

Table III. Information Useful for the Assessment of Nutritional Status of Communities (World Health Organization, 1963)

Sources of information	Nature of information obtained	Nutritional implications
1. Agricultural data Food balance sheets	Gross estimates of agricultural production Agricultrual methods Soil fertility Predominance of cash crops Overproduction of staples Food imports and exports	Approximate availability of food supplies to a population
2. Socioeconomic data Information on marketing, distribution and storage	Purchasing power Distribution and storage of foodstuffs	Unequal distribution of available foods between the socioeconomic groups in the community and within the family
3. Food consumption patterns Cultural–anthropological data	Lack of knowledge, erroneous beliefs and prejudices, indifference	
4. Dietary surveys	Food consumption Distribution within the family	Low, excessive, or unbalanced nutrient intake
5. Special studies on foods	Biological value of diets Presence of interfering factors (e.g., goitrogens) Effects of food processing	Special problems related to nutrient intake
6. Vital and health statistics	Morbidity and mortality data	Extent of risk to community Identification of high-risk groups
7. Anthropometric studies	Physical development	Effect of nutrition on physical development
8. Clinical nutritional surveys	Physical signs	Deviation from health due to malnutrition
9. Biochemical studies	Levels of nutrients, metabolites, and other components of body tissues and fluids	Nutrient supplies in the body Impairment of biochemical function
10. Additional medical information	Prevalent disease patterns, including infections and infestations	Interrelationships of state of nutrition and disease

that the information is out of date by the time it has been completed. Indeed, in many parts of the world, surplus information of megasurveys is currently lying unused on paper or tapes.

Selection of data to be included is a key issue, although always difficult. Even with limited investigators, funds, and all other resources, however, considerable information on prevalence, together with limited information on etiology, can be acquired by means of rapid surveys (especially using anthropometry to assess protein–calorie malnutrition of early childhood), combined with information from hospitals and health centers (and from their adjacent defined areas when such exist). Such studies are easier in poorer nutritional circumstances, as, for example, in Haiti

(Jelliffe and Jelliffe, 1960). More recently, simplified surveys of young children have been carried out in various countires, such as Liberia, using basic anthropometry (weight, height, arm circumference), hemoglobin, stool examination, and 24-hr dietary recall (Zerfas and Shorr, 1976).

The community nutrition-level equation approach mentioned earlier is, in essence, a practical guide from which to glean the major causative factors in the local ecology on which a practical program can be based. While this equation will be concerned with the whole community, it will inevitably relate in larger measure to the main vulnerable groups: young children, pregnant and lactating women, the elderly, and other culturally, socially, or economically disadvantaged groups.

Within the range of possible data, selection is always necessary, depending on the expected nutritional situation and resources (funds, trained staff, laboratory facilities, etc.). A limited range of areas often needing to be considered is given in Table III.

3. Nutritional Surveillance and Conclusions

The initial process in nutritional surveillance is essentially an epidemiological assessment of the nature of the nutritional problems (including the groups at risk), and the identification of casual factors and the formulation of a working hypothesis.

Following this, appropriate indicators will need to be defined with special reference to ease of measurement, speed and frequency of data availability, and cost. Indicators to be used in nutritional surveillance include data on the agricultural and socioeconomic situation as well as health and dietary indicators.

Data collected for nutritional surveillance must, as far as possible, be from existing channels, such as information already being collected by the statistical and meteorological agencies, and the agricultural and health services. In addition, other information will be required and will have to be obtained from sample surveys. This will certainly include the prevalence of protein–calorie malnutrition in young children, judged clinically and by growth failure.

As with all epidemiological investigations, the range of data that can be collected can vary greatly in complexity. The collection of such data is often mainly an organizational–managerial problem, needing a central collating unit for collection, analysis, interpretation, and dissemination.

Nutritional surveillance is, in fact, epidemiology in motion over time. Its objectives are the early detection of community nutrition problems and their ecological causes, and the formulation of appropriate policy, leading to the planning and evaluation of action programs for both national developmental and emergency situations as they affect human growth.

The significance of this "moving-picture" approach to epidemiology will undoubtedly be tested and refined in the next decade. It is part of a worldwide need to detect early, to analyze etiologically, and to deploy limited resources imaginatively and where most needed.

4. References

American Journal of Clinical Nutrition, 1977, At-risk factors and young child health and nutrition, **30:**242.

Bryant, J., 1969, *Health and the Developing World,* Cornell University Press, Ithaca, New York.

Falkner, F., and Roche, A., 1972, Creation of growth standards: A committee report, *Am. J. Clin. Nutr.* **25**:218.

Falkner, F., and Roche, A., 1974, Measurement of nutritional status, *Am. J. Clin. Nutr.* **27**:1259.

Gordon, G., 1976, Epidemiology applied to nutrition, in: Nutition in Preventive Medicine (G. H. Beaton and J. M. Bengoa, eds.), WHO Monograph No. 62, Geneva.

Jelliffe, D. B., 1957, Cultural blocks and protein malnutrition in West Bengal, *Pediatrics* **20**:128.

Jelliffe, D. B., 1966, Assessment of Nutritional Status of the Community, WHO Monograph No. 53, Geneva.

Jelliffe, D. B., 1968*a*, Community forces and child nutrition, *J. Trop. Pediatr. Env. Child Health* **14**:47.

Jelliffe, D. B., 1968*b*, Infant Nutrition in the Subtropics and Tropics, WHO Monograph No. 29, Geneva.

Jelliffe, D. B., 1974, Relative values of longitudinal, cross-sectional and mixed data collection in community studies, in: *Nutrition and Malnutrition* (A. Roche and F. Falkner, eds.), p. 309, Plenum Press, New York.

Jelliffe, D. B., and Jelliffe, E. F. P., 1960, The prevalence of protein calorie malnutrition in early childhood in Haiti, *Am. J. Publ. Health* **50**:1355.

Jelliffe, D. B., and Jelliffe, E. F. P., 1977, *Human Milk in the Modern World,* Oxford University Press, Oxford.

Jelliffe, E. F. P., and Jelliffe, D. B., 1969, The arm circumference as a public health indicator of protein–calorie malnutrition of early childhood, *J. Trop. Pediatr.* **15**:253.

King, K. R., Fougere, W., Webb, R. E., Berggren, G., Berggren, W., and Hillaire, A., 1978, Preventive and therapeutic benefits in relation to cost: Performance of over 10 years of mothercraft centers in Haiti (in press).

Morley, D., 1973, *Paediatric Priorities in the Developing World,* Butterworths, London.

Puffer, R. R., and Serrano, C. V., 1973, *Patterns of Mortality in Childhood,* Pan American Health Organization, Washington.

Waterlow, J., 1974, Some aspects of childhood malnutrition as a public health problem, *Br. Med. J.* **4**:88.

Waterlow, J., 1978, Expert committee on Nutritional Screening, WHO Bulletin (in press).

World Health Organization, 1963, WHO Expert Committee on Medical Assessment of Nutrition Status, Geneva.

World Health Organization, 1976, Methodology of Nutritional Surveillance, Tech. Rept. Ser. No. 593, Geneva.

Zerfas, A., 1975, The insertion tape, *Am. J. Clin. Nutr.* **28**:782.

Zerfas, A., and Shorr, I., 1977, A nutrition survey of Liberia, unpublished data.

Zerfas, A., Shorr, I., Jelliffe, D. B., and Jelliffe, E. F. P., 1977, Selected indicators of nutritional status obtained by direct examination, unpublished data.

14

Obesity

GEORGE F. CAHILL, JR., and
ALDO A. ROSSINI

1. Introduction

To the physiologist, obesity is an enigma; in fact, it also poses problems to many other disciplines including those dealing with its pathogenesis, with its physical and emotional sequelae, with its therapy, and even with its definition (Albrink, 1976; Apfelbaum, 1973; Bray, 1973, 1976; Bray and Bethune, 1974; Burland *et al.,* 1974; Garrow, 1974; Howard, 1975; James and Trayhurn, 1976; Jeanrenaud and Hepp, 1970; Jequier, 1975; Mann, 1974; Winick, 1975). To the Polynesian boarding his canoe prior to a several months' voyage, obesity was both critical and probably the norm; to the desert nomad who survived by agility and speed, it was an obvious disadvantage. In the former, by present-day criteria, namely survival, obesity was certainly healthy; in the latter, an obvious and possibly lethal handicap. Current standards for adult obesity are based on calculations from actuarial data obtained by insurance companies, and they give population statistics for longevity in middle- and upper-middle-class Americans or northern Europeans as a function of height, weight, and age (Dublin, 1953). Clearly, the single variable, mean longevity, is what interests insurance underwriters the most, and by these current standards, mortality doubles with moderate, and trebles with marked obesity (Bray, 1976).

Most statistics show that obesity is correlated with many morbid sequelae, especially the major disorders of hypertension, cardiovascular disease, sudden death, diabetes, and degenerative arthritis. It is also associated with other problems such as thrombophlebitis, pulmonary and gallbladder diseases, hernia, accidents, and dermatitis (Bray, 1976). Equally debilitating, but more difficult to measure, are the social and emotional morbidities, including prejudicial treatment of the very obese in almost all interpersonal contacts such as interviews, promotions, hirings, firings, courting, and many others. Thus today, in American and northern European cultures, obesity is usually but not exclusively, detrimental to both the individual's physical and social well-being and to those who are dependent on the individual, not

GEORGE F. CAHILL, JR., and ALDO A. ROSSINI • Joslin Diabetes Foundation, Inc., The Peter Bent Brigham Hospital and Harvard Medical School, Boston, Massachusetts.

only as a medical problem, but also as a problem with ramifications extending to financial security and to emotional and societal success.

Although a topic of some medical concern for centuries, obesity has gradually progressed from being an amusing curiosity to a major public health issue as well as a theme for sophisticated physiologic and behavioral research. This brief chapter will review the basic physiology, not as a detailed critique but rather as an attempt to integrate present knowledge into an understandable and logical basis on which pathophysiological processes can be constructed and therapy rationally applied. Other sections of this volume deal with adipose cell kinetics (see Chapters 2 and 11, Volume 2), with the genetics of body size (Chapter 11, Volume 1), and with the development of infant feeding behavior (Chapter 9, this volume).

2. Adipose Tissue

Vertebrates, especially mammals, have selected triglyceride as their form of stored energy. The teleology is obvious; triglyceride is nonionic, extra-aqueous, and yields 9.4 kcal/g, close to the theoretical maximal energy that might be provided by a saturated long-chain hydrocarbon. With adipose tissue being 70–90% triglyceride, it yields 6–8 kcal/g, which is many times more economical than glycogen storage. Glycogen, because of its water solubility, is accumulated as 1 g glycogen to 1–4 g water, yielding only approximately 1 kcal or less per gram of tissue. Thus mobile organisms, especially those working against gravity, store their extra calories as triglyceride once they have replenished their carbohydrate reserves as glycogen. An average, recently fed 70-kg adult man may have 75 g of liver and 200–400 g of muscle glycogen for a total of about 2000 kcal storage; however, in his 10–15 kg of adipose triglyceride, he has 100,000 kcal stored. To complete the caloric tally, his muscle mass may contain 25,000 kcal as protein. However, the body's machinery sets up metabolic defenses against the utilization of this protein during starvation. Instead, the body draws selectively on adipose tissue triglyceride (Cahill, 1971).

3. The Fat Cell

Although small amounts of triglyceride are found in all cells, the adipose tissue is unique for this purpose. Its metabolism is such that it is exquisitely insulin sensitive (with one or two exceptions such as retroorbital fat pads and intra-articular fat, and perhaps fat in the bone marrow), and the effect of insulin is to augment uptake of all three nutrients (carbohydrate, amino acids, and triglycerides) into and conversion to and storage into the triglyceride vacuole in the center of the fat cell. Thus the excess ingested calories, no matter what their source, end up as fat, but only after the glycogen and protein reserves in other tissues have been appropriately repleted.

When insulin levels are low, as between meals, the stored triglyceride inside the fat cell is hydrolyzed to free fatty acids to serve as fuel for most organs and tissues in the body, and the glycerol is taken up by liver and kidney to be made into glucose. The rate of release of free fatty acids is a function of two opposing processes: inhibition of lipolysis by the low level of insulin and augmentation by norepinephrine released from sympathetic nerve endings directly in the adipose tissue. With stress or with exercise, free fatty acid production is even greater thanks to the increase in sympathetic nerve discharge, and also, to a further decrease in insulin levels, since sympathetic nerve activity also decreases insulin release from the beta cells (Smith and Porte, 1976). Thus the fat cell serves as the body's caloric buffer, or in physical terms, its caloric capacitor. As such, it should take and store any caloric overload, and conversely, contribute its triglyceride to maintain homeostasis in times of deficit. Is it, however, purely a passive caloric warehouse or is there some accounting system whereby some central office is able to take inventory and make appropriate adjustments? Before leaving the fat cell and trying to answer this question, a few more words should be directed to the most important issue of whether a cell can monitor its own lipid reserve.

4. The Fat Mass

The liver cell is a site of active traffic in fat metabolism. One role is to remove fatty acids derived from the diet and which are not of the pattern to be integrated into man's fat depots in adipose tissue. These fatty acids are catabolized or remodeled into human type fatty acid and exported into blood as triglyceride packaged as a unique particle made up of triglyceride, cholesterol, phospholipid, and protein (very-low-density lipoprotein—VLDL, endogenous triglyceride, pre-beta lipoprotein—all synonyms) to be incorporated subsequently into adipose tissue. Also, during fasting and especially during stress or exercise, more fatty acids are released from adipose tissue than consumed, and the liver removes the excess and incorporates it also into VLDL as above. Finally, liver synthesizes much fat itself, and this is the third and probably the most significant precursor (in amount) to VLDL. The normal liver cell contains only about 3% triglyceride. This may be doubled during times of heavy fatty acid and triglyceride flux, but the question is how does it monitor its own mass of triglyceride, which is scattered in small fat vacuoles inside the hepatocyte? With chemical, nutritional, or infective damage, this sensitive monitor fails and a fatty liver ensues. Thus the liver cell must be able to recognize the surface area or volume of the lipid vacuoles, and as this increases, it must increase the formation of VLDL for subsequent export to adipose tissue. The important point is, however, that the liver cell recognizes and monitors its lipid content. The adipocyte, on the contrary, does not have the luxury of unloading its fat when accumulating too much, if not for its own well-being, certainly for the well-being of the organism. Instead, it can only swell further as calories accumulate. This may not be wholly true, since basal levels of circulating free fatty acids are higher in obese individuals, and the greater tendency of obese individuals to develop fatty livers at the slightest provocation, such as infection, malnutrition, or even anesthesia, is well documented. This suggests there may be some "backing up," but the total proportion is small relative to the immense mass of fat stored in adipose tissue. The only significant defense of the adipocytes is to signal the host somehow to decrease its caloric intake in order to prevent overload or to signal other primordial cells to help take on the burden and to become adipocytes *de novo* (Ashwell and Garrow 1973).

Returning, however, to the overall inventory of fat mass, the overwhelming bulk of data in both normal man (Campbell *et al.*,1971; Friedman and Stricker, 1976) and experimental animals (Hervey, 1959, 1969; Herberg and Coleman, 1977) strongly suggest a connection between total lipid storage and caloric ingestion; in other words, the fat mass does "speak" to the brain, but how? The simple fact that man's gut can process five times more calories per unit time, perhaps even 10 times more than needed basally, and that in general, his adipose stores change minimally with changes in caloric expenditure, attest to an accurate feedback process between caloric ingestion and expenditure. A young healthy adult may consume over 1,000,000 kcal annually with little or no change in body lipid stores. With age, this servomechanism appears to become less dominant as food becomes displaced from being a biologic necessity into being a source of emotional comfort, and the food itself is eaten more on the basis of flavor and appearance as well as the surrounding environment and the mood of the eater rather than on biologic need. More of this later.

5. Adipose Tissue Distribution

Since adipose tissue may serve secondary roles other than calorie storage, it is distributed throughout the body in various locations, some with and some without obvious purpose. In mammalia its insulatory role places it subcutaneously, and this process reaches its maximum in the marine mammals. It is interesting, however, that in desert mammals, such as the camel or the fat-tailed sheep (Young, 1976), excess fat may be accumulated not subcutaneously but in special organs such as the hump and the tail, respectively, not only to provide reserve calories but especially reserve water obtained from its subsequent metabolism. Another role of fat is mechanical, the buccal cheek pad and the buttocks serving as the main examples. Cosmetic roles have also developed, the steatopygia of the Hottentot bushman and the mammophilia of the Western cultures, including both white and black, as evidenced by both current modes as well as ancient images and drawings. Thus in the process of evolution and selection, culture has also played an apparently important role in man in fat accumulation and distribution.

In the process of differentiation in man, as genetically programmed, mesenchymal cells are distributed throughout the body in special areas: subcutaneous, omental, perirenal, to name the major sites, and also in many other areas, such as between muscle bundles, in the pericardium, around the spleen and the adrenals, and elsewhere. The major question today is how much of this distribution and mass and number is genetically determined and how much can be modified (Ashwell and Garrow, 1973; Ashwell *et al.*, 1975; Widdowson and Shaw, 1973; Kirtland *et al.*, 1975; Häger *et al.*, 1977), especially by under nutrition and overnutrition through the formative years (Knittle and Hirsch, 1968; Björntorp and Sjöstrom, 1971; Brook, 1972; Knittle, 1972), or even *in utero*. The fundamental approach of Hirsch and colleagues (Hirsch and Han, 1969; Hirsch, 1976; Hirsch and Batchelor, 1976), whereby number and size of fat cells can be determined has been of much help, but new concepts and problems have arisen, and the issue of fat cell kinetics, especially in man, is far from settled (Sjöstrom and Björntorp, 1974). The reader is referred for further discussion to the chapters by Brook (Chapter 2, Volume 2) and by Knittle (Chapter 11, Volume 2).

6. Adipocyte Number

That the human at birth has less cells that can be identified as adipocytes than the adult is beyond doubt. Whether a finite number is genetically determined of which some can be filled, or whether others can be recruited or perhaps even induced to become adipocytes *de novo* at a later time cannot as yet be determined (Widdowson and Shaw, 1973). One problem is that the method of Hirsch and Gallian necessitates a given amount of lipid in a cell to make it float so it can be counted, so cells with small amounts of lipid are not included in their census. Nevertheless the adipocyte number in adult rats is clearly a result of their early nutrition. The secular trend of increased growth with good nutrition (Chapter 16, Volume 2), whereby larger body size (for combat) and the fuel needed to keep the large size going and the increased fertility (Chapter 8, Volume 2) would all be ecological advantages in times of plenty, suggests a direct nutritional role on body caloric mass. The time constant for genetic drift and remodeling is so large that a certain amount of environmental modeling of the genetic information given to the phenotype certainly provides survival benefits to a given population, and this may express itself in secular variations in sizing and fuel storage.

The question still remains, however, whether adipocyte number is influenced at all and, if so, how much by nutrition in the formative years in man.

7. Nutrition and Adipocyte Number and Size

Adult man has approximately $3\text{--}5 \times 10^{10}$ fat cells with about 0.4 μg lipid/cell, giving an adipose triglyceride mass of 12–20 kg. With obesity, the lipid content/cell may double or even treble; however, with the more severe degrees of obesity, the increased lipid mass appears to be due to adipose hypercellularity in addition to hypertrophy (Salans *et al.,* 1973; Sjöstrom and Björntorp, 1974; Hirsch and Batchelor, 1976). In fact, once fat cells accumulate about 1.0 μg lipid, they appear not to enlarge further, and the increased adiposity appears to be provided by three to four times as many cells or even more.

As reviewed by Brook and Knittle in Chapters 2 and 11 of the second volume of this series, man appears to have two phases of adipose cell neogenesis. One is in infancy and the second just prior to adolescence. It has been generally accepted that in obesity starting after adolescence, especially as evidenced by the prospective studies of force-feeding of normal volunteers by Sims and colleagues (Sims *et al.,* 1973), an increase in weight is matched by increases in cell lipid content (Salans *et al.,* 1971). Thus middle-aged "spread" is hypertrophic obesity and juvenile obesity is more hyperplastic. The most recent data of Hirsch and Batchelor (1976), however, show much overlap, and the evidence in other species that latent adipocytes containing little fat can be recruited into adipocytes with weight gain makes one question whether some degree of neoformation of fat cells can occur in postadolescent obesity. One author is aware of a documented instance whereby a disturbed 33-year-old housewife, weighing approximately 120 pounds, fattened up to 350 pounds in two years. Half of this tenfold increase in lipid mass could have been provided by doubling the lipid mass/cell, but the remainder must have been new cell formation. Unfortunately she was not studied by morphometric techniques.

This still leaves the question of how much the adipose organ in development is a function of heredity and how much a function of nutrition. In the 1940s and 1950s, emphasis was placed on heredity as being dominant. Then came the studies of Hirsch and colleagues in rats showing a correlation in weanlings between nutrition and subsequent adipose cellularity (Hirsch and Han, 1969). Several recent reports have suggested that in man both genes and environment play roles. Significant correlations between chubby babies and fat adults have been made (Charney *et al.,* 1975), but infants in Holland born during an episode of severe caloric restriction have less obesity in later life than infants born before or afterward (Ravelli *et al.*, 1976), suggesting a strong environmental input. Yet intrapair differences between obesity in identical and nonidentical twins strongly suggest a marked genetic factor (Börjeson, 1976), as do differences in correlations of weight in adopted children with weight of their foster and biological parents. Infants of diabetic mothers have macrosomia in addition to an increase in adipose tissue at birth (Whitelaw, 1977), but the fatness is a function of increased cell size and not number (Björntorp *et al.*, 1974), and this argues against *in utero* overfeeding as being very significant, since in subsequent life they appear to show no greater predelection toward obesity. It is probably a safe assumption that both nutrition and inherited factors do contribute to cell number and the subsequent propensity to obesity, but these interrelationships are far from clear.

8. Treatment

Obesity is one of the most common medical disorders observed by the practicing physician, be he internist, generalist, surgeon, or pediatrician. It is a problem which has been reported to have a prevalence of 25–45% in the adult American population over 30 years of age and a childhood incidence ranging from 2 to 15% (Bray, 1973, 1976). Furthermore, the incidence of obesity, like cigarette smoking, in the adolescent population appears to be increasing (Bray, 1973).

This is not the place to detail the treatment of obesity; suffice it to say that the recent primary advance in the treatment of adult obesity has been based on the observations that the increased caloric intake is due to social and emotional use of food instead of a metabolic or biologic drive. The approach is to change the emotional and environmental cues which lead to the increased food intake, a process called "behavior modification" (Levitz and Stunkard, 1974; Stuart and Davis, 1972; Stunkard and Mahoney, 1976).

In the obese child, the problem is much more difficult, since the effort must be directed to those controlling the child's environment, namely, the parents. Since most obese children come from homes with fat parents, therapy of the entire family unit, directed to modifying the parents' behavior, is the only course of action. Attempts to help the child by the usual restrictive diet and augmented exercise is futile in a setting in which the parents do not comply. The same holds true for smoking and alcohol. In fact, the therapy of obesity, using behavior modification, self-help group support, and attempts to establish motivation, parallels closely that of therapy for alcoholism.

Pharmacologic agents, especially the amphetamines, have been used to treat obesity. However, patients rapidly become tolerant and even addicted, and their use is thus limited. Fenfluramine is an amphetamine analog with allegedly more anorexigenic and less analeptic qualities, and its eventual usefulness, especially in the young, is awaiting characterization (Bray, 1973, 1976). Finally, surgical bypass procedures, such as jejunoileostomy, have no place in the treatment of the young (Faloon, 1977) and variable degrees of success in the adult.

9. References

Albrink, M. J. (ed.), 1976, *Obesity; Clinics in Endocrinology and Metabolism,* Vol. 5, No. 2, W.B. Saunders, Philadelphia.
Apfelbaum, M. (ed.), 1973, *Energy Balance in Man,* Masson, Paris.

Ashwell, M., and Garrow, J. S., 1973, Full and empty fat cells, *Lancet* **2:**1036–1037.

Ashwell, M. A., Priest, P., and Sowter, C., 1975, Importance of fixed sections in the study of adipose tissue cellularity, *Nature* **256:**724–725.

Björntorp, P., and Sjöström, L., 1971, Number and size of adipose tissue fat cells in relation to metabolism in human obesity, *Metabolism* **20:**703–713.

Björntorp, P., Enzi, G., Karlsson, K., Krotkiewski, M., Sjöström, L., and Smith, U., 1974, The effect of maternal diabetes on adipose tissue cellularity in man and rat, *Diabetologia* **10:**205–209.

Börjeson, M., 1976, The aetiology of obesity in children: A study of 101 twin pairs, *Acta Pediatr. Scand.* **65:**279–287.

Bray, G. A. (ed.), 1973, Obesity in perspective. Fogarty International Center Series on Preventive Medicine, Vol. 2, Parts 1 & 2, DHEW Publication No. (NIH) 75-708. Bethesda, Maryland.

Bray, G. A., 1976, The obese patient, in: *Major Problems in Internal Medicine,* Vol. IX, W.B. Saunders, Philadelphia.

Bray, G. A., and Bethune, J. E., 1974, *Treatment and Management of Obesity,* Harper and Row, New York.

Brook, C. G. D., 1972, Evidence for a sensitive period in adipose-cell replication in man, *Lancet* **2:**624–627.

Burland, W. L., Samuel, P. D., and Yudkin, J. (eds.), 1974, *Obesity Symposium: Proceedings of a Servier Research Institute Symposium* held in December, 1973, Churchill Livingstone, Edinburgh.

Cahill, G. F., Jr., 1971, Physiology of insulin in man, *Diabetes* **20:**785.

Campbell, R. G., Hashim, S. A., and Van Itallie, T. B., 1971, Studies of food-intake regulation in man: Responses to variations in nutritive density in lean and obese subjects. *N. Engl. J. Med.* **285:**1402–1407.

Charney, E., Goodman, H. C., McBride, M., Lyon, B., and Pratt, R., 1975, Childhood antecedents of adult obesity. Do chubby infants become obese adults? *N. Engl. J. Med.* **295:**6–9.

Dublin, L. I., 1953, Relation of obesity to longevity, *N. Engl. J. Med.* **248:**971–974.

Faloon, W., 1977, Symposium on jejunoileostomy for obesity, *Am. J. Clin. Nutr.* **30:**1–129.

Friedman, M. I., and Stricker, E. M., 1976, The physiological psychology of hunger: A physiological perspective, *Psychol. Rev.* **83**(6):409–431.

Garrow, J. S., 1974, *Energy Balance and Obesity in Man,* American Elsevier, New York.

Häger, A., Sjöström, L., Arvidsson, B., Björntorp, P., and Smith, U., 1977, Body fat and adipose tissue cellularity in infants: A logitudinal study, *Metabolism* **6:**607–614.

Herberg, L., and Coleman, D. L., 1977, Laboratory animals exhibiting obesity and diabetes syndromes, *Metabolism* **26:**59–100.

Hervey, G. R., 1959, The effects of lesions in the hypothalamus in parabiotic rats, *J. Physiol.* **145:**336–352.

Hervey, G. R., 1969, Regulation of energy balance, *Nature* **222:**629–631.

Hirsch, J., 1976, The adipose-cell hypothesis, *N. Engl. J. Med.* **295:**389–390.

Hirsch, J., and Batchelor, B., 1976, Adipose tissue cellularity in human obesity, *Clin. Endocrinol. Metab.* **5:**299–311.

Hirsch, J., and Han, P. W., 1969, Cellularity of rat adipose tissue: Effects of growth, starvation and obesity, *J. Lipid Res.* **10:**77–82.

Howard, A. (ed.), 1975, *Recent Advances in Obesity Research: Proceedings of the 1st International Congress on Obesity,* Newman, London.

James, W. P. T., and Trayhurn, P., 1976, An integrated view of the metabolic and genetic basis for obesity, *Lancet* **2:**770–772.

Jeanrenaud, B., and Hepp, D. (eds.), 1970, Adipose tissue: Regulation and metabolic functions, *Horm. Metab. Res.* Suppl. 2.

Jequier, E. (ed.), 1975, *Regulation of Energy Balance in Man.* 2nd Congress, Editions Médicine et Hygiène, Geneva.

Kirtland, J., Gurr, M. I., Saville, G., and Widdowson, E. M., 1975, Occurrence of "pockets" of very small cells in adipose tissue of the guinea pig, *Nature* **256:**723–724.

Knittle, J. L., 1972, Obesity in childhood: A problem in adipose cellular development, *J. Pediatr.* **81:**1048–1059.

Knittle, J. L., and Hirsch, J., 1968, Effect of early nutrition on the development of rat epididymal fat pads: Cellularity and metabolism, *J. Clin. Invest.* **47:**2091–2098.

Levitz, L., and Stunkard, A. J., 1974, A therapeutic coalition for obesity. Behavior modification and patient self-help, *Am. J. Psychiatry* **131:**423.

Mann, G. V., 1974, The influence of obesity on health, *N. Engl. J. Med.* **291:**178–185, 226–232.

Ravelli, G. P., Stein, Z. A., and Susser, M. W., 1976, Obesity in young men after famine exposure in utero and early infancy, *N. Engl. J. Med.* **295:**349–353.

GEORGE F. CAHILL, JR.,
and ALDO A. ROSSINI

Salans, L. B., Horton, E. S., and Sims, E. A. H., 1971, Experimental obesity in man. Cellular characters of adipose tissue, *J. Clin. Invest.* **50**:1005–1011.

Salans, L. B., Cushman, S. W., and Weismann, R. E., 1973, Studies of human adipose tissue: Adipose cell size and number in nonobese and obese patients, *J. Clin. Invest.* **52**:929–941.

Sims, E. A. H., Danforth, E., Jr., Horton, E. S., Bray, G. A., Glennon, J. A., and Salans, L. B., 1973, Endocrine and metabolic effects of experimental obesity in men, *Rec. Progr. Horm. Res.* **29**:457–496.

Sjöström, L., and Björntorp, P., 1974, Body composition and adipose cellularity in human obesity, *Acta Med. Scand.* **195**:201–211.

Smith, P., and Porte, J. D., 1976, Neuropharmacology of the pancreatic islets, *Annu. Rev. Pharmacol.* **16**:209.

Stuart, R. B., and Davis, B., 1972, *Slim Chance in a Fat World Behavioral Control of Obesity,* Research Press, Champagne, Illinois.

Stunkard, A. J., and Mahoney, M., 1976, Behavioral treatment of the eating disorders, in: *Handbook of Behavior Modification and Behavior Therapy* (H. Leitenberg, ed.), pp. 45–73, Prentice Hall, Englewood Cliffs, New Jersey.

Whitelaw, A., 1977, Subcutaneous fat in newborn infants of diabetic mothers: An indication of quality of diabetic control, *Lancet* **1**:15–18.

Widdowson, E. M., and Shaw, W. T., 1973, Full and empty fat cells, *Lancet* **2**:905.

Winick, M., 1975, *Childhood Obesity,* Wiley, New York.

Young, R. A., 1976, Fat, energy and mammalian survival, *Ann. Zool.* **16**:699–710.

General Reference

United States Dept. of Health, Education and Welfare, 1977, *Obesity and Health; A Source Book of Current Information for Professional Health Personnel,* PHS Publication #1485, Washington, D.C.

15

Nutritional Deficiencies and Brain Development

ROBERT BALÁZS, PAUL D. LEWIS,
and AMBRISH J. PATEL

1. Introduction

Severe undernutrition and malnutrition are formidable problems affecting a
depressingly great proportion of the world's population. Undernutrition has many
medical consequences, among which those arising from the vulnerability of the
immature brain to metabolic imbalance loom large. This vulnerability results from a
complex, programmed chronology of development, interference with any part of
which may have irreversible effects on structure and function. In man, malnutrition
is usually associated with conditions in which social and psychological development
is adversely affected, and there is good evidence indicating that these factors may
also lead to impairment of brain function. Studies on animal models permit experiments in which the major variable—although by no means the only one—is
nutrition; moreover the consequences of this on the physical development of the
brain can be followed in depth. In this review we will consider primarily the
influence of nutritional factors (mainly in terms of their deficiency) on brain
development in experimental animals: The emphasis will be on structural and
biochemical alterations. Functional changes including behavioral alterations in man
thought to result from impaired nutrition will be dealt with in Chapter 16 of this
volume. Many reviews and symposia have been already devoted to various aspects
of the topic of nutrition and brain development, and these can be consulted for more
detailed information (e.g., Scrimshaw and Gordon, 1968; Symposium, 1970; Altman
et al., 1970a; Stanbury and Kroc, 1972; Pan American Health Organization, 1972;

ROBERT BALÁZS and AMBRISH J. PATEL • MRC Developmental Neurobiology Unit, Institute of
Neurology, London, England. PAUL D. LEWIS • Department of Histopathology, Royal Postgraduate Medical School, Hammersmith Hospital, London, England.

Gardner and Amacher, 1973; Elliott and Knight, 1972; Cravioto *et al.*, 1974; Dobbing and Smart, 1974; Brazier, 1975; Prescott *et al.*, 1975; Pampiglione *et al.*, 1975; Dodge *et al.*, 1975; Winick, 1976).

2. Some Comments on Methods

Systematic investigation of the effects of undernutrition on the central nervous system began late in the nineteenth century with the work of Donaldson and his colleagues at the Wistar Institute. Their studies have been recently summarized in the comprehensive treatise of Dodge *et al.* (1975). The importance of the timing of the nutritional insult emerged from this group in the studies of Sugita (1918), who devised some of the techniques still employed today for effecting nutritional deprivation during the period of most rapid brain development in the rat, which is probably the most extensively studied experimental species. These include methods aimed at reducing the amount of milk available to the young, either directly—by separating the young from the nursing mother for part of each day or by increasing the size of the litter—or indirectly—by decreasing milk production, feeding the dams either with a restricted amount of normal diet or with an unrestricted diet deficient in certain ingredients, such as protein, amino acids, and trace elements. An attractive hypothesis is that growth restriction and its timing are of primary importance rather than how the growth retardation is achieved (Dobbing, 1974). Nevertheless, although their effects are similar, the different modes of nutritional deprivation during development may not produce precisely the same results.

In this review we shall refer to the various types of nutritional deprivation as undernutrition, although it must be realized that even a simple reduction in the quantity of a balanced diet may lead to some dietary imbalance. Since we shall rely heavily at certain points in this presentation on our own experiments, it is appropriate to mention here that the technique we have used involves the underfeeding of the mother rats by providing approximately half of the normal food supply during a period in pregnancy (from the 6th day) and lactation. By comparing different methods of undernutrition Altman *et al.* (1971) observed that the most consistent developmental retardation of the young is brought about by this treatment. The fetus is relatively spared under these conditions at the expense of the mother (e.g., Smart and Dobbing, 1971; Balázs and Patel, 1973). However, there are limits to the protection offered, and more severe nutritional deficiency leads to retarded fetal growth affecting also the brain (e.g., Zamenhof *et al.*, 1971*a,b*). It is evident that consistent growth retardation is only one of the criteria which must be considered in the experimental design, and the choice of the technique of undernutrition is determined by the experimental aims: for example, by considering corresponding periods in brain development of humans and rats (e.g., Dobbing, 1974), Smart *et al.* (1976) have undernourished rats until the 5th postnatal day as an experimental model of small-for-date babies.

3. Effect of Undernutrition on Cell Acquisition in the Brain

Implementation of nutritional deprivation which during pregnancy interferes only slightly with fetal growth results, during lactation, in a severe retardation of the growth of the infant rats (Table I). This apparent difference is related to the

Table I.   Effect of Undernutrition during Pregnancy and Lactation[a]

(a) Body Weight of Mother Rats (g)

Treatment (No. of animals)	Day of pregnancy			Day of parturition
	6–7	13–14	20–21	
Control (12)	327 ± 8.4	341 ± 8.1	402 ± 13.8	324 ± 8.1
Undernourished (11)	323 ± 9.8	304 ± 10.6[b]	314 ± 9.0[b]	260 ± 9.9[b]

(b) Newborns

Treatment (No. of litters)	Litter size (no.)	Male (no.)	Female (no.)	Body weight (g)
Control (12)	11.0 ± 0.8	5.8 ± 0.6	5.2 ± 0.4	5.73 ± 0.13
Undernourished (11)	10.5 ± 0.5	5.7 ± 0.3	4.8 ± 0.3	4.87 ± 0.22[b]

(c) Postnatal Growth of the Young (g)

Treatment (No. of young)	Age (days)			
	6	10	14	21
Control (12)	12.2 ± 0.2	22.9 ± 0.2	34.9 ± 0.3	50.2 ± 0.4
Undernourished (12)	6.5 ± 0.3[b]	10.8 ± 0.2[b]	12.6 ± 0.4[b]	18.1 ± 0.2[b]

[a]Rats were undernourished from the 6th day of pregnancy throughout lactation by providing approximately half their normal diet (Breeding Diet for Rats and Mice, Oxoid Ltd; 10 g daily before parturition and 15, 20, and 25 g daily during the first, second, and third weeks of lactation respectively). Controls were fed the same food *ad libitum*. Water was always available to all rats. Values are means ± S.E. (From Patel and Balázs, 1973, and unpublished observations.)
[b]Significant differences between control and undernourished groups ($P < 0.05$).

nutritional requirement necessary to sustain the relatively great increase in the body weight of the young during the early postnatal period. In comparison with body weight, brain growth is much less affected by postnatal undernutrition (Sugita, 1918). Nevertheless, brain weight is significantly decreased, and this, in part, is the consequence of reduced cell numbers (Howard, 1965; Winick and Noble, 1966). Undernutrition interferes with cell proliferation throughout the body including the brain, and it is the mechanism of this effect that will now be considered.

3.1. Postnatal Cell Formation

In the rat brain vigorous cell proliferation proceeds during the early postnatal period and accounts for almost 50% and 97% of the final cell number in the forebrain and the cerebellum, respectively (Figure 1). There are certain characteristic features of postnatal cell formation in the rat brain which to some extent also apply to many other species including humans:

1. Cells are mainly formed at circumscribed sites, notably the subependymal layer lining the forebrain ventricles and the external granular layer covering the surface of the cerebellum. However, there are also progenitor cells, probably glial precursors, which are dispersed throughout the brain.

2. The period of extensive cell proliferation is limited; it is completed about 3 weeks after birth in the rat (Figure 1) and 1.5–2 years of age in man (Dobbing and Sands, 1973) with the dissolution of the germinal sites, such as the cerebellar external granular layer. However, fragments of the subependymal layer persist into later life in both rodents and primates, indicating that a fraction of the total cell

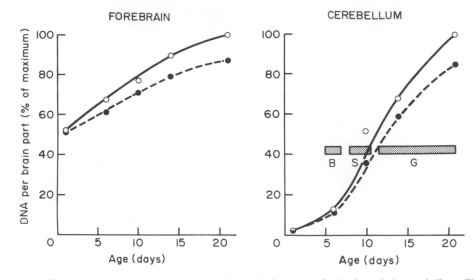

Fig. 1. Effect of undernutrition on postnatal cell acquisition in the forebrain and the cerebellum. The results are expressed as a percentage of the adult DNA content. The age curves for controls (○—○) differ significantly from those for undernourished (●---●) suckling rats ($P < 0.01$). The results are taken from Patel *et al.* (1973). The "birthdays" of the postnatal cerebellar interneurons are also indicated. The horizontal bars show the period when the formation of basket cells (B) and stellate cells (S) is maximal, and the period when about 50% of the granule cells (G) are formed. The data are taken from Altman (1969, 1972*a,c*).

population is continuously renewed even after the period of extensive cell formation (Smart and Leblond, 1961; Lewis, 1968*a,b;* Haas *et al.,* 1970).

3. Predominantly glial cells are formed in the forebrain during the postnatal period. However, extensive neurogenesis persists in the fascia dentata of the hippocampus and the olfactory lobes (Altman, 1969), and a few neuronal precursors continue to divide in the nucleus septalis accumbens in the first few days after birth (Creps, 1974). In this respect the behavior of the cerebellum is noteworthy; here most of the nerve cells are formed after birth. Only the Purkinje cells and the large neurons in the deep cerebellar nuclei as well as a fraction of the Golgi cells are formed prenatally. In general, the late-generated nerve cells are interneurons. They play an important role in the structural and functional organization of the central nervous system, and it has been suggested that, by virtue of their relatively late appearance, these cells may be targets of environmental influences leading to modification of the neuronal networks that they complete with the prenatally formed long-axoned neurons (Altman, 1970). There is evidence that this late wave of neurogenesis, at least in one part of the brain, the cerebellum, may proceed even in man postnatally. The cerebellar interneurons (with the exception of the Golgi cells) are generated in the external granular layer, and this layer is prominent in man for several months after birth; migrating cells in the cerebellar cortex disappear only 1.5 years after birth (Larroche, 1966). It has also been shown recently that the number of granule cells, the most abundant interneurons in the cerebellum, increases very substantially during the first two postnatal years in man (Gadsdon and Emery, 1976).

4. The formation of the different nerve cell types occurs, as in embryogenesis, in a well-defined time sequence in the cerebellum (Fujita, 1969; Altman, 1969, 1972*a–c*).

The relatively advanced knowledge of the characteristics of postnatal cell formation is of great advantage in studying the effect of metabolic imbalance on the developing rat brain. Since cell formation is restricted to a great extent to the germinal sites, autoradiographic studies are well suited to collect information on the possible interference of undernutrition with cell kinetics. This approach can be made even more telling by studying germinal sites which generate particular cell types, such as glial cells in the anterior part of the lateral ventricular subependymal layer or interneurons in the cerebellar external granular layer. The quantitative aspects of postnatal cell formation can be best tackled by complementing these studies with biochemical investigations, which permit the evaluation of the overall rate of cell acquisition and of DNA synthesis.

As shown in Figure 1 undernourishing the pregnant mother rat had no significant effect on the acquisition of cells in the brain of the fetus, for the amount of DNA was normal in the newborn (Patel *et al.*, 1973). In contrast, during the suckling period the rate of cell acquisition was depressed throughout the brain, and although the animals were rehabilitated from day 21, the moderate depression in cell numbers (10–20% depending on brain part) persisted. Other studies, in which different modes of undernutrition were employed, including food deprivation only during the suckling period, and in which the rehabilitation after weaning was longer, have also shown that the deficit in brain cell numbers incurred in early life is permanent (Howard and Granoff, 1968; Dobbing *et al.*, 1971; for reviews see Dobbing, 1974; Winick, 1976).

By fitting logistic curves to the data in Figure 1, it was found that in the control at the time of maximal cell acquisition the daily increase in cell number was 4.8×10^6 cells in the forebrain (day 2.7) and 17.9×10^6 cells in the cerebellum (day 12.8) (Patel *et al.*, 1973). These values represent 3% and 8% of the cells in the adult forebrain and cerebellum, respectively. It would appear that during the postnatal period the cerebellum is selectively sensitive to the maintenance of the metabolic balance. In comparison with other parts of the brain, different metabolic insults cause a greater deficit in cell number (Howard and Granoff, 1968; Fish and Winick, 1969; Chase *et al.*, 1969; Prensky *et al.*, 1971; Balázs, 1972a; Dobbing, 1974). However, these findings do not mean that the cells in the cerebellum are selectively vulnerable, rather they are related to the fact that here at birth cell number is only 3% of the final value in contrast to more than 50% in the cerebrum.

In an attempt to understand better the effect of undernutrition on cell acquisition in the brain the *in vivo* rate of DNA synthesis was also studied (Patel *et al.*, 1973). The results were unexpected. In contrast to the relatively mild effects on the cell number in the brain (approximately 15% final deficit), undernutrition caused a very severe depression of DNA synthesis rate, which at its nadir was only about 20% (forebrain) to 30% (cerebellum) of the control values (Figure 2). There are several factors which may account for this discrepancy.

1. [14C]Thymidine is a remote precursor of DNA; thus any alteration in its access to the cells in the brain and its conversion rate into [14C]thymidine nucleotides would result in an apparent decrease in [14C]-DNA formation. The kinetics of [14C]thymidine metabolism were therefore investigated in 12-day-old rats when the depression of DNA synthesis rate in both brain parts was great (Figure 3) (Patel *et al.*, 1978a). Although the concentration of [14C]thymidine was elevated in the undernourished brain, the decay of [14C]thymidine did not differ significantly from

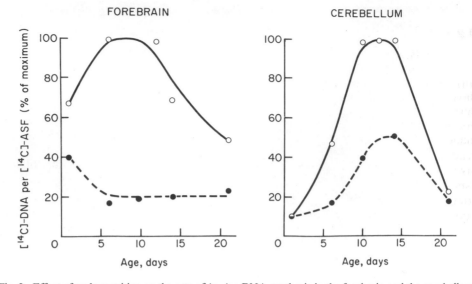

Fig. 2. Effect of undernutrition on the rate of *in vivo* DNA synthesis in the forebrain and the cerebellum. The incorporation of ^{14}C into DNA per brain part was expressed in terms of the concentration of acid-soluble [14]C 30 min after the subcutaneous injection of 20 μCi/100 g body wt of [2-^{14}C]thymidine. The results are expressed as a percentage of the maximum values. The age curves for the undernourished animals (●---●) differ significantly from controls (○—○) ($P < 0.01$). (From Patel *et al.*, 1973, 1976.)

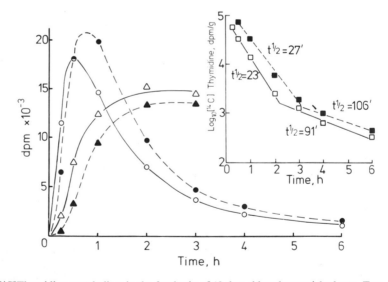

Fig. 3. [14]C]Thymidine metabolism in the forebrain of 12-day-old undernourished rats. Two control and experimental animals were killed by immersion into -150° Freon at the times indicated after a subcutaneous injection of [2-^{14}C]thymidine (15 μCi/100 g body wt); the constituents were isolated as described by Patel *et al.* (1976). The semilogarithmic plot of ^{14}C]thymidine decay is shown in the inset: The half-lives ($t\frac{1}{2}$) of the fast and the slow decay components were calculated from the slopes of the regression lines. The ^{14}C content is expressed in terms of g fresh wt for thymidine and the sum of thymidine nucleotides, whereas ^{14}C]-DNA refers to the whole forebrain. Control: [14]C]thymidine, □—□; ^{14}C]thymidine nucleotides, ○—○; ^{14}C]-DNA, Δ—Δ. Undernourished rats: [14]C]thymidine, ■---■; ^{14}C]thymidine nucleotides, ●---●; ^{14}C]-DNA, ▲---▲. (From Patel *et al.*, 1978*a*.)

controls and showed a rapid and a slow component (inset in Figure 3). In comparison with controls, the rate of [^{14}C]thymidine conversion into [^{14}C]thymidine nucleotides was slightly slower: the peak of [^{14}C]thymidine nucleotide concentration was reached about 15 min later, but the ^{14}C content of these compounds was even higher in the undernourished brain after 30 min. The rate of [^{14}C]-DNA formation was linear only for a limited time (about 0.5 hr in the control and 1 hr in experimental animals). The initial rate of [^{14}C]-DNA formation was severely depressed in undernutrition; when the [^{14}C]-DNA content was corrected on the basis of the concentration of [^{14}C]thymidine nucleotides, the rates were only 40% of the control values at 0.5 hr after the administration of the precursor. Thus the depressed incorporation of [^{14}C]thymidine into DNA reflects a genuine reduction in the rate of DNA synthesis and is not the consequence of an impairment in the availability of the proper labeled precursor.

2. The rate of cell acquisition is the result of the balance of new cell formation and cell death. A possible explanation for the discrepancy between the small decrease in cell number accompanied by a severe depression of DNA synthesis rate is a reduction of the normal rate of cell loss compensating for the lowered mitotic activity. However, it was observed that the proportion of degenerating cells was in fact elevated in the germinal sites of the undernourished brain at least at day 12 (Table II). In recent studies we have observed that the pyknotic index is also significantly increased in the forebrain subependymal layer of adult food-deprived animals (Lewis *et al.*, 1977*b*).

3. When a relatively long period, such as the first three weeks of life, is considered in an asynchronously replicating cell population, assuming a growth fraction of about 1, the rate of cell acquisition is related to the turnover time of the proliferating cells, whereas the rate of incorporation of a precursor into DNA is primarily a function of the length of the DNA synthesis phase (S phase) of the cell cycle. Thus, if the S phase were prolonged by undernutrition to a greater extent than the turnover time, the rate of DNA synthesis would be more depressed than the rate of cell acquisition. This hypothesis can be tested by studying cell kinetics in the germinal zones of undernourished animals (Lewis *et al.*, 1975).

Table II. Effect of Undernutrition (UN) on Cell Formation and Cell Degeneration[a]

Age (days)	Mitotic index (%) Control	UN	Labeling index (%) Control	UN	Turnover time (hr) Control	UN	Pyknotic index (%) Control	UN
Forebrain lateral ventricular subependymal layer								
1	1.34	0.99	12.6	14.7	79	83	0.22	0.14
6	0.95	0.82	9.7	14.7	111	98	0.45	0.78
12	1.15	0.90	9.3	14.5	118	140	0.54	1.41[b]
21	0.87	0.83	11.8	13.9	105	101	0.56	0.63
Cerebellar external granular layer								
1	2.72	2.29	25.7	32.8	36	43	0.30	0.62
6	2.31	2.12	27.2	25.6	35	54	0.39	0.25
12	1.82	1.75	21.9	31.1	49	61	0.58	1.30[b]
21	0.69	1.29	17.2	20.3	56	61	0.36	0.38

[a] Mitotic and pyknotic indices were obtained from counts of 1000 nuclei in sections from 10 animals in each group. Labeling indices were derived from counts of 1000 nuclei in animals killed 1 hr after injection of [^{3}H]thymidine (see Figure 4). The turnover time was calculated according to the formula: turnover time = 100 × length of S phase/labeling index. (From Lewis *et al.*, 1975; Lewis, 1975.)
[b] Significant differences between control and undernourished groups: $P < 0.01$

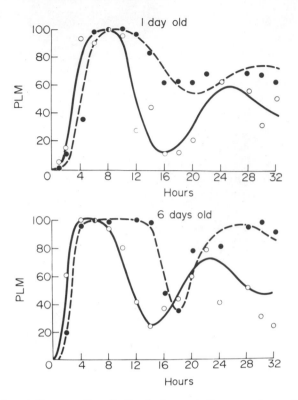

Fig. 4. Effect of undernutrition on cell replication in the cerebellar external granular layer. Computer-generated curves are superimposed on the data of percentage labeled mitoses (PLM) observed 1–32 hr after injection of [³H]thymidine (250 μCi/100 g body wt) at the different ages indicated. Controls, ○—○; undernourished, ●---●. Estimates of cell cycle parameters are given in Table III. (From Lewis *et al.,* 1975.)

Table III. *Effect of Undernutrition on Cell Cycle Parameters in the Germinal Sites of Developing Rat Brain*[a]

Phase	Treatment	Age (days)			
		1–2	6–7	12–13	21–22
S	Control (hr)	9.6	10.2	10.8	11.1
	Undernutrition (%)	136[b]	138[b]	182[b]	119
G₁	Control (hr)	5.7	3.4	2.9	5.0
	Undernutrition (%)	23[b]	6[b]	7[b]	114
G₂	Control (hr)	3.3	2.1	1.5	2.1
	Undernutrition (%)	130	133	240[b]	114
Total cell cycle	Control (hr)	18.7	16.8	16.1	18.1
	Undernutrition (%)	107	104	148[b]	117

[a]The estimates of the length of the cell cycle phases were obtained by fitting computer-generated curves to the data representing the percentage labeled mitoses at 1–32 hr after injection of [³H]thymidine (see Figure 4). Since the estimates in the subependymal layer in the forebrain and the EGL in the cerebellum did not differ significantly, the data for the two germinal sites were combined. The results for controls are given in hours, and the values in the undernourished animals are expressed as a percentage of controls. (From Lewis *et al.,* 1975.)
[b]Significant differences between control and undernourished groups: $P < 0.05$.

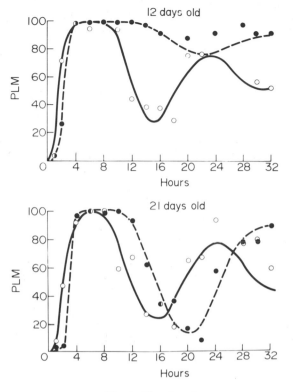

Fig. 4. (*Continued*)

Figure 4 shows computer–generated curves superimposed on the data of percentage labeled mitoses observed in the cerebellar external granular layer 1–32 hr after injection of [³H]thymidine. Estimates of the length of the cell cycle phases were obtained from these curves, and they are displayed in Table III. The results showed that, especially during the first two postnatal weeks when cell proliferation is most extensive, the S phase was prolonged much more than the generation time of the germinal cells in the undernourished brain. The relatively small effect of the treatment on cell cycle time in spite of the marked lengthening of the DNA synthesis phase was a consequence of a severe curtailment of the length of the G_1 phase of the cell cycle.

Mitotic and labeling indices and turnover times are given in Table II. The results showed that, in contrast to the pronounced prolongation of the S phase, undernutrition had relatively little effect on the turnover time, and thus the hypothesis which these experiments were set out to test has been substantiated.

However, some additional factors are probably also contributing to the effect. A slowing of DNA acquisition in the undernourished brain (Figure 1) implies, in the face of more or less unchanged germinal cell turnover times, that germinal cell numbers are reduced. Using both the biochemical and autoradiographic data, it was possible to make an approximate estimation of that fraction of the cells in the brain which belong to the germinal layers (Lewis *et al.,* 1975). For the subependymal layer in the control rats aged 1–12 days the calculated value was 2–3% of the total cell population in the forebrain; the estimate for the undernourished animals was 0.9–1.4%. For the cerebellum a less circumstantial estimate of germinal cell numbers can be derived from the cross-sectional area of the whole cerebellum and the external granular layer, together with the packing density of cells in that layer. The computed deficit in germinal cells in the undernourished cerebellum was almost 20%.

There is evidence indicating that the activity of certain enzymes associated with DNA synthesis, such as thymidine kinase, is closely linked to the life cycle of cells (Cleaver, 1967; Baril *et al.*, 1974). In synchronized cells in culture it has been found that thymidine kinase activity rises when the cells pass from G_1 to the S phase, reaches its maximum several hours after DNA synthesis was initiated, remaining high in the G_2 phase, and drops sharply as the cells finish mitosis (Stubblefield and Murphree, 1967; Gelbard *et al.*, 1971). Similar behavior has been reported for ribonucleotide reductase (Murphree *et al.*, 1969), thymidylate kinase (Johnson and Schmidt, 1966), deoxycytidine deaminase (Gelbard *et al.*, 1969), deoxycytidylate synthetase (Adams *et al.*, 1966), and DNA polymerase II (Baril *et al.*, 1974). The estimation of the activities of these enzymes may therefore provide useful information on the proliferative potential of the tissue. Table IV shows that the activity of thymidine kinase during the first month after birth is consistent with our knowledge of postnatal cell proliferation in the brain (compare, e.g., with Figure 2). In undernutrition, thymidine kinase activity per forebrain or cerebellum was significantly less than in controls at days 6 and 12. However, in the cerebellum the precipitous fall in the enzyme activity after the second week of life was somewhat retarded in undernutrition (see values at day 22). In a previous report Weichsel and Dawson (1976) have observed that the effects on the cerebellar thymidine kinase activity (per mg supernatant protein) also depended on the degree of undernutrition; e.g., at day 12 animals which were less growth retarded showed a significant decrease in activity, while there was an increase in the most affected group. Giuffrida *et al.* (1975) have interpreted the effects of undernutrition on the activity of cerebellar thymidine kinase as indicating a delay in the normal developmental changes. This can be especially well seen when the results are expressed in terms of unit wet weight rather than whole cerebellum.

In summary, the main factors involved in the effect of undernutrition on cell

Table IV.   Effect of Undernutrition on the Activity of Thymidine Kinase in the Developing Rat Brain[a]

	Age (days)				
	1	6	12	22	29
Forebrain					
		As a percentage of the value at 12 days of age			
Control	54	80	100	40	42
		As a percentage of control values			
Undernourished	120	72[b]	70[b]	95	83
Cerebellum					
		As a percentage of the value at 12 days of age			
Control	2	36	100	5	3
		As a percentage of control values			
Undernourished	118	44[b]	69[b]	211[b]	90

[a]The results are expressed as $p$mol/min per brain region; the values for 12-day-old controls are 497 and 1153 for the forebrain and the cerebellum, respectively. (From Patel *et al.*, 1978a.)
[b]Significant differences between control and undernourished groups: $P < 0.05$.

acquisition in the developing brain are as follows: (1) The generation cycle is distorted; the S phase is substantially prolonged, whereas the G_1 phase is severely curtailed; (2) in the first two weeks after birth the germinal cell population is decreased; (3) during the second week of life increased numbers of degenerate postmitotic cells are found in the germinal zones; and (4) a tendency to compensate for the reduced cell acquisition is detected in the slightly prolonged persistence of the cerebellar external granular layer (see also pp. 432 and 439).

The derangement of the generation cycle is unique in undernutrition to the developing brain. Food deprivation in tissues which in the adult contain rapidly proliferating cells results in a marked prolongation of the cell cycle time including both the S phase and the G_1 phase (Wiebecke *et al.*, 1969; Rose *et al.*, 1971; Deo *et al.*, 1975). In developing organs the reduction in cell acquisition is usually more severe than that in the brain, implying that in these organs undernutrition effected a significant lengthening of the cell cycle time.

The uniqueness of the response to undernutrition of the cell cycle in dividing cells in the brain during its major growth period is underlined by recent experiments on animals aged 6 weeks, in which the subependymal layer was studied autoradiographically (Lewis *et al.*, 1977*b*). Groups of rats were either chronically undernourished or acutely food-deprived. With both treatments the generation cycle in subependymal cells was markedly prolonged compared with controls, and both S phase and G_1 phase were lengthened, as occurs in other tissues in undernutrition. The duration of mitosis in these young adult animals also appeared to be increased.

3.2.1. Possible Mechanisms Involved in the Increase in the Length of the DNA Synthesis Phase of the Cell Cycle

There is little information about the possible cause of the marked prolongation of the S phase produced by undernutrition in the germinal cells in the developing brain. An observation resembling this situation has been reported for hepatoma cells cultured in the presence of dibutyryl cyclic adenylic acid (cAMP) (van Wijk *et al.*, 1973). However, here the effect was due to a severe reduction in the conversion of thymidine to thymidine nucleotides, which is not the case in the brain *in vivo* (Figure 3).

Although the regulation of DNA replication in eukaryotes has been extensively studied, our understanding of the mechanisms involved is still very limited. Current knowledge has been lucidly summarized in a recent review (Prescott, 1976). It is now well established that the reason why the replication of the eukaryote DNA, which is huge in comparison with the bacterial genome, lasts such a relatively short time is that there are large numbers of replicating units per genome. Autoradiographic studies on pulse-labeled purified DNA have indicated that the average replicating unit is about 30 μm long (6×10^7 daltons; Huberman and Riggs, 1968; Lark *et al.*, 1971). Since the DNA in a haploid set of mammalian chromosomes is about 90 cm long (3 pg or 1.8×10^{12} daltons) the number of replicating units per haploid genome is about 30,000. It seems that in these units DNA replication proceeds from a single origin in two directions with an average speed of almost 1 μm/min. Thus about 15 min is needed for the duplication of the average mammalian replicating unit. Many replicating units scattered throughout the whole genome initiate replication simultaneously. These groups are defined as banks, and a haploid mammalian cell would contain a minimum of 25 banks. Thus the S phase is characterized by a cascade of initiation and replication of many thousands of replicating units in a single mammalian cell nucleus. The nature of the initiation signal is unknown, but it is usually assumed that protein synthesis is involved. Food deprivation is associated with a reduction in the rate of protein synthesis (e.g., Symposium, 1970) and may influence, in particular, the synthesis of protein species involved in the program of initiation of DNA synthesis.

It may be important for the understanding of the effect of undernutrition on the length of the S phase in the replicating cells of the developing brain that great variations in the duration of this period of the cell cycle have been observed in different developmental states within a single species. In *Triturus* the length of the S phase is about 10 days immediately preceding male meiosis, approximately half that time in the somatic cells, and it is almost 50 times faster in the replicating embryonic cells (Callan, 1972). Variations in S-phase length depending on the developmental state have also been reported in *Drosophila* (Blumenthal *et al.,* 1974; Kriegstein and Hogness, 1974; Wolstenholme, 1973), while lengthening of the duration of DNA synthesis with advancing development has been repeatedly shown in avian muscle and mammalian brain (Schultze *et al.,* 1974; Lewis, 1978). It has been suggested that the great differences in *Drosophila* and *Triturus* may result from changes in the number of initiation points and thus in the size of the replicating units, rather than from alterations in the rate of copying the DNA segments (Callan, 1972; Prescott, 1976).

3.2.2. Curtailment of G_1 Phase: Possible Functional Consequences

Within the cell cycle the G_1 phase usually shows the greatest variation in length (Cleaver, 1967). During early embryogenesis the G_1 period is very short or even absent, and it tends to lengthen with advancing age (see Prescott, 1976; Lewis, 1978). It would appear, therefore, that the germinal cells in the developing brain of undernourished rats in the postnatal period reverted, at least in this context, to an earlier pattern of replication. Although the G_1 phase may be an expendable period in the cell cycle, there are indications that certain processes occurring during a limited period in the G_1 phase are critical in terms of the full expression of the normal differentiated functions of some cells (Vonderhaar and Toper, 1974). Thus, while the virtual elimination of the G_1 phase is an effective mechanism compensating for the prolongation of the S phase the germinal cells of the undernourished brain, it may have adverse effects on the progeny of cells which are near to their terminal division in the postnatal period.

3.2.3. Errors in Cell Replication

The increased pyknotic index observed in the germinal zones in the undernourished brain gives only a crude estimate of the lethally damaged newly formed cells. The number of cells with sublethal but functionally significant damage probably exceeds this value. A pertinent observation in this context is the report that the incidence of chromosomal anomalies in lymphocytes from malnourished patients is more frequent than that observed in healthy controls (Armendares *et al.,* 1971; Khouri and McLaren, 1973; Betancourt *et al.,* 1974). However, further studies are needed to exclude the effects of associated mutagenic agents such as toxic factors and viral infections (Thornburn *et al.,* 1972; Khouri and McLaren, 1973).

3.3.1. Microtubules and Microfilaments

A possible mechanism underlying the adverse effect of undernutrition on cell division is suggested by the influence of glucose deficiency on the formation of microtubules in the Langerhans islets of the pancreas. Microtubules are long tubelike subcellular organelles in which an electron-dense annulus of 250 Å diameter surrounds an electron-translucent core (diameter about 150 Å). The dense annulus is composed of a number of laterally associating protofilaments each comprising globular subunits. The building blocks of microtubules are polymers of tubulin (molecular weight about 100,000) which are dimers of nonidentical subunits (α and β). Microtubules are believed to be involved in the transmission of signals from the cell surface to the internal machinery associated with cell replication, cellular mobility, and interactions (e.g., Symposium, 1974; Soifer, 1975; Marx, 1976). Normal cells in culture have an extensive microtubular network which extends from the cell center to the periphery and terminates just under the cell membrane (Fonte and Porter, 1974; Brinkley *et al.*, 1975). During mitosis this complex disappears. At least in lymphocytes, which have been most extensively studied, the capacity to divide appears to be influenced by alterations in the mobility of certain membrane constituents which in turn is believed to be controlled by microtubules and microfilaments are not fully understood, but the involvement of stimulation or inhibition of cell division may thus depend on the induction of different states of the assembly or disassembly of microtubules by different membrane signals. That these mechanisms, rather than the concentration of tubulin, are involved in the regulation has been suggested by observations on transformed cells which lack the normal control of cell division (Fonte and Porter, 1974; Brinkley *et al.*, 1975). Further, in an inherited disorder, the Chediak–Higashi syndrome, which is characterized by a high susceptibility to infection, the leukocytes resemble normal cells in which microtubules are experimentally disrupted (e.g., with colchicine) (Oliver *et al.*, 1975). These cells make tubulin, but the defect is in the mechanism controlling microtubule assembly involving cyclic guanylic acid (cGMP). This nucleotide promotes the polymerization of microtubules. It is thought that microfilaments, which presumably contain actin and are long, linear structures of 40- to 60-Å diameter, serve as connecting links between membrane components and microtubules (Edelman, 1976). The factors controlling the state of assembly of microtubules and microfilaments are not fully understood, but the involvement of Ca^{2+} ions seems now to be established (e.g., Weisenberg, 1972).

In the pancreas, glucose stimulates the replication of the islet cells both *in vivo* (Brosky *et al.*, 1972) and in tissue culture (Chick *et al.*, 1973; Andersson, 1975), and it also enhances the synthesis of tubulin and its polymerization to microtubules (Pipeleers *et al.*, 1976). The effect of glucose on the pancreas islets may be mediated through mechanisms functioning with cAMP systems: Tubulin synthesis is stimulated by cAMP (Pipeleers *et al.*, 1976), the concentration of which is decreased in starvation (Selawry *et al.*, 1973), while after glucose feeding the activity of adenylate cyclase is significantly enhanced (Howell *et al.*, 1973). It is worth noting here that the effect of food deprivation varies according to the metabolic properties of the different tissues. Starvation is known to cause an increase in the cAMP concentration of other organs, such as the liver, kidney, and adipose tissue, as a result of the response of adenylate cyclase in these organs to the elevated blood level of certain hormones (such as adrenaline and glucagon; Selawry *et al.*, 1973).

Tubulin is a major protein constituent of the brain; it accounts, depending on the developmental stage, for over 20% of the soluble brain proteins in the chick (Wilson *et al.*, 1974) and almost 15% in the rat (Feit *et al.*, 1971): It is plainly involved in processes other than those associated with cell replication. Its role in neuronal function is emphasized by the observation that the concentration and the rate of synthesis of soluble tubulin are markedly increased in the rat visual cortex at about the time of eye-opening (Cronly-Dillon and Perry, 1976). Moreover, it has been reported that about 30% of the soluble protein in synaptosomal preparations is tubulin, which is claimed to be a significant constituent of the membranes in the isolated postsynaptic densities (Banker *et al.*, 1974; Matus *et al.*, 1975; Walters and Matus, 1975).* Besides, microtubule formation is believed to be involved in the function and growth of nerve fibers (Yamada *et al.*, 1970; Daniels, 1972), in axonal flow (Schmitt, 1968) and, as claimed recently, in the delivery and discharge of neurotransmitters at the synapse (Gray, 1975). In the control of microtubule formation Ca^{2+} ions (Weisenberg, 1972) and a specific protein, τ factor, seem to play important roles (Weingarten *et al.*, 1975): This factor may be rate-limiting in tubulin polymerization during early development (Fellous *et al.*, 1976). It would appear, therefore, that the study of the metabolism and polymerization of tubulin and of the interaction of fibrillary proteins with cell membrane constituents may advance the better understanding of the molecular processes underlying normal brain development (see, e.g., Edelman, 1976), as well as the effects of undernutrition on the developing brain.

Besides the observations on the Langerhans islets, there are also other results indicating that glucose (calorie) deficiency interferes selectively with cell proliferation in certain tissues. Cheek and Hill (1970) have found that muscle cell multiplication is retarded in rats fed on a diet containing adequate protein but low in calories. These authors believe that the effect is mediated by changes in the endocrine balance, since growth hormone seems to be important in the muscle for the increase in cell number, while insulin is important for cytoplasmic growth. It has been reported that while the mitotic index is severely reduced in the Lieberkühn crypts of fasted animals, the effect is slight when a protein-deficient isocaloric diet is given (Mönckeberg, 1975). It has also been claimed that the trend is similar in man: In the crypts in the jejunum the mitotic index is markedly depressed in marasmic patients, while little change is observed in kwashiorkor (Mönckeberg, 1975). Because of difficulties in experimental design, it is not yet known whether deficiency in calories or in proteins is mainly responsible for the depressed mitotic activity in the brain of suckling rats reared by undernourished dams. It seems that dietary manipulation of the mother rats affects the quantity of the milk rather than its composition (Mueller and Cox, 1937; see also Shoemaker and Wurtman, 1973).

3.3.2. Nutrients as Growth Factors

In model systems, it has been shown that many growth factors (about 28 were identified in 1965) are required to maintain growth of mammalian cells *in vitro:* By controlling the availability of one or more of these factors the average generation

---

*It has been found recently that although the isolated postsynaptic densities contain tubulin and proteins of the cytoskeletal network, there is a predominant protein constituent which does not correspond to any known fibrous protein (Cotman and Nadler, 1978).

time can be raised from less than a day to many days (Eagle, 1965). For example when the intracellular pool of a single amino acid falls below a critical concentration, usually $10–40$ μM, there is no demonstrable protein synthesis or cellular growth, and cells are arrested in the early G_1 phase of the cell cycle (Holly and Kiernan, 1974). Raising the concentration of the limiting nutrient usually leads to a disproportionate increase in cell proliferation. It has been suggested that, besides serum factors of protein or polypeptide nature and hormones, low-molecular-weight nutrients are also involved in the regulation of DNA synthesis and growth (Holley, 1972; Holley and Trowbridge, 1975).

The involvement of amino acids in the initiation of DNA synthesis in a resting cell population, such as liver parenchymal cells in the adult animal, has recently been studied by Short *et al.* (1974) *in vivo*. These authors have found that blood-borne factors play a crucial role in inducing replication of resting hepatocytes. The critical factors included triiodothyronine (T_3), glucagon, certain essential amino acids and heparin (Short *et al.,* 1972). A wave of mitotic activity in the liver can also be triggered within a day of shifting intact mice or rat from low- to high-protein diet (Leduc, 1949; Short *et al.,* 1974). Critical events that prepare the cells for the protein stimulus take place during the period of protein deprivation (Bailey *et al.,* 1976). The nucleolus seems to be involved in these events: By five days on a protein-free diet its size is doubled with parallel changes in the activity of RNA polymerase I. Supplementing the protein-free diet with individual amino acids, Bailey *et al.* (1976) have found that the increase in nucleolar size does not occur when methionine is given; tryptophan and threonine have some, but less, influence. Similarly when methionine is included in the protein free meal, the rise in DNA synthesis rate does not occur after shifting to high-protein diet. The means by which methionine deficiency leads to an increase in nucleolar size and prepares the hepatocytes for DNA synthesis is not known, but it is noteworthy that ribosome maturation is blocked through reduced methylation of rRNA precursors in the absence of methionine (Vaughan *et al.,* 1967).

3.3.3. RNA and Protein Metabolism

It seems therefore that RNA and presumably protein metabolism are affected by the nutritional factors prior to the initiation of DNA synthesis. Thus it is appropriate to consider now experiments, in which the influence of amino acid supply on these aspects of cellular metabolism has been investigated. The evidence mainly derived from studies on the liver and various mammalian cell lines in culture indicates that nucleic acid metabolism is sensitive to amino acid supply. Cell proliferation and DNA replication is an extensive process in the liver of young rats, leading to a 36% increase in cell number and significant polyploidy during a 4-week period after weaning (Mariani *et al.,* 1966): These processes are severely retarded when rats are weaned to a protein-deficient diet (Wannemacher *et al.,* 1968). The effects are associated with alterations in RNA metabolism which are similar to those seen in adult animals. The RNA/DNA ratio decreases exponentially during food deprivation, and the fractional renewal rate of ribosomes is reduced from about 14% per day to 9% (Hirsch and Hiatt, 1966). It has also been found that the synthesis of ribosomal precursor RNA (45 S RNA) is retarded under these conditions (Rickwood and Klemperer, 1970). In attempting to elucidate the effect, the activities of DNA-dependent RNA polymerases have been investigated in a number of studies. The influence of food deprivation on the activity of these enzymes is still controversial: Some investigators have reported an increase (Mandel and Quirin-

Stricker, 1967; Shaw and Fillios, 1968; Clark and Jacob, 1972), others a decrease (Rickwood and Klemperer, 1970; Henderson, 1970; Von der Decken and Andersson, 1972), and yet others no significant change in cells in tissue culture (Smulson, 1970). The early studies have suffered from the disadvantage that crude RNA polymerizing activity was assayed so that alterations in RNA synthesis could not be attributed with certainty to changes in enzyme or template. In a recent investigation, Andersson and Von der Decken (1975) have established that the activities of both the isolated RNA polymerase I and II, which are involved in the synthesis of ribosomal and nucleoplasmic RNA, respectively, are reduced after 30-day-old animals were kept on a low-protein diet. The results have always been consistent concerning the effect of supplying a balanced amino acid mixture or protein to deficient rats: RNA polymerase activity shows an immediate sharp rise (Henderson, 1970; Barbiroli *et al.*, 1975). This seems to be related to both *de novo* synthesis of the enzyme protein (Henderson, 1970) and structural modifications as well as changes in template availability (Barbiroli *et al.*, 1975). Besides affecting DNA transcription, the amino acid supply also influences the processing of newly formed RNA. One of the enzymes involved is poly(adenylic acid) polymerase which attaches 100–250 adenylic acid residues to the end of the newly synthesized mRNA molecules (Lim and Canellakis, 1970). Jacob *et al.* (1976) have found that starvation results in a very severe depression in the activity of both the purified nuclear and mitochondrial enzymes. Refeeding the animals with a complete amino acid mixture increased the activity of the enzyme from the nuclei and mitochondria 2-fold and 63-fold, respectively, within three hours.

The provision of adequate amino acid supply also accelerates the transfer of mRNA into the cytoplasm. Murty and Sidransky (1972) have observed that one hour after giving tryptophan to fasting rats the amount of mRNA in liver cytoplasm is markedly elevated. This amino acid seems to have a unique role in controlling anabolic reactions: In mice, although not in rats, it limits protein synthesis in the liver even under normal conditions (Sidransky *et al.*, 1968). In the rat a tryptophan-deficient diet results in a decrease in the protein-synthesizing competence of microsomal preparations and in the disaggregation of polyribosomes (Fleck *et al.*, 1965). Both under these conditions and in protein deficiency, there is a marked increase in single ribosomes and ribosomal subunits: After providing the missing amino acids there is a rapid appearance of newly formed mRNA in the cytoplasm associated with a recycling of the ribosomes into active units and the establishment of the normal polyribosomal profile (Gaetani *et al.*, 1972). Besides the availability of mRNA, another factor which seems to contribute to the disaggregation of polyribosomes in protein deficiency is the level of charged tRNAs. In bacteria, RNA synthesis is stringently controlled by the availability of amino acids. The control may involve tRNA charging levels, for it has been shown that an amino acid must be activated in order to promote RNA synthesis but that translation need not occur (Fangman and Neidhardt, 1964). It would appear that, in contrast to bacteria, in mammalian cells the control of transcription by amino acids is not stringent; it has been reported that in HeLa cells the rate of synthesis of protein is much more severely depressed than that of RNA in the absence of tryptophan (Smulson, 1970), and in ascites tumor cells the lack of amino acids causes a substantial decrease in the rate of synthesis of protein without affecting that of RNA, although ribosomal subunit maturation is retarded (Shields and Korner, 1970). Allen *et al.* (1969) have shown that feeding of amino acid mixtures lacking certain amino acids caused the level of the corresponding tRNA in the charged form to fall considerably below that seen in fasting rats. Tryptophan deficiency produced the most marked decrease in the corresponding charged tRNA. These authors have suggested that the great sensitivity to the availability of tryptophan of charging tRNA may account in part for the peculiar deleterious effect of tryptophan deficiency on hepatic polyribosome aggregation and activity.

Whereas it seems that in response to amino acid supply there is a coordinated increase in different components of the protein-synthesizing machinery in certain cells, various compensating mechanisms come into operation throughout the body when the food supply is restricted or is unbalanced. In the liver, the rates of protein synthesis and the catabolism of amino acids are greatly reduced, and enhanced breakdown of protein in certain organs, such as pancreas, liver, intestinal mucosa, and muscle, provides an endogenous supply of amino acids (see Symposium, 1970). As a result certain organs, such as the brain, are relatively little affected by food deprivation in the adult. The remarkable resistance of the chemical composition of the brain after prolonged starvation has been known for a long time (Addis *et al.*, 1936; Mandel *et al.,* 1950; Lehr and Gayet, 1967). In contrast to the liver, the DNA-dependent RNA polymerase activity is normal in the brain of protein-deficient adult rats and the *in vitro* protein-synthesizing activity is also less influenced (Von der Decken and Wronski, 1971; Von der Decken and Andersson, 1972). Such negative observations are probably responsible for the fact that cerebral RNA and protein metabolism in developing undernourished animals has hitherto attracted relatively little attention. Lee (1970) has studied, after two weeks of rehabilitation, the effect of giving to mice either an unbalanced protein diet or half of the normal diet from the 14th day of pregnancy throughout lactation. The incorporation of labeled amino acids into protein or orotic acid into RNA was unaffected with the first treatment, whereas undernutrition caused about 25% reduction in the specific activity of RNA only. However, the labeling was only studied at 24 hr after the injection of the precursors, and thus the relatively small effects may have been due to factors unrelated to the biosynthesis of RNA and protein. Lee (1970) has also claimed remarkable qualitative changes in the chromatographic and electrophoretic patterns of the soluble proteins extracted from the brain of the experimental animals. These results still await confirmation. Reid *et al.* (1970) have observed that acute food deprivation, which in the liver caused a marked depression of *in vitro* amino acid incorporating activity associated with disaggregation of polyribosomes, had no significant effect in the brain of 10- and 35-day-old rats. De Guglielmone *et al.* (1974) have studied the effect of undernutrition (by increasing the litter size) on the labeling of forebrain RNA 1 hr after intracerebral injection of [^{3}H]orotic acid at the age of 10 and 20 days, as well as 10 days after rehabilitation, at day 30. In order to take into account the variation in brain radioactivity content, which is unavoidable after intracerebral injections, these authors corrected the incorporation values on the basis of the total radioactivity in the brain. However, [^{3}H]orotic acid is metabolized in the whole body and the tritiated water formed enters the brain freely; thus the total radioactivity content is not an estimate of the availability of the precursor in the brain. Although the relative specific radioactivity values they have reported are unacceptable, the comparison of the estimates obtained for RNA in the various subcellular fraction suggests that the labeling of RNA in the microsomal fraction may have been less than in controls during the whole experimental period. However, this comparison is very sensitive to experimental errors, since at 1 hr after the administration of the precursor, the amount of radioactivity in cytoplasmic RNA relative to that in nuclear RNA is very small. De Guglielmone *et al.* (1974) have also estimated the activity of Mg^{2+}-activated (ribosomal) DNA-dependent RNA polymerase in isolated brain nuclei and observed a significant decrease in the undernourished infant rats. The activity was restored to normal after the relatively short rehabilitation period. In a recent study, Patel *et al.* (1975) have observed that the rate of incorporation of labeled leucine into protein is significantly depressed in the brain of suckling rats in undernutrition, but the effect on RNA metabolism was not investigated.

3.3.4. Endocrine Systems

Observations of Short *et al.* (1972) that T_3 is the most important single constituent contributing to the stimulation of hepatic DNA synthesis in intact rats raises the possibility that thyroid deficiency associated with undernutrition may play a part in the depressed mitotic activity. Thyroid hypofunction has indeed been documented in severe infantile malnutrition by the findings of low basal metabolic rate, low serum levels of thyroxine (T_4), and reduced radioiodine uptake by the thyroid gland (e.g., Pimstone *et al.*, 1973*b*). The decrease in the total serum T_4 level is not a good index of thyroid hypofunction since the physiologically active hormone is the free T_4 which is not reduced in protein–calorie malnutrition: There is a depression in the concentration of T_4-binding globulin under these conditions (Graham and Blizzard, 1973). The results hitherto obtained on the basal circulating level of thyroid-stimulating hormone (TSH) are conflicting (depression, Harland and Parkin, 1972; no change in kwashiorkor and slight decrease in marasmus, Godard, 1973; no change or elevation, Pimstone *et al.*, 1973*b*). However, the regulation of thyroid function seems to be more or less normal: TSH concentration in the blood is increased after the administration of TSH-releasing hormone (TRH) to malnourished infants (although the response is more sustained than in controls), and the basal level of TSH was decreased after T_3 was given (Pimstone *et al.*, 1973*b*). It has recently been claimed that the hypothyroidism accompanying neonatal food deprivation is of hypothalamic origin mediated through TRH deficiency (Shambaugh and Wilber, 1974). However, overt thyroid dysfunction in malnutrition has not been noted by other investigators (e.g., Meites and Wolterink, 1950; Sobotka *et al.*, 1974; see also various contributions in Gardner and Amacher, 1973).

The comparison of the effect of thyroid deficiency and undernutrition on cell acquisition in the brain provides convincing evidence that the observed changes are genuinely due to undernutrition (Patel *et al.*, 1973, 1976; Lewis *et al.*, 1975, 1976; Balázs, 1977). The major differences between the two conditions are as follows: (1) In undernutrition the rate of cell acquisition is depressed throughout the brain during the whole postnatal period. On the other hand, the effect of thyroid deficiency seems to be confined to those parts of the brain where neurogenesis is significant after birth (such as the cerebellum and olfactory lobes). (2) The rate of *in vivo* DNA synthesis is severly inhibited in undernutrition during most of the postnatal period, whereas in thyroid deficiency it is mainly reduced in the cerebellum and in the second week of life. (3) There is a slight tendency in undernutrition toward compensation for the deficit in cell numbers by abnormally high mitotic activity at the time, at about the end of the third week, when cell proliferation sharply decreases in the cerebellum (see also Rebière and Legrand, 1972; Barnes and Altman, 1973*b*; Gopinath *et al.*, 1976; Clos *et al.*, 1977). This effect is very marked in thyroid deficiency (see also Legrand, 1967; Nicholson and Altman, 1972*a*): Cell proliferation, in terms of DNA synthesis, is even greater at day 21 than at the time, day 12, when in the cerebellum of untreated rats cell division reaches its maximum rate. In contrast to undernutrition, cell replication carries on for a relatively long time in the cerebellar external granular layer of hypothyroid rats, resulting in full restoration of normal cell numbers. (4) An abnormally high rate of cell degeneration occurs in undernutrition in the germinal layers, whereas in thyroid deficiency the differentiated granule cells in the internal granular layer of the cerebellum are selectively affected (see also Rabié *et al.*, 1977). (5) Finally the most prominent difference between these two conditions rests in their effect on the generation cycle of the dividing cells. In undernutrition the S phase is markedly prolonged, while the G_1 phase is drastically curtailed. In contrast the generation cycle appears to be normal in the cerebral germinal cells of the thyroid-deficient animals.

The effect of growth hormone on cell acquisition in the brain of experimental animals has also been studied. Zamenhof *et al.* (1966) have claimed that treatment with growth hormone during gestation results in an increase in the number of cells, especially of neurons, in the brain. Sara and Lazarus (1975) have found a marked increase in the labeled DNA content of the brain of 7-day-old rats which had been given a single injection of [^{3}H]thymidine during the period of growth hormone treatment at the 20th day of gestation. However, it has not been ascertained whether the latter findings reflect a genuine increase in cell proliferation, or result from changes in the availability of [^{3}H]thymidine to the embryo. The claims that growth hormone promotes cell proliferation in the fetal brain are not substantiated by the histological studies of Clendinnen and Eayrs (1961) and by biochemical investigations. Cotterrell (1971) has observed that the DNA content was normal in the brain of rats which were exposed to growth hormone treatment during gestation. Zamenhof *et al.* (1971*a*) have also reported similar negative results, although claiming that growth hormone can prevent the depression in cell acquisition caused by nutritional deprivation of pregnant rats. These findings would be consistent with the view prominent a few decades ago that pituitary hypofunction is associated with severe undernutrition (Mulinos and Pomerantz, 1941). However, in the meantime evidence has accumulated that severely challenges this hypothesis. The basal level of circulating growth hormone has been found to be elevated in kwashiorkor, although some of the normal stimuli fail to trigger further increases in the hormone concentration (Pimstone *et al.*, 1973*a*). Nevertheless certain biological effects of growth hormone may be impaired in spite of elevated plasma levels: The production of somatomedin is markedly depressed in protein–calorie malnutrition (Pimstone *et al.*, 1973*a*).

That endocrine functions are not necessarily depressed as a result of nutritional deprivation is also indicated by observations showing that, compared to controls, the fasting plasma cortisol levels are elevated in both marasmic and kwashiorkor patients (Alleyne and Young, 1967; Rao *et al.*, 1968). Furthermore the proportion of cortisol in the free form is abnormally great (Leonard, 1973). The concentrations of both specific hormone-binding proteins (such as T_4-binding globulin and transcortin) and the nonspecifically binding albumin are markedly depressed in severe undernutrition. Similar findings have been obtained in developing undernourished rats (Adlard and Smart, 1972). However, when these animals grew up on unrestricted food after weaning, they showed depressed adrenocortical response to stress. It has been proposed that increased levels of circulating corticosteroids in the undernourished infant rats may permanently modify certain behavioral responses of these animals (see also Dobbing and Smart, 1974). It is of interest that the young of adrenalectomized mothers are also undernourished; nevertheless their stress response is permanently augmented, i.e., it is just the opposite to that seen after "uncomplicated" early undernutrition (Levine, 1974).

4. Effect of Undernutrition on Brain Maturation

Cellular differentiation in the central nervous system is manifested in characteristic structural and biochemical changes. Some of the parameters that reflect the progress of maturation in the rat brain are shown in Figure 5, where the results are expressed as a percentage of the adult values. The inherent limitations in collecting these data must be remembered when comparing the various estimates. The biochemical data usually refer to the whole forebrain or cerebellum. Although technological refinements in the last few years have permitted the biochemical investigation of relatively small and well-defined regions (e.g., Otsuka, 1972; Palkovits, 1973; Fonnum, 1973; Kizer *et al.,* 1974), even these data often represent mean, pooled estimates of complex structures, the constituents of which are

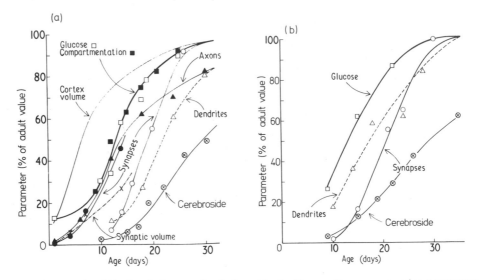

Fig. 5. Structural and biochemical maturation in the rat brain. The results are expressed as a percentage of the adult values. The number of synapses and the presynaptic volume in the cerebral cortex were computed from the results given per unit volume and the volume of the cortex; thus the estimates are rough approximations since the calculation implies the assumption that the values per unit volume are uniform throughout the cortex. (a) Forebrain. Volume of cerebral cortex (Sugita, 1918), ——; maturation of glucose metabolism in terms of conversion of glucose carbon into amino acids (Gaitonde and Richter, 1966; Cocks *et al.,* 1970), □; development of metabolic compartmentation, in terms of glutamate/glutamine specific radioactivity ratio after administration of [U-^{14}C]leucine (Patel and Balázs, 1971), ■; axonal density in the sensory-motor cortex (area 2) (Eayrs and Goodhead, 1959), ▲; number of synapses in the superficial motor cortex (layers I and II) (osmium impregnated preparations), ●: and presynaptic volume as a fraction of cortical volume (Armstrong-James and Johnson, 1970),⊗; number of synapses in the molecular layer of the parietal cortex (phosphotungstic acid preparations) (Aghajanian and Bloom, 1967), ○; development of the dendritic arborization of pyramidal cells in layer 5b of the sensory-motor cortex (Eayrs and Goodhead, 1959), Δ; cerebroside content (Balázs *et al.,* 1971; Hauser, 1968),⊗. (b) Cerebellum. Maturation of glucose metabolism (Patel *et al.,* 1974), □; total number of synapses in the molecular layer of the pyramis (Nicholson and Altman, 1972*b*), ○; dendritic arborization of Purkinje cells in the monticulus of the vermis (Legrand, 1967), Δ; cerebroside content (Balázs *et al.,* 1971),⊗.

usually sampled separately in morphological studies. The latter include, for example, observations on a single layer in the cerebral or cerebellar cortex or on the changes in a specific cell type (pyramidal cells or Purkinje cells) in one of the layers. However, extrapolation from these results, and especially from the ultrastructural observations, to the response to metabolic disturbance of the whole structure is liable to considerable error. Thus these graphs must be considered chiefly as *qualitative* guides to brain development; they may also be helpful in appreciating the importance of the time factor in the effects of early undernutrition, often described in the following sections only in comparison to controls.

In spite of these important limitations, a pattern seems to emerge from the results. The expansion of the cerebral cortex occurs earlier than the rapid phase of those maturational processes that are intimately associated with neuronal differentiation (Figure 5a). Further, it seems that the biochemical reflections of differentiation, such as the development of glucose metabolism and metabolic compartmentation of glutamate, which mature simultaneously, precede the morphological signs of differentiation, for which there is a chronological sequence. First fibers penetrate into the cerebral cortex, then there is a parallel or consecutive increase in the number of nerve terminals, and this is followed by the arborization of dendrites. It is of interest that the timing of the biochemical maturation in the cerebellum is similar to that in the forebrain (Figure 5b), although the assembly of the neuronal population is postnatal rather than prenatal. However, in contrast to the cerebral cortex, dendritic arborization in the cerebellar molecular layer is advanced compared with synaptogenesis. This may be related, in part, to the characteristic development of the cerebellar cortex, where the "oldest" neurons are the Purkinje cells, but the most numerous synapses are those of the granule cells, which are the "youngest" cells. It seems that the age curves obtained for synaptogenesis differ according to the techniques used for detecting synapses. In comparison with observations on osmicated preparations, results from the use of the phosphotungstic acid method give an age curve for synaptogenesis which is displaced to the right on the time axis [compare Aghajanian and Bloom (1967) with Armstrong-James and Johnson (1970). Dyson and Jones (1976) used both methods in the same study and noted that immature synapses can be detected earlier in the osmium-impregnated preparations than in those stained with phosphotungstic acid. Incidentally, this is probably the reason why they found that the ethanolic phosphotungstic acid method indicates a more severe reduction in nerve terminal numbers in the 10-day-old undernourished cerebral cortex in comparison with the estimates on osmicated preparations.

4.1. Undernutrition and the Structural Development of the Brain

Morphological changes in the brains of animals undernourished during development have been sought and recorded for many years. With the development of refined staining and of quantitative techniques in neuroanatomy, emphasis has shifted in recent years from the simple description of changes to their measurement, with the result that alterations that hitherto were only suspected to occur are now evident. A sizable volume of data relating to experimentally undernourished animals has accumulated in print; some of this is summarized in Tables V and VI. Such is the diversity of experimental procedure, timing of nutritional insult, species and focus of observation that the neuropathological picture of the effects of undernutrition on the developing brain is still far from complete. It should be emphasized that our knowledge of the morphological alterations in the undernourished brain comes almost exclusively from experimental animal work. As noted by Trowell *et al.* (1954), the changes in the human infant brain in undernutrition "have not been made the subject of special study and it is not known if there is any histological counterpart to the striking mental changes." Although over 20 years have elapsed since this comment, the position is unaltered. Reduced head size (Stoch and Smythe, 1963,1967,1976), brain weight (Brown, 1966), brain DNA, RNA, and protein content (Winick and Rosso, 1969*a,b*; Dobbing, 1974) and brain lipids (Fishman *et al.*, 1969) have been noted in severely undernourished infants, as in animals, but the contribution of morphological studies (e.g., Udani, 1960) to the description of the effect of undernutrition in man is small.

A review of animal experimental work should begin with Sugita (1918). His morphological observations have been amply confirmed: Undernutrition in early life results in a deficit in brain weight associated with a reduction in the thickness of the cerebral cortex in young rats. Cragg (1972) has found that the cortex (about 1.5 mm thick) was slightly thinner (0.2 mm) in rats undernourished throughout 50 days of life, and an approximately 10% deficit has been reported for animals rehabilitated for 27 weeks after food deprivation during the first 3 weeks only (Dobbing *et al.*, 1971). However, most investigators have found no qualitative abnormalities in the light microscopic appearance of the cerebral cortex in rats either continuously underfed or rehabilitated after weaning (Cragg, 1972; Dobbing *et al.*, 1971),

Table V. Some Reported Effects of Undernutrition on the Morphology of the Cerebral Cortex[a]

	Undernourished[b]	Rehabilitated[b]
Whole forebrain		
Weight	↓ [1]	↓ [2]
Cell numbers	↓ [3, 4]	↓ [2]
Cerebral cortex		
Volume	↓ [1]	
Thickness	↓ [5]	↓ [6], 0[7]
Nerve cells		
Numbers	0[1]	↓ [6]
Packing density	↑ [1,5,8,9]	0[9], ↓ [6,10]
Fiber density	↓ [11]	0[9,c]
Dendritic spines on pyramidal cells	↓ [12,13]	0[10]
Synaptic density	0[5], ↓ [8,14]	0[14]
Synapses/neuron	↓ [5]	
Presynaptic thickness	↓ [15]	↓ [15]
Postsynaptic thickness	0[15]	↓ [15]
Glial cells		
Numbers	↓ [5,7]	↓ [6,7,16]
Myelination	↓ [7,17]	↓ [7,17]

[a] Most of the results refer to rats which were undernourished during the period of brain development (mainly during the suckling period) although in some experiments the food restriction was continued into the postweaning period. To study the reversibility of the effects of undernutrition, the animals were nutritionally rehabilitated after weaning. The results indicate, in comparison with controls, retardation, ↓; no effect, 0; or increase, ↑.

[b] References: (1) Sugita, 1918; (2) Dobbing, 1974; (3) Patel *et al.*, 1973; (4) Winick, 1976; (5) Cragg, 1972 (rats were undernourished till 24 or 50 days of age); (6) Dobbing *et al.*, 1971; (7) Bass *et al.*, 1970*a;* (8) Gambetti *et al.*, 1974; (9) Horn, 1955; (10) Escobar, 1974; (11) Eayrs and Horn, 1955; (12) Salas *et al.*, 1974; (13) West and Kemper, 1976; (14) Dyson and Jones, 1976; (15) Jones and Dyson, 1976; (16) Siassi and Siassi, 1973; (17) Benton *et al.*, 1966.

[c] The fiber density was restored to normal even in animals which were undernourished throughout their lives, until 60 days of age (Horn, 1955).

Table VI. *Some Reported Effects of Undernutrition on the Morphology of the Cerebellar Cortex*[a]

	Undernourished[b]	Rehabilitated[b]
Whole cerebellum		
Weight	↓ [1,2,3]	↓ [1,4]
Cell numbers	↓ [2]	↓ [1]
Cerebellar cortex		
Area	↓ [3,5]	↓ [4,6,c]
Nerve cell numbers		
Purkinje cells	0[5], ↓ [4,d]	↓ [4,6]
Basket cells	↓ [5]	
Stellate cells	0[5]	
Golgi cells	0[5]	
Granule cells	↓ [3,5]	↓ [1,4]
Purkinje cells		
Dendritic arborization	↓ [7,8,9,e]	0[4,10]
Dendritic spines	0[7]	
Length of dendritic segments	↑ [8]	
Synaptic density in molecular layer	0[10]	
Granule cells		
Dendritic number	0[11]	
Development of neuronal circuits		
Climbing fiber–Purkinje cell circuit	0[12]	
Moss fiber-granule cell–Purkinje cell circuit	0[12,13,14]	
Glial cell numbers		
Bergmann glia	↓ [5]	
In molecular layer	↓ [5]	
Astrocytes in internal granular layer	↓ [5]	
Glial spaces in molecular layer	↓ [15]	

[a]Most of the results refer to rats which were undernourished during the period of brain development (mainly during the suckling period), although in some experiments the food restriction was continued in the postweaning period (e.g., Clos *et al.*, 1977, till day 35). To study the reversibility of the effects of undernutrition, the animals were nutritionally rehabilitated after weaning. The results indicate, in comparison with controls, retardation, ↓ ; no effect, 0; or increase, ↑ (slight effects, ↑).

[b](1) Dobbing, 1974; (2) Patel *et al.*, 1973; (3) Barnes and Altman, 1973*a;* (4) Barnes and Altman, 1973*b;* (5) Clos *et al.*, 1977; (6) Dobbing *et al.*, 1971; (7) Rebière and Legrand, 1972; (8) McConnell and Berry, 1978; (9) Sima and Persson, 1975; (10) Rebière, 1973; (11) West and Kemper, 1976; (12) Hajós, F., Patel, A. J., and Balázs, R., unpublished observations; (13) Legrand, 1967; (14) Neville and Chase, 1971; (15) Clos *et al.*, 1973.

[c]Barnes and Altman (1973*b*) have found that after rehabilitation the area of the cortex is reduced till the age of 90 days in rats which were exposed to severe food restriction before weaning, but not in mildly undernourished animals. However, at the age of 120 days the deficit in the area was not significant even in the animals severely undernourished before weaning.

[d]This is surmised from the estimates given by Barnes and Altman (1973*b*) for the 30-day-old rats undernourished till 21 days of age.

[e]There is only a retardation in dendritic arborization.

although Stewart *et al.* (1974) reported some delay in the development of cortical layering. In contrast, Bass *et al.* (1970*a*) claim that undernutrition during the suckling period results in relatively severe morphological changes which persist in the cerebral cortex even after rehabilitation: The six-layered stratification of the cortex is poorly defined, neuronal processes stained with silver are reduced in number, and glial migration is markedly retarded. Nevertheless, these authors have not observed a significant reduction in the thickness of the cortex after rehabilitation.

While animals are undernourished the packing density of neurons in the cortex seems to be higher than in controls, indicating a reduction in the development of the neuropil and in the glial compartment (e.g., Cragg, 1972; Stewart *et al.,* 1974; Gambetti *et al.,* 1974). However, in the nutritionally rehabilitated rats Dobbing *et al.* (1971) have observed a reduction in the packing density of cells. Especially in the deeper cortical areas, the decrease in nerve cells was more marked than in glial cells, suggesting a selective neuronal loss under these conditions. Nerve cell loss, especially in cortical layer VI, has also been reported by Escobar (1974). Since the full neuronal complement of the cerebral neocortex is believed to be generated in the rat before birth (Altman, 1969), i.e., before the major impact of the nutritional insult on the developing brain, these results would suggest that normal food supply is required not only for the formation of cells, but also for their maintenance.

In Section 3.2 the evidence was summarized indicating that undernutrition interferes with cell proliferation in the brain. Here we want to look at the consequences of this interference in terms of the cellular composition of a part of the brain, the cerebellum, which acquires a great fraction of its neurons during the early postnatal period (Table VI). Moreover, studies on the cerebellum may further our insight into the functional consequences, and the underlying structural alterations, of early metabolic insult: Knowledge of the role of the various nerve cells in the neuronal circuits is relatively advanced (Eccles *et al.,* 1967; Palay and Chan-Palay, 1974) and methods of testing motor coordination, in which the cerebellum plays an important role, are well developed (e.g., Altman *et al.,* 1970*b*; Lynch *et al.,* 1975).

Purkinje cells are formed exclusively before birth, but their spectacular differentiation in terms of the development of the extensive dendritic arborization characteristic of these cells takes place mainly in the first 3–4 postnatal weeks. Dobbing *et al.* (1971) have reported a mean decrease of about 20% in Purkinje cell numbers and a 12% reduction in the ratio of granule cells to Purkinje cells in the vermis. Barnes and Altman (1973*b*) have observed a persistent deficit in Purkinje cells during the period 30–120 days in animals rehabilitated after a severe gestational–lactational undernutrition: At 120 days the deficit was only 12% and the granule cell/Purkinje cell ratio was normal. They also found that milder undernutrition (the mother rats receiving 75% versus 50% of the normal diet) had no influence on the final number of Purkinje cells. Using a regime similar to the severe undernutrition of Barnes and Altman (1973*a,b*), Clos *et al.* (1977) observed no change in Purkinje cell numbers at the age 35 days. Further, the results of Neville and Chase (1971) are consistent with the absence of deficit in the Purkinje cell number in rehabilitated rats which were undernourished during the suckling period (experimental conditions similar to those used by Dobbing *et al.,* 1971). These authors reported a 14% increase in the number of Purkinje cells per mm in the vermis; this implies no significant change in Purkinje cell number, since the weight of the vermis was comparably reduced. Thus the results obtained hitherto do not support generalizations concerning the vulnerability of established neuronal cell populations to nutritional insult, especially since Dobbing *et al.* (1971), who reported the greatest deficit in Purkinje cell number, make it clear that their observations must be considered preliminary. (Only two rats were studied, and there was an overlap in the estimates between control and experimental animals.)

The other cell types for which estimates from different laboratories are available are the granule cells. These are the most abundant and the only excitatory interneurons in the cerebellar cortex, and about half their total number is generated after 10 days of age (Figure 1). All observations have shown that undernutrition results in an ultimate decrease in granule cell numbers (Dobbing *et al.,* 1971; Neville and Chase, 1971; Barnes and Altman, 1973*b*; Clos *et al.,* 1977). It would appear, however, from the results of Barnes and Altman (1973*b*) that some degree of restoration of the granule cell population is effected by rehabilitation after weaning. At the age of 20 days, the total number of granule cells in a sagittal section of the pyramid was about 10,000 in the undernourished rats (Barnes and Altman, 1973*a*). After 10 days of rehabilitation the value was about 13,000 and remained more or less constant during the rest of the experimental period till day 120. In contrast, granule cell numbers in the control (about 14,000) increased only marginally between 20 and 30 days of age. Barnes and Altman (1973*b*) have suggested that the compensation is due to the prolonged existence of the external granular layer in the cerebellum of the undernourished rats. This has also been observed by others (Rebière and Legrand, 1972; Lewis *et al.*, 1975; Clos *et al.*, 1977; Gopinath *et al.*, 1976), but the persistence of this germinal layer is not as prolonged as in thyroid deficiency. Granule cells constitute a great proportion, almost half of the total cells in the cerebellum (Balázs *et al.*, 1977). However, estimates on total cell numbers (e.g., Dobbing *et al.*, 1971) do not seem to support the claim that the deficit incurred during the period of neonatal food deprivation is effectively restored by postweaning rehabilitation.

The influence of undernutrition on the other cell types in the cerebellum has been studied only by Clos *et al.* (1977). Cell counts were made in sections of the culmen and declive in rats undernourished until 35 days of age. Clos *et al.* (1977) found no significant effect on the number of either the perinatally formed Golgi cells or the stellate cells ("birthday" in the second week of life; Altman, 1969). In contrast there was a marked deficit, about 35%, in the number of basket cells. These inhibitory interneurons are formed in the first postnatal week (Altman, 1969, 1972*b*), and they make powerful synaptic contact on the lower part of the perikarya and on the initial axon segments of the Purkinje cells. It is intriguing that the basket cells seem to be the cell type most vulnerable to metabolic insult of various kinds: Their number is also irreversibly and markedly depressed in rats made thyroid deficient or hyperthyroid during the neonatal period (Nicholson and Altman, 1972*a*; Clos and Legrand, 1973).

The functional plan of the cerebellar cortex is based on the Purkinje cells, which are the only efferent cells from the cortex, activated by two major afferent systems (climbing and mossy fibers) and inhibited by the noradrenergic input from the locus coeruleus (Eccles *et al.*, 1967; Palay and Chan-Palay, 1974). The functioning of both the excitatory input and the output systems is modulated by internal circuits. In the adult, the climbing fiber synaptic contact is on the Purkinje cell dendrites, whereas the mossy fiber input is transmitted first to the granule cells, which make synaptic contact with the spiny branchlets of the Purkinje cell dendrites through their axons, the parallel fibers. In the modulation of both circuits, the inhibitory interneurons (Golgi, basket, and stellate cells) play an important role. The development of the neural circuits is associated with structural changes which can be followed by light and electron microscopy. In the first four days after birth the Purkinje cells align in a row, and then start to develop a massive dendritic tree which becomes the skeleton of the molecular layer. Initially the climbing fibers make synaptic contact on the cell bodies of the Purkinje cells. These axosomatic synapses disappear by the end of the second week. The granule cells start to appear in great numbers in the internal granular layer after the first postnatal week and make synaptic contact with the terminals of the mossy fibers in the cerebellar glomeruli. The ultrastructure of the glomeruli undergoes developmental changes affecting not only the size of the mossy fiber terminal, but also the granule cell dendrites. These are initially electron-dense and make multiple synaptic contacts with the terminal, but by the end of the third week they are electron-translucent and digitated, each digit making a synaptic contact with the mossy fiber rosette.

Table VI shows that undernutrition has only a minor adverse influence on the development of neuronal circuits in the cerebellum. The maturation of the Purkinje cell is only slightly delayed (Rebière and Legrand, 1972; Sima and Persson, 1975) and, following rehabilitation after early undernutrition, the Purkinje cell domain, which is a reflection of the dendritic arborization of these cells, is grossly unaffected (Neville and Chase, 191; Barnes and Altman, 1973b). However, in 30-day-old undernourished or malnourished rats the total area of the vermis, and especially the area of the molecular layer is markedly reduced (McConnell and Berry, 1978; West and Kemper, 1976). Since the dendritic tree of the Purkinje cells and the ascending and descending vertical dendrites of the basket cells extend across the entire thickness of the molecular layer, they are reduced in proportion to the decrease in the width of this layer. The length of the dendrites of the granule cells and of the oblique branches of the basket cells are also reduced, but the number of the granule cell dendrites is normal (West and Kemper, 1976). McConnell and Berry (1978) have observed that while the overall size of the Purkinje cell dendritic tree is depressed, due to a reduction in total number of dendritic segments and in the length of distal segments, the network develops as in controls by terminal branching with only minor deviation from the usual purely random branching pattern.

The most abundant synapses in the cerebellar cortex are on the Purkinje cell dendrites in the molecular layer. Some observations suggest that the number of both the postsynaptic specializations, in terms of dendritic spines on the Purkinje cells, and the presynaptic structures are unaffected by undernutrition (Rebière and Legrand, 1972; Rebière, 1973). Systematic ultrastructural studies have not been conducted yet on the developing neuronal circuits, mainly because preliminary investigations did not indicate marked deviations from the normal pattern (Legrand, 1967; Neville and Chase, 1971; unpublished observations by Hajós, Patel, and Balázs from our laboratory). The failure of undernutrition to produce a marked influence on the structural manifestations of the development of the neural circuits in the cerebellum is very strong evidence against the involvement of thyroid deficiency as an important factor in the neurological effects of food deprivation (see also Section 3.3.4), since Hajós *et al.* (1973) observed that the development of neural circuits in terms of ultrastructural appearance is severely retarded in neonatal thyroid deficiency. The relatively mild effect of undernutrition in this respect is intriguing, as motor performance is significantly affected (Altman *et al.,* 1970b; Lynch *et al.,* 1975).

The effect of undernutrition on the development of neuronal processes has also been studied in the cerebral cortex (Table V). However, most investigations have concentrated on the influence of the metabolic insult during the period of undernutrition, which in many of these studies was before weaning when the development of neuronal processes is not yet completed (see Figure 5). Bearing these limitations in mind it seems to be established that the area occupied in the cerebral cortex by the neuropil is decreased in undernourished animals (Gambetti *et al.,* 1974). In the developing experimental animal all components of the neuropil seem to be affected. In comparison with controls, the fiber density is markedly reduced at 24 days (Eayrs and Horn, 1955), as is the basilar dendritic network of the large pyramidal cells in the 7- to 15-day-old undernourished rats (Salas *et al.,* 1974). However, the length of the oblique and basal dendrites of the pyramidal cells in layer IIIb is normal in the

30-day-old young of protein-deficient mother rats (West and Kemper, 1976). The results also indicate that the development of neuronal interconnections is retarded. Although the distribution of various forms of spines, which are believed to be postsynaptic specializations, on the dendrites of pyramidal cells is normal, the number and density of the spines are significantly reduced in undernutrition (Salas *et al.*, 1974; West and Kemper, 1976).

Further, the calculated value of the number of nerve terminals per nerve cell in the cerebral cortex is severely depressed (about 60% of control; Cragg, 1972). As pointed out by Cragg (1972) this estimate limits the complexity of the circuits that neurons can form and is perhaps the structural parameter most closely related to mental performance that can be measured at present. Because of the great importance of these results, it is necessary to analyze them in more detail. Certain limitations in the approach have been emphasized by Cragg himself (1970, 1972). The calculated number of nerve terminals per neuron is formidably great (about 10^4), and this makes it very difficult to appreciate the consequences, in terms of normal interconnections, of even a statistically significant reduction. Moreover, because of serious technical difficulties, the estimate has been obtained by counting nerve terminals in one layer of the cortex and dividing these values by the packing density of neurons throughout the cortex. Thus calculations are based on the assumption that results on nerve terminal density in one layer are representative of the whole cortex. However, in considering possible functional consequences of the structural changes it is most important that the animals investigated in this study were undernourished throughout life after birth (24 or 50 days). Therefore it is not known whether or not the insult has caused irreversible structural alterations. Furthermore, the severe depression in nerve terminals per neuron in chronically undernourished animals was due to an increased packing density of cells, rather than alterations in the density of synapses. It has been observed before that while undernutrition for a comparable length of time results in a marked increase in the packing density of neurons in the cortex, the difference compared to controls is reduced to insignificant levels in the rehabilitated animal (Horn, 1955). If so, would rehabilitation also restore the nerve terminal/neuron ratio to normal? It is worth noting that fiber density, which is significantly reduced in the cerebral cortex of undernourished rats at the age of 24 days, has been reported to be normal at 60 days in spite of continuous food deprivation during the whole experimental period (Horn, 1955). The remarkable plasticity of the central nervous system is indicated by recent observations on the reoccupation of postsynaptic sites vacated after denervation in certain parts of the brain (e.g., Raisman, 1976; Matthews *et al.*, 1976).

Thus, although most of the evidence is consistent with the view that the development of neuronal interconnections is retarded in undernutrition, the question of reversibility is still open. A potential for some degree of reversibility is indicated by the observations of Dyson and Jones (1976). These authors found, using the ethanolic phosphotungstic acid technique for staining synaptic specializations, that at 20 days of age the synaptic density in the cerebral cortex was only 60% of the control value (see p. 435). However, after nutritional rehabilitation the synaptic density was restored to normal. Dyson and Jones (1976) also found that the structural maturation of the synapses was retarded in the 20-day-old undernourished rats, but this was in part rectified after rehabilitation. Nevertheless, certain anomalies persisted; for example the widths of both the pre- and the postsynaptic specializations were reduced (Jones and Dyson, 1976).

It should be noted that the structural alterations induced by undernutrition are not exclusively negative ones. Cravioto *et al.* (1976) have observed that there is a persistent 75% increase in synaptic density in the midbrain reticular formation of mice rehabilitated after early life undernutrition. It is suggested by these authors that the morphological changes may be the structural correlates of the functional alterations in arousal in the undernourished and rehabilitated animals (Randt and Derby, 1973; Dobbing and Smart, 1974).

4.2. Effects on the Development of Glia

Many observations indicate that the maturation of glial cells is severely affected by undernutrition. Bass *et al.* (1970*a,b*) have reported that both the formation and the migration of glial cells to the cerebral cortex is severely retarded, leading to lasting effects. Even after rehabilitation the content of constituents characteristic of the myelin membrane was only about half normal in the somato-sensory cortex and the subcortical white matter, and this correlated with decreased numbers of mature oligodendroglia and poorly stained myelin. Dobbing *et al.* (1971) also observed a significant reduction in glial cell numbers throughout the cerebral cortex of undernourished rats rehabilitated after weaning.

The cerebellar cortex offers certain advantages for studying the relative vulnerability of different cell types to food deprivation, since here the formation of neurons and glial cells proceeds on a large scale during the early postnatal period. Quantitative histological studies in rats undernourished till 35 days of age have shown that the deficit in total number of neurons is only about 10% while that of glial cells is approximately 40% (Clos *et al.*, 1977). The severe interference with the formation and/or maintenance of glial cells affects all types of cells: The number of cells, as a percentage of control, was Bergmann glia 80%, astrocytes in the internal granular layer 76%, and glial cells in the molecular layer 36%. The decrease in glial cell numbers evidently contributes to the reduction of the weight of the cerebellum and of the cross-sectional area of the vermis. In addition there is a tendency for a reduction of the size of the glial cells which occupy a smaller fraction of the area of the cerebellar cortex in the molecular layer than in controls (Clos *et al.*, 1973). It is of interest to note that, in contrast to undernutrition, thyroid deficiency results in a marked increase in the fractional area occupied by glia.

Evidence is accumulating that undernutrition may have different effects on the various neuroglial subpopulations not only in the cerebellum, but throughout the brain. In rats and mice, it is apparent that the formation of astrocytes, part of which occurs in late fetal life, precedes that of oligodendroglia, the extensive generation of which only begins toward the end of the first week after birth (Del Cerro and Swarz, 1976; Skoff *et al.*, 1976; Lewis *et al.*, 1977*a*). Undernutrition, unless it is so severe as to interfere with the survival of both fetus and mother, will have its effect on cell proliferation in the early postnatal period and after a significant proportion of the astrocyte subpopulation has been acquired. The studies of Sturrock *et al.* (1978) in fact show reduction in the number of oligodendroglia in the white matter of mice undernourished during the suckling period and subsequently rehabilitated without diminution of astrocyte numbers in a gray region.

The oligodendroglia manufacture the myelin sheath enwrapping the axons. The expression of this differentiated function can be followed quantitatively by biochemical methods. That there was a retardation in myelin production was originally inferred from studies on the cholesterol content of the undernourished brain (Dobbing, 1964). Soon after, it was reported that cerebrosides were affected more than most of the other groups of lipids (Culley and Mertz, 1965). By that time it had been firmly established that certain lipid classes are more closely associated with myelin than others (Brante, 1949; Folch, 1955). These include, besides glycolipids, proteolipid proteins and plasmalogens which are relatively severely affected by neonatal undernutrition (Benton *et al.*, 1966; Chase *et al.*, 1967; Culley and Lineberger, 1968). Fishman *et al.*, (1971) have directly isolated myelin and reported that in the 21- and 53-day-old undernourished animals its weight was, respectively, 87% and 71% of that obtained from control brains. Morphological investigations have also shown that neonatal undernutrition results in retarded myelination. The observations of Clos and Legrand (1969, 1970) suggest that myelination is defective under these conditions. In 12-day-old undernourished rats the normal relationship between axon diameter and number of myelin lamellae deposited is distorted and in the periphery the ultrastructural features of Schwann cell differentiation are impaired. Hedley-Whyte (1973) has confirmed that in the sciatic nerve undernutrition during the suckling period slows down the acquisition of myelin sheath more than it retards axonal expansion (see also the results of Krigman and Hogan, 1976, for the pyramidal tract), but she did not observe structural abnormalities in the Schwann cells. She also noticed that the effect of food deprivation limited to the early part of the suckling period was reversed after three weeks of rehabilitation, but effects brought about by prolonging undernutrition until near the end of the suckling period were only partially reversed by optimal feeding up to 46 days of age. Sturrock *et al.* (1976) studied the effect of neonatal undernutrition in mice after 19 weeks of rehabilitation and observed that in this species the relationship between axon diameter and number of myelin lamellae was normal in the anterior commissure.

Characteristic changes occur in myelin membranes during development. It has been suggested that myelin resembles glial plasma membranes when initially formed, and is converted during maturation into compact myelin by an increase in the content of glycolipids at the expense of phospholipids (Davison *et al.*, 1966; Agrawal *et al.*, 1970). That undernutrition leads to a retardation in myelination is indicated by many observations including those showing that in comparison with controls, promyelinating fibers constitute a greater proportion of myelinated fibers in the pyramidal tract at 30 days of age (about 5% vs. 0.3%) (Krigman and Hogan, 1976). Other changes associated with the maturation of the myelin membrane include a rise in the fractional content of proteolipids and an increase in the proportion of phospholipids accounted for by sphingomyelins (see, e.g., Fishman *et al.*, 1971). The fatty acid composition also changes; there is a decrease in the proportion of $C_{18:0}$ fatty acids in the most abundant phospholipid classes, phosphatidylethanolamine and phosphatidyl choline, with a concomitant increase in $C_{18:1}$ fatty acids. In glycolipids and proportion of C_{18}, C_{20} and C_{22} fatty acids is reduced, while that of C_{23} and especially $C_{24:1}$ (nervonic acid) is increased. Fishman *et al.* (1971) observed that, with minor exceptions, the maturational changes in the myelin isolated from the brain of undernourished animals were normal. Plasmalogens accounted for a smaller fraction of total phosphatidylethanolamine, and $C_{24:1}$ as well as $C_{24(hydroxy):1}$ fatty acids constituted a greater proportion of the glycolipid fatty acids. These changes were partially reversible on rehabilitation.

The results of most investigations are consistent with the view that undernutrition after weaning has little effect on the concentration and composition of brain myelin (e.g., Donaldson, 1911). However, deficient nutrition in infancy results in a decrease in the amount of myelin deposited, although its composition is more or less normal. Brain tissue retains the capability, even after a nutritional insult during the neonatal period, for further deposition of myelin: in some studies rehabilitation led to a complete restoration to normal of the concentration of myelin lipids (e.g., Benton *et al.*, 1966), in others the restitution was only partial (e.g., Dobbing, 1968; Fishman *et al.*, 1971; Rajalakshmi *et al.*, 1974*a*). It would seem that the degree of the restitution depends on the severity and duration of the metabolic insult (for reviews see, e.g., Dobbing, 1974; Rajalakshmi, 1975; Dodge *et al.*, 1975). It is worth noting here the recent studies of Wiggins *et al.* (1976) on the reversibility of undernutrition-induced deficits in myelination. In this work (see also Krigman and Hogan, 1976) a combination of morphological and biochemical methods was used, including a sophisticated labeling technique of proteins and lipids separated from subcellular fractions of undernourished, rehabilitated, and control animals. The food deprivation was very severe in these experiments since the yield of myelin was only about 25% of the control values (see Fishman *et al.*, 1971, p. 443). A delay in myelination was indicated by the finding that the proportion of basic and proteolipid proteins was significantly reduced at 15 and 20 days of age, although by 30 days the protein composition of myelin was normal. By studying the effect of rehabilitation from 20 days to 26 days of age on protein and lipid synthesis using a double labeling technique, Wiggins *et al.* (1976) came to the conclusion that under their experimental conditions undernutrition resulted in an irreversible deficit in myelination. However, in view of the time course of myelination (see Figure 5), the rehabilitation period was too short to permit such a definitive conclusion. Furthermore, although the experimental design made it possible to pinpoint myelin as the subcellular fraction most affected by undernutrition, the results did not distinguish between an influence on the amount or synthesis rate of myelin constituents. In these experiments ^{3}H- or ^{14}C-labeled precursors (e.g., amino acids) were given to undernourished or control animals, which were killed 3 hr later. The brains were then combined and subcellular fractionation carried out. The isotope ratio (undernourished/control) of proteins in each subcellular fraction was determined: The ratios obtained in the various subcellular fractions were expressed as a percentage of the value in the microsomal fraction. In contrast to the approximately 100% values in the other subcellular fractions, myelin gave only 34% at day 20 (or 75% at day 26) indicating, according to Wiggins *et al.* (1976), that the rate of myelin protein synthesis in the 20-day-old undernourished animal was about a third of the synthetic rate in controls, assuming that the incorporation rates in the microsomal fractions were comparable. However, the same results would be obtained if, in comparison with controls, the rate of myelin protein synthesis were the same in undernourished animals, but the amount of myelin protein were only a third. A rehabilitation period of 6 days only cannot be expected to restore to normal a 75% deficit in the amount of myelin, thus the relative incorporation rate of 90% obtained under these conditions is consistent with a synthesis rate of myelin proteins which cannot be much less, and is probably even greater, than in controls.

4.3. Neurotransmitter Systems

Neuronal differentiation can also be followed by using biochemical markers. There is good evidence indicating that critical enzymes in neurotransmitter synthesis are concentrated in nerve terminals in the brain (e.g., Whittaker and Barker, 1972; DeRobertis and Rodríguez DeLores Arnaiz, 1969; McLaughlin *et al.*, 1974; Hökfelt, 1974). During development the elaboration of neuronal processes is associ-

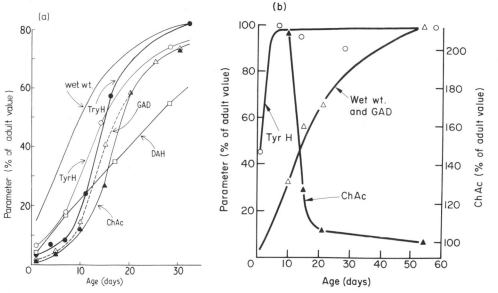

Fig. 6. Developmental changes in the activity of neurotransmitter enzymes in the forebrain (a) and the cerebellum (b). Tissue weights and enzyme activities per g wet wt are expressed as a percentage of the adult values. Organ weight, ———; tyrosine hydroxylase (TyrH), ○; and dopamine β-hydroxylase (DAH), □ (Coyle and Axelrod, 1972a,b); tryptophan hydroxylase (TryH) (Schmidt and Sanders-Busch, 1971), ●; glutamate decarboxylase (GAD) [in (a) Van den Berg et al., 1965, and in (b) Patel et al., 1978b], △; choline acetyltransferase (ChAC) [in (a) Ladinsky et al., 1972, and in (b) Patel et al., 1978b], △.

ated with an increase in the activities of these enzymes (Figure 6), which is also accompanied by a rise in the activities of various transmitter receptors in the different brain regions (e.g., Enna *et al.,* 1976). Figure 6b is presented in order to draw attention to regional differences in the development of various transmitter systems, and thus to the need to study brain parts separately. In comparison with the whole brain, tyrosine hydroxylase activity (expressed as a percentage of adult values) is relatively great in the cerebellum at birth and also reaches its final level much earlier (by the end of the first week). The main targets of the noradrenergic innervation in the cerebellum are the Purkinje cells which are formed during gestation and start to differentiate relatively early. In contrast to the other transmitter enzymes, choline acetyltransferase (ChAc) activity in the cerebellum declines sharply during postnatal development. This is due to the concentration of cholinergic systems in the rat archicerebellum which is the earliest-developing part of the cerebellum. Thus the enzyme activity per unit weight decreases with the rapid postnatal growth of the other, principally noncholinergic, parts of the cerebellum which constitute a very great proportion of the tissue. Among the transmitter enzymes considered, only glutamate decarboxylase (GAD) seems to show a developmental chronology similar to that in the forebrain.

Neonatal undernutrition results in a decrease in the activity of brain glutamate decarboxylase, the enzyme forming γ-aminobutyric acid (GABA), an important central inhibitory transmitter (Rajalakshmi *et al.,* 1974b). However, after rehabilitation the enzyme activity per whole brain is restored to normal and there is even a slight increase in comparison with controls, when the activity is expressed in terms of unit weight (Adlard *et al.,* 1972). Although it is not an unequivocally reliable marker, acetylcholine esterase (AChE) activity has been determined in several studies to monitor the development of the cholinergic system. Sereni *et al.* (1966) noted a transient reduction in the concentration of this enzyme in the brain of neonatally undernourished animals, an observation confirmed by detailed studies of Adlard and Dobbing (1971*a,b*; 1972), who also found that the activity of this enzyme, in terms of unit weight but not of the whole brain, was significantly greater than in controls when the animals were kept undernourished or rehabilitated after weaning. Im *et al.* (1973) have found that in undernourished pigs the specific activity of ChE was elevated in the forebrain, but it was normal in the brain stem and cerebellum.

It would appear than ChAc is a more satisfactory index of the cholinergic system than AchE activity (e.g., Silver, 1974). Further, estimates on whole brain tissue may mask changes occurring in certain brain parts. Thus, in a recent study we investigated the effect of undernutrition on the development of ChAc in different parts of the brain (Patel *et al.*, 1978*b*). To evaluate the behavior of another neurotransmitter system, the activity of GAD was also determined. Undernutrition was effected by halving the diet of mother rats from the 6th day of gestation throughout lactation, and the young were given food *ad libitum* after weaning. The results showed that in the different parts of the undernourished brain the normal developmental rise in both enzyme activities per gram wet weight were markedly retarded. Thus, when the estimates were expressed as a percentage of the control values, a significant reduction was observed which, however, was diminishing with age even during the suckling period, and completely disappeared after four weeks of rehabilitation (Figure 7). The findings thus suggest that undernutrition caused, in the development of the cholinergic and GABA-ergic systems, a retardation rather than an irreversible depression. In the cerebellum the effects on ChAc, but not on GAD, were apparently different from those in the rest of the brain. Here the activity per unit weight was not lower than in controls at 10 and 15 days of age, although the total activity per cerebellum was significantly reduced. The unique development of ChAc activity in the rat cerebellum described above is relevant for the understanding of this effect. Since the growth and development of the cerebellum are markedly retarded in undernutrition, the "dilution" of the cholinergic system by other structures is expected to be less pronounced than in controls.

It is possible, on the basis of these results, to reconsider the question whether the retardation in synaptic development in the brain of neonatally undernourished animals leads to a persistent deficit which cannot be rectified by rehabilitation (see also Section 4.1). As mentioned before, no conclusive morphological observations are yet available concerning this question. However, quantitative electron-microscopic studies, even if they would be available, have the limitation that generalizations must be made from results in extremely small samples. Biochemical estimations, although they have their own limitations and can only be considered as markers of certain structures, offer here certain advantages. The results in Figure 7 showed that the deficit in the concentration of the two transmitter enzymes completely disappeared throughout the brain after four weeks of rehabilitation. Further, the enzyme activities per brain part were also restored to normal, with the exception of the cerebral cortex where a small but significant deficit persisted (as a percentage of control, GAD was 88% and ChAc 78%). Previous studies indicated that the number of nerve cells in the cerebral cortex of animals rehabilitated after neonatal undernutrition is somewhat reduced (Dobbing *et al.*, 1971; Escobar, 1974). Thus the morphological and biochemical results are consistent with the view that

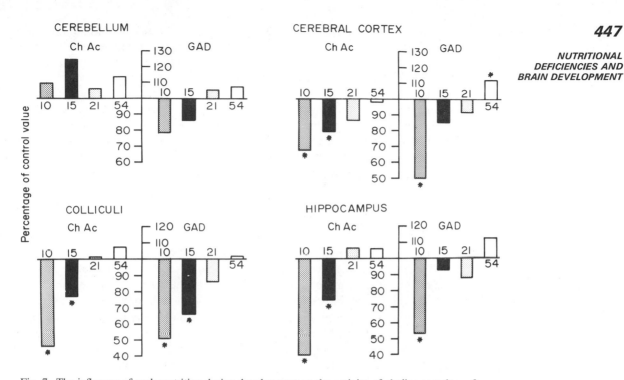

Fig. 7. The influence of undernutrition during development on the activity of choline acetyltransferase (ChAc) and glutamate decarboxylase (GAD) in different brain parts: effect of nutritional rehabilitation. The young of undernourished mother rats were studied at the ages indicated in days on the abscissa: They were weaned to unrestricted food at day 22 and killed at day 54 (nutritionally rehabilitated group). The enzyme activities per g wet wt were expressed as a percentage of the control values at the ages indicated and are displayed on the ordinate (for normal developmental changes see Fig. 6). Significant differences between experimental and control rats ($P < 0.01$) are indicated by asterisks. ▨, 10-day-old; ■, 15-day-old; ▨, 21-day-old undernourished rats; and □, 54-day-old rats rehabilitated from day 22. (From Patel *et al.*, 1978*b*.)

under these conditions there is no marked reduction in the association of, at least, the above two transmitter systems with cortical neurons, and they suggest that the severe decrease in the calculated association of nerve terminals with cortical neurons (Cragg, 1972) may reflect the effect of chronic undernutrition.

The influence of undernutrition on the development of central monaminergic systems has also been studied. In early investigations Sereni *et al.* (1966) determined the concentration of catecholamines in neonatally undernourished rats. Rats are born with only 30% and 15% of the adult amounts of brain dopamine and noradrenaline, respectively (Agrawal *et al.*, 1966). Sereni *et al.* (1966) reported that in undernutrition the brain concentration of catecholamines was reduced during the early neonatal period, although it was restored to normal even after prolonging undernutrition until 35 days of age. On the other hand, Stern *et al.* (1975*a*) have observed, in the brain of protein-deficient rats studied from birth to 300 days of age, a trend for increased noradrenaline concentrations with the most consistent changes recorded in the midbrain, pons, and medulla. A reduction in the total amount, but not in concentration, of brain catecholamines has been found by Shoemaker and Wurtman (1971, 1973) in 24-day-old rats reared by protein-deficient mothers. Since the accumulation in the brain of intracisternally administered labeled noradrenaline was normal, these authors argued that the reduction in catecholamine content was not due to a decrease in catecholaminergic terminals, which they assumed to be the principal sites of uptake, but to the amount of the transmitter in the neurons. The results also suggested that noradrenaline turnover was decreased in the brain of undernourished rats. On the other hand, Shoemaker and Wurtman (1971) have claimed that the activity of partially purified tyrosine hydroxylase— believed to be the rate-limiting enzyme of catecholamine synthesis—was above the control levels. In contrast, Lee and Dubos (1972) have reported that tyrosine hydroxylase levels are depressed in the brain of undernourished mice, although they also found a reduction in catecholamine content and in noradrenaline turnover. Estimates on transmitter content are not sufficient information to assess whether the size of the relevant neuronal population, including the proliferation of axons and terminals, is affected by adverse conditions. Nevertheless it is worth noting that the brain noradrenaline levels have been reported consistently to be restored to normal after nutritional rehabilitation (Sereni *et al.*, 1966; Wiener, 1972).

It has been proposed recently that nutritional factors may influence initial monoamine metabolism by affecting the availability of the precursor amino acids to the brain (for review see Fernstrom, 1976). In contrast to previous views that tyrosine hydroxylase, the rate-limiting enzyme of catecholamine synthesis, is saturated with the amino acid substrate (e.g., Carlsson *et al.*, 1972), it has been observed that the rate of catechol synthesis is increased after a rise in serum tyrosine levels and it is decreased when the serum concentration of other large neutral amino acids that compete with tyrosine for brain uptake is elevated (Wurtman and Fernstrom, 1975).

The results concerning the effects of undernutrition on 5-hydroxytryptamine (5HT) content in the developing brain are rather controversial. Sereni *et al.* (1966) have reported a transient decrease in the concentration of 5HT in the brain of infant undernourished rats. On the other hand, Stern *et al.* (1975*a*) have observed elevated values of 5HT and its major metabolite 5-hydroxyindole acetic acid (5HIAA) in the brain of the young of protein-deficient mother rats. Significant increases in the concentration of both 5HT and 5HIAA have been found in the brain stem (where the 5HT cell bodies are localized) with relatively smaller changes in the telencephalon (where the 5HT cells project) of malnourished infant rats (Sobotka *et al.*, 1974; Stern *et al.*, 1975*a*). However, Dickerson and Pao (1975) have reported that the brain concentration of 5HT was normal both at birth and day 21 in the young of protein-deficient mothers: 5HIAA was not determined in this study.

The precursor of 5HT is tryptophan, an amino acid which has been claimed to have a unique role in the regulation of protein synthesis in the body (see Section 3.3.3). It seems that tryptophan has also a regulatory role in 5HT synthesis in the brain (Fernstrom and Wurtman, 1971). The cerebral concentration of this amino acid is relatively low (about 30 μM) in comparison with the kinetic properties of the enzyme hydroxylating tryptophan (apparent affinity constant, K_m 50 μM; Friedman *et al.*, 1972). As in the case of catecholamines, nutritional factors have an influence on brain 5HT synthesis by affecting the serum concentration of the precursor and other amino acids competing with tryptophan for cerebral uptake. However, it is believed that the availability of tryptophan to the brain is also influenced by a factor that is specific to this amino acid. Tryptophan is the only amino acid which is stereospecifically bound to albumin in the plasma (McMenamy and Oncley, 1958). It has been proposed that the uptake of tryptophan and consequently the 5HT synthesis rate in the brain relate to the concentration of the free rather than total tryptophan in the blood (e.g., Knott and Curzon, 1972). Food deprivation may lead to accelerated 5HT metabolism in the brain (e.g., Stern *et al., 1975a*) because of an elevation in the plasma concentration of free tryptophan (Tagliamonte *et al.,* 1973) resulting from a reduction in serum albumin content (Dodge *et al.,* 1975) and an increase in circulating free fatty acids which compete with tryptophan for the albumin binding sites (McMenamy and Oncley, 1958). However, the hypothesis that changes in brain tryptophan level can be predicted from alterations in the size of the free tryptophan pool in the serum has been challenged recently (Madras *et al.,* 1974). Irrespective of the mechanisms involved, the evidence currently available is in favor of the observations indicating an increase in 5HT metabolism in the brain of undernourished developing animals, although it is not yet known whether this occurs in the functionally relevant compartments (e.g., Grahame-Smith, 1971).

Besides effects manifested during the period of undernutrition, there is also evidence suggesting that nutritional deficiency during brain development results in permanent alterations in certain brain functions involving monoamine transmitter systems. Stern *et al.* (1975*b*) reported that rats brought up by protein deficient dams and receiving low-protein diet throughout life showed a significant depletion of 5HT and noradrenaline in the midbrain and pons–medulla regions when exposed up to 90 min to electrical foot-shock, which in controls or in normal rats switched in adulthood to low-protein diet had only minimal effect. Behavioral measures of reactivity to foot-shock stress, however, did not reveal marked differences related to nutrition during development. Smart *et al.* (1976) found that 5HT turnover is markedly increased (about 170% of control) in the hippocampus, although not in other parts of the brain studied, in adult rats rehabilitated after food restriction in early life. Serotoninergic systems have been implicated in the regulation of emotional behavior (Costa *et al.,* 1974), which seems to be permanently altered by early undernutrition (Hanson and Simonson, 1971; Sobotka *et al.,* 1974; Smart *et al.,* 1976). Smart *et al.* (1976) have rightly pointed out that changes in neurotransmitter metabolism might have been mediated by nonnutritional factors, such as hormonal imbalance. The blood corticosteroid levels are high in undernourished infant rats (see Section 3.3.4). It seems that the hippocampus is a major target for circulating corticosteroids (McEwen *et al.,* 1972), which may influence 5HT metabolism through an induction of tryptophan hydroxylase (Azmitia and McEwen, 1974).

4.4. Effects of Undernutrition on Glucose Metabolism and Metabolic Compartmentation

The biochemical reflections of brain maturation involve quantitative changes, e.g., in enzyme activities and steady-state concentration of constituents. The effect of undernutrition on some of these developmental changes has been described above and further information can be found in recent reviews (e.g., Balázs, 1972*b*; Dobbing and Smart, 1974; Dodge *et al.,* 1975; Winick, 1976). Here we want to consider certain qualitative changes associated with biochemical differentiation in the brain as manifested by following the development of glucose metabolism and of metabolic compartmentation (for review see Patel and Balázs, 1975).

It seems that the fate of labeled glucose in the brain reflects the coordination of relevant metabolic pathways in the tissue during development (Gaitonde and Richter, 1966; Cocks *et al.,* 1970). Figure 8 shows that the rate of conversion of glucose carbon into amino acids associated with the tricarboxylic acid cycle ("cycle" amino acids) is low up to about 10 days after birth, and it increases sharply during the period of functional maturation of the rat brain. This is the result, in part, of quantitative biochemical changes; the concentration of "cycle" amino acids and the flux through the tricarboxylic acid cycle are more than doubled. However, qualitative changes are also involved, as highlighted by the replacement of a wide range of substrates by glucose as the major fuel in the tissue. These biochemical changes occur during the period of extensive neuronal differentiation, and there is independent evidence for their interrelationship (for references see Cocks *et al.*, 1970). As a result of the development of the metabolic compartments associated with neuronal processes, the apparently homogeneous metabolic pattern is replaced by the heterogeneous pattern characteristic of the adult brain. Most studies have concentrated hitherto on metabolic compartmentation of the tricarboxylic acid cycle and associated amino acids in the brain, and a high ratio of glutamine–glutamate specific radioactivity (ratio >1) has been used as an index of metabolic compartmentation. In the adult this ratio is high when labeled amino acids and fatty acids are oxidized in the brain. The ratio is low (<1) in the immature brain, and the development of metabolic compartmentation of glutamate parallels the extensive conversion of glucose carbon into amino acids.

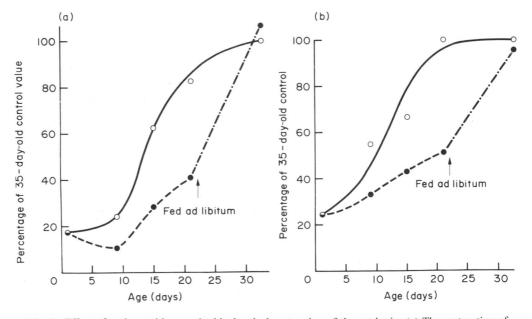

Fig. 8. Effect of undernutrition on the biochemical maturation of the rat brain. (a) The maturation of glucose metabolism in terms of conversion of glucose carbon into amino acids: The ^{14}C content of the amino acid fraction was expressed as a percentage of the acid-soluble ^{14}C at 10 min after a subcutaneous administration of [U-^{14}C]glucose. (b) Development of metabolic compartmentation of glutamate, in terms of the glutamate/glutamine specific radioactivity ratio at 10 min after a subcutaneous injection of [U-^{14}C]leucine. The results in both (a) and (b) are expressed as a percentage of 35-day-old control values (which are comparable with estimates obtained in adults). The young of undernourished mother rats were weaned to unrestricted food at day 21 and the effect of nutritional rehabilitation was studied at day 35. Control, ○—○; and undernourished rats, ●---●. During the suckling period undernutrition resulted in a significant depression ($P < 0.01$) in the estimates in both (a) and (b) with the exception of the values at day 1; there was no significant difference ($P > 0.05$) between control and rehabilitated rats. The results were taken for (a) from Balázs and Patel (1973), and for (b) from Patel *et al.* (1975).

Balázs and Patel (1973) observed that brain maturation, assessed by the age curve of conversion of glucose carbon into amino acids is severely retarded in undernutrition: The experimental animals at day 21 had only reached a developmental stage similar to 12-day-old normal rats (Figure 8). Detailed studies were then conducted on animals at 22 days of age, and the results provided some insight into the effect of undernutrition on the major developmental features underlying the age-dependent changes in the fate of [^{14}C]glucose in the brain.

Cycle Amino Acids. In contrast to most essential amino acids, the concentration of cycle amino acids increases during brain maturation. Since the rate of exchange transamination is great relative to the flux in the tricarboxylic acid cycle (Haslam and Krebs, 1963; Balázs and Haslam, 1965), these amino acids, which are present at much higher concentration than the cycle intermediates, trap a progressively increasing proportion of the glucose carbon entering the cycle. In contrast to a preliminary report (Balázs and Patel, 1973), we have now established that the developmental rise in the concentration of glutamate and aspartate is significantly retarded in the brain of undernourished rats, while the concentration of GABA and glutamine is not influenced significantly (Patel *et al.*, 1975). (For other results concerning the effect of undernutrition on brain amino acids, see Mandel and Mark, 1965; Rajalakshmi *et al.*, 1967; Mourek *et al.*, 1970; Reddy *et al.*, 1971; Thurston *et al.*, 1971; Wapnir, 1973; Dickerson and Pao, 1975.) However, since in comparison with controls the decrease in the concentration of cycle amino acids is only of the magnitude of 10%, this effect cannot account completely for the depressed conversion of glucose carbon into amino acids.

Contribution of Substrates Other Than Glucose to Overall Oxidation. This seems to be higher in the immature than in the adult brain (Patel and Balázs, 1970; van den Berg, 1970; Cremer, 1972; Williamson and Buckley, 1973; DeVivo *et al.*, 1975). Among the alternative substrates ketone bodies, which are present at relatively high concentration in the blood of infant rats and show significant increase in food deprivation (e.g., Dahlquist and Persson, 1976), are especially important; their contribution to overall oxidation in the brain is a function of their blood concentration (Hawkins *et al.*, 1971; Cremer and Heath, 1974). The nearly 60% rise in circulating ketone body concentration (Figure 9) would increase the contribution to brain oxidation from about 17% in the controls to 28% in undernutrition, thus accounting for a fraction (about 12%) of the observed depression in the incorporation of glucose carbon into cerebral amino acids.

Glucose Flux. This is more than doubled during the early postnatal period (Lowry *et al.*, 1964). To investigate systematically the effect of undernutrition on glucose flux, the fate of [^{14}C]glucose was studied between 2 and 60 min after injection (Patel and Balázs, 1975) (Table VII). In comparison with controls, the specific radioactivity of glucose reached a maximum at about the same time (10 min), but declined much more slowly, so that by 1 hr it was more than double. On the other hand, the specific radioactivities of lactate and of the cycle amino acids in particular were markedly reduced in the undernourished animal. The results indicated that glycolytic flux was severely depressed in the undernourished brain since the alanine/glucose (or lactate/glucose) specific radioactivity ratios were 40–60% lower than the control values. The data also suggested a decrease in the cycle flux since, in comparison with controls, the glutamate/alanine (or lactate) specific radioactivity ratios were reduced, although a marked depression was only detectable at

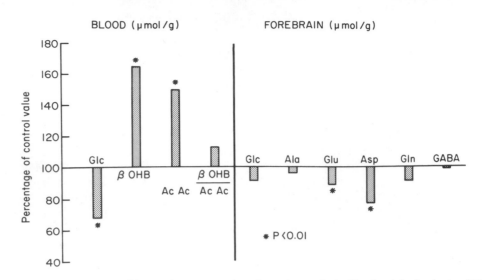

Fig. 9. Effect of undernutrition on the concentration of constituents in the blood and the forebrain of 22-day-old rats. The results are expressed as a percentage of the control values. Significant differences ($P <$ 0.01) between control and experimental animals are indicated by asterisks. Glc, glucose; β OHB, β-hydroxybutyrate; AcAc, acetoacetate; Ala, alanine; Glu, glutamate; Asp, aspartate; Gln, glutamine; GABA, γ-aminobutyrate. The results were taken from Patel *et al.* (1975) and Patel and Balázs (1975).

40–60 min. However, quantitation of the effect is complicated by the enhanced oxidation of ketone bodies causing a dilution of cycle intermediates.

The aspartate/glutamate specific radioactivity ratio was significantly higher than in controls (Table VII). These results indicate an increase in the rate of CO_2 fixation, introducing labeled pyruvate at the level of dicarboxylic acids without prior dilution by intermediates associated with the cycle. An increase in CO_2 fixation has previously been observed in the liver and muscle of starved animals, but this has not been noted in the mature brain (Freedman *et al.,* 1960; O'Neal and Koeppe, 1966). However, Lehr and Gayet (1967) have also found an increase in the aspartate/glutamate specific radioactivity ratio after giving labeled glucose to undernourished adult animals, and Patel (1974) showed that $^{14}CO_2$ fixation is stimulated by β-hydroxybutyrate in brain mitochondrial preparations incubated in the presence of pyruvate.

Summary of Effects of Undernutrition. It would appear therefore that the depression of the conversion of glucose carbon into amino acids in the developing undernourished brain is a composite effect resulting from (1) a retardation in the normal increase in the concentration of cycle amino acids, (2) an elevation in the supply of ketone bodies for oxidation, and (3) a decrease in glucose utilization flux in the brain. However, the effect is completely reversible: The conversion of glucose carbon into brain amino acids is restored to normal by two weeks after weaning the undernourished rats to unrestricted diet (Figure 8).

It is important to note here that undernutrition resulted in no decrease in the energy reserve, concentrations of ATP, phosphocreatine, glycogen, and glucose in the brain (Thurston *et al.,* 1971; Chase *et al.,* 1976). The normal steady-state level of these constituents is, therefore, maintained in spite of the reduction in glucose catabolism, including probably a depression in the cycle flux. The results imply that the energy-utilizing processes must have been adjusted to a lower than normal level.

This view is consistent with the general observation of retarded structural, biochemical, and functional development of the brain in undernutrition.

In a recent study Chase *et al.* (1976) have reported that the labeling of amino acids is substantially depressed in the brain of 10-day-old undernourished rats 1 hr after the injection of [U-^{14}C]glucose, although significant differences between control and experimental animals have not been noted at 17 and 24 days. A comparison of these results with those of Patel and Balázs (1975) is hampered by the lack of kinetic studies on the fate of labeled glucose and of information on the total radioactivity content of the acid-soluble fraction and the specific radioactivity of glucose. As shown in Table VII the decay of the specific radioactivity of glucose is slower in the undernourished than in the control animals. The total radioactivity content of the amino acids depends not only on the rate of glucose breakdown, but also on the total amount of [^{14}C]glucose available to the brain during the whole experimental period, which in the studies of Chase *et al.* (1976) was relatively long (1 hr). Dickerson and Pao (1975) have reported that the ^{14}C content of the cycle amino acids is not reduced in the brain of 21-day-old malnourished rats 30 min after the injection of [U-^{14}C]glucose. However, it can be calculated from their results that an unusually small fraction of the radioactivity content of the tissue was recovered in the cycle amino acids (less than 20% vs. over 50% found in other studies). Further, Dickerson and Pao (1975) separated the amino acids by paper chromatography without prior removal of neutral and carboxylic acid constituents, which may have caused contaminations.

The conversion of glucose carbon into lipids and proteins has also been investigated in the brain of undernourished developing rats (Agrawal *et al.*, 1971; Chase *et al.*, 1976). There is a marked reduction in the ^{14}C content combined in both constituents. Among the different lipid classes the greatest depression has been observed in glycolipids: This is consistent with the results of other studies showing a severe depression in the incorporation of different precursors into glycolipids and a more severe retardation, in comparison with general lipid constituents of cells

Table VII.   Effect of Undernutrition on [2-^{14}C] Glucose Metabolism in the Forebrain of 22-day-Old Rats[a]

	Time after injection (min)					
	2	5	10	20	40	60
Control rats						
Specific radioactivity of glucose (dpm × 10^{-2}/μmol)	120	237	311	294	179	76
Undernourished rats						
Specific radioactivity of glucose (% of control)	136	124	106	110	145[b]	222[b]
Specific radioactivity ratios (% of control)						
Alanine/glucose	44[b]	48[b]	53[b]	62[b]	65[b]	76
Glutamate/alanine	83	84	100	79	41[b]	35[b]
Aspartate/glutamate	102	138[b]	123[b]	128[b]	134[b]	113

[a]Mother rats were undernourished from the sixth day of gestation throughout the suckling period. The 22-day-old young of undernourished or control mothers received subcutaneous injection of [2-^{14}C]glucose (10 μCi/100 g body wt), and they were killed 2–60 min later by immersion into −150°C Freon. The resuts of the undernourished rats are expressed as a percentage of the control values. (From Patel and Balázs, 1975.)
[b]Significant differences between control and undernourished groups: $P < 0.05$.

(e.g., phospholipids), in the accumulation of those lipid classes which are closely associated with myelin (see Section 4.2). Agrawal *et al.* (1971) have observed a significant decrease with incorporation of ^{14}C into protein from [U-^{14}C]glucose, but not from [1-^{14}C]leucine, in the brain of 28-day-old undernourished rats at 6 hr after the administration of the precursor. They interpreted their results as indicating a decrease in the undernourished brain in the conversion of [^{14}C]glucose into [^{14}C]amino acids used for protein synthesis.

Although this conclusion is correct, it is difficult to infer from the results whether or not protein synthesis proper is also affected by undernutrition. By day 28 the period of relatively high protein synthesis rate in the developing brain is long over. Moreover, the incorporation of [^{14}C]leucine into protein was only studied at 6 hr after the injection. Thus it is possible that the over-all supply of the precursor to the brain of undernourished rats, whose protein synthesizing rate is depressed throughout the body, may have been increased in comparison with controls. In that case, a genuine reduction in protein synthesis rate would be masked by the increase in the specific radioactivity of the precursor. This explanation is consistent with the findings of Patel *et al.* (1975), who studied [U-^{14}C]leucine metabolism in undernourished rats during the period 1–21 days. It was observed that 10 min after the administration of the precursor the specific radioactivity of protein relative to that of the free leucine was substantially less than in controls. The conversion of leucine carbon into brain lipids was also severely reduced: The ^{14}C content per gram wet weight corrected for the specific radioactivity of free leucine was only about a third of the control value at 15 and 21 days of age.

4.4.2. Metabolic Compartmentation of Glutamate and Undernutrition

[U-^{14}C]Leucine is also a good precursor to study metabolic compartmentation of glutamate in the brain. It has been observed that the development of metabolic compartmentation, in terms of the age-dependent rise in the glutamine/glutamate specific radioactivity ratio, is severely retarded in the brain of undernourished rats (Patel and Balázs, 1973; Roach *et al.,* 1974: Patel *et al.,* 1975). However, this compartmentation index is restored to normal as a result of nutritional rehabilitation after weaning (Figure 8). As mentioned above, the normal pattern of [^{14}C]glucose metabolism is also restored after rehabilitating food-deprived animals. Therefore, neonatal undernutrition did not inflict detectable irreversible damage on the biochemical processes underlying the changes in these two indices of brain maturation. Furthermore, after rehabilitation the rate of leucine incorporation into brain proteins became also similar to controls.

Although the results of Roach *et al.* (1974) and Patel *et al.* (1975) are similar, they also show an interesting difference. The reduction, in comparison with controls, in the value of the glutamine/glutamate specific radioactivity ratio is mainly due to an increase in the conversion of leucine carbon into glutamate in the experiments of Roach *et al.* (1974). In this respect the results are very similar to those produced by neonatal thyroid deficiency (Patel and Balázs, 1971). A detailed analysis of these data in conjunction with other evidence led to the conclusion that the changes in leucine metabolic pattern result from a severe retardation of the development of the neuronal compartment in the thyroid deficient brain. On the other hand, in the experiments of Patel *et al.* (1975) the decrease in the compartmentation index in the undernourished brain was mainly the consequence of a reduction of specific radioactivity of glutamine. Considering the overall leucine metabolic pattern, Patel *et al.* (1975) have proposed that the results are consistent with retarded development of both the neuronal and glial compartment under these conditions. There is an important difference in the experimental conditions between these two studies: Undernutrition was produced by giving to the mother rats, during a period in pregnancy and throughout lactation, an isocaloric, low-protein diet in the studies of Roach *et al.* (1974), in contrast to a reduced amount of normal diet in the experiments of Patel *et al.* (1975). Thus a possible explanation would be that more pronounced thyroid dysfunction is associated with protein deficiency than with overall undernutrition.

5. Effects of Vitamin Deficiencies on Brain Growth

In this chapter we have dealt hitherto with the effects of protein and calorie deficiencies on brain development. These deficiencies, especially in humans, are often associated with deficits in minor food constituents such as vitamins and trace elements. We propose to examine here the influence of these substances on brain development so as to indicate the way in which their deficiency might contribute to disturbances in the growth and maturation of the nervous system. It must be stated, first, that this is only a small part of the study of vitamins and secondly, that the information available concerning the role of vitamins in brain growth has been gathered to a great extent from areas different from those largely examined so far in this chapter—pediatrics, human neuropathology, teratology, and behavioral studies.

There are many comprehensive reviews dealing with the effects of vitamin deficiencies on the nervous system (for neurological manifestations of vitamin deficiencies see Pallis and Lewis, 1974; Cruickshank, 1976; Still, 1976; Weiner and Klawans, 1976; Victor, 1976). However, not all of the vitamin-deficiency-induced changes in the mature nervous system are paralleled by comparable disturbances in the developing brain. Deficiency of vitamins C and D during brain growth seems to have no irreversible structural or functional consequences. The effects of riboflavin and nicotinamide deficiency on the maturing human brain are ill-defined, although electroencephalographic abnormalities are said to occur (Arakawa *et al.,* 1968; Srikantia *et al.,* 1968). However, deficiency of some other B vitamins in early life is clearly related to behavioral, electrophysiological, and other functional changes in man and in experimental animals; these include thiamine and pyridoxine which will be discussed here. Experimental deficiency of vitamins A and E also results in changes in the developing brain, and these will also be considered; the induced abnormalities hitherto documented are mainly structural, and their relationship to the known biochemical actions of these vitamins is not yet clear. Folic acid and its relationship to brain development is a controversial field. Whether folate deficiency *per se* is ever a cause of neurological symptoms and signs in the adult human is contentious (see, e.g., Pallis and Lewis, 1974; Kanig, 1976). Certain observations suggest that it *may* have effects on the developing nervous system. However, the experimental studies of Stempak (1965), in which pregnant rats were treated with folic acid antagonists and produced hydrocephalic offspring, and of Haltia (1970), whose folate-deficient chicks became severely undernourished and showed reduced Purkinje cell RNA content, present complex situations, the interpretation of which is difficult. Equally, the neurological abnormalities seen early in life in patients with congenital specific folate malabsorption (Luhby *et al.,* 1961; Lanzkowsky, 1970) cannot unequivocally be interpreted as due to folate deficiency and may in fact progress despite adequate treatment with folic acid.

The role of trace elements in brain growth would also merit consideration here, but is felt to be beyond the scope of the present contribution. At the speculative level, it is possible that "nutritional deficiencies (even marginal or short-term) superimposed upon a particularly susceptible genetic constitution, with perhaps a drug, toxin, or other environmental factor interacting, could cause aberrations of developmental processes with deleterious effects in the perinatal period" (Hurley, 1976). However, deficiencies of only a few trace elements are clearly related to disturbances of brain development. One of the most important of these is probably iodine. Iodine deficiency has a powerful influence in disturbing the development of the brain, but it has not yet been firmly established whether or not the effects are exclusively due to the resulting thyroid deficiency (see Stanbury and Kroc, 1972; Balázs *et al.*, 1975). As well as iodine, data are currently available about the effects of some other elemental deficiencies on brain growth. Deficiencies which appear to produce structural and/or functional abnormalities in developing brain are those of copper (for review, see O'Dell, 1976) and zinc (Halas *et al.*, 1976); these will not be discussed further.

5.1. B-Complex Vitamins

Although dietary deficiencies of various water-soluble vitamins have been associated with neurological disease in man, this vast subject can be touched on here. In the first place, the role of some isolated vitamin deficiencies in producing symptoms and signs in the nervous system is uncertain, and secondly most of the available data on specific deficiency states pertain to the adult. There is in addition a substantial literature on vitamin *dependency* in a number of metabolic disorders (e.g., aminoacidurias), in which neurological manifestations may occur; this is well discussed by Dodge *et al.* (1975) and will not be mentioned further.

Besides biochemical investigation of the role of B vitamins in brain development, which will be discussed in this chapter, there is a body of literature indicating the occurrence of behavioral and/or electrophysiological abnormalities in deficiency states. Much of the older data, reviewed by Brožek and Vaes (1961), is difficult to evaluate for a variety of methodological reasons, including the use of only semipure diets; however, some of the relatively recent studies strongly suggest that deficiency of some B-complex vitamins in early life can have marked effects on functional brain development. The evidence is strongest for pyridoxine. Deficiency of this vitamin in suckling rats (as in human infants, see below) produces seizures and EEG changes (Stephens *et al.*, 1971). In postweaning animals, deficits in avoidance learning have also been documented (Stewart *et al.*, 1968), together with increased latency of visually evoked cortical responses (Stewart *et al.*, 1973) and reduced activity and curiosity scores (Driskell and Foshee, 1974). Thiamine deficiency in early life is also believed to produce behavioral disturbances, and these may precede other clinical manifestations of the deficiency syndrome (Peskin *et al.*, 1967).

5.1.1. Thiamine

The clinical and pathological features of the Wernicke-Korsakoff syndrome have been authoritatively reviewed by many workers, most recently by Victor *et al.*

(1971). The relationship of this syndrome to thiamine deficiency has long been known, but whether it occurs in children—and if it does, the role of vitamin deficiency—is uncertain. A number of case reports (e.g., Tanaka, 1934; Guerrero, 1949) indicate that the clinical and/or pathological features of Wernicke's encephalopathy may occur in infants who have more general evidence of thiamine deficiency, i.e., beri-beri. In the absence of clear-cut evidence of infantile beri-beri, interpretation of seemingly classical neurological and pathological findings as being due to thiamine deficiency is fraught with difficulty, since Leigh's encephalopathy, an etiologically unrelated condition, may show identical lesions in the brain (see David *et al.*, 1976).

Infantile beri-beri is more often associated with a distinct and unusual neurological syndrome, including increased irritability, anorexia, vomiting, neck stiffness, spasticity of limbs, paroxysms of muscular paralysis, and aphonia, possibly due to a laryngeal paresis of vagal origin (Hirota, 1898; Andrews, 1912; Haridas, 1937). The mother herself has signs of thiamine deficiency; thus the milk is deficient in this vitamin, and the infant's vomiting exacerbates the deficiency. Systemic and neurological response to thiamine treatment is said to be generally rapid, only the presumptive vagal lesion being slow to recover.

Biochemically, thiamine in the form of thiamine pyrophosphate serves as a cofactor in reactions catalyzed by transketolase, pyruvate dehydrogenase, and 2-oxoglutarate dehydrogenase. The changes in carbohydrate metabolism attributed to thiamine deficiency have been extensively studied, but the precise molecular events underlying the neurological disorder are still poorly understood.

In the CNS, transketolase (which is involved in the pentose phosphate pathway) is the enzyme that seems to be consistently affected by thiamine deficiency (Sie *et al.*, 1961; McCandless and Schenker, 1968; however, see Gaitonde *et al.*, 1974, for the *in vivo* activity of this enzyme). Developing rats seem to be more susceptible to thiamine depletion than adults, yet unexpectedly there is no evidence that in the young animal the vitamin deficiency would affect the availability of NADPH for lipid synthesis believed to be one of the main functions of the pentose phosphate pathway (Geel and Dreyfus, 1974, 1975).

The influence of thiamine deficiency on pyruvate dehydrogenase activity in the brain is still equivocal: Some investigators have reported a decrease (Dreyfus and Hauser, 1965; McCandless and Schenker, 1968; Reinauer *et al.*, 1968), whereas others found normal activity (Gubler, 1961; Koeppe *et al.*, 1964; Heinrich *et al.*, 1973). Only in the case of deficiency produced by pyrithiamine, an antagonist of thiamine, has a marked reduction in pyruvate dehydrogenase activity been consistently observed (Gubler, 1961, 1968; Bennett *et al.*, 1966; Holowach *et al.*, 1968). Thus, it is not surprising that large variations in the effect of thiamine deficiency on the acetyl CoA and acetylcholine levels of the brain have been reported. Hosein *et al.* (1966), Speeg *et al.* (1970), and Stern and Igic (1970) did not observe any change in the levels of acetylcholine in the brain of thiamine-deficient animals. On the other hand, Gubler (1968) and Cheney *et al.* (1969) found a reduction of acetylcholine content. Heinrich *et al.* (1973) reinvestigated this problem and reported a marked decrease in the concentrations of acetyl CoA and acetylcholine without any appreciable change in choline acetyltransferase and acetylcholine esterase activities in thiamine deficient brains. These results were again contradicted by Reynolds and Blass (1975).

Thiamine deficiency is thought to have no effect on the activity of 2-oxoglutarate dehydrogenase in the brain (Gubler, 1961; Koeppe *et al.*, 1964) although an increase in the concentration of 2-oxoglutarate has been reported (Gubler *et al.,* 1974). By studying the incorporation of ^{14}C into certain brain metabolites at different times after the injection of [U-^{14}C]glucose or [U-^{14}C]ribose into rats maintained on thiamine-deficient diet, Gaitonde *et al.* (1974) concluded that the activities *in vivo* of the thiamine pyrophosphate requiring enzymes (pyruvate dehydrogenase, 2-oxoglutarate dehydrogenase, and transketolase) were similar to that of control rats.

The concentration of some amino acids seems to be altered in the brain of thiamine-deficient animals: It has been reported that the concentration of glutamate, aspartate, and threonine is decreased, while that of glycine is increased (Ferrari, 1957; Gaitonde *et al.,* 1974; Gaitonde, 1975). Gubler *et al.* (1974) observed a significant reduction in the levels of GABA and glutamate, whereas the activities of enzymes associated with the GABA bypath were not affected. The decrease in brain concentration of threonine, an essential amino acid, is noteworthy. Increased catabolism of [U-^{14}C]threonine to [^{14}C]lactate through the amino acetone pathway (catalyzed by threonine dehydrogenase), and to succinate via propionate by the α-oxobutyrate pathway (catalyzed by threonine dehydrates) has been reported in the brain of thiamine-deficient rats (Gaitonde, 1975).

5.1.2. Pyridoxine

Dietary deficiency of pyridoxine was shown to affect the postnatally developing human nervous system by Snyderman *et al.* in 1953. Soon afterwards, an outbreak of seizures accompanied by irritability occurred in bottle-fed children. Pyridoxine was found to be deficient in the artificial milk preparation used and both the clinical and electrophysiological abnormalities were corrected when given by injection (Coursin, 1954; Moloney and Parmalee, 1954). Response to anticonvulsants was incomplete; in contrast, the therapeutic effect of the vitamin was complete and rapid, suggesting that the syndrome was due to lack of adequate amounts of pyridoxal phosphate in the brain.

A dietary-induced vitamin-B_6 deficiency in man produces a marked effect on the metabolism of tryptophan and of methionine (Shin and Linkswiler, 1974). The increased urinary excretion of certain tryptophan metabolites of the kynurenine pathway following loading of tryptophan and of cystathionine following loading with methionine has been observed in pyridoxine-deficient patients. In experimental animals the concentration of GABA is markedly decreased in the pyridoxine-deficient brain (Tews and Lovell, 1967; Tews, 1969; Stephens *et al.,* 1971; Bayoumi *et al.,* 1972). The conversion of glutamic acid to GABA and the subsequent formation to succinic semialdehyde are catalyzed by GAD and GABA-transaminase, respectively, and both of these enzymes require pyridoxal phosphate as coenzyme (Awapara *et al.,* 1950; Roberts *et al.,* 1951; Baxter and Roberts, 1958). In the adult B_6-deficient brain the activity of GAD, when assayed without added coenzyme, is reduced to a greater extent than GABA-transaminase, presumably because of a greater degree of dissociation of pyridoxal phosphate from the enzyme protein (Massieu *et al.,* 1962). However, the GAD apoenzyme content, which is assayed with an excess of the exogenous coenzyme, is unaffected under these conditions (Bayoumi and Smith, 1973). In contrast, when B_6-deficiency is induced during brain development, there is a tendency for a reduction in the GABA-transaminase activity and for an increase in the apparent GAD apoenzyme content (Stephens *et al.,* 1971; Bayoumi and Smith, 1972). The degree of the increase in the reported GAD values shows great variation above the control level (e.g., even from the same group mean values for the whole brain ranging from a 24% to a 140% rise have been reported in different communications, Bayoumi and Smith, 1972, 1973), and it has not yet been established whether the effect is due to changes in kinetic properties or in the content of the enzyme protein. A regulation of the biosynthesis and/or the catabolism of the GAD apoenzyme either by pyridoxal phosphate (Bayoumi and Smith, 1973), or by GABA, the product of the enzyme reaction (Sze, 1970), the concentration of which is also depressed (see above), has been suggested.

In the developing brain, pyridoxine deficiency seems to affect specifically the GAD apoenzyme content: the level of the following enzymes has been found unaltered—aromatic L-amino acid decarboxylase, 5-hydroxytryptamine decarboxylase, GABA transaminase, aspartate aminotransferase, alanine aminotransferase, and tyrosine aminotransferase (Eberle and Eiduson, 1968; Eiduson *et al.*, 1972; Bayoumi and Smith, 1972, 1973).

Pyridoxine deficiency also affects the concentration of serine and glycine in the adult brain (Tews, 1969). The effect on serine levels can result from pyridoxal phosphate requirement for both the synthesis and the catabolism of this amino acid, whereas the changes in glycine concentration may be related to the relative proportion of serine which is converted to glycine and to pyruvate.

Pyridoxine deficiency also influences lipid metabolism in the developing brain. It has been reported that under these conditions the central concentrations of sphingomyelin and cerebrosides are reduced, in comparison with controls, in 21-day-old rats (Williamson and Coniglio, 1971) and that during brain development the decrease in sphingolipids is relatively greater than that in glycerophosphatides and plasmalogens which are also reduced (Kurtz *et al.*, 1972). Vitamin B_6 is involved in the biosynthesis of sphingosine and is implicated in the conversion of phosphatidyl-serine to phosphotidylethanolamine (Brady *et al.*, 1958; Braun and Snell, 1968; Borkenhagen *et al.*, 1961). The activity of 3-ketodihydrosphingosine synthetase is severely impaired in B_6-deficient rats (Kurtz and Kanfer, 1973). These authors have also observed that the amount of myelin is reduced in vitamin-B_6-deficient animals. However, the lipid composition of myelin was similar to that in controls except for a marked reduction of polyunsaturated fatty acids in phospholipids.

The young of pyridoxine-deficient mothers are relatively small, their motor development is retarded, and they show spontaneous convulsions (Alton-Mackey and Walker, 1973; Dakshinamurti and Stephens, 1969). Digital defects, cleft palate, omphalocoele, micrognathia, exencephaly, and hypoplasia of the spleen and thymus (Davis *et al.*, 1970; Miller, 1972) are among the congenital abnormalities reported.

Certain systemic changes induced by pyridoxine deficiency could have repercussions on brain growth. Thus vitamin B_6 deficiency has been shown to inhibit both cellular and humoral immunity and results in a reduction of the rate of synthesis of liver, spleen, and serum proteins (Axelrod and Trakatellis, 1964; Davis, 1974).

5.1.3. Vitamin B_{12}

The neurology of vitamin B_{12} deficiency in man has been reviewed recently (Pallis and Lewis, 1974). It would appear that in two of the conditions in which this deficiency may occur early in life, i.e., congenital intrinsic factor deficiency and congenital malabsorption of vitamin B_{12}, the developing brain may be affected.

Cases of congenital intrinsic factor deficiency described by Reisner *et al.* (1951) and Pearson *et al.* (1964) were infants who developed severe and progressive neurological disease which improved dramatically after treatment with injections containing vitamin B_{12}. Central nervous system involvement in congenital malabsorption of vitamin B_{12} has been reported by Francois *et al.* (1967) and by Brizard and Fraisse (1973); these children are of an older age group and have clinical features more like those of subacute combined degeneration in adults.

Dietary deficiency of vitamin B_{12} affecting the nervous system may occur in infants. Jadhav *et al.* (1962) reported from Vellore six cases of an unusual syndrome seen in breast-fed infants of vitamin-B_{12}-deficient mothers. The milk of such mothers is vitamin B_{12} deficient, and the infants show apathy, mucosal pallor, hyperpigmentation of the extremities, developmental regression, and complex involuntary movements of head, trunk, and limbs, The infants, all less than 12 months old, were well nourished but had megaloblastic bone marrow, low serum vitamin B_{12} levels, and normal B_{12} absorption. Response to small oral doses of B_{12} was good.

Apart from any specific effects it may have on the developing brain, vitamin B_{12} has a more general role in development. The studies of Neuberne and Young (1973) have shown that in mother rats made marginally B_{12} deficient by the prevention of coprophagy, a high dietary B_{12} intake produces increased birth weight and greater capacity to withstand infections.

5.2. Fat-Soluble Vitamins

5.2.1. Vitamin A

Deficiency of vitamin A is usually associated with protein deficiency (McLaren, 1970); the vitamin in its active form is found only in food of animal origin and, as its β-carotene precursor, in a wide variety of green vegetables. Both types of food are lacking in the impoverished, predominantly carbohydrate diet of much of the Third World. Vitamin A plays an important role in retinal physiology, being responsible for sensitivity to light and in dark adaptation. Jeghers (1937) showed that night blindness was the earliest and most constant manifestation of vitamin A deficiency; this—even without causing severe eye disease—may thus affect the normal behavioral development of children by restricting input from their environment.

The vitamin also appears to be necessary for the normal anatomical development of the nervous system. Deficiency has been implicated as a cause of mental retardation and raised intracranial pressure in man (Cornfeld and Cooke, 1952; Bass and Caplan, 1955; Bass and Fisch, 1961).

In pigs a variety of degenerative nerve fiber lesions have been described in the peripheral and central nervous system. The early literature in this field has been reviewed by Moore (1957), whose conclusion is that these lesions can usually, if not always, be attributed to skeletal abnormalities induced by hypovitaminosis A. Certainly malformations of the skull have been described in puppies brought up on vitamin-A-deficient diets, and these may have adverse effects on the developing brain by physically restricting its growth. Mellanby (1947) dramatically described these blind, deaf, and anosmic animals, in which "the skull and vertebrae had lost their primary function of protecting the nervous system and had become destructive agents." Mellanby's view that these neurological changes were entirely consequences of increased bone formation has been supported by the work of Howell and

Thompson (1967), although an effect of vitamin A deficiency on production, circulation, and absorption of CSF in the developing brain—which seems probable from the human cases of raised intracranial pressure—has not been adequately investigated. In rabbits, hydrocephalus was produced by feeding the mothers a diet containing insufficient vitamin A during pregnancy (Millen *et al.*, 1953). Here the cause was neither bony abnormality of the skull nor aqueduct stenosis, and it was concluded that excessive production of CSF from the choroid plexus was the primary event.

Vitamin A excess is a well-known cause of raised intracranial pressure in man, and there is evidence that this condition is associated with altered CSF hydrodynamics. Maddux *et al.* (1974) showed in rats overdosed with the vitamin that capillary permeability in the brain was altered, producing cerebral edema, and that increased CSF outflow occurred. These changes were related by these authors to the *in vitro* action of the vitamin on cell and organelle membranes (Dingle, 1968). It should also be noted that both hydrocephalus and cranial abnormalities have been reported as occurring in animals after large doses of vitamin A (Calhoun *et al.*, 1965), while overdosage during gestation may result in gross malformations of the nervous system (Cohlan, 1953, 1954). Langman and Welch (1966) gave pregnant rats excess of the vitamin on days 8–10 of gestation, i.e., prior to closure of the neural tube, and found a high incidence of anencephaly and exencephaly. The only cellular abnormalities seen were a disturbance of mitotic activity and the appearance of luminal cytoplasmic protrusions in the neuroepithelium. In a subsequent study (Langman and Welch, 1967), mice were treated with large doses of the vitamin on day 14–15 of gestation, and analysis of the cell cycle of the forebrain neuroepithelial cells was made by the percentage labeled mitoses method. Although treated animals were reported as showing prolongation of cell cycle time, duration of mitosis, and length of DNA synthesis, scrutiny of the published data suggests that the rather small changes observed between control and treated animals represent no more than normal interexperiment differences.

5.2.2. Vitamin E

Although deficiency of vitamin E (α-tocopherol) has been regarded as responsible for central nervous changes in developing and mature animals, it is still uncertain whether such a phenomenon occurs in man.

In experimental animals, vitamin E deficiency is well known to produce both a myopathy (which will not be discussed further) and degeneration in the central nervous system. In 1935 Ringsted first described a nervous disorder in rats resulting from a chronic deficiency. Clinical and pathological features were comprehensively reviewed by Einarson and Ringsted (1938), and many confirmatory reports have since appeared (Luttrell and Mason, 1949; Malamud *et al.*, 1949). The essential signs, which evolve over 6–15 months, are a progressive ataxic paraparesis with muscle wasting, hypalgesia, and urinary incontinence, while the underlying pathology is a combined neural and muscular one, atrophy of spinal origin supervening on an initially myopathic process. The earliest central nervous lesions are degeneration of posterior spinal roots and posterior columns with demyelination, loss of axons, and gliosis. Anterior horn cells subsequently show chromatolysis and lipochrome accumulation, and later disappear. Degeneration in the pyramidal tract can also be found.

As well as being responsible for lesions in the already formed nervous system, vitamin E deficiency, when induced in the latter half of pregnancy, may be implicated in such changes in the developing rat brain as encephalocoele and obstructive hydrocephalus (Verma and Wei King, 1967).

Pentschew and Schwarz (1962) and Lampert *et al.* (1964) have found microscopic extracellular eosinophilic bodies in the gracile and cuneate nuclei of rats deprived of dietary vitamin E for months. These bodies appear before obvious dorsal column degeneration occurring distally in long neuronal processes. Similar bodies have been found in children with cystic fibrosis (Sung, 1964) and biliary atresia (Sung and Stadlan, 1966), and it has been argued that since vitamin E deficiency is present in these states (Nitowsky *et al.,* 1956), distal neuronal degeneration of this type is related to hypovitaminosis E. However, such bodies have little specificity, occurring in a variety of spontaneous and induced diseases of the nervous system, and their relationship to vitamin deficiency must be regarded as unproven.

Attempts have been made to link the neurological features of the Bassen-Kornzweig syndrome (acanthocytosis, abetalipoproteinemia) with vitamin E deficiency (for review of this disease, see Pallis and Lewis, 1974). An inherited disorder, with steatorrhea from infancy and clinical involvement of the nervous system from 2 years onwards, the Bassen-Kornzweig syndrome has a neurological component resembling Freidreich's ataxia. Blood levels of vitamin E are too low to be measurable (Kayden *et al.,* 1965). While certain aspects of the disease (increased susceptibility of erythrocytes to autohemolysis when exposed to hydrogen peroxide; abnormal jejunal epithelial ultrastructure) can be related to deficiency of this vitamin, the discrepancy between the neuropathology of the Bassen-Kornzweig syndrome and the experimental findings in dietarily deprived animals suggests that such a relationship does not hold for its central nervous component.

6. Functional Consequences

In this chapter we have summarized the evidence showing that nutritional deficiency can affect the structural and biochemical development of the brain. Some, but by no means all indices of the physical development of the brain point to a reversible influence. However, because of the importance of programmed chronology in development in relation to the ultimate integrative organization of the central nervous system, the apparently reversible nature of some of the alterations produced by undernutrition does not preclude distortions which may lead to abnormalities in brain function (Dobbing, 1974; Altman *et al.,* 1970*a*). Studies on experimental animals have indeed provided evidence for certain aspects of behavior being permanently altered after nutritional insults during the period of brain development (see, e.g., Dobbing and Smart, 1974; Dodge *et al.,* 1975). Research workers in developmental psychology originally turned to animal models for the study of the effects of early undernutrition upon learning ability and behavior because it was thought that by using experimental animals the confounding variables associated with human undernutrition—a combination of social and environmental factors—could be controlled. However, it is now apparent that the interpretation of the findings of these investigations, and their extrapolation to the human situation, is problematic (see, e.g., Crnic, 1976). The procedures currently used to produce nutritional deficiency, such as separation of the young from the mother and increased litter size, also induce major changes in the suckling rats' social and environmental conditions, and these changes may play an important part in the determination of later behavior (see, e.g., Frankova, 1974). Early undernutrition and the variables compounded with it also affect subsequent size, motor ability, behavior toward food, ability to absorb and use food, and to adjust to food deprivation, and responsiveness to environmental stimuli. Thus, the influence of early undernutrition may involve effects upon these other variables which can affect learning performance, rather than upon learning capacity *per se*. It is worth noting here that the effects of undernutrition are not exclusively negative. Some of the permanent behavioral alterations seen in animals undernourished in early life seem to be consistent with a lower threshold for arousal (Smart, 1971; Randt and Derby, 1973), which can possibly be related to structural and biochemical changes (see p. 441 and p. 449).

The effects of early undernutrition on mental functions in man are considered in detail in Chapter 16 of this volume. It is sufficient to mention here that although controlled studies on the effect of undernutrition on the structural and biochemical development of the brain belong mainly to the province of animal experimentation, efforts are now made in several centers to obtain information directly on the human brain and to relate this to other data, such as head circumference and electrophysiology, which have been collected over many years. The limited information currently available indicates that the reduced head circumference after early undernutrition is accompanied by a decrease in brain weight (e.g., Pretorius and Novis, 1965; Garrow *et al.*, 1965; Brown, 1966), which can be linked to a reduction in both the number and size of cells in the brain (for reviews see Dobbing, 1974; Dodge *et al.*, 1975; Winick, 1976). Recent progress in biochemical studies on human cadaver brains may provide important new information on the effects of early undernutrition on brain development in man.

Human undernutrition is only one aspect of the culture of poverty, and its effect on brain development is thus inextricably intertwined with other adverse environmental factors. However, data from numerous research workers in several parts of the world and accumulated over a period of years strongly suggest that, while other factors are also involved, severe nutritional deprivation *per se* in early life impairs human mental development substantially and permanently (see, e.g., Klein *et al.*, 1971; Cravioto and DeLicardie, 1972; Birch, 1972; Hertzig *et al.*, 1972; Pan American Health Organization, 1972; Richardson *et al.*, 1973; Tizard, 1974; Hoorweg and Stanfield, 1976; Stoch and Smythe, 1976). The presentation and analysis of these data are outside the scope of this chapter, in which we have set out to examine only the structural and biochemical disturbances which may underlie the adverse effect of nutritional deprivation on brain function. The findings we have reviewed suggest that undernutrition is a very important nongenetic factor influencing the developing central nervous system and thus ultimately intellectual performance. This conclusion has an important bearing on preventive medicine, but it should also provide the scientific foundation for action at the social and political level for the eradication of a major, and often man-made, handicap.

ACKNOWLEDGMENTS

We thank Drs. M. Berry and R. G. Whitehead for allowing us access to unpublished material.

7. References

Adams, R. L. P., Abrams, R., and Lieberman, I., 1966, Deoxycytidylate synthesis and entry into the period of deoxyribonucleic acid replication in rabbit kidney cells, *J. Biol. Chem.* **241**:903–905.

Addis, T., Poo, L. J., and Lew, W., 1936, The quantities of protein lost by the various organs and tissues of the body during a fast, *J. Biol Chem.* **115**:111–116.

Adlard, B. P. F., and Dobbing, J., 1971a, Vulnerability of developing brain. III. Development of four enzymes in the brains of normal and undernourished rats, *Brain Res.* **28**:97–107.

Adlard, B. P. F., and Dobbing, J., 1971b, Elevated acetylcholinesterase activity in adult rat brain after undernutrition in early life, *Brain Res.* **30**:198–199.

Adlard, B. P. F., and Dobbing, J., 1972, Vulnerability of developing brain. VIII. Regional acetylcholinesterase activity in the brains of adult rats undernourished in early life, *Br. J. Nutr.* **28**:139–143.

Adlard, B. P. F., and Smart, J. L., 1972, Adrenocortical function in rats subjected to nutritional deprivation in early life, *J. Endocrinol.* **54**:99–105.

Adlard, B. P. F., Dobbing, J., Lynch, A., Balázs, R., and Reynolds, A. P., 1972, Effect of undernutrition in early life on glutamate decarboxylase activity in the adult brain, *Biochem. J.* **130**:12P.

Aghajanian, G. K., and Bloom, F. E., 1967, The formation of synaptic junctions in developing rat brain: A quantitative electron microscopic study, *Brain Res.* **6**:716–727.

Agrawal, H. C., Glisson, S. N., and Himwich, W. A., 1966, Changes in monoamines of rat brain during postnatal ontogeny, *Biochim. Biophys. Acta* **130**:511–513.

Agrawal, H. C., Banik, N. L., Bone, A. H., Davison, A. N., Mitchell, R. F., and Spohn, M., 1970, The identity of a myelin-like fraction isolated from developing brain, *Biochem. J.* **120**:635–642.

Agrawal, H. C., Fishman, M. A., and Prensky, A. L., 1971, A possible block in the intermediary metabolism of glucose into proteins and lipids in the brains of undernourished rats, *Lipids* **6**:431–433.

Allen, R. E., Raines, P. L., and Regen, D. M., 1969, Regulatory significance of transfer RNA charging levels. I. Measurements of charging levels in livers of chow-fed rats, fasting rats, and rats fed balanced or imbalanced mixtures of amino acids, *Biochim. Biophys. Acta* **190**:323–336.

Alleyne, G. A. O., and Young, V. H., 1967, Adrenocortical function in children with severe protein–calorie malnutrition, *Clin. Sci.* **33**:189–200.

Altman, J., 1969, DNA metabolism and cell proliferation, in: *Handbook of Neurochemistry,* Vol. 2 (A. Lajtha, ed.), pp. 137–182, Plenum Press, New York.

Altman, J., 1970, Postnatal neurogenesis and the problem of neural plasticity, in: *Developmental Neurobiology* (W. Himwich, ed.), pp. 197–237, Charles C Thomas, Springfield, Illinois.

Altman, J., 1972a, Postnatal development of the cerebellar cortex in the rat. I. The external germinal layer and the transitional molecular layer, *J. Comp. Neurol.* **145**:353–397.

Altman, J., 1972b, Postnatal development of the cerebellar cortex in the rat. II. Phases in the maturation of Purkinje cells and of the molecular layer, *J. Comp. Neurol.* **145**:399–463.

Altman, J., 1972c, Postnatal development of the cerebellar cortex in the rat. III. Maturation of the components of the granular layer, *J. Comp. Neurol.* **145**:465–513.

Altman, J., Das, G. D., and Sudarshan, K, 1970a, The influence of nutrition on neural and behavioural development. I. Critical review of some data on the growth of the body and the brain following dietary deprivation during gestation and lactation, *Dev. Psychobiol.* **3**:281–301.

Altman, J., Sudarshan, K., Das, G. D., McCormick, N., and Barnes, D., 1970b, The influence of nutrition on neural and behavioural development. III. Development of some motor, particularly locomotor patterns during infancy, *Dev. Psychobiol.* **4**:97–114.

Altman, J., Das, G. D., Sudarshan, K., and Anderson, J. B., 1971, The influence of nutrition on neural and behavioural development. II. Growth of body and brain in infant rats using different techniques of undernutrition, *Dev. Psychobiol.* **4**:55–70.

Alton-Mackey, M. G., and Walker, B. L., 1973, Graded levels of pyridoxine in the rat diet during gestation and the physical and neuromotor development of offspring, *Am. J. Clin. Nutr.* **26**:420–428.

Andersson, A., 1975, Synthesis of DNA in isolated pancreatic islets maintained in tissue culture, *Endocrinology* **96**:1051–1054.

Andersson, G. M., and Von der Decken, A., 1975, Deoxyribonucleic acid-dependent ribonucleic acid polymerase activity in rat liver after protein restriction, *Biochem. J.* **148**:49–56.

Andrews, V. L., 1912, Infantile beriberi, *Philipp. J. Sci.* **7B**:67–89.

Arakawa, T., Mizuno, T., Chiba, F., Sakai, K., Watanabe, S., Tamura, T., Tatsumi, S., and Coursin, D. B., 1968, Frequency analysis of electroencephalograms and latency of photically induced average evoked responses in children with ariboflavinosis (preliminary report), *Tohoku J. Exp. Med.* **94**:327–335.

Armendares, S., Salamanca, F., and Frenk, S., 1971, Chromosome abnormalities in severe protein calorie malnutrition, *Nature* **232:**271–273.

Armstrong-James, M., and Johnson, R., 1970, Quantitative studies of postnatal changes in synapses in rat superficial motor cerebral cortex, *Z. Zellforsch.* **110:**559–568.

Awapara, J., Landua, A. J., Fuerst, R., and Seale, B., 1950, Free γ-aminobutyric acid in brain, *J. Biol. Chem.* **187:**35–39.

Axelrod, A. E., and Trakatellis, A. C., 1964, Relationship of pyridoxine to immunological phenomena, *Vitam. Horm.* **22:**591–607.

Azmitia, E. C., Jr., and McEwen, B. S., 1974, Adrenalcortical influence on rat brain tryptophan hydroxylase activity, *Brain Res.* **78:**291–302.

Bailey, R. P., Vrooman, M. J., Sawai, Y., Tsukada, K., Short, J., and Lieberman, I., 1976, Amino acids and control of nucleolar size, the activity of RNA polymerase I, and DNA synthesis in liver, *Proc. Natl. Acad. Sci. U.S.A.* **73:**3201–3205.

Balázs, R., 1972*a*, Hormonal aspects of brain development, in: *The Brain in Unclassified Mental Retardation* (J. B. Cavanagh, ed.), IRMR Study Group No. 3, pp. 61–72, Churchill Livingstone, London.

Balázs, R. 1972*b*, Effects of hormones and nutrition on brain development, in: *Human Development and the Thyroid Gland: Relation to Endemic Cretinism, Advances in Experimental Medicine and Biology,* Vol. 30 (J. B. Stanbury and R. L. Kroc, eds.), pp. 385–415, Plenum Press, New York.

Balázs, R., 1977, Effect of thyroid hormone and undernutrition on cell acquisition in the rat brain, in: *Thyroid Hormone and Brain Development* (G. D. Grave, ed.), pp. 288–298, Raven Press, New York.

Balázs, R., and Haslam, R. J., 1965, Exchange transamination and the metabolism of glutamate in brain, *Biochem. J.* **94:**131–142.

Balázs, R., and Patel, A. J., 1973, Factors affecting the biochemical maturation of the brain. Effect of undernutrition during early life, in: *Neurobiological Aspects of Maturation and Ageing, Progress in Brain Research,* Vol. 40 (D. H. Ford, ed.), pp. 115–128, Elsevier, Amsterdam.

Balázs, R., Brooksbank, B. W. L., Patel, A. J., Johnson, A. L., and Wilson, D. A., 1971, Incorporation of [^{35}S]sulphate into brain constituents during development and the effects of thyroid hormone on myelination, *Brain Res.* **30:**273–293.

Balázs, R., Lewis, P. D., and Patel, A. J. 1975, Effects of metabolic factors on brain development, in: *Growth and Development of the Brain* (M. B. Brazier, ed.), IBRO Monograph Series, Vol. 1, pp. 83–115, Raven Press, New York.

Balázs, R., Patel, A. J., and Lewis, P. D., 1977, Metabolic influences on cell proliferation in the brain, in: *Biochemical Correlates of Brain Structure and Function* (A. N. Davison, ed.), pp. 43–83, Academic Press, New York.

Banker, G., Churchill, L., and Cotman, C. W., 1974, Proteins of the post-synaptic density, *J. Cell Biol.* **63:**456–465.

Barbiroli, B., Tadolini, B., Moruzzi, M. S., and Monti, M. G., 1975, Modification of the template capacity of liver chromatin for form-B RNA polymerase by food-intake in rats under controlled feeding schedules, *Biochem. J.* **146:**687–696.

Baril, E., Baril, B., Elford, H., and Luftig, R. B. 1974, DNA polymerases and a possible multienzyme complex for DNA biosynthesis in eukaryotes, in: *Mechanism and Regulation of DNA Replication* (A. R. Kolber and M. Kohiyama, eds.), pp. 275–291, Plenum Press, New York.

Barnes, D., and Altman, J. 1973*a*, Effects of different schedules of early undernutrition on the preweaning growth of the rat cerebellum, *Exp. Neurol.* **38:**406–419.

Barnes, D., and Altman, J., 1973*b*, Effects of two levels of gestational–lactational undernutrition on the postweaning growth of the rat cerebellum, *Exp. Neurol.* **38:**420–428.

Bass, M. H., and Caplan, J., 1955, Vitamin A deficiency in infancy, *J. Pediatr.* **47:**690–695.

Bass, M. H., and Fisch, G. R., 1961, Increased intracranial pressure with bulging fontenel. A symptom of vitamin A deficiency in infants, *Neurology* **11:**1091–1094.

Bass, N. H., Netsky, M. G., and Young, E., 1970*a*, Effect of neonatal malnutrition on developing cerebrum. I. Microchemical and histologic study of cellular differentiation in the rat, *Arch. Neurol.* **23:**289–302.

Bass, N. H., Netsky, M. G., and Young, E., 1970*b*, Effect of neonatal malnutrition on developing cerebrum. II. Microchemical and histologic study of myelin formation in the rat, *Arch. Neurol.* **23:**303–313.

Baxter, C. F., and Roberts, E., 1958, The γ-aminobutyric acid–α-ketoglutaric acid transaminase of beef brain, *J. Biol. Chem.* **233:**1135–1139.

Bayoumi, R. A., and Smith W. R. D., 1972, Some effects of dietary vitamin B$_6$ deficiency on γ-aminobutyric acid metabolism in developing rat brain, *J. Neurochem.* **19:**1883–1897.

Bayoumi, R. A., and Smith, W. R. D., 1973, Regional distribution of glutamic acid decarboxylase in the developing brain of the pyridoxine-deficient rat, *J. Neurochem.* **21**:603–613.

Bayoumi, R. A., Kirwan, J. R., and Smith, W. R. D., 1972, Some effects of dietary vitamin B₆ deficiency and 4 deoxypyridoxine γ-aminobutyric acid metabolism in rat brain, *J Neurochem.* **19**:569–576.

Bennett, C. D., Jones, J. H., and Nelson, J., 1966, The effects of thiamine deficiency on the metabolism of the brain. I. Oxidation of various substrates *in vitro* by the liver and brain of normal and pyrithiamine-fed rats, *J. Neurochem.* **13**:449–459.

Benton, J. W., Moser, H. W., Dodge, P. R., and Carr, S., 1966, Modification of the schedule of myelination in the rat by early nutritional deprivation, *Pediatrics* **38**:801–807.

Betancourt, M., de la Roca, J. M., Sáenz, M. E., Diaz, R., and Cravioto, J., 1974, Chromosome aberrations in protein–calorie malnutrition, *Lancet* **1**:168.

Birch, H. G., 1972, Malnutrition, learning and intelligence, *Am. J. Public Health* **62**:773–784.

Blumenthal, A. B., Kriegstein, H. J., and Hogness, D. J., 1974, The units of DNA replication in Drosophila melanogaster chromosomes, *Cold Spring Harbor Symp. Quant. Biol.* **38**:205–223.

Borkenhagen, L. F., Kennedy, E. P., and Fielding, L., 1961, Enzymatic formation and decarboxylation of phosphatidylserine, *J. Biol. Chem.* **236**:PC28–PC30.

Brady, R. O., Formica, J. V., and Koval, G. J., 1958, The enzymatic synthesis of sphingosine. II. Further studies on the mechanism of the reaction, *J. Biol. Chem.* **233**:1072–1076.

Brante, G., 1949, Studies on lipids in the nervous system with special reference to quantitative chemical determination and topical distribution, *Acta Physiol. Scand.* **18**(Suppl. 63):1–189.

Braun, P. E., and Snell, E. E., 1968, Biosynthesis of sphingolipid bases II. Keto intermediates in synthesis of sphingosine and dihydrosphingosine by cell-free extracts of *Hansenula ciferri*, *J. Biol. Chem.* **243**:3775–3783.

Brazier, M. A. B., ed., 1975, *Growth and Development of the Brain*, IBRO Monograph Series, Vol. 1, Raven Press, New York.

Brinkley, B. R., Fuller, G. M., and Highfield, D. P., 1975, Cytoplasmic microtubules in normal and transformed cells in culture: Analysis by tubulin antibody immunofluorescence, *Proc. Natl. Acad. Sci. U.S.A.* **72**:4981–4985.

Brizard, C. P., and Fraisse, H., 1973, Maladie d'Imerslund-Najman-Grasbeck avec syndrome neuroanémique, *Lyon Med.* **229**:59–63.

Brosky, G. M., Kern, H. F., and Logothetopoulos, J., 1972, Ultrastructure, mitotic activity and insulin biosynthesis of pancreatic beta cells stimulated by glucose infusions, *Fed. Proc.* **31**:256 (abstract).

Brown, R. E., 1966, Organ weight in malnutrition with special reference to brain weight, *Dev. Med. Child Neurol.* **8**:512–522.

Brožek, J., and Vaes, G., 1961, Experimental investigations on the effects of dietary deficiencies on animal and human behavior, *Vitam. Horm.* **19**:43–94.

Calhoun, M. C., Rousseau, J. E., Jr., Hall, R. C., Jr., Eaton, H. D., Nelson, S. W., and Lucas, J. J., 1965, Cisternal cerebrospinal fluid pressure during development of chronic bovine hypervitaminosis, *J. Dairy Sci.* **48**:729–732.

Callan, H. G., 1972, Replication of DNA in the chromosome of eukaryotes, *Proc. R. Soc. London, Ser. B* **181**:19–41.

Carlson, A., Davis, J. N., Kehr, W., Lindquist, M., and Atack, C. V., 1972, Simultaneous measurement of tyrosine and tryptophan hydroxylase activities in brain *in vivo* using an inhibitor of the aromatic amino acid decarboxylase, *Arch. Pharmacol.* **275**:153–168.

Chase, H. P., Dorsey, J., and McKhann, G. M., 1967, The effect of malnutrition on the synthesis of a myelin lipid, *Pediatrics* **40**:551–559.

Chase, H. P., Lindsley, W. F. B., and O'Brien, D., 1969, Undernutrition and cerebellar development, *Nature* **221**:554–555.

Chase, H. P., Rodgerson, D. O., Lindsley, W., Jr., Thorne, T., and Cheung, G., 1976, Brain glucose utilization in undernourished rats, *Pediatr. Res.* **10**:102–107.

Cheek, D. B., and Hill, D. E., 1970, Muscle and liver cell growth: Role of hormones and nutritional factors, *Fed. Proc.* **29**:1503–1509.

Cheney, D. L., Gubler, C. J., and Jaussi, A. W., 1969, Production of acetylcholine in rat brain following thiamine deprivation and treatment with thiamine antagonists, *J. Neurochem.* **16**:1283–1291.

Chick, W. L., Lauris, V., Flewelling, J. H., Andrews, K. A., and Woodruff, J. M., 1973, Effects of glucose on beta cells in pancreatic monolayer cultures, *Endocrinology* **92**:212–218.

Clark, A. J., and Jacob, M., 1972, Effect of diet on RNA polymerase activity in rats, *Life Sci.* **11**:(Part 1):1147–1154.

Cleaver, J. E., 1967, *Thymidine Metabolism and Cell Kinetics*, North-Holland, Amsterdam.

Clendinnen, B. G., and Eayrs, J. T., 1961, The anatomical and physiological effects of prenatally administered somatotrophin on cerebral development in rats, *J. Endocrinol.* **22**:183–193.

Clos, J., and Legrand, J., 1969, Influence de la déficience thyroidienne et de la sous-alimentation sur la croissance et la myélinisation des fibres nerveuses de la moelle cervicale et du nerf sciatique chez le jeune rat blanc, *Arch. Anat. Micro. Morphol. Exp.* **58:**339–354.

Clos, J., and Legrand, J., 1970, Influence de la déficience thyroidienne et de la sous-alimentation sur la croissance et la myélinisation des fibres nerveuses du nerf sciatique chez le jeune rat blanc. Etude au microscope électronique, *Brain Res.* **22:**285–297.

Clos, J., and Legrand, J., 1973, Effects of thyroid deficiency on the different cell population of the cerebellum in the young rat, *Brain Res.* **63:**450–455.

Clos, J., Rebière, A., and Legrand, J., 1973, Differential effects of hypothyroidism and undernutrition on the development of glia in the rat cerebellum, *Brain Res.* **63:**445–449.

Clos, J., Favre, C., Selme-Matrat, M., and Legrand, J., 1977, Effects of undernutrition on cell formation in the rat brain and specially on cellular composition of the cerebellum, *Brain Res.* **123:**13–26.

Cocks, J. A., Balázs, R., Johnson, A. L., and Eayrs, J. T., 1970, Effect of thyroid hormone on the biochemical maturation of rat brain: Conversion of glucose-carbon into amino acids, *J. Neurochem.* **17:**1275–1285.

Cohlan, S. Q., 1953, Excessive intake of vitamin A as a cause of congenital anomalies in the rat, *Science* **117:**535–536.

Cohlan, S. Q., 1954, Congenital anomalies in the rat produced by excessive intake of vitamin A during pregnancy, *Pediatrics* **13:**556–567.

Cornfield, D., and Cooke, R. E., 1952, Vitamin A deficiency: Case report, *Pediatrics* **10:**33–39.

Costa, E., Gessa, G. L., and Sandler, M., eds., 1974, Serotonin—new vistas: Biochemistry and behavioral and clinical studies, *Advances in Biochemistry and Psychopharmacology,* Vol. 11, Raven Press, New York.

Cotman, C. W., and Nadler, J. V., 1978, Reactive synaptogenesis in the hippocampus, in: *Neuronal Plasticity* (C. W. Cotman, ed.), pp. 227–271, Raven Press, New York.

Cotterell, M., 1971, The effects of growth hormone and corticosteroid treatment on the biochemical maturation of rat brain, MSc thesis, Council for National Academic Awards, U.K.

Coursin, D. B., 1954, Convulsive seizures in infants with pyridoxine-deficient diet, *JAMA* **154:**406–408.

Coyle, J. T., and Axelrod, J., 1972a, Dopamine-β-hydroxylase in the rat brain: Developmental characteristics, *J. Neurochem.* **19:**449–459.

Coyle, J. T., and Axelrod, J., 1972b, Tyrosine hydroxylase in rat brain: Developmental characteristics, *J. Neurochem.* **19:**1117–1123.

Cragg, B. G., 1970, Synapses and membranous bodies in experimental hypothyroidism, *Brain Res.* **18:**297–307.

Cragg, B. G., 1972, The development of cortical synapses during starvation in the rat, *Brain* **95:**143–150.

Cravioto, J., and DeLicardie, E. R., 1972, Environmental correlates of severe clinical malnutrition and language development in survivors from kwashiorkor or marasmus, in: Pan American Health Organization, *Nutrition, the Nervous System and Behaviour,* pp. 73–94, Scientific Publ. No. 251, Washington, D.C.

Cravioto, J., Hambraens, L., and Vahlquist, B., eds., 1974, *Early Malnutrition and Mental Development,* Almqvist and Wiksell, Uppsala.

Cravioto, H. M., Randt, C. T., Derby, B. M., and Diaz, A., 1976, A quantitative ultrastructural study of synapses in the brains of mice following early life undernutrition, *Brain Res.* **118:**304–306.

Cremer, J. E., 1972, Changes in enzyme activity and the developing brain, in: *The Brain in Unclassified Mental Retardation* (J. B. Cavanagh, ed.), pp. 77–87, Churchill Livingstone, London.

Cremer, J. E., and Heath, D. F., 1974, The estimation of rates of utilization of glucose and ketone bodies in the brain of the suckling rat using compartmental analysis of isotopic data, *Biochem. J.* **142:**527–544.

Creps, E. S., 1974, Time of neuron origin in preoptic and septal areas of the mouse: An autoradiographic study, *J. Comp. Neurol.* **157:**161–243.

Crnic, L. S., 1976, Effects of infantile undernutrition on adult learning in rats: Methodological and design problems, *Psychol. Bull.* **83:**715–728.

Cronly-Dillon, J. R., and Perry, G. W., 1976, Tubulin synthesis in developing rat visual cortex, *Nature* **261:**581–583.

Cruickshank, E. K., 1976, Effects of malnutrition on the central nervous system and the nerves, in: *Handbook of Chemical Neurology* (P. J. Vinken and G. W. Bruyn, eds.), Vol. 28, Metabolic and Deficiency Diseases of the Nervous System, Part II, pp. 1–41, North-Holland, Amsterdam.

Culley, W. J., and Mertz, E. T., 1965, Effect of restricted food intake on growth and composition of preweanling rat brain, *Proc. Soc. Exp. Biol. Med.* **118:**233–235.

Culley, W. J., and Lineberger, R. O., 1968, Effect of undernutrition on the size and composition of the rat brain, *J. Nutr.* **96:**375–381.

Dahlquist, G., and Persson, B., 1976, The rate of cerebral utilization of glucose, ketone bodies, and oxygen: A comparative *in vivo* study of infant and adult rats, *Pediatr. Res.* **10:**910–917.

Dakshinamurti, K., and Stephens, M. C., 1969, Pyridoxine deficiency in the neonatal rat, *J. Neurochem.* **16:**1515–1522.

Daniels, M. P., 1972, Colchicine inhibition of nerve fiber formation *in vitro, J. Cell Biol.* **53:**164–176.

David, R. B., Mamunes, P., and Rodsenblum, I., 1976, Necrotizing encephalomyelopathy (Leigh), in: *Handbook of Chemical Neurology* (P. J. Vinken and G. W. Bruyn, eds.), Vol. 28, Metabolic and Deficiency Diseases of the Nervous System. Part II, pp. 349–363, North-Holland, Amsterdam.

Davis, S. D., 1974, Immunodeficiency and runting syndrome in rats from congenital pyridoxine deficiency, *Nature* **251:**548–550.

Davis, S. D., Nelson, T., and Shepard, T. H., 1970, Teratogenicity of vitamin B_6 deficiency: Omphalocele, skeletal and neural defects, and splenic hypoplasia, *Science* **169:**1329–1330.

Davison, A. N., Cuzner, M. L., Banik, N. L., and Oxberry, J., 1966, Myelinogenesis in the rat brain, *Nature* **212:**1373–1374.

De Guglielmone, A. E. R., Soto, A. M., and Duvilanski, B. H., 1974, Neonatal undernutrition and RNA synthesis in developing rat brain, *J. Neurochem.* **22:**529–533.

Del Cerro, M., and Swarz, J. R., 1976, Prenatal development of Bergmann glial fibres in rodent cerebellum, *J. Neurocytol.* **5:**669–676.

Deo, M. G., Bijlani, V., and Ramalingaswami, V., 1975, Nutrition and cellular growth and differentiation, in: *Growth and Development of the Brain,* IBRO Monograph Series, Vol. 1 (M. A. B. Brazier, ed.), pp. 1–16, Raven Press, New York.

DeRobertis, E., and Rodríguez DeLores Arnaiz, G., 1969, Structural components of the synaptic region, in: *Handbook of Neurochemistry,* Vol. 2 (A. Lajtha, ed.), pp. 365–392, Plenum Press, New York.

DeVivo, D. C., Leckie, M. P., and Agrawal, H. C., 1975, D-β-Hydroxybutyrate: A major precursor of amino acids in developing rat brain, *J. Neurochem.* **25:**161–170.

Dickerson, J. W. T., and Pao, S.-K., 1975, Effect of pre- and post-natal maternal protein deficiency on free amino acids and amines of rat brain, *Biol. Neonat.* **25:**114–124.

Dingle, J. T., 1968, Vacuoles, vesicles and lysosomes, *Br. Med. Bull.* **24:**141–145.

Dobbing, J., 1964, The influence of early nutrition on the development and myelination of the brain, *Proc. R. Soc. London, Ser. B* **159:**503–509.

Dobbing, J., 1968, Effects of experimental undernutrition on development of the nervous system, in: *Malnutrition, Learning and Behaviour* (N. S. Scrimshaw and J. E. Gordon, eds.), pp. 181–202, The MIT Press, Cambridge, Massachusetts.

Dobbing, J., 1974, The later development of the brain and its vulnerability, in: *Scientific Foundations of Pediatrics* (J. A. Davis and J. Dobbing, eds.), pp. 565–577, Heinemann, London.

Dobbing, J., and Sands, J., 1973, The quantitative growth and development of human brain, *Arch. Dis. Child.* **48:**757–767.

Dobbing, J., and Smart, J., 1974, Vulnerability of developing brain and behaviour, *Br. Med. Bull.* **30:**164–168.

Dobbing, J., Hopewell, J. W., and Lynch, A., 1971, Vulnerability of developing brain: VII. Permanent deficit of neurons in cerebral and cerebellar cortex following early mild undernutrition, *Exp. Neurol.* **32:**439–447.

Dodge, P. R., Prensky, A. L., and Feigin, R. D., 1975, *Nutrition and the Developing Nervous System,* C. V. Mosby, St. Louis, Missouri.

Donaldson, H. H., 1911, The effect of underfeeding on the percentage of water, on the ether–alcohol extract, and on medullation in the central nervous system of the albino rat, *J. Comp. Neurol.* **21:**139–145.

Dreyfus, P. M., and Hauser, G., 1965, The effect of thiamine deficiency on the pyruvate decarboxylase system of the central nervous system, *Biochim. Biophys. Acta* **104:**78–84.

Driskell, J. A., and Foshee, D. P., 1974, Behavioral patterns and brain nucleic acid and pyridoxal phosphate contents of male rat progeny from vitamin B_6-repleted dams, *J. Nutr.* **104:**810–818.

Dyson, S. E., and Jones, D. G., 1976, Some effects of undernutrition on synaptic development—a quantitative ultrastructural study, *Brain Res.* **114:**365–378.

Eagle, H., 1965, Metabolic controls in cultured mammalian cells, *Science* **148:**42–51.

Eayrs, J. T., and Goodhead, B., 1959, Postnatal development of the cerebral cortex in the rat, *J. Anat.* **93:**385–402.

Eayrs, J. T., and Horn, G., 1955, The development of cerebral cortex in hypothyroid and starved rats, *Anat. Rec.* **121:**53–61.

Eberle, E. D., and Eiduson, S., 1968, Effect of pyridoxine deficiency on aromatic L-amino acid decarboxylase in the developing rat liver and brain. *J. Neurochem.* **15:**1071–1083.

Eccles, J. C., Ito, M., and Szentagothai, J., 1967, *The Cerebellum as a Neuronal Machine,* Springer, Berlin.

Edelman, G. M., 1976, Surface modulation in cell recognition and cell growth, *Science* **192**:218–226.

Eiduson, S., Yuwiler, A., and Eberle, E. D., 1972, The effect of pyridoxine deficiency on L-aromatic amino acid decarboxylase and tyrosine aminotransferase in developing brain, *Adv. Biochem. Psychopharmacol.* **4**:63–80.

Einarson, L., and Ringsted, A., 1938, *Effect of Chronic Vitamin E Deficiency on the Nervous System and the Skeletal Musculature in Adult Rats: A Neurotropic Factor in Wheat Germ Oil,* Munksgaard, Copenhagen.

Elliott, K., and Knight, J., eds., 1972, *Lipids, Malnutrition and the Developing Brain,* CIBA Symposium, Series 3, Elsevier, Amsterdam.

Enna, S. J., Yamamura, H. I., and Snyder, S. H., 1976, Development of muscarinic cholinergic and GABA receptor binding in chick embryo brain, *Brain Res.* **101**:177–183.

Escobar, A., 1974, Cytoarchitectonic derangement in the cerebral cortex of the undernourished rat, in: *Early Malnutrition and Mental Development* (J. Cravioto, L. Hambraens, and B. Vahlquist, eds.), pp. 55–59, Almqvist and Wiksell, Uppsala.

Fangman, W. L., and Neidhardt, F. C., 1964, Protein and ribonucleic acid synthesis in a mutant of *Escherichia coli* with an altered aminoacyl ribonucleic acid synthetase, *J. Biol. Chem.* **239**:1844–1847.

Feit, H., Dutton, G. R., Barondes, S. H., and Shelanski, M. L., 1971, Microtubule protein. Identification in and transport to nerve endings, *J. Cell Biol.* **51**:138–147.

Fellous, A., Francon, J., Lennon, A. M., Nunez, J., Osty, J., and Chantoux, F., 1976, Initiation of neurotubulin polymerisation and rat brain development, *FEBS Lett.* **64**:400–403.

Fernstrom, J. D., 1976, The effect of nutritional factors on brain amino acid levels and monoamine synthesis, *Fed. Proc.* **35**:1151–1156.

Fernstrom, J. D., and Wurtman, R. J., 1971, Brain serotonin content: Physiological dependence on plasma tryptophan levels, *Science* **173**:149–152.

Ferrari, V., 1957, The metabolic changes in thiamine deficiency as reflected on the individual free amino acids in tissues. I. Observations on brain, *Acta Vitam. Milano* **11**:53–66.

Fish, I., and Winick, M., 1969, Cellular growth in various regions of the developing rat brain, *Pediatr. Res.* **3**:407–412.

Fishman, M. A., Prensky, A. L., and Dodge, P. R., 1969, Low content of cerebral lipids in infants suffering from malnutrition, *Nature* **221**:552–553.

Fishman, M. A., Madyastha, P., and Prensky, A. L., 1971, The effect of undernutrition on the development of myelin in the rat central nervous system, *Lipids* **6**:458–465.

Fleck, A., Shepherd, J., and Munro, H. N., 1965, Protein synthesis in rat liver: Influence of amino acids in diet on microsomes and polysomes, *Science* **150**:628–629.

Folch, P. J., 1955, Composition of the brain in relation to maturation, in: *Biochemistry of the Developing Nervous System* (H. Waelsch, ed.), pp. 121–136, Academic Press, New York.

Fonnum, F., 1973, Localization of cholinergic and γ-aminobutyric acid containing pathways in brain, in: *Metabolic Compartmentation in the Brain* (R. Balázs and J. E. Cremer, eds.), pp. 245–257, Macmillan, London.

Fonte, V., and Porter, K. R., 1974, Topographic changes associated with the viral transformation of normal cells to tumorigenicity, in: *8th International Conference on Electron Microscopy* (J. V. Sanders and D. J. Goodchild, eds.), Vol. 2, pp. 334–335, Australian Academy of Science, Canberra.

Francois, P., Revol, L., Germain, D., and Bourlier, V., 1967, Le syndrome d'Imerslund, *Ann. Pediatr. (Sem. Hop. de Paris)* **43**:490–503.

Frankova, S., 1974, Interaction between early malnutrition and stimulation in animals, in: *Early Malnutrition and Mental Development* (J. Cravioto, L. Hambraeus, and B. Vahlquist, eds.), *XII Symp. Swedish Nutrition Foundation,* pp. 203–209, Almqvist and Wiksell, Uppsala.

Freedman, A. D., Rumsey, P., and Graff, S., 1960, The metabolism of pyruvate in the tricarboxylic acid cycle, *J. Biol. Chem.* **235**:1854–1855.

Friedman, P. A., Kappelman, A. H., and Kaufman, S., 1972, Partial purification and characterization of tryptophan hydroxylase from rabbit hindbrain, *J. Biol. Chem.* **247**:4165–4173.

Fujita, S., 1969, Autoradiographic studies on histogenesis of the cerebellar cortex, in: *Neurobiology of Cerebellar Evolution and Development* (R. Llinas, ed.), pp. 743–747, AMA Education and Research Foundation, Chicago.

Gaetani, S., Mengheri, E., and Spadoni, M. A., 1972, Characteristics of newly formed cytoplasmic RNA in liver of protein depleted amino acid refed rats. *Nutr. Rep. Int.* **6**:75–82.

Gadsdon, D. R., and Emery, J. L., 1976, Some quantitative morphological aspects of post-natal human cerebellar growth, *J. Neurol. Sci.* **29**:137–148.

Gaitonde, M. K., 1975, Conversion of [U-^{14}C]threonine into ^{14}C-labelled amino acids in the brain of thiamin-deficient rats, *Biochem. J.* **150**:285–295.

Gaitonde, M. K., and Richter, D., 1966, Changes with age in the utilization of glucose carbon in liver and brain, *J. Neurochem.* **13**:1309–1316.

Gaitonde, M. K., Nixey, R. W. K., and Sharman, I. M., 1974, The effect of deficiency of thiamine on the metabolism of [U-^{14}C]glucose and [U-^{14}C]ribose and the levels of amino acids in rat brain, *J. Neurochem.* **22**:53–61.

Gambetti, P., Autilio-Gambetti, L., Rizzuto, N., Shafer, B., and Pfaff, L., 1974, Synapses and malnutrition: Quantitative ultrastructural study of rat cerebral cortex, *Exp. Neurol.* **43**:464–473.

Gardner, L. I., and Amacher, P., eds., 1973, *Endocrine Aspects of Malnutrition,* Kroc Foundation Symposium No. 1, The Kroc Foundation, Santa Ynez, California.

Garrow, J. S., Fletcher, K., and Halliday, D., 1965, Body composition in severe infantile malnutrition, *J. Clin. Invest.* **44**:417–425.

Geel, S. E., and Dreyfus, P. M., 1974, Thiamine deficiency encephalopathy in the developing rat, *Brain Res.* **76**:435–445.

Geel, S. E., and Dreyfus, P. M. 1975, Brain lipid composition of immature thiamine-deficient and undernourished rats, *J. Neurochem.* **24**:353–360.

Gelbard, A. S., Kim, J. H., and Perez, A. G., 1969, Fluctuations in deoxycytidine monophosphate deaminase activity during the cell cycle in synchronous populations of HeLa cells, *Biochim. Biophys. Acta* **182**:564–566.

Gelbard, A. S., Perez, A. G., Kim, J. H., and Djordjevic, B., 1971, Effect of x-irradiation on thymidine kinase activity in synchronous populations of HeLa cells, *Radiat. Res.* **46**:334–342.

Giuffrida, A. M., Cambria, A., Vanella, A., and Serra, I., 1975, DNA polymerase and thymidine kinase activity in different regions of rat brain during postnatal development: Effect of undernutrition. Abstracts of 10th FEBS Meeting, Paris, Abstract No. 1321, p. 189.

Godard, C., 1973, Plasma thyrotropin levels in severe infantile malnutrition, in: *Endocrine Aspects of Malnutrition* (L. I. Gardner and P. Amacher, eds.), pp. 221–227, The Kroc Foundation, Santa Ynez, California.

Gopinath, G., Bijlani, V., and Deo, M. G., 1976, Undernutrition and the developing cerebellar cortex in the rat, *J. Neuropathol. Exp. Neurol.* **35**:125–135.

Graham, G. G., and Blizzard, R. M., 1973, Thyroid hormone studies in severely malnourished Peruvian infants and small children, in: *Endocrine Aspects of Malnutrition* (L. I. Gardner and P. Amacher, eds.), pp. 205–219, The Kroc Foundation, Santa Ynez, California.

Grahame-Smith, D. G., 1971, Studies *in vivo* on the relationship between brain tryptophan, brain 5-HT synthesis and hyperactivity in rats treated with a monoamine oxidase inhibitor and L-trytophan, *J. Neurochem.* **18**:1053–1066.

Gray, E. G., 1975, Presynaptic microtubules and their association with synaptic vesicles, *Proc. R. Soc. London, Ser. B* **190**:369–372.

Gubler, C. J. 1961, Studies on the physiological functions of thiamine. I. The effects of thiamine deficiency and thiamine antagonists on the oxidation of α-keto acids by rat tissues, *J. Biol. Chem.* **236**:3112–3120.

Gubler, C. J., 1968, Enzyme studies in thiamine deficiency *Z. Vitaminforsch.* **38**:287–303.

Gubler, C. J., Adams, B. L., Hammond, B., Yuan, E. C., Guo, S. M., and Bennion, M., 1974, Effect of thiamine deprivation and thiamine antagonists on the level of γ-aminobutyric acid and on 2-oxoglutarate metabolism in rat brain, *J. Neurochem.* **22**:831–836.

Guerrero, R. M., 1949, Wernicke's syndrome due to vitamin B deficiency. Report of the disease in two infants, *Am. J. Dis. Child.* **78**:88–96.

Haas, R. J., Werner, J., and Fliedner, T. M., 1970, Cytokinetics of neonatal brain cell development in rats as studied by the 'complete ^{3}H-thymidine labelling' method, *J. Anat.* **107**:421–437.

Hajós, F., Patel, A. J., and Balázs, R., 1973, Effect of thyroid deficiency on the synaptic organization of the rat cerebellar cortex, *Brain Res.* **50**:387–401.

Halas, E. S., Rowe, M. C., Johnson, D. R., McKenzie, J. M., and Sandstead, H. H., 1976, Effects of intrauterine zinc deficiency on subsequent behaviour, in: *Trace Elements in Human Health and Disease,* Vol. 1, Zinc and Copper (A. S. Prada and D. Oberleas, eds.), pp. 327–343, Academic Press, New York.

Haltia, M., 1970, The effect of folate deficiency on neuronal RNA content. A quantitative cytochemical study, *Br. J. Exp. Pathol.* **51**:191–196.

Hanson, H. M., and Simonson, M., 1971, Effects of fetal undernourishment on experimental anxiety, *Nutr. Rep. Int.* **4**:307–314.

Haridas, G., 1937, Infantile beri-beri, *J. Malaya Branch, Br. Med. Assoc.* **1**:26–37.

Harland, P. S. E. G., and Parkin, J. M., 1972, TSH levels in severe malnutrition, *Lancet* **2:**1145.

Haslam, R. J., and Krebs, H. A., 1963, The metabolism of glutamate in homogenates and slices of brain cortex, *Biochem. J.* **88:**566–578.

Hauser, G., 1968, Cerebroside and sulphatide levels in developing rat brain, *J. Neurochem.* **15:**1237–1238.

Hawkins, R. A., Williamson, D. H., and Krebs, H. A., 1971, Ketone-body utilization by adult and suckling rat brain *in vivo, Biochem. J.* **122:**13–18.

Hedley-Whyte, E. T., 1973, Myelination of rat sciatic nerve: Comparison of undernutrition and cholesterol biosynthesis inhibition, *J. Neuropathol. Exp. Neurol.* **32:**284–302.

Heinrich, C. P., Stadler, H., and Weiser, H., 1973, The effect of thiamine deficiency on the acetylcoenzymea and acetylcholine levels in the rat brain, *J. Neurochem.* **21:**1273–1281.

Henderson, A. R., 1970, The effect of feeding with a tryptophan-free amino acid mixture on rat liver magnesium ion-activated deoxyribonucleic acid-dependent ribonucleic acid polymerase, *Biochem. J.* **120:**205–214.

Hertzig, M. E., Birch, H. G., Richardson, S. A., and Tizard, J., 1972, Intellectual levels of school children severely malnourished during the first two years of life, *Pediatrics* **49:**814–824.

Hirota, Z., 1898, Ueber die durch die Milch der an Kakke (Beriberi) leidenden Frauen verursachte Krankheit der Sauglinge, *Zentralbl. Inn Med.* **19:**385–392.

Hirsch, C. A., and Hiatt, H. H., 1966, Turnover of liver ribosomes in fed and fasted rats, *J. Biol. Chem.* **241:**5936–5940.

Hökfelt, T., 1974, Morphological contributions to monoamine pharmacology, *Fed. Proc.* **33:**2177–2186.

Holley, R. W., 1972, A unifying hypothesis concerning the nature of malignant growth, *Proc. Natl. Acad. Sci. U.S.A.* **69:**2840–2841.

Holley, R. W., and Kiernan, J. A., 1974, Control of the initiation of DNA synthesis in 3T3 cells: Low-molecular-weight nutrients, *Proc. Natl. Acad. Sci. U.S.A.* **71:**2942–2945.

Holley, R. W., and Trowbridge, I. S., 1975, Control of growth of mammalian cells, in: *Molecular Approaches to Immunology* (E. E. Smith and D. W. Ribbons, eds.), pp. 289–301, Academic Press, New York.

Holowach, J., Kauffman, F., Ikossi, M. G., Thomas, C., and McDougal, D. B., Jr., 1968, The effects of a thiamine antagonist, pyrithiamine, on levels of selected metabolic intermediates and on activities of thiamine-dependent enzymes in brain and liver, *J. Neurochem.* **15:**621–631.

Hoorweg, J., and Stanfield, J. P., 1976, The effects of protein energy malnutrition on intellectual and motor abilities in later childhood and adolescence, *Dev. Med. Child Neurol.* **18:**330–350.

Horn, G., 1955, Thyroid deficiency and inanition: The effects of replacement therapy on the development of the cerebral cortex of young albino rats, *Anat. Rec.* **121:**63–79.

Hosein, E. A., Chabrol, J. G., and Freedman, G., 1966, The effect of thiamine deficiency in rats and pigeons on the content of materials with acetylcholine-like activity in brain, *Rev. Can. Biol.* **25:**129–134.

Howard, E., 1965, Effects of corticosterone and food restriction on growth and on DNA, RNA and cholesterol contents of the brain and liver in infant mice, *J. Neurochem.* **12:**181–191.

Howard, E., and Granoff, D. M., 1968, Effect of neonatal food restriction in mice on brain growth, DNA and cholesterol, and on adult delayed response learning, *J. Nutr.* **95:**111–121.

Howell, J. McC., and Thompson, J. N., 1967, Lesions associated with the development of ataxia in vitamin A-deficient chicks, *Br. J. Nutr.* **21:**741–750.

Howell, S. L., Green, I. C., and Montague, W., 1973, A possible role of adenylate cyclase in the long-term dietary regulation of insulin secretion from rat islets of Langerhans, *Biochem. J.* **136:**343–349.

Huberman, J. A., and Riggs, A. D., 1968, On the mechanism of DNA replication in mammalian chromosomes, *J. Mol. Biol.* **32:**327–341.

Hurley, L. S., 1976, Perinatal effects of trace element deficiencies, in: *Trace Elements in Health and Disease* (A. S. Prasad, ed.), Vol. II, Essential and Toxic Elements, pp. 301–314, Academic Press, New York.

Im, H. S., Barnes, R. H., Levitsky, D. A., and Pond, W. G., 1973, Postnatal malnutrition and regional cholinesterase activities in brain of pigs, *Brain Res.* **63:**461–465.

Jacob, S. T., Rose, K. M., and Munro, H. N., 1976, Response of poly(adenylic acid)polymerase in rat liver nuclei and mitochondria to starvation and refeeding with amino acids, *Biochem. J.* **158:**161–167.

Jadhav, M., Webb, J. K. G., Vaishnava, S., and Baker, S. J., 1962, Vitamin-B$_{12}$ deficiency in Indian infants. A clinical syndrome, *Lancet* **2:**903–907.

Jeghers, H., 1937, Night blindness as a criterion of vitamin A deficiency: Review of the literature with preliminary observations of the degree and prevalence of vitamin A deficiency among adults in both health and disease, *Ann. Intern. Med.* **10**:1304–1334.

Johnson, R. A., and Schmidt, R. R., 1966, Enzymic control of nucleic acid synthesis during synchronous growth of *Chlorella pyrenoidosa*. I. Deoxythymidine monophosphate kinase, *Biochim. Biophys. Acta* **129**:140–144.

Jones, D. G., and Dyson, S. E., 1976, Synaptic junctions in undernourished rat brain—an ultrastructural investigation, *Exp. Neurol.* **51**:529–535.

Kanig, K., 1976, Other deficiencies and toxicities of water-soluble vitamins, in: *Handbook of Chemical Neurology* (P. J. Vinken and G. W. Bruyn, eds.), Vol. 28, Metabolic and Deficiency Diseases of the Nervous System, Part II, pp. 199–223, North-Holland, Amsterdam.

Kayden, H. J., Silber, R., and Kossmann, C. E., 1965, The role of vitamin E deficiency in the abnormal autohemolysis of acanthocytosis, *Trans. Assoc. Am. Physicians* **78**:334–342.

Khouri, F. P., and McLaren, D. S., 1973, Cytogenetic studies in protein–calorie malnutrition, *Am. J. Hum. Genet.* **25**:465–470.

Kizer, J. S., Palkovits, M., Zivin, J., Brownstein, M., Saavedra, J. M., and Kopin, I. J., 1974, The effect of endocrinological manipulations on tyrosine hydroxylase and dopamine-β-hydroxylase activities in individual hypothalamic nuclei of the adult male rat, *Endocrinology* **95**:799–812.

Klein, R. E., Habicht, J. P., and Yarbrough, C., 1971, Effects of protein-calorie malnutrition on mental development, *Ad. Pediatr.* **18**:75–91.

Knott, P. J., and Curzon, G., 1972, Free tryptophan in plasma and brain tryptophan metabolism, *Nature* **239**:452–453.

Koeppe, R. E., O'Neal, R. M., and Hahn, C. H., 1964, Pyruvate decarboxylation in thiamine deficient brain, *J. Neurochem.* **11**:695–699.

Kriegstein, H. J., and Hogness, D. S., 1974, Mechanism of DNA replication in *Drosophila* chromosomes: Structure of replication forks and evidence for bidirectionality, *Proc. Natl. Acad. Sci. U.S.A.* **71**:135–139.

Krigman, M. R., and Hogan, E. L., 1976, Undernutrition in the developing rat: Effect upon myelination, *Brain Res.* **107**:239–255.

Kurtz, D. J., and Kanfer, J. N., 1973, Composition of myelin lipids and synthesis of 3-ketodihydrosphingosine in the vitamin B_6-deficient developing rat, *J. Neurochem.* **20**:963–968.

Kurtz, D. L., Levy, H., and Kanfer, J. N., 1972, Cerebral lipids and amino acids in the vitamin B_6 deficient suckling rats, *J. Nutr.* **102**:291–298.

Ladinsky, H., Consolo, S., Peri, G., and Garattini, S., 1972, Acetylcholine, choline and choline acetyltransferase activity in the developing brain of normal and hypothyroid rats, *J. Neurochem.* **19**:1947–1952.

Lampert, P., Blumberg, J. M., and Pentschew, A., 1964, An electron microscopic study of dystrophic axons in the gracile and cuneate nuclei of vitamin E-deficient rats, *J. Neuropathol. Exp. Neurol.* **23**:60–77.

Langman, J., and Welch, G. W., 1966, Effect of vitamin A on development of the central nervous system, *J. Comp. Neurol.* **128**:1–15.

Langman, J., and Welch, G. W., 1967, Excess vitamin A and development of the cerebral cortex, *J. Comp. Neurol.* **131**:15–25.

Lanzkowsky, P., 1970, Congenital malabsorption of folate, *Am. J. Med.* **48**:580–583.

Lark, K. G., Consigli, R., and Toliver, A., 1971, DNA replication in Chinese hamster cells: Evidence for a single replication fork per replicon, *J. Mol. Biol.* **58**:873–875.

Larroche, J.-C., 1966, The development of the central nervous system during intrauterine life, in: *Human Development* (F. Falkner, ed.), pp. 257–276, W. B. Saunders, Philadelphia.

Leduc, E. H., 1949, Mitotic activity in the liver of the mouse during inanition followed by refeeding with different levels of protein, *Am. J. Anat.* **81**:397–429.

Lee, C.-J., 1970, Biosynthesis and characteristics of brain protein and ribonucleic acid in mice subjected to neonatal infection or undernutrition, *J. Biol. Chem.* **245**:1998–2004.

Lee, C.-J., and Dubos, R., 1972, Lasting biological effects of early environmental influences. VIII. Effects of neonatal infection, perinatal malnutrition, and crowding on catecholamine metabolism of brain, *J. Exp. Med.* **136**:1031–1042.

Legrand, J., 1967, Analyse de l'action morphogénétique des hormones thyroidiennes sur le cervelet du jeune rat, *Arch. Anat. Microsc. Morphol. Exp.* **56**:205–244.

Lehr, P., and Gayet, J., 1967, Response of the cerebral cortex of the rat to prolonged protein depletion. III. Entry of glucose carbon into free amino acids *in vivo*, *J. Neurochem.* **14**:927–936.

Leonard, P. J., 1973, Cortisol binding in serum of kwashiorkor: East African studies, in: *Endocrine*

Aspects of Malnutrition (L. I. Gardner and P. Amacher, eds.), pp. 355–362, The Kroc Foundation, Santa Ynez, California

Levine, S., 1974, Malnutrition and neuroendocrine development, in: *Early Malnutrition and Mental Development* (J. Cravioto, L. Hambreus, and B. Vahlquist, eds.), pp. 90–95, Almqvist and Wiksell, Stockholm.

Lewis, P. D., 1968a, The fate of the subependymal cell in the adult rat brain, with a note on the origin of microglia, *Brain* **91:**721–736.

Lewis, P. D., 1968b, Mitotic activity in the primate subependymal layer and the genesis of gliomas, *Nature* **217:**974–975.

Lewis, P. D., 1975, Cell death in the germinal layers of the postnatal rat brain, *Neuropathol. Appl. Neurobiol.* **1:**1–9.

Lewis, P. D., 1978, The application of cell turnover studies to neuropathology, in: *Recent Advances in Neuropathology* (W. T. Smith, ed.), Churchill Livingstone, London (in press).

Lewis, P. D., Balázs, R., Patel, A. J., and Johnson, A. L., 1975, The effect of undernutrition in early life on cell generation in the rat brain, *Brain Res.* **83:**235–247.

Lewis, P. D., Patel, A. J., Johnson, A. L., and Balázs, R., 1976, Effect of thyroid deficiency on cell acquisition in the postnatal rat brain: A quantitative histological study, *Brain Res.* **104:**49–62.

Lewis, P. D., Fülöp, Z., Hajós, F., Balázs, R., and Woodhams, P. L., 1977a, Neuroglia in the internal granular layer of the developing rat cerebellar cortex, *Neuropathol. Appl. Neurobiol.* **3:**183–190.

Lewis, P. D., Patel, A. J. and Balázs, R., 1977b, Effect of undernutrition on cell generation in the adult rat brain, *Brain Res.* **138:**511–519.

Lim, L., and Canellakis, E. S., 1970, Adenine-rich polymer associated with rabbit reticulocyte messenger RNA, *Nature* **227:**710–712.

Lowry, O. H., Passonneau, J. V., Hasselberger, F. X., and Schulz, D. W., 1964, Effect of ischemia on known substrates and cofactors of the glycolytic pathway in brain, *J. Biol. Chem.* **239:**18–30.

Luhby, A. L., Eagle, F. J., Roth, E., and Cooperman, J. M., 1961, Relapsing megaloblastic anemia in an infant due to a specific defect in gastrointestinal absorption of folic acid, *Am. J. Dis. Child.* **102:**482–483.

Luttrell, C. N., and Mason, K. E., 1949, Vitamin E deficiency, dietary fat, and spinal cord lesions in rat, *Ann. N.Y. Acad. Sci.* **52:**113–120.

Lynch, A., Smart, J. L., and Dobbing, J., 1975, Motor coordination and cerebellar size in adult rats undernourished in early life, *Brain Res.* **83:**249–259.

Maddux, G. W., Foltz, F. M., and Nelson S. R., 1974, Effect of vitamin A intoxication on intracranial pressure and brain water in rats, *J. Nutr.* **104:**478–482.

Madras, B. K., Cohen, E. L., Messing, R. L., Munro, H. N., and Wurtman, R. J., 1974, Relevance of free tryptophan in serum tissue tryptophan concentrations, *Metabolism* **23:**1107–1116.

Malamud, N., Nelson, M. M., and Evans, H. M., 1949, Effect of chronic vitamin E deficiency on nervous system in rats, *Ann. N.Y. Acad. Sci.* **52:**135–138.

Mandel, P., and Mark, J., 1965, The influence of nitrogen deprivation on free amino acids in rat brain, *J. Neurochem.* **12:**987–992.

Mandel, P., and Quirin-Stricker, C., 1967, Effect of protein deprivation on soluble and "aggregate" RNA polymerase in rat liver, *Life Sci.* **6:**1299–1303.

Mandel, P., Jacob, M., and Mandel, L., 1950, Etude sur le métabolisme des acides nucléiques. I. Action du jeûne protéique prolongé sur les deux acides nucléiques du foie, du rein et du cerveau, *Bull. Soc. Chim. Biol. (Paris)* **32:**80–88.

Mariani, A., Migliaccio, P. A., Spadoni, M. A., and Ticca, M., 1966, Amino acid activation in the liver of growing rats maintained with normal and with protein-deficient diets, *J. Nutr.* **90:**25–30.

Marx, J. L., 1976, Cell biology: Cell surfaces and the regulation of mitosis, *Science* **192:**455–457.

Massieu, G. H., Ortega, B. G., Syrquin, A., and Tuena, M., 1962, Free amino acids in brain and liver of deoxypyridoxine-treated mice subjected to insulin shock, *J. Neurochem.* **9:**143–151.

Matthews, D. A., Cotman, C., and Lynch, G., 1976, An electron microscopic study of lesion-induced synaptogenesis in the dentate gyrus of the adult rat. II. Reappearance of morphologically normal synaptic contacts, *Brain Res.* **115:**23–41.

Matus, A. I., Walters, B. B., and Mughal, S., 1975, Immunohistochemical demonstration of tubulin associated with microtubules and synaptic junctions in mammalian brain, *J. Neurocytol.* **4:**733–755.

McCandless, D. W., and Schenker, S., 1968, Encephalopathy or thiamine deficiency: Studies of intracerebral mechanisms, *J. Clin. Invest.* **47:**2268–2280.

McConnell, P., and Berry, M., 1978, The effects of undernutrition on Purkinje cell dendritic growth in the rat, *J. Comp. Neurol.* **178:**759–772.

McEwen, B. S., Zigmond, R. E., and Gerlach, J. L., 1972, Sites of steroid binding and action in the brain, in: *The Structure and Function of Nervous Tissue,* Vol. 5 (G. H. Bourne, ed.), pp. 205–291, Academic Press, New York.

McLaren, D. S., 1970, Effects of vitamin A deficiency in man, in: *Vitamins, Chemistry, Physiology, Pathology, Methods* (W. H. Sebrell and R. S. Harris, eds.), Vol. 1, pp. 267–280, Academic Press, New York.

McLaughlin, B. J., Wood, J. G., Saito, K., Barber, R., Vaughn, J. E., Roberts, E., and Wu, J.-Y., 1974, The fine structural localization of glutamate decarboxylase in synaptic terminals of rodent cerebellum, *Brain Res.* **76:**377–391.

McMenamy, R. H., and Oncley, J. L., 1958, The specific binding of L-tryptophan to serum albumin, *J. Biol. Chem.* **233:**1436–1447.

Meites, J., and Wolterink, L. F., 1950, Uptake of radioactive iodine by the thyroids of underfed rats, *Science* **111:**175–176.

Mellanby, E., 1947, Vitamin A and bone growth: The reversibility of vitamin A-deficiency changes, *J. Physiol.* **105:**382–399.

Millen, J. W., Woollam, D. H. M., and Lamming, G. E., 1953, Hydrocephalus associated with deficiency of vitamin A, *Lancet* **2:**1234–1236.

Miller, T. J., 1972, Cleft palate formation: A role for pyridoxine in the closure of the secondary palate in mice, *Teratology* **6:**351–356.

Moloney, C. J., and Parmalee, A. H., 1954, Convulsions in young infants as a result of pyridoxine (vitamin B_6) deficiency, *J. Am. Med. Assoc.* **154:**405–406.

Mönckeberg, F., 1975, The effect of malnutrition on physical growth and brain development, in: *Brain Function and Malnutrition. Neuropsychological Methods of Assessment* (J. W. Prescott, M. S. Read, and D. B. Coursin, eds.), pp. 15–36, Wiley, New York.

Moore, T., 1957, *Vitamin A,* Elsevier, London.

Mourek, J., Agrawal, H. C., Davis, J. M., and Himwich, W. A., 1970, The effects of short-term starvation on amino acid content in rat brain during ontogeny, *Brain Res.* **19:**229–237.

Mueller, A. J., and Cox, W. M., Jr., 1937, The effect of changes in diet on the volume and composition of rat milk. II. *J. Biol. Chem.* **119:**LXXII.

Mulinos, M. G., and Pomerantz, L., 1941, Pituitary replacement therapy in pseudo-hypophysectomy. Effect of pituitary implants upon organ weights of starved and underfed rats, *Endocrinology* **29:**558–563.

Murphree, S., Stubblefield, E., and Moore, E. C., 1969, Synchronized mammalian cell cultures. III. Variation of ribonucleotide reductase activity during the replication cycle of Chinese hamster fibroblasts, *Exp. Cell Res.* **58:**118–124.

Murty, C. N., and Sidransky, H., 1972, The effect of tryptophan on messenger RNA of the livers of fasted mice, *Biochim. Biophys. Acta* **262:**328–335.

Neuberne, P. M., and Young, V. R., 1973, Marginal vitamin B_{12} intake during gestation in the rat has long term effects on the offspring, *Nature* **242:**263–265.

Neville, H. E., and Chase, H. P., 1971, Undernutrition and cerebellar development, *Exp. Neurol.* **33:**485–497.

Nicholson, J. L., and Altman, J., 1972*a,* The effects of early hypo- and hyperthyroidism on the development of rat cerebellar cortex. I. Cell proliferation and differentiation, *Brain Res.* **44:**13–23.

Nicholson, J. L., and Altman, J., 1972*b,* Synaptogenesis in the rat cerebellum: Effects of early hypo- and hyperthyroidism, *Science* **176:**530–532.

Nitowsky, H. M., Gordon, H. H., and Tildon, J. T., 1956, Studies of tocopherol deficiency in infants and children. IV. The effect of alpha tocopherol on creatinuria in patients with cystic fibrosis of the pancreas and biliary atresia, *Bull. Johns Hopkins Hosp.* **98:**361–371.

O'Dell, B. L., 1976, Biochemistry and physiology of copper in vertebrates, in: *Trace Elements in Human Health and Disease,* Vol. 1, Zinc and Copper (A. S. Prasad and D. Oberleas, eds.), pp. 399–413, Academic Press, New York.

Oliver, J. M., Zurier, R. B., and Berlin, R. D., 1975, Concanavalin A cap formation on polymorphonuclear leukocytes of normal and beige (Chediak-Higashi) mice, *Nature* **253:**471–473.

O'Neal, R. M., and Koeppe, R. E., 1966, Precursors *in vivo* of glutamate, aspartate and their derivatives of rat brain, *J. Neurochem.* **13:**835–847.

Otsuka, M., 1972, γ-Aminobutyric acid in the nervous system, in: *The Structure and Function of Nervous Tissue,* Vol. IV (G. H. Bourne, ed.), pp. 249–289, Academic Press, New York.

Palay, S. L., and Chan-Palay, V., 1974, *Cerebellar Cortex, Cytology and Organization,* Springer, Berlin.

Palkovits, M., 1973, Isolated removal of hypothalamic or other brain nuclei of the rat, *Brain Res.* **59**:449–450.

Pallis, C. A., and Lewis, P. D., 1974, *The Neurology of Gastrointestinal Disease*, Saunders, London.

Pampiglione, G., 1975, Electroencephalographic studies in animals with experimental malnutrition, in: *Brain Function and Malnutrition: Neurophysiological Methods of Assessment* (J. W. Prescott, M. S. Read, and D. B. Coursin, eds.), pp. 233–246, Wiley, New York.

Pan American Health Organization, 1972, Nutrition, the Nervous System and Behaviour, Scientific Publication No. 251, Pan American Health Organization, Washington, D.C.

Patel, M. S., 1974, The effect of ketone bodies on pyruvate carboxylation by rat brain mitochondria, *J. Neurochem.* **23**:865–867.

Patel, A. J., and Balázs, R., 1970, Manifestation of metabolic compartmentation during the maturation of the rat brain, *J. Neurochem.* **17**:955–971.

Patel, A. J., and Balázs, R., 1971, Effect of thyroid hormone on metabolic compartmentation in the developing rat brain, *Biochem. J.* **121**:469–481.

Patel, A. J., and Balázs, R., 1973, Effect of undernutrition on the biochemical development of the brain, *Proc. 4th Int. Meet. Int. Soc. Neurochem.*, p. 436, Tokyo.

Patel, A. J., and Balázs, R., 1975, Factors affecting the development of metabolic compartmentation in the brain, in: *Metabolic Compartmentation and Neurotransmission, Relation to Brain Structure and Function* (S. Berl and D. Schneider, eds.), pp. 385–395, Plenum Press, New York.

Patel, A. J., Balázs, R., and Johnson, A. L., 1973, Effect of undernutrition on cell formation in the rat brain, *J. Neurochem.* **20**:1151–1165.

Patel, A. J., Michaelson, I. A., Cremer, J. E., and Balázs, R., 1974, The metabolism of [^{14}C]glucose by the brains of suckling rats intoxicated with inorganic lead, *J. Neurochem.* **22**:581–590.

Patel, A. J., Atkinson, D. J., and Balázs, R., 1975, Effect of undernutrition on metabolic compartmentation of glutamate and on the incorporation of [^{14}C]leucine into protein in the developing rat brain, *Dev. Psychobiol.* **8**:453–464.

Patel, A. J., Rabié, A., Lewis, P. D., and Balázs, R., 1976, Effect of thyroid deficiency on postnatal cell formation in the rat brain: A biochemical investigation, *Brain Res.* **104**:33–48.

Patel, A. J., Hunt, A., Lewis, P. D., and Balázs, R., 1978a, Effect of undernutrition on thymidine metabolism in the developing rat brain (in preparation).

Patel, A. J., Del Vecchio, M., and Atkinson, D. J., 1978b, Effect of undernutrition on the regional development of transmitter enzymes: Glutamate decarboxylase and choline acetyltransferase, *Develop. Neurosci.* **1**:41–53.

Pearson, H. A., Vinson, R., and Smith R. T., 1964, Pernicious anemia with neurologic involvement in childhood, *J. Pediatr.* **65**:334–339.

Pentschew, A., and Schwarz, K., 1962, Systemic axonal dystrophy in vitamin E deficient adult rats, *Acta Neuropathol.* **1**:313–334.

Peskin, M. R., Newton, G., and Brin, M., 1967, Thiamine deficiency, infantile manipulation and startle response in rats, *J. Nutr.* **91**:20–24.

Pimstone, B., Becker, D., and Hausen, J. D. L., 1973a, Human growth hormone and sulphation factor in protein–calorie malnutrition, in: *Endocrine Aspects of Malnutrition* (L. I. Gardner and P. Amacher, eds.), pp. 73–90, The Kroc Foundation, Santa Ynez, California.

Pimstone, B., Becker, D., and Hendricks, S., 1973b, TSH response to synthetic TRH in human protein–calorie malnutrition, in: *Endocrine Aspects of Malnutrition* (L. I. Gardner and P. Amacher, eds.), pp. 243–255, The Kroc Foundation, Santa Ynez, California.

Pipeleers, D. G., Pipeleers-Marichal, M. A., and Kipnis, D. M., 1976, Regulation of tubulin synthesis in islets of Langerhans, *Proc. Natl. Acad. Sci. U.S.A.* **73**:3188–3191.

Prensky, A. L., Fishman, M. A., and Daftari, B., 1971, Differential effects of hyperphenylalaninemia on the development of the brain in the rat, *Brain Res.* **33**:181–191.

Prescott, D. M., 1976, The cell cycle and the control of cellular replication, *Adv. Genet.* **18**:100–177.

Prescott, J. W., Read, M. S., and Coursin, D. B., eds., 1975, *Brain Function and Malnutrition: Neurophysiological Methods of Assessment*, Wiley, New York.

Pretorius, P. J., and Novis, H., 1965, Nutritional marasmus in Bantu infants in the Pretoria area. II. Clinical and pathological aspects, *S. Afr. Med. J.* **39**:501–505.

Rabié, A., Favre, C., Clavel, M. C., and Legrand, J., 1977, Effects of thyroid dysfunction on the development of the rat cerebellum, with special reference to cell death within the internal granular layer, *Brain Res.* **120**:521–531.

Raisman, G., 1976, The reaction of synaptogenesis in the central and peripheral nervous system of the adult rat, in: *Neural Mechanisms of Learning and Memory* (M. R. Rozenweig and E. L. Bennett, eds.), pp. 348–351, MIT Press, Cambridge, Massachusetts.

Rajalakshmi, R., 1975, Effects of nutritional deficiencies on the composition and metabolic activity of the brain in the rat, in: *Growth and Development of the Brain*, IBRO Monograph Series, Vol. 1 (M. A. B. Brazier), pp. 139–156, Raven Press, New York.

Rajalakshmi, R., Ali, S. Z., and Ramakrishnan, C. V., 1967, Effect of inanition during the neonatal period on discrimination learning and brain biochemistry in the albino rat, *J. Neurochem.* **14:**29–34.

Rajalakshmi, R., Nakhasi, H. L., and Ramakrishnan, C. V., 1974*a*, Effects of preweaning and post-weaning deficiencies on the composition of brain lipids in rats, *Ind. J. Biochem. Biophys.* **11:**57–60.

Rajalakshmi, R., Parameswaran, M., Telang, S. D., and Ramakrishnan, C. V., 1974*b*, Effects of undernutrition and protein deficiency on glutamate dehydrogenase and decarboxylase in rat brain, *J. Neurochem.* **23:**129–133.

Randt, C. T., and Derby, B. M., 1973, Behavioral and brain correlations in early life nutritional deprivation, *Arch. Neurol. (Chicago)* **28:**167–172.

Rao, K. S. J., Srikantia, S. G., and Gopalan, C., 1968, Plasma cortisol levels in protein-calorie malnutrition, *Arch. Dis. Child.* **43:**365–367.

Rebière, A., 1973, Aspects quantitatifs de la synaptogenèse dans le cervelet du rat sous-alimenté dès la naissance. Comparaison avec l'animal hypothyroïdien. *C. R. Acad. Sci. (Paris)* **276:**2317–2320.

Rebière, A., and Legrand, J., 1972, Effects comparés de la sous-alimentation, de l'hypothyroïdisme et de l'hyperthyroïdisme sur la maturation histologique de la zone moléculaire du cortex cérébelleux chez le jeune rat, *Arch. Anat. Microsc. Morphol. Exp.* **61:**105–106.

Reddy, B. S., Pleasants, J. R., and Wostmann, B. S., 1971, Effect of protein-calorie restriction on brain amino acid pool in neonatal rats. *Proc. Soc. Exp. Biol. Med.* **136:**949–953.

Reid, I. M., Verney, E., and Sidransky, H., 1970, Influence of acute nutritional stress on polyribosomes and protein synthesis in brain and liver of young rats, *J. Nutr.* **100:**1149–1156.

Reinauer, V. H., Grassow, G., and Hollmann, S., 1968, Aktivitäsänderungen der Pyruvatdehydrogenase im Thiaminmangel, *Hoppe-Seyler's Z. Physiol. Chem.* **349:**969–978.

Reisner, E. H., Wolff, J. A., McKay, R. J., and Doyle, E. F., 1951, Juvenile pernicious anemia, *Pediatrics* **8:**88–106.

Reynolds, S. F., and Blass, J. P., 1975, Normal levels of acetyl coenzyme A and of acetylcholine in the brains of thiamin-deficient rats, *J. Neurochem.* **24:**185–186.

Richardson, S. A., Birch, H. G., and Hertzig, M. E., 1973, School performance of children who were severely malnourished in infancy, *Am. J. Men. Defic.* **77:**623–632.

Rickwood, D., and Klemperer, H. G., 1970, Decreased ribonucleic acid synthesis in isolated rat liver nuclei during starvation, *Biochem. J.* **120:**381–384.

Ringsted, A., 1935, A preliminary note on the appearance of paresis in adult rats suffering from chronic avitaminosis E, *Biochem. J.* **29:**788–795.

Roach, M. K., Corbin, J., and Pennington, W., 1974, Effect of undernutrition on amino acid compartmentation in the developing rat brain, *J. Neurochem.* **22:**521–528.

Roberts, E., Younger, F., and Frankel, S., 1951, Influence of dietary pyridoxine on glutamic decarboxylase activity of brain, *J. Biol. Chem.* **191:**277–285.

Rose, P. M., Hopper, A. F., and Wannemacher, R. W., Jr., 1971, Cell population changes in the intestinal mucosa of protein-depleted or starved rats. I. Changes in mitotic cycle time, *J. Cell Biol.* **50:**887–892.

Salas, M., Díaz, S., and Nieto, A., 1974, Effects of neonatal food deprivation on cortical spines and dendritic development of the rat, *Brain Res.* **73:**139–144.

Sara, V. R., and Lazarus, L., 1975, Maternal growth hormone and growth and function, *Dev. Psychobiol.* **8:**489–502.

Schmidt, M. J., and Sanders-Busch, E., 1971, Tryptophan hydroxylase activity in developing rat brain, *J. Neurochem.* **18:**2549–2551.

Schmitt, F. O., 1968, Fibrous proteins—neuronal organelles, *Proc. Natl. Acad. Sci. U.S.A.* **60:**1092–1101.

Schultze, B., Nowak, B., and Maurer, W., 1974, Cycle tissues of neuronal epithelial cells of various types of neuron in the rat. An autoradiographic study, *J. Comp. Neurol.* **158:**207–218.

Scrimshaw, N. S., and Gordon, J. E., eds., 1968, *Malnutrition, Learning and Behaviour*, MIT Press, Cambridge, Massachusetts.

Selawry, H., Gutman, R., Fink, G., and Recant, L., 1973, The effect of starvation on tissue adenosine 3'-5' monophosphate levels, *Biochem. Biophys. Res. Commun.* **51:**198–204.

Sereni, F., Principi, N., Perletti, L., and Sereni, L. P., 1966, Undernutrition and the developing rat brain.

I. Influence on acetylcholinesterase and succinic acid dehydrogenase activities and on norepineph-rine and 5-OH-tryptamine tissue concentration, *Biol. Neonat.* **10**:254–265.

Shambaugh, G. E., III, and Wilber, J. F., 1974, The effect caloric deprivation upon thyroid function in the neonatal rat, *Endocrinology* **94**:1145–1149.

Shaw, C., and Fillios, L. C., 1968, RNA polymerase activities and other aspects of hepatic protein synthesis during early protein depletion in the rat, *J. Nutr.* **96**:327–336.

Shields, R., and Korner, A., 1970, Regulation of mammalian ribosome synthesis by amino acids, *Biochim. Biophys. Acta* **204**:521–530.

Shin, H. K., and Linkswiler, H. M., 1974, Tryptophan and methionine metabolism of adult females as affected by vitamin B-6 deficiency, *J. Nutr.* **104**:1348–1355.

Shoemaker, W. J., and Wurtman, R. J., 1971, Perinatal undernutrition: Accumulation of catecholamines in rat brain, *Science* **171**:1017–1019.

Shoemaker, W. J., and Wurtman, R. J., 1973, Effect of perinatal undernutrition on the metabolism of catecholamines in the rat brain, *J. Nutr.* **103**:1537–1547.

Short, J., Brown, R. F., Husakova, A., Gilbertson, J. R., Zemel, R., and Lieberman, I., 1972, Induction of deoxyribonucleic acid synthesis in the liver of the intact animal, *J. Biol. Chem.* **247**:1757–1766.

Short, J., Armstrong, N. B., Kolitzky, M. A., Mitchell, R. A., Zemel, R., and Lieberman, I., 1974, Amino acids and the control of nuclear DNA replication in liver, in: *Control of Proliferation in Animal Cells* (B. Clarkson and R. Baserga, eds.), pp. 37–48, Cold Spring Harbor Laboratory, New York.

Siassi, F., and Siassi, B., 1973, Differential effects of protein–calorie restriction and subsequent repletion on neuronal and nonneuronal components of cerebral cortex in newborn rats. *J. Nutr.* **103**:1625–1633.

Sidransky, H., Sarma, D. S. R., Bongiorno, M., and Verney, E., 1968, Effect of dietary tryptophan on hepatic polyribosomes and protein synthesis in fasted mice, *J. Biol. Chem.* **243**:1123–1132.

Sie, H. G., Nigam, V. N., and Fishman, W. H., 1961, Capacity of rat tissues to form heptulose phosphates in thiamine deficiency, *Biochim. Biophys. Acta* **50**:277–286.

Silver, A., 1974, *The Biology of Cholinesterases,* North-Holland, Amsterdam.

Sima, A., and Persson, L., 1975, The effect of pre- and postnatal undernutrition on the development of the rat cerebellar cortex. I. Morphological observations, *Neurobiology* **5**:23–34.

Skoff, R. P., Price, D. L., and Stocks, A., 1976, Electron microscopic autoradiographic studies of gliogenesis in rat optic nerve. II. Time of origin, *J. Comp. Neurol.* **169**:313–333.

Smart, I., and Leblond, C. P., 1961, Evidence for division and transformations of neuroglia cells in the mouse brain, as derived from radioautography after injection of thymidine-H³, *J. Comp. Neurol.* **116**:349–367.

Smart, J. L., 1971, An experimental investigation of the effect of early nutritional deprivation on behavior, in: *Proc. XIIIth Congress of Pediatrics,* Vienna, Vol. 2, Wischen. Prog. Kom. Wien Verlag der Wiener Med. Akad., Vienna, p. 65.

Smart, J. L., and Dobbing, J., 1971, Vulnerability of developing brain. II. Effects of early nutritional deprivation on reflex ontogeny and development of behaviour in the rat, *Brain Res.* **28**:85–95.

Smart, J. L., Fricklebank, M. D., Adlard, B. P. F., and Dobbing, J., 1976, Nutritionally small-for-date rats: Their subsequent growth, regional brain 5-hydroxytryptamine turnover, and behavior, *Pediatr. Res.* **10**:807–811.

Smulson, M., 1970, Amino acid deprivation of human cells: Effects on RNA synthesis, RNA polymerase, and ribonucleoside phosphorylation, *Biochim. Biophys. Acta* **199**:537–540.

Snyderman, S. E., Holt, L. E., Carretero, R., and Jacobs, K., 1953, Pyridoxine deficiency in the human infant, *J. Clin. Nutr.* **1**:200–207.

Sobotka, T. J., Cook, M. P., and Brodie, R. E., 1974, Neonatal malnutrition: Neurochemical, hormonal and behavioral manifestations, *Brain Res.* **65**:443–457.

Soifer, D., ed., 1975, The biology of cytoplasmic microtubules, *Ann. N.Y. Acad. Sci.* **253**:1–802.

Speeg, K. V., Jr., Chen, D., McCandless, D. W., and Schenker, S., 1970, Cerebral acetylcholine in thiamine deficiency, *Proc. Soc. Exp. Biol. Med.* **134**:1005–1009.

Srikantia, S. G., Reddy, M. V., and Krishnaswamy, K., 1968, Electroencephalographic patterns in pellagra, *Electroencephalogr. Clin. Neurophysiol.* **25**:386–388.

Stanbury, J. B., and Kroc, R. L., eds., 1972, *Human Development and the Thyroid Gland—Relation to Endemic Cretinism,* Advances in Experimental Medicine and Biology, Vol. 30, Plenum Press, New York.

Stempak, J. G., 1965, Etiology of antenatal hydrocephalus induced by folic acid deficiency in the albino rat, *Anat. Rec.* **151**:287–296.

Stephens, M. C., Havlíček, V., and Dakshinamurti, K., 1971, Pyridoxine deficiency and development of the central nervous system in the rat, *J. Neurochem.* **18:**2407–2416.

Stern, P., and Igic, R., 1970, The content of material with acetylcholine-like activity in the brain of animals following thiamine deprivation and treatment with pyrithiamine, in: *Drugs and Cholinergic Mechanisms in the CNS* (E. Hailbronn and A. Winter, eds.), pp. 419–427, Research Institute for National Defense, Stockholm.

Stern, W. C., Miller, M., Forbes, W. B., Morgane, P. J., and Resnick, O., 1975*a*, Ontogeny of the levels of biogenic amines in various parts of the brain and in peripheral tissues in normal and protein malnourished rats, *Exp. Neurol.* **49:**314–326.

Stern, W. C., Morgane, P. J., Miller, M., and Resnick, O., 1975*b*, Protein malnutrition in rats: Response of brain amines and behavior to foot shock stress, *Exp. Neurol.* **47:**56–67.

Stewart, C. N., Bhagavan, H. N., and Coursin, D. B., 1968, Some behavioral consequences of pyridoxine deficiency in rats, in: *International Symposium on Pyridoxal Enzymes* (L. Yamada, N. Katunuma, and H. Wada, eds.), pp. 181–183, Maruzen Co., Tokyo.

Stewart, C. N., Coursin, D. B., and Bhagavan, H. N., 1973, Cortical-evoked responses in pyridoxine-deficient rats, *J. Nutr.* **103:**462–467.

Stewart, R. J., Merat, A., and Dickerson, J. W., 1974, Effect of low protein diet in mother rats on the structure of the brain of the offspring, *Biol. Neonat.* **25:**125–134.

Still, C. N., 1976, Nicotinic acid and nicotinamide deficiency: Pellagra and related disorders of the nervous system, in: *Metabolic and Deficiency Diseases of the Nervous System, Part II* (P. J. Vinken and G. W. Bruyn, eds.), pp. 59–104, Handbook of Clinical Neurology, Vol. 28, North-Holland, Amsterdam.

Stoch, M. B., and Smythe, P. M., 1963, Does undernutrition during infancy inhibit brain growth and subsequent intellectual development?, *Arch. Dis. Child.* **38:**546–552.

Stoch, M. B., and Smythe, P. M., 1967, The effect of undernutrition during infancy on subsequent brain growth and intellectual development, *S. Afr. Med. J.* **41:**1027–1030.

Stoch, M. B., and Smythe, P. M., 1976, 15-year developmental study on effects of severe undernutrition during infancy on subsequent physical growth and intellectual functioning, *Arch. Dis. Child.* **51:**327–336.

Stubblefield, E., and Murphree, S., 1967, Synchronized mammalian cell cultures. II. Thymidine kinase activity in colcemid synchronized fibroblasts, *Exp. Cell Res.* **48:**652–656.

Sturrock, R. R., Smart, J. L., and Dobbing, J., 1976, Effects of undernutrition during the suckling period on growth of the anterior and posterior limbs of the mouse anterior commissure, *Neuropathol. Appl. Neurobiol.* **2:**411–419.

Sturrock, R. R., Smart, J. L., and Dobbing, J., 1978, A quantitative light microscopic study of the mouse brain following undernutrition during the suckling period, *Neuropathol. Appl. Neurobiol.* **4:**(in press).

Sugita, N., 1918, Comparative studies on the growth of the cerebral cortex. VII. On the influence of starvation at an early age upon the development of the cerebral cortex. Albino rat, *J. Comp. Neurol.* **29:**177–240.

Sung, J. H., 1964, Neuroaxonal dystrophy in mucoviscidosis, *J. Neuropathol. Exp. Neurol.* **23:**567–583.

Sung, J. H., and Stadlan, E. M., 1966, Neuroaxonal dystrophy in congenital biliary atresia, *J. Neuropathol. Exp. Neurol.* **25:**341–361.

Symposium: Nutrition and Cell Development, 1970, *Fed. Proc.* **29:**1489–1521.

Symposium: Pharmacological and Biochemical Properties of Microtubule Proteins, 1974, *Fed. Proc.* **33:**151–174.

Sze, P. Y., 1970, Possible repression of L-glutamic acid decarboxylase by gamma-aminobutyric acid in developing mouse brain, *Brain Res.* **19:**322–325.

Tagliamonte, A., Biggio, G., Vargiu, L., and Gessa, G. L., 1973, Free tryptophan in serum controls brain tryptophan level and serotonin synthesis, *Life Sci.* **12**(Part II):277–287.

Tanaka, T., 1934, So-called breast milk intoxication, *Am. J. Dis. Child.* **47:**1286–1298.

Tews, J. K., 1969, Pyridoxine deficiency and brain amino acids, *Ann. N.Y. Acad. Sci.* **166:**74–82.

Tews, J. K., and Lovell, R. A., 1967, The effect of a nutritional pyridoxine deficiency on free amino acids and related substances in mouse brain, *J. Neurochem.* **14:**1–7.

Thorburn, M. J., Hutchinson, S., and Alleyne, G. A. O., 1972, Chromosome abnormalities in malnourished children, *Lancet* **1:**591.

Thurston, J. H., Prensky, A. L., Warren, S. K., and Albone, K.-R., 1971, The effects of undernutrition upon the energy reserve of the brain and upon other selected metabolic intermediates in brains and livers of infant rats, *J. Neurochem.* **18:**161–166.

Tizard, J., 1974, Early malnutrition, growth and mental development in man, *Br. Med. Bull.* **30:**169–174.

Trowell, H. C., Davies, J. N. P., and Dean, R. F. A., 1954, *Kwashiorkor,* Arnold, London.

Udani, P. M., 1960, Neurological manifestations of kwashiorkor, *Indian J. Child Health* 9:103–112.

Van den Berg, C. J., 1970, Compartmentation of glutamate metabolism in the developing brain: experiments with labelled glucose, acetate, phenylalanine, tyrosine and proline, *J. Neurochem.* 17:973–983.

Van den Berg, C. J., Van Kemper, G. M. J., Schadé, J. P., and Veldstra, H., 1965, Levels and intracellular localization of glutamate decarboxylase and γ-aminobutyrate transaminase and other enzymes during the development of the brain, *J. Neurochem.* 12:863–869.

Van Wijk, R., Wicks, W. D., Bevers, M. M., and Van Rijn, J., 1973, Rapid arrest of DNA synthesis by N^6, $O^{2'}$-dibutyryl cyclic adenosine 3′, 5′-monophosphate in cultured hepatoma cells, *Cancer Res.* 33:1331–1338.

Vaughan, M. H., Jr., Soeiro, R., Warner, J. R., and Darnell, J. E., Jr., 1967, The effects of methionine deprivation on ribosome synthesis in HeLa cells, *Proc. Natl. Acad. Sci. U.S.A.* 58:1527–1534.

Verma, K., and Wei King, D., 1967, Disorders of the developing nervous system of vitamin E-deficient rats, *Acta Anat.* 67:623–635.

Victor, M., 1976, The Wernicke-Korsakoff syndrome, in: *Metabolic and Deficiency Diseases of the Nervous System, Part II* (P. J. Vinken and G. W. Bruyn, eds.), pp. 243–270, Handbook of Clinical Neurology, Vol. 28, North-Holland, Amsterdam.

Victor, M., Adams, R. D., and Collins, G. H., 1971, The Wernicke-Korsakoff syndrome. A clinical and pathological study of 245 patients, 82 with post-mortem examinations, in: *Contemporary Neurology Series,* Vol. 7 (F. Plum and F. H. McDonnell, eds.), Davis, Philadelphia.

Von der Decken, A., and Andersson, M. G., 1972, Effect of protein intake on DNA-dependent RNA polymerase activity and protein synthesis *in vitro* in rat liver and brain, *Nutr. Rep. Int.* 5:413–419.

Von der Decken, A., and Wronski, A., 1971, Protein synthesis *in vitro* in rat brain after short-term protein starvation and refeeding, *J. Neurochem.* 18:2383–2388.

Vonderhaar, B. K., and Topper, Y. J., 1974, A role of the cell cycle in hormone-dependent differentiation, *J. Cell Biol.* 63:707–712.

Walters, B. B., and Matus, A. I., 1975, Tubulin in postsynaptic junctional lattice, *Nature* 257:496–498.

Wannemacher, R. W., Jr., Cooper, W. K., and Yatvin, M. B., 1968, The regulation of protein synthesis in the liver of rats. Mechanisms of dietary amino acid control in the immature animal, *Biochem. J.* 107:615–623.

Wapnir, R. A., 1973, Rat brain and blood gluconeogenic metabolic changes in neonatal starvation, *Brain Res.* 57:187–195.

Weichsel, M. E., Jr., and Dawson, L., 1976, Effects of hypothyroidism and undernutrition on DNA content and thymidine kinase activity during cerebellar development in the rat, *J. Neurochem.* 26:675–681.

Weiner, W. J., and Klawans, H. L., 1976, Vitamin B_6, in: *Metabolic and Deficiency Diseases of the Nervous System, Part II* (P. J. Vinken and G. W. Bruyn, eds.), pp. 105–139, Handbook of Clinical Neurology, Vol. 28, North-Holland, Amsterdam.

Weingarten, M. D., Lockwood, A. H., Hwo, S.-Y., and Kirschner, M. W., 1975, A protein factor essential for microtubule assembly, *Proc. Natl. Acad. Sci. U.S.A.* 72:1858–1862.

Weisenberg, R. C., 1972, Microtubule formation *in vitro* in solutions containing low calcium concentrations, *Science* 177:1104–1105.

West, C. D., and Kemper, T. L., 1976, The effect of a low protein diet on the anatomical development of the rat brain, *Brain Res.* 107:221–237.

Whittaker, V. P., and Barker, L. A., 1972, The subcellular fractionation of brain tissue with special reference to the preparation of synaptosomes and their component organelles, in: *Methods in Neurochemistry,* Vol. 2 (R. Fried, ed.), pp. 1–52, Marcel Dekker, New York.

Wiebecke, B., Heybowitz, R., Löhrs, U., and Eder, M., 1969, The effect of starvation on the proliferation kinetics of the mucosa of the small and large bowel of the mouse. *Virchows Arch. B* 4:164–175.

Wiener, S. G., 1972, Post-weaning rehabilitation of catecholamine levels in the rat brain and heart after perinatal undernutrition, MA thesis, MIT, Cambridge, Massachusetts.

Wiggins, R. C., Miller, S. L., Benjamins, J. A., Krigman, M. R., and Morell, P., 1976, Myelin synthesis during postnatal nutritional deprivation and subsequent rehabilitation, *Brain Res.* 107:257–273.

Williamson, D. H., and Buckley, B. M., 1973, The role of ketone bodies in brain development, in: *Inborn Errors of Metabolism* (F. A. Hommes and C. J. Van den Berg, eds.), pp. 81–96, Academic Press, London.

Williamson, B., and Coniglio, J. G., 1971, The effects of pyridoxine deficiency and of caloric restriction on lipids in the developing rat brain, *J. Neurochem.* 18:267–276.

Wilson, L., Bamburg, J. R., Mizel, S. B., Grisham, L. M., and Cresswell, K. M., 1974, Interaction of drugs with microtubule proteins, *Fed. Proc.* **33:**158–166.

Winick, M., 1976, *Malnutrition and Brain Development,* Oxford University Press, London.

Winick, M., and Noble, A., 1966, Cellular response in rats during malnutrition at various ages, *J. Nutr.* **89:**300–306.

Winick, M., and Rosso, P., 1969a, The effect of severe early malnutrition on cellular growth of human brain, *Pediatr. Res.* **3:**181–184.

Winick, M., and Rosso, P., 1969b, Head circumference and cellular growth of the brain in normal and marasmic children, *J. Pediatr.* **74:**774–778.

Wolstenholme, D. R., 1973, Replicating DNA molecules from eggs of Drosophila melansoaster, *Chromosoma* **43:**1–18.

Wurtman, R. J., and Fernstrom, J. D., 1975, Control of brain monoamine synthesis by diet and plasma amino acids, *Am. J. Clin. Nutr.* **28:**638–647.

Yahara, I., and Edelman, G. M., 1975, Modulation of lymphocyte receptor mobility by concanavalin A and colchicine, *Ann. N.Y. Acad. Sci.* **253:**455–469.

Yamada, K. M., Spooner, B. S., and Wessells, N. K., 1970, Axon growth: Roles of microfilaments and microtubules, *Proc. Natl. Acad. Sci. U.S.A.* **66:**1206–1212.

Zamenhof, S., Mosley, J., and Schuller, E., 1966, Stimulation of the proliferation of cortical neurons by prenatal treatment with growth hormone, *Science* **152:**1396–1397.

Zamenhof, S., van Marthens, E., and Grauel, L., 1971a, Prenatal cerebral development: Effect of restricted diet, reversal by growth hormone, *Science* **174:**954–955.

Zamenhof, S., van Marthens, E., and Grauel, L., 1971b, DNA (cell number) and protein in neonatal rat brain: Alteration by timing of maternal dietary protein restriction, *J. Nutr.* **101:**1265–1270.

16

Nutrition, Mental Development, and Learning

JOAQUÍN CRAVIOTO and
ELSA R. DELICARDIE

1. Introduction

In recent years there has been an increasing concern both with the health of peoples and with the requirements of national development and public policy. This has led to the consideration of nutritional inadequacy as an important factor in the life of an individual from the time of gestation to the time of his acceptance of full responsibility as a socially functioning adult. This concern has not been limited to the developing countries. Industrial–advanced nations have also had to take nutrition into account in order to understand better both the health and educational performances of the marginal and/or minority segments of their populations. From a historical viewpoint, perhaps the greatest impetus to the study of the relation of nutrition to health and to human development was given by Orr (1936) who, in his classical *Food, Health and Income,* analyzed the relationships among nutrition, development, and health performance in the population of a technological–advanced country. Leitch (1959) summarized these relationships by stating that "whether the picture is of the broad differences between rich productive countries and the underdeveloped regions, or between social classes in this or in any other country, there is always a gradient with health in quantity and quality of diet associated with parallel gradients in rate of growth and adult stature, physical performance–mental ability and resistance to disease."

Because of the significant association between nutritional status and income level, particularly in preindustrial societies, individuals who have been at greatest

___

JOAQUÍN CRAVIOTO • Instituto Nacional de Ciencias y Tecnología de la Salud del Niño, Sistema Nacional para el Desarrollo Integral de la Familia (DIF), Mexico City, Mexico.   *ELSA R. DELICAR-DIE* • Rural Research and Training Center, Instituto Nacional de Ciencias y Tecnología de la Salud del Niño, Sistema Nacional para el Desarrollo Integral de la Familia (DIF), Mexico City, Mexico.

JOAQUÍN CRAVIOTO
and ELSA R.
DELICARDIE

and most persistent nutritional risk tend to cluster in the lowest socioeconomic segments. Such segments of the population differ from the remainder in a host of other variables. They tend to have poorer housing, higher morbidity rates, lower levels of formal education, greater degrees of attachment to outmoded patterns of child care, and in general to live in circumstances which are less conducive for the development of technologic and educational competence. Among these segments of society, cognition is seldom identified as an active powerful tool of individual achievement. This is most often seen during infancy and the preschool years. In the presence of low purchasing power, resulting directly from the lack of modern technology and factual information, parents are preoccupied with the more pressing needs of life. Problems related to housing, sufficient food, employment, transportation, disease, physical energy, family conflict, and economic and physical safety take the highest priority. Under this load, frequently neglected are activities of the infant and preschool child in manipulating and exploring his physical environment and in being introduced through play to auditory, visual, and tactile stimuli that constitute the precursors of symbols. Seldom is there time for the adult to play with, talk to, or read to a child. In many of these families there is a lack of awareness of the importance of these activities for the child's development. Moreover, the effects of these circumstances may be intergenerational, and they appear, in animal experiments, as a facsimile of familial or hereditary processes (Hsueh *et al.,* 1967).

The presence of these associations makes it inevitable that any consequences for mental development and learning deriving from nutritional conditions of risk be associated with social status and the variables attaching to it. There has been some tendency to view this relationship circularly and to conclude that the abnormal outcomes in development and learning can be accounted for by the social status per se. This is unfortunate since it substitutes a truism for an analysis. Given the associations between lower socioeconomic standing and undesirable mental outcomes, the task of the analysis is to identify the effective variables which mediate these outcomes. If intelligence is operationally defined as the process through which the child learns the use of the tools of his culture in order to know and to manipulate the environment, it is easy to accept that at each stage of development intelligence will be directly associated with both the genetic endowment of the individual and with the several environments in which the child has lived so far. Mental growth is modified to the degree to which conditions of life associated with depressed social position function directly to modify the growth and differentiation of the central nervous system, and indirectly to affect the opportunities for obtaining and the motives for profiting from experience. Similarly social class as such does not determine physical stature. Rather individuals are stunted when their social positions provide an environment in terms of nutrition morbidity, habits, and housing which influence the biological processes involved in growing in length.

Since the factors cited are present in various degrees in different communities, and also vary within families in the same community, it is possible to identify several patterns of life style, nutrition, health, and child care among the underprivileged members of the society. The study of child development, behavior, and learning across the gradient of disadvantage has permitted the assessment of the effects of environment, and particularly malnutrition, on intelligence performance, learning, and behavior.

In this chapter, through a review of selected investigations, we have attempted to present in summary form the current status of our knowledge on the relationship between malnutrition and mental development and learning.

2. Some Problems Involved in the Assessment of the Role of Malnutrition

In attempting to assess the role that malnutrition may have on mental functioning of the human, it becomes essential to consider the meaning of food and feeding in at least three dimensions. The first one is the physiological dimension which has as a unit of measurement the nutrient or the joule and whose function is to provide chemical substances to the organism for purposes of growth, maintenance, and metabolic regulation.

The second dimension of food may be considered as psychophysical. Its unit of measurement would be the foodstuff which, through its organoleptic characteristics, would provide the organism with a variety of stimuli (texture, color, aroma, taste, temperature, etc.). In this context, a foodstuff presented at the table as two different kitchen preparations having the same nutrient and energy content would, in fact, appear as if two different foods were offered to the individual.

Finally, the third food dimension may be considered as psychosocial in nature. Its unit of measurement would be the mealtime. The functions of food along this line are, on the one hand, to aid in symbol formation through the value family and society attach to food, such as a form of reward or punishment, as an experience attached to a gratifying person, or as an identifying characteristic of an ethnic or subcultural group. On the other hand, the mealtime provides opportunities to demonstrate, clarify, and practice role and status at the family and at the community level. Who is waited on first? Who sits at the place of honor at the table? Who receives the best part of a dish? Who moderates conversation at the table? These are some examples of the way in which this food dimension is expressed.

It seems easy to visualize that food deprivation in young children represents not only a shortage of nutrients necessary for the increase in mass of the infant, but also a deprivation of sensory stimuli and of social experiences.

Malnutrition at the community level can be considered as a man-made disorder characteristic of the underprivileged members of society. This is particularly true of the preindustrial societies where the social system, consciously or unconsciously, produces malnourished individuals generation after generation through a series of social mechanisms, among which limited access to goods and services, limited social mobility, and restricted opportunities for social stimuli and experiences at crucial times in life play a major role. This makes it difficult to determine the contribution which malnutrition, *vis à vis* all the other potential conditions of risk for maldevelopment, may be making to the inhibition of growth in mental abilities. The problem is made more complex because of the varying degrees to which the central nervous system is vulnerable to insult at different ages (Chase *et al.,* 1967; Benton *et al.,* 1966; Davison and Dobbing, 1966; Dobbing, 1963). These investigators have reported species differences in the age of most rapid myelination and, for the species exposed to food restriction, have documented the relation of abnormal outcomes in central nervous system structure and composition to the conjunction of malnutrition and the time of most rapid growth and differentiation of the nervous system. Accordingly, not only the presence of malnutrition but the time of life at which it is experienced must be taken into account, if the relation between exposure and mental performance or learning adequacy is to be properly assessed.

The time factor has another important bearing on the problem of malnutrition and development. The assessment of the developmental consequences of malnutrition suffered during infancy or the preschool years cannot be fully evaluated at the time of the insult. This is particularly true when one wishes to quantify the effects of malnutrition on such functions as intellectual abilities, school achievement, and later social and economic competence. For the assessment of these late effects, a considerable time must elapse between the period of exposure and the period in which the functions to be evaluated can be examined meaningfully. It is apparent that this gap in time allows the potential intervention of many environmental factors, other than the fact of prior malnutrition, which may influence the course of the child's developing competence. Thus, for the effect of the nutritional condition to be assessed, the influence of these operating factors must be accounted for.

Nonetheless, the time between the exposure to malnutrition and the assessment of complex nervous system activities is an asset in itself. Clearly, the effect of malnutrition on the developing nervous system is unlikely, except in rare cases, to be fully manifested by changes in simple reflex and adaptive behaviors. Many years ago, Lashley (1929) demonstrated that as much as 20% of the rat cerebral cortex could be extirpated without demonstrable consequences for simple maze learning. However, Maier (1932), using the same species, showed that as little as 3% of destroyed tissue demonstrably affected more complex learning performances. Thus lags in developmental differentiation, or even distributed lesions, prodiced in the nervous system by malnutrition would be expected to manifest themselves at varying distances in time from the age of the primary insult. At these latter ages, when more complex demands for integration are made, opportunities for increased sensitivity in the assessment of consequences exist. A time gap in the assessment of consequences therefore is essential if the full force of the potential insult is to be measured. However, the existence of a time gap does make it necessary to know and to account for the influence of social and other variables on the course of development during the intervening period.

In dealing with the problem of protein–calorie malnutrition in humans and its effects on mental function, investigators have only rarely used a neurophysiological approach. Rather than explore the question of the impairment of mechanisms of brain functioning which may result from deficient nutritive intake, they have attempted to answer the practical question of whether the more readily noted reductions in somatic growth and biochemical maturation are associated with reduced mental performance. Further, they have wished to know whether such mental lags, when they have been found, represent permanent changes in functional effectiveness or are merely transient phenomena which disappear with nutritional recovery.

3. Sensory–Motor Development in Malnourished Infants

In several regions of the world where malnutrition in early infancy is highly prevalent, a direct association between deficits in height and weight of severely malnourished preschool children and lags in psychomotor, adaptive, language, and social–personal behavior, as measured by the Gesell, Catell, or Bayley methods, has been reported by Geber and Dean (1956) in Ugandan children, Robles *et al*. (1959) and Cravioto and Robles (1962) in Mexican children, and Barrera-Moncada (1963) in Venezuelan children.

Serial studies on sensory–motor development in kwashiorkor patients have shown that as recovery from malnutrition takes place, developmental quotients, which are much lower than those obtained in nonmalnourished children of similar age and social class, increase in most patients, and the gap between normal age expectation and the actual performance of the child progressively diminishes for all except infants whose age on admission for treatment is less than 6 months. When developmental quotients are plotted against days of hospitalization, it is apparent that the rate of behavioral recovery varies in direct relation to age at which malnutrition occurred. The older the group the greater the value of the slope (Cravioto and Robles, 1963, 1965).

Research conducted in infants recovering from nutritional marasmus has also disclosed that intelligence and psychomotor activity, as judged by the Bayley scales, remain severely retarded despite a clear somatic recovery (Pollit and Granoff, 1967). The results of these studies in kwashiorkor and marasmic patients have been confirmed in South Africa by Stoch and Smythe (1963), in Chile by Monckeberg (1968) and Kardonsky *et al.* (1971), in Lebanon by Botha-Antoun *et al.* (1968) and Yatkin and McLaren (1970), in the U.S.A. by Chase and Martin (1970), and in Brazil by Marcondes *et al.* (1969). All these reports point out the fact that children affected with either marasmus or kwashiorkor exhibit marked retardation in sensory–motor development which is still present even after physical and biochemical rehabilitation have occurred.

It is not only in general measures of mental development that malnourished children show a poorer performance. Brockman and Ricciuti (1971) have examined a more specific cognitive function, categorization behavior, in relation to malnutrition in 20 severely marasmic children and in 19 control children matched for age and sex, without a history of malnutrition and with heights above the 10th percentile of the Boston norms. Using simple sorting tasks, it was found that the total test scores on categorization behavior of the malnourished children were significantly lower than those obtained in the control children. On retest after 12 weeks of treatment the malnourished children showed no significant increase in scores. Analysis of the ten individual sorting tasks differences disclosed not only lower performance levels for the malnourished children than for the controls, but also appreciably less variation among task differences. Children with a longer period of successful treatment showed a greater gain in body length and head circumference and, with a higher clinical rating of nutritional recovery, tended to perform better on the cognitive tasks.

All the above-mentioned studies make it clear that even after a period of several months children who had made a successful nutritional recovery of severe malnutrition and are medically considered as cured still show developmental lags not only in psychomotor behavior but also in several other areas, including hearing and speech, social–personal behavior, problem-solving ability, eye–hand coordination, and categorization skills.

4. Early Malnutrition and Intelligence at School Age

Follow-up studies of children, who as infants have suffered from severe malnutrition requiring hospitalization, constitute one of the model systems through which research workers have assessed the degree to which severe malnutrition in early life may affect subsequent levels of intelligence. In order to minimize the possibility that the period of hospitalization might itself have continued to exert a depressing effect on performance, the assessment of level of functioning must be conducted in survivors several years after discharge from the hospital. Within this model two main strategies have been used. The first one compares the intelligence-test performance of children with documented episodes of severe malnutrition with the performance of children living in the same community, but without a history of severe malnutrition. Since malnutrition develops in environments conducive, in many ways, to lower performance, investigators have tried to match survivors of early malnutrition with control children for those nonnutritional variables considered as capable of playing an important role.

The second strategy involves the comparison of children who have experienced early severe malnutrition with their own sibs raised in the same family environment but not experiencing the same severity of nutritional insult. The assumption behind this strategy is that siblings as controls cancel out the majority of the demographic or macroenvironmental variables, leaving those related to the specific microenvironment of each child within his or her own family to be accounted for by other means or study.

Table I shows the main characteristics and results of the studies which have contrasted survivors of severe malnutrition and non-severely malnourished children drawn from the general population among which the cases of malnutrition have contemporaneously occurred. As may be seen these reports deal, in total, with children living in markedly different cultural settings, with geographical representations from Europe (Yugoslavia), Asia (India, Indonesia, and the Philippines), Africa (Uganda), and Latin America (El Salvador). Survivors of malnutrition were tested several years after they were considered to be cured. Intelligence-test performance was assessed by a variety of tests, either standardized for the country or specially designed for the particular study. The comparison groups were drawn from the population at risk. In the majority of the studies survivors and controls were only matched for age and general socioeconomic status at the time of retesting: Two exceptions are the study of Champakam *et al.* (1968) in which the groups were matched for age, sex, religion, caste, socioeconomic status, family size, birth order, and educational background of parents; and the study of Hoorveg and Standfield (1972) in which the control group was made up of children with records of good nutrition and growth during their first two years of life.

Srikantia *et al.* (1975) have reported on the follow-up of the group of survivors of kwashiorkor whose mental performance at school age had been previously described by Champakan *et al.* (1968). Eleven years after discharge from the hospital the means in the Wechsler test were 49.9 for the experimental boys and 60.5 for the control boys; for girls the scores were 35.7 and 50.0, respectively, for experimental and control groups. In the 2-year period that followed, the control boys had improved their score by 17.5 ± 4.05 points while the experimental boys had improved theirs by 22.0 ± 2.94 points. Similarly, control girls had improved their performance by 14.2 ± 1.4 points and the experimental by 18.1 ± 4.39 points. The differences in mean increment between the groups were not statistically significant, although both at the beginning and at the end of the observation time period the absolute scores were higher in the control groups, and more markedly so in boys. In other words, differences in IQ initially present in survivors of kwashiorkor and in control children persisted after a lapse of over 13 years, but the early episode of severe clinical malnutrition did not alter significantly the subsequent rate of growth in intellectual performance.

The results of the six reported studies show that the presence of an episode of severe malnutrition occurring early in life and of enough severity to force the family to take the child to a hospital significantly increases the chances of the survivors,

Table I. Main Characteristics and Results of Studies on the Mental Performance of Survivors of Malnutrition

Authors	Country	Children's age when malnourished	Children's age when tested	Tests administered	Comparison group	Results
Cabak and Najdavic (1965)	Yugoslavia	4–24 months	7–14 years	Stepanovic's adaptation of Binet-Simon	Children of unskilled workers	Intensity and duration of malnutrition in early life appeared to be related to magnitude of intellectual deficit after rehabilitation
Liang *et al.* (1967)	Indonesia	2–4 years	5–12 years	Goodenough Wechsler	Age-mates whose nutritional status at age 2–4 years was known	Previously malnourished children had significantly lower scores than the children who were regarded as nonmalnourished during the 2- to 4-year age period
Champakam *et al.* (1968)	India	18–36 months	8–11 years	Specially designed tests, standardized in a comparable population	Children matched for: age, sex, religion, caste, socioeconomic status, family size, birth order, parent's education, locality, and school	Lower intellectual performance of previously malnourished children, differences more marked in the younger age group
Guthrie *et al.* (1969)	Philippines	Presumably at preschool age	School age	Philippine nonverbal intelligence test	Height gradient within the sample studied	Taller children at all ages performed better on intelligence than shorter children
Guillen-Alvarez (1971)	El Salvador, C. A.	3–9 months	10–12 years	Raven, Koch, Goodenough	Rural children, same age and socioeconomic status	Survivors of malnutrition clustered on the lowest region of the IQ score distribution
Hoorveg and Standfield (1972)	Uganda	Below 27 months	11–17 years	Raven, Wechsler, WAIS, Porteus, Memory, Knox, Lincoln-Oseretky	Children with records of good nutrition and growth during first 2 years of life	Except for verbal abilities, a maze test, and a short memory test, survivors of malnutrition had lower performance levels in all the other tests; in general those who suffered malnutrition at a younger age did poorer, particularly in several aspects of memory and learning

years after discharge, scoring at values much lower than those obtained in the children of the control groups.

Two studies on the effect of early malnutrition on intellectual performance at later ages have used the sibling strategy. In the first one, conducted in Mexico (Birch *et al.,* 1971), measured intelligence at school age was compared in 37 survivors of severe malnutrition and in 37 of their siblings closest in age. The malnourished children all had been hospitalized for severe chronic malnutrition of the kwashiorkor type when they were between 6 and 30 months of age. No cases of marasmus were included, and the group was relatively homogeneous both for severity of illness and for type. All children had been discharged to the family following nutritional rehabilitation. The average time of hospitalization was 6 weeks, with a range of 4–8 weeks. During the hospital stay children were visited by their mothers for a 3- to 4-hr period every other day. The ratio of nursing staff to children was high, with one nurse available for every three children. In general, care in hospital was good and considerate, but no special stimulation procedures were applied.

No detailed data were available with respect to the quality of the diets received by the children after their discharge from the hospital. Although it would be appropriate to assume from knowledge of the social circumstances in which the children lived that such diets were suboptimal, in no cases were any of the children ever readmitted to any hospital for severe malnutrition.

At follow-up the children were between 5 and 13 years of age. In all cases intellectual evaluation was carried out at least three years after discharge from the hospital. The child chosen as the control was the one in the surviving sibship nearest in age to the index case and without any prior history of an incident of severe malnutrition. The age distributions of the index cases and of the sibling controls were very similar. This was the consequence of the sibling closest in age tending to be randomly older or younger than the index case. Sex distribution was somewhat divergent with relatively more females in the control cases. Difference in sex ratio, however, was not significant (chi square = 1.95; $P > 0.10$).

At follow-up each child was individually examined to determine intellectual level. The test used was the Wechsler Intelligence Scale for Children (WISC) in its Spanish adaptation. Although it was recognized that this test lacks sensitivity for the youngest children studied, it was judged desirable to use it for all children rather than to substitute a different and probably noncomparable test for the youngest age group.

Full-scale WISC IQ of the index cases (sexes combined) was 68.5 and of the controls 81.5. Verbal and performance differences were of similar magnitude and in the same direction. All mean IQ differences were significant at less than the 5% level of confidence. If an IQ score of below 70 is considered a customary cut-off point for the definition of mental retardation, then twice as many of the previously malnourished children as their siblings functioned below this level; 18 malnourished, compared with nine of the control subjects, had an IQ below 70. Moreover, of those with a low IQ, ten of the index cases were below 60, contrasted with only two of the control subjects. At the other extreme of the distribution, in the conventionally normal range, the reverse picture obtains: Ten of the control subjects, compared with four of the index cases, had an IQ of 90 or higher. Differences in age between the survivors of malnutrition and their siblings did not affect the differences in intellectual test performance between the two groups. No significant sex difference in IQ was found among the rehabilitated children. The mean full-scale IQ for boys and girls, respectively, was 70.7 ± 14.08 and 68.6 ± 13.83.

The situation in the sibs group was different. In these children there was a significant difference between boys and girls, with the boys having significantly higher full-scale and verbal IQ than the girls. When patients and control subjects were compared separately by sex, it was found that the previously malnourished boys differed from the control boys significantly, with control subjects having a full-scale IQ approximately 12 points higher than patients. Although control girls also had somewhat higher IQ than girls who were previously malnourished, the size of the difference was insufficient to result in statistical significance. This pattern of finding was entirely the result of the depressed level of IQ in the girls in the control group. No reasons for this depression could be found in differing educational experience since there were no sex differences in school attendance. A depressed level of IQ in girls relative to boys, however, has been repeatedly found in children of school age in the social groups considered and probably derives from the far lower value attached to the education of girls in this particular subculture (Cravioto *et al.*, 1971).

When the scaled scores on the 11 subtests of the WISC obtained in the survivors of malnutrition were compared with the values found in the siblings, the analysis of the subtest profiles suggested a relative uniform depression of performance in the index cases with respect to a variety of cognitive demands.

In the second study using siblings, Hertzig *et al.* (1972) decided to investigate the degree to which children malnourished before two years of age differ from their sibs and classmates in intellectual competence at school age. The index cases were 74 Jamaican boys who had been treated in a hospital for severe malnutrition before the age of two years. The three main clinical types of severe protein–calorie malnutrition (kwashiorkor, marasmus, and marasmic-kwashiorkor) were represented. On the average, children were hospitalized for a period of 8 weeks. Follow-up visits in the homes by nurses were carried out for 2 years after discharge. At the time of the intelligence testing, the children's age ranged from 6 to 10 years. This age range was selected in order to be far enough removed from the time of hospitalization and for the children to be at an age where intelligence test performance has predictive value.

A sib and a classmate or yardmate were selected as controls for each index case. Two comparison groups were thus included, the first one made up of male sibs. For a sib to be selected he had to be between 6 and 12 years old, nearest in age to his malnourished brother, and without a history of severe malnutrition. The second comparison group was made up of classmates or neighbors of the index cases. For index children attending school, two male classmates closest in age to the index case were selected. If the first comparison child was not available for examination, the second comparison was used. Some of the index boys, although of school age, were not going to school. For these cases, a comparison case was chosen by finding the nearest neighboring child who was not a relative and who was of an age within 6 months of the index case. For some index cases at small schools, no classmate was within 6 months of age. For these cases neighbor children were also used as comparisons. As would be expected from the method of selecting comparison children, index and comparison children lived in the same general neighborhood from which the school drew its pupils.

Each child's intellectual level was individually evaluated by means of the

Wechsler Intelligence Scale for Children. All children were examined without the examiners being aware of the group to which the child belonged.

Results showed that mean verbal, performance, and total (full-scale) intellectual quotients were lower in the survivors of severe malnutrition, with sibs scoring at an intermediate level, and classmates and yardmates having the mean highest scores.

Since the WISC test has a floor of 46 points for both the full scale and the verbal scale, and of 44 points for the performance scale, it is apparent that even if a child's responses were all wrong his minimal intellectual quotient would be the floor value, and therefore a comparison of mean IQs would be inappropriate. With this in mind, Hertzig *et al.* compared the number of survivors of severe malnutrition and of control children who scored at the lowest level of measurable performance. Twenty-three percent of the survivors against 7% of the controls were at the floor level for the full-scale IQ. Similar results were obtained for the performance scale, with no significant difference in the verbal scale. This analysis clearly shows the very low levels of IQ present in the survivors of severe malnutrition. In interpreting the data, it is of importance to note that no age trend for IQ was found in the study groups, nor was ordinal position for sibs responsible for the differences obtained.

The use of siblings as comparison subjects has certain implications for the interpretation of findings. Such comparisons are, of course, advantageous in that the children who are contrasted come from the same families and tend to share a common experiential ambience. Demographic data, however, strongly point out that having a child hospitalized for severe clinical malnutrition in fact identifies a family in which all children are at risk for significant mild–moderate malnutrition on a chronic basis. Therefore, index cases and sibs are similar in sharing a common exposure to subnutrition on a lifelong basis and differ, in this nutritional background, only in that the index cases have a superimposed episode of acute exacerbation. Consequently, the comparison of sibs and index children does not provide a full picture of the overall effects of nutritional inadequacy on intelligence performance. Rather, it indicates the additional consequence for maldevelopment which may attach to the superimposed episode of acute exacerbation. The use of sibs as controls also means that the children compared have shared a generally disadvantageous social and family environment, several features of which can in themselves significantly contribute to the depression of intellectual performance levels. This factor, too, should result in the minimizing of differences between survivors of severe malnutrition and their sibs and provides further support to the significance of the difference found on measured intelligence.

In spite of their limitations, all the reviewed studies clearly show that the environment in which children at risk of malnutrition live is highly negative in its effects on mental development. Irrespective of the presence or absence of a previous admission to a hospital because of severe malnutrition, children developing in this milieu have a high probability of showing poor performance on intelligence testing. The presence of a superimposed episode of malnutrition occurring early in life, and of enough severity to force the child into a hospital, increases the chance of scoring at values even lower than those characteristic of the poor environment.

It must be emphasized that the finding of an association between early malnutrition and lower mental development is by no means evidence that the insufficient intake of nutrients and energy per se affects intellectual competence in man.

Data from comparative psychology (Maier and Schneirla, 1935; Birch, 1954) and evolutionary physiology (Voronin and Guselnikov, 1963) have suggested that the emergence of complex adaptive capacities is underlain by the growth of increasing liaison and interdependence among the separate sense systems. In addition a variety of studies (Birch and Bitterman, 1949, 1951) have indicated that the basic mechanisms involved in primary learning (i.e., the formation of conditioned reflexes) are probably the effective establishment and patterning of intersensory integration. Cravioto *et al.* in Guatemala and Mexico (1966, 1967*a,b*), Champakam *et al.* in India (1968), and Wray in Thailand (1975), in trying further to explore the mental functioning of school-age children who had been at various degrees of risk of malnutrition in the preschool years, have measured the development of intersensory integration in the kinesthetic–visual, kinesthetic–haptic, and haptic–visual modalities. The selection of these intersensory modalities was made on the basis of the work of Birch and Lefford (1963), who found that in normal children between the ages of 6 and 12 years the interrelations among three sense systems—touch, vision, and kinesthesis—improved in an age-specific manner and resulted in developmental curves that were as regular as those for skeletal growth.

In the Indian study, children rehabilitated from kwashiorkor, suffered between the ages of 18 and 36 months, were compared at the age of 8–11 years with matched controls. Intersensory organization was poorer in the index cases than in the control subjects, with highly significant differences. When these children were again tested 5 years later, the differences between the two groups had considerably decreased. Although the survivors of severe protein–calorie malnutrition still made more errors than the control children, the difference was not statistically significant. After another follow-up period of two years, the performance of both groups of children was free of errors.

The apparent catch-up of the nutrition-rehabilitated group has to be taken cautiously. When applying a test with a clearly developmental course, the point at which the asymptotic performance is reached might be the only difference between a group with normal development and a group with a developmental lag. If a child has already completed the maximal level of performance and another child obtains that same level later in time, although both children now have the same score it cannot properly be said that the second child caught up with the first. Moreover, in societies where the demands are chronological-age specific, the importance of a lag in development might be fundamental for the future role and status of those affected in spite of the fact that later in life, such as in adulthood, the test performance of these individuals may not differ at all from that obtained by their more fortunate mates.

In the Mexican, Guatemalan, and Thailand studies, school-age children aged 6 through 11 years were tested. According to the physical growth achievement at each age level, the lower quartile of height distribution (most stunted children) and the upper quartile of the distribution (most fully grown children) were contrasted in their intersensory abilities. Differences in neurointegrative skills were manifested in all combinations of intersensory integrations examined. Tall children, particularly in the younger age groups, performed at a higher level of competence than stunted children. Figures 1 to 3 illustrate the results of the Guatemalan study. Not only were mean differences significant, individual variability in performance also tended to be

JOAQUÍN CRAVIOTO
and ELSA R.
DELICARDIE

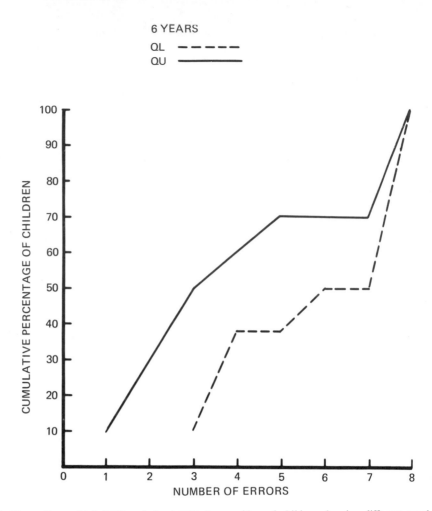

Fig. 1. Proportions of tall (QU) and short (QL) 6-year-old rural children showing different number of errors in the visual–kinesthetic judgment of nonidentical forms.

greater in the shorter children. Obviously, when height is used as an index of risk of exposure to prior malnutrition, at least three contaminating variables must be controlled in interpreting its association with levels of performance. The first is that height differences should not merely be the reflection of familial differences in stature; the second, that short stature should not be just another manifestation of a general developmental lag; and the third, that shorter children must not come from familial environments at significantly lower sociocultural levels.

These nonnutritional factors were ruled out as main contributions to the results obtained since: (1) height of parents and children were not significantly correlated; (2) no significant association was found between the height of the fathers and the level of intersensory competence achieved by the child; (3) tall and short children of the same age in populations of children without antecedent conditions of nutritional risk did not exhibit differences in their levels of intersensory adequacy; and (4) by

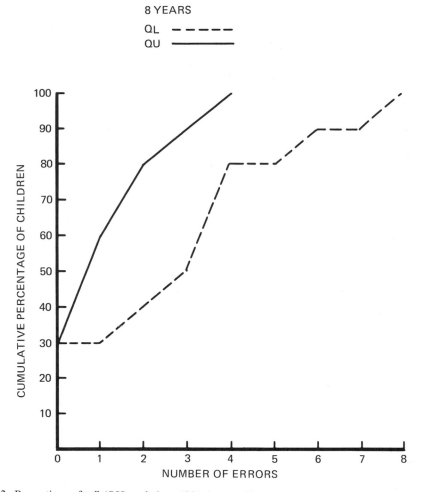

Fig. 2. Proportions of tall (QU) and short (QL) 8-year-old rural children showing different number of errors in the haptic–kinesthetic judgment of nonidentical forms.

the lack of correlation between height and income, occupation, housing, personal cleanliness, presence of sanitary facilities in the home, and contact of parents with mass communication media. Educational level could not be eliminated as an important intervening variable since mother's level of formal education was significantly associated with her child's height. Accordingly, the Guatemalan, Mexican, and Thailand results could be interpreted in the sense that the inadequacy in intersensory development could represent the effects of earlier malnutrition in association with more general subcultural differences between the tall and short children.

A more detailed definition of the roles of malnutrition and cultural factors obviously requires a prospective study of a community of children with varying quantified risks of malnutrition in infancy and during the preschool years.

More recently, we have had the opportunity of a preliminary analysis of the

JOAQUÍN CRAVIOTO
and ELSA R.
DELICARDIE

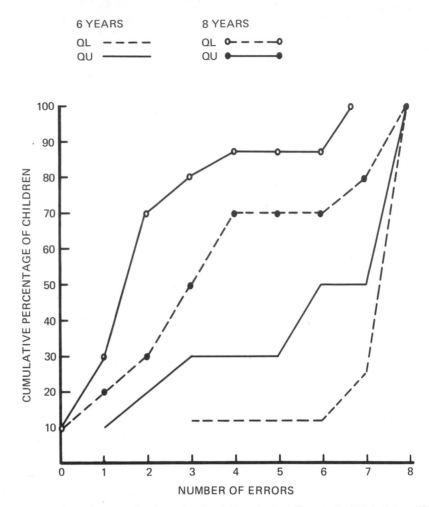

Fig. 3. Changes in performance levels on the visual–kinesthetic ability of tall (QU) and short (QL) rural children.

visual–kinesthetic development in children with known nutritional histories during their first 7 years of life (Cravioto *et al.*, 1969*a,b,* 1970). As may be seen in Figure 4, the competence level at 78 months of age in survivors of severe malnutrition suffered before 3 years of age is markedly lower than performance obtained in children from the same birth cohort who never had severe malnutrition. When the survivors of malnutrition were matched with nonmalnourished controls for total stimulation available in the home, as measured by the inventory of Bettye Caldwell (1967), Figure 5 shows the persistence of the lower levels of competence in the children with a previous history of severe malnutrition.

The reports of neurointegrative adequacy in four different cultural settings (India, Guatemala, Thailand, and Mexico) may be significant because they suggest that functional lags could occur at mild to moderate degrees of protein–calorie malnutrition and may not be limited to the extremely severe cases represented by kwashiorkor and marasmus.

Trying to extend our inquiry to other types of cross-modal integrations, two studies of auditory–visual equivalence were carried out. The first involved schoolchildren of a Mexican rural area where malnutrition in early life is highly prevalent (Cravioto *et al.,* 1967*a*). All children aged 7–12 were weighed and measured, and on the basis of height achievement the lower quartile and the upper quartile of the distribution at each age were compared in their auditory–visual performance. The children's ability to integrate auditory and visual stimuli was individually tested by a method of equivalence in which the children were asked to identify visual dot patterns corresponding to rhythmic auditory patterns (Birch and Belmont, 1964).

At each age level the mean performance of the taller group was higher than that of the shorter. The difference was most striking at age 12, when 42% of the taller children were making eight or more correct judgments, with 30% achieving a perfect score of ten. In contrast, only 9% of the shorter children in this age group achieved scores of eight or greater, with none making a perfect score.

In the second study (Cravioto, 1971) the developmental course of auditory–visual equivalence was studied in 39 school-age children who had suffered severe clinical malnutrition before the age of 30 months, and in 39 siblings of similar sex and age. When the performance of siblings and index cases was contrasted age by

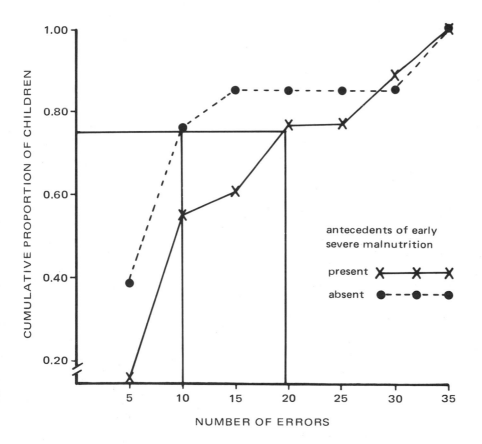

Fig. 4. Visual–kinesthetic intersensory development of 78-month-old children with and without antecedents of early severe malnutrition (nonidentical forms).

**JOAQUÍN CRAVIOTO
and ELSA R.
DELICARDIE**

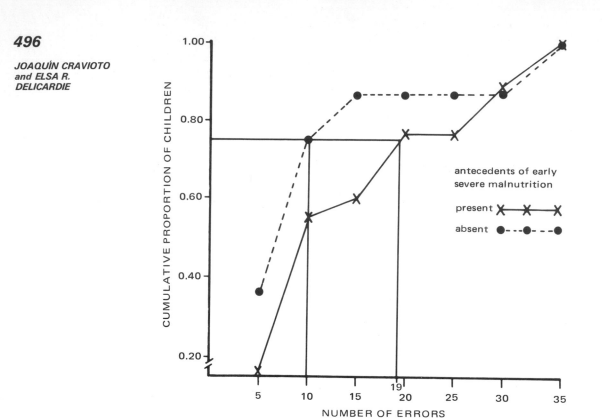

Fig. 5. Visual–kinesthetic performance (nonidentical forms) obtained by 78-month-old children with and without antecedents of early severe malnutrition matched for total scores on home stimulation.

age, the survivors of severe malnutrition were well below their siblings in auditory–visual competence.

Independent of whether intersensory organization is disrupted by malnutrition per se or through one or more of the variables which produce or accompany malnutrition, neurointegrative development is delayed in those children at nutritional risk who have grown poorly. It is, therefore important to consider the potential significance of this type of developmental lag in so primary a process as intersensory organization for more complex psychological functioning. In this connection it is of interest to consider two significant features: learning-conditioned reflex formation and acquisition of academic skills.

In most conditioning situations what is being demanded is the integration between two stimuli, each of which belongs to a different sensory modality. Thus, in classical salivary conditioning or in the conditioning of a leg withdrawal, a gustatory or a tactile stimulus is in effect being linked to an auditory or visual one. The process of conditioning, when effective, therefore involves the establishment of intimate equivalences between initially nonequivalent stimuli in different sensory modalities. If interrelations among sensory modalities are inadequately established the possibility exists that conditioning will either be delayed in its occurrence or that the pairing of stimuli will be ineffective in producing conditioned reflexes. Therefore, failure for intersensory integration to occur at normal age-specific points can contribute to inadequate primary learning at the given age level.

Evidence already exists that the lag in the development of certain varieties of intersensory integrations have a high correlation with backwardness in learning to read or to write. In this respect, Birch and Belmont (1964, 1965) and Kahn (1965) have data strongly suggesting that backwardness in reading is significantly associated with inadequacy in auditory–visual integration. Data are also available that point out the dependence of visual–motor control in design-copying on visual–kinesthetic intersensory organization (Birch and Lefford, 1967). Since the time of Baldwin (1897) it has been recognized that such visual–motor control is essential for learning to write, therefore, the lag in intersensory organization can interfere with a second primary educational skill—learning to write.

Thus, inadequacies of intersensory development can place the child at risk of failing to establish an ordinary normal background of conditionings in his preschool years and at risk of failing to profit from educational experience in the school years.

6. Visual Perception

An essential prerequisite for learning to read is the child's ability to distinguish simple visually presented figures and to respond to more differentiated aspects of the figural percept such as angular properties and spatial orientation. Failure to respond to these aspects can result in the child confusing the letters which in the Roman alphabet are identical in form but distinguishable by their spatial positioning. Letters such as lower-case b and d, p and q, or capital N and Z, W and M, all represent equivalent shapes with the distinction among them depending upon the child's ability to respond simultaneously to shape and to orientation in visual space.

Using the visual discrimination task developed by Birch and Lefford for their studies on voluntary motor control (1967), visual perception of forms was explored in 39 schoolchildren who had been treated for severe malnutrition at ages below 3 years and in an equal number of their nearest-age and same-sex siblings. Age by age, the levels of performance of recognition of geometric forms, although low for both groups of children, were significantly lower for the survivors of severe malnutrition until age 9 years. By this age both siblings and previously malnourished children achieved similar levels of competence; however, this asymptotic performance level had been reached by the siblings at age 7 years.

When the children were tested for their ability to analyze geometric forms, the mean number of errors committed also decreased as age advanced. Once again a marked difference was found in favor of the siblings. Thus, survivors of severe malnutrition present a lag in readiness to read which is even lower than the readiness characteristic of their low socioeconomic class (Cravioto et al., 1969c).

7. Language Development

In the course of a longitudinal ecological study of a total cohort of children born during one calendar year in a rural village of central Mexico, of the 334 infants included 14 girls and 8 boys developed clinically severe protein–calorie malnutrition before the age of 5 years. Of the 22 patients, 15 developed kwashiorkor and the other 7 marasmus. Ten children, 6 with kwashiorkor and 4 with marasmus, were treated at home, while 9 children with kwashiorkor and 3 with marasmus were treated in a pediatric hospital. No deaths occurred in the latter group, but 3 of the 10 children treated at home died.

These cases occurred despite the fact that all children in the cohort studied were medically examined on a biweekly basis. Children who failed to grow normally were identified, their infectious illnesses were treated, and their parents were given advice (which they did not follow) on the appropriate feeding and management of children who failed to thrive. In contrast to its lack of influence on the incidence of clinically severe PCM, this medical attention decreased the infant mortality rate from a figure of 96/1000 to 46, and reduced the preschool mortality of the cohort by half. Incidentally, these data point out, once more, that traditional medical care can strongly influence mortality with minimal or no effect on morbidity.

Since one of the aspects under investigation in the cohort was language and communication, it was possible to analyze its development prior to, during, and after the episode of severe malnutrition. In a first report, Cravioto and DeLicardie (1972) have described language development and appearance of bipolar concepts in the group of survivors of severe malnutrition in comparison with a group of children selected from the same birth cohort who were never diagnosed as severely malnourished and who were matched, case by case at birth, for sex, gestational age, season of birth, body weight, body length, and performance on the Gesell test.

During the first 3 years of life, language acquisition was evaluated by means of a technique similar to the Gesell test (Gesell and Amatruda, 1947). Mean language development was very similar in index cases and controls during the first year of life when only one case of severe malnutrition had been diagnosed. As time elapsed and more children became ill with severe malnutrition, a difference in language performance favorable to the matched controls became evident. The difference was more pronounced at each successive age tested. Not only were mean values significantly lower in the index cases, the distribution of individual scores was also markedly different. At 1080 days of life 11 of the 19 control children had language scores above 1021 day-equivalents. In contrast, while none of the children in the malnourished group scored above 960 day-equivalents, 12 of the 19 cases had values below 720 days, with 3 of these children exhibiting language development inferior by more than 6 months to that observed in the control children who had the lowest scores.

Since examinations of bipolar concept acquisition were done in the total cohort at 26, 31, 34, 38, 46, 52, and 58 months of age, it was possible to compare the patterns of competence of this function in the group of children who developed severe malnutrition and in their selected controls, both at the time some of the children were suffering from severe malnutrition and after they had been nutritionally rehabilitated.

Data were obtained by Rojano (1970) in the total cohort of Mexican children tested at successive ages with the series of bipolar concepts, developed by Francis H. Palmer at the Institute for Child Development and Experimental Education of the City University of New York, U.S.A., for the systematic training and enrichment of language experience. They documented a clearly developmental course of competence in response to tasks involving the utilization of such bipolar concepts as a scale of progressive difficulty. The test could be used with young children less than 3 years of age because most of the items included only required the selection by the child of an object representing a given pole (e.g., big–little: long–short; in–out; black–white) from two objects differing only with respect to their position on one of the concepts continua. The items are grouped into two forms so that each form contains items covering both poles of each concept. The forms differ only with respect to the setting in which the concepts are placed. The score derived from the test provides a measure of the child's knowledge of various categories that are commonly used in organizing sensory experience.

Since examinations of bipolar concept acquisition were done in the total cohort at 26, 31, 34, 38, 46, 52, and 58 months of age, it was possible to compare the patterns of competence of this function in the group of children who developed severe malnutrition and in their selected controls, both at the time some of the children were suffering from severe malnutrition and after they had been nutritionally rehabilitated.

The mean number of bipolar concepts present in the children with present or past severe malnutrition was significantly lower than the mean number of concepts shown by the control group. The differences at all ages but age 31 months were statistically significant. Not only were mean values significantly higher in the control children, the proportion of better performers was also higher. Thus, for example, at 26 months of age while almost half the control children had five or six concepts, none of the malnourished subjects had attained this level of competence. On the other hand, one out of every three malnourished children had a maximum of one concept while seven out of every ten controls are at the level of two or more concepts. Even after nutritional rehabilitation the survivors continued to lag behind the controls. Thus, at age 58 months four out of every ten survivors had not more than 17 concepts, while the minimum number shown by the control children was 18 concepts. Similarly, none of the previously malnourished children had more than 20 concepts while two out of every ten controls were above this level. A study of increments at successive ages showed that between ages 26 and 38 months, the period in which the last cases of severe malnutrition appeared, increments in the control group were far greater than those in the malnourished group. During rehabilitation the survivors of severe malnutrition had an increment twice as big as that of the controls; unfortunately, at later ages the increments tended to be lower, and there was not enough time for the previously malnourished children to catch up. In other words, the lag in language development present when the children were actually malnourished continued to persist after clinical recovery had taken place.

Since the group of malnourished children had significantly lower scores in home stimulation than the control children, an attempt was made to separate the possible influences of stimuli deprivation from those of malnutrition and an analysis of the interrelations among these two variables and the number of bipolar concepts was carried out. As a first approach to this issue a technique of partial correlation was employed in order to estimate the degree of association between two of the variables "holding constant" the influence of the third variable. Since the number of survivors of severe clinical malnutrition is rather small, it was decided to test for the interrelations in the total birth cohort assuming that body height is a good indicator, in the community under study, of the risk of chronic malnutrition.

The coefficients of correlation (product × moment) among home stimulation scores, total body height, and number of bipolar concepts present at 46 months of age in the total cohort were all significant at $P < 0.05$. When the relation between home stimulation and number of bipolar concepts was partialed out for body height, the coefficient of correlation dropped from 0.20 to 0.15. When the relation between body height and number of bipolar concepts was partialed out for home stimulation, the coefficient changed from 0.26 to 0.23; finally when the number of bipolar concepts was held constant, the coefficient of correlation between home stimulation and body height changed from 0.23 to 0.19.

These results suggest that the association between home stimulation and number of bipolar concepts is mediated, to a good extent, through body height which in turn holds a significant degree of association with the number of bipolar concepts independent, to a large extent, of home stimulation. Within the limits of the probabilities given by the magnitude of the coefficients, home stimulation contributes relatively more to body height than to number of bipolar concepts while body height contributes more than home stimulation to the variance of bipolar concepts.

8. Styles of Response to Cognitive Demands

Since the studies on the possible effects of malnutrition on mental development and learning in humans have, for the most part, focused on achievement without considering behavioral style, DeLicardie and Cravioto (1973) analyzed, as a part of their longitudinal study on the effects of environment on mental development, the modification that an episode of severe malnutrition might make on the behavioral style of lower-class children. In other words, they were concerned not with *how well* but with *how* the children with antecedents of severe malnutrition behaved. To answer this question the strategy for the analysis focused on the comparison of responsiveness to cognitive demands in 14 survivors of severe clinical malnutrition suffered before 38 months of age, and in two groups of children selected from the same birth cohort who never showed signs or symptoms of severe malnutrition and who were matched, case by case, with the survivors. In one comparison group the matching was done for sex, gestational age, season of birth, body weight, total body length, and organization of the central nervous system as determined by the Gesell method. The second comparison group included 14 children, full-term and healthy at delivery, who were matched for sex and IQ (WPPSI) at 5 years of age with the survivors of severe malnutrition.

At the age of 5 years, all children were individually examined using an adapted version of the Wechsler Preschool Primary Scale of Intelligence (WPPSI). The tasks of this test were used to obtain behavioral information on response style to a standardly presented set of cognitive demands. During administration of the test, a detailed protocol of the child's behavior and verbalization was recorded by an independent observer.

The analysis of response styles followed the logic tree developed by Hertzig *et al*. (1968). When confronted with a demand, a child can respond either by working or not working. This initial choice can be expressed through either verbalization or motor action. If the choice was to work, whether verbally or nonverbally, the response can involve an expression of spontaneous ideational extension or it can be delimited and restricted to the defined requirements of the task. In relation to nonwork responses the different styles can be expressed as simple negation, substitutive behavior, request for assistance, or as a rationalization with respect to competence when the starting point is verbal. If the nonwork response is nonverbal, the subdivisions are simple negation, substitutive behavior, request for assistance, or passive unresponsiveness. All categories, in terms of which response styles to cognitive demands are classified, are objective generalizations about observed behavior. As Hertzig and co-workers have emphasized, *the categories are not, nor are they intended to be, inferences about the underlying reasons for the expression of the observed behavioral patterns.*

Since the logic classification of styles is done mainly through a series of dichotomies, the way for identifying individual differences in the responsiveness to cognitive demands consisted of ascribing a child to a particular style when his proportion of responses of that style was equal to or greater than 0.75. Table II shows the main style of response in survivors of malnutrition and controls.

It is apparent that while survivors and controls for IQ and sex tend to be similar in their styles of response, the children in the group of controls matched at birth showed a different style. Verbal behavior predominated among this latter group, and the children who were classified as of the ''not-work'' type had verbal competence as their main style of response. Conversely, among the survivors and the controls matched for IQ and sex, all the not-work children were of the not-work–nonverbal type, unresponsiveness being the typical behavior of these not-work children. Moreover, controls for size at birth, when they give not-work verbal responses, express them mainly in terms of competence rationalizations.

In the presence of these findings, one might tend to consider that the differences in patterns of response between the controls matched at birth and the survivors of severe malnutrition could be explained on the basis of differences in intellectual performance. The observation of similar behavioral patterns in children with and without antecedents of severe malnutrition but with equally low intellectual performance would be in favor of this explanation. Moreover, Hertzig *et al.* (1968), in their study of middle-class American and working-class Puerto Rican

Table II.   Main Style of Response Obtained in Survivors of Early Severe Malnutrition and Controls for Size at Birth, and Sex and IQ at 5 Years of Age, in the "Land of the White Dust"

Behavioral style	Proportion of		
	Survivors	Controls matched for IQ and sex	Controls for size at birth
Work	0.57	0.71	0.85
Not-work	0.43	0.29	0.15
Work verbal	0.62	0.80	0.75
Work nonverbal	0.38	0.20	0.25
Not-work verbal	0.00	0.00	1.00
Competence	0.00	0.00	1.00
Not-work nonverbal	1.00	1.00	0.00
Passivity	1.00	1.00	0.00

JOAQUÍN CRAVIOTO
and ELSA R.
DELICARDIE

children, found that differences in IQ affected the proportion of demands that was met by a work response, the proportion of total responses that were verbally expressed, and the style of verbal not-work. However, the difference in styles observed between middle-class and Puerto Rican children persisted at all IQ levels. Lugo (1971), in a study of urban Mexican children of three different social classes, also found a quantitative difference in responsiveness as a function of IQ level, but the differences in response style persisted across the socioeconomic levels in the presence of a common IQ range. Accordingly, one cannot accept differences in IQ as the reason for differences in behavioral patterns of response.

Besides the IQ difference between survivors of malnutrition and controls matched for size at birth, home stimulation scores, as assessed by the inventory of Caldwell (1967), were significantly higher in the controls matched at birth, with similar low scores in both survivors and controls matched for IQ and sex. This difference in stimulation could significantly contribute to the development of dissimilar patterns of responsiveness, particularly in relation to the amount and type of verbalization.

In attempting to tease out the effects that may be due to malnutrition from those which may be due to stimulus deprivation, survivors of severe malnutrition and controls with equal scores in home stimulation were identified. Ten controls for IQ and sex, seven survivors of severe malnutrition, and six controls for size at birth met this requirement. When the styles of response of these children were compared, no difference was found among the three groups in the mean number of total responses, in the proportion of work responses, in the proportion of total responses verbally expressed, and in the styles of nonverbal not-work responses. On the other hand, the proportion of verbal not-work responses observed in controls matched at birth (0.87) was almost three times the proportion found in either survivors (0.30) or controls matched for IQ and sex (0.26). The difference between controls matched at birth and the other two groups was significant at the level of confidence of 0.001.

When the styles of verbal not-work responses were compared, it was observed that the three groups were markedly different. Controls matched at birth expressed their verbal not-work responses in terms of rationalization of competence, survivors expressed their responses mainly as request for aid, and controls matched for IQ and sex had similar proportions of styles of competence, requests for aid, and substitution.

The finding of unresponsiveness as a characteristic style in children who, with or without antecedents of severe malnutrition, have as a common background a low level of home stimulation leads one to consider that, regardless of a physiological component that may be present when the children were malnourished, the passive behavior of survivors of severe malnutrition is probably linked to stimuli deprivation.

The differences observed between survivors of severe malnutrition and controls with equal scores of home stimulation seem to indicate that, besides the effect of stimulus deprivation on the styles of response, the antecedent of clinical severe malnutrition appears to be another influential factor.

It has been suggested by Canosa et al. (1973) that differences in performance between well-nourished and malnourished children may be due to a lower ability of the malnourished subjects to attend to or concentrate on the given task. However, since these children have a high frequency of passive responses when confronted with a demand, failure to perform efficiently could be a consequence of this style of behavior. For example, any task requiring systematic scanning or elaboration of information by progressively more complicated steps requires the child to engage actively in the pursuit of an answer. Unresponsiveness as a style of behavior may lead to a quick answer without regard for its accuracy. Similarly children with passive patterns of behavior may give low scores on tasks that are penalized for time.

9. Influence of Age and Duration of Malnutrition on Later Performance

Dobbing (1976), through a masterly series of studies, has documented his concept of the developing brain's vulnerability. According to this concept there is a pathology of brain development in which the timing of etiological factors in relation to the growth spurt sequence is of even more significance than their duration and their severity. Moderate growth restriction, which before the brain growth spurt or afterwards does not leave traces of pathology, results in permanent structural and functional alterations when it occurs at the time when the brain is passing through the period of most rapid increase in weight.

In accordance with this concept, age at the time of insult and duration of the period of malnutrition, or rather of growth restriction, becomes of the greatest importance in relation to the prognosis. The general hypothesis would be that the effect of severe postnatal malnutrition, capable of significantly restricting somatic growth, would vary as a function of the period of life at which the lag in growth was experienced.

To test this hypothesis Cravioto and Robles (1965) examined the mental recovery of 20 Mexican infants hospitalized for kwashiorkor. On admission for treatment, 6 infants were below 6 months of age, 9 between 15 and 29 months, and 5 between the ages of 37 and 42 months. All children were tested with the Gesell method at regular intervals of 2 weeks. As recovery from malnutrition occurred, developmental quotients increased in most of the patients and the gap between normal age expectation and the actual performance of the child progressively diminished for all except those in the group whose age on admission was less than 6 months. These younger malnourished infants showed no tendency to catch up and increased in developmental age only by a figure equal to the number of months they remained in the hospital. Serial data on performance plotted against days of hospitalization showed that the older the group, the greater the value of the slope.

Hoorveg and Standfield (1972) selected from their records in Uganda three groups of children aged 11–17 years who had been treated for severe clinical malnutrition suffered before 27 months of age. Each group consisted of 20 former patients whose ages when admitted for treatment were less than 16 months for group one, between 16 and 21 months for group two, and between 22 and 27 months for those included in group three. General intelligence, verbal abilities, spatial and perceptual abilities, visual memory, short-term memory, learning and incidental learning, and motor development were evaluated. Significant differences among groups were found in memory for designs, in which group three performed better than groups one and two, and in the learning task, in which group two did better than group one. In two other tests, incidental learning and short-term memory, the differences as a function of age at which malnutrition was treated reached a significant level of confidence of 0.10. The findings are in the expected direction in which those children who suffered younger do more poorly.

Intelligence quotients, derived from a battery of tests that included Raven Progressive Matrices, the Koch test, and the Goodenough Draw-a-man test, were found significantly associated with the age on admission for treatment of severe malnutrition in a group of 14 survivors examined 9–11 years after discharge from a hospital in El Salvador, Central America (Guillen-Alvarez, 1971).

Chase and Martin (1970), in the U.S.A., did a follow-up study of 19 infants who were hospitalized at less than one year of age because of severe undernutrition. Forty-two months after discharge from the hospital, the mean developmental quotient obtained with the Yale Revised Developmental Examination was significantly lower than the quotient found in a comparison group of children matched for birth date, weight, sex, race, and socioeconomic status. Children admitted to the hospital with severe malnutrition lasting longer than the first 4 months of life gave the lowest quotients on the follow-up. All the children admitted in the first 4 months of their lives had, 3½ years later, developmental quotients above 80, whereas 9 of 10 children who had suffered malnutrition for periods longer than their first 4 months of life had developmental quotients below 80. These findings are in line with a report of DeLicardie *et al.* (1971), who found that a group of infants who, for no detectable reason, had body weights at 15 days of life below their birth weight, continued to weigh less than their matched controls throughout the first year of life and lag behind in total body length, head circumference, chest circumference, arm circumference, and brachial skinfold thickness without exhibiting a significant difference in their mental development, assessed by the Gesell technique. Along this same line of age and duration, Stein and co-workers (1975), in their brilliant epidemiologic study of the Dutch famine, did not find a relation between the mental performance of young men at the time of their military induction and their prenatal exposure to famine. Since no evidence could be found of interaction of prenatal famine exposure with indicators of social environment that might have influenced compensatory learning opportunities and promoted better subsequent mental development, and also no evidence was found that selected survival might have masked or distorted an association of prenatal famine exposure with mental performance at the age of military induction, it appears as if deficient food intake confined to the prenatal period may be either of too short duration to produce a detectable effect on mental performance on the young adult or the insufficient nutrient intake period was out of phase with the brain's growth spurt, which for the human starts around mid-gestation but has its maximal rate postnatally and extends its duration at least well into the second year of life (Dobbing, 1974).

Hertzig and associates (1972), in their study of Jamaican school-age boys who had suffered from severe malnutrition before age 2, did not find a significant correlation between age of the child at the time of admittance for treatment and Wechsler intelligence quotient. When the IQs of the 74 children were divided into those belonging to the children admitted for treatment before 8 months of age, between 8 and 12 months, and between 13 and 24 months, an analysis of variance showed that the means for the subgroups were not statistically different at the 5% level of confidence. Hertzig's data show very clearly that IQ mean values in the survivors of malnutrition are misleading due to the significant number of children who scored at the floor level of the Wechsler. Hence, before accepting the author's conclusion of no relationship between time at which severe malnutrition occurs and severity of mental outcome, it might be convenient to test for performance differences in their age groups using nonparametric statistical techniques.

Klein *et al.* (1975) studied 50 children, 44 boys and 6 girls, aged 5–14 years, who had been treated for congenital pyloric stenosis. They estimated the severity of starvation as the percentage difference of infant's weight on admission for treatment and expected weight for age, extrapolated from birth weight. General intelligence, measured by the Peabody Picture Vocabulary test and the Raven Progressive Matrices test, showed a significant correlation with the estimated degree of starvation ($r = -0.323$, $P < 0.05$). The scores on a scale that measures the parental evaluation of the child's intellectual development and expected educational potential also correlated significantly with the severity of starvation ($r = -0.367$; $P < 0.01$). These data apparently are at variance with the study of Berglund and Rabo (1973), who did a follow-up of 174 Swedish boys who suffered from inanition, starting at ages 6–20 days, because of pyloric stenosis. After a variable period of starvation, during which body weight became lower than birth weight, the patients recovered. A significant correlation between height at adulthood and weight loss and duration of starvation in infancy was found, but the performance on an intelligence test, administered at the time of registration for military service, did not correlate with the history of malnutrition in infancy.

Even considering the limitations inherent to the types of studies reported, it seems apparent that severity, duration, and particularly the time at which the malnutrition insult occurs are relevant factors in determining the permanency or transiency of the physical and mental aftereffects. It must be remembered at all times that nutritional deprivation may constitute just another of the numerous factors present in the impoverished environment in which infants and children develop malnutrition.

10. Mechanisms of Action of Malnutrition

It must be emphasized that the finding of an association between early malnutrition and lags in mental development and learning and a distortion in behavior is by no means evidence that the insufficient intake of nutrients and calories per se affects intellectual competence, learning, and behavior in man. In attempting to explain the effect of malnutrition on intellectual competence and learning at least two possibilities can be entertained. The first and simplest hypothesis would be that the nutrient deficiency affects mental functioning by directly modifying the anatomical and/or biochemical structure of the central nervous system. In favor of this explanation it is relevant to remember that increase of cell cytoplasm, with extension of axons and dendrites (one of the processes associated with the growth of the brain in early life), is largely a process of protein synthesis. From the microspectrographic investigation of the regenerating nerve fibers, it has been estimated that protein substance increases more than 200 times as the apolar neuroblast matures into the young anterior horn cell. In experimental animals specific amino acid deficiencies can cause structural and functional lesions of the central nervous system (Scott, 1964). In mice, inhibition of protein synthesis in the brain produced by puromycin is accompanied by loss of memory (Flexner *et al.,* 1962). Delays in myelination and reduction in cell number and in cell distribution in the brain caused by interference with adequate nutrition in early age have been amply documented (Dobbing, 1968, 1974; Winick and Rosso, 1969; Chase *et al.,* 1967; Benton *et al.,* 1966; Davison and Dobbing, 1966; Guthrie and Brown, 1968; Kerr *et al.,* 1976; Winick, 1969; Winick and Noble, 1966). Preliminary findings of reduction in brain size and even in cell number in children who died with severe malnutrition have been reported from Mexico (Ambrosius, 1961), Uganda (Brown, 1965), and Chile (Winick, 1969; Winick *et al.,* 1970). In this respect Stein *et al.* (1975) have rightly pointed out that even autopsy at death cannot prove the irreversibility of the brain cell depletion reported in fatal cases of early malnutrition nor indeed can it be inferred from these studies whether the depletion is compatible with life and even to be found in survivors.

In regard to functional changes linked to inadequate diets, Barnes (1968; Barnes *et al.,* 1970) and Frankova and Barnes (1968) have documented a series of disorders, including learning disability, in rats and pigs, and Platt *et al.* (1964) have reported electrophysiological changes in various animal species.

The second hypothesis considers that malnutrition in human children does not need to produce structural lesions of the central nervous system to affect intellectual competence, behavior, and learning. Three possible indirect mechanisms can be postulated:

Loss of Learning Time. Since the child was less responsive to the environment when malnourished, at the very least the child had less time in which to learn and had lost a certain number of months of experience. On the simplest basis, therefore, the child would be expected to show some developmental lags.

Interference with Learning during Critical Periods of Development. A considerable body of evidence exists indicating that interference with the learning process at specific times during its course may result in disturbances in function that are both profound and of long-term significance. Such disturbance is not merely a function of the length of time the organism is deprived of the opportunities for learning. Rather, what appears to be important is the correlation of the experiential opportunity with a given stage of development. It is possible that exposure to malnutrition at particular ages may in fact interfere with development at critical points in the child's growth course and so provide either abnormalities in the sequential emergence of competence or a redirection of the developmental course in undesired directions.

Motivation and Personality Changes. It should be recognized that the mother's response to the infant is to a considerable degree a function of the child's own characteristics of reactivity. One of the first effects of malnutrition is a reduction in the child's responsiveness to stimulation and the emergence of various degrees of apathy. Apathetic behavior, in its turn, can function to reduce the value of the child as a stimulus and to diminish the adult's responsiveness to the child. Thus, apathy can provoke apathy and so contribute to a cumulative pattern of reduced adult–child interaction. If this occurs, it can have consequences for stimulation, learning, maturation, and interpersonal relations, the end result being significant backwardness in performance on later more complex learning tasks. It has been reported in experimental animals that small but statistically significant differences in the size of the cerebral cortex can be obtained by manipulation of the stimulatory aspects of the environment (Diamond *et al.,* 1966). Recently we have reported (Castilla-Serna *et al.,* 1973) the synergistic effects of malnutrition and stimulus deprivation on the biochemical structure of the brain, confirming and extending the findings of Levitsky and Barnes (1972) on the effects of nutrition and isolation on animal behavior. Cravioto and DeLicardie, in a study on the environmental correlates of severe malnutrition (1972), found that none of the characteristics of parents (biological, social, or cultural) or family circumstances (including income per capita, main source of income, and family size) were significantly associated with the presence or absence of severe malnourishment. On the other hand, mother's listening to the radio and level of stimulation available in the home help substantially to differentiate families with or without severely malnourished children. These microenvironmental characteristics, as well as an inadequate mother–child interaction (Cravioto and DeLicardie, 1974), were present a long time before the appearance of the severe episode of malnutrition.

Klein and co-workers (1972) have shown that in children rehabilitated from malnutrition body height and head circumference increase the level of prediction of cognitive function over and above the prediction level given by sociocultural factors such as quality of house, father's education, mother's dress, mother's personal cleanliness, task instruction, and social contacts. Richardson (1976) has reported that the IQs of Jamaican school-age children who suffered severe clinical malnutrition before age 2 years were significantly associated with the presence or absence of severe malnutrition as such, with total body length at the time of intelligence testing, and with a measure of the child's social background. These studies point out the importance of the environment as a synergistic factor both in the development of early malnutrition and its consequences.

Research work in experimental animals leads in the same direction. Thus, Barnes and his group, on the basis of their results from a long series of animal experiments (Barnes, 1968; Frankova and Barnes, 1968; Barnes *et al.,* 1970; Levitsky and Barnes, 1973), prefer to speak of the interaction between malnutrition and environmental stimulation. The similarity of the biochemical changes produced in the brain by nutrition or by stimulation have led them to consider that the physiological mechanisms which may be responsible for the long-term effects of early stimulation may not be operative if a concurrent state of malnutrition is present during a critical period of development. Malnutrition may thus change the experience of perception of the environment by physiologically rendering the animal less capable of receiving or integrating, or both, information about the environment. These authors have also considered that even in the absence of biochemical alterations of the brain, malnutrition may elicit behavior that is incompatible with the incorporation of environmental information necessary for optimum cognitive development. Behavior primarily food-oriented and behavior expressed as apathy and social withdrawal are two examples of the kind of behaviors exhibited with a very high frequency by malnourished subjects.

At present we are just beginning to have enough data to tease out the specific contributions of a lack of nutrients, inadequate stimulation, and diminished experiential opportunities to defective cognitive function. It is most probable that all factors are interdependent and interactive (Barnes, 1976; Cravioto *et al.,* 1966; Pollitt, 1973; Richardson, 1976). However, the data reviewed leave no doubt that survivors of severe malnutrition show, during quite a long time after rehabilitation, decreased measured intelligence together with developmental lags in tasks related to the learning of language and of certain basic academic skills. The most important questions to be answered, hopefully by the on-going longitudinal growth studies, are related to a documentation of the quantitative effects of deficient nutritional status on mental development and to the mechanisms of action of malnutrition either alone or in conjunction with the other features of the unfavorable macro- and microenvironments in which malnutrition flourishes.

ACKNOWLEDGMENTS

The Nutrition Foundation, Inc., The Foundation for Child Development (formerly Association for the Aid of Crippled Children), The Van Ameringen Foundation, The Von Monell Foundation, and The Mexican National System for the Integral Development of the Family (DIF) have financially supported this work.

11. References

Ambrosius, K., 1961, El comportamiento del peso de algunos órganos en niños con desnutrición de tercer grado, *Bol. Med. Hosp. Infant. (Mexico)* **18**:47.

Baldwin, J. M., 1897, *Mental Development in the Child and the Race,* Macmillan, New York.

Barnes, R. H., 1968, Behavioral changes caused by malnutrition in the rat and the pig, in: *Environmental Influences* (H. Glass, ed.), pp. 62–60, Rockefeller University Press and Russell Sage Foundation, New York.

Barnes, R. H., 1976, Dual role of environmental deprivation and malnutrition in retarding intellectual development, *Am. J. Clin. Nutr.* **29**:912.

Barnes, R. H., Moore, A. V., and Pond, W. G., 1970, Behavioral abnormalities in young adult pigs caused by malnutrition in early life, *J. Nutr.* **100**:14.

Barrera-Moncada, G., 1963, *Estudios Sobre el Crecimiento y Desarrollo Psicológico del Síndrome Pluricarencial (Kwashiorkor),* Grafos, Caracas.

Bennett, E. L., Diamond, M. C., Krech, D., and Rosenzweig, M. R., 1964, Chemical and anatomical plasticity of brain, *Science* **146**:610.

Benton, J. W., Moser, H. W., Dodge, P. R., and Carr, S., 1966, Modification of the schedule of myelination in the rat by early nutritional deprivation, *Pediatrics* **38**:801.

Berlund, G., and Rabo, E., 1973, A long-term follow-up investigation of patients with hypertrophic pyloric stenosis with special reference to the physical and mental development, *Acta Paediatr. Scand.* **62**:125.

Birch, H. G., 1954, Comparative psychology, in: *Areas of Psychology* (F. Marcuse, ed.), pp. 446–477, Harper, New York.

Birch, H. G., and Belmont, L., 1964, Auditory-visual integration in normal and retarded readers, *Am. J. Orthopsychiatry* **34**:852.

Birch, H. G., and Belmont, L., 1965, Auditory visual integration, intelligence and reading ability in school children, *Percept. Mot. Skills* **20**:295.

Birch, H. G., and Bitterman, M. E., 1949, Reinforcement and learning. The process of sensory integration, *Psychol. Rev.* **65**:292.

Birch, H. G., and Bitterman, M. E., 1951, Sensory integration and cognitive theory, *Psychol. Rev.* **58**:355.

Birch, H. G., and Lefford, A., 1963, Intersensory development in children, *Monogr. Soc. Res. Child Dev.* **28**:1.

Birch, H. G., and Lefford, A., 1967, Visual differentiation, intersensory integration and voluntary motor control, *Monogr. Soc. Res. Child Dev.* **32**:1.

Birch, H. G., Piñeiro, C., Alcalde, E., Toca, T., and Cravioto, J., 1971, Relation of Kwashiorkor in early childhood and intelligence at school age, *Pediatr. Res.* **5**:579.

Botha-Antoun, E., Babayan, S., and Harfouche, J., 1968, Intellectual development relating to nutritional status, *J. Trop. Pediatr.* **14**:112.

Brockman, L. M., and Ricciuti, H. N., 1971, Severe protein–calorie malnutrition in infancy and childhood, *Dev. Psychol.* **4:**312.

Brown, R. E., 1965, Decreased brain weight in malnutrition and its implications, *East Afr. Med. J.* **42:**584.

Cabak, V., and Najdavic, R., 1965, Effect of undernutrition in early life on physical and mental development, *Arch. Dis. Child.* **40:**532.

Caldwell, B. M., 1967, Descriptive evaluations of child development and of developmental settings, *Pediatrics* **40:**46.

Canosa, C. A., Solomon, R. L., and Klein, R. E., 1973, The intervention approach: The Guatemalan study, in: *Nutrition, Growth and Development of the North American Indian Children* (W. M. Moore, M. M. Silver, and M. S. Read, eds.), pp.185–199, Publication No. NIH 72, U.S. Govt. Printing Office, Washington, D.C.

Castilla-Serna, L., Cravioto, R. A., and Cravioto, J., 1973, "Interacción de la Estimulación y la Nutrición Sobre el Desarrollo Bioquímico del Sistema Nervioso Central (Informe Preliminar)," Proc. XXXVII Meeting Mexican Pediatric Society, San José Vista Hermosa, Mor., Mexico, December 7–8.

Champakam, S., Srikantia, S. G., and Gopalan, C., 1968, Kwashiorkor and mental development, *Am. J. Clin. Nutr.* **21:**844.

Chase, P. H., and Martin, H. P., 1970, Undernutrition and child development, *N. Engl. J. Med.* **282:**933.

Chase, P. H., Dorsey, J., and McKhann, G. M., 1967, The effect of malnutrition of the synthesis of myelin lipid, *Pediatrics* **40:**551.

Cravioto, J., 1971, Infant malnutrition and later learning, in: *Progress in Human Nutrition* (S. Margen and N. L. Wilson, eds.), pp. 80–96, Avi, Westport, Connecticut.

Cravioto, J., and DeLicardie, E. R., 1972, Environmental correlates of severe clinical malnutrition and language development in survivors from Kwashiorkor or marasmus, in: *Nutrition, The Nervous System and Behavior,* pp. 73–94, Panamerican Health Organization, Scientific Publication No. 251.

Cravioto, J., and DeLicardie, E. R., 1974, Mother–infant relationship prior to the development of clinical severe malnutrition in the child, pp. 126–137, *Proc. IV-Western Hemisphere, Nutrition Congress* (P. L. White and N. Selvey, eds.), pp. 126–137, Bal Harbour Florida, August 19–22.

Cravioto, J., and Robles, B., 1963, The influence of protein-calorie malnutrition on psychological test behavior, in: *First Symposium of the Swedish Nutrition Foundation, Mild–Moderate Forms of Protein-Calorie Malnutrition* (G. Blix, ed.), p. 115, Almqvist and Wiksell, Uppsala.

Cravioto, J., and Robles, B., 1965, Evolution of adaptive and motor behavior during rehabilitation from kwashiorkor, *Am. J. Orthopsychiatry* **35:**449.

Cravioto, J., DeLicardie, E. R., and Birch, H. G., 1966, Nutrition, growth and neurointegrative development. An experimental and ecologic study, *Pediatrics* **38:**319.

Cravioto, J., Gaona-Espinosa, C., and Birch, H. G., 1967*a,* Early malnutrition and auditory-visual integration in school-age children, *J. Spec. Educ.* **2:**75.

Cravioto, J., Birch, H. G., and DeLicardie, E. R., 1967*b,* Influencia de la desnutrición sobre la capacidad de aprendizaje del niño escolar, *Biol. Med. Hosp. Infant. Mexico* **24:**217.

Cravioto, J., DeLicardie, E. R., Rosales, L., and Vega, L., 1969*a,* The ecology of growth and development in Mexican preindustrial community, 1. Methods and findings from birth to one month of age, *Monogr. Soc. Res. Child Dev.* **34:**129.

Cravioto, J., DeLicardie, E. R., Piñeiro, C., and Alcalde, E., 1969*b,* Neurointegrative Development and Intelligence in School Children Recovered From Malnutrition in Infancy, Seminar on Effects of Malnutrition on Growth and Development. Memoirs Golden Jubilee Nutrition Research Laboratories of India, Hyderabad, India.

Cravioto, J., Piñeiro, C., Arroyo, M., and Alcalde, E., 1970, Mental performance of school children who suffered malnutrition in early age, in: *7th Symposium Swedish Nutrition Foundation, Nutrition in Preschool and School Age* (G. Blix, ed.), pp. 85–91, Almqvist and Wiksell, Uppsala.

Cravioto, J., Lindoro, M., and Birch, H. G., 1971, Sex differences in I.Q. Pattern of children with congenital heart defects, *Science* **174:**1042.

Davison, A. N., and Dobbing, J., 1966, Myelination as a vulnerable period in brain development, *Br. Med. Bull.* **22:**40.

DeLicardie, E. R., and Cravioto, J., 1974, Behavioral responsiveness of survivors of clinical severe malnutrition to cognitive demands, in: *Early Malnutrition and Mental Development. Twelfth Symposium of the Swedish Nutrition Foundation* (J. Cravioto, L. Harnbreus, and P. Vahlquist, eds.), pp. 134–153, August 20–22.

DeLicardie, E. R., Vega, L., Birch, H. G., and Cravioto, J., 1971, The effect of weight loss from birth to fifteen days on growth and development in their first year, *Biol. Neonat.* **17:**249.

Diamond, M. C., Law, F., Rhodes, H., Lindner, B., Rosenzweig, M. R., Krech, D., and Bennett, E. L.,

1966, Increases in cortical depth and glia number in rats subjected to enriched environment, *J. Comp. Neurol.* **128**:117.

Dobbing, J., 1963, The influence of early nutrition on the development and myelination of the brain, *Proc. R. Soc. Med.* **159**:503.

Dobbing, J., 1968, Vulnerable periods in developing brain, in: *Applied Neurochemistry* (A. N. Davison and J. Dobbing, eds.), pp. 287–316, Blackwell, Oxford.

Dobbing, J., 1974, The later development of the central nervous system and its vulnerability, in: *Scientific Foundations of Pediatrics* (J. A. Davis and J. Dobbing, eds.), pp. 1–6, Heinemann, London.

Dobbing, J., 1976, Vulnerable periods in brain growth, in: *The Biology of Human Fetal Growth* (D. F. Roberts and A. M. Thomson, eds.), Taylor and Francis, London.

Flexner, L. B., Stellar, E., de la Haba, G., and Roberts, R. B., 1962, Inhibition of protein synthesis in brain and learning and memory following puromycin, *J. Neurochem.* **5**:595.

Frankova, S., and Barnes, R. H., 1968, Effect of malnutrition in early life on avoidance conditioning and behavior of adult rats, *J. Nutr.* **96**:485.

Geber, M., and Dean, R. F. A., 1956, The psychological changes accompanying kwashiorkor, *Courrier* **6**:3.

Gesell, A., and Amatruda, C., 1947, *Developmental Diagnosis: Normal and Abnormal Child Development,* Hoeber, New York.

Guillen-Alvarez, G., 1971, Influence of severe marasmic malnutrition in early infancy on mental development at school age, *Proceedings of the Twelfth International Congress of Pediatrics,* Vienna, Austria, August 29–September 4, Wiener Medizinischen Akademic.

Guthrie, H. A., and Brown, M. L., 1968, Effect of severe undernutrition in early life on growth, brain size and composition in adult rats, *J. Nutr.* **94**:419.

Guthrie, H. A., Guthrie, G. M., and Tayag, A., 1969, Nutritional status and intellectual performance in a rural Philippine community, *Philipp. J. Nutr.* **22**:2.

Hertzig, M., Birch, H. G., Thomas, A., and Arán-Méndez, O. A. 1968, Class and ethnic differences in the responsiveness of preschool children to cognitive demands, *Monogr. Soc. Res. Child Dev.* **33**: (117).

Hertzig, M. E., Birch, H. G., Richardson, S. A., and Tizard, J., 1972, Intellectual levels of school age children severely malnourished during the first two years of life, *Pediatrics* **49**:814.

Hoorveg, J., and Standfield, P., 1972, The influence of malnutrition on psychologic and neurologic development. Preliminary communication, in: *Nutrition, the Nervous System and Behavior,* Panamerican Health Organization. Scientific Publication No. 251.

Hsueh, A. M., Agustin, C. E., and Chow, B. C., 1976, Growth of young rats after differential manipulation of maternal diet, *J. Nutr.* **91**:195.

Kahn, D., 1965, A developmental study of the relationship between auditory–visual integration and reading achievement in boys, PhD dissertation, Teachers College, Columbia University.

Kardonsky, V., Alvarado, M., Undurraga, O., Manterola, A., and Segure, T., 1971, Desarrollo Intelectual y Físico en el Niño Desnutrido, unpublished manuscript, University of Chile, Department of Psychology, Santiago.

Kerr, G. R., Helmuth, R., Campbell, J. V., and El Lozy, M., 1976, Malnutrition studies in Macaca mulatta. V. Effect on biochemical and cytochemical composition of major organs, *Am. J. Clin. Nutr.* **29**:868.

Klein, R. E., Lester, B. M., Yarbrough, C., and Habitch, J. P., 1972, On malnutrition and mental development: Some preliminary findings, Proceedings of the Ninth International Congress of Nutrition. Mexico City, Mexico, September 2–9.

Klein, P. S., Forbes, G. B., and Nader, P. R., 1975, Effects of starvation in infancy (pyloric stenosis) on subsequent learning abilities, *Pediatrics* **87**:8.

Lashley, K. S., 1929, *Brain Mechanisms and Intelligence,* University of Chicago Press, Chicago.

Leitch, I., 1959, Growth, heredity and nutrition, *Eugen. Rev.* **51**:155.

Levitsky, D. A., and Barnes, R. H., 1972, Nutritional and environmental interactions in the behavioral development of the rat: Long term effects, *Science* **176**:68.

Levitsky, D. A., and Barnes, R. H., 1973, Malnutrition and animal behavior, in: *Nutrition, Development and Social Behavior* (D. J. Kallen, ed.), Publ. No. NIH 73-242, U.S. Govt. Printing Office, Washington, D.C.

Liang, P. H., Hie, T. T., Jan, O. H., and Giok, L. T., 1967, Evaluation of mental development in relation to nutrition, *Am. J. Clin. Nutr.* **20**:1290.

Lugo, G., 1971, Influencia de la Clase Social Sobre el Estilo de Respuesta Ante una Demanda Cognoscitiva, Thesis, Facultad de Filosofía y Letras, UNAM, Mexico, D.F.

Maier, N. R. F., 1932, The effect of cerebral destruction on reasoning and learning in the rat, *J. Comp. Neurol.* **54:**45.

Maier, N. R. F., and Schneirla, T. C., 1935, *Principles of Animal Behavior,* McGraw-Hill, New York.

Marcondes, E., Lefevre, A. B., and Machado, D. V., 1969, Desenvolvimiento neuropsicomotor de Crianca Desnutrida, *Rev. Brasil. Psiquiatr.* **3:**173.

Monckeberg, F., 1968, Effect of early marasmic malnutrition on subsequent physical and psychological development, in: *Malnutrition, Learning and Behavior* (N. E. Scrimshaw and J. E. Gordon, eds.), MIT Press, Cambridge, Massachusetts.

Orr, J. B., 1936, *Food, Health and Income,* London, Macmillan.

Platt, B. S., Heard, C. R. C., and Steward, R. J. C., 1964, Experimental protein–calorie deficiency, in: *Mammalian Protein Metabolism* (M. Munro and A. Allison, eds.), Academic Press, New York.

Pollitt, E., 1973, Behavior of infant in causation of nutritional marasmus, *Am. J. Clin. Nutr.* **26:**264.

Pollitt, E., and Granoff, D., 1968, Mental and motor development of Peruvian children treated for severe malnutrition, *Rev. Interam. Psicol.* **1:**93.

Richardson, S. A., 1976, The relation of severe malnutrition in infancy to the intelligence of school children with differing life histories, *Pediatr. Res.* **10:**57.

Robles, R., Cravioto, J., Rivera, L., Vilches, A., Santibañez, E., Vega, L., and Pérez-Navarrete, J. L., 1959, Influencia de ciertos factores ecológicos sobre la conducta del niño rural mexicano, in: Proceedings IX Biannual Meeting Mexican Society of Pediatric Research, Mexico.

Rojano, M. E., 1970, Desarrollo de conceptos antitéticos en el niño como prueba para evaluar el lenguaje oral en la edad preescolar temprana, Master's dissertation, Universidad Autónoma de México.

Scott, E. B., 1964, Histopathology of amino acid deficiencies. VII Valine, *J. Exp. Mol. Pathol.* **3:**10.

Stein, Z., Susser, M. W., Saenger, G., and Marolla, F., 1975, *Famine and Human Development: The Dutch Hunger Winter,* London, Oxford University Press.

Stoch, M. B., and Smythe, P. M., 1963, Does undernutrition during infancy inhibit brain growth and subsequent intellectual development?, *Arch. Dis. Child.* **38:**546.

Srikantia, S. G., Sastry, C. Y., and Naidu, A. N., 1975, Malnutrition and mental function, Proceedings Xth International Congress of Nutrition, Kyoto, Japan, August 3–9.

Voronin, L. G., and Guselnikov, V. I., 1963, On the phylogenesis of internal mechanisms of the analytic and synthetic activity of the brain, *Pavlov J. Higher Nerv. Act.* **13:**193.

Winick, M., 1969, Malnutrition and brain development, *J. Pediatr.* **74:**667.

Winick, M., and Noble, A., 1966, Cellular response in rats during malnutrition at various ages, *J. Nutr.* **89:**300.

Winick, M., and Rosso, P., 1969, The effect of severe early malnutrition on cellular growth of the human brain, *Pediatr. Res.* **3:**181.

Winick, M., Rosso, P., and Waterlow, J., 1970, Cellular growth of the cerebrum, cerebellum and brain stem in normal and marasmic children, *Exp. Neurol.* **26:**393.

Wray, J., 1975, Intersensory development in school-age children at a high risk of severe malnutrition during the preschool years, unpublished manuscript.

Yatkin, U. S., and McLaren, D. S., 1970, The behavioral development of infants recovering from severe malnutrition, *J. Ment. Defic. Res.* **14:**25.

VIII

History of Growth Studies

17

A Concise History of Growth Studies from Buffon to Boas

J. M. TANNER

1. Buffon and the First Growth Study

Although the space available for this essay precludes the traditional start with Aristotle, father of scientific biology, happily our subject matter leads us directly to his greatest successor, Buffon. The one wrote almost as much as the other; and more, if small matters of politics, ethics, logic, and philosophy be excluded. Buffon himself would have welcomed the comparison. When in 1739 he was appointed Director of the Royal Medicinal Herb Garden in Paris and directed to prepare a catalog of the Royal Collection, he went one better, and declared his intention of making a catalog of all Nature, a work to be comparable with that of Aristotle, only larger.

This aim he and his successors, chiefly Lamarck and Cuvier, actually accomplished. Buffon's capacity for work was phenomenal; throughout each day he read and wrote in his library, emerging only in the evening for dinner. He revised continually, setting aside a manuscript until he had forgotten its contents, then having it read aloud by a person who had no technical knowledge of the subject; everything that person failed to understand was rewritten (Flourens, 1860). The prospectus for the Catalog of Nature foresaw 15 volumes covering all the animal and vegetable kingdoms, together with rocks, fossils, and minerals. In the event the task proved even greater than it looked. Instead of 15 volumes, Buffon had published 35, with the 36th in press, at the time of his death, and covered only mammals, quadrupeds, birds, minerals, and earth history. But the series continued, called *Suites à Buffon,* until the definitive description of the whole animal kingdom was completed.

This chapter is a condensed version of the middle section of *A History of the Study of Human Growth,* shortly to be published by Cambridge University Press.

J. M. TANNER • Department of Growth and Development, Institute of Child Health, University of London, London, England.

The first volume of the *Natural History* appeared in 1749, three thousand copies being sold in 6 weeks. The fifteenth volume, the last of the original series, was published in 1767. The natural history of man occupies parts of Volumes 2 and 3; both were released in 1749. A second set, of 7 volumes, called *Supplements to the Natural History,* was published between 1774 and 1789, and it is in the fourth of these, which appeared in 1777, that the first growth study ever made was published (Figure 1). The study was carried out, undoubtedly at Buffon's behest, by Montbeillard, his closest family friend. Count Philibert Guéneau de Montbeillard (1720–1788) was a landowner and physician and a devoted admirer of Buffon, helping him for many years in an unacknowledged capacity. He drafted much of the *Natural History of Birds,* but his collaboration was kept secret until the appearance of the third volume, as Buffon, and it seems Montbeillard too, feared that no hand other than Buffon's could write, and sell, so well. Montbeillard was quiet but courageous; he was the first person in France to carry out a vaccination, or more accurately a variolation, for this was much before the time of Jenner, and pus from subjects with the disease was inoculated. On May 7, 1766, he inoculated his son, with, says his memoire to the Dijon Academy, "the trembling hand of a father" (Flourens, 1860). The inoculation was successful (or at least, not disastrous) and advanced greatly the cause of smallpox prevention in France. The success did not prevent the boy later falling victim to the guillotine under Robespierre, however (a fate shared also by

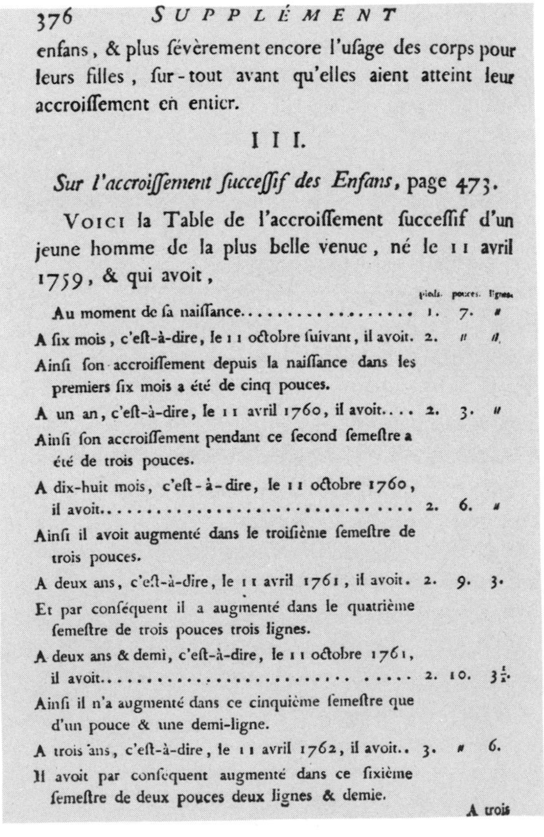

376 SUPPLÉMENT

enfans, & plus ſévèrement encore l'uſage des corps pour leurs filles, ſur-tout avant qu'elles aient atteint leur accroiſſement en entier.

III.

Sur l'accroiſſement ſucceſſif des Enfans, page 473.

VOICI la Table de l'accroiſſement ſucceſſif d'un jeune homme de la plus belle venue, né le 11 avril 1759, & qui avoit,

	pieds.	pouces.	lignes.
Au moment de ſa naiſſance..................	1.	7.	"
A ſix mois, c'eſt-à-dire, le 11 octobre ſuivant, il avoit.	2.	"	"
Ainſi ſon accroiſſement depuis la naiſſance dans les premiers ſix mois a été de cinq pouces.			
A un an, c'eſt-à-dire, le 11 avril 1760, il avoit....	2.	3.	"
Ainſi ſon accroiſſement pendant ce ſecond ſemeſtre a été de trois pouces.			
A dix-huit mois, c'eſt-à-dire, le 11 octobre 1760, il avoit..........................	2.	6.	"
Ainſi il avoit augmenté dans le troiſième ſemeſtre de trois pouces.			
A deux ans, c'eſt-à-dire, le 11 avril 1761, il avoit.	2.	9.	3.
Et par conſéquent il a augmenté dans le quatrième ſemeſtre de trois pouces trois lignes.			
A deux ans & demi, c'eſt-à-dire, le 11 octobre 1761, il avoit..........................	2.	10.	$3\frac{1}{2}$.
Ainſi il n'a augmenté dans ce cinquième ſemeſtre que d'un pouce & une demi-ligne.			
A trois ans, c'eſt-à-dire, le 11 avril 1762, il avoit..	3.	"	6.
Il avoit par conſequent augmenté dans ce ſixième ſemeſtre de deux pouces deux lignes & demie.			

A trois

Fig. 1. A page from Buffon's original publication in 1777 of the measurements of Montbeillard's son, in *Supplements à l'Histoire Naturelle,* Volume 4.

Buffon's son). Montbeillard and Buffon were evidently very different sorts of men: Buffon indefatigable, at work all day, but given to familiarity and even boorishness in the evenings; Montbeillard "of a delicate constitution" and starting each day with family madrigals.

Montbeillard's growth study was well known to scientists in the nineteenth century, being quoted by Quetelet in 1835, and later by Roberts and Bowditch in the 1870s. But then it became neglected and passed into modern books only through the agency of Richard Scammon, professor of anatomy at the University of Minnesota. Scammon was the author of a celebrated book on fetal measurements (Scammon and Calkins, 1929) and of numerous pregnant and tantalizing abstracts referring to work never written up in full. He also began to write a history of growth studies and had evidently assembled a vast bibliography—partly analyzed in 1927—when he was overtaken by illness (Scammon, 1927*b*). In 1927 he reprinted Montbeillard's measurements from the Cuvier edition (Volume 4 of the *Oeuvres Complètes*, Buffon, 1836; Scammon, 1927*a*), converted them to the metric system, and plotted them as a graph of height against age. Later D'Arcy Thompson calculated the increments and plotted the height velocity curve in the revised edition of *Growth and Form* (1942). Tanner then used the combination of height-for-age and height velocity curves as the opening figure in *Growth at Adolescence* (Tanner, 1955, 1962). This figure has been recopied many times since, and the growth of Montbeillard's son is one of the best-known illustrations in human auxology.

The graphs are reproduced in Figure 2. In the velocity chart "rolling" yearly velocities are plotted, that is to say velocities calculated successively, e.g., from 0.0 to 1.0 years plotted at 0.5; 0.5 to 1.5 years plotted at 1.0, and so on. In this way seasonal effects are removed. Where the incremental periods are uneven, the appropriate adjustment has been made to obtain true velocities and each velocity has been plotted at the actual midpoint of the interval concerned. There is a gap in the measurements from 10.0 to 11.5 years for reasons not disclosed in the report nor in the voluminous correspondence between Buffon and Montbeillard. Unfortunately nothing is said as to how the measurements were taken, except that all were done by Montbeillard himself, with his son barefoot. At first the boy must have been lying down, but the point at which the changeover to standing took place is not indicated. Since standing height is usually about 1 cm less than supine length, one would expect a particularly low increment at the point of changeover, and there is indeed one such at the interval 3.0–3.5 years (1.6 cm in 0.5 years). If the first standing height was really taken at 3.5 years, then the two velocities 2.5–3.5 and 3.0–4.0 are artificially reduced and should be augmented by about 1 cm/year each. The tentatively "corrected" points are shown in brackets in the velocity plot. The "correction" does make the velocity 3.5–4.5 look more consonant with the rest of the curve. Both distance and velocity plots show all the features of modern curves obtained in the same way, that is by a single measurer observing at six-monthly intervals, and indeed Montbeillard's record has never been surpassed, and seldom equaled, in elegance and presumed accuracy.

Montbeillard's son was very tall, ending above the 97th centile for modern data. He also had a marked adolescent growth spurt (or pubertal growth spurt, the terms are used here synonymously), with a whole-year peak velocity of 12.1 cm/year. This is well above the present-day average. The age of peak velocity is about 14.3 years (smoothing by eye) which is entirely in line with modern data, being about 0.3 years later than the modern mean. The whole curve shows a velocity falling rather strongly from birth to about age 4.5; then much more slowly (or not at

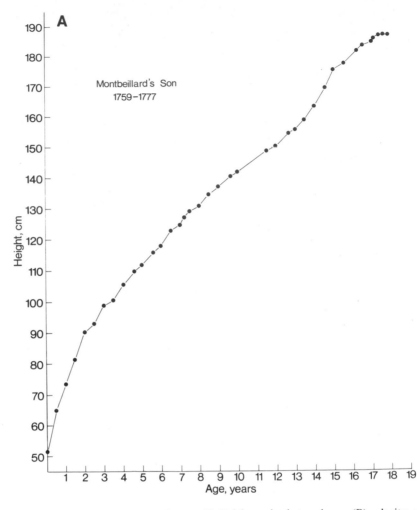

Fig. 2. Growth in height of Montbeillard's son. (A) Height attained at each age; (B) velocity, or rate of growth, calculated in successive whole-year periods, e.g., 0.0–1.0 years plotted at 0.5 years; 0.5–1.5 years plotted at 1.0 years. Points in brackets are "corrected" velocities allowing for presumed change from supine length to standing height.

all) till about 8.5. Between 8.5 and the beginning of the spurt at 13 the velocity declines noticeably.

Buffon did not remark on any of these features, however, in the *Supplement* where the record appeared. He did report Montbeillard's observation that his son, when approaching maturity, suffered an apparent decrease of height when measured the morning after an all-night dance, a decrease that had disappeared by the following morning. He also remarked on the seasonal effect on growth rate, total growth in the age period 5–10 years being 7 *pouces* 1 *ligne* in the summer months (April to October) and only 4 *pouces* 1½ *lignes* in the winter months. These relative values agree closely with the modern ones reported by W. A. Marshall (1971).

Buffon's lack of specific comment on the marked and beautiful growth spurt is easier to understand when we read what he had already written on growth and puberty in the first edition of the *Histoire Naturelle* (1749). "There is a quite remarkable thing about the growth of the human. The fetus . . . grows always more

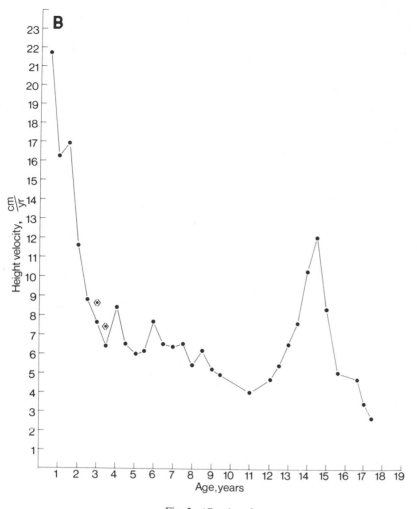

Fig. 2. *(Continued)*

and more (rapidly) up to the moment of birth [Buffon was wrong here, while his contemporary, Haller, was right]; in contrast the child grows always less and less up to the age of puberty, when he grows, one might say, in a bound (*tout à coup*) and arrives in very little time at the height that he has for always'' (p. 472). Thus Buffon was fully aware of the adolescent spurt, partly from his reading perhaps, but partly, one may be sure, from personal observation. Although others had more or less vaguely described the adolescent spurt before him, it is he who gave the first adequate and proper description.

Buffon made other remarks on puberty: ''In the whole human species females arrive at puberty earlier than males, but the age of puberty differs in different people, and seems to depend partly on the temperature and climate and on the quality of the food; in the towns and amongst people who are well-off, children accustomed to succulent and abundant food arrive earlier at that state, while in the country and amongst poor people children take two or three years longer because they are nourished poorly and too little. In all the southern parts of Europe and in the towns the majority of girls have puberty (*sont pubères*) at 12 years and the boys at 14, but in the north and in the countryside the girls scarcely reach it by 14 or the boys by 16" (p. 489). (Presumably by *pubère* Buffon means the traditional first appearance of pubic hair.) All this is a usual view and exactly echoes Guarinoni (1610), a physician and commentator of the previous century.

These opinions continued unchanged for a further hundred years. Dictionaries and encyclopedias supplied much of the demand for medical textbooks in the nineteenth century, especially in France. The most frequently quoted article that concerns growth is that by Virey (1775–1846), published in 1816. Invited to contribute an article on giants, Virey managed to squeeze 15 pages about factors which determine the height of ordinary folk between a brief definition of gigantism and a coda concerned with the question of whether giants seven to eight feet tall existed as a race in antiquity. As to this latter, he thinks it quite possible indeed, especially in the light of the then current version of the abominable snowman, whose gigantic and human footprints had been found in 1815 on the banks of the Swan River in western Australia.

Virey is still firmly in the humoral era: There is little sign (except, perhaps, in the elegant and beautiful style) that he is writing after Buffon. "Nourish a man or an animal parsimoniously, with dry and hard foods, smoked, salted, spiced or sharp and astringent; permit him to drink only a little and then a sharp and sour wine such as tartarous vin rouge, give him above all acid and bitter things which harden and contract the fibres, it is very obvious that such a person will become thin, short, compact in all his organs. In contrast, stuff a child with very soggy foods, get him used to taking milk and gruel and dough, to slimey drinks like beer, mead, whey, and oily-chocolate, to warm and dilute liquids, cram him with all the foods necessary to distend and enlarge him, in the way one fattens geese and pigs, he is able to become colossal and gigantic in stature relative to a person nourished in the opposite way" (p. 553).

This is Guarinoni, too, more forcefully expressed. Jean-Jacques Rousseau and some philosophers, says Virey scornfully, may think savages are tall and strong and beautiful, but the truth, easy to see, is otherwise. The bad climate and lack of nourishment obtainable in the humid and dark forests or on the barren lands makes their inhabitants unhealthy and small. It is civilized Europeans who are the tall ones. Or at least they used to be, in the times of the German tribes described by Tacitus (in his work *Germani*, written in A.D. 98) largely as a criticism of the degenerate Rome of his day. In a passage that would have done credit to Baden-Powell he continues, "Nowadays we are soft and effeminate. The age of puberty is advanced because of a precocious awareness (*précocité du moral*), because of the pernicious solitary pleasures which bring on prematurely the sexual organs and exhaust youth. This turning the greater part of nutrition towards the excretion of sperm stops growth and people stay short in stature. The promiscuity in towns and amongst the rich makes them feeble" (p. 560). The German tribes were strong and brave because of sexual abstinence, for Caesar said they thought it disgraceful to approach a woman under the age of 20. As for us, we live in the height of immoral and enervating luxury and no good can come of it, not even so far as stature is concerned.

2. Schiller and the Carlschule

There is a second source of individual longitudinal growth records in the eighteenth century, although it remained unknown until 1970. This is the collection of measurements made on the pupils of the Carlschule in Stuttgart from 1772 to

1794. The records, now in the Stuttgart City Archives, were discovered by Dr. Robert Uhland (1953), the state archivist and historian of the Carlschule, and Professor Wilhelm Theopold (1967), professor of pediatrics in Stuttgart and a medical historian. Much of the material was published in the thesis of one of Professor Theopold's students (Hartmann, 1970).

The Carlschule was an institution unique in its time (Wagner, 1856–1858). It was the beloved brain-child of Carl Eugen (1728–1794), Duke of Württemberg, a man who in his maturity attended in exemplary fashion to the welfare and especially to the education of his subjects (it was quite otherwise in his youth). The great high school began as an institution for teaching the sons of soldiers how to be gardeners, who were needed in the grounds of the Duke's newly built and favorite hunting lodge, "Solitude." Soon the school was turning out plasterers as well, and from that it was a short step to training musicians and dancers, also in demand at the lodge. At the same time Carl Eugen was making plans for a military orphanage to train a variety of artisans, and the two institutions were merged to form a military academy, which was opened in 1770. It became the Duke's favorite project, and he himself acted as headmaster, dining with the pupils and taking the school assemblies. He called the pupils his dear sons and placed himself firmly *in loco parentis,* which was perhaps as well, since for many years neither pupils nor staff had any holidays whatsoever in which to visit their families; even later there were only two periods of one week each per year.

The school provided such an excellent education for the period that officers, officials, and even people of standing at court sent their sons there. Thus the original character changed. In 1775 the school moved into Stuttgart to be housed in a former barracks behind the castle. By this time it had three divisions: a basic stage into which boys were admitted at 7 years old, a gymnasium, and an upper section which, despite the opposition of the neighboring University of Tübingen, was soon declared of university status by the emperor. The Carlschule included a medical school from 1775, and indeed boasted a wider selection of faculties than any German university of the time. Educational methods were advanced, and this attracted some of the best teachers in Germany. Sons of nobility and of ordinary folk were taught together and were subject to the same discipline, although each had their own dormitories and ate at separate tables in the dining hall (albeit the same food). Students wore uniforms and used military ceremony and discipline. At the end of the yearly examination there was a major assembly and prize-giving.

During its whole existence the school graduated 2211 pupils. The majority were from Württemberg, but such was the reputation of the place that others were sent from all over Europe. The graduates, not unlike the later polytechnicians of Napoleon's school, formed an elite that ruled Württemberg and spread far and wide beyond it. From the start the school was menaced by epidemics; the worst was typhus in 1784, when many parents took their sons away. The deficit was made up by day-boys from the town, and the school continued till Carl Eugen's death in 1794. But then it proved too much his own plaything and, with his death, it dissolved.

There were a number of famous people who were pupils at the Carlschule. Friedrich Schiller (1759–1805) was one and Cuvier another. Schiller graduated in jurisprudence and medicine, and at the prize-giving ceremony in 1779 saw Goethe for the first time, standing behind the young Duke of Saxe-Weimar who was a guest that year. Schiller thoroughly disliked Carl Eugen and hated the school, military discipline not forming the most suitable environment for the new leader of the romantic movement. He was overcome with humiliation at being seen there by the

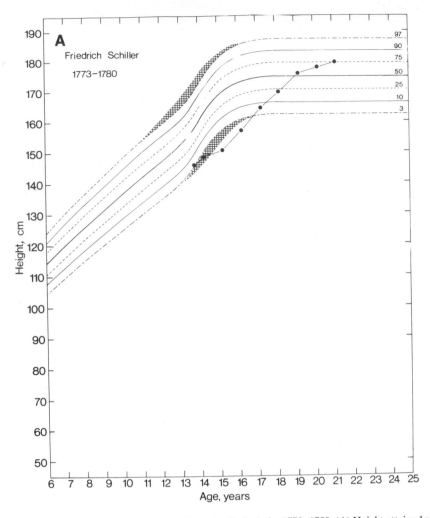

Fig. 3. Growth curve of Friedrich Schiller while at the Carlschule, 1773–1780. (A) Height attained at each age; (B) height velocity calculated in successive whole-year periods, e.g., 14.2–15.2 plotted at 14.7, 15.4–16.2 plotted at 15.7, etc. (except for first point, representing 0.5 years only). Plotted on British standards from Tanner and Whitehouse (1976). (Data from Hartmann, 1970.)

author of the *Sorrows of Young Werther* (albeit now quite a mature functionary). His humiliation was worse because while prizewinners among the nobles kissed the Duke's hand on receiving their prizes, sons of the bourgeois, among whom was Schiller, whose father was an ordinary army officer, were only allowed a nibble at the hem of the Ducal robe (Theopold, 1967, p. 31).

Schiller's discontent and rebelliousness at the Carlschule may have been fed from another source, shared by so many thousands of adolescents after him. He was a late maturer, with peak height velocity only at about 16.7 years, some nine months later than the average bourgeois and over a year later than the average noble. He was a very tall man; indeed, both before and after puberty, he was the tallest but one among the bourgeois whose records Hartmann reproduces. At the frequent Carlschule parades the boys were lined up strictly according to height and Schiller must have started further up the line than all his bourgeois contemporaries.

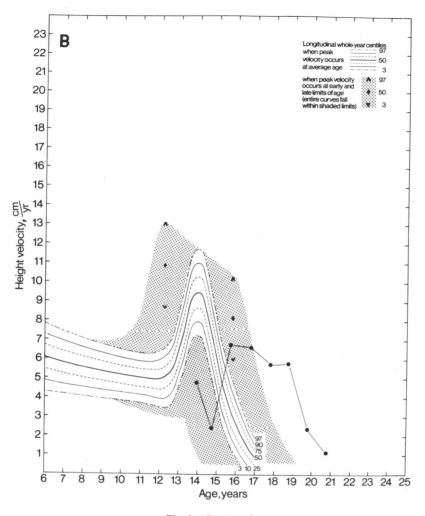

Fig. 3. *(Continued)*

But he sank progressively down during his fifteenth and sixteenth years; when 15.0 he was only slightly above the bourgeois average and actually below the average of the nobles. With puberty, but only then, he regained his place. His height and height velocity curves are shown in Figure 3.

About one third of the records are of the nobility and two-thirds of the bourgeois. The number of students measured rises from 92 at age 8, to 442 at age 15, then diminishes to 155 at age 21. The series then, is a typical mixed longitudinal one (Tanner, 1962), just like the Harpenden Growth Study of the present century. Hartmann gives distance and velocity curves for 60 nobles and 60 bourgeois

All pupils were measured at regular intervals, twice and sometimes three times a year. This continued from 1772 to the time of the epidemics of 1782–1784 when the frequency of measuring diminished. At each measuring session the previous, as well as the present, measurement was recorded, and in later years the increment between the two occasions was calculated and written in also. Evidently Carl Eugen understood that velocity told him more about the pupil's health than did "distance" (that is, height attained).

students; most have 10–12 measurements and show quite typical adolescent growth spurts. Between the nobility and the bourgeois there was a difference in height, despite the similarity of environment and nutrition at the school. The difference relates almost entirely to advancement in growth, however, and not to final height (see Hartmann, Figure 33). At ages 10 and 11 the nobles were on average 2.5 cm taller, and at age 15 nearly 7 cm taller; by 20 and 21, however, the difference was down to 1 cm, a quite insignificant amount. (The means at 21 are, respectively, 168.8 cm and 167.6 cm and seem to represent nearly fully mature height.) Peak height velocity was reached on average at a little under 16.0 in the bourgeois and probably earlier in the nobles, although a curious chance of sampling makes the figure for the nobles uncertain. Thus the bourgeois, at least, grew up 18 months to 2 years later than their equivalents nowadays. (Peak height velocity is at approximately 14.0 in the British population as a whole, with the best-off section some 3–6 months earlier.) The mature height of the Carlschule students was less than nowadays, by about 6–7 cm.

3. Goethe and the Recruiting Officers

Besides Buffon and the Carlschule there is another source of growth records in the eighteenth century. This was the recruiting sergeant.

Goethe, much to his disgust, was involved in recruiting as part of his duties for the Duke of Saxe-Weimar, and he drew the sardonic little pen, pencil, and wash drawing reproduced in Figure 4. To the present-day auxologist the sketch is technically revealing, for it shows an altogether modern approach to measurement. The recruit is being made to stand up straight, aided by upward pressure after the manner of R. H. Whitehouse; his head is held correctly, and a correctly aligned block is being slid down the wall in contact with a fixed rule. There is even the requisite recorder; and the shoes have been removed, as shown by the figure in the foreground. To the literary historian the sketch is revealing too. In the decoration over the door of the room that the successful recruit is entering, with a soldierly pat on the back from the recruiting sergeant, Goethe has replaced what was presumably the regimental badge by a gallows, which neatly picks up the line of the stadiometer. The sketch was done, it is believed, in 1779, the year in which Goethe was in charge of recruiting. The gallows clearly reflects Goethe's feelings about the military life. The motto under the badge represents Goethe's comment, too, rather than the reality. It says *Thor des Ruhms*. At first glance this means "gateway to glory." But Goethe evidently intended a pun, for "thor" has two meanings, depending on its gender; the second meaning is fool. Furthermore Goethe wrote *Thor* des *Ruhms* and not *Thor* zum *Ruhm;* fool/gateway *of* glory, not *to* glory. (The motto of the Duke of Saxe-Weimar was naturally a little different and said *Pro Fide, Rege et Lege.* One wonders where Goethe kept his subversive material.)

The earliest military data which have been retrieved and analyzed come from Norway, and go back to 1741 (Kiil, 1939; Udjus, 1964). Minimum standards, however, go back earlier and in England a minimum standard for the height of a recruit was laid down in the Recruiting Act of 1708 as 5 ft 5 inches (Scouller, 1966). Wasse (Wace), the Rector of Aynho in Northamptonshire, wrote in a letter which was communicated to the Royal Society in 1724 that he had observed "several soldiers discharged from being a little under the Standard and helped by telling the officer of the difference between morning and later." "Since that time" he contin-

Fig. 4. Goethe's drawing of the measurement of recruits in Weimar in 1779. Reproduced from Bruford (1962); original in the Goethe Museum, Weimar.

ued "I have measured Sir H. A., Mr. C. and a great many sedentary people and Day-Labourers of all ages and shapes and find the difference to be near an inch." He then quotes experiments on himself, apparently in relation to sitting height rather than stature. On August 21, 1723, at 11 AM he sat down and fixed an iron pin "so as to touch it, and that but barely." After half an hour rolling his rectory lawn he was $5/10$ inches below the mark. At 6 AM the following morning he touched it fully, but after riding 4 miles he was $6/10$ inches short. "If I study closely, though I never stir from my writing desk, yet in 5 or 6 hours, I lose one inch. All the difference I find between labourers and sedentary people is that the former are longer in losing their Morning Height and sink rather less in the whole than the latter . . . I cannot perceive that when height is lost it can be regained by any rest that day or by the use of the cold bath . . . the alteration in the human stature, I imagine, proceeds from the yielding of the cartilage between the vertebrae." A lively discussion followed this first demonstration of the diminution of stature during the day, with many words expended on arguing that since it *could* indeed be possible according to current theory, then it probably *had* indeed occurred. Isaac Newton, who was in the chair, happily, came out strongly on Wasse's side.

4. Quetelet and the Mathematics of Growth

J. M. TANNER

Adolphe Quetelet (1796–1874) was one of the major figures in science and public health in the first half of the nineteenth century. By training a mathematician, he was the creator and director of the Brussels Observatory, where he lived from 1832 until his death. He contributed greatly to meteorology and earth science, was permanent secretary, from 1834 till his death, of the Brussels Royal Academy of Science and Letters, later (1845) the Belgian Royal Academy of Science, Letters and Fine Art, and held a quite extraordinary place in Brussels society in the 1840s and 1850s. He was in many ways the founder of modern statistics. It was he who introduced into practical work the Normal curve, discovered in another context by Laplace and Gauss. It was he who in 1833, as delegate from newly independent Belgium, urged the members of the third meeting of the British Association for the Advancement of Science at Cambridge to found a statistical section. This fathered the Statistical (later: Royal Statistical) Society; Malthus, Babbage, Farr, Jones, and Herschel were among its luminaries. Florence Nightingale was Quetelet's devoted admirer, and Metternich was among the many expatriates of all countries and persuasions who sought solace and advice at the Brussels Observatory. Quetelet made an extensive study of the growth of children and is often, but wrongly, taken as the father of modern work on growth. That honor, for reasons that will become clear as we proceed, must be reserved for the advent, 50 years later, of Franz Boas.

But for all this, Quetelet was at heart an artist; Goethe and Gauss were his intellectual parents, with Goethe dominant. Quetelet was born in Ghent in 1796 and appointed teacher of mathematics at the Ghent College (high school) on his nineteenth birthday, just four months before the battle of Waterloo. He obtained his doctorate in mathematics in 1819 with a thesis on a new family of curves derived from conic sections, and was soon called to be professor of mathematics at the Athénée in Brussels. But he had earlier thought of being a painter. "I finished my schooling" he wrote "at the time of the events of 1814 which separated the Netherlands and France. I was dividing my time between art and science. My taste for art had already led to my being a student in a painter's studio, which I abandoned later when I accepted the chair of mathematics . . . . The taste for art continued to be associated with that for science in my moments of leisure" (1870, p. 6, footnote). Paul Levy (1974), in a most perceptive essay contributed to Quetelet's memorial at the centenary of his death, wrote "It was the research of Beauty which obsessed him. It was that which directed him towards the mean (moyenne), which necessitated measurement, and from the mean towards the conception of the laws of social physics (physique sociale). Running through all his work one discerns a constant preoccupation with discovering, under common appearances, the intention of the Supreme Intelligence who had conceived it all. How to determine the model of human beauty? Surely with a synthesis, in which all accidents would cancel out."

Quetelet's meeting with Goethe, whom he greatly admired, must have strengthened, if it did not arouse, his search for the ideal or representative type. In 1823 Quetelet, then aged 27, was given the job of organizing the observatory to be set up in Brussels, and during the next few years, he traveled extensively in Europe assessing astronomical equipment and methods. He spent the winter of 1823–1824 in Paris learning how to make astronomical observations, and the brilliance of his doctoral thesis procured meetings with Laplace, Gauss, and Fourier. Then or a little later, Fourier introduced him to the great French epidemiologist Louis-René Villermé, with results we shall see later. In 1827 Quetelet toured Great Britain and met Babbage and Herschel, the Astronomer Royal, who became a lifelong friend. The following year he was in Germany and went to Weimar with his wife with the intention of spending a couple of days at the celebrations associated with Goethe's eightieth birthday (Collard, 1934). The old man was impressed and attracted by Quetelet, had him come every day for a week to his very select evening receptions, gave him a copy of his famous book *Zür Naturwissenshaft überhaupt besonderers zür Morphologie*, and asked him to send details of the proceedings of the scientific congress in Heidelberg to which Quetelet was en route. Goethe, we now sometimes forget, rated his scientific works more highly than his poetry, felt his views on optics to have been unjustly neglected, and begged Quetelet to promote them, if possible, in Heidelberg. Quetelet, for his part, shared Goethe's preoccupation with Beauty.

Quetelet's visit to Paris a few years earlier had also been of immense importance to him. From the mathematicians he obtained his central overriding idea that is his lasting claim to statistical fame, that is, the applicability of the curve derived from the Law of Errors to describe the distributions of all kinds of things from children's heights to the ages of murderers. From Villermé (1782–1863), the chief founder of public health in France, he obtained a powerful impulse to apply his ideas to the benefit of society as well as to the perfection of man. Villermé was 14 years older than Quetelet, and a very different man. Brought up in a small village outside Paris, his studies were interrupted in 1804 by entering the army of the Napoleonic wars as a surgical assistant (Astruc, 1933). He remained in the army for 10 years, working for the most part under atrocious conditions. His thesis for the full doctorate had to wait until the fall of Napoleon; it was entitled "Effects of famine on health in places which are theatres of war." Experience of war and famine had affected him deeply; his thesis describes the Spanish campaign in Estremadura and the effects of starvation on people's behavior in precisely the terms of Goya's "Horrors of the War." In later life he avoided talking of his war years, but after only four years in medical practice he used his small family fortune as support while he devoted himself to examination of the plight of the unfortunates in post-Napoleonic France; the weavers and cottonspinners, the silk workers, the persons in prisons. We shall follow this part of his career in more detail later.

Villermé was a humble man, very conscious that his education was far inferior to Quetelet's (although his experience of fieldwork was incomparably greater). He was an admirer and encourager of the younger man's work and "if he disagreed he said so with a straight-forwardness and honesty for which he was known amongst his contemporaries" (Ackerknecht, 1952; the best paper on Villermé). Over the period of 1823–1863 Villermé, a poor, almost negligible correspondent, wrote 86 letters to Quetelet, including the last one of his life. Quetelet, in his turn, was always prompt to acknowledge the debt he owed to Villermé, referring to him in the second edition of *Physique Sociale* (1869) no less than 48 times, twice as often as to any other author.

Quetelet and his wife were in Italy at a congress when the revolt of 1830 that separated Belgium from Holland broke out. The half-completed observatory building was turned into a fortress, and it was not until 1832 that Quetelet was able to move in and begin his astronomical and meteorological work. In the meanwhile he carried out the first large-scale cross-sectional survey of children ever made. Altogether he did two separate surveys in 1831–1832, the first of height only, the second of weight and height. Probably he was urged to do it by Villermé, who had only two years before published his classic analysis of the effects of poverty on the height of French recruits (see below). At all events Quetelet never returned again to the study of children in a similar context; thereafter births, deaths, marriages, criminology, and the search for ideal proportion occupied that portion of his mind which was concerned with statistics.

The results of the surveys were first reported to the Belgian Academy (Quetelet, 1831) and subsequently reproduced in Quetelet's first and most important book, *Sur l'homme et le développement de ses facultés,* published in 1835 in Paris and seen through the press there by Villermé. The book won immediate acclaim; a German translation quickly followed in 1838 and a British one, made in Edinburgh by Dr. R. Knox, lecturer in anatomy, in 1842. Quetelet always referred to the book as *Physique Sociale* or *Social Physics* and the second, enlarged edition (1869), had this as its main title. Florence Nightingale thought it should be the basis of an honours course at Oxford for future administrators and members of Parliament.

Before we turn to the somewhat complex task of unraveling the results of the surveys and placing them in a modern context, it is as well to continue the consideration of Quetelet's underlying, more philosophical thoughts. As Hilts (1973) says, in a valuable essay contrasting especially the attitudes of Quetelet and Galton, "Quetelet's lasting reputation is based upon his discovery that human stature is distributed in many populations according to the law of error. This was the first indication that the error distribution which had been used by Laplace (in 1812–1814) and Gauss (contemporaneously with Laplace) in connection with errors of astronomical observations had a much more general application. However, Quetelet never broke with the traditional concept that the law of error really is a law of error and not a more general law of distribution." This was the great difference between Quetelet and Galton. Hilts attributes Quetelet's myopia in this respect to the hold that his concept of *l'homme moyen* achieved over him. Certainly it was powerful, almost mesmeric. "Of the admirable laws which Nature attaches to the preservation of the species, I think I may put in the first rank that of maintaining the type. . . . The human type, for men of the same race and of the same age, is so well established that the difference between the results of observations and of calculation, notwithstanding the numerous accidental causes which might induce or exaggerate them, scarcely exceed those which unskillfulness may produce in a series of measurements taken on an individual" (Quetelet, cited in Hilts). This curiously phrased statement seems to imply that errors of fit of an empirically found distribution to the Normal curve are not greater than measuring error. What Quetelet does not grasp is that the two are related only if sampling error is taken into consideration. The errors of fit depend on the numbers entering into the empirical distribution; only if the number at each point is infinite are the errors of fit reduced to errors of measurement. Quetelet never distinguished sampling from distributional errors; at that time what we call the standard deviation was called error and the, to us elementary, distinction between the standard error of the mean and the standard deviation of the distribution did not occur to him. It was Galton who, realizing the difference, felt uncomfortable with the existing terminology and Karl Pearson who in 1894 actually introduced the term "standard deviation" (see the excellent account of the history of statistics by Walker, 1929).

Quetelet's discoveries about the applicability of the Normal curve were expressed in his second important book *Lettres . . . sur la théorie des probabilités,* published in Brussels in 1846 and in translation in London in 1849. This small book, with another *"Du système social et des lois qui le régissent* (1848), grew out of the lessons he gave as private tutor to the two Princes of Saxe-Coburg in the period around 1837. He dedicated the first to the older brother Ernest, by that time Duke of Saxe-Coburg himself, and the second to Albert, become Prince Consort to Victoria. It is uncertain to what extent the government of the Duchy was affected by Quetelet's book celebrating the Rule of Number; but one may well imagine that the idea of the *homme moyen* as *centrum mundi* struck an answering chord in Albert's breast.

The *homme moyen* was, of course, that fictive individual who had the average value. In *Sur l'homme* Quetelet wrote "If the average man were completely determined, we might consider him as the type of perfection; and everything differing from his proportions would constitute deformity and disease; everything found dissimilar, not only as regards proportion and form but as exceeding the observed limits, would constitute monstrosity (Quetelet, 1842, p. 99). . . . An individual who combined in his own person all the qualities of the average man would . . . represent all which is grand, beautiful and excellent" (p. 100). (Herschel, in a note added to his review of the book, remarked simply that if this were the case the highest degree of beauty would constitute the most common category of people, a conclusion absolutely contrary to experience.)

In 1855, when Quetelet was at the height of his powers, shining with a brilliance unequaled in Brussels for decades to come, he suffered a severe stroke. His biographer Mailly (1875) was present when it happened and describes the immediate and long-term effects with sympathetic honesty. Quetelet turned over the directorship of the observatory to his son Ernest; he continued writing and was punctilious in his duties as secretary of the academy till his death 21 years later. But his memory was damaged and his assistants had to help with his articles. The scientific and sociological world continued to heap honors upon him, and he recovered sufficiently to enjoy them, in particular a visit to St. Petersburg in 1872 where he met Farr. He died in 1874.

We now turn to Quetelet's measurements of children. The original series was collected in 1830–1831. Quetelet had the lengths of 50 male and 50 female newborns at the Foundling Hospital in Brussels measured, using "M. Chaussier's méconometre." In *Sur l'homme* he gives frequency distributions, and remarks, for the first time so far as can be seen, that boy babies are longer than girls "by a trifle less than half an inch." He continues "By adding these numbers to those which have been obtained in the junior schools of Brussels, the Orphan Hospital, boarding-houses and in public life, in respect of persons of different classes, I have been able to construct the following table, comprising the rate of growth from birth to 20 years; the height of the shoe is not included" (1842, p. 58). He added that "apparently but little interest is attached to the determination of the stature and weight of man or to his physical development at different ages. . . . I do not think before Buffon any inquiries had been made to determine the rate of human growth successively from birth to maturity; and even this celebrated naturalist cites only a single particular example" (1842, pp. 57, 58).

The actual numbers of children that he measured in constructing his table (p. 58 of the 1842 edition, reproduced here in Table I) are never stated. Judging by the

Table I. Quetelet's First Series of Measurements of Height in Boys and Girls[a]

Age	Height (cm)		Yearly velocity (calculated, cm/year)	
	Boys	Girls	Boys	Girls
0	50.0	49.0		
1	69.8		(13.2)	
2	79.6	78.0	9.8	
3	86.7	85.3	7.1	7.3
4	93.0	91.3	6.3	6.0
5	98.6	97.8	5.6	6.5
6	104.5	103.5	5.9	5.7
7		109.1		5.6
8	116.0	115.4		6.3
9	122.1	120.5	6.1	5.1
10	128.0	125.6	5.9	5.1
11	133.4	128.6	5.4	3.0
12	138.4	134.0	5.0	5.4
13	143.1	141.7	4.7	7.7
14	148.9	147.5	5.8	5.8
15	154.9	149.6	6.0	2.1
16	160.0	151.8	5.1	2.2
17	164.0	155.3	4.0	3.5
18		156.4		1.1
19	166.5	157.0		0.6
20		157.4		0.4
"Terminated"	168.4	157.9		0.5

[a]From p. 58 in the English edition of *Treatise on Man* (1842). Velocities calculated by writer and placed at age center on unverifiable assumption that, e.g., "age 2" refers to age range 2.0 to 2.9 (except at 0.0).

fluctuations of the means and the absence of any measurements at all of boys aged seven, they were not large. A footnote acknowledges the help of three people, perhaps the measurers. There is also something very odd, indeed unique, about this first Quetelet series: At no age are girls on average taller than boys. At ages 13 and 14 there is merely a diminution of the boys' larger difference to 1.4 cm compared with 4.8 and 4.4 cm at 11 and 12, and 5.3, 8.2, and 8.7 cm at ages 15, 16, and 17. The oddity can scarcely be ascribed to the poverty of the children, for Cowell and Stanway's contemporaneous sample of working children in England (see below) shows the girls' ascendancy with the usual clarity, at ages 12, 13, and 14. Something has gone wrong, it seems, with Quetelet's sampling. He does mention in his first report on the series (Quetelet, 1831) that the girls from 7 to 20 were all from the Hospice des Orphalines and also warns that as their numbers were less than those of the boys, less confidence should be placed in them. As to the velocities, the girls have a fairly clear peak of 7.7 cm, in the 12–13 interval. The boys' peak is scarcely to be seen, however; the increments are successively 5.4, 5.0, 4.7, 5.8, 6.0, 5.1, and 4.0 in the intervals 10–11, 11–12, 12–13, 13–14, 14–15, 15–16, and 16–17. The small and uncertain peak is located in the 13–14 and 14–15 intervals. Quetelet says absolutely nothing about these increases in rate at puberty, despite his acquaintance with Buffon's work. He merely observes that the average rate of growth of girls from 5 to 15 is 5.2 cm/year, lower than that of boys which is 5.6 cm/year.

Furthermore, in determining for our own comparative purposes the age at peak height velocity a difficulty arises, often met with in the past and sometimes even now (see Eveleth and Tanner, 1976, p. 7). It is unclear whether Quetelet's ages refer to age at last birthday, or whether the age marked 13 includes children from 12.5 to 13.5. One presumes the former, with centers of age classes as 13.5, etc. until in the appendix on page 115 (1842 edition), in comparing English children reported in half-yearly age groups with Belgians reported in yearly ones, Quetelet writes "In order to get the height of a child of 9 years of age we have taken the mean of the child's height in the age between 8.5 and 9.0 years, and the height of the age of 9.0 and 9.5." It is, therefore, impossible to be sure whether the modest peak height velocity shown by Quetelet's first series of boys is located at 15.0 (on the first assumption) or 14.5 (on the second). Either is relatively early in comparison with schoolboys of later in the century (see below). The girls' peak, equally, may be at 13.0 or perhaps 12.5 years. On either reading the usual 2-year gap is seen, although Quetelet is naturally silent about this, since he is not taking notice of the spurts, now or, as it turns out, ever.

Quetelet goes on to discuss the possible differences in growth between urban and rural children. "Already Dr. Villermé has proved, contrary to the generally received notion, that the inhabitants of towns are taller than those of the country. I have arrived at the same conclusion in respect of the inhabitants of Brabant" (1842, p. 59). This time he gives the (substantial) numbers, the subjects being 19-year-old conscripts. "It could still happen," he very correctly remarks, "that the inhabitants of the country might attain a greater eventual height than the inhabitants of the town if those in the town were closer to ending their growth at 19." Villermé was of the same opinion about slowness of development in the country, and Quetelet quotes him on poverty causing delay in, as well as stunting of, growth (the quotation is given below on p. 548). To determine when growth in stature in men ceased, Quetelet (1830) examined the Brussels conscript registers made about 1815 and found the mean height of 19-year-olds was 166.5 cm, of 25-year-olds 167.5 cm, and of 30-year-olds 168.4 cm. He concluded that growth did not terminate in Brussels "even at the age of 25, which is very much opposed to general opinion" (1842, p. 59).

After quoting the figures for Manchester and Stockport children that Cowell sent him (see below), which show, incidentally, an absolutely clear spurt in boys at the 14–15 interval and a less clear one in girls in the 13–14 interval, Quetelet fits a curve to his succession of means. This was the first time a mathematical expression had been used to express growth data." I have endeavoured," he says, "to render the preceding results *sensible* [his italics] by the construction of a line which indicates the growth at different ages. Changing Quetelet's terminology to one consistent with our usage elsewhere the formula reads:

$$h + \frac{h}{100(h_{max} - h)} = at + \frac{h_0 + t}{1 + 4t/3}$$

where h is height in meters, h_0 height at birth, h_{max} height at maturity, a a constant, and t age in years. (For Brussels males Quetelet had $h_0 = 0.500$, $h_{max} = 1.684$, and $a = 0.545$. For women $h_0 = 0.49$, $h_{max} = 1.579$, and $a = 0.520$.) This is a curve which gives always a falling velocity of growth although of varying amount (see Figure 5 below). Using the fitted curve, he gave a *Table of Growth*. Thus Quetelet enshrined in a rule the monotonically descending velocity of growth, obliterating the adolescent spurt, and confusing a number of later workers, some of great distinction in fields other than human auxology. D'Arcy Thompson used the table in the first edition of his famous *Growth and Form* (1917) but in the revised edition (1942) substituted Quetelet's original empirical means, showing the spurt, presumably because in the meantime Scammon had resurrected Montbeillard's curve. As late as 1945 Brody, whose book is a mine of information on the growth of farm animals, wrote "The prepubertal acceleration, so conspicuous in the literature on growth of children, is usually found only in the curves of poorly nourished children" (legend to Figure 16.24, p. 511; also p. 539). Although Brody's figure shows

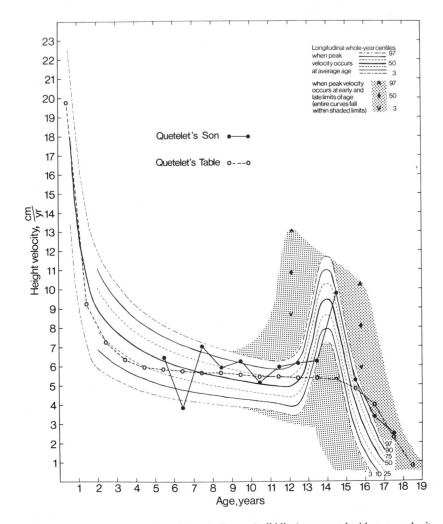

Fig. 5. Velocity of height growth of Quetelet's son Ernest (solid line) compared with mean velocity curve (dashed line) given in Quetelet's smoothed *Table of Growth* (p. 65, *Treatise on Man,* 1842). Son's measurements from *Anthropométrie* (1870, p. 185). Plotted on British standards from Tanner and Whitehouse (1976).

children studied at the Horace Mann School by Baldwin (1921), this opinion certainly does not derive from Baldwin, but is practically word for word Quetelet, who, however, is not actually quoted in Brody's book. Those who themselves measured children, however, soon realized something was wrong. Bowditch (see p. 558) in the 1880s rejected Quetelet's smoothed growth tables, and Stratz (1915) wrote, "Like Weissenberg (1911) I have come to the conclusion that Quetelet's tables are so far removed from actual findings that one cannot consider them any longer as authoritative."

Quetelet and his associates made a second survey, in which both heights and weights were obtained. Again the sample sizes are unstated, but it seems the source of subjects was the same up to age 12 (schools of Brussels and orphan hospital). From 12 onwards most of the subjects were from the better-off sections of society, and it is to this that Quetelet attributes the fact that the heights of boys from 15 upwards were greater than in the first survey. In giving a general curve for weight he wished to eliminate this bias, and thought to do so by observing that "for individuals of the same age, the weight may be considered as having a pretty constant relation to the size of the body (i.e., height). It will be sufficient then, to know the ratios . . . and to have a good general table of growth, to deduce the corresponding table of weight" (p. 64, 1842). For the "general table of growths" he takes the *fitted* height values previously obtained and applies to them the values at each age of the weight/height ratio obtained in this new survey, and thus obtains his weight table. Thus the weight tables as well as the height tables on his p. 65 are smoothed. Later workers wishing to compare Brussels weights with modern values should use the empirical means found in his p. 64 tables, bearing in mind the selection (and that he used clothed weight in these empirical tables, as did Cowell). The heights of this second series, incidently, show an adolescent spurt equally distributed in the 12–13 and 13–14 intervals in girls and a much clearer spurt in boys than in the first series, in the intervals 13–14 and 14–15. Once again Quetelet ignored this. (The intervals again are of uncertain boundary.)

Quetelet also carried out some determinations of strength of hand grip and pulling strength ("lumbar power"), these being the first such investigations to be made in children. He took 10 subjects of each sex for each year of age from 6 upwards. His means show a considerable adolescent spurt in the 14- to 15-year interval in boys' grips; a less clear spurt in girls a year earlier.

In his later book *Anthropométrie* (1870), published four years before his death, he describes a third series of measurements, this time made purely to establish the laws of development of human proportions. Thus Quetelet returned to his first and constant love. Using only 10 subjects in each yearly age group, he took no less than 80 measurements, 18 being of the face and head. He gives tables of means of the actual measurements and of each as a percentage of height. This, therefore, was the first series of measurements of children concerning more than height, weight, and one or two other dimensions. Nothing is said about selection (except that he only took children of regular build), and as one looks at the means it becomes clear that the omission is important. The biacromial diameters of the boys have the following increments from 10–11 to 19–20: 1.1, 1.1, 1.1, 1.1, 1.0, 1.1, 0.9, 0.9, 0.9, 0.6 cm. No ordinary series of biacromial diameters could so almost entirely fail to show the spurt. Indeed increments based on means from 10 randomly selected subjects at each age could scarcely be so regular. Either the sample is very special or else Quetelet has smoothed the means, employing something like his height growth function to do so. The height increments over the same ages are 5.2, 5.0, 4.8, 4.6, 4.4, 4.1, 4.0, 3.6, 2.5, 1.4 cm. The girls' height increments show the same type of curve.

Quetelet senses there is something wrong, or something missing. "At this time of puberty . . . there is a considerable change in the development of the human body" he writes (1870, p. 176). "This change comes about by quick sudden and unforeseen transformations which do not take place at exactly the same age in all individuals, which is what makes difficult the appreciation of the size of the changes." He sticks to his point though and in summarizing says, "Growth becomes less rapid the further one moves away from the time of birth" (p. 180). "In considering a particular individual . . . there are nearly always periods of arrest . . . as also of more or less rapid growth. These anomalies are observed around the age of puberty and above all following illnesses. . . . When one works on a larger number of persons these little anomalies disappear in the general mean and what lacks in the developmental of the one is compensated by an excess of growth in another" (p. 183). In his first article Quetelet (1831) actually says that Montbeillard's son's slowing down at 10–12 followed by a reprise must be due to a disorder (dérangement) of growth, frequent at puberty and indeed more frequent in girls than boys (p. 112).

There is perhaps, a further or personal reason for Quetelet's unease. Partly because of Buffon, and partly perhaps because Dr. Eduard Mallet, a distinguished analyzer of the Geneva archives of births, deaths, and marriages, in reviewing the 1832 edition of *Sur l'homme* had criticized him for not taking the measurements at successive ages in the same subjects, Quetelet made measurements of his own son and daughter from age 5 to maturity, as well as of two other daughters of a friend. These measurements are reproduced here (Table II) just as he presented them (in his original, alongside those of Montbeillard's son; his p. 185). In Figure 5 Quetelet's son's velocity curve is plotted, together with the velocity curve for boys from

Table II. Quetelet's Measurements of Height (cm) on His Son and Daughter and Two Daughters of Friends[a]

Age	Ernest Q.	Isaure Q.	Antoinette	Amélie
4		90.3		
5	102.5	94.6		
6	109.0	99.6		
7	112.9	108.3		
8	120.0	116.7		
9	126.0	121.2		
10	132.3	128.4	126.0	
11	137.5	133.2	131.0	
12	143.5	141.8	137.5	129.8
13	149.7	146.0	143.5	132.0
14	156.0	153.5	152.0	
15	165.8	159.0	157.4	
16	171.1	159.5	159.4	152.4
17	174.5		161.0	155.0
18	177.0			
25	179.5			

[a]From *Anthropométrie* (1870), p. 185. Age presumed at birthday.

Quetelet's fitted *Table of Growth,* against the background of present-day British standards. Ernest Quetelet shows a clear adolescent spurt at the expected age and of the expected amplitude. Quetelet's daughter Isaure's curve is less regular (although with an identifiable spurt at 13–14) but Antoinette, the friend, has an absolutely regular velocity curve, with the peak at 13–14, amplitude 8.5 cm/year. Yet even with these data in front of him, the oversimplified dream refused to fade; *l'homme moyen,* at least, should have a growth curve of perfect regularity and impeccable form.

5. Chadwick, Horner, and Villermé: The Beginning of Auxological Epidemiology

In the early part of the nineteenth century a school of growth studies grew up, born not of science but of the reaction of humanitarians to the appalling conditions of the poor and their children. This school developed among the family of Factory Legislation, Poor Law Commission Reports, and Sanitation and Housing Acts which constituted the new and powerful practice of public health. In England its progenitors were such men as John Fielding, Jeremy Bentham, Robert Owen, John Howard, and William Wilberforce; its pioneers Chadwick, Senior, and Farr. Their activities initiated the studies of growth which have continued in an unbroken, although often tenuous, chain through the measurements in schools and infant welfare clinics, to the London County Council Surveys of 1905–1965, modern British growth standards, and the current National Survey of Health and Growth. We may call this activity auxological epidemiology, the use of growth data to search out, and later to define, suboptimal conditions of health.

There was another, less humanitarian, side to this development. As a result of the continual recurrence of European wars and the disruption and epidemics that they entailed, statesmen in the eighteenth century became very much occupied with what we would call the manpower situation and the human resources problem. Men were needed to fight wars, and at the beginning of the eighteenth century no country really knew how many men it had. Furthermore, men who were physically fit were needed, and as the censuses began to be done the authorities were appalled to discover how many men were undersized, consumptive, and crippled, in exactly the same way as the British and American authorities were appalled in World War I (Anon., 1918; Ireland *et al.,* 1919; Davenport and Love, 1920). Such men might give rise, it was thought, to inferior stock while the fit men were away engaging in more military pursuits. Indeed such men might give rise to no stock at all, and who would then fight the next war?

These worries gave considerable impulsion to the development of numerical methods of investigation and gradually turned men's minds toward the idea that "political arithmetic," as it was named by William Petty (1623–1687), could be useful. All the same, the idea took nearly a century to become fully accepted (see Rosen's valuable article, 1955). Gradually, however, the necessary mathematical bases were established. In the *Ars Conjectandi,* published posthumously in 1713, the Swiss mathematician Jacques Bernoulli (1654–1705) laid the foundations of probability theory, with the theorem that an event occurring with a certain probability appears with a frequency approaching more closely that probability as the number of observations increases. In 1733 the Huguenot refugee De Moivre (1667–1754), working on games of chance in Soho, published a technical paper that

Walker (1929) calls "The fons et origo of the Normal Curve." Gottfried Achenwall (1719–1772), professor at Göttingen, replaced "political arithmetic" by the term "statistics" in 1749. In 1786 Laplace in Paris developed methods to solve the currently very important problem of whether inoculation for smallpox (begun in France, it will be remembered, by Montbeillard in 1766) actually was or was not effective in diminishing mortality, and he brilliantly used Bernoulli's theorem to deduce, from a study of the sex ratios of newborn infants in Paris and in the rest of France, that parents in the countryside sent a higher proportion of girls than boys to the Paris Hospital for Foundlings, where the majority perished (Hilts, 1973). Finally came the discovery by Laplace and Gauss, around 1810, of the curve of errors, and the extension of the curve to describe population distributions by Quetelet in the 1830s. Thus, just as a century before the iatro-mathematicians had paved the way for the emergence of modern physiology, so now the statistical mathematicians paved the way for the emergence of public health.

In eighteenth-century Europe the condition of the poor, and especially of their children, was dreadful in the extreme. Descriptions of London, such as that given in so vivid and careful detail by George (1965), call to mind exactly the cities of the backward countries nowadays. Around a central urban core there was a shanty-town of persons living in broken-down sheds, with few clothes, little food, and less work. Abandoned and vagrant children were everywhere, and their mortality was such that, when from 1756 to 1760 the Foundling Hospital in Coram's Fields opened its doors to all homeless infants and children in London, out of 14,934 admitted just 4400 survived to be apprenticed. In the Foundling Hospital in Paris the survival rate was about the same; in Dublin it was said to be 10%. In several London workhouses the survivors were even fewer (Pinchbeck and Hewitt, 1969–1973, p. 181). In London, however, things began to improve about 1750 (George, 1965). In the country as a whole the population began to rise, slowly at first and rapidly after about 1780. The revolution in manufacturing techniques consequent on the discovery and adoption of steam power concentrated working people in new towns, built around factories and populated by immigrants from the countryside. Although hideous and cramped, these towns at least provided wages for the working population.

In the agricultural areas the economic situation was quite simply that of a backward country, and it remained so well into the nineteenth century, with much disguised underemployment, particularly in the southern counties. From 1795 to 1834 the wages of agricultural workers, in Blaug's (1964) very informed opinion, were below subsistence level, to a degree that probably caused low production consequent on actual undernutrition. Therefore, when we see that the growth of children of the manual laboring classes in England in the 1830s was even more depressed than that of the poorer groups in underdeveloped countries nowadays (see Figure 8 below) we should not be too surprised. Indeed recent research (Trussell and Steckel, 1978) has shown that children of slaves in the plantations of the southern states of America at this time were taller than contemporary children of manual laborers in England, whereas nowadays heights of children of European and African descent living under similar circumstances are very nearly the same (see Eveleth and Tanner, 1976). This also would have occasioned no surprise to the social critics in the 1830s. Writing about factory labor, Southey remarked in 1833, that "the slave trade is mercy compared to it," and Gibbins commented "The spectacle of England buying the freedom of the black slaves by riches drawn from the labour of her white ones affords an interesting study for the cynical philosopher'' (Pinchbeck and Hewitt, 1969–1973, p. 407). The growth curves fully bear out this comment.

In the eighteenth century children worked, and worked far longer hours than any trade union nowadays would think permissible for adults. At the beginning of the century, ''Public opinion,'' wrote Marshall (1926) in her classic study of the English poor, ''strongly approved of the employment of children, and the ideas of a later age on the subject of education and the need to foster a child's self-development would have been met with blank incomprehension'' (p. 24). Pinchbeck and Hewitt (1969–1973) in their history of *Children in English Society* confirm this. ''From the days of Elizabeth onwards labour of children was a social ideal explicitly encouraged both by the provisions for parish apprenticeships in the Poor Laws and also by the Statute of Artificers of 1563; an ideal of which the extensive use of child labour in all types of industry was evidence'' (p. 98). By the eighteenth century considerations of class had considerably diluted those of conscience. ''Few children,'' wrote Mandeville in 1723, ''make any progress at school but at the same time are capable of being employed in some business or other, so that every hour of those poor people spent at their books is so much time lost to Society. Going to school in comparison to working is Idleness and the longer boys continue in this easy sort of life, the more unfit they will be, when grown up, for downright labour, both as to Strength and Inclination. Men who are to remain and end their days in a laborious, tiresome and painful Station in life, the sooner they are put upon it at first, the more patiently they will submit to it for ever after'' (p. 328). Considerations of economics were also becoming paramount. ''The eighteenth century accounts of the children's workhouse movement bear eloquent testimony to the deterioration in social attitudes towards the children of the poor since the sixteenth century. Elizabethan reformers had set out in a burst of generosity and idealism to end poverty by education and training (in the Poor Law Acts of 1572, 1576 and 1597). . . . When these early hopes failed, the mid-seventeenth century produced the children's workhouses, where the emphasis was to be on work, with only a modicum of education. Here too, the original ideal of giving children some kind of skill was soon abandoned in favour of the principle of using the labour of children to reduce the 'Poor Rates.' And when this could not be achieved because their earnings (from work done in the workhouse) were so small, children were left untrained and in idleness in the interests of economy. From the end of the seventeenth century onwards, economy in relation to the children of the poor became the dominant idea in the minds of most administrators'' (Pinchbeck and Hewitt, 1969–1973, p. 175). Furthermore, confusion in the public mind between the problems of vagabond children and of able-bodied poor adults resulted in the low work output of children being ascribed to idleness and general incorrigibility. ''The opinion that the problem of training the young in the habits of industry must be solved, if the Poor Rates were to be reduced, was shared by all writers'' says Marshall (1926, p. 25).

By the end of the century this attitude had been greatly modified. It was the time of Rousseau's *Social Contract,* of Howard's advocation of prison reform, of Wilberforce's campaign to abolish the slave trade and, a little later, of the social and educational projects of Robert Owen and George Birkbeck. Children had their champion, too, in the person of Jonas Hanway (1712–1786), who from 1750 to 1770 continually complained to the House of Commons about the state of children in workhouses which he called ''the greatest sink of mortality in these kingdoms.'' In the end his strictures led to complete discredit of the workhouse as an institution.

Children, whether parish or not, sometimes began work in factories and even the mines at the age of 5 or less, and entry at age 8 was the usual thing. Even tiny children had for centuries worked at their homes in such things as cotton manufacture and, "when, due to later inventions, the textile industry moved from the home to the factory, the children went with it. Small children of only three or four years of age were employed to pick up cotton waste, creeping under unguarded machines where bigger people could not go. The older children worked for fifteen hours a day, and on night work too, under conditions which were often enforced by fear and brutality" (Pinchbeck and Hewitt, 1969–1973, p. 354). We should remember too, in passing, that conditions of agricultural labor for children were scarcely better; indeed the physical demands made, especially in the "gangs" system, passes present-day belief (see Pinchbeck and Hewitt, 1969–1973, p. 391ff). Gradually a series of Acts were passed which mitigated these conditions. It was in connection with these Acts that the first large-scale studies of children's heights and weights were made.

The most important document in this development was the *Report of the Commissioners on the Employment of Children in Factories* (*English Historical Documents*, 1956; Vol. 12, pp. 934–949). The report resulted from an inquiry into "the whole subject of the labour of children, as now enforced in the various Mills and Factories or places of work throughout the country" (*Parliamentary Papers*, 1833, p. 79). Edwin Chadwick (1800–1890), in his youth secretary to Bentham and the chief architect of public health reform in England, was one of its signatories. With the *Instructions to Members of the Commission* we are quite suddenly in a new and wholly up-to-date world. Not only is the inquiry thorough-going, requesting data on morbidity, mortality, stillbirth and illegitimacy rates, accidents, and so forth; it lays great emphasis on investigating a suitable control population and on sampling by domiciliary visits to whole streets. "A given amount of evil is experienced by a class placed under peculiar circumstances; a large portion of that evil is shared by other classes not under these peculiar circumstances; to attribute the whole of the evil experienced by the first class to those peculiar circumstances is obviously fallacious" (*Report of the Commissioners*, p. 84). In each area a medical commissioner was appointed and, among other things, he was specifically required to ascertain the stature of the children to see "whether there be any difference at any age, and what age, and in either sex, between persons brought up from an early period in a Factory and persons of the same age and sex and station not brought up in a Factory."

There is no indication of the antecedents of this request for height measurement, but presumably Chadwick was in contact with Villermé and his associates in France and knew of the 1829 memoir on adult height and the conclusions Villermé had drawn from it. At any rate, the instructions were too advanced for the medical commissioners; all, that is, except a certain Samuel Stanway, who worked in the area centered in Manchester, under the supervision of Assistant Commissioner J. W. Cowell. All other commissioners ignored, it seems, the question about stature; but Stanway and Cowell visited the Bennett Street and St. Augustine's Sunday Schools in Manchester and two schools in Stockport, and measured heights and weights of a total of 1933 children aged 9–18. Some of these children, the majority, worked in the factories; others did not. Their results on the factory children are reproduced, in the original measurements, in Table III. They do not state their technique, but Quetelet (1835) says that boots were kept on, and in comparing them with Belgian children, subtracted what he thought was a suitable amount (although

Table III. Stanway's Measurements of Heights and Weights of Children Working in Factories in the Manchester Area, 1833, with Factory Assistant Commissioner Cowell[a]

Age	Boys			Girls		
	N	Height (inches)	Weight (lb)	N	Height (inches)	Weight (lb)
9+	17	48.14	51.76	30	47.97	51.13
10+	48	49.79	57.00	41	49.62	54.80
11+	53	51.26	61.84	53	51.16	59.69
12+	42	53.38	65.97	80	53.70	66.08
13+	45	54.48	72.11	63	55.64	73.25
14+	61	56.59	77.09	80	57.75	83.41
15+	54	59.64	88.35	81	58.50	87.86
16+	52	61.60	98.00	83	59.81	96.22
17+	26	62.67	104.46	75	60.41	100.21
18+	22	63.31	106.13	65	62.72[b]	106.35
	420			651		

[a] Probably with indoor clothes and boots. From *Parliamentary Papers* (1833), *Reports from Commissioners,* Vol. 4, First report. . . . into the Employment of Children in Factories, pp. 697, 698 (D. 1, p. 87).
[b] Presumably a misprint.

his Edinburgh translator wonders, as well he might, whether Lancashire children did not wear wooden clogs even to Sunday school).

The Stanway and Cowell figures, in contrast to Quetelet's, show just the sort of results one might expect. Among factory children, the numerically larger group, girls' height exceeds boys' at ages 12, 13, and 14. The boys have a clear-cut adolescent spurt with a peak at 15.0 years (see Figures 6 and 7); a year later than nowadays. The girls have a less clear-cut spurt centering probably at about 13.5 years. The nonfactory, or control, children were taller, but here the so-far very modern statistical technique breaks down. In calculating average heights of the two groups Stanway fails to weight his means according to the numbers in each age group. The nonfactory children were considerably younger, so the differences he gives (0.3 inches for boys and a negative value for girls) are misleading. The average mean differences calculated within years of age and weighted according to numbers comes to approximately 0.25 inches (about 0.5 cm.) both for girls (aged 9–15) and boys (aged 9–17).

Chadwick's 1842 *Report on the Sanitary Condition of the Labouring Population of Great Britain,* made for the Poor Law Commissioners, is justly a famous document, but the earlier *Report on the Employment of Children in Factories* is just as detailed, thorough, and well-organized, a model in all respects. As a result of the report, the Factories Regulation Act (1833) was passed which prohibited children under 9 years of age (i.e., before the ninth birthday) from working in various types of textile factories, and stipulated that children between the ninth and thirteenth birthdays should have 1.5 hours a day for meals and rest. Such children had to have a certificate "from a surgeon or medical man resident in the township . . . who shall certify on inspection of the child that he believes it to be of the full growth and usual condition of a child of the age prescribed . . . and fit for employment in a manufactory. This certificate should be given in the presence of a magistrate by whom it should be countersigned, provided he also were satisfied" (*English Historical Documents,* 1956, p. 944). Age is not enough, however: "The physical condition alone is the proper qualification for employment, and unless a discretion of this nature were given to the parties certifying, they might feel themselves bound to

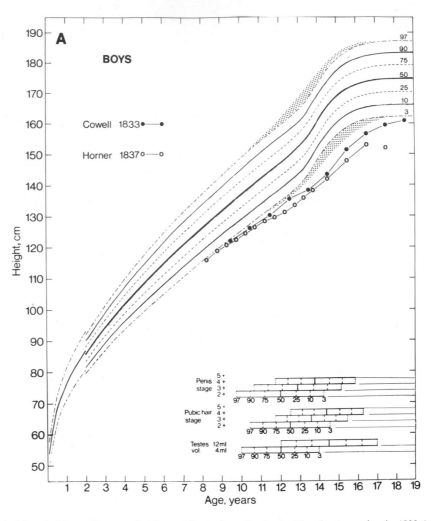

Fig. 6. Mean heights of boys and girls working in factories in the Manchester region in 1833 (●—●), measured by Stanway and Cowell (see Table III), and in the Lancashire area in 1837 (○---○) measured at Horner's request (see Table IV). Plotted on British standards from Tanner and Whitehouse (1976). Peak height velocity (PHV) approximately 15.0 in both series of boys. Two probably erroneous points are left unjoined to others.

certify to the age of a child on the production of copies of the baptismal registers, which are easily forged, or on the evidence of parents who would be under temptation to perjure themselves" (p. 944). The flavor of the underdeveloped country is very strong. The *Report* also adds that the local doctor's certificate would not be much use as "his practice is liable to be dependent on the labouring classes" (thus prefiguring the present-day certificates for sore throats and back-ache). Thus a more stringent certificate given by an inspector must be produced before entry to full-time work at 13.

It was the belief that ages were being falsified which led directly to the second and much larger survey of childrens' heights and weights, organized by Cowell's colleague Horner. Leonard Horner (1785–1864) was for 25 years one of the four government inspectors of factories. He was a man of wide interests and influence,

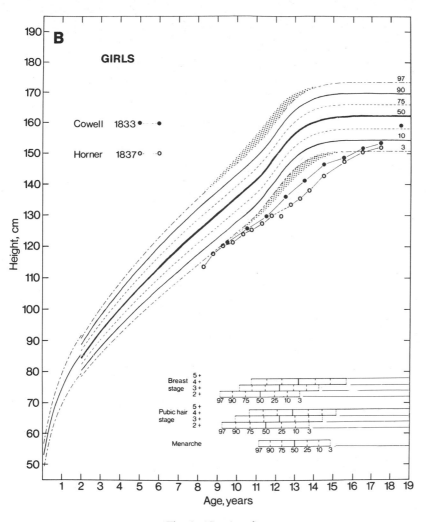

Fig. 6. *(Continued)*

especially among the scientists of the time. The son of a well-to-do wholesale linen merchant in Edinburgh, he took up geology at Edinburgh University and although he went into the family business, geology remained a passion throughout his life. He was twice president of the Geological Society of London, a fellow of the Royal Society, a close friend of Francis Galton's father Tertius, of Babbage, and of Hooker, and well-known to Charles Darwin, who rather charmingly wrote to him in 1860 "I believe variations arise . . . accidently or spontaneously and these are naturally selected or preserved, from being beneficial to the successive individual animals in their struggle for life. I do not know whether I make myself clear" (Lyell, 1890, Vol. 2, p. 300). Horner's second passion was the education of working men and children. In 1821 he founded the first of the numerous Mechanics Institutions which sprang up in Great Britain in the 1820s and 1830s. This was the Edinburgh School of Arts, a "College for the better education of the Mechanics of Edinburgh." He also founded, jointly with Lord Cockburn, the Edinburgh Academy. Small wonder then that in 1826 he was invited to be the first warden (equivalent of

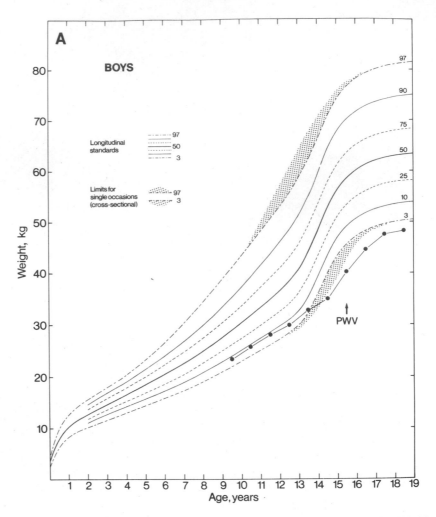

Fig. 7. Mean weights of boys and girls working in factories in the Manchester region in 1833. Measurements by Stanway and Cowell (Parliamentary Papers, 1833). Plotted on British standards from Tanner and Whitehouse (1976). Peak weight velocity (PWV) for boys approximately 15.5.

principal) of the new University of London. He retired from business, and after relinquishing the wardenship in 1830 was appointed to the Factory Commission in 1833.

Horner took up office on April 30, visited factories in May, reported (with Cowell and two other colleagues) in June, and got the act passed on August 29. He continued as factory inspector, at first in the Scottish region and from 1836 in the Lancashire area, until he was over 70. In 1837 he got 27 surgeons appointed to factories in his district to measure 8469 boys and 7933 girls aged 8–14 inclusive in the towns of Manchester, Stockport, Bolton, Preston, Leeds, Halifax, Rochdale, Huddersfield, and Skipton, and the neighboring rural districts. Details of selection and technique were never published. The results were printed as Appendix XI in the English version of Quetelet's (1842) *Treatise on Man.* Horner's family letters (Lyell, 1890) unhappily make no reference to the circumstances of this large-scale survey, nor indeed do they mention Quetelet; Horner's international acquaintance was large, but for the most part geological. However, there is a letter extant from

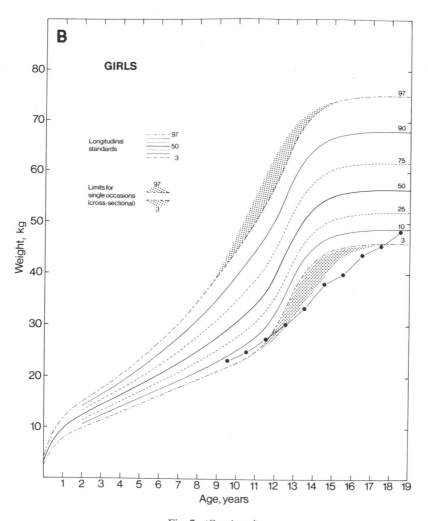

Fig. 7. *(Continued)*

Horner to N. W. Senior, chief of inspectorate, in which, among other critical remarks, Horner writes "From the great imperfections of the Act in all that relates to the determination of the age of the children, it is impossible for the Inspector to check the most palpable frauds, and to prevent the admission of children full-time long before they are 13 years of age. I have tried various checks—with but partial success—and I am pursuaded that fully one-half of the children now working under surgeons certificates of 13 are in fact not more than 12, many not more than 11 years" (Horner, 1837, p. 34). The reason for the alleged deception was that parents were paid according to the hours worked and at age 13 adult hours were permissible.

The Factory Act of 1833 by no means satisfied Horner. He campaigned continually for better factory conditions, especially as regards accident prevention, and in a small book that still makes red-hot reading he urged reform in other areas besides textile manufacture. "The inhumanity, injustice and impolicy of extorting labour from children unsuitable to their age and strength—of subjecting them, in truth, to the hardships of slavery (for they are not free agents)—have been condemned by the public voice in other countries [also]" he wrote (Horner, 1840). He goes on to cite the law in other countries and adds a (rather complacent) letter from Horace Mann (1840), the secretary of the Board of Education of the state of Massachusetts. In 1840 Horner and three other commissioners reported on the situation of children working in mines. The report (*English Historical Documents*, 1956, pp. 963–968) created such a scandal that it led to a total ban on women and children in mines. Some children of 8 or even younger were spending 12 hours a day standing in solitude and darkness (owing to the particular task they did). The commissioners reported that "The employment in these mines generally produces in the first instance an extraordinary degree of muscular development, accompanied by a corresponding degree of muscular strength; this preternatural development and strength being acquired at the expense of other organs, as is shown by the general stunted growth of the body" (p. 976). Unfortunately, there seems to be no evidence that the heights of mine children were actually measured.

Horner's results are given in Table IV and plotted on the standard 1965 British height, weight, and height velocity charts (Tanner *et al.*, 1966; Tanner and White-house, 1976) in Figure 6 for comparison with Cowell's. The numbers in Horner's survey are substantial up to and including age 14. At 15, 16, and 17 the means are based on smaller numbers of children, measured by one particular surgeon, Harrison, in Preston. Figures 6 and 7 show how extraordinarily small these children were compared with those of the present day. At age 8–10 the means of both 1833 and 1837 series are at about the third centile of the present-day standards, and at older ages they are well below the third centile until 18 in the boys and 16 in the girls. The mature height seems to be at about the present British tenth centile for girls and third centile for boys. Horner's series has lower means than Cowell's, especially in the case of girls. One has to remember, however, that Horner thought his factory children were dissembling their ages; if his 13-year-olds and perhaps his 12-year-

Table IV. *Horner's Measurements of Heights of Children Working in Factories in the Lancashire Area in 1837*[a]

Age	Boys		Girls	
	N	Height (inches)	*N*	Height (inches)
8.0+	327	45.62	267	44.69
8.5+	339	47.00	272	46.37
9.0+	527	47.62	438	47.37
9.5+	418	48.12	375	48.00
10.0+	574	49.00	506	49.00
10.5+	550	49.87	421	49.62
11.0+	664	50.37	577	50.06
11.5+	559	51.06	478	51.25
12.0+	767	51.75	712	51.06[b]
12.5+	660	52.25	618	52.75
13.0+	1269	53.50	1260	53.33
13.5+	864	54.37	980	54.50
14.0+	951	55.75	1029	56.00
14.0+	117	56.25	140	57.00
15.0+	82	58.50	106	58.75
16.0+	43	60.50	90	59.50
17.0+	47	60.00	112	60.00

[a]Taken from Appendix XI of Quetelet's *Treatise on Man* (1842).
[b]Presumably a misprint for 52.06.

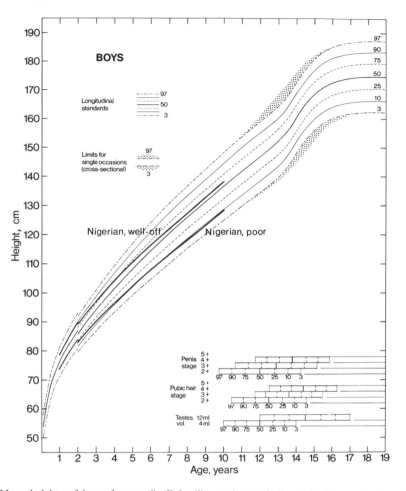

Fig. 8. Mean heights of boys from well-off families and poor indigent families in Ibadan, Nigeria. Longitudinal data from Janes (1975); figure from Tanner (1978). Plotted on British standards from Tanner and Whitehouse (1976).

olds also were plotted as if really 6–9 months younger they would much more nearly approach the values obtained by Cowell. Cowell's children did not have the same reason to falsify their ages, because they were already working full-time in factories, before the law was passed. Secondly, Cowell's children were selected by being those who attended Sunday school, and since this was not compulsory they would presumably be the children of the more intelligent and caring parents. Horner's children (one assumes) were measured for the most part in factories and were not subject to this selection. Cowell's children wore Sunday shoes and indoor clothing; Horner's we are uncertain about, but often children went without shoes in the factories, so it seems likely that all were measured barefoot.

These series represent the first properly documented cross-sectional surveys of children's heights known to the writer. It is instructive and surprising—and perhaps encouraging—to compare the results with those for a typical underdeveloped country nowadays. In Ibadan, Nigeria, Dr. Margaret Janes (1975) has for many years been engaged in a longitudinal study of two groups of children, one rich, the other very poor, being dwellers in a slum near the central marketplace. Figure 8 shows that while the rich in Ibadan are exactly comparable with present-day British standards for height, the poor are considerably taller than Manchester factory children in the 1830s, their means being at the British tenth centile.

Scrutiny of the velocity curves of Cowell's and Horner's children shows a growth spurt delayed compared with nowadays, but by no means so delayed as by itself to account for the smallness. The girls' ascendancy (girls taller than boys) is at 12, 13, and 14; nowadays it is at 11, 12, and 13. The peak height velocity seems to be at about 15.0 for boys and 13.5 for girls, which is only 1.0–1.5 years later than nowadays. (There is no indication of puberty status in either series.) Most of the smallness of the children, then, is persisting smallness, not growth delay. The remarks of Horner and his colleagues about children in the mines, given above, seem very pertinent.

In light of modern knowledge the causes of such short stature, persisting into adult life and not to a very great extent accompanied by retardation of growth, have to be sought in early childhood and even the fetal period (see, e.g., Tanner, 1978). Severe malnutrition of the pregnant mother followed by chronic and severe under-nutrition of the infant could cause this result, but more likely is a low birthweight and/or a low weight gain in infancy caused by injurious substances breathed or eaten by the pregnant mother at work, and by the newborn child. The remarks of Pinchbeck and Hewitt seem very relevant: "In the nineteenth century, where mothers were much employed from home in mills, workshops and factories, the infant mortality rates were inflated by the deaths of babies ill-fed and often ill-used by those in whose care they were left by their mothers. Some starved to death; others died from being fed totally unsuitable food (patent baby foods did not appear in the UK till 1867); many more were the victims of the reckless use of the narcotics—opium, laudanum, morphia—which were the major ingredients of God-frey's Cordial, Atkinson's Royal Infants Preservative and Mrs. Wilkinson's Sooth-ing Syrup, administered to calm children and which in many cases established a calm that was but a prelude to a deeper quiet. There was nothing unusual in this. Whenever mothers of young children were fully employed, whether in the fields, the cottage, or the factory, the administration of drugs to keep children quiet was, and as far as we know, always had been, a common phenomenon. Among embroi-derers for example, 'the practice, which is most common, usually is begun when the child is three or four weeks old, but Mr. Brown, the coroner of Nottingham states that he knows Godfrey's Cordial is given on the day of birth, and that it is even prepared in readiness for the event . . . the result is that a great number of infants perish . . . those who escape with life become pale and sickly, often half idiotic and always with a ruined constitution.' Here as in so many other respects, the experi-ence of factory industry, far from being unique, was in fact the experience of cottage and workshop industry, writ large for all to see" (p. 406).

Another contributory cause of the shortness might have been the undernutri-tion of the mothers much earlier, when they themselves were children, causing stunting of size and in consequence the production of small babies. The mothers had grown up in the days of the Napoleonic Wars, when conditions in the countryside were at about their worst. Some of the mothers, too, must have been immigrants from Ireland, an area Horner reports as greatly depressed even compared with northwest England.

This remarkable situation was not confined to the factory towns of the indus-trialized countries. In a masterly and most unjustly neglected study, Kiil (1939) has analyzed with great care a Norwegian archive on the heights of young men that stretches back, effectively, to 1761. In Norway nearly every young man over the age of 17 had to attend an annual enrollment muster, from which the standing militia was selected. At this muster, age was recorded and height measured, in stockinged or bare feet. Since many men came back on successive occasions longitudinal records are abstractable, although with much difficulty, from this archive. Kiil gives the means of two groups of Romsdal peasants in northwest Norway measured during the years 1826–1837, just about the time of the Factory Commission surveys. One group consisted of 238 men measured each year from age 17.3 to 22.3, the other of 133 men measured each year from age 16.3 to 22.3. (Kiil, 1939, Tables 56 and 57, p. 118). The mean heights of each pure longitudinal group are plotted against the modern British standards in Figure 9. The modern average Norwegian, it must be remembered, is a little taller than the average Briton, with height at about the British 60th centile. The mature height of the Romsdal men is some 5–6 cm less than that of comparable Norwegians nowadays, and it was not reached till age 25 or later. It is not possible to locate the age at peak height velocity with accuracy from

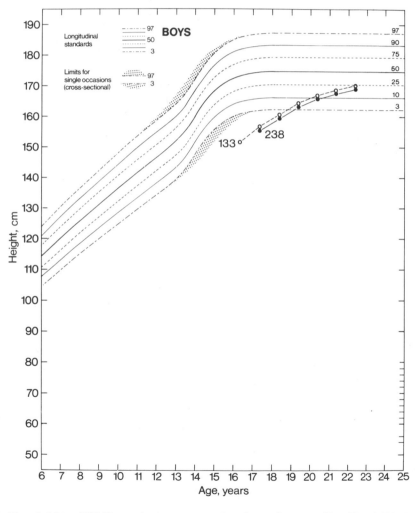

Fig. 9. Mean heights of 238 Norwegian men measured each year from age 17 to 22 and 133 Norwegian men measured each year from age 16 to 22 in years 1826–1837. Data from Kiil (1939, p. 118). Plotted on British standards from Tanner and Whitehouse (1976).

Kiil's data; cross-sectional material for 1795 indicates a peak at around 17.0 (p. 90) whereas the 1830s data seems to indicate a peak at about 16.0. In both cases the velocities are very low compared to modern ones, suggesting a long drawn-out puberty. Although these country dwellers could scarcely have been exposed to the adulterated atmosphere, food, or even cordials of the urban factory workers, undernutrition was fairly severe (until, Kiil thinks, the general introduction of the potato in the 1830s). This undernutrition produced, as we would expect, a greater delay than that seen in the factory children, but a smaller diminution of mature height. These data provide some corroboration then, for the interpretation given above.

Horner's was the first large-scale survey of the heights of children of which we have any record. It is perhaps a rather baleful reflection that just as the first surveys of adult height were made to see what steps could be taken to obtain better soldiers, so this survey of childrens' heights was made to see whether parents were cheating the law in the exploitation of their children.

5.1. France: Villermé

In France the same sort of developments had been taking place, with Villermé in the position of Chadwick. Villermé's first large memoir (1828) was on the mortality of rich and poor in France. The memoir begins by saying that the matter is disputed: Some say the poor die more because of their privation, but others that the rich die more because of their "luxury, passions and excesses of all kinds." Villermé established that in Paris the rate of mortality over all ages combined was nearly twice as great in the poorest as in the richest arrondisement; he concluded that "the mortality is greater in the poor to a degree of which we had no inkling" (p. 75)—"easy circumstances prolong life, poverty shortens it" (*L'aisance de fortune conserve notre vie: la misère l'abrège*) (p. 81). Villermé (1829) went on to examine adult stature in relation to the same factors. He analyzed the results of a large government enquiry made in 1812–1813 about the heights of conscripts, and the proportion of conscripts rejected for small size, illness, or deformities. In Bouches-de-la-Meuse, later part of Holland, the mean height of conscripts of 1808–1810 was 167.7 cm and 66 out of 1000 were rejected. In Chiavani, in the Apennine Mountains, mean height was 156.0 cm and 300 out of 2100 were rejected. The prefect of Puy-de-Dóme, remarked, "The young people develop late and many of them finish their growth at 24 years. . . . This slowness of development relates in part to race but more to poverty (la misère), to the smallness of the care that the poor give to the education of the children, to the precocious work which is necessary, to lack of healthy and abundant food and to the type of industry here." (p. 363). In the arrondisements of Paris there was an association between the percentage of persons owning houses and the mean stature of the district. Poverty, said Villermé, arguing against the still-circulating views of Hippocrates in *Airs, Waters and Places,* was much more important than climate in influencing growth. "Human height," he said, "becomes greater and growth takes place more rapidly, other things being equal, in proportion as the country is richer, the comfort more general, houses, clothes and nourishment better, and labour, fatigue and privation during infancy and youth less; in other words, the circumstances accompanying poverty delay the age at which complete stature is reached and stunt adult height."

Villermé had absolutely no word to say about the effects of heredity on stature, and his views on poverty did not go for long unchallenged. In France, Dr. M. Boudin (1806–1867), a distinguished army surgeon, who in 1859 was a founder-member of the Societé d'Anthropologie, published in the year of Villermé's death a massive account of recruitment procedures in various countries, and said the view that differences in heights between populations were due to differences in nutrition were quite mistaken; altitude, climate, and race were the important factors (Perier, 1967). Indeed before Boudin, Dr. Knox, the translator of Quetelet, had exploded in a footnote over Quetelet's quote of Villermé's findings that the men in the Meuse region were taller than those in the Apennines (Quetelet, 1842, p. 60). "The translator is persuaded," said Knox, "that Dr. Villermé and M. Quetelet have failed to detect the real cause of differences in stature in those two Departments: it is a question purely of race and not of feeding or locality. The latter conscripts were Saxons . . . the shorter, found in the Apennines . . . were the descendents of the Ancient Celtic population of that country." In light of modern research (see Eveleth and Tanner, 1976) the truth lies between these two extremes of opinion, but Villermé's view is the closer to it.

The condition of the factory workers and the poor was even worse in France after the Napoleonic wars than it was in England and, in the 1830s and 1840s, Villermé played the part of Farr (1807–1883), a man 20 years younger, and Chadwick. In 1832 the Académie des Sciences Morales et Politiques was reinstated in France after its suppression 30 years earlier by Napoleon, and it set up the same sort of inquiry among the textile workers in France that the commission of 1833 had made in England. Villermé undertook the inquiry; ten years in the army had accustomed him to fieldwork and he did all the investigations himself (see Villermé, 1840, introduction), traveling from place to place and writing to Quetelet in June 1835, "You ask me if I am satisfied with my tour; yes, in some respects, no in others. I was at Lyon, Avignon, Nismes, Montpellier, Marseilles, Toulon, Geneva, Lausanne, Berne, Zurich, Basle, Mulhouse and the neighbourhoods. . . . If following my tour a maximum duration of work for children in factories should be adopted I would certainly be well rewarded for all my trouble. This law, which would only be a copy of one passed in England not long ago, is absolutely demanded by conscience and humanity" (Ackerknecht, 1952). His work was issued in a report entitled *Tableau de l'état physique et morale des ouvriers* (Villermé, 1840; Deslandres and Michelin, 1938), which was successful in obtaining a law the following year forbidding children aged 8–12 working more than 8 hours a day or doing any night work, and making schoolwork mandatory until the age of 12.

6. Roberts, Galton, and Bowditch: Social Class and Family Likeness

The Factory Acts continued to provide increasing protection to working children as the century progressed, and conditions in the textile factories changed, both because of this and because of the introduction of new machinery. Not all the changes were for the better; the machines were more complicated and ran faster, and the work was more monotonous. Children of 8 still left home without breakfast at 5:30 on a winter's morning to begin work in the factory at 6, and pregnant women still stayed at their jobs till 2 weeks or even 2 days before confinement. Many then returned within a month postpartum, either leaving the infant in the care of some frequently ill-paid childminder or bringing the baby to the factory where it could not be suckled sufficiently, for the mother could only break her work at 5- or 6-hour intervals.

In 1872 a parliamentary commission was set up, under a doctor, J. H. Bridges, and a factory inspector, T. Holmes, to inquire into all aspects of conditions of the work of women and children in textile factories (Parliamentary Papers, 1873). They took evidence from workers, employers, and the factory surgeons. These latter stressed particularly the plight of pregnant and postpartum women. They remarked that growth of the fetus *in utero* was affected by the conditions of work, although they gave no evidence in the form of statistics of birth weight. In fact, as we have discussed above, a low birth weight, especially if combined with undernutrition in the first few months after birth, would provide a good deal more convincing explanation of the shape of the childhood growth curves of 1833 and 1873 (see below) than any simple childhood deprivation. The curves show height greatly reduced at all ages, including maturity, with only a relatively minor degree of retardation. Such curves are seen nowadays in babies with low birth weight (see Tanner *et al.,* 1975) due to disorders of fetus or placenta. If the factory mothers did indeed produce such babies, the cause would more likely have lain in the poisonous atmosphere and the frequently adulterated food than in simple undernutrition (see Tanner, 1978), although undernutrition in the first months after birth might perhaps produce a not dissimilar picture, and this must have been general.

In addition to taking verbal evidence, the commission instituted "a careful and systematic examination of children upon an extensive scale, in a great variety of areas . . . registering their height, weight and dimensions of the chest and recording all instances of malformation or disease . . . our object was to compare and contrast children employed in factories first, with children inhabiting factory districts but not employed in factories; secondly, with children from adjacent districts where no factories were situated. It was also necessary to distinguish between factories situated in large towns and those in suburban or semi-rural districts, to meet the obvious objections that whatever results were observed might be attributable to the child's locality rather than to its occupation. Lastly, the occupation . . . of the child's parents seemed to us an element which it was essential to take into account" (Parliamentary Papers, 1873, pp. 846–847). There were five doctors employed to do the examinations and take the measurements. During the winter of 1872–1873 they visited "a large number of schools in Lancashire, Cheshire and the West Riding of Yorkshire" and examined nearly 10,000 "factory and other children of the working classes." All children in a given school were taken without selection. Height was measured in bare feet, weight in indoor clothes, and chest circumference with boys stripped to the waist. Bradford, Halifax, Preston, Rochdale, Stockport, and Macclesfield were some of the urban factory districts; York, Kendal, Sedbergh, Lancaster, Chester, and Keswick some of the nonfactory towns. "The general impression made by the factory children was in many respects not unfavourable. As compared with the children of the East of London, or of the poorer parts of Liverpool, they were markedly superior. They did not appear to be more liable than ordinary country children to rickets or scrofula . . . ."; however, "very few were free from vermin" (Parliamentary Papers, 1873, p. 848), and most were flea bitten. The "personal habits of cleanliness of the factory children compared very unfavourably with those in the agricultural districts" (*Parliamentary Papers,* 1876, Appendix D).

The tables in the report show that urban and suburban children who worked in the factories, and had parents doing the same, were about 3 cm shorter at ages 8–11 than children who lived in nonfactory rural and suburban districts. At 12 the difference was 4 cm in height, and in weight about 2 kg. Boys and girls showed similar differences. The same factory children were smaller by almost as much when compared with children living in the factory towns but not working in factories, nor having parents so working (although all were of the manual laborer class). The differences in this comparison averaged 2–3 cm. Hence simply living in the factory town was insufficient to cause the shortness; the social environment that went with factory work was the important thing. Factory children in the big towns were slightly smaller than factory children in the suburban areas, by about 1 cm. This report was soon followed by the Factory Act of 1874 which raised the minimum age of employment—in the textile factories only—from 8 to 10 and the full age of employment from 13 to 14, unless a child of 13 could pass a standard of education. Meal and rest times were increased to 2 hours out of the working day, and schools, already by law attached to factories, had to be certified as efficient by examiners from the Department of Education.

One of the five doctors who did the inspections and measurements was a man whose interest in anthropometry after, if not before, this experience was profound. Charles Roberts (died 1901) was a Yorkshireman who studied medicine at St. George's Hospital, London. He and Bridges were fellow students and qualified in the same year, 1859. He practiced for a time in York and became surgeon to North Riding Prison, but then returned to London following Bridges' appointment as Medical Inspector to the Local Government Board there. Roberts continued his surgical career in London, but he was a man of independent means, and devoted much time to his chief interests, anthropology and natural history. (He for some years edited the *Selborne Society's Magazine*.) Roberts read and admired Quetelet's *Anthropométrie* (1870) and set his heart on writing a book to be entitled *Physical Development and Proportions of the Human Body*. He never succeeded in this, but he did publish a shorter work, *A Manual of Anthropometry* (1878), which has much of Quetelet in it, including a section on body proportions addressed particulary to art students and using Quetelet's own measurements of body segments. About Quetelet's survey of height, and weights, however, Roberts was critical.

Roberts' place in the history of growth studies rests on two papers, published between 1874 and 1876. He resembles, therefore, his contemporary and transatlantic collaborator, Henry Bowditch, rather than the later figure of Franz Boas, with his dozens of papers and decades of work. Roberts' (1876) report to the Statistical Society deals with the old question of how large and strong a child should be before he was permitted to work in a factory. The provisions of the 1833 Act were still in force, and had been added to, so that in 1872 all children under the age of 16 applying for employment in a factory had to produce a certificate from a certifying surgeon saying they were not incapacitated by illness and had "the ordinary strength and appearance" of a child of at least 8 years or a young person of at least 13 years, according to which category they entered, that is part time (5 hours a day) or full time (10 hours a day). A Royal Commission, appointed to inquire into the working of the Factory and Workshops Act, asked Roberts to establish whether the medical examination was really necessary, given a valid certificate (*Parliamentary Papers,* 1876). Roberts examined the large series of measurements of height, weight, and chest circumference taken in the 1872–1873 survey and established, first, that "there are no physical qualities sufficiently distinct and constant to indicate the age of a child within two and often three years of its actual birthday, and that the certificate of birth is the only evidence that can be relied on"; and, second, "that age is not an indication of any constant physical qualities, and a certificate of birth is not, therefore, sufficient evidence of fitness for factory work, and cannot supply the place of a medical examination" (1876, p. 682). He examined whether or not the state of the eruption of the teeth could be used to validate age but, although agreeing teeth were a better guide than body size, he wrote, "I am convinced that order of appearance of teeth varies so widely in different individuals and different classes of society that it cannot be trusted as a test of age" (p. 684). He then proposed using height, weight, chest circumference, and weight-for-height as criteria for fitness for employment, excluding the children at the lower ends of the frequency distributions of each of the variates. The actual values suggested as lower limits for height and weight are interesting: at age 8, 42 inches and 45 lb; at age 9, 44 inches and 57 lb; at age 12, 49 inches and 59 lb; and at age 13, 50 inches and 65 lb. Children as small as this would be quite impossible to find nowadays, except in cases of specific pathological disorders; the height limits are below our 1 in 1000 line. The weights, even when clothes are allowed for, are a little higher, and about at our first centile.

The survey of 1872–1873 was a considerable affair. However, it concerned only children aged 8–12 inclusive (and alleged), so Roberts, in addition, extracted measurements from the records of the Chelsea Royal Military Asylum, a boarding school for sons of soldiers, and from the Royal Hospital School, Greenwich, a boarding school for sons of naval personnel, largely coast guards in rural areas. He thus extended the age range for boys to 13+. He divided the children into those living in rural and in urban areas, with each subdivided into those working in factories and those not working in factories. Means and modes are given (Roberts calls the modern mean the "average" in his tables and the "mathematical mean" at one place in his text; in all his papers the word "mean" in his tables indicates the modern "mode").

There was a clear difference between the urban (factory and nonfactory) and rural boys, with the rural about 1.5 cm taller at 8–11, 2.0 cm taller at 12, and 3.0 cm taller at 13. This is the reverse of the pattern seen in most modern statistics (see Eveleth and Tanner, 1976). The girls showed a similar but smaller difference, amounting to about 1.5 cm at 8 and 9 and less at later ages.

Roberts compares his factory children's weights at ages 9, 10, 11, and 12 with those given by Cowell in 1833. The differences were 6.8, 4.5, 4.8, and 4.6 lb. "From these figures," he wrote, "it will be seen that a Factory child (he meant boy) of the present-day of the age of 9 years, weighs as much as one of 10 did in 1833, one of 10 now as much as one of 11 then, and one of 11 now as much as one of 12 then; each age has gained one year in 40 years" (Roberts, 1876, p. 691). However, he says nothing about heights. A comparison of his 1873 urban factory boys' heights with the 1833 factory boys of Cowell shows only very small differences: at 9, 1.0 cm; at 10, 0.5 cm; at 11, 0.6 cm; and at 12, nil. Even when all the "labouring class" children, urban and rural, factory and nonfactory, are pooled together, as in Table 1 in Roberts' other paper of 1874–1876 (see below), the boys' heights exceed those of Cowell's boys only by 1.8, 1.8, 1.5, −0.7, 0.7, and 0.5 cm at ages 9–14 (at 15 and above Roberts' "labouring class" is selected and invalid for comparison). It seems that between 1833 and 1873 the heights of children changed very little indeed, at least in the age range 9–14. Thus we have to question Roberts' interpretation of his results on weight—or at least the present writer's earlier interpretation of them (Tanner, 1962). It looks as if no secular change in height has occurred, only in weight-for-height. Some of the 4- to 6-lb difference may even be due to differences in the weights of clothing included (Roberts added 10 lb to the nude weights of the Chelsea and Woolwich boys to bring them equal to the rest, so that is his estimate of the weight of winter indoor clothing; Cowell's children were "lightly clothed and had nothing in their pockets," and Quetelet suggests an allowance of only 5–6 lb for their clothing). Cowell's children had their height measured in shoes, it is true, but this is unlikely to have added more than 1 or maximally 2 cm. Furthermore, the same lack of difference is seen between the 1873 figures for factory children of factory parents and those in Horner's survey of 1837. The 1873 factory boys aged 8–12 average a mere 0.6 cm taller, the girls 1.2 cm.

Certainly the position of factory children was still deplorable. Roberts (1874–1876) reported that the prevalence of severe flat feet reached 5% in urban factory children aged 9 and 10 and 10% and over in those aged 11 and 12. In country towns and agricultural districts the figures were between 1 and 2%, as they were in the factory children on entry at age 8. This, however, was the only difference between factory and nonfactory children that Roberts thought was due to the work in the factory itself. "Nearly the whole of the disadvantages the factory children labour under are to be attributed to social causes," he wrote, "rather than factory work" (*Parliamentary Papers,* 1876, Appendix D, p. 180).

Much of Robert's other large paper (1874–1876) is an exposition of Quetelet, although not an uncritical one. Quetelet had measured 30 individuals of the same age, divided them into three groups of ten each and observed that the means of each group were nearly the same; hence, he said, a group of ten persons was sufficient to characterize typical proportions. "In this respect," says Roberts (p. 15), "my tables will show M. Quetelet was in error, and will account for the differences which occur between his results and mine."

Modes, means, and frequency distributions of height and weight are given for each year, effectively from age 5 till 25. Roberts pools all his 1872–1873 values for the "labouring class," which now includes the schoolboys of Chelsea and Greenwich, and boys in rural and urban areas, whether working in factories or not: All, however, belong to the families of artisans or laborers, the manually employed. The numbers at age 5–15 years were 175, 327, 581, 1071, 1252, 1669, 1246, 940, 735, 869, and 409; over age 15 the sample is biased, because it consists only of boys accepted for entrance to the army and navy and there was a lower limit for height. This bias is nicely shown in the relative progression of means and modes. Up to age 16 there is very little difference between mean and mode, but at age 16 and upwards the mean becomes considerably greater. In consequence, in Figure 10 and in the interpretations below, the modes have been used for the age groups 16 and over. Viewed this way the adolescent spurt is clear, with a peak height velocity located at about 15.5 years, that is 1.5 years later than today. Due to the selection it is not really possible to say what the final height of the manual worker group was, or when it was reached, although the modes terminate at 168 cm (5 ft 6 inches).

The table also contains the heights of boys of the nonlaboring (i.e., nonmanual, white-collar) class, mostly collected by Francis Galton and his associates at almost exactly the same time (see below). These were boys at public (i.e., private middleclass) schools, naval officer colleges, and universities. Fairly substantial numbers were again involved, the figures in successive age groups from 9 to 24 years being: 62, 235, 430, 745, 969, 990, 819, 462, 313, 300, 344, 262, 471, 600, 660, and 689. This time no differences between mean and mode are seen in the older age groups, as no selection was operative. The means have been plotted in Figure 10. Peak height velocity is reached at about 15.0 years, 0.5 years earlier than in the manual class. Final height is about 173 cm (5 ft 8 inches) which is some 5 cm shorter than the average for a similar social class nowadays. It was reached by the average boy at about age 20. The difference between manual and nonmanual classes was about 6

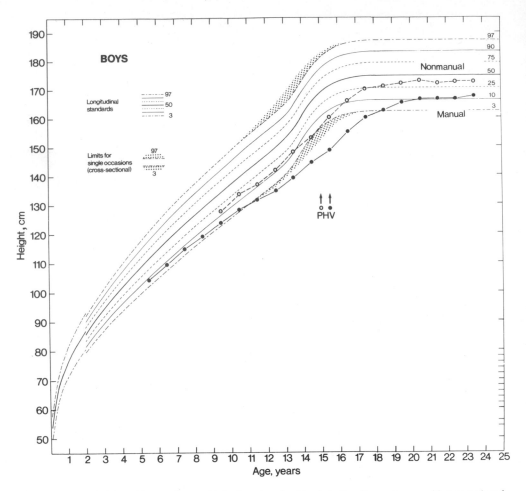

Fig. 10. Mean heights of boys of manual laboring class families (whether themselves working or not) and boys of families in the nonmanual classes, in England in 1872–1874. Data from Roberts (1874–1876; Table 1). Plotted on British standards from Tanner and Whitehouse (1976). Peak height velocities (PHV) 15.0 for nonmanual, 15.5 for manual.

cm at ages 9–11, 7 cm at 12, 9 cm at 13 and 14, and 11 cm at 15 and 16. It then gradually reduced to a final figure of between 5 and 6 cm.

Roberts noticed and, in contrast to Quetelet, accepted that "with the accession of puberty there is, according to my observations, an increased rate of growth" (p. 22). He saw this chiefly in the nonmanual class where it was more decided and called it "pre-pubertic growth."

Roberts, in contrast to Quetelet and indeed all his predecessors, laid great emphasis on *variation* around the mean. "A glance at the Table will show how impossible it is to study the progressive development of man, and the causes which check or promote it, by the use of mere averages. . . . The difference in height of the average public school boys of the ages of 13.5 and 14.5 is about 2 inches but the difference between the tallest and shortest boys of either of these ages is 20 inches . . . it is for this reason that I have given the whole range of heights which I have found to occur at each age" (p. 20). Roberts gave also tables of weight against heights, which, simultaneously with those of Bowditch, were the first bivariate tables of measurements ever to appear.

A few years later Roberts published *A Manual of Anthropometry* (1878). For the most part this merely repeats the material of his two big papers; but the tables he gives of growth of children of the manual and nonmanual classes have different numbers and somewhat different means. He prints also Bowditch's figures for Boston children, saying that "Professor Bowditch has, at my suggestion, tabulated his statistics in the manner I am now advocating, which I first adopted in my paper on Factory children. . . . I have already insisted on the absolute necessity of making statistical tables complete in every detail" (p. 87). Roberts maintained his interest in growth and later wrote a memorandum on physical education in schools for a Royal Commission on secondary education (1895) and a note commenting on sex differences in growth (1890). He retired at the turn of the century and died shortly after.

6.1. Galton and Family Likeness

Roberts' insistence on the importance of variation certainly echoed, if it did not indeed stimulate, one of Francis Galton's many preoccupations. Galton (1822–1911, see Forrest, 1974) had a hand in everything, and growth was no exception. In 1873 he put up a proposal to the recently formed Anthropological Institute asking them to sponsor a program of body measurements in schools. Since schoolmasters were "trustworthy and intelligent in no small degree," the statistics would be reliable, and the boys measured would grow up with favorable recollections of the procedure and hence would willingly submit to further measurement when grown up, and working in universities, factories, etc. (Galton clearly had in mind his family record study, to be initiated some ten years later.) There is an echo of Villermé about the opening words of the proposal: "We do not know whether the general physique of the nation remains year after year at the same level, or whether it is distinctly deteriorating or advancing in any respects. Still less are we able to ascertain how we stand in comparison with other nations, because the necessary statistical facts are, speaking generally, as deficient with them as with ourselves" (Galton, 1873–1874). Galton made clear that he wanted to recruit all classes of schools "from Public schools to schools for pauper children" and then to obtain an idea of the overall situation in the whole country by weighting the different social classes according to their numbers in the census. This must have been one of the first proposals for the use of stratified sampling.

In the event only a number of Public schools cooperated: they were Marlborough, Clifton, Haileybury, Wellington, and Eton College in the country; and City of London School, Christ's Hospital, King Edwards School, Birmingham, and Liverpool College in the big towns. Marlborough College, a school then some 30 years old, attracting primarily the sons of parsons, doctors, officers, and lawyers living in the southwest of England, was the first to get going, in 1874. This was because the headmaster, Dr. Farrer, was "honourably known to the scientific world by his early and outspoken advocacy of the introduction of science teaching into schools" (Galton, 1874). In consequence the college had already a full-time biology master, G. F. Rodwell, and in addition the first resident doctor to be appointed to an English school (in 1849), Dr. Walter Fergus (see Fergus, 1887). Fergus and Rodwell (1874) submitted their own report to the Institute, where it was read and commented on by Galton (1874). They measured height, weight, chest circumference, upper arm circumference with the arm flexed, calf circumference, and head circumference in 550 boys aged 10–19. They constructed a height measurer much superior to any used before; or perhaps Galton did, for whereas they say they used a "vertical board provided with a sliding square at right angles to it," Galton added blandly in a footnote, "A bracket sliding between vertical guides, and balanced by a counterweight acting over two pulleys, as in the Figure [reproduced as Figure 11 here] will be found easy, quick and sure in action. The vertical board and foot-piece may be dispensed with, if the guides can be nailed to the wall" (1874, p. 126). He thus anticipated R. H. Whitehouse's modern Harpenden stadiometer in everything except digital read-out. Height was taken without boots, and weight clothed and with boots. Frequency distributions were given for each year of age for height, weight, and head circumference, and Galton added the "probable error" values for each distribution, having remarked first that most of the distributions approximated to the law of error described in Quetelet's *Letters on the Theory of Probability*. This is one of the first occasions on which the probable error appeared; it was later (1888) renamed "probable deviation" and in 1894 replaced by Karl Pearson's "standard deviation." In the Marlborough figures there is a clear spurt in height, with a peak a little before 15.5, say 15.2 years. The head circumference figures show evidence for

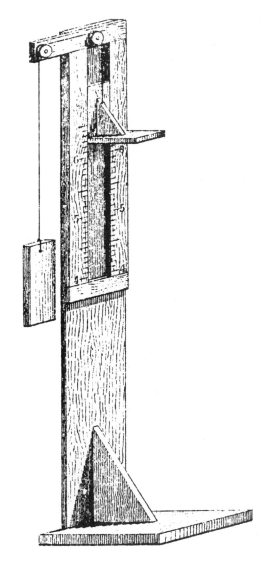

Fig. 11. Galton's stadiometer, 1874. Reproduced from Fergus and Rodwell (1874; Fig. 3).

a small spurt also, although Fergus and Rodwell do not remark on this, saying only that head circumference is not related in any obvious way to school ability.

Results from other schools came in more slowly and were never published in detail. However, Galton (1875–1876) did contrast the heights of 14-year-old boys in the country schools with those in the city schools in a paper chiefly remarkable for introducing the use of percentiles for the first time, a system invented by Galton himself in 1875. The country boys were some 3.5 cm taller; however, Galton does not mention that the country schools were also those with pupils from generally richer families. In none of these publications is Roberts mentioned, nor are the Commissioners or the Factory Acts, although Galton must have been generally aware of the earlier studies of the 1830s for at that time, as a child, he was a frequent visitor to Horner's house. It seems that Roberts sought Galton out when he heard what was going on (through Dr. Fergus, he implies) and got permission to use the entire statistical archive accumulated by the Anthropological Institute. The records of these schools appear as the "non-labouring" children in his 1874–1876 paper and again in his 1878 book.

Galton later turned his attentions to the totally neglected subject of the inheritance of height (Galton, 1885–1886, 1886). "I had to collect all my data myself," he says, "as nothing existed, so far as I know, that would satisfy even my primary requirement. This was to obtain records of at least two successive generations of some population of considerable size" (Galton, 1889). He collected 150 family records of height simply by offering publicly a prize to the best ones sent in. He obtained records of 205 couples, with 930 children; but the data were poor, the heights taken some with and some without shoes; some were only estimated. By personal correspondence, he obtained rather better data on 783 brothers from 295 families, and he was able also to make use of the measurements of nearly 10,000 unrelated persons measured in his Anthropometric Laboratory at the International Health Exhibition held in London in 1884. Percentiles for these latter were published in 1885. The 5th, 50th, and 95th for men's and women's heights (in inches) were 63.2, 67.9 (172 cm), 72.4; 58.8, 63.3 (161 cm), and 67.3; for sitting heights 33.6, 36.0, 38.2; 31.8, 33.9, and 36.0; and for arm span 65.0, 69.9, 74.8; 58.6, 63.0, and 68.0. In this paper (1884–1885) incidentally, Galton mentions, it seems for the first time, the difficulty occasioned by not knowing whether an observer has read the measuring instrument to the nearest unit or to the last completed unit (it making a difference of half a unit in the means).

It was in studying these family data that Galton introduced the regression coefficient and the coefficient of correlation (in 1885 and 1888, respectively, but the better known work is *Natural Inheritance,* 1889). Both with his emphasis on the inherited element in stature and his invention of the most important statistical tools since Quetelet's recognition of the general applicability of the law of Laplace and Gauss, Galton transformed the study of growth.

6.2. Bowditch

This new burst of activity in England was matched, if not yet exceeded, in the United States. Galton (1883) had written, "The records of growth of numerous persons from childhood to age are required before it can be possible to rightly appraise the effect of external conditions upon development, and records of this kind are at present non-existant" (p. 41). He was wrong; besides the records of Montbeillard's son and Quetelet's children, and, had anyone known it, of the

Carlschule, there were the slightly more numerous records kept by an extended New England family on the backs of its doors.

At a meeting of the Boston Society of Medical Science in 1872, Henry Pickering Bowditch (1840–1911), then lately returned from postgraduate training in Germany to take over the teaching of physiology at Harvard Medical School, presented a paper that was the beginning of all the North American work on human growth. The paper was never published except in summary. "Dr. Bowditch exhibited a diagram showing the rate of growth in height of the two sexes. . . . The curves represented the average measurements on 13 individuals of the female and 12 individuals of the male sex. The measurements were all taken annually during the last 25 years, and the individuals were all nearly related to each other" (Bowditch, 1872, 1877, p. 275). Sadly, we do not know the name of the Galtonian spirit in Boston who was responsible. It seems quite likely it was someone in Bowditch's own family, which was certainly sufficiently scientific and perhaps sufficiently large, for he himself was one of six siblings. Bowditch's paternal grandfather translated Laplace's *Mécanique Celeste* and wrote a classic seafarer's manual called the *Practical Navigator*. His father, a merchant, maintained a strong scientific interest and edited new editions of the *Practical Navigator* as they were needed. Perhaps this man and his brothers and sisters produced between them the 25 children measured.

The most interesting aspect of the inquiry, however, is what struck Bowditch about the curves. He noted that "The boys are taller than the girls from early childhood till 12, but then at about 12.5 years of age the girls begin to grow faster than boys and during the 14th year are about 1 inch taller than boys of the same age. At 14.5 boys are again taller, girls having at this period very nearly completed their growth while boys continue to grow rapidly till 19 years of age. The tables and curves of growth given by Quetelet show that in Belgium girls are at no period of their lives taller than boys of the same age" (1877, p. 275). Bowditch was intrigued, perhaps disquieted. He notes that Cowell's figures for working children in Manchester and Stockport also showed girls taller at 13 and 14 but ends "It would be interesting to determine in what races and under what climatic conditions the growth of girls, at about the period of puberty, is the most rapid. It is possible that in this way facts may be discovered bearing upon the alleged inferiority in physique of American Women." Quetelet's error had borne dialectical fruit.

Bowditch was a prominent Bostonian who went to school with Oliver Wendell Holmes, studied with Claude Bernard and Carl Ludwig and was the teacher of William James, Stanley Hall, Charles Minot, and Walter B. Cannon. He was the creator, practically speaking, of experimental medicine in North America, cofounder and second president (from 1888–1895) of the American Physiological Society and dean of Harvard Medical School from 1883 to 1893. He was the driving force behind the building of the new Harvard Medical School which was opening in 1906, but by this time he had fallen victim to parkinsonism; he retired after the dedication ceremony and died in 1911. He left a succession of devoted students. Prominent among these was his successor and memoirist W. B. Cannon (1924; see also the memoir of his youngest son, Bowditch, 1958).

The source of Bowditch's interest in growth is not clear. He was certainly greatly beholden to Charles Roberts, as he was careful to say, for introducing him to frequency distributions, and also to Francis Galton for demonstrating how to calculate percentiles. But his first work was done quite independently, and in the same year (1872) as the Bridger–Holmes Factory Survey. His later contact with

Charles Roberts was perhaps made through Galton or through Michael Foster, the founder (1887) of the *Journal of Physiology* of which Bowditch was American editor. Roberts and Bowditch exchanged unpublished data across the Atlantic and joined hands in rebutting Quetelet's ideas about the growth curve. Both were active in growth work during exactly the same span of dates, 1872–1891. But whereas Roberts had no successor, Bowditch was the instigator of a considerable New England school, which included W. T. Porter, and fractionally later, Franz Boas.

In 1875, Bowditch persuaded the Boston School Committee to do a growth survey of pupils in the public (i.e., state) schools and the survey was extended into some of the private schools as well. In all 24,500 children were measured, more than twice the number in the 1872–1873 factory survey. Heights were taken without shoes, weight in indoor clothes. The birthplaces of the parents were recorded and their occupations obtained from school records. Bowditch's training as a meticulous and quantitative physiologist appears at once. He was the first person to go on record as editing growth data. "In the progress of the work," he wrote (which was placed, incidentally "in the hands of professional accountants"), "many cases were met with of heights and weights differing so widely from the average measurement to which they belonged as to excite a suspicion of error in the observation" (p. 281). Forty errors were discovered and corrected by remeasuring the children concerned.

Bowditch's main contributions to growth were published in the 8th, 10th, 21st, and 22nd annual reports (1877, 1879, 1890, 1891) of the Board of Health of the State of Massachusetts (which he chaired in the late 1870s). All are of the highest quality, and that of the 8th Annual Report (1877) is a classic of the international growth literature.

The age range covered in the survey was 5–18 years, and although Bowditch illustrates only what we call nowadays distance curves, he gives tables of mean increments from year to year, discusses whether this is legitimate as an estimate of what should really be means of successive measurements on the same children, and points out that the center of the mean increment from age class 5 to age class 6 is at 6.0 years. His tables of increments show the adolescent growth spurts in both sexes very clearly. The children were divided into those of American-born parents and those of Irish-born parents. The boys of the American-born parents had a peak height velocity at age 14.0, the girl's, a peak at 12.0. These values are barely 6 months later than present-day North American ones. Bowditch spelled out the relation of the spurt and puberty more clearly than did Roberts (indeed the spurt was often referred to as "Bowditch's Law of Growth" by American writers in the period 1880–1910). "The age at which the rate of growth attains its maximum in the two sexes," Bowditch wrote, "suggests a connection with the period of puberty which presents a similar difference in the time of its occurrence." He quotes recall data from 575 American-born women attending the Boston City Hospital which put the average age of menarche at about 14.5 years; two years, as he says, after the maximum rate of growth (the modern figure is about 1.5 years).

By the time the survey data were accumulated, Bowditch was quite sure of his ground against Quetelet and leveled at him much the same criticisms as we have done above. He sums up the answer to his original query—as to the existence of the girls-taller phase—by saying that undoubtedly Quetelet would have obtained the same results in Belgium as Cowell and Roberts in England and he himself in Boston, if only he had used proper methods.

The parents of the majority of the Boston children were born in either America or Ireland. Bowditch analyzed the difference in growth between the two groups, showing that children with American parents tended to be larger. In his second report (1879) he analyzed this difference further and incorporated a test of whether it was but a reflection of the differences Roberts had demonstrated in England between children whose parents (rather than themselves) were manual or nonmanual workers. The boys of nonmanual American-born parents were indeed taller than those of the manual American group by about 1.5 cm at 6–11 years, 3 cm at 12, and 2 cm at 18. The girls showed a smaller difference, with the nonmanual only fractionally larger. The sons of Irish-born parents showed a similar occupational differential, which reached 2 cm at 12 and 3 cm at 13, above which age there were no data for nonmanual parents. The nonmanual Irish girls averaged between 1 and 2 cm taller, ending 2 cm taller at age 16 years.

When the American and Irish were pooled, the nonmanual boys were taller than the manual by about 1 cm at 8–10, 2 cm at 11–12, 3 cm at 13–14, and 1 cm at 18. The nonmanual girls were taller than the manual by about 1 cm at 13 but ended practically the same. These were smaller occupational class differences than Roberts had obtained; but of course Roberts was comparing not only occupation of parents, but groups of children themselves working or not working. Citing both Villermé and Boudin and siding with the former, Bowditch interprets the American–Irish difference as being largely due to the higher standard of living of the American group. He pointed out that the American group was about the same height as the English Public school group of Galton's survey, but that a similar American private school group abstracted from his material had heights averaging a little more. Both Roberts and Bowditch plotted bivariate distributions of weight on height, although neither could analyze them further until Galton invented regression. Both realized, however, that the manual groups had a higher weight for height, a finding consistently found in all industrialized countries since (see Eveleth and Tanner, 1976). "Deprivation of the comforts of life," wrote Bowditch, "has a greater tendency to diminish the stature than the weight of a growing child" (1877).

Bowditch summed up the sex difference very clearly: "Until the age of 11 or 12 years boys are both taller and heavier than girls of the same age. At this period of life girls begin to grow very rapidly and for the next two or three years surpass boys of the same age in both height and weight. Boys then acquire and retain a size superior to that of girls who have now nearly completed their full growth. This statement is based on observations on several different races in various conditions of life" (1877). What neither Roberts nor Bowditch realized was the part played by early or late maturing in creating differences between occupational or national groups. Roberts partly saw the importance of this in relation to individuals of the same sex, and Bowditch went so far as to query the wisdom of coeducation in schools in the light of the different timing of the acceleration of growth in boys and girls (1877). But the real recognition of differences in tempo of growth had to wait for Boas.

Bowditch (1891) did point out an interesting feature of his centile curves for height, which had an influence on Boas' first analyses of growth data. The curves were symmetrically distributed only up till 10 in girls and 12 in boys. Then the upper centiles moved further away, making the distribution asymmetric. At 12 in boys and 14 in girls exactly the reverse asymmetry occurred. Bowditch concluded that "The pre-pubertal period of growth acceleration . . . which is such a distinct phenomenon in the growth of children, occurs at an earlier age in large than in small children" (p. 502). In thus interpreting cross-sectional distributions he was confounding them with longitudinally derived data (which was Boas' point later). Bowditch did not realize that earlier maturers on average were naturally found at higher centiles in the early pubertal years and that it was these growth spurts which pulled out the centiles in a skew.

7. Educational Auxology: School Surveys and School Surveillance

Roberts and Bowditch were not alone in their investigations; soon school commissioners got to work in various parts of Europe, and by the end of the century growth curves, of better or worse provenance, were available for many countries. Before this, however, was the remarkable pioneer study of Pagliani (1875–1876; also 1879), an assistant in the Physiological Institute of Turin. In 1872 an agricultural boarding school for orphaned and abandoned boys called the Institute Bonafous was set up, and Pagliani followed for three years the height, weight, chest circumference, vital capacity, and muscular strength of all the boys admitted in the initial year. This, therefore, was the earliest short-term mixed longitudinal study, except for the unpublished one made in the Carlschule. The numbers were not large; 6 boys were followed from 10 to 13; 10 from 11 to 14; 8 from 12 to 15; 5 from 13 to 16, with more over shorter periods. In all a little over 200 boys, aged 10–19, were measured.

Pagliani does not merely give means cross-sectionally at each year of age and mean increments derived from them. He starts this way, but then he calculates the individual increments, takes their means for each year, and constructs a growth curve of height by adding these mean increments successively onto the cross-sectional mean value of the 10-year-olds. He was thus the first to use a method for getting rid of sampling bias introduced, or reintroduced, by Tanner (1965; Tanner and Gupta, 1968) 80 years later. Pagliani notes the very considerable difference this makes to the increment curves. His figures for height increments are reproduced in Table V. The left-hand column gives the differences between yearly means cross-sectionally calculated, and the right-hand column the mean increments calculated longitudinally. Peak height velocity is reached at about 16.0 years.

Pagliani's purpose was to show that an improvement in the conditions of life of the orphans caused improved growth, and indeed he interpreted his data as showing some degree of catch-up in the first year after admission to the school (although he

Table V. Pagliani's Demonstration That in a Mixed–Longitudinal Growth Study Means of Individual Yearly Increments Are More Precise Than Differences between Successive Yearly Means (Boys) (Pagliani, 1875)

Age (center)	Differences between successive means (cm)	Means of individual yearly increments (cm)
11.0	1.8	2.8
12.0	4.0	3.7
13.0	5.4	4.4
14.0	2.5	5.9
15.0	8.6	6.2
16.0	2.6	7.4
17.0	0.2	6.9
18.0	2.9	4.8

did not use that word or its Italian equivalent). The big school surveys were made with the same general objective: to see if the conditions of the schools were such as to encourage healthy growth.

At the center of the early school survey work was the journal *Zeitschrift für Schulgesundheitspflege*, begun in Hamburg in 1888. In the foreword to the first issue, the editor, L. Kotelmann, declares its purpose. To date, he says, there is no journal devoted to studies of bodily development or training (*körperliche Ausbildung*). The journal will answer this need. It will also publish studies concerning all matters pertaining to school buildings and classroom and furniture design; and concerning the "hygiene of teaching" (*Hygiene des Unterrichts*) under which heading come problems of time of first lessons in the day, intervals between lessons, age at starting homework, and so forth. Health and disease are also the journal's concern. "By thus uniting engineers, doctors and teachers in joint work *made from the standpoint of exact science and careful measurement* [his italics] we may hope to contribute to the attainment of the educational ideal . . . and by applying the principles not only of pedagogy but of physiology and hygiene to bring [the pupil] to a harmonious proportion [*Gestaltung*]." Such an attitude became widespread in the last years of the nineteenth century. Maria Montessori is remembered still in educational circles, but scarcely in medical ones. Yet she was a medical doctor, and her main work was called *Pedagogical Anthropology* (Montessori, 1913).

Kotelmann himself published, in 1879, a paper that was a pioneer work in the analysis of changes in body composition with age. He measured arm and calf circumferences, the thickness of a fold of subcutaneous fat pinched up over the biceps, this last with the regular curved caliper used by anthropologists to measure head breadth. His subjects were 515 Hamburg Gymnasium boys mostly aged 10–17.

In the first volume of the *Zeitschrift* was an account of the first large-scale study in continental Europe. In Denmark a school commission was set up in 1882 to investigate, throughout the whole country, the conditions of children in all types of schools, including elementary and village schools. The main questions derived directly from the earlier studies of working children. The prevalence of illnesses of all sorts was inquired into, and also the question of "overpressure," the school equivalent to overwork (Hertel, 1885). Some 17,600 boys and 11,600 girls were investigated, with results given by Hertel (1888), the community medicine doctor in Copenhagen. The prevalence of ill health was astonishingly high; 30% of boys and 40% of girls were noted as having some illness, often scrofula, anemia, or nervousness. A controversy arose as to the age at which the prevalence was greatest: Hertel and also Axel Key (1889, 1891; see below) thought the period of slow growth just prior to puberty was the worst for disease; Combe and some later authors thought the prevalence rose at puberty. Burk (1898) quotes Key (1889) as saying, "The curve of disease, in boys, reaches its first summit directly before, or more correctly at the beginning of, the pubertal development. But as soon as the development sets in forcibly, the curve sinks year by year, so long as that accelerated growth continues. . . . Directly at the conclusion of the pubertal period, when the yearly increases in height and weight hastily decrease, the curve of disease . . . jumps to its second summit, which it reaches in the 18th and 19th year." As to growth in the Danish children, the boys' peak velocity of height was at 15.0 years and the girls' just 2 years earlier. Boys aged 11–14 in gymnasia averaged about 3 cm taller than boys in the realschulen, and 5–6 cm taller than boys in the bürgerschulen and the volkschulen.

A similar Swedish school commission of 1883 which, however, only investigated pupils in secondary schools, was chaired by Axel Key, professor of pathological anatomy at the Karolinska Hospital and friend of the famous Retzius. Key's report (1885), now a rare work, is commented on in a modern paper by Ljung *et al.* (1974), who reproduce his mean values for height and weight. Many more boys (16,637) than girls (3165) were measured, and substantial numbers were present only from age 11 upwards. All Sweden is represented, but since only secondary schools were taken, the better-off class was overrepresented, as it tended to be in most early school statistics. Peak height velocity was reached in boys at 15.5 and in girls apparently as early as 12.5; but as peak weight velocity in girls was at 15.0, it seems that the earliness of height is probably due to chance or to poor measurements. Key's Swedish boys averaged a little less in height than those in Danish gymnasia and realschulen boys and were only about 1 cm taller than the general mean of all Danish schoolboys.

In Norway school measurements were made in 1891–1892, but in secondary schools only and chiefly in Oslo (Faye and Hald, 1896). The results are illustrated in Kiil (1939, p. 27); boys had their peak height velocity at 15.5; girls at 13.75. The schoolchild-measuring tradition was signally revived in Norway later, in particular by Schiötz, the chief medical officer for schools in Oslo in the 1920s. In 1888 a similar survey was carried out in the schools of the Moscow region by Erismann (1888), professor of hygiene at Moscow University and, a little later, Sack (1893) measured some 6600 Moscow boys in middle schools (gymnasia and realschulen).

In 1888 Geissler and Uhlitzsch (1888) reported a study of Freiberg children which was noteworthy in two respects. To begin with, it was probably the first study made specifically to enable proper-sized school desks to be designed; it is thus the source-paper of auxological ergonomics. Also it seems to have been the first paper to describe the distributions of heights and weights in terms of "probable deviations" since Galton's original introduction of the method in 1874 (see p. 556). During the next 20 years many similar school surveys were made, chiefly in Germany, but also in Warsaw and in Paris. Among them Schmidt-Monnard's (1901) paper is of special interest, for he alone covers the whole age range. Indeed he has more preschool children than children of school age. This, therefore, is a pioneer report; before this time a number of infants had been measured during the first year (Camerer, 1893, 1901) but practically no children aged 2–6, who are still the most difficult to study en masse. Schmidt-Monnard simply gives yearly increments, calculated from cross-sectional means, and displays them in a way still used by many modern authors. Vertical bars for each year of age are drawn to represent the increments, thus giving the appearance of a sort of histogram in time. His figure is reproduced as Figure 12.

In the United Kingdom work on growth slowed down after the early beginnings. Not until 1904 did the successor to Galton's and Robert's school studies appear; and it was a pretty meager affair. Thorne (1904) and Berry (1904) measured the heights and weights (both in boots) of boys and girls aged 11–15 admitted to the London County Council schools by competitive scholarships from poor homes. At the same time a survey was made of a number of schools in Glasgow (Kay, 1904–1905); this was chiefly remarkable as the forerunner of the first survey in the U.K. covering a whole school population, which was done in Glasgow in 1905–1906. The first analysis of the Glasgow survey contained errors of computation, corrected in the later report of Elderton (1914). In this survey there were some 66,000 children

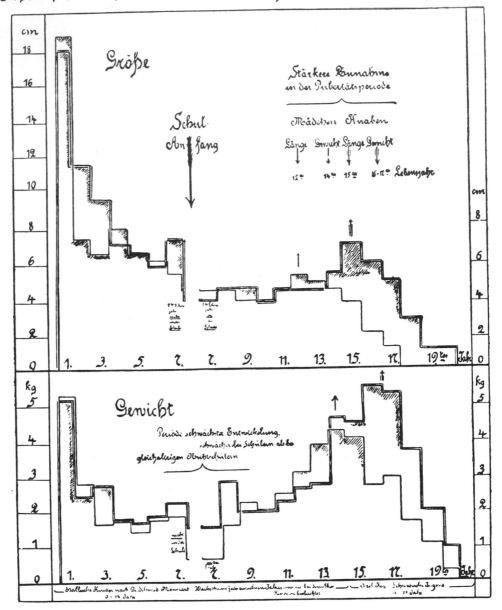

Fig. 12. Schmidt-Monnard's representation of mean yearly growth velocities by height of vertical bars. Reproduced from Schmidt-Monnard (1901, p. 54).

aged 6–14 divided by social class according to the area of the city in which the school was located. The boys from schools in good areas averaged about 5 cm taller than those from schools in bad areas; the girls' differences were only a trifle less.

School children in Edinburgh were surveyed in the following year (City of Edinburgh, 1906) and those in Birmingham in 1910 (Auden, 1910). Differences between children attending a school in a poor part of Edinburgh and those attending a school for relatively well-off nonmanual workers averaged about 8 cm at ages 10–14 (City of Edinburgh, 1906). In Birmingham children in the best-off of the five city wards, as judged by infant mortality, averaged 5 cm taller from age 5 to 8 than those from the worst-off ward. The mean heights of 7-year-olds in the best ward were the present-day 10th centile, those of the worst at the present-day 3rd. School statistics for 1909–1910 covering the whole of England were collected from school medical officers by Tuxford and Clegg (1911); mean heights are at about the modern 5th centile. The rural children were still taller than the urban children by about 1 cm, and children from the south were about 1 cm taller than those from the north.

A particularly interesting paper is that by McGregor (1908), since it deals largely with preschool children, always a difficult group to study. All children aged 1–10 admitted to a Glasgow fever hospital with scarlet fever, measles, and whooping cough in 1907–1908 were measured. A marked difference in height and weight was found between children whose parents had three rooms for their family to live in and those from families living in two rooms, or with a still larger difference, in one room. Thus differences in height of the order of 5 cm were already established by age 3 between children living in different conditions (all of them poor by comparison with modern housing).

Meanwhile the examinations of Boer War recruits, which revealed a disastrous level of disease and malfunction, were having a delayed effect; an interdepartmental committee on "physical deterioration" was set up and recommended regular and permanent anthropometric surveys of the population, especially of young people in schools and factories. Little was done, however, and in 1977, two wars and three quarters of a century later, a government committee, known appropriately as COMA (Committee on the Medical Applications of Food Policy) found itself recommending exactly the same thing. However, the London County Council did initiate a survey in 1904–1905 and measured 3500 schoolchildren. In 1906 this was extended to 18,000. The results were circulated in a report (London County Council, 1908, cited in Cameron, 1977; Weir, 1952). In 1907 an Education (Administrative Provisions) Act was passed which made the examination of elementary schoolchildren mandatory over the whole of England and Wales. The results of the examinations, when forwarded to the Chief Medical Officer of the Board of Education, caused him to say that "Of the six million children registered on the books of the public elementary schools of England about 10% suffer from a serious defect of vision, 3 to 5% from defective hearing, 1 to 3% have suppurating ears, 6 to 8% adenoids or enlarged tonsils of sufficient degree to obstruct the nose or throat and to require surgical treatment, about 40% suffer from extensive and injurious decay of the teeth, about 30 to 40% have unclean heads or bodies, about 1% suffer from ringworm, 1% from tuberculosis in readily recognizable form, from 1 to 2% are afflicted with heart disease and a considerable percentage of children are suffering from a greater or less degree of malnutrition." "Such a statement," says R. H. Tawney in his introduction to Arthur Greenwood's *Health and Physique of School Children* (1915), "is likely to be read in the future with the sensation aroused today by a study of the reports of the early Commissioners of Child Labour in Factories and Mines'' (p. xii). Greenwood, later a well-known Labour government minister, collected the country-wide statistics for 1908, 1909, and 1910. The total number of children exceeded 800,000. Although the curves of the means are not altogether easy to interpret, presumably because of the varying methods and care of the collectors, both boys' and girls' means are at the modern 3rd centile for ages 5–10, and drop to the first centile at ages 12–14. Rural children are again clearly taller and heavier than children living in large towns, by about 1.5–2 cm and about 1.0–1.5 kg.

Before turning to examine the situation in the United States, we may summarize at least one aspect of the results of the European studies of 1880–1910. The mean age at peak height velocity in Norway in 1795 was about 17.0 years, the latest recorded for Europe. At the Carlschule, at much the same time, age at peak was about 16.0 in the bourgeois. In Germany in the 1880s and 1890s the means of different series mostly varied between 15.0 and 15.5, with Russian textile workers and, oddly, Hamburg gymnasium students at 16.0. The means of peak height velocity for girls varied mostly between 13.0 and 14.0, with Russian textile workers at 15.0. Thus on the average children were experiencing their adolescent growth spurt about 1.0–1.5 years later than nowadays. This agrees well with the data on menarche (see Tanner, 1978). Average (prepubertal) heights were very low compared with nowadays; relatively well-off children were around the modern 10th centile, and poor children at or below the modern 3rd.

7.1. W. T. Porter and the Relation between Size and Ability

When Bowditch persuaded the Boston School Committee in 1875 to undertake a survey of the growth of Boston schoolchildren, he saw it as part of a larger undertaking spread over the whole of the United States. At the meeting of the Social Science Association in Detroit that year he read an account of his own plans and urged others to make similar studies. The only immediate response came from the town of Milwaukee, Wisconsin, in the person of G. W. Peckham (1882), the teacher of biology in the Milwaukee high school. In 1881 Peckham organized a survey of the heights, weights, and sitting heights of 10,000 children in the city, aged 4–18, copying Bowditch's protocol exactly. German immigrants, however, substituted for Irish ones in the comparison between childen born to native Americans and those born to immigrant parents. Peckham gives excellent graphs of the growth of Milwaukee children; girls are equal to boys in height at age 11 and greater at 12, 13, and 14; they reach peak height velocity at about 13.5 and have ended growth by 17, with a final height of 158 cm (present 25th centile). Boys reach peak height velocity at about 16.0. As in Boston, children of immigrants were a little smaller than children of native-born Americans.

The most important of the American studies of this time, however, was that made by W. T. Porter in the schools of St. Louis. [An excellent review of all American studies up to 1898 is given by Burk (1898), who worked in Stanley Hall's department (see below) at Clark University, and published in the journal that Stanley Hall edited.]

William Townsend Porter (1862–1949; see Carlson, 1949; Landis, 1949) became, like Bowditch, one of the founding fathers of American physiology. Born in Ohio, he graduated in 1885 from what was then the St. Louis Medical College, now Washington University, and was appointed professor of the new science, physiology, only three years later. He spent the year 1887 in Breslau, Kiel, and Berlin, where he evidently had contact with the physical anthropologists, to judge from his introduction of head measurements into his school survey and from his using an article by Stieda in the *Archiv für Anthropologie* of 1882 as his source of all statistical formulae. In Breslau he presumably became familiar with Carstädt's (1888) growth study in the high school, which was in progress while he was there. His appointment in St. Louis to the physiological department must have at once brought him into contact with Bowditch in Boston. He was friendly too, with Franz Boas at Clark University, for Boas supplied him with a sketch of a head with measurement landmarks, to put in one of his papers (Porter, 1894). In 1891 Porter sought permission to make a growth study in the public schools of St. Louis, and in the following year this was carried out. Height, weight, sitting height, span, head and face measurements, vital capacity, and grip strength were measured in 33,500 children aged 6–15. Porter had scarcely had time to complete this work when Bowditch called him to Harvard to be his assistant professor of physiology. He never returned to the study of growth; with Bowditch he founded the *American Journal of Physiology,* and he guided it as managing editor as well as financial guarantor from its beginning in 1898 until 1914. Like Bowditch, he was a master craftsman and loved to build the sort of equipment that permitted clear-cut experimentation. In 1901 he founded the Harvard Apparatus Company, an organization which supplied equipment for physiological teaching throughout the world and whose profits went to support research fellowships in physiology in American universities. Porter retired in 1928 and died in 1949. In his long obituaries, no one thought of mentioning his work on human growth.

What distinguishes Porter's survey from others is the relating of body size to apparent ability at school. In St. Louis the schools were organized in grades, and a pupil moved up a grade when he had successfully completed the work of the previous grade, irrespective of his age. Porter found that the pupils in the higher grades were heavier than the pupils of the same age in the lower grades. Figure 13 is redrawn from the first of Porter's papers (1893*a*, plate 2). It shows the median weights of boys and girls divided into those in above-average grades for their age and those in below-average grades. In Figure 13 the difference in weight between the two sets of pupils is clear to see and has been confirmed in hundreds of investigations since Porter's day (see Rietz, 1906; Tanner, 1966). For height Porter gives means (his "average") and modes (his "mean") at age 10 only in boys and 12 only in girls. But the same distinction holds; 10-year-olds in grade 1 averaged 5 cm shorter than those in grade 4; 12-year-olds in grade 2 averaged 5 cm shorter than those in grade 6.

Porter's findings raised incredulous but ill-documented opposition then, as sometimes now. Porter thought that physical strength, which he equated with height and weight, conditioned the amount of mental effort a child could make, and was much concerned, like all his contemporaries, with the question of school "overpressure." Hertel's *Overpressure in High Schools of Denmark* had been published in translation in America in 1885; and in 1887 in Breslau, Berlin, and Kiel, Porter must have felt the influence of the group connected with the nascent *Zeitschrift für Schulgesundheitspflege.* At the end of his paper, Porter wrote: "No child whose weight is below the average of its age should be permitted to enter a school grade beyond the average of its age, except after such a physical examination as shall make it probable that the child's strength shall be equal to the strain." This is pure Factory Commission concern, almost word for word.

Porter, however, went further. He pointed out that the growth curves of the two groups of children (he referred to one as "precocious," but meant "bright,"

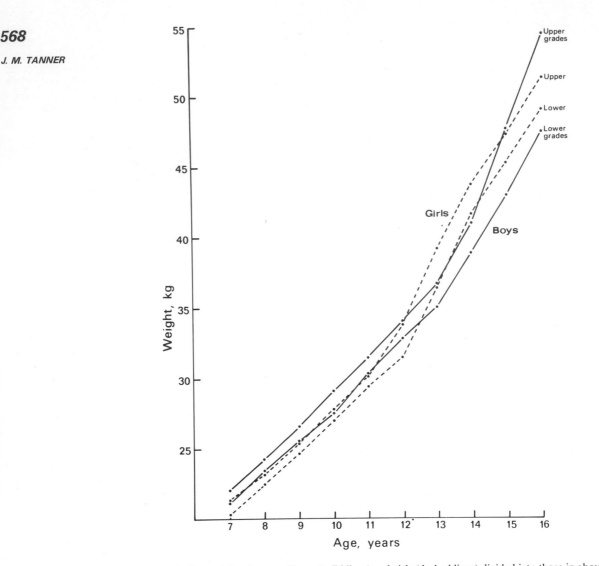

Fig. 13. Median weights for age of boys (solid lines) and girls (dashed lines) divided into those in above-average grades for their age and those in below-average grades in schools in St. Louis in 1892. Data from Porter (1893); figure reproduced from Tanner (1966).

and to the other as "dull") followed parallel courses, with the adolescent increases in weight occurring at about the same time in both. He thus indirectly inferred (although never unequivocally claimed) that these differences of size and mental ability would persist into adult life. To this conclusion Franz Boas (1895*a*) took strong exception. "I should prefer to call the less favourably developed grade of children *retarded* [his italics] not dull," he wrote, "and these terms are by no means equivalent, as a retarded child may develop and become quite bright . . . furthermore I do not believe that the facts found by Dr. Porter establish a basis for precocity and dullness, but only that precocious children are at the same time better developed physically. . . . Dr. Porter has shown that mental and physical growth are correlated, or depend upon common causes; not that mental development depends on physical growth." Boas, as we shall see, was the first man really to appreciate that children grow at very different rates; thus naturally he saw the relation between ability and size as probably caused by differences in rate of development, some children being advanced both physically and mentally, others delayed. If so, then by adulthood the delayed child would have caught up with the advanced, and there would no longer be any difference between them either in size or in ability. After nearly a century the controversy continues, and the data needed to resolve it are only beginning to be accumulated. Suffice it to say that in most countries, although perhaps not all, a very small positive relation between height and ability does indeed continue to exist in adulthood, probably generated by differential social mobility linked to size as well as intelligence, and also perhaps by early environmental influences.

Porter's next paper on his growth study (1893*b*) was only the third paper to appear in which the "probable deviations" of the distributions of heights and weights at each age were given. (The standard deviation did not appear till the following year.) Porter starts by invoking Quetelet, but whereas Quetelet was confused about the difference between error and distribution, Porter was perfectly clear and at once splits up the variance, saying, "The Probable Deviation contains the Error of Observation, as well as the Physiological Difference of the Individual from the Type." He then shows that the error of observation is relatively small; one boy aged 17 was measured 78 times (whether by one or more measurers is unclear) and the probable deviation of the 78 measurements was ± 0.24 cm (giving a SD of $0.24 \div 0.6745 = 0.35$ cm, a value in line with modern work, if we assume that several different measurers were involved). The probable deviation of 78 boys each measured once, however, was 5.15 cm. Porter then goes on to show, for the first time, that the probable deviation, when considered as an absolute value, increases with age. However, when considered as a percentage of the mean value at the same age (i.e., as the equivalent of the modern coefficient of variation), it changes little from 6 until puberty, when it rises; at maturity it falls to a lower value than during growth. This coincides with the increases of height velocities, the ages of peak being 13.0 in girls and 15.0 in boys. (Porter, like the early Glasgow survey of 1905, uses age at nearest birthday not last birthday, a potential source of confusion.) Porter concludes: "The Physiological Difference between the individual children in an anthropometric series and the Physical Type of the series is directly related to the Quickness of Growth" (p. 247). In this generalization he missed the point, for it seems he was really thinking about *individual's* rates of growth; he failed to realize that the probable deviation increase was more due to differences in tempo *between* individuals. This is where we must proceed to the work of Boas, for this is one of the two points from which Boas started. Before considering the work of Boas, however, we will pause for a short intermezzo.

Porter took the whole question of school health very seriously and the third of his 1893 papers (1893*c*) is an interesting early attempt at a theory of growth standards. If we have sufficient data obtained by the "individualizing method" (i.e., longitudinal studies), Porter says, "the deviation of children from the laws of normal growth could be quickly recognised and by timely treatment largely overcome; the evil effects of over-study could be watched and intelligently combatted and systems of education, no longer exacting from all that which should only be exacted from the mean, could be rationally adapted to the special needs of the exceptionally weak and the exceptionally strong" (p. 578). Data (or standards) obtained by the generalizing method (i.e., cross-sectional studies) could not do this because (to translate Porter into modern language) they could not give standards for velocity of height, and that was what was needed. However, it would take a generation to produce such standards, and in their absence Porter suggested using standards of weight for height (irrespective of age) as the best measure of health. "Overwork," he writes, "may cause a temporary or a permanent deviation in these (growth) curves . . . a prolonged strain in a growing child harms for life and leaves a mark which can never be effaced. The danger is greatest in the periods of quickest development, and particularly great in the prepubertal period. The child should be guarded against the possibility of harm. The anthropometrical system proposed in this article offers a means of doing so. It infallibly discovers those whose physical development is below the standard for their age" (p. 158). Such enthusiasm, scarcely shared by modern and more cautious standards-constructors, is worthy of Godin, whose work we now briefly consider.

7.2. Paul Godin and Growth Surveillance

The large-scale surveys of heights and weights in schools were usually made by public health doctors, remote from the pupils themselves, in order to establish large-scale general facts, for example, that children from overcrowded homes grew less well as a class than children from better homes. The action that followed, if the persons concerned were fortunate, was action directed to ameliorate the environment, to improve the homes, the conditions of work, or the low incomes. Surveys of this sort continue to be made and continue, alas, to be useful.

But growth measurements can also be used to follow the health of individual children, for, as Porter saw, one of the surest signs of ill health is a gradual falling off in growth rate. This use of measurements for individual surveillance is the province of general practitioners, school doctors, or the teachers themselves. It was adopted with enthusiasm at the beginning of the century and then fell into almost total disuse, largely because later generations of school and family doctors had little training in how to interpret the measurements, failed to plot them or to calculate the essential increments, and usually thought their stethoscopes more practical. Recently, with the emphasis turning to prevention and family medicine and developmental assessment, growth surveillance is coming back, and child clinics, general practitioners, and schools are beginning again to keep serial records of individual children.

One of the pioneers of such surveillance was a French physician called Paul Godin (ca. 1860–1935). Godin was an army doctor and his publications include works such as *l'Earth système* (1901), which deals, after the manner of army doctors, with experimental researches on the earth latrine. But Godin was an educationist from the start, and the thesis for his doctorate was entitled "Physical Education in the Family and the School." Like Maria Montessori, he was also greatly interested in physical anthropology. From 1891 to 1900 he worked in schools for children of army personnel, and in 1893 he studied physical anthropology in Paris, with the celebrated Professor Manouvrier (Godin, 1914). In 1895 he was appointed doctor to the military preparatory school for sons of noncommissioned officers at St. Hippolyte-du-Fort on the Mediterranean coast, where he had 400 boys aged 13 upwards in his care. He was already a convinced longitudinalist, declaring that only a study of "la marche de la croissance individuelle" would reveal the laws of growth, and deploring the fact that, Montbeillard apart, nobody had yet concerned themselves with this.

Accordingly, Godin took two consecutive entry classes, each comprised of 115 boys aged 13 or 14, and followed the boys until they left the school, most of them after 4 years, some, regrettably, sooner. Each boy was measured every 6 months. Godin was only able to measure 3 boys each day, because, trained by the Paris school of physical anthropology of the 1890s, he took 129 measurements (32 heights from the ground, 9 transverse diameters, 27 circumferences, 23 head and face measurements, and 32 measurements on tracings of the hands and feet). In addition he made 46 ratings, for example of pubic and facial hair, skin color, eye form, etc. He did this for 5 successive years, something which seems to have astonished even him slightly. Godin is a man after the heart of anyone who has personally made longitudinal studies on children. In the book which reports the study (Godin, 1903; reprinted with addition, 1935), he writes that although such activity demands a certain effort, "I never yielded to physical lassitude, never allowed anyone to substitute for me; there is not a rating that I have not personally dictated to my corporal assistant; not a height above the ground, a diameter or a circumference which has been measured by anyone but myself" (p. 6). He is a real operator too; he understands exactly. Arm span is always the worst measurement, but in the tyro's hands the circumferences are the next worst, despite their apparent simplicity. Godin knows why: "The uninitiated, even if a doctor, always has a feeling of respect, whether he knows it or not, for the circular and sliding compasses. Already the stadiometer (toise) seems less imposing; despite its special shape and the disposition of its scale it is an instrument he has seen before . . . but the uninitiated does not even pause to glance at the tape, the familiar object which is in everybody's hand and is used by all." As to his devotion to the measurements, it has only been equaled by that of R. H. Whitehouse, also, and perhaps significantly, a man with a military training as background. Godin took 129 measurements on what must have been about 1000 boy-occasions, a total of about 1.3×10^5 measurements. At the Harpenden Growth Study over the years 1949–1970, Whitehouse took 15 measurements on approximately 9000 child-occasions. The totals are about the same, but at least Whitehouse had time to put down his instruments.

Sadly, when it came to working up his data, Godin was less inspired. Despite his absolutely correct insistence on obtaining longitudinal data, his text, and above all his results, make it fairly clear that he worked everything out cross-sectionally. Instead of taking increments for individuals, he took 100 subjects at each 6 months of age from 13.5 to 17.5 years, calculated the means at each age and subtracted one from the next to get the mean increments. Had the series been pure longitudinal, with the same 100 boys at every age, this would not have mattered, but Godin says many boys left the school in the middle of the study; furthermore the entry classes were not comprised of boys all of the same age. Thus his increment curves have irregularities, which, it seems, must be due to sampling bias.

Godin produced the following values for 6-monthly increments of thigh and calf length: thigh 9, 4, 8, 12, 2, 3, 3, 0; calf 2, 9, 0, 6, 4, 10, 3, 1. Inspection of the values seems to reveal an inverse relation; 9 with 2, followed by 4 with 9, 8 with 0, 12 with 6, etc. His figures on upper arm and forearm lengths, calf length, and calf circumferences appeared to alternate also. He therefore enunciated the *Loi des Alternances*: "The growth of the long bones of the limbs proceeds by alternating periods of activity and rest, which follow each other regularly. These periods are in opposition in the two body segments of the same limb" (p. 98; and see Godin, 1914). There was also an opposition between length and breadth growth in the same limb segment. Clearly such a generalization goes far beyond the evidence. Like most other doctors of his time, Godin was evidently quite ignorant of statistics. His alternations between limb segments seem exceedingly unlikely to be real; but it has to be said that no modern test of his idea has ever been made; judgment should yet be suspended.

It was Godin who introduced the terms "auxology" and "auxological" into human growth. They were not used in the 1903 edition of his *Recherches* and seem to appear for the first time in 1919 in an article entitled *La Méthode Auxologique*. He seems to have taken it over from the plant physiologist G. Bonnier, who wrote of *"Croissance (auxologie) des végétaux"* in Richet's *Dictionaire de Physiologie* (1895). Plant growth hormones were already called *auxins* at that time so the Greek root *auxein,* to increase, had been already adopted. The word is convenient because it yields an elegant adjective and adverb, which the word growth cannot do.

Godin gradually elaborated a whole system for following the growth of individual children. In 1914 he retired from the French army, and although he continued to live at Saint-Raphael and then Nice, he took up an appointment in Geneva at the already famous Institut Jean-Jacques Rousseau des Sciences de l'Education, founded by Claparède and Bovet and later to be graced by the lifetime's work of Piaget (see Godin, 1921). In 1913 he published a book on the educational applications of the study of growth and in 1919 a *Manual of Pedagogical Anthropology* (1919*a*) in which his full system is described. He begins the manual by saying that he has evolved "The individual formula for the child . . . something wished for by all the great educators from Montaigne to Montessori and Baden-Powell." His system consists essentially in obtaining a physiological age (something already popularized by Boas and others) by relating one body measurement or block of body measurements to others. The system owes a great deal to the mental-age concept of his countryman Binet, and a good deal to the once-famous but now almost entirely forgotten Italian school of constitutional medicine, represented by Di Giovanni (1904–1909) and Viola (1932). Indices of trunk volume, limb lengths, and head size are obtained, and the ratios of each to the other compared with tables for age. Godin stresses that it is changes in the ratios which matter more than their absolute values, but his use of this imperfect instrument is somewhat hair-raising. He gives several examples of his advice to parents; thus when a child was below par in the motor development ratio, the parents were advised to take him out of boarding school and send him back to the country air.

However this may be, Godin envisaged a whole new and useful area of preventive pedagogical medicine. In discussing a boy with retardation of his indices, he writes "The child is not sick, and that removes the usefulness of a clinical doctor; but it does not remove the usefulness of an educational doctor (*médicin éducateur*)." Godin's work and even his name are quite forgotten nowadays. He was an enthusiast, perhaps an overenthusiast, some might say an uncritical quack. But he saw further than most into the future of medicine in childhood, and although his methods are long since dead, his approach is very much alive. He deserves a more enduring memory.

7.3. Springing Up and Filling Out

Godin was not the only doctor to describe supposed alternations in growth between different parts of the body. This particular hare was started, about the time Godin was making his myriad measurements, by Winfield Hall, a doctor who had worked in Sargent's laboratory at Harvard and who, from 1889 to 1893, was medical examiner at Haverford College and in three Friends' schools in the Philadelphia region. Hall took 25 anthropometric measurements and measured vital capacity and the strength of the back, legs, and arms. The age range covered was 9–23 years and some 2400 boys were measured, once each. When Hall regarded the tables of medians at each year of age, he was struck by the fact that the greatest increment of height was from 12 to 13. The increments for 11–12 and 13–14 were less, but that for 14–15 was again large. Body circumferences, however, showed no such thing and indeed had rather larger increases in the 13–14 interval. The actual yearly increments, from 10–11 on were: for height 3.0, 3.9, 10.6, 3.8, 7.5, 3.0, 1.7, and 0.5 cm, and for the sum of knee, ankle, elbow, and wrist circumferences 2.6, 3.0, 3.5, 4.0, 2.7, 1.9, 0.9, and 2.1 cm. Thus the girths show a perfectly ordinary curve with peak velocity at 14.0 as expected, but the heights are quite peculiar, the third (12–13) increment being enormously high and the subsequent increment conspicuously low. The fifth value, 7.5 cm/year, may represent the true peak height velocity, centered at 15.0. It seems likely that the 13-year-old median height has fallen victim to arithmetic error; but if so, so have those of knee height and pubic bone height, which show the same pattern. Perhaps some maverick or misread value consistent over all three measurements got in at 13. However this may be, to the modern eye the supposed alternation of "springing-up" and "filling-out" periods is most easily explained as artifact, statistical or otherwise. But Dr. Hall, like Godin, was an enthusiast. He enunciated at once and in italics a *Law of Growth* (thus beating Godin to it). "When the vertical dimension of the human body is undergoing an acceleration of its rate of growth, the horizontal dimensions undergo a retardation; and conversely" (Hall, 1896).

Hall was probably influenced in this by the work of Malling-Hansen (1883) in the school for deaf and dumb children in Copenhagen. Malling-Hansen measured the heights and weights of 70 boys each month for two years and amply confirmed the findings of Montbeillard that height velocity is greater in the spring (maximum: March to August) than in the autumn (minimum: August to November). Weight, on the contrary, increased maximally in autumn (August to December) and minimally in late spring (April to July). These observations have been frequently confirmed since; most recently, and in a detail which bears out Malling-Hansen's quantitative as well as qualitative findings, by W. A. Marshall (1971). Malling-Hansen wrote that "In the maximal period of lengthening, the thickening of the body is at its minimum, and vice versa, the thickening has its maximum in the time of minimal lengthening" (cited in Burk, 1898). However, the seasonal periodicities to which he referred were, of course, quite different from the supposed periodicities, extending over a whole year or more, postulated by Hall.

The most popular and persistent version of the filling-out and springing-up description, however, was that promulgated by C. H. Stratz, professor at the Hague and prolific writer of textbooks and reviews. In 1909 he proposed the scheme of growth shown in Figure 14. In Stratz' hands the periods depended on the relative increments of weight and height. In the light of modern data it is clear that his first *fülle* reflects the increase of fat in the first year and his first *streckung* its loss in the period 5–8. The second *fülle* reflects the gradual increase of fat before puberty and the second *streckung* the loss of fat and increase in height at puberty. Stratz set great store by his scheme and reacted sharply (1915) to the very lukewarm reception accorded it in the first more or less modern textbook of growth, by Weissenberg (1911). However, textbooks continued for many years to talk of the alternating springing-up and filling-out periods without further explanation (or in some cases

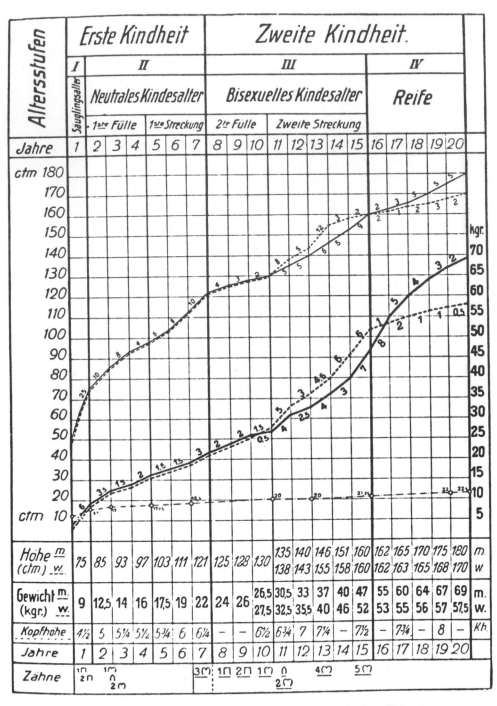

Größen- und Gewichtszunahme nach der Geburt.

Fig. 14. Stratz' scheme of human growth. Reproduced from Stratz (1909).

with explanations relevant to Hall's periods). Dr. Howard Meredith, the great anthropometrist from Iowa, cited Hall's work and its subsequent history as a classical example of how a spurious generalization gets into the textbooks and takes years of effort and numerous editions to be got out (Meredith, 1950).

Stratz' 1909 paper also contained a review of fetal growth which was authoritative and apposite. His two diagrams (Figure 15) showing changes of proportion have been reproduced in various versions innumerable times.

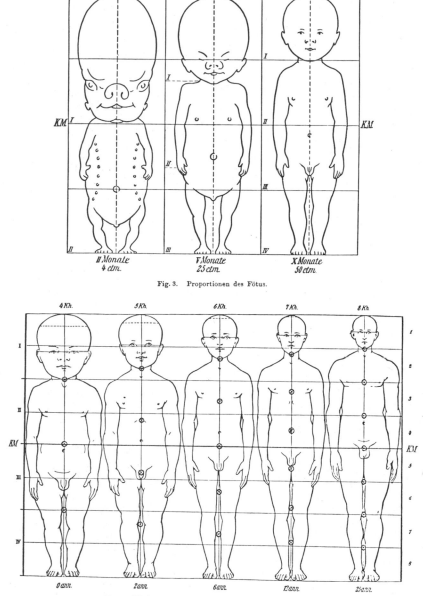

Fig. 15. Stratz' diagrams of changes of body proportions in the fetal and postnatal periods. Reproduced from Stratz (1909).

During the late nineteenth century, several physicians took the trouble, like Montbeillard and Quetelet, actually to measure their own children throughout their growth. Professor C. Wiener, of Karlsruhe, had four sons, the first three by his first wife, the fourth by his second. Each boy was measured on or near his birthday. Wiener published the results in 1890 in a small volume now very hard to find, but the results are readily available, as tables of raw measurements and as graphs, in Burk (1898). All four boys had entirely typical growth curves; the peak height velocities of the first two were during the year 13.0–14.0 with whole-year values of 9.5 and 8.2 cm/year; the third boy matured a year later, with a whole-year peak of 9.4 cm/year; the last son was a strikingly early maturer, having a whole-year peak velocity of 9.9 cm/year at age 12.0–13.0. Wiener can scarcely have started his observations later than 1860, which was ten years before Quetelet published the results of his own family. Perhaps there was a connection between Wiener and the Boston family measurers whose results were already available to Bowditch in 1872. Or, more likely, we are seeing only the survivors (Ford, 1958, reports one) of an army of Victorian child-measurers, whose records, like those of the eighteenth-century military recruits, were pulped into Sunday newspapers or lie moldering in the attics of half-demolished houses.

8. Franz Boas and D'Arcy Thompson

Tradition has it that a railway train bore Franz Boas (1858–1942) into the field of human growth. True or apocryphal, the story is that Boas, en route to the Cleveland meeting of the American Association for the Advancement of Science in 1888, became engaged in conversation with the man sitting next to him. As the train drew into Cleveland the man was revealed as G. Stanley Hall, a redoubtable figure in contemporary American academic life, who had just become first president, and professor of psychology and pedagogy, at a brand-new university named Clark at Worcester, Massachusetts. Stanley Hall, "the father of child study in America" (Grinder, 1969; and see Wallin, 1968) and the founder of the American Psychological Association, is now chiefly remembered for his remarkably polysyllabic and mid-Victorian book on *Adolescence,* which, although published in 1904, is full of phrases like "the secret vice" and "enfeebled heredity" and "the evolution of the soul." But he was a brilliant teacher and a far-sighted man, and he ended that railway journey by asking Boas to join him as head of a division of anthropology in his own department of psychology.

This was in 1888 and Boas was 30. He was already becoming known as an anthropologist interested especially in language and folklore, and formidably armed with a knowledge of scientific method. He was born in Minden, Germany, and studied at Heidelberg, Bonn, and Kiel (Herskovits, 1953). Although destined originally for medicine, he had always been fascinated by strange lands and places, and he finally took his doctorate in geography. In 1883–1884 he went on his first field expedition, to study the Baffinland Eskimo, and then settled in the United States. Despite his already wide interests and growing reputation in quite other branches of anthropology, Boas accepted Hall's offer and embarked on the study of human growth. Boas' first paper on growth appeared in 1892, his last in 1941. He was largely responsible for the discovery that some individuals are, throughout their childhood, more advanced on the road to maturity than others, and for the introduction of the concept of physiological or developmental age. It was his studies which established growth and development securely as an item in the practice and teaching of physical anthropology in North America. Indeed, the first PhD ever given in anthropology by a North American university was to his student and successor A. F. Chamberlain in 1892 for a study on the heights and weights of Worcester children.

Boas approached the problem of *tempo of growth*—his own exact and expressive phrase—in a very characterisitc way. He was a contemporary of Karl Pearson and for many years in the forefront of those who were developing the new techniques of biometry. Bowditch (1891; sec p. 558) found that the distribution of heights became skewed as the pubertal acceleration occurred at an earlier age in large than in small children. Boas (1892a) at once saw another and more likely explanation. Suppose there is an underlying variable of physiological status which is Normally distributed around an average value at any age. Then when the average rate of growth is not changing, the individuals who are advanced in physiological status and occupy on average the upper part of the height distribution gain the same amount as those who are retarded and occupy on average the lower part of the height distribution. Thus the distribution curve of height remains symmetrical, like the distribution curve of physiological status itself. But when the pubertal acceleration occurs, it happens first to those who are physiologically advanced. They therefore gain more than those who are physiologically retarded and thus pull out the height distribution into a positively skewed shape. Later they grow less than the retarded, so the skew reverses. Thus (although Boas in 1892 was in no position to put it this way) in early puberty there is a positive correlation between height and height gain and in late puberty a corresponding negative correlation. By 1906 he was able to show that such changes in correlation did actually occur (see below).

Boas realized—and he was the first to do so—that variation between individuals in tempo of growth, combined with the fact of an adolescent growth acceleration, must mean that most children did not stay in their percentile channels during this phase of growth. This was contrary to what Bowditch had supposed. "Consider for a moment," says Boas (1892b), "all those children separately who will, as adults, have a certain percentile rank, and investigate their position during the period of rapidly decreasing growth, during adolescence. It seems reasonable to assume that the average individual (not the average of all individuals) will retain its percentile grade throughout life." (The content of the parentheses is Boas'.) "At seventeen, say, some of these individuals (let us say of the mature 80th centile) are advanced, others retarded. As the amount of growth is decreasing rapidly at this period, the number of retarded individuals will have a greater effect on the average than individuals of accelerated growth . . . thus the average of all observed values will be lower than the value belonging to the average boy of seventeen years of age." This sounds a bit obscure, even to the professional auxologist, because in 1892 Boas had not yet fully realized the very large implications of what he was struggling to say. He was not yet thinking in terms of velocity of growth. Thus the explanation that follows carries us beyond his position in 1892 to the threshold of his second period of papers upon growth. Boas gave a somewhat clearer explanation (1895a), when he commented on Porter's (1893a) identical results on the skew, and in explaining his own results (Boas, 1897) on Worcester children. Not till 1930, however, had he fully absorbed the implications of his discovery, and it was he who then initiated the use of the "maximum growth age" (Boas, 1930), his second great contribution to the methodology and theory of human growth.

Figure 16A explains the effect of differences in tempo of growth in mass

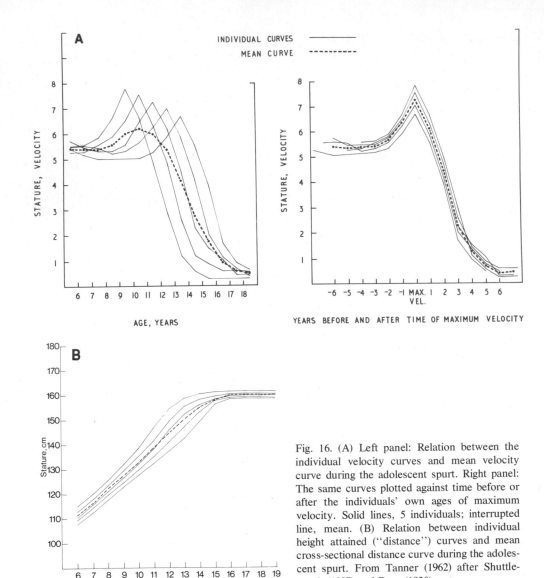

Fig. 16. (A) Left panel: Relation between the individual velocity curves and mean velocity curve during the adolescent spurt. Right panel: The same curves plotted against time before or after the individuals' own ages of maximum velocity. Solid lines, 5 individuals; interrupted line, mean. (B) Relation between individual height attained ("distance") curves and mean cross-sectional distance curve during the adolescent spurt. From Tanner (1962) after Shuttleworth (1937) and Boas (1930).

"velocity" curves. It shows the velocity curves from 5 to 18 years of 5 individuals, each one of whom starts his spurt at a different time. The mean of these curves, taking the average value at each age, is shown by the heavy interrupted line. It is obvious that this line characterizes the average velocity curve very poorly; in fact, it makes a travesty of it, smoothing out the adolescent spurt and spreading it along the time axis ("the average of all observed values will be lower than the value belonging to the average boy"). Figure 16B shows the same thing in terms of height for age. Two things are noticeable. Not only does the early developer rise through the centiles at first and drop back later (while the late developer does the reverse); even the average individual (shown by the heavy interrupted line) departs from the average centile if the centiles are calculated cross-sectionally, from mass data, for the reasons made clear in Figure 16A.

It was the realization of this effect which prompted Boas in 1892 to insist on the importance of following individuals. Nevertheless, his advice was ignored by nearly all students of growth until revived and clarified by Boas himself in 1930, and restated by Davenport in 1931 and Shuttleworth in 1937. Even then standards of growth for individuals continued to be based on cross-sectional data, despite efforts by Bayley (1956) and Bayer and Bayley (1959) to provide separate curves for early and late maturers. It was not till 1966 that Tanner, Whitehouse, and Takaishi introduced longitudinally based standards in which the boy of average height and average tempo actually followed the average curve (see also Tanner and White-house, 1976).

8.1. The First American Longitudinal Growth Study

In May 1891 Boas began a long-term longitudinal study of Worcester school-children. In addition to height and weight, sitting height, forearm length, hand breadth, and head length and breadth were taken, and later reported by his assistant F. M. West (1893, 1894) and by Boas himself (Boas and Wissler, 1906). The study was ill-fated, however. Just over a year after it started, Boas resigned from Clark University to go to the Chicago Columbian Exposition. When he wrote up his results in 1897, Boas used his scanty material to excellent effect. The experience of actually handling his own longitudinal data seems to have clarified considerably his concepts, and in this paper one feels he has really got a grip on the acceleration/retardation problem. He gives means and standard deviations of the yearly increments, and comments that "young children grow more uniformly than older children. The increase in variability (of growth rate) is very great during the years of adolescence . . . this increase must be considered due to the effects of retardation and acceleration." He then divides his data into halves, one containing all the children who are shorter than average for their age and the other all the children who are taller. He found the shorter children grew less than the taller in the years before adolescence. But during adolescence the shorter grew more; that is, they continued to grow at a time when the taller were stopping. He thus demonstrated for the first time the relative independence of final adult height and the speed with which it is reached. He overplayed his hand, perhaps, in concluding that "small children are throughout their period of growth retarded in development, and smallness at any given period as compared to the average must in most cases be interpreted as due to slowness of development." He had a special reason for doing this, in the shape of his controversy with Porter (see p. 568). In line with his position in this controversy he added also: "The differences in development between social classes are to a great extent the results of acceleration and retardation of growth" (1897). Boas did not challenge Porter's figures; indeed he himself got similar results in his Worcester children (Boas and Wissler, 1906). But he felt himself to be championing the cause of the late developer, and he never lost interest in the subject, returning to it in his very last paper on growth (Boas, 1941).

In the 1897 paper he also extended the application of tempo differences to explain differences in height between different social and ethnic groups. Returning to Bowditch's Boston data, he took the 13 different classes constituted by five different nationalities of parents and four different occupational classes in Americans and Irish and showed that differences between them reached their maximum at the fourteenth year in boys and the twelfth year in girls. "The figures prove," he wrote, "that the differences in development between various social classes are, to a great extent, results of acceleration and retardation of growth, which act in such a way that the social groups which show higher values of measurements do so on account of accelerated growth and that they cease to grow earlier than those whose growth is in the beginning less rapid, so that there is a tendency to decreasing differences between these groups during the last years of growth."

8.2. Standards of Height and Weight for American Children

Besides starting what was to have been the first substantial longitudinal growth study, Boas produced also the first national standards for height and weight of North American children. In 1891 various school authorities were approached and those in Oakland (California), and in Toronto agreed to contribute figures to add to those already collected by Bowditch in Boston, Peckham in Milwaukee, Boas and West in Worcester, and Porter in St. Louis. A total of nearly 90,000 children between 5 and 18 years old were involved. Boas pooled the results from these six cities to produce the composite North American standards, remarking that, although he would have liked to weight each series in accordance with the percentage of the total North American population it sampled, the lack of accurate census data prevented his doing so.

Although larger than most contemporary European children, these subjects were still small by modern standards. Both boys' and girls' average heights correspond to the modern British 20th centile at prepubertal ages, drop to the 5th centile at midpuberty, and end at the British 30th with growth complete or nearly so. In relation to modern North American standards they would be about 5 centile points lower. Peak height velocity is at 15.0 in boys and 13.0 in girls, around 15 months later than in the United States nowadays. Oakland children were taller and heavier than children in any other of the towns (Barnes, 1892–1893). Boas (1895b) used the Oakland and Toronto data to show that there was a relation between birth rank and height and weight at all ages in boys from 6 to 16 (where the data terminated) and in girls from 6 to 18; the first born was largest, and others successively smaller. This seems to have been the first demonstration of this sibling number and birth rank effect: Boas treated it only as a difference between firstborn and all later-born, but his data show a graded regression.

Boas left the Chicago Exposition in 1894 and in 1896 settled in New York, at the American Museum and Columbia University, where he was to spend the rest of his life. We may regard this move as bringing to a close his first period of the study of growth. His interest had been increasingly engaged by the problems of the effect of heredity and environment on body form, and in 1908 he began a famous series of studies on immigrants. Until 1930, when he embarked on a second period of the study of human growth curves, his interest in children's measurements relates rather to their family relationships than to the process of growth itself. During the first years of the century, however, a further report of the commissioner of education was issued (Boas and Wissler, 1906) as well as a paper (Boas, 1912a) assimilating his thoughts on the new methods for measuring developmental age then being developed.

The 1906 paper, with Wissler, is chiefly remarkable for presenting what seem to be the first correlation coefficients calculated upon human growth data. Boas, as we have seen already, was in the forefront of the biometrical advance; all his life he taught and encouraged anthropologists to use statistical methods. As early as 1894 he wrote a paper (Boas, 1894), expository in character and placed in the *American Anthropologist,* describing the correlations between two anthropometric measurements. Galton (whom Boas does not mention in this article, but whose work must certainly have inspired it) had only introduced the correlation coefficient in December 1888, and the book that brought Karl Pearson to him, *Natural Inheritance,* appeared in 1889. Pearson's method for calculating the correlation coefficient, which replaced Galton's, was published in 1896. So Boas came to the biometric fold as early as the great K. P. himself. In this 1894 paper Boas presents no coefficients of correlation, but gives diagrams of the regression of two anthropometric measurements, each on the other, in Sioux and Crow Indians. He remarks that the greater closeness of the two lines for stature and span than those for head length and head breadth indicates that the former pair of variables is the more closely associated. He also understood clearly the reduction in variance of one measurement when standardized for a fixed value of a correlated measurement. He remarked that the discovery of ultimate causes of correlations was a futile endeavor and wrote, nevertheless, that "the results of anthropometric statistics are a means of describing in exact terms a certain variety (of man) and its variability . . . it is clear that a *biometric* [his italics] method would undoubtedly open new ways of attacking problems of variation."

The Worcester longitudinal data are the ones used for the correlational analysis of 1906, supplemented by new data from the file of the Newark Academy, a private school near New York. Boas calculated the correlations of stature and other measurements at year t with their respective increments from t to $t + 1$. He demonstrates in these terms what he had really already shown by his division into short and tall groups in the 1897 analysis, that "the expected results [occurred], that is a maximum of positive correlations during the periods of most rapid growth, and a sudden drop to negative correlations when growth is nearly completed." For the Newark Academy data, which encompassed longitudinal series of 70 boys from 12 to 17 years, he calculated the correlations between the increments $t, t + 1$ and the increment $t + 1, t + 2$, and so on, showing them to be positive until 14–15 and negative thereafter.

It was just about this time that Pryor (1905) and Rotch (1910) demonstrated that children of the same chronological age showed considerable differences in the degree of ossification of the bones of the hand and that girls were ahead of boys in their ossification at all ages. It is not now clear whether these researches into developmental age were stimulated by contact with Boas and his ideas on tempo of growth, but it seems very likely. The other advance in developmental age measurement, that made by Crampton, was certainly stimulated, although not actually begun, by Boas himself. Crampton was a school physician in New York who felt the need of some more realistic criterion than chronological age for classifying his adolescent school boys for fitness to take part in athletic programs. About 1901, under the influence of Boas and Stanley Hall, he began studying the growth of pubic hair as providing such a criterion, and published a paper (Crampton, 1908) in which he described a threefold classification according to absence, initial appearance, and full development of the pubic hair. Godin, it will be remembered, had already been using a fivefold classification of pubic hair since 1895 and published an account of it in 1903. It is not clear if Crampton (or more likely, Stanley Hall) knew of this: Godin's work was known in Italy and Spain, but very little, it seems, in Germany and the English-speaking countries.

Boas welcomed these advances and additions to his basic conceptions and in his 1912 article wrote: "A study of the eruption of the teeth which I made a number of years ago, and the more recent interesting investigations by Rotch and Pryor on the ossification of the carpus, show that the difference in physiological development between the two sexes begins at a very early time and that in the fifth year it has already reached a value of more than a year and a half. . . . The conditions of the bones . . . gives us a better insight into the physiological development of the individual than his actual chronological age and may therefore be advantageously used for the regulation of child labour and school entrance, as Rotch and Crampton advocate'' (1912*a*).

8.2.1. Second Growth Period: 1930–1932

From 1912 to 1930 Boas published little on growth itself, but in 1930, at the age of 72, he produced the first of two papers embodying a second major contribution to the understanding of human growth. A number of individual growth curves had accumulated in the literature since 1906. In particular there was the work of Baldwin, an educational psychologist, who reported in 1914 the second series of longitudinal growth data systematically collected in the United States and became, in 1917, the first director of the Iowa Child Welfare Research Station. His monograph in 1921, justly famous for its annotated bibliography of studies of growth in all countries up to that time, contains a rather half-hearted analysis of the growth of 60 boys and 60 girls followed in the Horace Mann School of Teachers College, Columbia University, over various periods from age 7 to 17. Baldwin had data on age at menarche in the girls and did indeed conclude that growth in stature ceases soon after menarche. But despite drawing what is probably the first graph of velocity of growth in height for different age-at-menarche groups, he failed to emphasize the points that Boas later seized upon.

Boas had pondered over these longitudinal data and come to realize that his own correlational methods of 1906 were insufficient to extract more than a fraction of the information contained in the growth curves of individuals followed over many years. Their inadequacy was felt especially in dealing with events of adolescence. Boas wrote (1930) that he ''demonstrated a number of years ago (in the 1906 paper) that . . . the actual individual growth rate must show a much more decided decline during childhood and a much more decided increase during adolescence, and that at the same time the period of increased rapidity of growth during adolescence must be much shorter than would appear from the generalized curve.'' He now wished to study the details of these individual curves more closely, and he wished, in particular, to see if final adult height was related in any way to the time of occurrence of the spurt, or to its rapidity or intensity. Initially he used the same longitudinal records as in 1906, from boys at Newark Academy, but this time supplemented, where possible, by records of their adult stature. For the later paper (chiefly published in 1932, but with continuations in 1933, and less importantly, 1935*a*), he added similar longitudinal although retrospective material from the Horace Mann School, the Ethical Culture Schools, and the City College of New York.

Boas used graphical methods, subdividing his subjects into groups on the basis of height at 14, age at maximum height velocity, or age at menarche, and then plotting the growth of the means of each of the groups. The method, in Boas' hands, was a powerful and productive one. It inspired Shuttleworth's more detailed but essentially very similar studies on the more plentiful Harvard Growth Study data some 5 years later (Shuttleworth, 1937, 1938, 1939). Nowadays, curve fitting to the individual's data would be the appropriate, and indeed the equivalent approach. But in 1930 computers were not available and efficient methods of curve fitting were in their infancy; Boas liked to handle his data himself and not leave them to multitudes of assistants working desk machines.

The essence of his findings is shown in Figure 17, redrawn from illustrations in the 1930 and 1932 papers. In Figure 17 (1930, Figure 4) the distance curves for stature are given for five groups of boys, the grouping being by age at maximum velocity of growth. This is the first introduction of this method of classification. Two effects are clearly demonstrated. The first was that, on average, those who had an early adolescent spurt were already taller by age 11, but not necessarily taller when adult stature was achieved. Secondly, Boas discovered from these curves that the earlier the spurt occurs, the higher was the peak reached. The 1932 paper documents these conclusions with more exactitude and detail. This time the data are mostly for girls, who are classified both in relation to age at maximum velocity as before, and age at menarche.

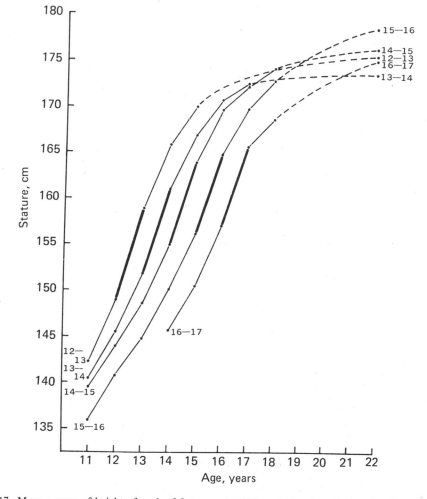

Fig. 17. Mean curves of height of each of five groups of Newark Academy boys, classified by age at maximum growth velocity. Group 1 contains all boys whose maximum velocity lies between ages 12.0 and 13.0; group 2, all boys whose maximum velocity lies between 13.0 and 14.0, and so on. (From Tanner, 1959; redrawn from Boas, 1930; Fig. 4.)

Later (1935*b*) Boas showed that the siblings he had studied in the Horace Mann School and the Hebrew Orphanage resembled each other in their times of maximum growth velocity. "These observations," he wrote, "may be summarized in the statement that each individual has by heredity a certain tempo of development that may be modified by outer conditions. The gross, generalized observations available at the present time suggest that in a socially uniform group the tempo of development may be considered as a hereditary characteristic of individuals" (1935*c*).

Such is a bare summary of Boas' contributions to the study of human growth. So much of his work has a contemporary ring that it is easy to forget the general state of growth studies in 1930. Of the later famous longitudinal studies, only the Harvard Growth Study (1922–1934), the Iowa Station, and the Merrill-Palmer School had been long in existence, and little of consequence for physical growth had yet been published from them except for Baldwin's graphs. The Brush Foundation Study, the studies at Berkeley, the Fels Research Institute, the Yale Normative Study, the Center for Research in Child Development at Harvard, the Child Research Council at Denver—all were nonexistent or only very lately begun. All built on the solid foundation laid by Boas' technical knowledge, critical insight, and personal wisdom.

8.2.2. Environment and Heredity

Between these two "growth study" periods, Boas' interest was engaged principally in the question of the relative influence of environment and heredity upon body size and form. (The portion of Boas' interest allotted to physical anthropology, that is, was so engaged. So adequate to one man's entire life study were Boas' contributions to this section of anthropology alone that it is almost impossible to bear in mind that he was at least equally active in two or three other fields, as witness the accounts in his memorial volume; Goldschmidt, 1959.) A full account of his contributions to the dispute between Pearson and the Mendelian School, and of his biometrical studies of "family lines" in physical characteristics will be found in Tanner's (1959) commemorative article, from which much of this account is taken. A word should be said, however, on his study of immigrants to America and their children, a study vast in scope, unexpected in outcome, and staggering in professional and public reception. Begun in 1908, it was made for the United States Immigrant Commission, whose object seems to have been to determine whether the influx of immigrants was resulting in deterioration of the physique of the American population. Boas, with some 13 assistants, prominent among whom was Crampton, measured the stature, head length and breadth, bizygomatic diameter, and in some cases the weight, of a total of nearly 18,000 immigrants and American-born children of immigrants in New York. About 5500 were aged 25 and older, and the remainder were children over the age of 4. The material was almost exclusively family groups, the main comparison being between children born before the parents immigrated ("foreign-born") and hence growing in Europe, and children born in America after the parents had come to New York.

The main report was issued in book form under the title *Changes in Bodily Form of Descendents of Immigrants* (Boas, 1912*b*). In the same year a summary of the main findings was published in the *American Anthropologist,* together with answers to some of the criticisms already encountered (Boas, 1912*c*). The criticisms were to continue, off and on, for years, and in 1928 the entire raw data were issued, constituting the largest collection of family measurements ever published. Boas' findings, that stature was a little greater in children growing up in America and that head length and breadth were a few millimeters more, would scarcely be regarded as surprising nowadays. But anthropologists at that time had an astonishing belief in the fixity of what they called human types or races, and when Boas showed that even the central tabneracle of the doctrine—the cephalic index—was built on sand, they chorused their displeasure and disbelief.

It is hard for us to realize the degree to which the anthropologists of the last century were obsessed with the constancy of the cephalic index. Head form appeared to them a method of tracing the origins and distributions of peoples; if it altered in response to environment, a major prop was kicked out from under the anthropological pile-dwelling. Long after the fuss had died down, Boas (1930) summarized the work as a whole in a statement with which nobody nowadays could quarrel:

> It has been known for a long time that the bulk of the body as expressed by stature and weight is easily modified by more or less favourable conditions of life. In Europe there has been a gradual increase in bulk between 1850 and 1914. Adult immigrants who came to America from south and east Europe have not taken part in the general increase . . . presumably because they were always selected from a body the social condition of which has not materially changed. Their children, however, born in America or who came here young, have participated in the general increase of stature of our native population. . . . With this go hand in hand appreciable differences in body form. . . . These changes do not obliterate the differences between genetic type but they show that the type as we see it contains elements that are not genetic but an expression of the influence of environment.

Boas was a remarkable innovator who nevertheless made few mistakes. He was a man of enormous breadth of interest, yet his writings in the small sector of auxology show no trace of dispersion of talent. On the contrary, most of his work makes the other writers of his time look like dilettantes; it is he who penetrates below the surface. To a large extent this is due to his biometrical skill and biological understanding. In a cogent address (Boas, 1904) he described anthropology as "partly a branch of biology, partly a branch of the mental sciences." A portion will split off as linguistics; a portion will remain as the study of cultural change; but anthropologists must always be familiar with the main lines of all the branches. On the physical side of anthropology he here, as elsewhere, declared an indebtedness and allegiance to Galton and Pearson, and before them to Quetelet. In the bridge between the older concepts of physical anthropology and the newer disciplines of human biology, Boas' work constitutes much of the framework.

When Crampton's paper (1908) on physiological age was reprinted in 1944, Crampton wrote a short postscript. It said, "The author [Crampton] is commonly given credit for originating the term 'physiological age.' This, I believe, properly belongs to the much respected and beloved Franz Boas." Many others might have written the same of their contributions to human biology in the first half of this century.

8.3. D'Arcy Thompson

No account of the history of growth studies could ignore the author of *Growth and Form*, and in placing D'Arcy Thompson (1860–1948) next to Boas there is a deeper logic than that of dates. D'Arcy Thompson and Boas were both giants, intellectually and personally, but giants of totally different shapes. Boas was a meticulous research worker, cast recognizably in the modern mold, who picked his way carefully forward with letters in *Science* and the theses of postgraduates. Thompson was a late flower of the Renaissance, a man of vast scholarship, whose footnotes alone are a delight and an education. *Growth and Form* (1917, 1942) is above all a Renaissance book, crammed with all sorts of considerations and written in a prose of such clarity and elegance that one reviewer compared him, and justly, with the author of the *Anatomy of Melancholy*.

D'Arcy, as he was universally called, was brought up in the classics, wrote a *Glossary of Greek Birds* and a *Glossary of Greek Fishes,* and translated Aristotle's *Historia Animalium* with an understanding only possible to a professional zoologist who was also president of the Classical Association of Scotland, England, and Wales. The scientific establishment, already grown unaccustomed to such scholarship, did not know what to make of him, and he spent his whole career rather peripheral to the main stream of scientific activity, as professor of natural history, first in Dundee (1884–1917) and then St. Andrews (1917–1948). *Growth and Form* is a massive book and its running page headings include "of modes of flight" and "the comparative anatomy of bridges" as well as "of growth in infancy." Two chapters only, out of 17, deal with matters of growth in the strict sense. One is entitled "On the Theory of Transformations, or the Comparisons of Related Forms"; in it appear the famous Dürer drawings of faces and the application of transformed coordinates to depict the evolution of a species or the relations between different genera. D'Arcy's method was graphical, and it was 50 years before the use of computers made analytical application of the technique generally possible; geographers and systematists in particular are now among its users.

The second chapter that concerns us is entitled "The Rate of Growth." Form is produced by growth of varying rates in different directions: "When in a two-dimensional diagram we represent a magnitude (for instance, length) in relation to time . . . we get that kind of vector diagram which is known as the *curve of growth.* We see that the phenomenon we are studying is a *velocity* . . . by measuring the slope or steepness of our curve of growth at successive epochs, we shall obtain a picture of the successive *velocities* or *growth rates*. Plotting these successive differences against time we obtain a curve each point on which represents a certain rate at a certain time; while the former curve (height against time) showed a continuous succession of varying *magnitudes,* this shows a succession of varying *velocities*. The mathematician calls it a curve of first differences; we may call it a curve of annual (or other) increments; but we shall not go wrong if we call it a curve of the *rate (or rates) of growth,* or still more simply a *velocity curve"* (1942, p. 95, his italics). Although others before D'Arcy Thompson had plotted annual increments (for example, Schmidt-Monnard), it was D'Arcy who emphasized the continuous nature of the *process* of velocity. It was Plato who defined growth as a form of motion, and D'Arcy as usual took his cue from the old philosophers; for him growth was something moving through time and for him "to say that children of a given age vary in the rate at which they are growing would seem to be a more fundamental statement than that they vary in the size to which they have grown" (1942, p. 128). These quotations are the origin of the term velocity curve, and of the emphasis laid on growth velocity by subsequent writers.

D'Arcy's biography has been well written by his daughter (R. D. Thompson, 1958), but *Growth and Form* will ever remain his epitaph. The generation who heard D'Arcy Thompson (which includes the writer as a schoolboy) witnessed not only an actor–scholar of superlative skill, but a sort of beneficent magus, large and bearded, surrounded always by the objects about which he was talking: dodecahedrons and examples of Aristotelian gnomons, in the writer's case, as well as shells, fishes, and in at least one instance a live chicken (Thompson, 1958, p. 213). D'Arcy was one of the few British scientists honored with a festschrift (Clark and Medawar, 1945) and when a symposium on aspects of growth was organized at the Royal Society as late as 1950, it was inevitable that its title should be *A Discussion on the Measurement of Growth and Form* (Zuckermann, 1950). Although he made no explicit contributions to the facts of human growth, D'Arcy Thompson, like Franz Boas, elevated the whole intellectual level of the field and broadened the outlook of all who work in it.

9. Coda

With Boas and D'Arcy Thompson we come to the present day. The scientific study of human growth began with Buffon and Montbeillard, its social application with Quetelet, Cowell, and Horner. Roberts, the factory doctor; Bowditch and Townsend, the physiologists; and Kotelmann and the other school physicians established growth firmly in science and medicine. Boas and D'Arcy Thompson complete the development, the one taking us squarely into the modern context, the other reminding us that the roots of knowledge and understanding are nourished through soil deposited in the past.

ACKNOWLEDGMENT

I wish to thank the staff of the Library of the Wellcome Institute for the History of Medicine and in particular Mr. H. R. Denham, for much expert bibliographic assistance; Dr. Noel Cameron and Mr. Ray Lunnon for help with the diagrams; and Mrs. Sara Batterley for her patience with innumerable typescripts.

10. References

Ackerknecht, E. H., 1952, Villermé and Quetelet, *Bull. Hist. Med.* **26:**317–329.

Anonymous, 1918, A physical census and its lesson, *Br. Med. J.* **2:**348–349.

Astruc, P., 1933, Villermé, Louis-Réné, 1782–1863, *Les Biographies Médicales,* Vol. 7, pp. 215–244, Baillière, Paris.

Auden, G. A., 1910, Heights and weights of Birmingham school children in relation to infant mortality, *Sch. Hyg.* **1:**290–291.

Baldwin, B. T., 1914, *Physical Growth and School Progress*, U.S. Bureau of Education, No. 10, Washington, 212 pp.

Baldwin, B. T., 1921, The physical growth of children from birth to maturity, *University of Iowa Studies in Child Welfare,* Vol. 1, 411 pp.

Barnes, E., 1892–1893, Physical development of Oakland children, *Oakland Sch. Rep.* **93:**38–44; cited in Baldwin (1921); not seen.

Bayer, L. M., and Bayley, N., 1959, *Growth Diagnosis,* University Press, Chicago.

Bayley, N., 1956, Growth curves of height and weight by age for boys and girls scaled according to physical maturity, *J. Paediatr.* **48:**187–194.

Bell, E. T., 1945, *The Development of Mathematics,* 2nd ed., Macmillan, New York.

Berry, F. M. D., 1904, On the physical examination of 1580 girls from the elementary schools of London; *Br. Med. J.* **1:**1248–1249.

Blaug, M., 1964, The poor law report re-examined, *J. Econ. Hist.* **24**:229–245.

Boas, F., 1892*a*, The growth of children, *Science* **19**:256–257; 281–282.

Boas, F., 1892*b*, The growth of children, *Science* **20**:351–352.

Boas, F., 1894, The correlation of anatomical or physiological measurements, *Am. Anthropol.* **7**:313–324.

Boas, F., 1895*a*, On Dr. William Townsend Porter's investigation of the growth of the school children of St. Louis, *Science* N. S. **1**:225–230.

Boas, F., 1895*b*, The growth of first-born children, *Science* N. S. **1**:402–404.

Boas, F., 1897, The growth of children, *Science* N.S. **5**:570–573.

Boas, F., 1904, The history of anthropology, *Science* N.S. **20**:513–524.

Boas, F., 1912*a*, The growth of children, *Science* N.S. **36**:815–818.

Boas, F., 1912*b*, *Changes in Bodily Form of Descendants of Immigrants,* Columbia Univ. Press, New York.

Boas, F., 1912*c*, Changes in the bodily form of descendants of immigrants, *Am. Anthropol.* N.S., **14**:530–562.

Boas, F., 1922, Review of R. M. Woodbury's "Statures and weights of children under six years of age," *Am. J. Phys. Anthropol.* **5**:279–282.

Boas, F., 1928, *Materials for the Study of Inheritance in Man,* Columbia Univ. Press, New York.

Boas, F., 1930, Observations on the growth of children, *Science* **72**:44–48.

Boas, F., 1932, Studies in growth, *Hum. Biol.* **4**:307–350.

Boas, F., 1933, Studies in growth II, *Hum. Biol.* **5**:429–444.

Boas, F., 1935*a*, Studies in growth III, *Hum. Biol.* **7**:303–318.

Boas, F., 1935*b*, The tempo of growth of fraternities, *Proc. Nat. Acad. Sci.* **21**:413–418. Reprinted in *Race, Language and Culture* (Boas, 1940).

Boas, F., 1935*c*, Conditions controlling the tempo of development and decay, *Assoc. Life Inst. Med. Dir. Am.* **22**:212–223.

Boas, F., 1936, Effects of American environment on the immigrants and their descendants, *Science* N.S. **84**:522–525.

Boas, F., 1940, *Race, Language and Culture,* Macmillan, New York.

Boas, F., 1941, The relation between physical and mental development, *Science* **93**:339–342.

Boas, F., and Wissler, C., 1906, Statistics of growth, *Report of U.S. Commissioner of Education for 1904,* pp. 25–132, Washington, D.C.

Bowditch, H. P., 1872, Comparative rate of growth in the two sexes, *Boston Med. Surg. J.* **10**:434 (old series **87**).

Bowditch, H. P., 1877, The growth of children, *8th Annual Report of State Board of Health of Massachusetts,* pp. 275–327, Wright, Boston.

Bowditch, H. P., 1879, Growth of children, *10th Annual Report of State Board of Health of Massachusetts,* pp. 33–62, Wright, Boston.

Bowditch, H. P., 1890, The physique of women in Massachusetts, *21st Annual Report of the State Board of Health of Massachusetts,* pp. 287–304, Wright & Potter, Boston.

Bowditch, H. P., 1891, The growth of children studied by Galton's percentile grades, *22nd Annual Report of the State Board of Health of Massachusetts,* pp. 479–525, Wright & Potter, Boston.

Bowditch, M., 1958, Henry Pickering Bowditch: An intimate memoir, *Physiologist* **1**:7–11.

Bruford, W. H., 1962, *Culture and Society in Classical Weimar, 1775–1806,* Cambridge University Press, London.

Buffon, G. L. L. de, 1749, *Histoire Naturelle,* Vols. 2 and 3, Imprimerie Royale, Paris.

Buffon, G. L. L. de, 1836, *Oeuvres complets avec les supplémens, augmenté de la classification de G. Cuvier,* Vol. 4, p. 70, Dumenil, Paris.

Burk, F., 1898, Growth of children in height and weight, *Am. J. Psychol.* **9**:253–326.

Camerer, W., 1893, Untersuchungen über Massenwachstum und Längenwachstum der Kinder, *Jahrb. Kinderheilk.* **36**:249–293.

Camerer, W., 1901, Das Gewichts und Längenwachstum, des Menschen, insbesondere im 1 Lebenjahr, *Jahrb. Kinderheilk.* **53**:381–446.

Cameron, N., 1977, An Analysis of the Growth of London Schoolchildren: The London County Council's 1966–67 Growth Survey, Ph.D. Thesis, University of London.

Cannon, W. B., 1924, Henry Pickering Bowditch, *Biographical Memoirs of National Academy of Sciences,* Vol. 17, pp. 183–196, Senate Documents of 68th Congress, U.S. Govt. Printing Office, Washington, D.C.

Carlson, A. J., 1949, William Townsend Porter, 1862–1949, *Science* **10**:111–112.

Carstädt, F., 1888, Über das Wachstum der Knaben vom 6 bis zum 16 lebensjahre, *Z. Schulgesund-heitspflege* **1**:65–69.

Chadwick, E., 1842, *Report on the Sanitary Condition of the Labouring Population of Great Britain. . . . from The Poor Law Commissioners,* Clowes, London.

City of Edinburgh Charity Organisation Society, 1906, *Report on the physical condition of 1400 school children in the city together with some account of their homes and surroundings,* King, London.

Clark, W. E. le Gros., and Medawar, P. B., Eds., 1945, *Essays on Growth and Form presented to D'Arcy Wentworth Thompson,* Clarendon, Oxford.

Collard, A., 1934, Goethe et Quetelet. Leurs relations de 1829 à 1832, *Isis* **20**:426–435.

Cowell, J. W., 1933, see *Parliamentary Papers* (1833).

Crampton, C. W., 1908, Physiological age, a fundamental principle, *Am. Phys. Educ. Rev.* **13**:(3–6). Reprinted in *Child Dev.* **15**:1–12 (1944).

Davenport, C. B., and Love, A. G., 1920, Report on defects found by draft boards in USA conscripts. *Sci. Month.* Jan., pp. 5–25; Feb., pp. 125–141.

De Giovanni, A., 1904–1909, *Mofologia del corpo umano,* 3 vols., Hoepli, Milan.

Deslandres, M., and Michelin, A., 1938, *Il y a Cent Ans. Etat Physique et Morale des Ouvriers au Temps du Libéralisme. Témoinage de Villermé,* Editions Spes, Paris.

Elderton, E. M., 1914, Height and weight of school children in Glasgow, *Biometrika* **10**:288–339.

English Historical Documents, 1883–1874, 1956 (G. M. Young and W. D. Hoardcock, eds.), Eyre & Spottiswoode, London.

Erismann, F., 1888, Schulhygiene auf der Jubiläumsausstellung der Gesellschaft für Beförderung der Arbeitsamkeit in Moskau, *Z. Schulgesundheitspflege* **1**:347–373, 393–419.

Eveleth, P. B., and Tanner, J. M., 1976, *Worldwide Variation in Human Growth,* University Press, Cambridge.

Faye, L., and Hald, 1896, Udersögelser om Sundhedstilstanden ved norske höiere Gutte-og Pigeskoler samnt Faellesskoler udförte i 1981 og 1892. Med un résumé en francais. *Kongeriget Norges 45 ordentlige Stortings Forhandlinger i Aaret 1896,* cited in Kiel (1939); not seen.

Fergus, W., 1887, Obituary notice, *Lancet* **1**:105.

Fergus, W., and Rodwell, G. F., 1874, On a series of measurements for statistical purposes recently made at Marlborough College, *J. Anthropol. Inst.* **4**:126–130.

Flourens, P., 1860, *Des Manuscripts de Buffon,* Garnier, Paris.

Ford, E. H. R., 1958, Growth in height of ten siblings, *Hum. Biol.* **30**:107–119.

Forrest, D. W., 1974, *Francis Galton. The Life and Work of a Victorian Genius,* Elek, London.

Galton, F., 1873–1874, Proposal to apply for anthropological statistics from schools, *J. Anthropol. Inst.* **3**:308–311.

Galton, F., 1874, Notes of the Marlborough School statistics, *J. Anthropol. Inst.* **4**:130–135.

Galton, F., 1875–1876, On the height and weight of boys ages 14 years in town and country public schools, *J. Anthropol. Inst.* **5**:174–181.

Galton, F., 1883, *Inquiries into Human Faculty and its Development,* Macmillan, London.

Galton, F., 1884–1885, Some results of the anthropometric laboratory, *J. Anthropol. Inst.* **14**:275–288.

Galton, F., 1885–1886, Hereditary stature, *J. Anthropol. Inst.* **14**:488–499.

Galton, F., 1886, Family likeness in stature, *Proc. R. Soc.* **40**:42–73.

Galton, F., 1889, *Natural Inheritance,* Macmillan, London.

Geissler, A., and Uhlitzsch, R., 1888, Die grössenverhältnisse der Schulkinder im Schulinspectionsbezirk Freiberg, *Zeitschrift des Königlichen Sächsischen Statistiken Bureaus* **34**:28–40, cited in Porter (1893*b*); not seen.

George, M. D., 1965, *London Life in the Eighteenth Century,* Capricorn, New York.

Godin, P., 1903, *Recherches anthropométriques sur la croissance des diverses parties du corps. Détermination de l'adolescent type aux différents âges pubertaires d'après 36,000 mensurations sur 100 sujets suivis individuellement de 13 à 18 ans,* Maloine, Paris.

Godin, P., 1914, Lois de croissance, *J. R. Anthropol. Inst.* **44**:295–301.

Godin, P., 1919*a, Manuel d'anthropologie pédagogique: basée sur l'anatomo-physiologie de la croissance méthode auxologique,* Delachaux et Niestle, Paris.

Godin, P., 1919*b,* La méthode auxologique, *Med. Fr.* March 15, cited in Godin (1935); not seen.

Godin, P., 1921, Mon enseignement à Genève 1912–1924, *Med. Fr.,* cited in Godin (1935); not seen.

Godin, P., 1935, *Recherches anthropométriques sur la croissance des diverses parties du corps,* 2nd ed., Legrand, Paris.

Goldschmidt, W., ed., 1959, *The anthropology of Franz Boas,* Chandler, San Francisco, and *Am. Anthropol.* **61**.

Greenwood, A., 1915, *The Health and Physique of School Children,* Ratan Tata Foundation, London, School of Economics.

Grinder, R. E., 1969, The concept of adolescence in the genetic psychology of G. Stanley Hall, *Child Dev.* **40:**355–369.

Guarinonius, H., 1610, *Die Grewel der Verwüstung menschlichen Geschlects,* Innsbruck.

Hall, W. S., 1896, Changes in the proportions of the human body during the period of growth, *J. Anthropol. Inst.* **25:**21–46.

Hartmann, W., 1970, *Beobachtungen zur Acceleration des Längenwachstums in der Zweten Hälfte des 18 Jahr-hunderts,* thesis, Frankfurt am Main.

Herskovits, M. J., 1953, *Franz Boas: The Science of Man in the Making,* Scribner, New York.

Hertel, A., 1885, *Overpressure in high schools of Denmark,* translated by G. Sorenson, Macmillan, New York, cited in Baldwin (1921); not seen.

Hertel, A., 1888, Neuere Untersuchungen über den allgemeinen Gesundheitszustand der Schüler und Schülerinnen, *Zeitschrift für Schulgesundheitspflege* **1:**167–183, 201–215.

Hilts, V. L., 1973, Statistics and social science, in: *Foundations of Scientific Method: The Nineteenth Century* (R. N. Gieve and R. S. Westfall, eds.), Indiana University Press, Bloomington.

Horner, L., 1837, *Letter to Mr. Senior of May 23rd, 1837,* in: *Letters on the factory Act as it affects cotton manufacture . . . by N. W. Senior . . . to which is appended a letter to Mr. Senior from Leonard Horner Esq,* pp. 30–42, Fellowes, London.

Horner, L., 1840, *On the employment of children in factories and other works in the United Kingdom and in some foreign countries,* Longman, London.

Ireland, M. W., Love, A. G., and Davenport, C. B., 1919, Physical examination of the first million draft recruits, *War Department Office of the Surgeon General. Bulletin No. II,* U.S. Govt. Printing Office, Washington, D.C.

Janes, M. D., 1975, Physical and psychological growth and development, *Environ. Child Health* **121:**26–30.

Kay, T., 1904–1905, Tables showing height, weight, mental capacity, condition of nutrition, teeth, etc., of Glasgow schoolchildren, *J. R. Sanit. Inst.* **26:**907–913.

Key, A., 1885, *Läroverkskomiténs underdaniga utlatande och förslag angaende organisationen af rikets allmanna läroverk och dermed samenhangende fragor,* Redogörelse, Stockholm.

Key, A., 1889, *Schulhygienische Untersuchungen,* Burgerstein, Leipzig, cited in Baldwin (1921); not seen.

Key, A., 1891, Die pubertätsentwicklung und das verhaltnis derselben zu den krankheitserscheinungen der Schøljugend. *Verhandlingen des 10 internationales medizinisches Kongresses,* Berlin, 1890, Vol. 1. Berlin: cited in Baldwin (1921); not seen.

Kiil, V., 1939, Stature and growth of Norwegian men during the past two hundred years, *Skrifter utgitt av det norske Videnscaps-Academi i Oslo I. Mat-Nat. Klasse,* No. 6.

Landis, E. M., 1949, William Townsend Porter, 1862–1949, *Am. J. Physiol.* **158:**v–vii.

Lange, E. von, 1903, Die Gesetzmassigkeit im Längenwachstum des Menschen, *Jahrbu. Kinderheilk.* **57:**261–324.

Ljung, B.-O., Bergsten-Brucefors, A., and Lindgren, G., 1974, The secular trend in physical growth in Sweden, *Ann. Hum. Biol.* **1:**245–256.

Lyell, K. M., 1890, *Memoir of Leonard Horner FRS, F.A.S., consisting of letters to his family and from some of his friends,* 2 vols., Women's Printing Society, London.

Mailly, N. C., 1875, *Essai sur la vie et les ouvrages de L.A.H. Quetelet,* Hayez, Brussels.

Malling-Hansen, P. R., 1883, Über periodizität im Gewicht der Kinder, Copenhagen, 35 pp.

Mandeville, B. de, 1723, *An Essay on Charity and Charity Schools,* 4th ed., Tonson, 1725, London, bound with *The Fable of the Bees,* by the same author.

Mann, H., 1840, *Letter to L. Horner,* quoted in Horner (1840), p. 107ff.

Marshall, D., 1926, *The English Poor in the Eighteenth Century,* Routledge, London.

Marshall, W. A., 1971, Evaluation of growth rate in height over periods of less than a year, *Arch. Dis. Child.* **46:**414–420.

McGregor, A. S. M., 1908, The physique of Glasgow children: An enquiry into the physical condition of children admitted to the City of Glasgow Fever Hospital, Belvidere during the years 1907–8, *Proc. R. Philos. Soc. Glasgow* **40:**156–176.

Meredith, H. V., 1950, The research worker's responsibility for generalization, *Phys. Educ.* **7:**47–48.

Montessori, M., 1913, *Pedagogical Anthropology,* translated from the Italian by F. T. Cooper, Heinemann, London.

Pagliani, L., 1875–1876, Sopra alcuni fattori dello sviluppo umano; richerche antropometriche, *Atti Accad. Sci. Torino* **11**:694–760.

Pagliani, L., 1879, *Lo svillupo umano per eta, sesso, condizione sociale ed etnica,* Civelli, Milan.

Parliamentary Papers, 1833, *Reports from Commissioners (4),* Vol. 20, First report . . . into the Employment of children in factories, H.M.S.O., London.

Parliamentary Papers, 1873, *Accounts and Papers (17),* Vol. 55, pp. 803–864, Report to the local Government Board on proposed changes in hours and age of employment in textile factories, H.M.S.O., London.

Parliamentary Papers, 1876, *Reports from Commissioners (15),* Vol. 29, Report of the commissioners on the working of the Factory and Workshops Act, with a view to their consolidation and amendment, H.M.S.O., London.

Peckman, A. W., 1882, The growth of children, *Report of State Board of Health, Wisconsin for 1881,* **6**:28–73.

Perier, J. A. N., 1867, Le docteur Boudin. Notice historique sur sa vie et ses travaux, *Recueil des mémoirs de medicine, de chirurgie et de pharmacie militaires,* 3rd series, Vol. 19, 39 pp., Rozier, Paris.

Pinchbeck, I., and Hewitt, M., 1969–1973, *Children in English Society,* 2 vols., Routledge & Kegan Paul, London.

Porter, W. T., 1893*a,* The physical basis of precocity and dullness, *Trans. St. Louis Acad. Sci.* **6**:161–181.

Porter, W. T., 1893*b,* The relation between the growth of children and their deviation from the physical type of their sex and age, *Trans. St. Louis Acad. Sci.* **6**:233–250.

Porter, W. T., 1893*c,* On the application to individual school children of the mean values derived from anthropological measurements by the generalizing method, *Q. Publ. Am. Stat. Assoc.* N.S. **3**:576–587.

Porter, W. T., 1894, The growth of St. Louis children, *Trans. St. Louis Acad.* **6**:263–380.

Pryor, J. W., 1905, Development of the bones of the hand as shown by the x-ray method, *Bull. State Coll. Kentucky* Series 2, No. 5.

Quetelet, A., 1830, Sur la taille moyenne de l'homme dans les villes et dans les campagnes, et sur l'âge ou la croissance est complètement achevée, *Ann. Hyg. Publique* **3**:24–26.

Quetelet, A., 1831, Recherches sur la loi de croissance de l'homme, *Ann. Hyg. Publique* **6**:89–113.

Quetelet, L. A. J., 1835, *Sur l'homme et le développement de ses facultés. Essai sur physique sociale,* 2 vols., Bachelier, Paris.

Quetelet, A., 1842, *A treatise on man and the development of his faculties,* translated into English by R. Knox, Chambers, Edinburgh.

Quetelet, A., 1846, *Lettres à SAR le duc regnant de Saxe-Coburg et Gotha, sur la théorie des probabilities, appliquée aux sciences morales et politiques,* Brussels.

Quetelet, A., 1848, *Du système social et des lois qui le régissent,* Paris.

Quetelet, A., 1869, *Physique sociale. Essai sur le développement des facultés de l'homme,* 2 vols., Muquardt, Brussels.

Quetelet, A., 1870, *Anthropométrie, ou mésure des différentes facultés de l'homme,* Muquardt, Brussels.

Rietz, E., 1906, Körperentwicklung und geistige Begabung, *Zeitschrift für Schulgesundheitspflege* **19**:65–98.

Roberts, C., 1874–1876, The physical development and the proportions of the human body, *St. George's Hosp. Rep.* **8**:1–48.

Roberts, C., 1876, The physical requirements of factory children, *J. Stat. Soc.* **39**:681–733.

Roberts, C., 1878, *A Manual of Anthropometry,* Churchill, London.

Roberts, C., 1890, Relative growth of boys and girls, *Nature* **42**:390.

Roberts, C., 1895, Memorandum on the medical inspection of, and physical education in, secondary schools, *Report, Royal Commission on Secondary Education of England,* Vol. V, pp. 352–374.

Roberts, C., 1901, Obituary, *Br. Med. J.* **1**:249.

Rosen, G., 1955, Problems in the application of statistical analysis to questions of health 1700–1800, *Bull. Hist. Med.* **29**:27–45.

Rotch, T. M., 1910, A comparison in boys and girls of height, weight and epiphyseal development, *Trans. Am. Paediatr. Soc.* **22**:36–38.

Sack, N., 1893, Über die körperliche entwicklung der knaben in der Mittelschulen Moskau's, *Zeitschrift für Schulgesundheitspflege* **6**:649–663.

Scammon, R. E., 1927a, The first seriatim study of human growth, *Amer. J. Phys. Anthropol.* **10**:329–336.

Scammon, R. E., 1927b, The literature on the growth and physical development of the fetus, infant, and child: A quantitative summary, *Anat. Rec.* **35**:241–267.

Scammon, R. E., and Calkins, L. A., 1929, *The Development and Growth of the External Dimensions of the Human Body in the Fetal Period,* University of Minnesota Press, Minneapolis.

Schmidt-Monnard, K., 1901, Über den werth von Körpermassen zur Beurhteilung des Körperzustandes bei kindern, *Jahrb. Kinderheilk.* **53**:50–58.

Scouller, R. E., 1966, *The Armies of Queen Anne,* Clarendon Press, Oxford.

Shuttleworth, F. K., 1937, Sexual maturation and the physical growth of girls age six to nineteen, *Monographs of the Society for Research in Child Development* **2**(5), 253 pp.

Shuttleworth F. K., 1938, Sexual maturation and the skeletal growth of girls age six to nineteen, *Monographs of the Society for Research in Child Development* **3**(3), 56 pp.

Shuttleworth, F. K., 1939, The physical and mental growth of girls and boys age six to nineteen in relation to age at maximum growth, *Monographs of the Society for Research in Child Development* **4**(3).

Stanway, S., 1833, See *Parliamentary Papers* (1833).

Stieda, L., 1882–1883, Über die Anwendung der Wahrscheinlichkeitsrechnung in der anthropologischen Statistik, *Arch. Anthropol.* **14**:167–182.

Stratz, C. H., 1909, Wachstum und Proportionen des Menschen vor und nach der Geburt, *Arch. Anthropol.* **8**:287–297.

Stratz, C. H., 1915, Betrachtungen über das Wachstum des Menschen, *Arch. Anthropol.* **14**:81–88.

Tanner, J. M., 1955, *Growth at Adolescence,* Blackwell Scientific Publications, Oxford.

Tanner, J. M., 1959, Boas' contributions to knowledge of human growth and form, in: *The Anthropology of Franz Boas* (W. Goldschmidt, ed.), Chandler, San Francisco, and *Am. Anthropol.* **61**:76–111.

Tanner, J. M., 1962, *Growth at Adolescence,* 2nd ed., Blackwell Scientific Publications, Oxford.

Tanner, J. M., 1965, Radiographic studies of body composition, in: *Body Composition* (J. Brozek, ed.), pp. 211–238, Pergamon, Oxford.

Tanner, J. M., 1966, Galtonian eugenics and the study of growth, *Eugen. Rev.* **58**:122–135.

Tanner, J. M., 1978, *Fetus into Man,* Open Books, London and Harvard University Press, Cambridge, Mass.

Tanner, J. M., and Gupta, D., 1968, A longitudinal study of the excretion of individual steroids in children from 8 to 12 years old, *J. Endocrinol.* **41**:139–156.

Tanner, J. M., and Whitehouse, R. H., 1976, Clinical longitudinal standards for height, weight, height velocity and weight velocity and the stages of puberty, *Arch. Dis. Child.* **51**:170–179.

Tanner, J. M., Whitehouse, R. H., and Takaishi, M., 1966, Standards from birth to maturity for height, weight, height velocity and weight velocity: British children. 1965, *Arch. Dis. Child.* **41**:454–471, 613–635.

Tanner, J. M., Lejarraga, H., and Cameron, N., 1975, The natural history of the Silver-Russell syndrome: A longitudinal study of thirty-nine cases, *Paediatr. Res.* **9**:611–623.

Theopold, W., 1967, *Der Herzog und die Heilkunst: die Medizin an der Hohen Carlschule zu Stuttgart,* Deutsche Arzte-Verlag, Cologne.

Thompson, D'A. W. , 1917, *On Growth and Form,* University Press, Cambridge.

Thompson, D'A. W., 1942, *On Growth and Form,* revised ed., University Press, Cambridge.

Thompson, R. D., 1958, *D'Arcy Wentworth Thompson: The Scholar–Naturalist, 1860–1948,* Oxford University Press, London.

Thorne, L. T., 1904, The physical development of the London schoolboy: 1890 examinations, *Br. Med. J.* **1**:829–831.

Trussell, J., and Steckel, R. H., 1978, The estimation of the mean age of female slaves at the time of menarche and their first birth, *J. Interdiscip. Hist.* **8** (in press), and personal communications.

Tuxford, A. W., and Glegg, R. A., 1911, The average height and weight of English schoolchildren, *Br. Med. J.* **1**:1423–1424.

Udjus, L. G., 1964, *Anthropometrical Changes in Norwegian Men in the Twentieth Century,* Universitetsforlaget. Oslo, 249 pp.

Uhland, R., 1953, *Geschichte der Hohen Karlschule in Stuttgart,* Kohlhammer, Stuttgart.

Villermé, L. R., 1828, Mémoire sur la mortalité en France dans la classe aisée et dans la classe indigente, *Mém. Acad. Méd.* **1**:51–98.

Villermé, L. R., 1829, Mémoire sur la taille de l'homme en France, *Ann. Hyg. Publique* **1**:551–599.

Villermé, L. R., 1840, *Tableau de l'état physique et moral des ouvriers employés dans les manufactures de coton, de laine et de soie,* 2 vols., Renouard, Paris.

Viola, G., 1932, *La Costituzione Individuale,* Cappelli, Bologna.

Virey, G., 1816, Géant, in: *Dictionaire des sciences médicales par une société de médicins et de chirurgiens,* Vol. 17, pp. 546–568, Panckouke, Paris.

Wagner, H., 1856–1858, *Geschichte der Hohen Carlschule,* 3 vols., Etlinger, Würzburg.

Walker, H. M., 1929, *Studies in the History of Statistical Method,* Williams and Wilkins, Baltimore.

Wallin, J. E. W., 1938, A tribute to G. Stanley Hall, *J. Genet. Psychol.* **113:**149–153.

Weir, J. B. de V., 1952, The assessment of the growth of schoolchildren with special reference to secular changes, *Brit. J. Nutr.* **6:**19–33.

Weissenberg, S., 1911, Das Wachstum des Menschen nach Alter, Geschlect und Rasse, in: *Studien und Forschungen zur Menschen und Volkerkunde* (G. Buschau, ed.), Strecker & Schrödner, Stuttgart.

West, G. M., 1893, Worcester school children: Growth of head, body and face, *Science* **21:**2–4.

West, G. M., 1894, Anthropometrische Untersuchungen über die Schulkinder in Worcester, Mass., *Arch. Anthropol.* **22:**13–48.

Zuckerman, S., ed., 1950, A discussion on the measurement of growth and form, *Proc. R. Soc. Ser. B* **137:**433–523.

Index